ELECTRON PARAMAGNETIC RESONANCE

THE WILEY BICENTENNIAL–KNOWLEDGE FOR GENERATIONS

*E*ach generation has its unique needs and aspirations. When Charles Wiley first opened his small printing shop in lower Manhattan in 1807, it was a generation of boundless potential searching for an identity. And we were there, helping to define a new American literary tradition. Over half a century later, in the midst of the Second Industrial Revolution, it was a generation focused on building the future. Once again, we were there, supplying the critical scientific, technical, and engineering knowledge that helped frame the world. Throughout the 20th Century, and into the new millennium, nations began to reach out beyond their own borders and a new international community was born. Wiley was there, expanding its operations around the world to enable a global exchange of ideas, opinions, and know-how.

For 200 years, Wiley has been an integral part of each generation's journey, enabling the flow of information and understanding necessary to meet their needs and fulfill their aspirations. Today, bold new technologies are changing the way we live and learn. Wiley will be there, providing you the must-have knowledge you need to imagine new worlds, new possibilities, and new opportunities.

Generations come and go, but you can always count on Wiley to provide you the knowledge you need, when and where you need it!

WILLIAM J. PESCE
PRESIDENT AND CHIEF EXECUTIVE OFFICER

PETER BOOTH WILEY
CHAIRMAN OF THE BOARD

ELECTRON PARAMAGNETIC RESONANCE
Elementary Theory and Practical Applications

Second Edition

JOHN A. WEIL
Department of Chemistry, University of Saskatchewan, Saskatoon,
Saskatchewan, S7N 0W0 Canada

JAMES R. BOLTON
Bolton Photosciences Inc., 628 Cheriton Cres. NW, Edmonton, AB T6R 2M5,
Canada

WILEY-INTERSCIENCE
A JOHN WILEY & SONS, INC., PUBLICATION

For general information on our other products and services or for technical support, please
contact our Customer Care Department within the United States at (800) 762-2974, outside
the United States at (317) 572-3993 or fax (317) 572-4002.

Wiley also publishes its books in a variety of electronic formats. Some content that appears in
print may not be available in electronic formats. For more information about Wiley products,
visit our web site at www.wiley.com.

Wiley Bicentennial Logo: Richard J. Pacifico

Library of Congress Cataloging-in-Publication Data:

Weil, John A. (John Ashley), 1929–
Electron paramagnetic resonance : elementary theory and practical
applications. –– 2nd ed./John A. Weil, James R. Bolton.
      p.cm.
Includes bibliographical references.
ISBN 978-0471-75496-1
1. Electron paramagnetic resonance spectroscopy. I. Bolton, James R.,
1937-II. Title.

QC763.W45 2007
543'.67--dc22   2006016130

10 9 8 7 6 5 4 3 2 1

Twinkle twinkle little Spin
Are you single or are you twin?
Are you real or are you false?
How I crave your resonant pulse

—JOHN A. WEIL

CONTENTS

10   RELAXATION TIMES, LINEWIDTHS AND SPIN KINETIC PHENOMENA   301

APPENDIX A   MATHEMATICAL OPERATIONS

APPENDIX D   PHOTONS   505

APPENDIX E   INSTRUMENTATION AND TECHNICAL PERFORMANCE   512

PREFACE

This book is intended to be an introduction to and a tutorial on electron paramagnetic resonance (EPR) spectroscopy. It has been written specifically for students at the senior undergraduate or graduate level, and can be used as either a textbook in a course or a self-study guide.

It would seem fair to demand of authors of a technical book that they enumerate in their first pages some of the benefits from its study. This is especially true if a clear understanding of the 'new' material requires an extensive investment of effort, say, in learning mathematical and quantum-mechanical techniques. We accept this challenge and enumerate the following benefits, which the diligent reader may expect to accrue:

1. Understanding of the fundamentals of EPR should be achieved even by readers having no previous training in quantum mechanics. In fact, the concepts in magnetic resonance provide an excellent tutorial path for reaching a deep understanding of this theory. Readers with some quantum-mechanical background may anticipate the acquisition of further mathematical and quantum-mechanical skills, plus powerful experimental and theoretical techniques; these will permit the interpretation of a wide range of EPR spectra.
2. The reader will be stimulated to consider the application of EPR techniques to the solution of problems of interest in the areas of organic, inorganic, biological or analytical chemistry; chemical physics; mineralogy; and geophysics. There is ample scope for this, since EPR is applicable to paramagnetic species in the solid, liquid and gaseous states.
3. The reader will have made considerable progress toward an understanding of exciting new EPR developments now in progress. An example of this is

time-resolved spectroscopy, which involves acquisition of EPR data for short-lived species. Thus one may hope more and more to extend the EPR technique to initially diamagnetic species that all can be excited to a paramagnetic state.

4. In listing possible benefits to the reader, one must include the acquisition of an historical perspective. We thus enumerate below a few of the successes of the EPR technique:

   a. EPR studies established that new absorption bands observed in optical spectra following excitation by light of certain molecules (e.g., naphthalene) arise from the temporary unpairing of two electron spins. These 'triplet' states had been proposed, but their existence had not been proved. Much of the information now available on triplet states has come from EPR studies.

   b. The mechanism of photosynthesis has been under study for decades. The primary donor in the photosynthetic process has been shown by EPR to be a chlorophyll free radical, and many other key intermediates in this reaction have similarly been identified by EPR.

   c. By appending a paramagnetic fragment (a 'spin label') to a molecule of biological importance, one in effect has acquired a 'transmitter' to supply data on the interactions of biological molecules. Very many systems of biomedical interest (e.g., oxygen carriers, various enzymes) have had their structure and function elucidated by application of the EPR technique.

   d. EPR has excelled over any other technique in the identification of paramagnetic species in insulators and semiconductors, and in describing their environment. Isotopic enrichment of samples has added many details to EPR interpretations. Such data have provided stringent tests of theory.

   e. EPR has allowed chemists to probe into the details of reaction mechanisms by using the technique of 'spin trapping' to identify reactive radical intermediates.

More than three decades have elapsed since the publication of the first edition of this book. The EPR spectra that it interpreted reflected the interests of early experimenters. Numerous free radicals in solution and transition-group ions in crystalline electric fields of high symmetry were cited as examples. We strive in this edition to demonstrate the similarities and the unity of approach (e.g., spin-hamiltonian analysis) possible for all the myriad of paramagnetic systems open to study by EPR. For pedagogical clarity we have continued to cite some early results. However, space limitations have forced us to choose between broad coverage of specific topics (e.g., biological applications) and detailed coverage of limited examples. We have usually chosen the latter. For example, the reader will not find a systematic examination of the various transition ions in this edition. Fortunately, many sources are now available (and cited) where the reader can find such material. Certainly it has become trivially easy to use the various computer-based search engines available to obtain even highly technical references and data.

Examination of the extensive EPR literature (e.g., one may scan the nearly 3000 EPR review papers published up to 2004, about 60% of which are in English) shows a pattern of increasing complexity in the systems studied and in the experimental equipment used. The free radicals of old have in part been supplanted by biological or mineralogical samples or complicated transition-ion complexes. Some of the spectrometers now in use operate not only in the classical continuous-wave mode but also in a pulsed mode. Magnetic fields as high as 100 T are now used in pulsed experiments. Computers direct the spectrometer control as well as data acquisition, analyze the spectra to yield the underlying parameters and produce accurate simulated spectra.

In this evolving environment, we continue to aspire to provide the reader with a respectable level of understanding, both of EPR experiments and equipment and of their theoretical background. We have striven both for clarity and for accuracy.

We emphasize that this book does not attempt a comprehensive coverage of the subject. Where appropriate, we refer to texts, reviews and key papers for further reading. Our primary purpose is to develop a sound base to enable the reader to go on to the specialized field of interest. In this regard we make liberal use of elementary interpretive theory (e.g., Hückel theory, molecular-orbital theory and crystal-field theory) as pedagogical tools, even though these theories have severe limitations in more advanced applications, and large-scale computer-based calculations for simulation and interpretation have now become routine.

Our primary focus is on *isolated* paramagnetic species. Later we deal fleetingly with interacting systems (e.g., electron exchange and transfer, polarization).

Finally, we have generated sets of problems of varying difficulty for the reader. The effort involved in their solution will probably be generously repaid by enhanced understanding of the subject material.

The EPR community is a compatible and congenial one. The International EPR Society (see IES Website http://www.ieprs.org/; ca. 1000 members in 2006) sponsors various meetings, globally, and produces a fine quarterly newsletter.

The senior author of this third edition takes most of the responsibility for its contents. Sadly, the senior author of the 1972 edition, John Wertz, was able to participate in only the early stages of formulating the 1994 edition, and passed away soon after that. The junior author (JRB) of the first editions evolved away from EPR in his research interests many years ago; although he played an important supportive role in preparing and editing the present manuscript, he did not contribute material for the 2007 edition.

The authors would greatly appreciate receiving from the reader any suggested improvements and corrections, via the available Website. This mechanism was most helpful in the writing of the present book.

REMARKS SPECIFIC TO THIS EDITION

EPR spectroscopy is now considered to be a 'mature' scientific field and, while still evolving with much work to be done, the fundamentals are in place. Thus a previous text on this topic need not be revised dramatically.

Advances in the field have been mostly in obtaining more sophisticated analyses of the theoretical background and of developing more modern electronics and magnetry, much of it on the road that the frequency domain is taking from the micro-wave region toward the infrared. The time scales open for examination of EPR phenomena have, of course, shrunk appreciably. The funds required to do modern work have not. It will be most interesting to learn how EPR imaging evolves in the next decade.

We thank the numerous readers who took the time to communicate with us about this book. What was strongly urged by readership and publishers alike was to keep the size and feeling of the book much the same. We have tried to heed this advice. Thus the (relatively few) errors in the previous edition have been attended to, clar-ifications have been added where suggested, and various new references have been built in. Much remains as it was. Thus tutorial 'gems' from the 1960s are deemed to retain their value, not to be exchanged for 2000s replacements merely for temporal reasons.

ACKNOWLEDGMENTS

We have received most valuable support from many persons in preparing this book. Students and postdoctoral scientists who have critically read parts of the volume and who have worked various problems include Philip Bailey, Ning Chen, Deok Choi, Raymond Dickson, Jana Dyck, David Howarth, Susanne Manley, Rod McEachern, Dennis McGavin, Renata Michaelis, Michael Mombourquette, Dagang Ni, Lizhong Sun, John Williams and Zbigniew Zimpel. We thank them all. We also acknowledge helpful comments received from Drs. Michael Bowman, Naresh Dalal, Jack Freed, Betty Gaffney, Ira Goldberg, Brian Hoffman, James Hyde, Janis Kliava, Olga Kryliouk, Alexander Maryasov, John Morton, Arnold Raitsimring, Balu Ramakrishna, Chary Rangacharyulu, Stefan Stoll and Craig Tennant, and to Keith Preston for his discerning critique. Very special gratitude goes to Gareth and Sandra Eaton, who provided extraordinary help with Appendices E and F. Particular thanks also go to Nancy Leppan, Barbara Nelson, Martina Steinmetz, and especially to Sarah Figley, David Howarth and Wayne McMillan, who all contributed substantially in preparing the manuscript, diagrams and figures. We are especially grateful to Ms. Xiong Yang, who carefully scanned the proofs for flaws, and found many. Many persons contributed comments and suggestions to the relevant Website.

Generous financial support received from the Universities of Saskatchewan and Western Ontario is gratefully acknowledged. Not least, the authors have enjoyed helpful and most pleasant relations with this book's publisher, as represented by Ms. Amy Byers, Shirley Thomas, and others.

This book would certainly not have reached completion without the loving encouragement and tolerance contributed over the many years by the authors' wives and other family members.

CHAPTER 1

BASIC PRINCIPLES OF PARAMAGNETIC RESONANCE

1.1 INTRODUCTION

The science of electron paramagnetic resonance (EPR) spectroscopy is very similar in concept to the more familiar nuclear magnetic resonance (NMR) technique. Both deal with the interaction between electromagnetic radiation and magnetic moments; in the case of EPR, the magnetic moments arise from electrons rather than nuclei. Whether or not the reader has an immediate interest in the multitude of systems to which EPR is applicable, the insights that it provides cannot be ignored. Furthermore, there is hardly another technique from which one can gain a clearer insight into many of the fundamental concepts of quantum mechanics.

Much of our knowledge of the structure of molecules has been obtained from the analysis of molecular absorption spectra. Such spectra are obtained by measuring the attenuation versus frequency (or wavelength) of a beam of electromagnetic radiation as it passes through a sample of matter. Lines or bands in a spectrum represent transitions between energy levels of the absorbing species. The frequency of each line or band measures the energy separation of two levels. Given enough data and some guidance from theory, one may construct an energy-level diagram from a spectrum. Comparison of an energy-level diagram and an observed spectrum shows clearly that, of the many transitions that may occur between the various levels, only a relatively few 'allowed' transitions are observed. Hence the prediction of transition intensities requires a knowledge of selection rules.

Electromagnetic radiation may be regarded classically as coupled electric ($\mathbf{E}_1$) and magnetic ($\mathbf{B}_1$) fields perpendicular to the direction of propagation (Fig. 1.1).

Electron Paramagnetic Resonance, Second Edition, by John A. Weil and James R. Bolton
Copyright © 2007 John Wiley & Sons, Inc.

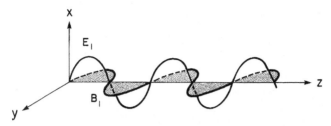

FIGURE 1.1 Instantaneous amplitudes of electric field ($\mathbf{E}_1$) and magnetic-field ($\mathbf{B}_1$) components in a propagating plane-polarized and monochromatic electromagnetic beam. We note that $\mathbf{E}_1$ is confined to plane xz, $\mathbf{B}_1$ is confined to plane yz, with wave propagation along $\mathbf{z}$.

Both oscillate at some frequency v, within the theoretical range 0 (DC) to infinity. For our purposes, in EPR, the commonly used frequency range is $10^9 - 10^{12}\ s^{-1}$ (1–1000 GHz).

We must also consider the particulate nature of electromagnetic radiation in that it can be represented as a stream of particles called *photons*. These have no mass or net electrical charge but are to be thought of as wave packets having electromagnetic fields and a type of spin angular momentum. Furthermore, photons travel in observable directions, always at the speed of light; that is, they constitute light. The electric ($\mathbf{E}_1$) and magnetic ($\mathbf{B}_1$) components of the fields associated with them (see Appendix D) are generally perpendicular to each other and to the direction of propagation and oscillate in a narrow range centered at frequency v.

The energy of any given photon is given by the quantity hv, where h is the famous Planck constant. When a photon is absorbed or emitted by an electron, atom or molecule, the energy and angular momentum of the combined (total) system must be conserved. For this reason, the direction of photon travel relative to the alignment of the photoactive chemical system is of crucial importance.

In most spectroscopic studies, other than magnetic resonance, it is the electric-field component of the radiation that interacts with molecules. For absorption to occur, two conditions must be fulfilled: (1) the energy (hv) of a quantum of radiation must correspond to the separation between certain energy levels in the molecule, and (2) the oscillating electric-field component $\mathbf{E}_1$ must be able to interact with an oscillating electric-dipole (or higher) moment. An example is gaseous HCl; molecular rotation of HCl creates the required fluctuation in the direction of the electric dipole along the bond. Likewise, infrared radiation interacts with the molecules in vibrational modes, dependent on the change in the electric-dipole moment magnitude with bond-length fluctuations. Similarly, a molecule containing a magnetic dipole might be expected to interact with the oscillating magnetic component $\mathbf{B}_1$ of electromagnetic radiation. This indeed is so and forms the basis for magnetic-resonance spectroscopy. Herein we are concerned with permanent dipole moments, that is, those that exist in the absence of external fields. However, in most magnetic-resonance experiments, a static magnetic field $\mathbf{B}$ is applied (in addition to $\mathbf{B}_1$) to align the moments and shift the energy levels to achieve conveniently measured splittings.

Each electron possesses an intrinsic magnetic-dipole moment that arises from its spin.[1] In most systems electrons occur in pairs such that the net moment is zero. Hence only species that contain one or more unpaired electrons possess the net spin moment necessary for suitable interaction with an electromagnetic field.

A magnetic-dipole moment in an atom or molecule (neutral or charged) may arise from unpaired electrons, as well as from magnetic nuclei. The magnetic-dipole moments of these particles in turn arise, respectively, from electronic or nuclear angular momenta. Hence one of the fundamental phenomena to be understood in EPR spectroscopy is the nature and quantization of angular momenta (see Section 1.6 and Appendix B).

1.2 HISTORICAL PERSPECTIVE

The technique of electron paramagnetic resonance spectroscopy may be regarded as a fascinating extension of the famed Stern-Gerlach experiment. In one of the most fundamental experiments on the structure of matter, Stern and Gerlach [3] in the 1920s showed that an electron magnetic moment in an atom can take on only discrete orientations in a magnetic field, despite the sphericity of the atom. Subsequently, Uhlenbeck and Goudsmit [4] (see also Ref. 5) linked the electron magnetic moment with the concept of electron spin angular momentum. In the hydrogen atom, one has additional angular momentum arising from the proton nucleus. Breit and Rabi [6] described the resultant energy levels of a hydrogen atom in a magnetic field. Rabi et al. [7] studied transitions between levels induced by an oscillating magnetic field. This experiment was the first observation of magnetic resonance.

The first observation of an electron paramagnetic resonance peak was made in 1945 when Zavoisky [8] detected a radiofrequency absorption line from a $CuCl_2 \cdot 2H_2O$ sample. He found a resonance at a magnetic field of 4.76 mT for a frequency of 133 MHz; in this case the electron Zeeman factor g is approximately 2 (Sections 1.7 and 1.8). Zavoisky's results were interpreted by Frenkel [9] as showing paramagnetic resonance absorption. Later experiments at higher (microwave) frequencies in magnetic fields of 100–300 mT showed the advantage of the use of high frequencies and fields.

Rapid exploitation of paramagnetic resonance after 1946 was catalyzed by the widespread availability of complete microwave systems following World War II. For example, equipment for the 9-GHz region had been extensively used for radar, and components were easily available at low cost. Almost simultaneously, EPR studies were undertaken in the United States (Cummerow and Halliday [10]) and in England (Bagguley and Griffiths [11]). Major contributions toward the interpretation of EPR spectra were made by many theorists. Important figures in this endeavor include Abragam, Bleaney, Pryce and Van Vleck. The early history of magnetic resonance has been summarized by Ramsey [12] and others.[2]

The state of the art has advanced on many fronts. In general, pulsed spin-excitation schemes and ultra-rapid-reaction techniques have now become not only feasible but almost commonplace. One remarkable accomplishment in recent years has been the observation of an EPR signal from a single electron held in space by a configuration of applied electric and magnetic fields (in a so-called Penning trap) [15].

1.3   A SIMPLE EPR SPECTROMETER

The use of a magnetic field is the unique aspect of magnetic-dipole spectroscopy. We shall illustrate the effect of the field and the components of a basic EPR spectrometer, but first we must consider the energy states of the chemical species being examined.

The simplest energy-level diagram for a particle of spin $\frac{1}{2}$ in a magnetic field is shown in Fig. 1.2. The levels are labeled with the symbols α and β, or with the numbers $M = \pm\frac{1}{2}$, to be defined. By varying the static field **B**, one may change the energy-level separation, as indicated. Resonant absorption occurs if the frequency is adjusted so that $\Delta U = h\nu$. Here ν is the center frequency of the source of incident radiant energy. The magnitude of the transition shown is the energy that must be absorbed from the oscillating **B$_1$** field to move from the lower state to the upper state. No numerical values appear on the qualitative diagram. We merely note that, for many simple unpaired-electron systems, resonance occurs at a field of about 0.3 T if ν is approximately 9 GHz. The variation of energy with magnetic field need not be linear, and more complex systems have additional pairs of energy levels.

The energies of the magnetic dipoles in a typical static magnetic field **B** are such that frequencies in the microwave region are required. A possible experimental arrangement for the detection of magnetic-dipole transitions is the microwave EPR spectrometer shown in Fig. 1.3a. An optical spectrometer is shown in Fig. 1.3b to suggest by analogy the function of components in the two spectrometers.

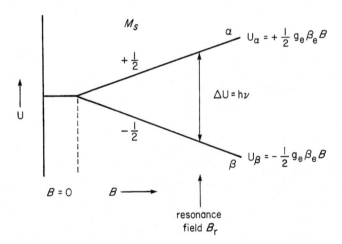

FIGURE 1.2   Energy-level scheme for the simplest system (e.g., free electron) as a function of applied magnetic field **B**, showing EPR absorption. U_α and U_β represent the energies of the $M = \pm\frac{1}{2}$ states. For electron spins, M is written as M_S. The constants g_e and β_e are defined in Section 1.7.

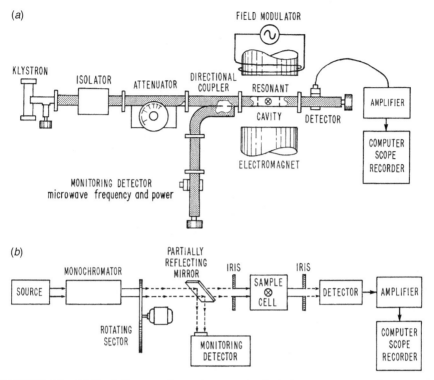

FIGURE 1.3 (*a*) Block diagram of a continuous-wave (cw) electron paramagnetic resonance (EPR) spectrometer; (*b*) block diagram of an optical spectrometer, where $\otimes$ denotes the sample. Note that there is a pair of irises in the end faces of the transmission cavity.

In either case, approximately monochromatic radiation falls on a sample in an appropriate cell, and one looks for changes in the intensity of the transmitted (or reflected) radiation by means of a suitable detector. Two primary classes of fixed-frequency spectrometers exist: either continuous or pulsed in the amplitude of $\mathbf{B}_1$. We shall now describe briefly the principal components of a simple EPR spectrometer. More details can be found in Appendix E.

Source. The frequency of radiant energy used in the majority of EPR spectrometers is approximately 9.5 GHz, in the medium-frequency microwave region. This frequency corresponds to a wavelength of about 32 mm. The microwave source is usually a klystron, which is a vacuum tube well known for its low-noise characteristics (see Appendix E). The field $\mathbf{B}_1$ is generated by oscillations within its own tunable cavity. In the range of about 1–100 GHz the mode of energy transmission is either by special coaxial cables or by waveguides. The latter are usually rectangular brass pipes, flanged to facilitate assembly of discrete components. In standard instruments, the microwave power is incident on the sample continuously (i.e., continuous wave, commonly abbreviated cw). Alternatively, in certain modern spectrometers, the power is pulsed.

In Fig. 1.3a and in Appendix E, in addition to the waveguide-connected klystron, there are other components; the most important are a resonator, a magnet and a detector. These components perform the following functions:

Resonator. This is most commonly a resonant cavity, which admits microwaves through an iris. The frequency of the source is tuned to the appropriate resonant frequency of the cavity. The corresponding resonant wavelengths are related to the dimensions of the cavity. One wishes to operate in a resonant mode that maximizes B_1 at the location of the sample. At resonance, the energy density in the resonator may be thousands of times that in the waveguide, which maximizes the opportunity to detect resonant absorption in a sample. A recently developed loop-gap resonator has been advocated as an alternative to the usual resonant cavities for energy-dissipative samples (Appendix E).

Figure 1.3a features a transmission cavity, with separate input and output irises. In practice, a reflection cavity, in which a single iris fulfills both functions, is usually used.

Magnetic Field. In magnetic resonance experiments, the static magnetic field **B** usually must be very well controlled and stable. Variations of this field are translated into corresponding variations in energy separation ΔU. The magnitude of B may be measured and controlled by a Hall-effect detector. Since every absorption line has a non-zero width, one finds it convenient to use a scanning unit to traverse the region of field B encompassing the line. Unless B is uniform over the sample volume, the observed spectral line is broadened.

Detector. Numerous types of solid-state diodes are sensitive to microwave energy. Absorption lines can be observed in the EPR spectrum when the separation of two energy levels is equal to (or very close to) the quantum energy $h\nu$ of an incident microwave photon. The absorption of such photons by the sample in Fig. 1.3a is indicated by a change in the detector current.

The direct detection of the absorption signal, as in Fig. 1.3a, is possible only for samples containing an unusually high concentration of unpaired electrons; noise components over a wide range of frequencies appear with the signal, making its detection difficult. In the optical spectrometer (Fig. 1.3b), the signal-to-noise ratio may be improved greatly by chopping the light beam at a preselected frequency. This permits narrow-band amplification of the detected signal; hence noise components are limited to those in a narrow band centered at the chopping frequency.

In a typical fixed-frequency magnetic-resonance spectrometer, the role of the light chopper is taken by a field modulator to impose an alternating component on the static magnetic field **B** (Appendix E). This results in an alternating signal at the microwave detector that can be amplified in a narrow-band amplifier. Typically, the resulting signal is rectified and takes on a B dependence that resembles the first derivative of an absorption line. The shape of the absorption line often is fitted to a functional formula (e.g., gaussian, lorentzian or elaboration thereof; Appendix F) approximating its field or frequency dependence.

An alternative to detection of magnetic resonance via energy absorption is measurement of the direct change in the angular momentum of the spin system occurring as a result of photon absorption [16]. Other means of detecting EPR lines continue to be developed but also remain unconventional. These include use of magnetic force microscopy [17], optical detection (e.g., of EPR absorption from a single molecule) [18] and use of a superconducting quantum interference device (SQUID) [19,20].

1.4 SCOPE OF THE EPR TECHNIQUE

In almost all cases encountered in EPR spectroscopy, the electron magnetic dipole arises from spin angular momentum with only a small contribution from orbital motion. Resonant absorption of electromagnetic radiation by such systems is variously called 'paramagnetic resonance', 'electron spin resonance' or 'electron paramagnetic resonance'. The term *resonance* is appropriate, since the well-defined separation of energy levels is matched to the energy of a quantum of incident monochromatic radiation. Resonant transitions between energy levels of nuclear dipoles are the subject of study in *nuclear magnetic resonance* (NMR) spectroscopy. The term *electron paramagnetic resonance* (EPR)[3] was introduced as a designation taking into account contributions from electron orbital as well as spin angular momentum. The term *electron spin resonance*[4] (ESR) has also been widely used because in most cases the absorption is linked primarily to the electron-spin angular momentum. Electron magnetic resonance (EMR) is an alternative. We note also that the term *paramagnetic resonance* was employed at the Clarendon Laboratory in Oxford, England, where much of the early inorganic EPR work was carried out. After considering the various options, we have decided to use the designation electron paramagnetic resonance since this encompasses all the phenomena observable by the technique.

In any given molecule or atom there exists literally an infinite set of electronic states that are of importance in optical spectroscopy. However, in EPR spectroscopy the energy of the photons is very low; hence one can ignore the multitude of electronic states except the ground state (plus perhaps a few very nearby states) of the species. The unique feature of EPR spectroscopy is that it is a technique applicable to systems in a paramagnetic state (or that can be placed in such a state), that is, a state having net electron angular momentum (usually spin angular momentum). The species exists either in a paramagnetic ground state or may be temporarily excited into a paramagnetic state, for instance, by irradiation. Thus, in principle, *all* atoms and molecules are amenable to study by EPR (see Section F.1). Typical systems that have been studied include

1. *Free Radicals in the Solid, Liquid or Gaseous Phases.* A free radical is herein defined as an atom, molecule or ion containing one unpaired electron. (Transition ions and 'point' defects in solids fitting this description are not normally called 'free radicals'.)

2. *Transition Ions Including Actinide Ions.* These routinely may have up to five or seven unpaired electrons (Chapter 8).

3. *Various 'Point' Defects (Localized Imperfections, with Electron Spin Distributed over Relatively Few Atoms) in Solids.* Best known in this class

is the *F* center (Fig. 4.2*a*), an electron trapped at a negative-ion vacancy in crystals and glasses. Deficiency of an electron (a 'positive hole') may also give rise to a paramagnetic entity.

4. *Systems with More than One Unpaired Electron.* Excluding ions in category 2, these include: (a) *Triplet-state systems.* Here the interaction between the two unpaired electrons is strong. Some of these systems are stable in a triplet ground state but most are unstable, requiring excitation, either thermal or usually optical, for their creation (Sections 6.3.4–6.3.6). (b) *Biradicals.* These systems contain two unpaired electrons that are sufficiently remote from one another so that interactions between them are very weak. Such a system behaves as two weakly interacting free radicals (Section 6.4). (c) *Multiradicals.* Such species (having more than two unpaired electrons) also exist.

5. *Systems with Conducting Electrons.* These (e.g., semiconductors and metals) are not treated extensively in this book.

EPR spectra may convey a remarkable wealth of significant chemical information. A brief summary of structural or kinetic information derivable from Figs. 1.4–1.6 foreshadows the diversity of the applications of the method. Each of these spectra is considered at a later point.

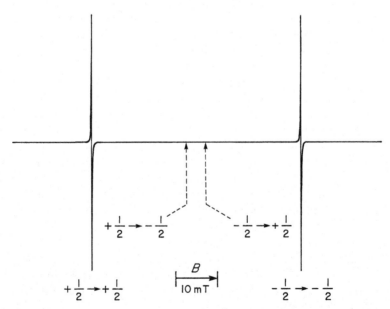

FIGURE 1.4 Simulated first-derivative EPR spectrum of a hydrogen atom (^{1}H) in the gas phase ($\mathbf{B} \perp \mathbf{B}_1$, $\nu = 10$ GHz). The quantum number M_I, denoting the EPR transitions, is defined in Chapter 2 and is consistent with $g_n > 0$. Note that two EPR transitions are allowed, occurring at $B = 329.554545$ and 380.495624 mT, and two transitions (dashed lines) are forbidden.

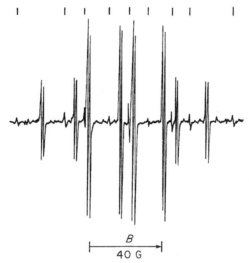

FIGURE 1.5 First-derivative EPR spectrum of the CH_3CHOH radical produced by continuous ultraviolet photolysis of a mixture of H_2O_2 and CH_3CH_2OH. The photolysis produces the OH radical, which then abstracts a hydrogen atom from the ethanol molecule. The weak lines, which are marked above the spectrum, arise from the radical CH_2CH_2OH. [After R. Livingston, H. Zeldes, *J. Chem. Phys.*, **44**, 1245 (1966).]

Figure 1.4 presents a gas-phase EPR spectrum of hydrogen atoms (^1H). This simplest atom, since it has only one electron, of necessity has electronic spin $S = \frac{1}{2}$. Here the atom can exist in any one of four spin energy levels. One can think of the system as being composed effectively of two chemical 'species' (the proton has a spin $I = \frac{1}{2}$

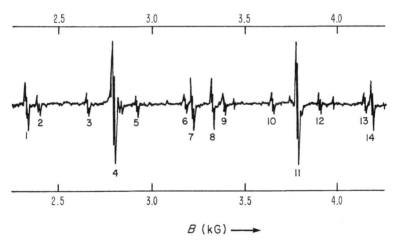

FIGURE 1.6 First-derivative EPR spectrum (9070 MHz) of XeF trapped in a single crystal of XeF_4 at 77 K. The numbered lines are examined in Problem 3.12. [After J. R. Morton, W. E. Falconer, *J. Chem. Phys.*, **39**, 427 (1963).]

and hence all atoms having the nuclear spin component $M_I = +\frac{1}{2}$ constitute one species and those with $M_I = -\frac{1}{2}$ constitute the other) giving rise to the two lines observed. As is usual, the lines are presented as first derivatives (dY/dB: see Sections E.1.6 and F.2.1) of the power absorbed by the spins. This system is treated extensively in Appendix C.

Figure 1.5 shows the liquid-phase EPR spectrum of the CH_3CHOH radical produced as a transient species via H-atom removal in the ultraviolet photolysis of a solution of H_2O_2 in ethanol. The photolysis produces the $\cdot OH$ radical, which then abstracts a hydrogen atom from the ethanol molecule. This is an excellent example of the use of EPR spectra in the identification of radical intermediates in chemical reactions.

Figure 1.6 shows an EPR spectrum of species formed by γ irradiation of a single crystal of XeF_4; again the number, spacing and intensity of the lines provide identification of one xenon atom and one fluorine atom, that is, the unstable XeF molecule. Here the positive identification of xenon comes from the observation of lines arising from several of its isotopes occurring in natural abundance.

The proper interpretation of EPR spectra requires some understanding of basic quantum mechanics, especially that associated with angular momentum. A full understanding is best obtained by reconstruction of the spectrum from the parameters of the quantum-mechanical treatment. To understand an EPR spectrum, it is desirable to have a working acquaintance with the following topics:

1. Mathematical techniques such as operator[5] methods, matrix algebra and matrix diagonalization (summarized in Appendix A). These are required for the solution of the Schrödinger equation, for the representation of angular momentum by quantum numbers, and for relating vectors (e.g., angular momentum and the magnetic moment) (Appendix B).
2. Familiarity with the operation of microwave magnetic-resonance spectrometers, including interfacing with computers (Appendixes E and F).

The elementary aspects of these topics are treated where needed in the text or in appendixes. *Even the reader who has had no previous training in quantum mechanics should be able to acquire considerable understanding of the fundamentals of electron paramagnetic resonance.* Indeed, we believe that this is a fine way to learn quantum mechanics! We shall undertake the development of the necessary background in a step-by-step fashion. Beyond this fundamental background, there are certain special areas of EPR that require particular background material:

1. Understanding of EPR requires an analysis of the energy levels of the system and of the influence of the surrounding environment on these levels. For example:
   a. The interpretation of EPR spectra of organic free radicals, π-electron free radicals, is aided by use of the elementary molecular-orbital approach due to Hückel (HMO approach; Chapter 9). In most cases more refined

theoretical treatments are necessary to obtain a completely satisfactory interpretation of the data.

b. Understanding of transition-ion spectra requires knowledge of the splitting of orbital and spin energy levels by local electric fields of various symmetries (Chapter 8).

2. The properties of some systems are independent of orientation in a magnetic field; that is, they are *isotropic*. Most systems are *anisotropic*, and thus their energy-level separations and the magnitude of the observable properties depend strongly on orientation in the applied magnetic field. The description of systems showing anisotropic behavior usually requires that each spectroscopic property be described by six independent parameters. It is convenient to order these parameters in a symmetric 3×3 array known as a *matrix*. Each such matrix can be considered in terms of the intrinsic information provided by its numerical components that define a set of spatial coordinate axes (its principal-axis system) and a set of three basic numerical parameters (its principal values). Simple examples of matrices are given in Appendix A, and numerous other examples are encountered in the text.

3. Time-dependent phenomena, such as the formation or decay of paramagnetic species, molecular motions (e.g., internal rotation or reorientation by discrete jumps), changes in the population polarization of spin states and chemical or electron exchange, can affect EPR spectra in many ways. An analysis of these effects leads to information about specific kinetic processes (both internal and external). These various phenomena are described in Chapter 10.

The last two points (2 and 3 above) are related in that most free radicals in fluid solution of low viscosity exhibit simplified EPR spectra with narrow lines. These are characterized by parameters arising from an effective averaging of the anisotropic interactions by the (sufficiently) rapid molecular tumbling. Thus such solutions effectively act as isotropic media. The key requirement is that the characteristic time, inverse of the tumbling rate, must be much less than the *time scale* appropriate for the EPR experiment (Chapter 10). Fortunately, this condition is easily met in most fluids at moderate temperatures.

The simple spectra we examine in the next two chapters are of systems that are either inherently isotropic (e.g., the hydrogen atom) or are effectively isotropic by virtue of rapid molecular tumbling.

1.5   ENERGY FLOW IN PARAMAGNETIC SYSTEMS

It is important at an early stage to note how the appearance of EPR spectral lines, or even the ability to detect them, is dependent on energy flows in the chemical sample. This is depicted in Fig. 1.7, which shows the net flow beginning at the excitation source (photons, B_1) and ending in the thermal motions of the atoms including the surroundings of the paramagnetic sample.

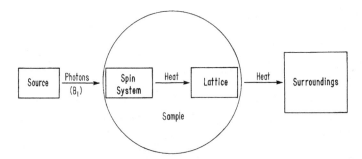

FIGURE 1.7   Energy flow in a magnetic-resonance experiment. The spin system is irradiated by a photon source (usually a microwave oscillator) at the frequency v of $\mathbf{B}_1$. The absorbed radiation is lost by energy diffusion to the lattice at an exponential rate, which allows continuing absorption of photons to occur. Energy ultimately passes from the sample to the surroundings.

The reader seeking to understand some aspects of EPR spectra is likely to have encountered closely analogous phenomena in nuclear magnetic resonance. In optical spectroscopy one may use intense sources to irradiate in absorption bands without causing a significant temperature rise of the sample. But in many NMR samples, even at low power levels, the NMR signal amplitude diminishes as the radiofrequency power level (i.e., B_1) is increased. The same is true in many EPR samples as the microwave power is increased. For these samples one speaks of power saturation or alternatively of heating the spin system.

This behavior results from a limited ability of the sample to dissipate energy from its spin system to its internal thermal motions. The surroundings of the spin are commonly referred to as the 'lattice', regardless of the sample's physical state. Samples differ widely in their ability to relax to the ground spin state after absorbing a quantum of energy.

The coupling between the spins and the lattice is measured by a characteristic spin-lattice relaxation time τ_1 (Chapter 10).[6] The same symbol is used extensively in NMR systems, for which it was first defined. Efficient relaxation implies a sufficiently small value of τ_1.

The magnitude of the observed EPR signal is proportional to the net resultant (polarization) of the spin orientations of the set of paramagnetic species. The system is said to be saturated when the rate of upward and of downward transitions is equalized; then no net energy is transferred between $\mathbf{B}_1$ and the spin system.

If the electron spin-lattice relaxation time τ_1 is very long, one may have to make observations at very low microwave power to avoid saturation. In the opposite case of very short τ_1, lifetime broadening (Chapter 10) may be so great that the line is broadened beyond detection. This is a difficulty frequently observed with transition ions (Chapter 8). It is usually dealt with by taking spectra at very low temperatures, since the value of τ_1 tends to increase dramatically with decreasing temperature.

In addition to τ_1, there are a number of other 'relaxation' times defined to describe the linewidth. These are dealt with at appropriate places (e.g., Chapters 10 and 11) in this book.

1.6 QUANTIZATION OF ANGULAR MOMENTA

In quantum mechanics the allowed values of the magnitude of any angular momentum arising from its operator $\hat{\mathbf{J}}$ (Appendix B) are given by $[J(J + 1)]$ where J is the primary angular-momentum quantum number ($J = 0, \frac{1}{2}, 1, \frac{3}{2}, \ldots$). We adopt the usual convention that all angular momenta and their components are given in units of $\hbar$. The allowed values of the component of vector $\hat{\mathbf{J}}$ along any selected direction are restricted to the quantum numbers M_J, which range in unit increments from $-J$ to $+J$, giving $2J + 1$ possible components along an arbitrary direction.

An example of the conditions described above is the spin angular momentum operator $\hat{\mathbf{S}}$ for a single electron that has a quantum number S with the value $\frac{1}{2}$. For systems of two or more unpaired electrons, S is $1, \frac{3}{2}, 2, \ldots$. The spin angular-momentum vectors and their projected components for $S = \frac{1}{2}, 1$ and $\frac{3}{2}$ are represented in Fig. 1.8. States with $S = \frac{1}{2}$ are referred to as *doublet states* since the multiplicity $2S + 1$ is equal to 2. This situation is certainly of most interest, since it includes free radicals. States with $S = 1$ are called *triplet states* (Chapter 6). For paramagnetic ions, especially those of the transition ions, states with $S > \frac{1}{2}$ are common. EPR transitions do not alter the value of S.

The nuclear-spin angular-momentum operator $\hat{\mathbf{I}}$ is quantized in an exactly analogous fashion. The nuclear-spin quantum number is I (a non-negative number, which may be integral or half-integral).[7]

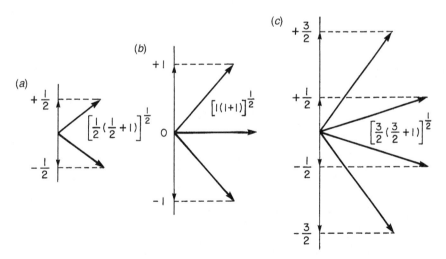

FIGURE 1.8 Allowed values (in units of $\hbar$) of the total spin angular momentum $[S(S + 1)]^{1/2}$ and of its component M_S along a fixed direction (vertical line, e.g., B) for (a) $S = \frac{1}{2}$, (b) $S = 1$, and (c) $S = \frac{3}{2}$.

Very often we must consider a whole set of spin-bearing nuclei. Parameters such as the nuclear Zeeman factor, hyperfine coupling, or quadrupole factors, that determine line positions are required for each. In addition, there are other parameters (e.g., relaxation times) to define the lineshapes and intensities. Finally, one often is interested in the quantitative analytical aspects of EPR spectroscopy [21].

For the sake of simplicity, we shall often discuss and give examples of single-nucleus systems. When dealing with more than one unpaired electron, because of their mobility and delocalization, it is often useful and correct to work with a single total electron spin operator $\hat{\mathbf{S}}$ and a single Zeeman parameter matrix $\mathbf{g}$ associated with it.

In certain cases there may exist non-zero electronic (orbital angular momentum, designated by the quantum number L, which is a non-negative integer). Usually electron-spin and orbital angular momenta initially can be considered separately, later introducing a small correction to account for the 'spin-orbit' interaction. For systems containing light atoms (such as free radicals) that have essentially zero orbital angular momentum, the spin-orbit interaction is usually very small; hence for most purposes, attention may be focused wholly on the spin angular momentum. However, spin-orbit interaction must necessarily be included in discussion of the EPR behavior of transition ions (Chapters 4 and 8). Further details about angular momentum are to be found in Appendix B.

The notation we shall use in dealing with angular momenta (i.e., $J =$ any of S, I, L, ...) is that when there are several particles of one type (electron, nuclei, etc.), we shall append a subindex indicating the individual particle being considered. When no subindex is present, then it is the *total* angular momentum that is at hand. Thus for the operators one has

$$\hat{\mathbf{J}} = \sum_{i=1}^{N} \hat{\mathbf{J}}_i \tag{1.1a}$$

and for the component values one has

$$M_J = \sum_{i=1}^{N} M_{J_i} \tag{1.1b}$$

When $N = 1$, the index is omitted. At times a pre-superscript t will be attached to $\hat{\mathbf{J}}$ and M_J to emphasize 'total'.

1.7 RELATION BETWEEN MAGNETIC MOMENTS AND ANGULAR MOMENTA

The magnetic moment and angular momentum are proportional to each other, in both classical and quantum mechanics. An analog of an orbital magnetic dipole is a classical particle of mass m and charge q, rotating with velocity $\mathbf{v}$ (speed v) in a circle of radius r, taken to lie in the xy plane. Associated with the circulating electric current i is a magnetic field equivalent to that produced by a point magnetic dipole. Such a dipole has a moment $i\mathcal{A}$ and is normal to the plane, where $\mathcal{A} = \pi r^2$ is the area of the circle. The effective electrical current i (charge flow per unit time) is $qv/2\pi r$. The magnetic moment points along the direction $\mathbf{z}$ perpendicular to the plane of the circle and is given by

$$\mu_z = i\mathcal{A} = \pm \frac{qv\pi r^2}{2\pi r} = \pm \frac{q}{2m} mvr = \frac{q}{2m} l_z \tag{1.2}$$

The sign choice depends on the direction of rotation of the particle. Here l_z is the orbital angular momentum of the particle about the axis z. The proportionality constant γ ($=q/2m$) is called the *magnetogyric ratio* (or sometimes the *gyromagnetic ratio*) and has units C kg^{-1} = s^{-1} T^{-1} (Section 1.8). Factor γ converts angular momentum to magnetic moment. More generally $\gamma = gq/2m$, where g is the Zeeman (correction) factor. Thus, quantum mechanically, each integral multiple $\hbar$ of orbital angular momentum has an associated orbital magnetic moment of magnitude $\beta = |q|\hbar/2m = |\gamma\hbar/g|$. The latter equality is valid for particles when they are free, but must be generalized further when electric fields are acting on them [22].

We now return specifically to the free electron. The component μ_z of electron-spin magnetic moment along the direction of the magnetic field $\mathbf{B}$ applied along the direction $\mathbf{z}$ is

$$\mu_z = \gamma_e \hbar M_S = -g_e \beta_e M_S \tag{1.3}$$

where g_e is the free-electron g factor. The negative sign arises because of the negative charge on the electron and the choices of β_e and g_e as positive quantities (Eqs. 1.10 and 1.11).

1.8 MAGNETIC FIELD QUANTITIES AND UNITS

In this book, we shall use the Système International (SI) [23] for units of all parameters. This comprises use of the meter, the kilogram, the second and the coulomb, that is, the rationalized mksC scheme of units. Among other benefits, this system offers a convenient and self-consistent way of checking equations. Especially with regard to the units for electromagnetic parameters, there has been much inconsistency and carelessness in the EPR literature. We shall attempt herein to encourage appropriate usage.

The two magnetic-field vectorial quantities[8] $\mathbf{B}$ and $\mathbf{H}$ are related to each other via

$$\mathbf{H} = \mathbf{B}/\mu_m \tag{1.4}$$

where the permeability

$$\mu_m = \kappa_m \mu_0 \tag{1.5}$$

is expressed in terms of the permeability μ_0 of the vacuum and κ_m is a dimensionless parameter (unity for the vacuum) describing the (isotropic) medium considered. Subindex m labels the medium. Here

$$\mu_0 = 4\pi \times 10^{-7} \, \text{J} \, \text{C}^{-2} \, \text{s}^2 \, \text{m}^{-1} (= \text{T}^2 \, \text{J}^{-1} \, \text{m}^3) \tag{1.6}$$

is, of course, a universal constant; J denotes the unit *joule* ($=\text{kg} \, \text{m}^2 \text{s}^{-2}$). The unit ampere A is just coulombs per second (C s^{-1}). We see that the magnetic flux density (alias *magnetic induction*) **B** has dimensions and units *different* from those of the magnetic 'field' **H**. Nevertheless, the term 'magnetic field' (meaning **B**) is in almost universal usage in magnetic resonance; hence we shall continue to use the term *magnetic field* for the quantity **B**.[9] Specifically, the more fundamental quantity **B** has the unit of tesla (T $= \text{kg} \, \text{s}^{-1} \, \text{C}^{-1}$), where

$$1 \text{ tesla (T)} = 1 \text{ kg } \text{s}^{-1} \, \text{C}^{-1}$$

$$= 1 \, \text{J} \, \text{C}^{-1} \, \text{m}^{-2} \, \text{s}$$

$$= 1 \times 10^4 \text{ gauss (G)} \tag{1.7}$$

We shall use for **B** either the unit of tesla or less frequently the unit of gauss. On the other hand, **H** has the derived unit of coulombs per meter per second (C m^{-1} s^{-1}, which is identical to J T^{-1} m^{-3}). One such unit is equal to $4\pi \times 10^{-3}$ oersted (Oe). The vector **H** measures the total magnetic field (externally applied from distant current-carrying conductors), plus contributions from any (almost) fixed and sufficiently close particles that may be present [24–26]. The vector **B** deals only with the former. It follows that $\mathbf{B} = \mu_0 \mathbf{H}$ when there are no neighbors (vacuum).

A very important quantity in this book is the magnetic-dipole moment μ, which has units of J T^{-1}. The classical dipole moment can be regarded as being the 'handle' by which each magnetic species can change its energy, that is, its orientation in an external field **B**, by reacting to external magnetic excitation. The macroscopic collection of N such dipoles in a given volume V has the resultant macroscopic moment

$$\mathbf{M} = \frac{1}{V} \sum_{i=1}^{N} \boldsymbol{\mu}_i \tag{1.8}$$

per unit volume, called the *magnetization* (Section 10.3), which has units of J T^{-1} m^{-3} (the same as for **H**). **M** is thus the net magnetic moment per unit volume.[10]

Since the magnetic moments of nuclei, atoms and molecules are proportional to the angular momenta of these species, it is convenient to write each such proportionality factor as a product of a dimensionless g factor and a dimensioned factor (a conglomeration of physical constants) called the *magneton*. Thus typically

$$\boldsymbol{\mu} = \alpha g \beta \mathbf{J} \tag{1.9}$$

where β has the same units as vector $\boldsymbol{\mu}$; g is the magnitude of the electron Zeeman factor for the species considered, and $\mathbf{J}$ is the general angular-momentum vector. This is taken to be dimensionless (in units of $\hbar$ $\hbar = h/2\pi$), as its units (J s) have been incorporated into β; the factor α ($= \pm 1$) is defined below. The circumflex ($\wedge$) is placed above symbols such as J and μ when it is desirable that these be interpreted as quantum-mechanical operators. Such operators can simultaneously be vectors.

For free electrons (i.e., single electrons *in vacuo*; see Eq. 1.3), $\alpha_e = -1$ and $\hat{\mathbf{J}}$ is the electron spin operator $\hat{\mathbf{S}}$, so that β becomes[11]

$$\beta_e \equiv \frac{|e|\hbar}{2\,m_e} = 9.27400949(80) \times 10^{-24} \text{ J T}^{-1} \tag{1.10}$$

which is called the *Bohr magneton*; e is the electronic charge, $2\pi\hbar = h$ is Planck's constant, and m_e is the mass of the electron (Table H.1). The Zeeman splitting constant (2006 measurement [28]) for the free-electron Zeeman factor

$$g_e = 2.0023193043617(15) \tag{1.11}$$

is one of the most accurately known of the physical constants. For those readers with masochistic tendencies, we furnish some references [29–31] to the quantum electrodynamic theory of the electron magnetic moment, which has been spectacularly successful in matching the observed value of g_e (Eq. 1.11), and continues to evolve with ever-increasing sophistication; the plain symbol g is utilized when electrons interact with other particles, in which case $g \neq g_e$.

There is an instructive area of EPR spectroscopy, or at least a close relative thereof, one that features electrons in a vacuum circulating normal to a large magnetic field: the electron beam in a synchrotron storage ring. Here, in applying the theory, the choice of coordinate system [fixed laboratory, or moving with the - electron(s)] is important, and the macroscopic orbital motion enters appreciably together with the spin dynamics, in setting up the observed g factor. The equilibrium magnetic polarization (distribution of spins among the two M_S states) is distinctive, and explains the continuous emission of spin-flip synchrotron radiation [32,33].

For nuclei, $\hat{\mathbf{J}}$ is the nuclear-spin operator $\hat{\mathbf{I}}$ and $\alpha_n = +1$. Here the *nuclear magneton* is defined (Table H.1) as

$$\beta_n \equiv \frac{|e|\hbar}{2m_p} = 5.05078343(43) \times 10^{-27} \text{ J T}^{-1} \tag{1.12}$$

where m_p is the mass of the proton (^{1}H). Values of nuclear g factors g_n are given in Table H.4.

We next consider the magnetic moment $\boldsymbol{\mu}$ in a magnetic field $\mathbf{B}$, where $\boldsymbol{\mu}$ may describe either a nuclear or an electron magnetic dipole. Its component μ_z along

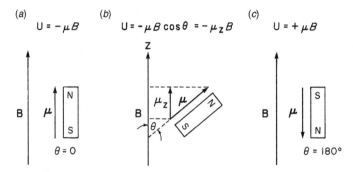

FIGURE 1.9   Energy of a classical magnetic dipole in a magnetic field as a function of the angle θ between the magnetic field and the axis of the dipole: (*a*) $\theta = 0$ (configuration of minimum energy); (*b*) arbitrary value of θ. (*c*) $\theta = 180°$ (maximum energy).

B (taken along **z**) is generally defined as

$$\mu_z = -\left.\frac{\partial U}{\partial B}\right|_{B=0} \tag{1.13}$$

Here $U(B)$ is the energy of a magnetic dipole of moment **μ** in a field **B**, and the use of the partial derivative symbols indicates that the only parameter to be varied is the field. In most situations, **μ** may be defined in terms of its scalar product with **B**

$$U = -\boldsymbol{\mu}^{\mathrm{T}} \cdot \mathbf{B} \tag{1.14a}$$

$$= -\mathbf{B}^{\mathrm{T}} \cdot \boldsymbol{\mu} \tag{1.14b}$$

$$= -|\mu B| \cos(\boldsymbol{\mu}, \mathbf{B}) \tag{1.14c}$$

where $(\boldsymbol{\mu}, \mathbf{B})$ represents the angle between μ and $\mathbf{B}$.[12] The form given in Eq. 1.14*b* proves to be advantageous in our future usage. For a given value of **B**, there is a minimum energy $-|\mu B|$, which occurs when $(\boldsymbol{\mu}, \mathbf{B})$ is equal to 0, that is, the dipole is parallel to the direction of **B** (Fig. 1.9*a*); the maximum energy $+|\mu B|$ occurs when $(\boldsymbol{\mu}, \mathbf{B}) = \pi$ (Fig. 1.9*c*); at intermediate angles, U lies between these two extremes (Fig. 1.9*b*).

1.9   BULK MAGNETIC PROPERTIES

Now consider a large ensemble of non-interacting (with each other) classical magnetic moments **μ** in a uniform magnetic field **B**. If the mean interaction energy $-\mathbf{B}^{\mathrm{T}} \cdot \boldsymbol{\mu}$ is large compared with the thermal energy $k_b T$ (e.g., in a field ~ 1 T and at 1 K), then practically all dipoles are aligned along the direction of **B** (corresponding to the case of minimum energy). Here k_b is Boltzmann's constant and T the absolute temperature. The resultant macroscopic magnetization **M** would be approximately equal to $N_V \boldsymbol{\mu}$, where N_V is the number of dipoles per unit volume.

However, $|\mu B / k_b T| \ll 1$ in almost all cases because the dipoles point in various directions. Thus the magnitude M is ordinarily several orders of magnitude smaller than $N_V \mu$, even for the relatively strong electronic magnetic dipoles.

Another equivalent approach to understanding these quantities is via the vector relation

$$\mathbf{H} = \mu_0^{-1} \mathbf{B} - \mathbf{M} \tag{1.15}$$

with the realization that $\mathbf{M} = \mathbf{M}(\mathbf{H})$ is dependent on the laboratory medium at hand, whereas $\mathbf{B}$ is based on the atom-free vacuum as the relevant medium.[13] Both $\mathbf{B}$ and $\mathbf{H}$ may be functions of the location of the observation point and/or of time. Usually, one can utilize the approximation that $\mathbf{M}$ is proportional to $\mathbf{H}$ (but not necessarily collinear with it), as seen in Eq. 1.15.

The magnetization $\mathbf{M}$ is related to the applied field $\mathbf{H}$ by a dimensionless proportionality factor χ_m, the rationalized volume magnetic susceptibility,[14] which can be evaluated by measuring the force on a macroscopic sample in an inhomogeneous static magnetic field [35,36]. The contribution to χ_m of a set of non-interacting magnetic dipoles in the simplest (isotropic) case is

$$\mathbf{M} = -\alpha[g/|g|] \chi_m \mathbf{H} \tag{1.16}$$

so that for electrons ($\alpha = -1$, $g > 0$), one has

$$\chi_m = \frac{M}{H} \tag{1.17a}$$

$$= \frac{M}{B/(\kappa_m \mu_0)} \tag{1.17b}$$

With assumption of equilibrium (i.e., Boltzmann distribution) [35, Sections 7.5 and 11.2; 37] and independent behavior of the members of the electron spin ensemble, this becomes

$$\chi_m = \frac{N_V \mu^2}{3 k_b T} \mu_0 \tag{1.17c}$$

$$\equiv \frac{C}{T} \geq 0 \tag{1.17d}$$

where $\mu^2 = g^2 \beta_e^2 S(S+1)$ and N_V is the number of magnetic species per unit volume. Here C is called the Curie law 'constant'. The quantity $\kappa_m = 1 + \chi_m$ is called the *relative permeability* (compared to free space). Typically, for Eq. 1.17c, $\chi_m \approx 10^{-6}$.

The literature abounds in the use of the next-best approximation, the Curie–Weiss law

$$\chi = \frac{C}{T - T_c} \tag{1.17e}$$

where T_c is a semi-empirical parameter giving a measure of spin-spin interaction (e.g., exchange) present.

In addition, there is a smaller additive but negative and (almost) temperature-independent contribution to χ_m, arising from the reaction in the motion and distribution of all electrons (and to a lesser extent of all nuclei) in the bulk sample to the applied field **B**. Note that, by definition, paramagnetic samples have $\chi_m > 0$ whereas diamagnetic samples have $\chi_m < 0$.

An example of the simplest paramagnetic case is a dilute ensemble of free radicals, each with one unpaired electron and having zero orbital angular momentum. The experimental determination of χ_m yields only the product $N_V \mu^2$; to obtain μ one must determine N_V from other data. EPR measurements allow N_V and μ to be determined independently.

Subindex m will generally be suppressed throughout the rest of this book.

1.10 MAGNETIC ENERGIES AND STATES

Since the individual-particle magnetic energy U is proportional to the magnetic moment (Eqs. 1.14), the quantization of spin angular momentum in a specified direction leads to the quantization of the energy levels of a magnetic dipole in a magnetic field. If the direction **z** is chosen to be along **B**, application of the expression $U = -\mu_z B$ to a 'spin-only' system and substitution of $-g_e \beta_e M_S$ for μ_z give a set of energies

$$U = g_e \beta_e B M_S \tag{1.18}$$

For a single unpaired electron, the possible values of M_S are $+\frac{1}{2}$ and $-\frac{1}{2}$. Hence the two possible values of μ_z are $\mp g_e \beta_e$ and the values of U are $\pm \frac{1}{2} g_e \beta_e B$ (Fig. 1.2). These are sometimes referred to as the *electronic Zeeman energies*.

Adjacent energy levels are separated by

$$\Delta U = U_{\text{upper}} - U_{\text{lower}} \tag{1.19a}$$
$$= g_e \beta_e B \tag{1.19b}$$
$$= -\gamma_e \hbar h$$

corresponding to $|\Delta M_S| = 1$. Note that, in this simplest case, Δ_U increases linearly with the magnetic field as shown in Fig. 1.2 (where now $M = M_S$).

The states of magnetic systems, as indicated earlier, are generally finite in number. If all the states in a set have the same energy, they are said to be *degenerate*. Each state is labeled with whatever set of quantum numbers is suitable. Thus, for an unpaired electron system, the quantum number M_S is required (Eq. 1.18). As we shall see in Chapter 2 (also Section A.5.4), the Dirac notation $|M_S\rangle$ (or $\langle M_S|$) is often used. For a single electron, since $M_S = +\frac{1}{2}$ or $-\frac{1}{2}$, the notation $|\alpha(e)\rangle$ and $|\beta(e)\rangle$ is found to be convenient. When there are several spin-bearing particles in a magnetic species, then quantum numbers for each particle may be needed to specify the spin state of the system. For every transition, both the initial state and the final state must be specified by sets of quantum numbers.

In atomic and molecular systems, no more than two electrons can occupy a given spatial orbital. This is expressed by the Pauli exclusion principle, which arises from the fact that electrons act quantum-statistically as fermions. When two electrons occupy any given orbital, their spin components (M_S) always have opposite sign, and their magnetic moments cancel each other. Thus filled orbitals are ineffective with respect to spin magnetism. An EPR signal will be observed only when at least one orbital in a chemical species contains a single electron, that is, is a semi-occupied atomic or molecular orbital (SOMO).

1.11 INTERACTION OF MAGNETIC DIPOLES WITH ELECTROMAGNETIC RADIATION

Transitions between the two electronic Zeeman levels may be induced by an electromagnetic field $\mathbf{B}_1$ of the appropriate frequency ν such that the photon energy $h\nu$ matches[15] the energy-level separation ΔU. Then from Eqs. 1.19 one has

$$\Delta U = h\nu = g_e \beta_e B \qquad (1.20)$$

where B designates the magnetic field that satisfies the resonance condition (Eq. 1.20). A more formal derivation of Eq. 1.20, valid for $S = \frac{1}{2}$, is deferred until Chapter 2. Even for systems with $S > \frac{1}{2}$, the conservation of angular momentum imposes a selection rule of $|\Delta M_S| = 1$ to such transitions because the photon has one unit ($\hbar h$) of angular momentum. Thus there is a second requirement, other than Eq. 1.20, that must be met for a transition to take place.

Let us briefly think in terms of absorption and emission of individual photons by our unpaired-electron system. The photon has its spin component ($\pm \hbar h$) along or opposed to its direction of motion [38]. This corresponds to right and left circular polarization see App. D. The photon has no magnetic moment. For absorption, depending on its direction of approach relative to the axis of the electron spin, it can deliver either energy $h\nu$ and angular momentum (photon type σ) or merely energy (photon type π). To meet the energy requirement of Eq. 1.20, several photons can cooperate, but only one of type σ can be involved, in order to match the condition of total (photon + electron) angular-momentum conservation. The situation is shown in Fig. 1.10. Such two-frequency EPR experiments are not routine but have been carried out, for instance, using the stable organic free radical DPPH (see Section F.2.2 and Ref. 39). In the vast majority of EPR experiments, only a single photon (of type σ) is involved in each transition excited. We shall now restrict ourselves to considerations of such transitions. However, in more recent EPR work, multi-quantum phenomena have become ever more evident and important. These effects (e.g., development of new EPR lines) appear as the excitation field intensity B_1 is increased.

It is of organizational value to distinguish between experimental techniques that provide EPR signal intensities that are linear in B_1, the usual case, and those that are

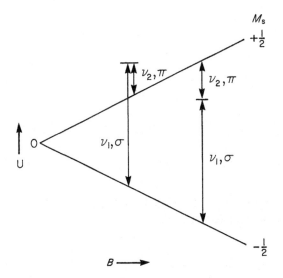

FIGURE 1.10 Energy levels for a $S = \frac{1}{2}$ system, as a function of applied magnetic field B, showing the (unusual) transitions induced when two excitation fields with two distinct frequencies are present. Photon types σ and π are discussed in the text.

not. The latter types include resonance line saturation, harmonic generation, multi-quantum transitions, spin decoupling, intermodulation and longitudinal detection (some of which will be discussed later in this book).

The transitions between the Zeeman levels require a change in the orientation of the electron magnetic moment. Hence transitions can occur only if the electromagnetic radiation can cause such a reorientation. To make transitions possible, the electromagnetic radiation must be polarized such that the oscillating magnetic field has a component *perpendicular* to the static magnetic field (justification for this statement is given in Section C.1.4). The requirement of a suitable oscillating perpendicular magnetic field (i.e., σ photons) is easily met at microwave frequencies. If we apply the electromagnetic radiation polarized such that its oscillating magnetic field $\mathbf{B_1}$ is oriented parallel to the static magnetic field $\mathbf{B}$, then the effect of the radiation would merely cause an oscillation at frequency ν in the *energies* of the Zeeman levels. Generally no *reorientation* of the electron magnetic moment would occur. In this case no transitions are possible, unless certain other conditions (to be discussed; e.g., in Appendix C) are met.

From Eq. 1.20 one may infer that there are two approaches to the detection of resonant energy absorption (or emission) by a paramagnetic sample. In the first case, the separation of the Zeeman levels is fixed by holding the magnetic field constant; the microwave frequency is then varied until a resonant absorption is found. In the second case the microwave frequency is fixed; the magnetic field is then varied. Until recently, the second method has been the one of choice, because experimentally it was relatively easy to vary the field B (i.e., the current in an electromagnet) but difficult and expensive to obtain microwave sources with wide frequency variability. The latter situation still holds; however, with the advent of pulsed microwave sources, it is now routine to work at fixed $\mathbf{B}$ and to utilize Fourier-transformation techniques to attain EPR spectra over modest frequency ranges. This subject is discussed in Section 11.4.

Everything that has been said about the electron-spin energy levels and transitions is also applicable to nuclear-spin systems. The nuclear Zeeman levels are given by an expression analogous to Eq. 1.18, namely, $U = -g_n \beta_n B M_I$; g_n is the nuclear g factor,[16] β_n is the nuclear magneton, and M_I is the component of the nuclear-spin angular-momentum vector in the z direction. In analogy to the electron-spin case, only dipolar transitions for which $|\Delta M_I| = 1$ (and I is unchanged) are allowed; hence

$$\Delta U = h\nu = |g_n|\beta_n B \qquad (1.21)$$

The corresponding spectroscopic phenomenon for nuclei is commonly referred to as *nuclear magnetic resonance* (NMR).

Nuclear spins and magnetic moments are very important in EPR studies; the interactions of the unpaired electron(s) with magnetic nuclei give rise to the rich hyperfine structure that characterizes many EPR spectra.

1.12 CHARACTERISTICS OF THE SPIN SYSTEMS

1.12.1 The *g* Factor

It should be noted that the actual field at each spin species is not necessarily only the magnetic field $\mathbf{B}_{ext}$ applied externally to the sample. In addition to this, there may exist local fields $\mathbf{B}_{local}$ that add vectorially to the external field to produce the total field $\mathbf{B}_{eff}$ effective at the electron being considered. Thus [35, Eq. 3.104]

$$\mathbf{B}_{eff} = \mathbf{B} + \mathbf{B}_{local} \qquad (1.22)$$

where the subindex ext has been suppressed (as is the usual practice). There are two types of local field: (1) those that are induced by $\mathbf{B}$, and hence have a magnitude dependent on $\mathbf{B}$; and (2) those that are permanent and independent of $\mathbf{B}$ except in their orientation.

For the moment consider only the first type, that is, the induced contribution to $\mathbf{B}_{local}$. Field B in Eq. 1.20 in principle should be replaced by B_{eff}; in practice it is much more convenient to retain the *external* magnetic field B. Then g_e must be replaced by a *variable g factor* (Section 4.8) that can and does deviate from g_e (according to the strength of $\mathbf{B}_{local}$). Thus we can write Eq. 1.22 as

$$\mathbf{B}_{eff} = (1 - \sigma)\mathbf{B} \qquad (1.23a)$$

$$= (g/g_e)\mathbf{B} \qquad (1.23b)$$

where σ is the EPR analog of the 'chemical shift' parameter σ_n used in NMR spectroscopy ($0 \leq \sigma^2 \ll 1$) and where g is the effective Zeeman factor used by EPR spectroscopists.[17] For now, we shall speak as if each magnetic species has a single unique g factor; however, we shall soon see that in fact each material exhibits a range of g factors. Many free radicals and some transition ions do have $g \approx g_e$, but there are many systems (e.g., many transition ions) that show marked deviations from this value (in some rare cases, g can be negative).

We note that incorporation of the generalized g factor into the magnetic moment (Eqs. 1.16–1.20) allows for a variable g to take account of field-induced local magnetic fields. For example, these local fields often arise from the orbital motion of the unpaired electron.

If it were not for the variation in g and the additional line structure contributed to $\mathbf{B}_{local}$ by various neighbor dipoles, EPR spectra would be very dull and uninteresting, consisting of a single line with $g = g_e$. In practice, these factors cause a multiplicity of fascinating and useful phenomena observable in EPR spectroscopy.

In most paramagnetic systems, there are so-called 'zero-field' terms in the energy that cause the resonance energy to be

$$h\nu = g\beta_e B + \text{terms} \tag{1.24}$$

At times, it is convenient to use an effective g parameter $g_{eff}(B)$ defined as

$$g_{eff} \equiv h\nu/(\beta_e B) \tag{1.25a}$$
$$= g + \text{terms}/(g\beta_e) \tag{1.25b}$$

We note that this type of g value (often found in the literature) is dependent on the magnetic field used (i.e., on the microwave frequency) and thus is far from being a constant (e.g., see Chapter 6).

There are many examples of systems for which the g factor is sufficiently distinctive to provide a reasonable identification of the paramagnetic species. Consider the spectrum of x-irradiated MgO (a cubic crystalline material) shown in Fig. 1.11 for a resonant frequency $\nu = 9.41756$ GHz. We seek to establish the origin of the very intense line to the right of the center of the spectrum. The weaker lines arise from Co^{2+} with effective spin $S' = \frac{1}{2}$, for which $g = 4.2785$ in this (isotropic) medium; the octet multiplicity of lines in this spectrum is due to magnetic (hyperfine) interaction with the ^{59}Co nucleus (100% natural abundance), which has spin $I = \frac{7}{2}$. This causes a type-2 contribution to $\mathbf{B}_{local}$. Substitution of the value 162.906 mT for the magnetic field B at the center of the intense line gives (using Table H.1) its electronic g factor as

$$g = \frac{h\nu}{\beta_e B} = \frac{(6.626069 \times 10^{-34} \text{ J s})(9.41756 \times 10^9 \text{ s}^{-1})}{(9.27401 \times 10^{-24} \text{ J T}^{-1})(0.162906 \text{ T})} = 4.1304 \tag{1.26}$$

where type-2 interactions are assumed (correctly) to be absent. A g factor of this size is unusual, and it gives an important clue as to the ion responsible for the intense

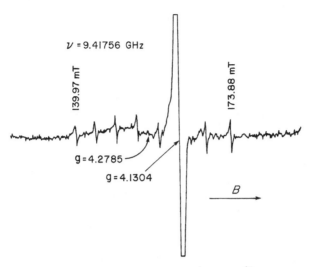

FIGURE 1.11   First-derivative EPR spectrum of Fe^+ and Co^{2+} in MgO at 4.2 K, with microwave frequency 9.41756 GHz. The Fe^+ spectrum consists of a single intense line at $g = 4.1304$ ($B = 162.906$ mT), while the Co^{2+} spectrum is an octet at $g = 4.2785$ arising from hyperfine splitting from the ^{59}Co nucleus, which has $I = \frac{7}{2}$. (Adapted from a spectrum supplied by Mr. F. Dravnieks.)

line. It is generally observed that isoelectronic ions (i.e., ions that have the same electronic configuration) in environments of similar symmetry have similar g factors. An ion that is isoelectronic with the $3d^7$ Co^{2+} ion is Fe^+. Considering the large deviation of both the g factors 4.2785 and 4.1304 from the free-electron g factor $g_e = 2.0023$, the two g factors may be considered similar enough to arise from isoelectronic ions. Hence the intense line is assigned to Fe^+. The disappearance of the Fe^+ line (but not the Co^{2+} spectrum) on heating the crystal to 400 K is consistent with expectation for an unstable oxidation state. It should be mentioned that EPR lines for both the Fe^+ and the Fe^{3+} ions may be observed in these crystals. It is typical of isoelectronic ions in an environment of similar symmetry that their EPR spectra are observable under comparable conditions. Neither Co^{2+} nor Fe^+ exhibit a resonance line at 77 K, yet one does find strong absorption for both at 20 K and lower. This similarity is confirmatory evidence for the identification of Fe^+. Inability to see lines at room temperatures or even at 77 K is shown in Chapter 10 to be due to excessive broadening of lines as a result of their very short relaxation times (τ_1). One of the joys of EPR spectroscopy is that advanced quantum theory can predict (usually after the fact) what is observed. This is so for the g values of the $3d^7$ ions just described [40].

Media yielding EPR spectra that are truly isotropic[18] are relatively rare. They do include all cubic crystalline materials not distorted by impurities or external forces. As stated above, dilute liquid solutions of low viscosity effectively act as magnetically isotropic systems. Their isotropic behavior is the result of rapid, random reorientation of the solute molecules. When such solutions are cooled sufficiently or even frozen, the EPR spectrum may consist of only a broad band. Such rigid solutions are 'isotropic', in that changing the sample orientation relative to the magnetic field **B** does not alter the EPR spectrum. However, the individual species responsible for it may well have anisotropic magnetic properties. It is with single-crystal systems that EPR reveals anisotropy, that is, dependence of the line positions and splittings on the crystal orientation relative to the magnetic field **B**.

It is *not* necessary to have an indefinitely large number of parameters to describe an anisotropic property quantitatively, in all directions. As stated earlier (Section 1.4), six parameters suffice. For our purposes, any physical system is deemed to have three mutually perpendicular inherent directions (principal axes) such that these, together with the results (principal values) measured along these directions, completely describe the anisotropic property. This is true for EPR line positions and splittings. Analogous statements may be made about other magnetic and optical properties (e.g., magnetic susceptibility, optical absorption behavior, or refractive index) of an anisotropic crystal.[19] The basic reason for this proliferation of parameters is that, for a general crystal orientation, the response is in a direction *different* from that of the applied stimulus.

Specifically, the simple resonance expression $B = h\nu/g\beta_e$ (Eqs. 1.20 and 1.24) with a single numerical value of parameter g is applicable only to systems that behave isotropically (and require no other types of energy terms). With anisotropic systems, variability of g with orientation relative to **B** is required. Thus (Eq. 1.23) the magnetic field $\mathbf{B}_{\mathrm{eff}}$ effective on $\hat{\mathbf{S}}$ generally differs in direction from that of **B**. Furthermore, the resonant field value is a function of the field orientation relative to the crystal (or molecular) axes. For some purposes, it is convenient to append subscripts on g to specify the field orientation defining it. If the principal axes of the paramagnetic entity are labeled X, Y and Z,[20] g_X is to be interpreted for our simple case as $h\nu/\beta_e B_X$, that is, the g factor for **B** along the X axis of the magnetic entity. A detailed treatment of anisotropy in EPR spectra is developed in Chapters 4–6.

A truly isotropic system is one for which

$$g_X = g_Y = g_Z \tag{1.27}$$

On the other hand, for paramagnetic species in a liquid system of low viscosity, the measured (apparently isotropic) g factor is to be regarded as an effective value averaged over all orientations.

It is important to distinguish between a *space-averaged* and a *time-averaged* quantity. In the case of the paramagnetic species in solution, *each* entity exhibits a time-averaged response, and hence the resultant spectral line is narrow. However, if the averaging is spatial, as would be the case if a crystal were ground into a powder, each center exhibits its own resonance position, depending on its orientation, and the resultant spectrum is broad since the resonance is an envelope representing a weighted distribution of all possible resonance fields.

We now comment on the functional dependences of parameters, such as g factors and hyperfine splitting factors a, which describe the paramagnetic species and that are needed to characterize EPR spectra. These parameters (often called 'constants') are functions of many factors (temperature, pressure, solvent or crystal surroundings; impurity content of the host; nature of the molecular or lattice vibrations in solid media, presence of any externally applied electric fields; etc.). These do not show any dependence on B for the usual magnetic fields applied. In principle, all these variables should be specified when values of parameters are reported. Further parameters, namely, those that describe the inherent lineshapes (lineshape function, linewidth and other 'moments') and the intensity of (area below) each line (which is proportional to the concentration of the paramagnetic species in the sample) must also be given. These, of course, depend in part on the instrumental settings.

The primary variables in EPR spectroscopy are either the magnetic field B or the frequency ν of the continuously applied exciting radiation.[21] When B is scanned at various fixed values of ν, the Zeeman terms (g factors) yield line positions proportional to B, whereas splittings of hyperfine multiplets tend to be independent of B.

Clearly, to obtain an EPR spectrum having appreciable intensity requires the presence of a large number of unpaired-electron species in the sample (Sections 4.6 and F.2.2). On the other hand, if the spin concentration in the sample is too high, the spins interact appreciably with each other, and this alters the nature of the EPR spectrum observed. The realm between these limits, which we term the 'magnetically dilute sample', is the one dealt with throughout most of this book. In other words, we consider each paramagnetic species to act independently of all others (but see Chapters 6 and 9).

1.12.2 Characteristics of Dipolar Interactions

As discussed earlier, if the interaction of unpaired electrons with externally applied homogeneous magnetic fields were the only effect operative, then all EPR spectra would consist of one line. The primary information to be garnered from these spectra would be the line positions, that is, the g factors. The EPR technique would thus provide rather limited information. Fortunately, other interactions can produce spectra rich in line components, offering a wealth of detailed information about the species studied.

Specifically, the magnetic-resonance spectrum of a dipole is very sensitive to the orientation of all other nearby magnetic dipoles (electronic or nuclear). These dipoles generate local magnetic fields that add vectorially and contribute to the local field $\mathbf{B}_{\text{local}}$ in Eq. 1.22. This local field is of the second type, that is, either independent of the applied field (but not its direction) or only weakly dependent on it. An important characteristic of these neighboring dipoles is that the magnitude and direction of the local-field contribution depend on the spin state of the center containing the dipole. Consequently, the EPR spectrum is split into a number of lines, each corresponding to a specific set of spin states.

In EPR the unpaired electron may interact with neighboring nuclear-dipole moments with a resulting splitting of the resonance. This interaction, and the resulting splitting, is called nuclear *hyperfine interaction* and *hyperfine splitting*. The term 'hyperfine splitting' was first used in atomic spectroscopy to designate the splitting of certain lines as a result of such an interaction with magnetic nuclei. The hyperfine interaction may be either isotropic (orientation-independent) or anisotropic (dependent on the orientation of **B** with respect to a molecular axis). As we shall see, an anisotropic hyperfine interaction can be accompanied by a significant isotropic component, and both are measurable. Hyperfine interactions with one or more magnetic nuclei are dealt with in Chapters 2, 3 and 5.

If there are two or more unpaired electrons in sufficiently close proximity, similar splittings (often called *fine structure*) may occur. This case is discussed in Chapter 6. A high concentration of species with one or more unpaired electrons results in intermolecular interactions of the dipoles that usually leads to line broadening.

Since the electron magnetic moment is much larger than that of nuclei, electron-electron dipolar interactions (when present) are usually very strong and dominate the spectral features. This leads to complications in the EPR spectra, discussion of which we defer until later (Chapter 6). For this reason, and also because the preponderance of EPR work has been carried out on species containing only one unpaired electron, we shall first treat those species in which the dominant feature is hyperfine interaction.

It should be noted that it is possible to observe EPR transitions at zero magnetic field because often energy-level splittings caused by local magnetic fields of type 2 are present. All the fine-structure and hyperfine parameters, but not the g factors, can be measured by zero-field EPR [41], but of course one has no control over the level splittings. Thus the frequency of the excitation field $\mathbf{B}_1$ must be scanned to find the transitions, and this can be technically problematic.[22]

At this point, we can discern and summarize the major use of EPR spectroscopy. By measuring the spectral parameters of any paramagnetic species encountered, we can expect (in due time) not only to identify it, but also to deduce details of its structure, to characterize its location, orientation and surroundings, as well as to measure its concentration. It is a primary goal of this book to train the reader in the art and science of this capability.

1.13 PARALLEL-FIELD EPR

We have seen that, usually, the condition $\mathbf{B}_1 \perp \mathbf{B}$ must be met to excite EPR transitions. However, the situation $\mathbf{B}_1 \parallel \mathbf{B}$ in certain circumstances also leads to appreciable EPR lines.[23] A prime example, certain transitions of the hydrogen atom, is discussed in Appendix C. Here the transition (labeled F and M_F) is $|1, 0\rangle \leftrightarrow |0, 0\rangle$ and can be thought of as involving simultaneous and opposite flips of the electronic and nuclear spins, so that no angular-momentum transfer with the radiation field ($\mathbf{B}_1$) occurs. Another example, involving triplet-state molecules and 'half-field' transitions of type $|\Delta M_S| = 2$ (quantum number appropriate in the high-field limit) is to be found in Section 6.3.2; no nuclear spins are involved here. Both examples involve single-photon transitions.

Since the mid-1990s, parallel-field EPR has become better known and exploited. Thus, high-spin electron species with large zero-field parameters (D and E: see Chapter 6) can exhibit ordinary EPR transitions only at quite high fields, whereas parallel-field transitions are readily accessible and their analysis yields all the needed spin-hamiltonian parameters. This is so for transition-ion clusters (Cr_{12}, $S = 6$ [43]), exchange-coupled high-spin Fe^{3+} and nearby Cu^{2+} in beef heart cytochrome c oxidase [44], high-spin biological FeS clusters ($S \leq \frac{9}{2}$ [45]), and Mn^{3+} in oxidized manganese superoxide dismutase ($S = 2$ [46]). Obviously, parallel-mode EPR is having substantial impact in the study of biomedical systems.

1.14 TIME-RESOLVED EPR

Time-resolved EPR refers to the research area dedicated to the detection of spectra from magnetic species as soon as possible after their creation, say, by flash photolysis or pulse radiolysis (e.g., electron beams) [47–49]. Their immediate subsequent behavior also has been a prime topic of interest. The time scale achieved has been down to 10^{-7} s. Here one cannot scan or modulate the B field, but must sample a sufficiently large set of such fixed fields. ESE techniques also have been widely used. Clearly, highly efficient computer-based digital data storage and processing is a crucial aspect of such endeavors, and special instrumentation is required.

1.15 COMPUTEROLOGY

The electronic computer has, of course, had a huge impact on magnetic resonance spectroscopy, as it has everywhere else.

Solid-state devices and printed circuits form the backbone of all modern spectrometers. In EPR, they control and stabilize and scan and measure all magnetic fields and the excitation electromagnetic sources—continuous wave and pulsed. They control and set the sample temperatures, and the sequence of experiments can be computer-controlled; for example, automatic variation of single-crystal orientation can be done in the absence of the scientist. All spectra are stored digitally, and are displayed and adjusted at the operator's will.

Furthermore, virtually all the mathematics relevant to analysis of magnetic resonance is programmed and enabled on computers, allowing best-fit attainment of the parameters by comparison of the actual and simulated spectra (see Appendix F). Review articles covering spectral simulation are at hand [50,51].

One very important feature of EPR spectroscopy is that all parameters obtained experimentally can be made available and published, allowing generation of the spectra (line positions and relative intensities) at will. The actual modeling of these spin-hamiltonian data, using the increasingly advanced techniques of molecular quantum mechanics, can be done separately and later.

One negative aspect of all this computerology is that the EPR user is tempted to use the programs and to bypass the understanding in depth of the mathematical background. Hopefully the present text will help to assuage this situation.

1.16 EPR IMAGING

We cannot devote much space herein to the topic of EPR imaging, which is a developing sister to EPR spectroscopy. Certainly in the case of NMR, there has been a revolution where MRI has become a dominant applied aspect of that technique. With EPR imaging, there has been slow steady development. Some relevant references can be found herein in Section 13.6. It is as yet nebulous as to what the future holds for the importance of this technique, but the EPR community is hopeful.

REFERENCES

1. H. C. Ohanian, *Am. J. Phys.*, **54**, 500 (1986).
2. A. O. Barut, N. Zanghi, *Phys. Rev. Lett.*, **52**, 2009 (1984).
3. W. Gerlach, O. Stern, *Z. Phys.*, **8**, 110 (1921); **9**, 349, 353 (1922).
4. G. E. Uhlenbeck, S. Goudsmit, *Naturwissenschaften*, **13**, 953 (1925).
5. A. Pais, *Phys. Today*, **42**(12), 34 (1989).
6. G. Breit, I. I. Rabi, *Phys. Rev.*, **38**, 2082 (1931).
7. I. I. Rabi, J. R. Zacharias, S. Millman, P. Kusch, *Phys. Rev.*, **53**, 318 (1938).
8. E. Zavoisky, *J. Phys. U.S.S.R.*, **9**, 211, 245 (1945); *ibid.*, **10**, 170 (1946).
9. J. Frenkel, *J. Phys. U.S.S.R.*, **9**, 299 (1945).
10. R. L. Cummerow, D. Halliday, *Phys. Rev.*, **70**, 433 (1946).
11. D. M. S. Bagguley, J. H. E. Griffiths, *Nature (London, U.K.)*, **160**, 532 (1947).
12. N. F. Ramsey, *Bull. Magn. Reson.*, **7**, 94 (1985).
13. A. Abragam, *Time Reversal*, Clarendon Press, Oxford, U.K., 1989.
14. B. I. Kochelaev, Y. V. Yablokov, *The Beginning of Paramagnetic Resonance*, World Scientific, Singapore, 1995.
15. H. Dehmelt, *Am. J. Phys.*, **58**, 17 (1990).
16. G. Alzetta, E. Arimondo, C. Ascoli, A. Gozzini, *Nuovo Cimento*, **52B**, 392 (1967).
17. R. Rugar, C. S. Yannoni, J. A. Sidles, *Nature (London)*, **360**, 563 (1992).
18. J. Wachtrup, C. von Borczyskowski, J. Bernard, M. Orrit, R. Brown, *Nature (London)*, **363**, 244 (1993).
19. B. Cage, S. Russek, *Rev. Sci. Instrum.*, **75**(11), 4401 (2004).
20. B. Cage, S. E. Russek, D. Zipse, N. S. Dalal, *J. Appl. Phys.*, **97**, 10M507 (2005).
21. For example, see I. B. Goldberg, A. J. Bard, in *Treatise on Analytical Chemistry*, P. J. Elving, Ed., Vol. 10, Wiley, New York, NY, U.S.A., 1983, Chapter 3.
22. M. Blume, S. Geschwind, Y. Yafet, *Phys. Rev.*, **181**, 478 (1969).
23. I. Mills, T. Cvitas, K. Homann, N. Kallay, K. Kuchitsu, *Quantities, Units and Symbols in Physical Chemistry*, 2nd ed., International Union of Pure and Applied Chemistry, Blackwell Scientific, Oxford, U.K., 1993.
24. J. D. Jackson, *Classical Electrodynamics*, 3rd ed., Wiley, New York, NY, U.S.A., 1999, pp. 174ff, 775ff.
25. J. H. Van Vleck, *The Theory of Electric and Magnetic Susceptibilities*, Oxford University Press, London, U.K., 1932, p. 3.

26. D. J. Griffiths, *Introduction to Electrodynamics*, 2nd ed., Prentice-Hall, Englewood Cliffs, NJ, U.S.A., 1989.

27. G. E. Pake, T. L. Estle, *The Physical Principles of Electron Paramagnetic Resonance*, 2nd ed., Benjamin, Reading, MA, U.S.A., 1973, pp. 7–9.

28. B. Odom, D. Hanneke, B. D'Urso, G. Gabrielse, *Phys. Rev. Lett.*, **97**, 030801 (2006).

29. J. Schwinger, *Particles, Sources and Fields*, Vol. III, Perseus Books, Reading, MA, U.S.A., 1998, pp. 189–251.

30. A. Das, S. Perez, *Phys. Lett. B*, **581**, 182 (2004).

31. T. Kinoshita, M. Nio, *Phys. Rev. D*, **73**, 013003 (2006).

32. J. D. Jackson, *Rev. Mod. Phys.*, **48**(3), 417 (1976).

33. P. Chen, Ed., *Quantum Aspects of Beam Physics*, Advanced ICFA Workshop, World Scientific, Singapore, 1998, pp. 19–33 (K.-J. Kim), 620–621 (D. P. Barber), 622–625 (J. D. Jackson), and other contributions.

34. M. B. Ferraro, T. E. Herr, P. Lazzeretti, M. Malagoli, R. Zanasi, *Phys. Rev. A*, **45**(2), 6272 (1992).

35. M. Gerloch, *Magnetism and Ligand-Field Analysis*, Cambridge University Press, Cambridge, U.K., 1983.

36. R. S. Drago, *Physical Methods in Chemistry*, 2nd ed., Saunders, Philadelphia, PA, U.S.A., 1992, Chapter 11.

37. C. A. Hutchison Jr., "Magnetic Susceptibilities", in *Determination of Organic Structures by Physical Methods*, E. A. Braude and F. C. Nachod, Eds., Academic Press, New York, NY, U.S.A., 1955, Chapter 7.

38. A. P. French, E. F. Taylor, *An Introduction to Quantum Physics*, Norton, New York, NY, U.S.A., 1978, Section 14.8.

39. J. Burget, M. Odehnal, V. Petříček, J. Šácha, L. Trlifaj, *Czech. J. Phys.*, **B11**, 719 (1961).

40. S.-Y. Wu, H.-N. Dong, *Z. Naturforsch.*, **60a**, 366 (2005).

41. R. Bramley, S. J. Strach, *Chem. Rev.*, **83**, 49 (1983).

42. E. Zavoiskii, *J. Exp. Theor. Phys. (U.S.S.R.)*, 15, 253 (1945) and *J. Phys. U.S.S.R.*, **8**, 377 (1944).

43. S. Piligkos, D. Collison, V. S. Oganesyan, G. Rajaraman, G. A. Timco, A. J. Thomson, R. E. P. Winpenny, E. J. L. McInnes, *Phys. Rev. B*, **69**, 134424 (2004).

44. D. J. B. Hunter, V. S. Oganesyan, J. C. Salerno, C. S. Butler, W. J. Ingledew, A. J. Thomson, *Biophys. J.*, **78**, 439 (2000).

45. S. J. Yoo, H. Angove, B. K. Burgess, M. P. Hendrich, E. Münck, *J. Am. Chem. Soc.*, **121**, 2534 (1999).

46. K. A. Campbell, E. Yikilmaz, C. V. Grant, W. Gregor, A.-F. Miller, R. D. Britt, *J. Am. Chem. Soc.*, **121**, 4714 (1999).

47. N. C. Verma, R. W. Fessenden, *J. Chem. Phys.*, **65**(6), 2139 (1976).

48. S. Basu, K. A. MacLauchlan, G. R. Sealy, *J. Phys. E: Sci. Instrum.*, **16**, 767 (1983).

49. A. D. Trifunac, M. C. Thurnauer, "Time-Resolved Electron Spin Resonance of Transient Radicals in Liquids", in *Time-Domain Electron Spin Resonance*, L. Kevan, R. N. Schwartz, Eds., Wiley, New York, NY, U.S.A., 1979, pp. 107 ff.

50. S. Brumby, *Magn. Reson. Rev.*, **8**, 1 (1983).

51. J. A. Weil, *Mol. Phys. Rep.*, **26**, 11 (1999).

NOTES

1. In truth, no one knows what an electron or a photon, and its spin, *really* is, but scientists and engineers can work wonderfully with these concepts. In relativistic quantum theory, it is postulated that the electron spin is a kind of orbital angular momentum associated with a very-high-frequency jitter motion (zitterbewegung) superimposed on its more classical 'time-averaged' trajectory [1,2]. The electron was discovered about 100 years ago, and has had a colorful history so far; for a good read, we recommend the book edited by Springford (see FURTHER READING).

2. An anectdotal colorful history of EPR has been included in an autobiography written by one of the prime sorcerers of its development [13] (see also Ref. 14).

3. The acronym EPR has more than 15 meanings. The predominant other use is the famed 'EPR paradox' of Einstein–Podolsky–Rosen (1935), which continues to be a source of fervent research regarding the root meaning of quantum mechanics. Often, the models used involve spatially separating two spin-paired electrons.

4. An example of a species showing transitions not appropriately described by the term *electron spin resonance* is O^{2-} in its ^1D state, which has two units of angular momentum about the internuclear axis but has zero spin angular momentum. Such entities may exhibit electron resonance in the gas phase (Chapter 7).

5. Mathematical operators are designated with a circumflex (e.g., $\hat{\mathcal{H}}$). A summary of the notation used herein for the symbols used can be found in Table I.1.

6. Many authors use the symbol T_1 for the spin-lattice relaxation time; we prefer the symbol τ_1 to avoid confusion with the symbol for temperature.

7. Note that $I = 0$ for all nuclei for which both the atomic mass number *and* the atomic number are *even*. If the atomic mass number is *even* and the atomic number *odd*, I is an integer $(0, 1, 2, \ldots)$; if the atomic mass number is *odd*, I is a half-integer $(\frac{1}{2}, \frac{3}{2}, \frac{5}{2}, \ldots)$.

8. When the directional aspects of quantities are of importance, we use vectors and designate these with **boldface** type. When only magnitudes or vector components are involved, we shall employ *italic* type. See Table I.1 herein.

9. Details about these quantities, as well as about the various systems of units used in the literature, may be found in the excellent treatise by Jackson [24]. The choice of which field quantity, **B** or **H**, is the more fundamental is a problematic and vexing one; see the classic book by Van Vleck [25].

10. In the quantum-mechanical treatment, **M** is the ensemble summation of the expectation value of the magnetic moment for each particle. Strictly speaking, Eq. 1.8 is not applicable as written, in that the right-hand sum should be replaced by the appropriate quantum-mechanical and statistical average of the operators μ_i [27].

11. The unfortunate IUPAC (International Union Pure and Applied Chemistry) recommendation of using the symbol μ_B for this quantity is not followed herein, since this latter symbol erroneously suggests the component of $\boldsymbol{\mu}$ along **B**. In cgs units, $\beta_e = |e|\hbar/2m_e c$.

12. The superscript "T" denotes taking the transpose of the vector; this operation is applicable when the vector specified is a row or column of components. The reader who is unfamiliar with these concepts or with scalar products is referred to Section A.4 and Table A.2.

13. For electric fields, the analogous equation $\mathbf{D} = \varepsilon_0 \mathbf{E} + \mathbf{P}$ is valid. Here **D** is the electrical displacement, **E** is the electric-field intensity (units of force/coulomb: $J\,C^-\,m^{-1} = m\,s^{-1}\,T$), and $\mathbf{P} = \mathbf{P(E)}$ is the electrical polarization. Electric susceptibilities are defined in exact analogy with the magnetic ones [25].

14. The volume magnetic susceptibility χ is an intensive (i.e., independent of sample amount) property and generally is anisotropic as well as dependent on the frequencies of any oscillating magnetic fields present (see Chapter 10). Here we have dropped the subscript m, which up to now has been used to indicate the medium. Scientists at times also use two related susceptibilities: $\chi_g \equiv \chi/\rho$ (where ρ is the mass density of the substance considered), and $\chi_M \equiv \chi_g m_M$ (where m_M denotes the molar mass of the substance considered). To convert from the older unrationalized cgs units to rationalized mksC (SI) units, multiply χ by 4π. Appropriate conversions between cgcC and mksC need, of course, also be done for all quantities. Susceptibility data in the literature should specify the type of χ at hand, whether it is rationalized or not, as well as (if relevant) the units used for ρ and m_M. Technically speaking, one should (in analogy with Eq. 1.13) utilize

$$\chi \equiv -\alpha \frac{g}{|g|} \frac{\partial M}{\partial H}\bigg|_{H=0} \tag{1.28}$$

and also take into account anisotropy, perhaps via a series expansion of vector $\mathbf{M}$ in terms of $\mathbf{H}$, where the first term (i.e., linear in H) is $\boldsymbol{\chi} \cdot \mathbf{H}$, with tensor $\boldsymbol{\chi}$ independent of field magnitude H. We recall that, for electrons, we have $\alpha = -1$ and $g > 0$. Clearly also, by substituting $N_v \mu$ for $\mathbf{M}$, one can define and work with a molecular rather than a bulk magnetic susceptibility (e.g., see Ref. 34).

15. The term *resonance condition* refers to the maximum in a spectral line. Strictly speaking, however, every system absorbs (and emits) electromagnetic radiation over the entire frequency range. Thus $g\beta_e B/h$ represents the peak of a line that (usually) drops off rapidly toward zero, as described by a lineshape function. For the same reason, no truly monochromatic source of radiation exists at a given frequency ν; that is, all sources emit over an infinite band.

16. In this book, we shall ignore the nuclear chemical shift σ_n as being negligible and take the effective nuclear Zeeman factor $g_n(1 - \sigma_n)$ to be simply g_n, i.e., that of the bare nucleus, as tabulated in Table H.4.

17. This effect can be viewed classically; magnetic moment $-g_e\beta_e\mathbf{S}$ induces a (usually) small magnetic moment in its surroundings.

18. By *isotropic*, we mean that reorienting the sample relative to $\mathbf{B}$ and $\mathbf{B}_1$ has no effect on the EPR spectrum.

19. Anisotropy causes recasting of Eq. 1.16 to become $\mathbf{M} = -\alpha(g/|g|)\boldsymbol{\chi}_m \cdot \mathbf{H}$. Here the magnetic susceptibility $\boldsymbol{\chi}_m$ is a 3×3 matrix, as is the relative permeability $\boldsymbol{\kappa}_m = \mathbf{1}_3 + \boldsymbol{\chi}_m$, where $\mathbf{1}_3$ is the 3×3 unit matrix.

20. Henceforth, x, y and z are used for laboratory-fixed axes and X, Y and Z for inherent axes fixed with respect to the paramagnetic species.

21. With pulsed sources of the stimulating electromagnetic radiation, it is *time* (length and spacing of pulses) that is important (Chapter 11). The time and frequency domains are interconvertible via Fourier transformation.

22. The time dependence of $\mathbf{B}_1$ usually is sinusoidal, at constant frequency, but in principle the frequency can be modulated or scanned linearly with time. In this book, we will not always state explicitly whether a given $\mathbf{B}_1$ (or its magnitude B_1) should be deemed to be time-dependent or just the constant amplitude. This aspect depends also on the coordinate system chosen; see discussion on 'rotating frame' in Section 10.3.3.

23. Parallel-field paramagnetic absorption has been investigated from the beginnings of EPR [42].

FURTHER READING

Appendix G contains a fairly complete list of books and monographs dealing with EPR and related topics. The texts in this field that appear below are especially recommended.

1. A. Abragam, B. Bleaney, *Electron Paramagnetic Resonance of Transition Ions*, Oxford University Press, Oxford, U.K., 1970.

2. R. S. Alger, *Electron Paramagnetic Resonance: Techniques and Applications*, Wiley, New York, NY, U.S.A., 1967.

3. S. A. Al'tshuler, B. M. Kozyrev, *Electron Paramagnetic Resonance*, Academic Press, New York, NY, U.S.A.,1964.

4. E. D. Becker, *High Resolution NMR—Theory and Chemical Applications*, 2nd ed., Academic press, New York, NY, U.S.A., 1980.

5. A. Carrington, A. D. McLachlan, *Introduction to Magnetic Resonance*, Harper & Row, New York, NY, U.S.A., 1967.

6. R. K. Harris, *Nuclear Magnetic Resonance Spectroscopy*, Longman, Harlow, U.K., 1986.

7. J. S. Hyde, S. S. Eaton, G. R. Eaton, "EPR at Work", *Concepts Magn. Reson.*, **28**A(1), 1–100 (2006) [a printing of the 40+ fine tutorial articles produced by Varian Assoc. (Palo Alto, CA, U.S.A.) during the years 1957–1965].

8. G. Jeschke, "Electron Paramagnetic Resonance: Recent Developments and Trends", *Curr. Opin. Solid State Mater. Sci.*, **7**, 181 (2003).

9. G. E. Pake, T. L. Estle, *The Physical Principles of Electron Paramagnetic Resonance*, 2nd ed., Benjamin, Reading, MA, U.S.A., 1973.

10. M. A. Plonus, *Applied Electromagnetics*, McGraw-Hill, New York, NY, U.S.A., 1978 (a good discussion of **B**, **H** and **M**).

11. C. P. Poole Jr., *Electron Spin Resonance: A Comprehensive Treatise on Experimental Techniques*, 2nd ed., Interscience, New York, NY, U.S.A., 1983.

12. J. A. Pople, W. G. Schneider, H. J. Bernstein, *High-Resolution Nuclear Magnetic Resonance*, McGraw-Hill, New York, NY, U.S.A., 1959.

13. C. P. Slichter, *Principles of Magnetic Resonance*, 3rd ed., Springer, New York, NY, U.S.A., 1989.

14. M. Springford, Ed., *The Electron—a Centenary Volume*, Cambridge University Press, Cambridge, U.K., 1997.

15. J. Talpe, *Theory of Experiments in EPR*, Pergamon Press, Oxford, U.K., 1971.

16. J. H. Van Vleck, *The Theory of Electric and Magnetic Susceptibilities*, Oxford University Press, London, U.K., 1932.

17. S. Weber, EPR Methods, in *Encyclopedia of Chemical Physics and Physical Chemistry*, Vol. II, J. H. Moore, N. D. Spencer, Eds., Inst. Phys., Bristol, U.K. 2001, Chapter 1.15, pp. 1351–1388.

18. W. Weltner Jr., *Magnetic Atoms and Molecules*, Van Nostrand Reinhold, New York, NY, U.S.A., 1983.

PROBLEMS

1.1 (*a*) Draw a continuous-wave (cw) NMR spectrometer analogous to the cw
EPR and optical spectrometers shown in Figs. 1.3*a* and 1.3*b*.

 (*b*) Describe the functioning of those NMR components that are significantly
different from their EPR analogs.

 (*c*) Contrast the frequency and field regions that are routinely accessible to the
two techniques. To be sure, cw NMR is close to obsolete, but nevertheless
is conceptually useful. It has been replaced by pulsed NMR, which techniques also exist for EPR (Chapter 11), but that have not been as widely
accepted and used in the latter.

1.2 What is the numerical value ℓ_z of the angular momentum of a classical electron
rotating in a fixed circular orbit of 1 Bohr radius ($r = 0.0529$ nm) with a
frequency ν of 10^{13} Hz? Note that the magnitude ν of the linear velocity of
a rotating particle is given by $\nu = r\omega$, where $\omega = 2\pi\nu$ is called the *angular
frequency*. Compare ℓ_z with $\hbar h$, the natural quantum-mechanical unit of
angular momentum. Discuss the difference.

1.3 The separation of two lines (splitting) in a free-radical EPR spectrum is given
as 75.0 MHz, and $g = 2.0050$. Express the splitting in mT and in cm^{-1}.

1.4 Is it possible to obtain EPR spectra with NMR equipment? Assuming
$g = 2.0050$, what magnetic field would be required to observe EPR at
$v = 400$ MHz?

1.5 A classical magnetic dipole placed in a static magnetic field precesses about
the magnetic-field direction with an angular frequency $\omega = 2\pi\nu$ given by
$\omega = \gamma B$. Consider the electron to be such a particle.

 (*a*) What is the magnetogyric ratio for a free electron?

 (*b*) At what frequency ν does this dipole 'precess' in a field $B = 350.0$ mT?

 (*c*) What would be the value of g for an electron trapped in a negative-ion
vacancy in KBr ($g = 1.985$)?

1.6 Calculate the ratio of the resonant frequencies of a free electron and a free
deuteron (^{2}H) in the same magnetic field.

1.7 Using the data in Table H.4, compute the NMR frequency for a proton at the
magnetic field used in X-band EPR (9.5 GHz) (this is the basis of a popular
gaussmeter for measuring magnetic-field strengths).

1.8 Explain why one might wish to perform an EPR experiment on an unpaired-
electron system as well as to determine separately its static magnetic susceptibility by force measurements; that is, what is the difference in information
provided by the two measurements?

CHAPTER 2

MAGNETIC INTERACTION BETWEEN PARTICLES

2.1 INTRODUCTION

The first dipolar interaction to be considered is that of the electron-spin magnetic dipole with that of nuclei in its vicinity. It was noted in Chapter 1 that some nuclei possess an intrinsic spin angular momentum. The spin quantum number I of these nuclei is found to have one of the values $\frac{1}{2}, 1, \frac{3}{2}, 2, \ldots$, with a corresponding multiplicity of nuclear-spin states given by $2I + 1$. Analogous to the electron case, there is a magnetic moment associated with the nuclear-spin angular momentum. The spins and magnetic moments of various nuclei are listed in Table H.4. For the present the discussion is restricted to species containing one unpaired electron $(S = \frac{1}{2})$, although much of this chapter applies equally well to species containing more than one unpaired electron $(S > \frac{1}{2})$.

The simplest system exhibiting nuclear hyperfine interaction is the hydrogen atom, which we first consider in a qualitative fashion. The details of the origin of the hyperfine interaction and the calculation of energy levels are discussed later in this chapter. The EPR spectrum of a hydrogen atom is shown in Fig. 1.4. As already mentioned in Section 1.12, instead of a single line characterized by $B = h\nu/g\beta_e$ with $g = 2.0022$, one observes a *pair* of lines,[1] which implies the presence of more than two spin energy levels. Since the proton has a spin $I = \frac{1}{2}$, M_I has two allowed values: $M_I = \pm\frac{1}{2}$. Hence at each position of the electron, there is one of two possible local fields (Section 1.12.1) at which resonance occurs, that is

$$B = B' - aM_I + \cdots = B' \mp \frac{a}{2} + \cdots \tag{2.1}$$

where aM_I is B_{local} at the electron, and where $B' = h\nu/g\beta_e$ is the would-be resonant field if the hyperfine parameter a were zero.[2] In addition, there is a series of terms (indicated by ... in Eq. 2.1) of the form a^q/B'^{q-1} with $q = 2, 3, \ldots$. These become less and less important as B' increases relative to a. These extra terms are considered later (Eq. 3.2; also Chapters 5 and 6, as well as Appendix C). For the free hydrogen atom, $a = 50.684$ mT, whereas at 9.5 GHz the spacing of the hyperfine doublet is 50.970 mT. We see here that, with such a large splitting, the above-mentioned correction is substantial. For most species (say, organic free radicals) the hyperfine parameters encountered are less than 1 mT (and $g \approx g_e$); hence the additional terms are sufficiently small that the spacing between hyperfine lines is well approximated by the parameter a.

The astute observer may notice that there are four possible EPR transitions in Fig. 2.1 for the one-nucleus $(I = \frac{1}{2})$ system. Two of the transitions involve simultaneous nuclear-spin flips (shown as dashed lines in Fig. 1.4). For the free hydrogen atom these have negligibly small EPR transition probabilities as compared to the

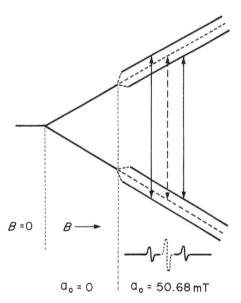

$B = 0$ $B \longrightarrow$

$a_0 = 0$ | $a_0 = 50.68$ mT

FIGURE 2.1 Energy levels of a system with one unpaired electron and one magnetic nucleus with $I = \frac{1}{2}$ (e.g., the free hydrogen atom) as a function of magnetic field. The dashed-line transition would be observed if a_0 were zero. The observed fixed-frequency spectrum (Section C.1.6 and Fig. 1.4) may be accounted for if the allowed transitions shown as solid lines are both drawn with the same length, since $h\nu$ is constant.

pure transitions (solid lines in Fig. 2.1). Here, then, only two EPR transitions are observed. In general, however, with other hydrogen-containing radicals, for which only relatively small hyperfine interactions are involved, all four EPR lines can be present. This subject is considered in Chapter 5 and Appendix C.

2.2 THEORETICAL CONSIDERATIONS OF THE HYPERFINE INTERACTION

If the electron and nuclear magnetic dipoles were to behave classically and a substantial externally applied static magnetic field **B** ($\|\mathbf{z}$) were present so as to align them, then the energy of dipole-dipole interaction (Section 5.2) between them would be given by the following approximate expression:[3]

$$U_{\text{dipolar}} = -\frac{\mu_0}{4\pi} \frac{3\cos^2\theta - 1}{r^3} \mu_{nz}\mu_{ez} = -B_{\text{local}}\mu_{ez} \tag{2.2}$$

Here the components of the electron and nuclear dipole moments along the applied magnetic field **B** are μ_{ez} and μ_{nz}. The dipoles are separated by the distance r, and θ is the angle between **B** and the line joining the two dipoles. This classical system is shown in Fig. 2.2. Depending on the value of θ, the local field B_{local} caused by the nucleus at the electron can either aid or oppose the external magnetic field. From Eq. 2.2 and Fig. 2.2 it is apparent that B_{local}, arising from the nucleus, depends markedly on the instantaneous value of θ (and of r).

It is clear from Eq. 2.2 that as the inter-particle distance r approaches zero, the interaction energy approaches infinity. This does not pose a problem, largely because the probability of this type of superposition of particles is suitably small. Further mathematical details of this pathological situation have been discussed by

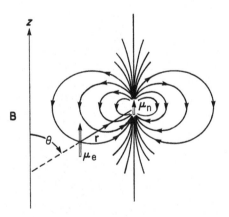

FIGURE 2.2   Interaction of the aligned magnetic dipoles $\boldsymbol{\mu}_e$ and $\boldsymbol{\mu}_n$ arising from an electronic spin and a nuclear spin. Vector $\boldsymbol{\mu}_e$ is indicated for the state $M_S = -\frac{1}{2}$ and vector $\boldsymbol{\mu}_n$ is indicated for the state $M_I = +\frac{1}{2}$. Angle θ is between the inter-dipole vector **r** and the applied field **B**.

Skinner and Weil [1]. This type of 'contact' interaction, to be explored in this section and the next, is important (but not huge) in most cases.

Since the electron is not localized at one position in space, the interaction energy U_{dipolar} must be averaged over the electron probability distribution function. If all regions of θ are equally probable (as for an electron in an s orbital centered on nucleus n), then the average local field at each r is obtained by inserting the value of $\cos^2\theta$ averaged over a sphere[4]

$$\langle \cos^2 \theta \rangle \; = \; \frac{\int_0^{2\pi} \int_0^{\pi} \cos^2 \theta \sin \theta \, d\theta \, d\phi}{\int_0^{2\pi} \int_0^{\pi} \sin \theta \, d\theta \, d\phi} \; = \; \frac{1}{3} \tag{2.3}$$

into Eq. 2.2. In spherical polar coordinates, $\sin \theta \, d\theta \, d\phi$ is the element of surface area on a sphere. Since $\langle \cos^2 \theta \rangle = \frac{1}{3}$, B_{local} in Eq. 2.2 vanishes. Consequently, the classical action-at-a-distance dipolar interaction *cannot* be the origin of the hyperfine splitting in the hydrogen atom since the electron distribution in a $1s$ orbital is spherically symmetric.

An understanding of the actual origin of the hyperfine interaction in the hydrogen atom may be obtained by examining the radial dependence of the hydrogen $1s$ orbital shown in Fig. 2.3. One notes that the $1s$ electron density at the nucleus (here taken to be a mathematical point) is non-zero;[5] it is precisely this non-zero density that gives rise to the hyperfine interaction. It is clear from Fig. 2.3 that *only electrons in s orbitals have a non-zero probability density at the nucleus*; $p, d, f, \ldots$ orbitals all have nodes at the nucleus. On the other hand, electrons in $2s, 3s, \ldots$ orbitals also have a non-zero electron density at the nucleus and give rise to such hyperfine interactions. From Table H.4, column 7, the hyperfine interactions for valence s electrons of some atoms are seen to attain very large values. By virtue of the spherical symmetry of s orbitals, the hyperfine interaction in these cases is, of course, *isotropic*.

Fermi [3] has shown that for systems with one electron the magnetic energy for the isotropic interaction is given approximately by

$$U_{\text{iso}} = -\frac{2\mu_0}{3} |\psi(0)|^2 \mu_{ez} \mu_{nz} \tag{2.4}$$

when the applied field **B** ($\|\mathbf{z}$) is sufficiently large. Here $\psi(0)$ represents the electron wavefunction evaluated at the point nucleus.[6] For example, the hydrogen atom ground-state wavefunction is given by

$$\psi_{1s}(r) = \left(\frac{1}{\pi r_b^3} \right)^{1/2} \exp\left(-\frac{r}{r_b} \right) \tag{2.5}$$

where r_b is the radius of the first Bohr orbit (52.9 pm). Using the probability density $|\psi_{1s}(0)|^2 = 1/\pi r_b^3$, one can then calculate a value of U_{iso} with the aid of Eq. 2.4. This calculation, which provides an excellent approximation to the actual value, is the subject of Problem 2.3. In Section 2.4, we relate U_{iso} to the hyperfine parameter a. Detailed consideration of hyperfine anisotropy is deferred to Chapter 5.

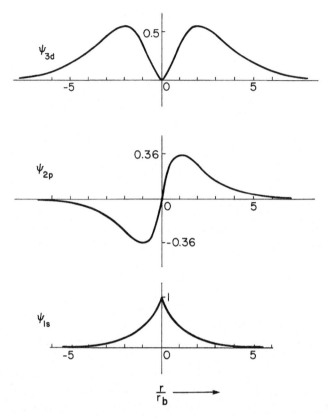

FIGURE 2.3   Radial dependence of the hydrogenic $1s$, $2p$ and $3d$ wavefunctions, showing the non-vanishing behavior at the point nucleus ($r = 0$) of s orbitals, as well as the change in sign of odd orbitals ($\ell = 1, 3, 5, \ldots$) in going through the origin. Here r_b is the Bohr radius, and the functions $(r/r_b)^{\ell}\exp(-r/nr_b)$ plotted are not normalized. Here n is the principal quantum number $(1, 2, 3, \ldots)$.

It is worthwhile noting that the above considerations hold for the hydrogen atom in its electronic ground state. EPR studies of the atom in any of its (infinite number of) excited states are, in principle, feasible and would yield different but analogous results. The same wealth of states is at hand for any of the molecular species treated in this book. Unless otherwise stated, only the electronic (and vibrational) ground state is dealt with. For examples of excited-state EPR, see Sections 6.3.4 and 6.3.5.

2.3   ANGULAR-MOMENTUM AND ENERGY OPERATORS

Before considering further details of the hyperfine interaction, it is instructive to introduce operator methods for determining the energies of a system of interest. In this chapter, we examine some relatively simple problems so that the reader may become familiar with these techniques, which are also applicable to more complicated systems. The algebraic manipulation of operators is briefly described in Section A.2. General properties of spin operators are given in Appendix B. Their application to the hydrogen atom and to radicals of type RH_2, which exhibits hyperfine interaction with two equivalent protons, is given in Appendix C.

2.3.1 Spin Operators and Hamiltonians

For a system having discrete energy levels described by well-defined quantum numbers, it is always possible to write an eigenvalue equation; that is, if $\hat{\Lambda}$ is the operator appropriate to the property under study, the eigenvalue equation (Eq. A.9) is

$$\hat{\Lambda}\psi_k = \lambda_k\psi_k \tag{2.6}$$

Here λ_k represents an eigenvalue of a state (labeled k) for which the eigenfunction is ψ_k.

The topic of primary interest in EPR is the quantization of spin angular momentum. Hence one seeks a *spin operator* that operates on a function describing a spin state, causing it to be multiplied by a constant characteristic of that state. For a system with electron spin $S = \frac{1}{2}$, the two states ($k = 1,2$) are characterized by the quantum numbers $M_S = \pm\frac{1}{2}$. These measure the components M_S of angular momentum along the direction $\mathbf{z}$ of the magnetic field, corresponding to the operator $\hat{S}_z$. Thus, if $\hat{\mathbf{S}}$ is the angular-momentum operator, then its z component obeys Eq. 2.6, written as

$$\hat{S}_z\phi_e = M_S\phi_e \tag{2.7}$$

For simplicity here and below we omit the index k. The factor M_S is called the *eigenvalue* of the operator $\hat{S}_z$ and $\phi_e(M_S)$ is the corresponding *eigenfunction*. We adopt the notation $\alpha(e) = \phi_e(M_S = +\frac{1}{2})$ and $\beta(e) = \phi_e(M_S = -\frac{1}{2})$, so that

$$\hat{S}_z\alpha(e) = +\frac{1}{2}\alpha(e) \tag{2.8a}$$

$$\hat{S}_z\beta(e) = -\frac{1}{2}\beta(e) \tag{2.8b}$$

Note that the angular momentum is taken in units of $\hbar$.

Similar expressions pertain to the nuclear-spin operator $\hat{I}_z$ for a nucleus with nuclear spin $I = \frac{1}{2}$ and z component M_I:

$$\hat{I}_z\alpha(n) = +\frac{1}{2}\alpha(n) \tag{2.9a}$$

$$\hat{I}_z\beta(n) = -\frac{1}{2}\beta(n) \tag{2.9b}$$

The symbolism for the representation of an eigenfunction can readily be simplified. Since the functions are distinguished by their quantum numbers, one may enclose these numbers in a distinctive way to represent the function. Dirac suggested the notation $|k\rangle$ for an eigenfunction ψ_k. (A function represented in such a way is called a 'ket'; see Sections A.5.4 and B.4.) Then Eqs. 2.8 and 2.9 may be rewritten as

$$\hat{S}_z|\alpha(e)\rangle = +\tfrac{1}{2}\,|\alpha(e)\rangle \tag{2.10a}$$

$$\hat{S}_z|\beta(e)\rangle = -\tfrac{1}{2}\,|\beta(e)\rangle \tag{2.10b}$$

and

$$\hat{I}_z|\alpha(n)\rangle = +\tfrac{1}{2}\,|\alpha(n)\rangle \tag{2.11a}$$

$$\hat{I}_z|\beta(n)\rangle = -\tfrac{1}{2}\,|\beta(n)\rangle \tag{2.11b}$$

The energies U_k of systems, for which M_S and M_I are precise measures of components of electronic and nuclear-spin angular momentum, are obtained from the time-independent Schrödinger equation

$$\hat{\mathcal{H}}_e\phi_{ek} = U_{ek}\phi_{ek} \tag{2.12}$$

$$\hat{\mathcal{H}}_n\phi_{nk} = U_{nk}\phi_{nk} \tag{2.13}$$

Here the hamiltonian operator $\mathcal{H}$ (which we consider commutes with $\hat{S}_z$ as well as with $\hat{I}_z$) is the operator for the total energy. The index k is any one of the labels of the eigenstates of the system. The importance of Eqs. 2.12 and 2.13, taken together with Eqs. 2.8 and 2.9, is that *the same ϕ_k is an eigenfunction of the z component of the spin angular momentum and of the energy*[7] (Section A.2.2).

Hence

$$\hat{\mathcal{H}}_e|\alpha(e)\rangle = U_{\alpha(e)}|\alpha(e)\rangle \tag{2.14a}$$

$$\hat{\mathcal{H}}_e|\beta(e)\rangle = U_{\beta(e)}|\beta(e)\rangle \tag{2.14b}$$

and

$$\hat{\mathcal{H}}_n|\alpha(n)\rangle = U_{\alpha(n)}|\alpha(n)\rangle \tag{2.15a}$$

$$\hat{\mathcal{H}}_n|\beta(n)\rangle = U_{\beta(n)}|\beta(n)\rangle \tag{2.15b}$$

It is often useful to express $\hat{\mathcal{H}}$ in a special reduced form. In general, the hamiltonian operator of a system is a function of the positions and momenta of all particles present (the spatial part), and of their intrinsic angular momenta (the spin part). Of necessity, since the hamiltonian contains spin operators, it is represented by a matrix (in quantum-mechanical state space) that is generated from angular-momentum matrices (Section B.5). Since the rules for setting these up are straightforward, it is possible to construct the matrix $\mathcal{H}$ for any system, as long as one knows which spins (electrons and nuclei) are present.

The energy eigenvalues are obtained by integrating over all spatial variables to yield numerical parameters, leaving the spin part of the hamiltonian operator intact. The resulting entity, consisting of parameters and spin operators, is called a *spin hamiltonian*. This approach has proved to be very valuable, in that it enables measurement by magnetic-resonance techniques of the parameters g, D, A, ... to be introduced later. These can be tabulated in the scientific literature and can be used to reproduce the original EPR spectra in detail. Theoretical analysis, to interpret the parameters in terms of the spatial behavior of the electrons and nuclei, can be carried out separately, possibly at a later date as appropriate mathematical tools evolve. Thus spin-hamiltonian parameter sets can be regarded as storehouses of quantitative information about atoms and molecules. We use the same symbol $\hat{\mathcal{H}}$ for the *hamiltonian* and the *spin hamiltonian*, but we take care with the explicit nomenclature to distinguish which of these is being considered.

As we shall see (Chapters 4–6), the spin-hamiltonian concept is especially suitable for description of EPR line positions and relative intensities of paramagnetic species in solids, but is also of major use in liquids. While originally developed for use with transition ions located in a symmetric environment (in certain salts), the spin hamiltonian is now utilized with all EPR-detectable species, inorganic and organic.

To obtain the energy values $U(B)$, in terms of the various parameters, one must solve the secular determinant of dimension $(2S+1)P_i(2I_i+1)$. For sufficiently simple small determinants, this can be done analytically to yield algebraic equations for the eigenvalues. (e.g., see the solution of the hamiltonian matrix $\mathcal{H}$ for the hydrogen atom in Appendix C.) Failing that, when the numerical parameters in $\mathcal{H}$ are available, one can always diagonalize it numerically by computer to obtain the possible values of $U(B)$.

2.3.2 Electronic and Nuclear Zeeman Interactions

The first problem we treat with spin-operator methods is the interaction of an electron or a nucleus with a static magnetic field taken along some direction z, that is, we re-derive the resonance equation (Eq. 1.19) for a system with $S = \frac{1}{2}$ and also for a system with $I = \frac{1}{2}$. In operator form, Eq. 1.14b becomes

$$\hat{\mathcal{H}} = -\mathbf{B}^{\mathrm{T}} \cdot \hat{\boldsymbol{\mu}} \tag{2.16a}$$

$$= -B\hat{\mu}_z \tag{2.16b}$$

where we have chosen axis **z** along **B**. We now use operator relations between the magnetic moment and the spin angular momentum. The electron magnetic-moment operator $\hat{\mu}_{ez}$ is proportional to the electron-spin operator $\hat{S}_z$ (Eq. 1.9). Similarly $\hat{\mu}_{nz}$ is proportional to the nuclear-spin operator $\hat{I}_z$. Thus

$$\hat{\mu}_{ez} = \gamma_e \hat{S}_z \hbar = -g\beta_e \hat{S}_z \tag{2.17}$$

$$\hat{\mu}_{nz} = \gamma_n \hat{I}_z \hbar = +g_n \beta_n \hat{I}_z \tag{2.18}$$

The extension of Eqs. 2.17 and 2.18 leads to the definition of the electron and nuclear *spin-hamiltonian operators*

$$\hat{\mathcal{H}}_e = +g\beta_e B\hat{S}_z \tag{2.19}$$

$$\hat{\mathcal{H}}_n = -g_n \beta_n B\hat{I}_z \tag{2.20}$$

Note that the only operators in spin hamiltonians are those of spin. We shall find this type of formulation very useful in more complex situations.[8] Now application of the spin hamiltonians of Eqs. 2.19 and 2.20 to the spin state functions (also called eigenfunctions) has the following results

$$\begin{aligned}\hat{\mathcal{H}}_e|\alpha(e)\rangle &= +g\beta_e B\hat{S}_z|\alpha(e)\rangle \\ &= +\tfrac{1}{2} g \beta_e B|\alpha(e)\rangle \end{aligned} \tag{2.21a}$$

and

$$\begin{aligned}\hat{\mathcal{H}}_e|\beta(e)\rangle &= +g\beta B_e \hat{S}_z|\beta(e)\rangle \\ &= -\tfrac{1}{2} g\beta_e B|\beta(e)\rangle \end{aligned} \tag{2.21b}$$

Similarly

$$\begin{aligned}\hat{\mathcal{H}}_n|\alpha(n)\rangle &= -g_n \beta_n B\hat{I}_z|\alpha(n)\rangle \\ &= -\tfrac{1}{2} g_n\beta_n B|\alpha(n)\rangle \end{aligned} \tag{2.22a}$$

and

$$\begin{aligned}\hat{\mathcal{H}}_n|\beta(n)\rangle &= -g_n \beta_n B\hat{I}_z|\beta(n)\rangle \\ &= +\tfrac{1}{2} g_n\beta_n B|\beta(n)\rangle \end{aligned} \tag{2.22b}$$

One may infer from Eqs. 2.21 and 2.22 that

$$U_{\alpha(e)} = +\tfrac{1}{2} g\beta_e B \tag{2.23a}$$

$$U_{\beta(e)} = -\tfrac{1}{2} g\beta_e B \tag{2.23b}$$

and

$$U_{\alpha(n)} = -\tfrac{1}{2} g_n\beta_n B \tag{2.24a}$$

$$U_{\beta(n)} = +\tfrac{1}{2} g_n\beta_n B \tag{2.24b}$$

Thus

$$\Delta U_e = U_{\alpha(e)} - U_{\beta(e)} = g\beta_e B = h\nu_e \tag{2.25}$$

$$\Delta U_n = U_{\beta(n)} - U_{\alpha(n)} = g_n\beta_n B = h\nu_n \tag{2.26}$$

The resonance equation (Eq. 2.25) corresponds to transitions between the states $|\beta(e)\rangle$ and $|\alpha(e)\rangle$ (an EPR transition) and the next resonance equation (Eq. 2.26) (considering $g_n > 0$) corresponds to transitions between the states $|\alpha(n)\rangle$ and $|\beta(n)\rangle$ (an NMR transition). Here $h\nu_e$ and $h\nu_n$ are the photon energies that stimulate the electronic and nuclear transitions.

A general procedure for determining the energy U from a given hamiltonian involves multiplication of both sides of Eq. 2.12 (and similarly Eq. 2.13) from the left by ϕ_k^{*}[9]

$$\begin{aligned} \phi_k^{*}\hat{\mathcal{H}}\phi_k &= \phi_k^{*} U_k \phi_k \\ &= U_k \phi_k^{*} \phi_k \quad \text{since } U_k \text{ is a constant} \end{aligned} \tag{2.27}$$

Multiplication of both sides by $d\tau$ (where τ represents one or more spatial variables of integration) and integration over the full range of the variable(s) τ give

$$\int_\tau \phi_k^{*} \hat{\mathcal{H}}\phi_k \, d\tau = U_k \int_\tau \phi_k^{*} \phi_k \, d\tau \tag{2.28}$$

Hence

$$U_k = \frac{\int_\tau \phi_k^{*} \hat{\mathcal{H}}\phi_k \, d\tau}{\int_\tau \phi_k^{*} \phi_k \, d\tau} \tag{2.29}$$

If the spatial functions ϕ_k are normalized, that is, if they satisfy the condition

$$\int_\tau \phi_k^{*} \phi_k \, d\tau = 1 \tag{2.30}$$

then

$$U_k = \int_\tau \phi_k^{*} \hat{\mathcal{H}}\phi_k \, d\tau \tag{2.31}$$

One can say that the expectation value $\langle\hat{\mathcal{H}}\rangle$ is U_k for the energy of the system in its kth state.

It is appropriate to rewrite Eqs. 2.27–2.31 in the Dirac notation used in Eqs. 2.14 and 2.15. The symbol appropriate to multiplication from the left by ϕ_k^{*} is $\langle\phi_k|$ (Dirac called this function a 'bra'). When $\langle\phi_k|$ is combined with the ket $|\phi_k\rangle$, integration over the full range of all variables is implied. Thus the combination $\langle\phi_k|\phi_k\rangle$, that is, *bra[c]ket*, suggests the origin of the notation. Then Eq. 2.28 becomes

$$\langle\phi_k|\hat{\mathcal{H}}|\phi_k\rangle = U_k\langle\phi_k|\phi_k\rangle \tag{2.32}$$

For normalized functions, Eq. 2.30 is

$$\langle \phi_k | \phi_k \rangle = 1 \tag{2.33}$$

and hence Eq. 2.31 is

$$U_k = \langle \phi_k | \hat{\mathcal{H}} | \phi_k \rangle \tag{2.34}$$

We see that the energy U_k is the kth diagonal element of matrix $\hat{\mathcal{H}}$. In dealing with spin hamiltonians, the bra-ket notation is the appropriate one, since there are *no* spatial variables to be integrated for spin states.

For the electronic and nuclear-spin states of our simple problem (Section 2.3.1), one writes

$$U_{\alpha(e)} = \langle \alpha(e) | \hat{\mathcal{H}}_e | \alpha(e) \rangle = +\tfrac{1}{2} g \beta_e B \tag{2.35a}$$

$$U_{\beta(e)} = \langle \beta(e) | \hat{\mathcal{H}}_e | \beta(e) \rangle = -\tfrac{1}{2} g \beta_e B \tag{2.35b}$$

and

$$U_{\alpha(n)} = \langle \alpha(n) | \hat{\mathcal{H}}_n | \alpha(n) \rangle = -\tfrac{1}{2} g_n \beta_n B \tag{2.36a}$$

$$U_{\beta(n)} = \langle \beta(n) | \hat{\mathcal{H}}_n | \beta(n) \rangle = +\tfrac{1}{2} g_n \beta_n B \tag{2.36b}$$

At this point, the reader who is interested in eigenfunctions and their manipulation may wish to turn to Section A.2.2, where the problem of a particle in a ring is considered in terms of both the angular-momentum operator and the energy operator.

2.3.3 Spin Hamiltonian Including Isotropic Hyperfine Interaction

Let us now consider the effects of the isotropic hyperfine interaction, deferring the anisotropic interaction until Chapter 5. The appropriate spin-hamiltonian operator may be obtained from Eq. 2.4 by replacing the classical magnetic moments by their corresponding operators (Eqs. 2.17 and 2.18). Following the same procedure as in the preceding section, one obtains the result

$$\hat{\mathcal{H}}_{\text{iso}} = \frac{2\mu_0}{3} g \beta_e \beta_n |\psi(0)|^2 \hat{S}_z \hat{I}_z \tag{2.37}$$

The factor multiplying $\hat{S}_z \hat{I}_z$ often is called the isotropic hyperfine *coupling* 'constant'

$$A_0 = \frac{2\mu_0}{3} g \beta_e \beta_n |\psi(0)|^2 \tag{2.38}$$

which measures the magnetic interaction energy (in joules) between the electron and the nucleus. Hence Eq. 2.37 becomes

$$\hat{\mathcal{H}}_{iso} = A_0\hat{S}_z\hat{I}_z \tag{2.39a}$$

When B is along an arbitrary direction (or absent), $\hat{\mathcal{H}}_{iso}$ should be expressed in its most general form

$$\hat{\mathcal{H}}_{iso} = A_0\hat{\mathbf{S}}^{\mathrm{T}} \cdot \hat{\mathbf{I}} \tag{2.39b}$$

Often the hyperfine coupling constant is given[10] as A_0/h in frequency units (MHz). It may also be expressed in magnetic-field units (Table H.4) and is then called the *hyperfine splitting* constant $a_0 = A_0/g_e\beta_e$. Strictly speaking, Eq. 2.37 should be written with the factor $\hat{\mathbf{S}}^{\mathrm{T}} \cdot \hat{\mathbf{I}}$ in place of $\hat{S}_z\hat{I}_z$; however, it is shown in Section C.1.7 that when the hyperfine interaction A_0 is small compared to the electron Zeeman interaction $g\beta_e B$, Eq. 2.37 is adequate.

The spin-hamiltonian operator for the hydrogen atom (and other isotropic systems with one electron and one nucleus with $I = \frac{1}{2}$) is obtained by adding Eqs. 2.19, 2.20 and 2.39a:

$$\hat{\mathcal{H}} = g\beta_e B\hat{S}_z - g_n\beta_n B\hat{I}_z + A_0\hat{I}_z\hat{S}_z \tag{2.40}$$

This is valid when B is sufficiently large. We note that the hyperfine term destroys the independence of the electron and nuclear spins. If more than one magnetic nucleus interacts with the electron, the terms in $\hat{I}_z$ are additive. Thus, summing over all nuclei yields

$$\hat{\mathcal{H}} = g\beta_e B\hat{S}_z - \sum_i g_{ni}\beta_n B\hat{I}_{iz} + \sum_i A_{0i}\hat{S}_z\hat{I}_{iz} \tag{2.41}$$

The nuclear Zeeman energy (the second term) has been included in Eqs. 2.40 and 2.41 but has little effect on the transition energies, since the contributions from this term tend to cancel when the hyperfine terms (third terms on the right) are relatively large (e.g., in the hydrogen atom). This is not the case when the second and third terms give contributions of similar magnitude in anisotropic systems (Section 5.3.2.1).

2.4 ENERGY LEVELS OF A SYSTEM WITH ONE UNPAIRED ELECTRON AND ONE NUCLEUS WITH $I = \frac{1}{2}$

This is the simplest case exhibiting hyperfine interaction, for which the hydrogen atom is the prototype.[11] Since the eigenvalues M_S of $\hat{S}_z$ are $+\frac{1}{2}$ and those of $\hat{I}_z$ are $M_I = +\frac{1}{2}$, there are four possible composite spin states; the kets are

$$|\alpha(e), \alpha(n)\rangle \quad |\alpha(e), \beta(n)\rangle \quad |\beta(e), \alpha(n)\rangle \quad |\beta(e), \beta(n)\rangle$$

Application of the spin operators $\hat{S}_z$ and $\hat{I}_z$ gives the following results

$$\hat{S}_z|\alpha(e), \alpha(n)\rangle = +\tfrac{1}{2}|\alpha(e), \alpha(n)\rangle \qquad (2.42a)$$

$$\hat{I}_z|\alpha(e), \alpha(n)\rangle = +\tfrac{1}{2}|\alpha(e), \alpha(n)\rangle \qquad (2.42b)$$

and so on for the other six combinations.

The energies of these states are obtained by evaluating expressions analogous to Eqs. 2.35 and 2.36; for example

$$
\begin{aligned}
U_{\alpha(e), \alpha(n)} &= \langle|\alpha(e), \alpha(n)|\hat{\mathcal{H}}|\alpha(e), \alpha(n)\rangle \\
&= \langle|\alpha(e), \alpha(n)|g\beta_e B\hat{S}_z - g_n\beta_n B\hat{I}_z + A_0\hat{I}_z\hat{S}_z + \cdots|\alpha(e), \alpha(n)\rangle \\
&= +\tfrac{1}{2}g\beta_e B - \tfrac{1}{2}g_n\beta_n B + \tfrac{1}{4}A_0 + \cdots
\end{aligned}
\qquad (2.43a)
$$

Similarly

$$U_{\alpha(e), \beta(n)} = +\tfrac{1}{2}g\beta_e B + \tfrac{1}{2}g_n\beta_n B - \tfrac{1}{4}A_0 + \cdots \qquad (2.43b)$$

$$U_{\beta(e), \alpha(n)} = -\tfrac{1}{2}g\beta_e B - \tfrac{1}{2}g_n\beta_n B - \tfrac{1}{4}A_0 + \cdots \qquad (2.43c)$$

$$U_{\beta(e), \beta(n)} = -\tfrac{1}{2}g\beta_e B + \tfrac{1}{2}g_n\beta_n B + \tfrac{1}{4}A_0 + \cdots \qquad (2.43d)$$

where the ellipses ($\cdots$) indicate terms that have been left implicit (Section 3.6). Neglecting these gives the so-called first-order energies. For the present case of $S = I = \tfrac{1}{2}$, the mathematical energy problem has been solved exactly, as a function of the field B, by Breit and Rabi [4]. The solution (Appendix C) can be expanded as an infinite series, the first terms of which are given explicitly in Eqs. 2.43. The energy levels are shown in Fig. 2.4a for a moderately high magnetic field, including the EPR transitions observable by scanning the frequency ν.

From these quantitative expressions for the energy levels, valid for sufficiently large B, we note the (near) equality of splitting of each nuclear doublet. Second, we note that the ordering (M_I) of the levels is reversed in the lower set of levels as compared with the upper set. Since the one unit ($\hbar$) of angular momentum in the absorbed photon is used to change the angular momentum of the electron, no change in the angular momentum of the nucleus is possible. However, in some cases it may be possible for more than one photon to be absorbed. We also note that, in the limit of $B = 0$, energy-level splittings arising from the hyperfine term remain, so that zero-field transitions at specific frequencies are observable when a suitable excitation magnetic field $\mathbf{B}_1$ is applied. A notable example of this is the 1420 MHz emission from atomic hydrogen in outer space (see Sections 7.8 and C.1 and Problem 2.3).

The energies of these two allowed EPR transitions are

$$\Delta U_1 = U_{\alpha(e), \alpha(n)} - U_{\beta(e), \alpha(n)} = g\beta_e B + \tfrac{1}{2}A_0 + \cdots \qquad (2.44a)$$

$$\Delta U_2 = U_{\alpha(e), \beta(n)} - U_{\beta(e), \beta(n)} = g\beta_e B - \tfrac{1}{2}A_0 + \cdots \qquad (2.44b)$$

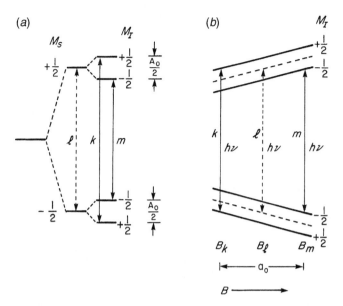

FIGURE 2.4 Energy levels of a system with one unpaired electron and one nucleus with $I = \frac{1}{2}$ (e.g., the hydrogen atom). (*a*) At a sufficiently high fixed magnetic field *B*. The dashed line would be the transition corresponding to $h\nu = g\beta_e B$ in the absence of hyperfine interaction (A_0). The solid lines marked *k* and *m* correspond to the allowed EPR transitions with hyperfine coupling operative. To first order, $h\nu = g\beta_e B \pm \frac{1}{2}A_0$, where A_0 is the isotropic hyperfine coupling constant. (*b*) As a function of an applied magnetic field. The dashed line corresponds to the transition in the hypothetical case of $A_0 = 0$. The solid lines *k* and *m* refer to transitions induced by a constant microwave quantum $h\nu$ of the same energy as for the transition ℓ. Here the resonant-field values corresponding to these two transitions are, to first order, given by $B = h\nu/g\beta_e \pm \frac{1}{2}(g_e/g)a_0$, so that $(g_e/g)a_0$ (measured in mT) is the hyperfine splitting constant given approximately by $B_m - B_k$. Note that these diagrams are specific to a nucleus with positive g_n and A_0 values, such as ^{1}H.

Note that the nuclear Zeeman terms cancel out. We examine these two EPR transitions under two conditions: constant magnetic field and constant frequency.

1. *Constant Magnetic Field B.* When the frequency is swept and $A_0 = 0$, a single transition occurs at a frequency $\nu = h^{-1}g\beta_e B$ (see the dashed transition mark in Fig. 2.4*a*). For non-zero hyperfine interaction, transitions occur at the two frequencies

$$\nu_k = h^{-1}\left(g\beta_e B + \tfrac{1}{2}A_0 + \cdots\right) \qquad \left(M_I = +\tfrac{1}{2}\right) \qquad (2.45a)$$

$$\nu_m = h^{-1}\left(g\beta_e B - \tfrac{1}{2}A_0 + \cdots\right) \qquad \left(M_I = -\tfrac{1}{2}\right) \qquad (2.45b)$$

(see transitions marked *k* and *m* in Fig. 2.4*a*). Note that each of the two transitions occurs between levels of identical M_I value. This corresponds to the selection rules $\Delta M_S = \pm 1$, $\Delta M_I = 0$ for EPR absorption.[12]

2. *Constant Microwave Frequency v.* Here the magnetic field is swept slowly. When $A_0 = 0$, a single transition occurs at the resonant magnetic field $B' = h\nu/g\beta_e$ (see the dashed transition in Fig. 2.4b). With $A_0 \neq 0$, EPR transitions occur at the two magnetic fields

$$B_k = h\nu/g\beta_e - A_0/2g\beta_e \cdots \qquad \left(M_I = +\tfrac{1}{2}\right) \qquad (2.46a)$$

$$B_m = h\nu/g\beta_e + A_0/2g\beta_e \cdots \qquad \left(M_I = -\tfrac{1}{2}\right) \qquad (2.46b)$$

(see transitions marked k and m in Fig. 2.4b).

The resonance equation becomes

$$h\nu = g\beta_e B + A_0 M_I + \cdots = g\beta_e[B + (g_e/g)a_0 M_I] + \cdots \qquad (2.47)$$

Here

$$a_0 = A_0/g_e\beta_e \qquad (2.48)$$

is the hyperfine *splitting* constant (in magnetic-field units), and the factor g_e/g represents the chemical shift correction (Eq. 1.22b). To first order, the hyperfine splitting is $(g_e/g)a_0$. For many free radicals, g is sufficiently close to g_e to allow neglect of the deviation from unity of the ratio g/g_e in Eq. 2.47.

Finally, we note that with this chemical system, the other type of magnetic-resonance transition (i.e., NMR) also occurs.[10] There are two pure NMR transitions, in which the electron-spin direction remains unaltered, but the nuclear-spin flips. Of more interest for our purposes are the electron-nuclear double-resonance (ENDOR) experiments, in which the two appropriate excitation magnetic fields are applied simultaneously (Chapter 12). A major advantage of this technique is the simplification of spectra, which facilitates analysis and measurement of spectral parameters for all unpaired-electron systems in which nuclear spins are present.

2.5 ENERGY LEVELS OF A SYSTEM WITH $S = \frac{1}{2}$ AND $I = 1$

The ^{2}H (deuterium) atom is a simple example of a system with $S = \frac{1}{2}$ and $I = 1$. As in Section 2.4, the energy levels are computed using the spin-hamiltonian operator $\hat{\mathcal{H}}$ (Eq. 2.40).[13] There are now six spin states, which are represented by $|M_S, M_I\rangle$

$$
\begin{array}{ll}
|+\tfrac{1}{2}, +1\rangle & |-\tfrac{1}{2}, -1\rangle \\
|+\tfrac{1}{2}, 0\rangle & |-\tfrac{1}{2}, 0\rangle \\
|+\tfrac{1}{2}, -1\rangle & |-\tfrac{1}{2}, +1\rangle
\end{array}
$$

These energies, given to first order by expressions analogous to Eqs. 2.43 (using the appropriate matrix elements—Section B.10), are

$$U_{+1/2, +1} = \frac{1}{2} g\beta_e B - g_n\beta_n B + \frac{1}{2} A_0 \qquad U_{-1/2, -1} = -\frac{1}{2} g\beta_e B + g_n\beta_n B + \frac{1}{2} A_0$$
$$U_{+1/2, 0} = \frac{1}{2} g\beta_e B \qquad\qquad\qquad U_{-1/2, 0} = -\frac{1}{2} g\beta_e B$$
$$U_{+1/2, -1} = \frac{1}{2} g\beta_e B + g_n\beta_n B - \frac{1}{2} A_0 \qquad U_{-1/2, +1} = -\frac{1}{2} g\beta_e B - g_n\beta_n B - \frac{1}{2} A_0$$

$$(2.49)$$

By virtue of the selection rules $\Delta M_S = \pm 1$ and $\Delta M_I = 0$, there are three allowed EPR transitions. These are depicted in Fig. 2.5a; a typical first-derivative spectrum in an increasing magnetic field is shown in Fig. 2.5b. A spectrum of the deuterium atom trapped in crystalline quartz is shown as the middle three lines in Fig. 2.6 [5]. Under conditions of constant microwave frequency, transitions to first order occur at the resonant fields

$$B_k = \frac{h\nu}{g\beta_e} - \frac{g_e}{g} a_0, \qquad B_l = \frac{h\nu}{g\beta_e}, \qquad B_m = \frac{h\nu}{g\beta_e} + \frac{g_e}{g} a_0 \qquad (2.50)$$

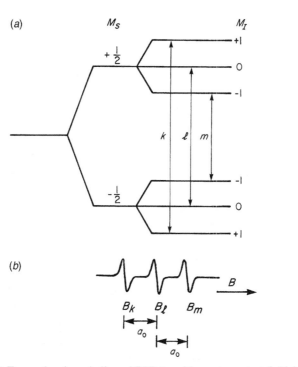

FIGURE 2.5   (a) Energy levels and allowed EPR transitions at constant field for an $S = \frac{1}{2}$, $I = 1$ atom (e.g., deuterium), for which $A_0 > 0$; (b) simulated constant-frequency spectrum.

for $M_I = +1$, 0 and -1. These lines are of equal intensity since there is no coincidence of states, that is, all states are non-degenerate.

The extension to systems with $S = \frac{1}{2}$ and $I > 1$ is straightforward. For $I = \frac{3}{2}$, four transitions of equal intensity are observed. In general, for a single nucleus interacting with one unpaired electron, there are $2I + 1$ lines of equal intensity; adjacent lines are separated by the hyperfine splitting a.

In this chapter expressions have been obtained for the energy levels of systems in which a single electron interacts with *one* magnetic nucleus. In most free radicals the unpaired electron interacts with a number of magnetic nuclei.

For instance, when a hydrogen atom is trapped in a crystal structure in which its surrounding atoms have nuclear spins, then *superhyperfine* structure is resolved. In the case of CaF_2, there are eight nearest-neighbor F^- ($I = \frac{1}{2}$) ions arranged at the corners of a cube, giving rise to such splittings (Fig. 2.7).

Examples of practical procedures for determining qualitative hyperfine splitting patterns, when more than one magnetic nucleus interacts with the unpaired electron, are given in the next chapter.

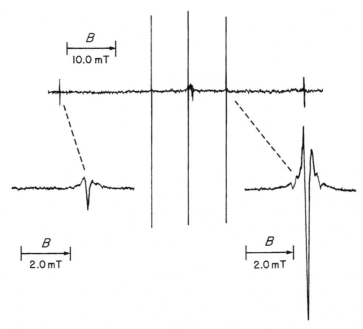

FIGURE 2.6 X-band EPR spectrum of isotopically enriched atomic hydrogen trapped in x-irradiated α-quartz at 95 K, presented in (almost pure) dispersion mode (Section F.3.5). The outer lines arise from ^1H (Fig. 1.4) and the three inner lines, from ^2H (Fig. 2.5). There are some unidentified lines present to the right of the central line. [After J. Isoya, J. A. Weil, P. H. Davis, *J. Phys. Chem. Solids*, **44**, 335 (1983).]

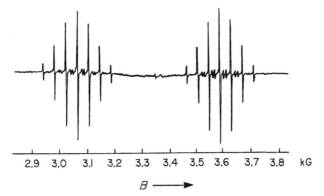

FIGURE 2.7 Room-temperature x-band EPR spectrum of interstitial hydrogen atoms in x-irradiated CaF_2. The weak lines barely visible are 'forbidden', analogous to transitions b and c of Fig. 5.4. [The small central line arises from DPPH used as a g marker (Section F.4).] [After J. L. Hall, R. T. Schumacher, *Phys. Rev.*, **127**, 1892 (1962).]

2.6 SIGNS OF ISOTROPIC HYPERFINE COUPLING CONSTANTS

The sign of the hyperfine coupling constant determines the energy order of the zero-field levels. Thus for $A_0 > 0$ (e.g., atomic hydrogen — Fig. C.1a), the triplet [$F = 1$ (Eq. C.9)] lies above the singlet ($F = 0$). For $A_0 < 0$, the opposite would be true. Here the EPR spectrum is unaffected by the sign of A_0. However, in principle, at a sufficiently low field ($B > 0$) and temperature, the NMR spectrum would reveal the sign of A_0, since one of the two NMR lines would be of lower intensity.

For one-electron atoms, the hyperfine coupling is given by Eq. 2.38. Thus, in this simple case, the sign of g_n determines that of A_0. Physically, the sign of A_0 indicates whether the magnetic moments of the electron and nucleus tend to align parallel or antiparallel. Note that A_0 is a property of the spin system considered and does not depend on the direction or magnitude of any external magnetic fields present.

More generally, for multielectron systems, there is another factor, which takes into account a mutual unpairing interaction between the electrons; that is, an outer unpaired electron may cause inner electron pairs to exhibit *spin polarization*, either parallel or antiparallel to it. In molecules, there may be regions with either polarization. The net electron-spin polarization around any nucleus determines the sign of its A_0.

We can quantify these ideas by using a generalized expression

$$A_0 = \frac{2\mu_0}{3} g\beta_e g_n\beta_n\langle\rho_s\rangle \tag{2.51}$$

for the isotropic hyperfine interaction parameter at nucleus n, appropriate for electronic state ψ of some atomic or molecular species. This allows us to take into account how each individual electron spin contributes at nucleus n, with its direction compared to the total electron-spin direction. In other words, expectation value $\langle \rho_s \rangle = \langle \psi | \hat{\rho}_s | \psi \rangle$ can contain both positive and negative contributions $|\phi_k|^2$ of individual electronic orbitals in ψ (Chapter 9); that is, it represents a competition at nucleus n between up-spin $|\alpha\rangle_k$ and down-spin $|\beta\rangle_k$ electrons. In the case of a single electron and nucleus, Eq. 2.51 reduces to Eq. 2.38. Finally, we note that $\langle \rho_s \rangle$ represents a true density, having dimensions of volume^{-1}. Thus $\langle \rho_s \rangle$ is called the *spin density*, and is itself a probability density (see Note 9.1). Further details concerning this concept, and the spin-density operator $\hat{\rho}_s$, are to be found in Chapters 5 and 9, as will the idea of unpaired-electron population.[14]

Ordinary first-order EPR spectra yield only $|A_0|$, since peak-position terms $A_0 M_I$ occur symmetrically with regard to $\pm |M_I|$ (Figs. 2.4 and 3.1). Higher-order hyperfine correction terms (Section 3.6) can yield relative signs (i.e., of A_{0i}/A_{0j}) when more than one nucleus (i.e., i and j) is present and both give sufficiently large splittings. Various other circumstances and special techniques yielding sign information are discussed later in this book (see Sections 3.6, 5.2, 5.3.2, 6.7, 9.2.4–9.2.7 and 10.5.5.1 and Problems 5.10 and 5.11).

2.7 DIPOLAR INTERACTIONS BETWEEN ELECTRONS

When two interacting spin moments arise from electrons, the ideas and theoretical aspects presented here remain valid. Of course, electrons are more mobile than nuclei and hence interact more readily so that exchange energy terms become important (Chapter 6). The magnitudes of the magnetic moments are about 2000 times (i.e., β_e/β_n) greater for electrons due to the smaller electron mass, so that the dipolar interaction energy at any given inter-particle distance r (Eq. 2.2) is greater by this factor. The detailed discussion of systems with more than one unpaired electron is deferred to Chapter 6, in view of the complications cited above.

REFERENCES

1. R. Skinner, J. A. Weil, *Am. J. Phys.*, **57**, 777 (1989).

2. J. E. Harriman, *Theoretical Foundations of Electron Spin Resonance*, Academic Press, New York, NY, U.S.A., 1978.

3. E. Fermi, *Z. Phys.*, **60**, 320 (1930).

4. G. Breit, I. I. Rabi, *Phys. Rev.*, **38**, 2082 (1931).

5. J. Isoya, J. A. Weil, P. H. Davis, *J. Phys. Chem. Solids*, **44**, 335 (1983).

6. A. Zheludev, V. Barone, M. Bonnet, B. Delley, A. Grand, E. Ressouche, P. Rey, R. Subra, J. Schweizer, *J. Am. Chem. Soc.*, **116**, 2019 (1994).

7. J. Schweizer, R. J. Papoular, E. Ressouche, F. Tasset, A. I. Zheludev, in *Electron, Spin and Momentum Densities and Chemical Reactivity*, P. G. Mezey, B. E. Robertson, Eds., Kluwer Academic, Dordrecht, Netherlands, 2000, pp. 37 ff.

8. Y. Pontillon, A. Caneschi, D. Gatteschi, E. Ressouche, F. Romero, J. Schweizer, R. Sessoli, R. Ziessel, *ibid.*, pp. 265 ff.

9. L. V. Bershov, *Geochem. Int.*, **10**, 853 (1970).

10. L. Essen, R. W. Donaldson, E. G. Hope, M. J. Bangham, *Metrologia*, **9**, 128 (1973).

11. J. S. Tiedeman, H. G. Robinson, *Phys. Rev. Lett.*, **39**, 602 (1977).

NOTES

1. The average position of these lines (Fig. 1.4) corresponds to $g \approx 2$. This averaging procedure for obtaining the g factor leads to an appreciable error when the separation of hyperfine lines is large, that is, >1 mT at $\nu = 10$ GHz (Section C.1.6).

2. Here we temporarily ignore (until Section 2.4) the small difference between g and g_e. Subindex 0 indicates that we are dealing with an isotropic parameter; that is, the free hydrogen atom basically is spherically symmetric (and we neglect small deviations from this induced by **B**).

3. This equation is valid only if the applied field is much greater than the magnetic fields present at the two particles as a result of the hyperfine interaction (Section 5.3.2).

4. The average (expectation) value of a quantity $g(q)$ weighted by a probability function $P(q)$ is given by

$$\langle g \rangle = \frac{\int gP \, dq}{\int P \, dq} \tag{2.52}$$

Each integration is taken over the allowed range of q.

5. In addition to this assumption, throughout this book, we utilize the concepts of electron distributions (orbitals) as derived from non-relativistic quantum mechanics. The changes in viewpoint and (in general) small corrections obtained from relativistic (Dirac) theory are beyond the scope of this book (see Ref. 2).

6. Bare protons and neutrons are not point particles, but rather have non-zero size (radius $\sim 0.7 \times 10^{-3}$ nm). Each contains a complex mixture of quarks, which yields a total spin of $I = \frac{1}{2}$, and a very appreciable nuclear magnetic moment (see Appendix H). Electrons can travel through these (and all) nuclides.

7. This is a general property of linear operators $\hat{A}$ and $\hat{B}$ for which the commutator $[\hat{A}, \hat{B}]_- = \hat{A}\hat{B} - \hat{B}\hat{A}$ is zero; that is, the operators commute (Section A.2).

8. It is often desirable to restructure the hamiltonian into a form that contains '*pseudo*-spin' operators, formally describing sets of states and their energy levels, even in the absence of actual spins. These are also called 'effective' or 'fictitious' spins.

9. $\phi_k{}^*$ is the complex conjugate of ϕ_k (Section A.1).

10. In this book we use the convention that all parameters (e.g., A_0) in a spin hamiltonian are assumed to be in energy units. Division by h is required to convert to frequency units and by $g_e\beta_e$ to convert to magnetic-field units (e.g., mT or G).

11. Hydrogen atoms and other free radicals (e.g., methyl) can be obtained by (say) γ-irradiation of certain solids, and in some circumstances are quite stable even at room temperature (e.g., see Ref. 9).

12. The selection rules $\Delta M_S = 0$, $\Delta M_I = \pm 1$ apply in the case of NMR spectroscopy, that is, when the system is irradiated only at the nuclear resonance frequency.

13. For $I > \frac{1}{2}$, another energy term should be added in Eq. 2.40, and represents the nuclear electric quadrupole interaction (see Section 5.6). However, this does not affect the EPR lines of isotropic systems.

14. Spin densities and population distributions can be measured quantitatively by polarized neutron diffraction. Such work for nitroxide free radicals[15] has been published by Schweizer and his group [6–8], and includes use of maximum-entropy reconstruction to generate projection maps.

15. Nitroxide is the name for species RR'N—O $\longleftrightarrow$ RR'N$^+$—O$^-$, also known as the aminoxyl free radical. These and related numerous species (including nitric oxide N—O itself) are of major importance now in biomedical areas.

FURTHER READING

1. D. J. Griffiths, *Hyperfine Splitting in the Ground State of Hydrogen*, Am. J. Phys., **50**, 698 (1982).

2. G. T. Rado, *Simple Derivation of the Electron-Nucleus Contact Hyperfine Interaction*, Am. J. Phys., **30**, 716 (1962).

PROBLEMS

2.1 (*a*) Carry out the integrations indicated in Eq. 2.3 and verify the result. (*b*) Compute $\langle \cos^2 \theta \rangle$ assuming that $\boldsymbol{\mu}_e$ and $\boldsymbol{\mu}_n$ are confined to a plane containing the magnetic field, for example, by setting $\phi = 0$. Assume that all values of θ are equally probable.

2.2 Taking the value of μ_p for the hydrogen nucleus (^{1}H) from Table H.1, compute the local magnetic field at an electron 0.2 nm from a proton when $\theta = 0°$ and again when $\theta = 90°$. What assumption is made in applying Eq. 2.2?

2.3 The experimental hyperfine coupling constant A_0/h for the free hydrogen atom in its electronic ground state is 1420.40575 MHz, and $g = 2.0022838$. [The data were taken from MASER (microwave amplification by stimulated emission of radiation) experiments [10,11].] Compare the value of A_0/h with that calculated using elementary quantum theory, that is, Eqs. 2.4 and 2.5. Can you give some reasons for the deviation (which can be represented by using a multiplicative correction factor $1 + \delta$ appended to Eq. 2.4)?

2.4 The sodium nucleus (^{23}Na) has $I = \frac{3}{2}$.

   (*a*) Specify the possible spin eigenfunctions for the sodium atom in its electronic ground state (^2S).

   (*b*) Use the spin hamiltonian of Eq. 2.40 (neglecting the nuclear Zeeman term $\hat{\mathcal{H}}_n$ therein) to derive expressions for the first-order energies of this spin system. What was assumed in using Eq. 2.40 and by neglecting $\hat{\mathcal{H}}_n$?

(*c*) Derive expressions for the possible EPR transitions and draw energy-level diagrams similar to Figs. 2.4 and 2.5. Use A_0 taken from Table H.4 for this purpose.

2.5 Calculate the energy levels of a two-proton radical RH_2 (Section C.2) at sufficiently high magnetic fields, using the spin hamiltonian of Eq. 2.41. R is any suitable molecular group. Write the spin eigenfunctions as $|M_S, {}^tM_I\rangle$, where ${}^tM_I = M_{I_1} + M_{I_2}$ (where M_{I_1} and M_{I_2} are the quantum numbers for the z components of the nuclear-spin angular momenta of protons 1 and 2). Plot the energy levels as a function of magnetic field (as in Problem 2.4) and indicate the allowed transitions and their relative intensities.

2.6 The EPR spectrum in Fig. 2.6 of hydrogen atoms trapped in a quartz crystal was taken at 9.94186 GHz. The ${}^1H$ doublet lines occur at 326.857 and 378.913 mT. Estimate the line positions of the ${}^2H$ triplet. Assume that $g = 2.002117$ for both species, and that high-field conditions pertain.

CHAPTER 3

ISOTROPIC HYPERFINE EFFECTS IN EPR SPECTRA

3.1 INTRODUCTION

In this chapter we continue to explore chemical species with a single unpaired electron ($S = \frac{1}{2}$). It was shown in Chapter 2 that the usual effect of the hyperfine interaction with a single proton ($I = \frac{1}{2}$) is to split each electron energy level into two, one pair for the $M_S = +\frac{1}{2}$ and one pair for the $M_S = -\frac{1}{2}$ states (Figs. 2.1 and 2.4). Interaction with a deuteron ($^2H = D, I = 1$) leads to splitting of each electron level into three (Fig. 2.5). The actual splitting observed is field-dependent at low fields but reaches a high-field limiting value of $A_0/2$, where A_0 is the isotropic hyperfine coupling constant. In general, if the nuclear spin is I, there are $2I + 1$ energy levels for each value of M_S.

Most free radicals contain several magnetic nuclei; in some molecules these may be grouped into magnetically equivalent sets. Usually the nuclei in a set are equivalent by virtue of the symmetry of the molecule; occasionally, the observed equivalence is accidental. The hyperfine spectra from radicals having numerous magnetic nuclei may give spectra rich in line components. The analysis of these spectra may be straightforward; more often a successful analysis requires some experience. This chapter presents a number of experimental spectra for radicals containing a single unpaired electron, ranging from the simple to the complex. The reader is urged to consider each spectrum carefully and to understand its analysis before proceeding to the next. Section 3.5 presents a number of rules that should aid in the analysis of complex spectra.

Electron Paramagnetic Resonance, Second Edition, by John A. Weil and James R. Bolton
Copyright © 2007 John Wiley & Sons, Inc.

Equivalent nuclei, each with spin I_i, can be treated by considering that they interact as *one* nucleus with a total composite nuclear spin ^tI equal to the sum $\Sigma_i I_i$ of the nuclear spins in the set and a corresponding total spin component $^tM_I = \Sigma_j M_{Ij}$; the number of levels is $2\,^tI + 1$ for each M_S value. This procedure gives the correct number of energy levels but does not account adequately for the degeneracy of some of the levels; this factor leads to variations in the relative intensities of the peaks. The degeneracies can be obtained from the simple rules given in Sections 3.2 and 3.6.

3.2 HYPERFINE SPLITTING FROM PROTONS

We first consider, for the sake of dealing with definite examples, the energy splittings and resultant spectral effects caused by the presence of protons, the most common nuclear-spin species in EPR. Thereafter we deal with other nuclides.

3.2.1 Single Set of Equivalent Protons

For a system of one unpaired electron interacting with two equivalent protons, it is possible to obtain the appropriate hyperfine energy levels by replacing the two nuclei with one nucleus having $I = 1$. The energy-level sequence is then the same as that in Fig. 2.5 for the deuterium atom; there the levels are labeled according to the value of M_I ($+1$, 0 or -1). Alternatively, the energy-level scheme may be obtained by successive splitting of levels as shown in Fig. 3.1a. Interaction with the first nucleus causes the $M_S = +\frac{1}{2}$ and $M_S = -\frac{1}{2}$ levels to be split by $A_0/2$; interaction with the second nucleus causes each level to be split again by $A_0/2$, since equivalence implies identity of hyperfine splitting constants. Figure 3.1a demonstrates that for the composite nuclear spin there is a coincidence of the intermediate levels ($M_I = 0$), in both the $M_S = +\frac{1}{2}$ and $M_S = -\frac{1}{2}$ groups. Note that the twofold degeneracy is associated with the two possible permutations of nuclear spins that give a net spin of zero (Fig. 3.1a). The factor of 2 in population of the $M_I = 0$ states as compared with the $M_I = +1$ or $M_I = -1$ states is reflected in the $1:2:1$ relative intensities of the allowed transitions. These intensities describe the system as long as $g\beta_e B \gg |A_0|$ and $k_b T \gg |A_0|$. Experimentally, these conditions almost always apply. The resulting simplest spectra are designated as 'first-order' spectra. These transitions are shown frequency-swept at constant field in Fig. 3.1a and field-swept at constant frequency in Fig. 3.1b. The selection rules are $\Delta M_S = +1$, $\Delta(^tM_I) = 0$, just as for the single-nucleus case.

For three equivalent nuclei, each with $I = \frac{1}{2}$, the repetitive splitting procedure leads to four levels for the $M_S = +\frac{1}{2}$ state and also for the $M_S = -\frac{1}{2}$ state.[1] The two inner levels each are three-fold degenerate, corresponding to the number of nuclear-spin states having $^tM_I = +\frac{1}{2}$ or $-\frac{1}{2}$. (Alternatively, the degeneracy of these levels may be viewed as a result of the fact that the $^tM_I = \pm\frac{1}{2}$ levels arise from the coincidence of a single level and a doubly degenerate level.) The normally applicable selection rules require that the allowed transitions occur between levels having the same

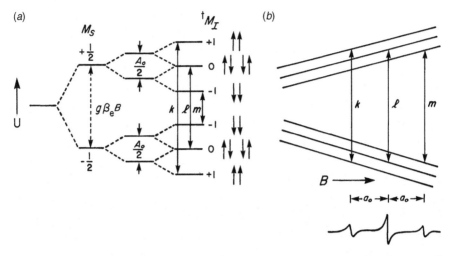

FIGURE 3.1 Energy levels and transitions for a system with one unpaired electron ($S = \frac{1}{2}$) and two equivalent nuclei with $I = \frac{1}{2}$. Here k, ℓ, and m denote the allowed transitions. Transition ℓ is twice as intense as k or m, since it occurs between doubly degenerate levels. (*a*) Constant-field conditions; the various possible configurations of the nuclear spins are shown at the right. (*b*) Constant-frequency conditions; this diagram assumes $A_0 > 0$. If $A_0 < 0$, the spectrum is unchanged; only the notation ^tM_I is altered ($\pm \rightarrow \mp$), where ^tM_I is given by Eq. 3.3.

value of ^tM_I and therefore having the same degeneracy. Hence the relative intensity of the observed lines is given by the ratio of the degeneracies of the levels between which transitions occur. Inspection of the intensity ratios $1:1$, $1:2:1$, $1:3:3:1$, and so on reveals that they are precisely the coefficients resulting from the binomial expansion[2] of $(1 + x)^n$, where n is the number of equivalent spin-$\frac{1}{2}$ nuclei in the set. The coefficient of the term x^x in the expansion represents the relative probability of occurrence of the state with spin component $^tM_I = \xi - {}^tI$, where $^tI = n/2$ is the total composite nuclear spin. The successive sets of coefficients for increasing n are readily found from Pascal's triangle (Fig. 3.2*a*, right-most column). Note that the sum of the values across any row is 2^n, which is the total number of energy levels (most of which are degenerate) for each value of M_S.

It is important for the reader to acquire and retain mental images of EPR spectra of radicals containing several equivalent protons. Figures 3.3*a*–*h* represent a collection of first-order spectra for radicals with up to eight equivalent protons. These have been drawn with the help of a computer so as to be both accurate and comparable. The spectrum of the benzene anion radical showing a septet of lines (Fig. 3.4) with the correct binomial intensities is a particularly important practical and historical example of this series.

It is true that the ideas described above, while aesthetically very pleasing, are only approximations, and that a closer look at reality reveals deviations from the simplest scheme (see Section 3.7).

(a)

$$I_i = 1/2 \quad (i = 1, 2, \ldots, n)$$

No. of nuclei n	No. of lines $2I + 1$	First-order line positions/a_0 M_I	Binomial intensity ratios
0	1	0	1
1	2	$\overline{1/2}$ 1/2	1 1
2	3	$\overline{1}$ 0 1	1 2 1
3	4	$\overline{3/2}$ $\overline{1/2}$ 1/2 3/2	1 3 3 1
4	5	$\overline{2}$ $\overline{1}$ 0 1 2	1 4 6 4 1
5	6	$\overline{5/2}$ $\overline{3/2}$ $\overline{1/2}$ 1/2 3/2 5/2	1 5 10 10 5 1
6	7	$\overline{3}$ $\overline{2}$ $\overline{1}$ 0 1 2 3	1 6 15 20 15 6 1
$\vdots$	$\vdots$	$\vdots$	$\vdots$

(b)

$$I_i = 1 \quad (i = 1, 2, \ldots, n)$$

No. of nuclei n	No. of lines $2I + 1$	First-order line positions/a_0 M_I	Multinomial intensity ratios
0	1	0	1
1	3	$\overline{1}$ 0 1	1 1 1
2	5	$\overline{2}$ $\overline{1}$ 0 1 2	1 2 3 2 1
3	7	$\overline{3}$ $\overline{2}$ $\overline{1}$ 0 1 2 3	1 3 6 7 6 3 1
4	9	$\overline{4}$ $\overline{3}$ $\overline{2}$ $\overline{1}$ 0 1 2 3 4	1 4 10 16 19 16 10 4 1
$\vdots$	$\vdots$	$\vdots$	$\vdots$

FIGURE 3.2 Triangles displaying the relative EPR line positions and intensities arising from the interaction of an electron-spin moment with the spin moments from n equivalent nuclei, each with spin I_i. The composite spin ^tI is a sum over all the individual spins I_i. (a) $I_i = \frac{1}{2}$; (b) $I_i = 1$. The right-hand triangles represent the coefficients in the expansion of $[1 + x + x^2 + \cdots + x^{2I_i}]^n$. The triangle for $I_i = \frac{1}{2}$ is usually attributed to Blaise Pascal. Note that the sum across any row is $(2I_i + 1)^n$ and that every non-peripheral integer is the sum of the $2I_i + 1$ integers closest above it.

As one encounters observed hyperfine splittings for a variety of radicals, including the hydrogen atom, one is struck by the marked smallness of the typical hyperfine splitting, as compared with the splitting of the hydrogen-atom doublet. Rationalization of this reduction is one of the important tasks of this book and is treated in Chapter 9.

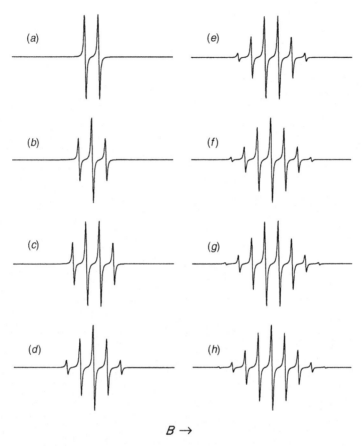

FIGURE 3.3 Computer simulations of EPR spectra for an unpaired electron interacting with (a) one, (b) two, (c) three, (d) four, (e) five, (f) six, (g) seven and (h) eight equivalent nuclei each with $I = \frac{1}{2}$. In all cases, $\nu = 9.50$ GHz, $B_r = 339$ mT, $a_0 = 0.50$ mT, and the lorentzian peak-to-peak linewidths ΔB_{pp} are 0.05 mT.

3.2.2 Multiple Sets of Equivalent Protons

As we have seen, the energy levels for hyperfine interaction with a single set of equivalent protons may be obtained by considering repetitive equal splitting of hyperfine levels of the $M_S = +\frac{1}{2}$ and $M_S = -\frac{1}{2}$ states. Chemically inequivalent protons, in general, have different splitting constants. Consider a radical containing two inequivalent protons, having hyperfine coupling constants A_1 and A_2, respectively, with $|A_1| \gg |A_2|$. The energy-level diagram may be constructed by representing the splitting $|A_1|/2$ arising from the first proton, and next taking each of the four resulting levels to be split into two levels separated by $|A_2|/2$. These energy levels are shown in Fig. 3.5. The typical allowed transitions are again those for which $\Delta M_S = \pm 1$ and $\Delta(^t M_I) = 0$. The spectrum shown in Fig. 3.5 is that of the HOCHCOOH radical. Here $g = 2.0038$, $|a_H(CH)| = 1.725$ mT and $|a_H(HOC)| = 0.255$ mT, at pH 1.3 [1].

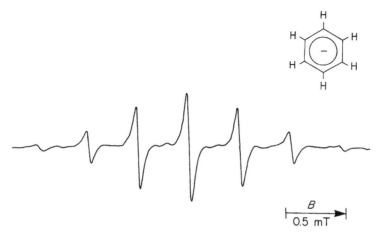

FIGURE 3.4 X-band EPR spectrum of the benzene anion radical in a solution of 2:1 tetrahydrofuran and dimethoxyethane at 173 K. Here the ^{13}C satellite lines are just barely visible. [After J. R. Bolton, *Mol. Phys.*, **6**, 219 (1963).]

No splitting from the acidic proton was observed. It was not possible to derive this assignment of the hyperfine splitting constants solely from an analysis of this spectrum; the assignment was made with the help of comparisons of these hyperfine splittings with those obtained from other similar radicals.

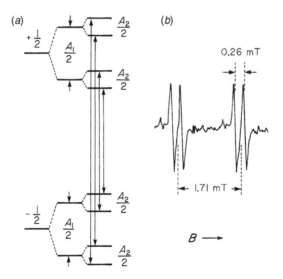

FIGURE 3.5 (*a*) Energy-level splitting by two inequivalent nuclei with $I = \frac{1}{2}$ in a given magnetic field. We have taken $A_1 > A_2 > 0$. (*b*) X-band EPR spectrum of the glycolic acid radical (HOCHCOOH) in aqueous solution at 298 K as an example of two inequivalent protons. The larger splitting arises from the CH proton and the smaller splitting from the nearest OH proton. (Spectrum taken by J. E. Wertz.)

The next case is that of a radical containing three protons, two of which are equivalent. Let A_1 be the hyperfine coupling constant for each proton of the equivalent pair and A_2 the coupling constant for the unique proton. Consider the case for $|A_1| \gg |A_2|$. One constructs the energy-level diagram by continuing the splitting process started in Fig. 3.1. Crossing of many levels during this construction may be avoided if the larger coupling constant is taken first. The final set of energy levels is independent of the order in which the splittings are considered. The CH$_2$OH radical (produced by photolysis of a methanol-H$_2$O$_2$ solution) is an example of this case. The spectrum is shown in Fig. 3.6. The smaller splitting is that of the OH proton: $|a(CH_2)| = 1.738$ mT and $|a(OH)| = 0.115$ mT at 26°C [2].

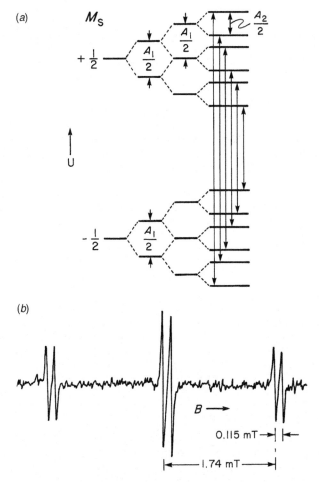

FIGURE 3.6 (*a*) Energy-level splitting by two equivalent nuclei plus another, all with $I = \frac{1}{2}$, in a given magnetic field ($A_1 > A_2 > 0$). (*b*) X-band EPR spectrum of the CH$_2$OH radical in methanol at 299 K. [After R. Livingston, H. Zeldes, *J. Chem. Phys.*, **44**, 1245 (1966).]

In the CH_2OH spectrum the six lines arise from a doubling by the unique proton of the three transitions expected for the two equivalent protons. In general, if there are sets of m and n equivalent protons in a molecule, then the maximum possible number of distinct lines in the spectrum is given by $(m + 1)(n + 1)$. Thus for an arbitrary number N of sets of such equivalent protons, the number of lines is given by $\Pi_j(n_j + 1)$, where Π_j indicates a product over all values of j $(= 1, 2, \ldots, N)$.

The 1,3-butadiene anion radical, $(H_2C\!=\!HC\!-\!CH\!=\!CH_2)^-$, is an example of a molecule with six protons, equivalent in sets of four and of two [3]. Whether the spectrum is seen to consist of three quintets or five triplets depends on the relative magnitude of the hyperfine splitting constants. These can be predicted with guidance from molecular-orbital theory, as discussed in Chapter 9. As is seen, the spectrum in Fig. 3.7a is readily interpreted in terms of five fully resolved groups of $1:2:1$ triplets. Here $|a_1| = |a_4| = 0.762$ mT; $|a_2| = |a_3| = 0.279$ mT. It is necessary to construct the set of energy levels for only one of the two M_S spin states, since the two sets of energy levels are mirror images.

When the energy levels are plotted to scale, the relative separation of levels corresponds to the separations of lines in the EPR spectrum. A set of lines is drawn with heights proportional to the degeneracy of the corresponding level. The relative amplitudes then correspond to the predicted relative intensities of the EPR lines. This 'stick-diagram' reconstruction of the spectrum is illustrated in Fig. 3.7b. The positions of the lines to first order in such a spectrum are a function of the proton hyperfine *splitting* constants a_j. The kth line in the spectrum is found at the field B_k given by

$$B_k = B' - \sum_{j=1}^{N} a_j{}^t M_{I(j)} \qquad (3.1)$$

where B' $(= h\nu/g\beta_e)$ is the magnetic field at the center of the spectrum, N is the number of different hyperfine splitting constants, and ${}^t M_{I(j)} = -{}^t I(j), -{}^t I(j) + 1, \ldots, +{}^t I(j) - 1, +{}^t I(j)$; here ${}^t I(j) = n_j/2$ is the total effective spin for the protons $(i = 1, 2, \ldots, n_j)$ of the jth set. In this example, $N = 2$, ${}^t I(1) = 2$ and ${}^t I(2) = 1$. As was discussed in Chapter 2, hyperfine splitting constants may have either positive or negative signs. In most cases the signs of the a_j parameters are unknown. In this situation, if we arbitrarily assign a positive sign to all a_j constants, then the negative ${}^t M_{I(j)}$ values label the high-field side of the spectrum (and vice versa).

We next consider the general case of a set of m protons and a set of n protons. The analysis of the spectrum is best carried out by beginning with the hyperfine pattern from the largest hyperfine splitting, taken to be a_n. The further splitting caused by the m protons (labeled $|a_m|$) is also shown. Consider the case $m = n = 4$ appropriate for the naphthalene anion radical (Fig. 3.8). The lines are labeled with the appropriate relative intensities. The reader may verify that (1) the central line of the final spectrum has an intensity 36 times that of the outermost components and (2) the sum of the relative intensities of all the lines is $2^8 = 256$. This is the number of energy levels for one value of M_S if eight protons are interacting with the unpaired electron.

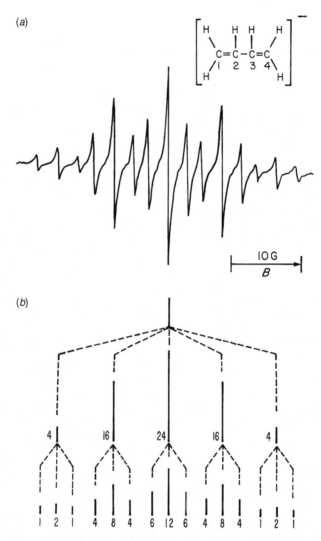

FIGURE 3.7 (*a*) X-band EPR spectrum of the 1,3-butadiene anion radical generated by electrolysis in liquid NH_3 at 195 K. [After D. H. Levy, R. J. Myers, *J. Chem. Phys.*, **41**, 1062 (1964).] (*b*) Reconstruction of this EPR spectrum, indicating relative intensities.

When $|a_n| > |a_m|$ but $|a_m|$ is sufficiently large, there is an intermingling of line groups. This is true for the naphthalene anion radical (Fig. 3.8), for which the splittings are given by $|a_1| = 0.495$ mT and $|a_2| = 0.187$ mT [4]. Here the analysis may not be immediately apparent. The separation of the outermost line from the next line is always the smallest hyperfine splitting. As an aid in the analysis, the degeneracy of the nuclear-spin states for each transition is given above the

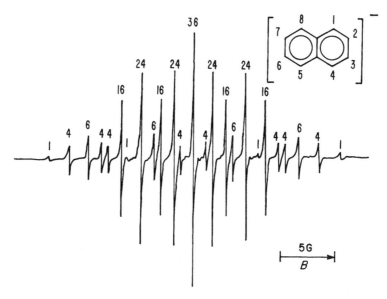

FIGURE 3.8 X-band EPR spectrum of the naphthalene anion radical in dimethoxyethane (K^+ is the counterion) at 298 K. The numbers above each line are the degeneracies of the corresponding nuclear-spin states. These numbers correspond approximately to the relative first-derivative amplitudes. (Spectrum taken by J. R. Bolton.)

corresponding line. The naphthalene anion is of special historical interest since it was the first radical for which proton hyperfine splitting was observed in solution [5].

If the difference between a_m and a_n is small, one may fail to see all the lines because of overlapping. Whenever large numbers of protons are involved, one must expect at least partial overlapping. If $a_n = ka_m$, where k is an integer, or the reciprocal of an integer, the spectrum has fewer than the expected number of lines, and the intensities do not follow a binomial distribution. There are numerous instances in the literature in which erroneous assignments have been made because of such accidental relations. When the *difference* between two splitting constants is exactly or nearly a multiple of another splitting constant, there is a further hazard of misassignment.

In some spectra one finds deviations from the binomial distribution of amplitudes. Such deviations are to be expected if the *linewidths* are different (see Appendix F for relations between amplitude and width). However, the integrated line *intensities* should follow the binomial distribution.

A simple example of a radical with three sets of symmetry-equivalent protons is that of the anthracene anion radical (Fig. 3.9). Here $|a_1| = 0.274$ mT, $|a_2| = 0.151$ mT and $|a_9| = 0.534$ mT [6]. The analysis of this spectrum is given as Problem 3.5.

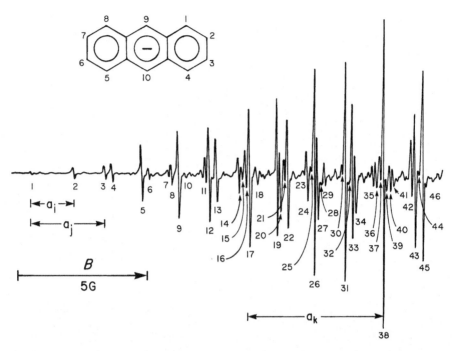

FIGURE 3.9 Low-field portion of the X-band EPR spectrum of the anthracene anion radical in dimethoxyethane at 295 K. Proton hyperfine lines are numbered; unnumbered lines arise from ^{13}C splittings. The three proton splitting constants are indicated. [After J. R. Bolton, G. K. Fraenkel, *J. Chem. Phys.*, **40**, 3307 (1964).]

3.3 HYPERFINE SPLITTINGS FROM OTHER NUCLEI WITH $I = \frac{1}{2}$

In organic radicals the most common nuclei with $I = \frac{1}{2}$ are ^{1}H, ^{13}C, ^{19}F and ^{31}P. Proton hyperfine splittings have already been discussed at length. Hyperfine splitting from ^{19}F or ^{31}P is usually indistinguishable from proton hyperfine splittings. It is an important characteristic of solution EPR spectra that an analysis usually yields only the *spin* of the interacting nucleus and the *hyperfine splitting*. Other evidence is required to identify the interacting nucleus. For this reason, everything that has been said about the analysis and reconstruction of the spectra involving proton splittings also applies to ^{19}F and ^{31}P. For ^{19}F, variations in linewidths across the hyperfine set can, in some instances (Chapter 10), be used to make an assignment [7].

Nuclides ^{19}F and ^{31}P occur with 100% natural abundance (Table H.4). For elements in which more than one isotope is present in significant amount, an assignment can usually be made by comparing intensities of hyperfine multiplets with known nuclear spins and isotopic abundances (Table H.4). As an example, ^{13}C splittings are considered shortly.

^{19}F hyperfine splittings have been observed in many organic radicals such as perfluoro-*p*-benzosemiquinone (**I**) [8].

(I) perfluoro-*p*-benzosemiquinone

An interesting example is the CF_3 radical [9], since its geometric configuration, planar or pyramidal, has been controversial. The ^{13}C splittings observed in this species, at low temperatures due to its instability, have been helpful in resolving this question in favor of the pyramidal structure (Section 9.3). The EPR spectrum is shown in Fig. 3.16, where the ^{19}F splitting is 14.45 mT.

PO_3^{2-} is an example of a radical showing ^{31}P hyperfine splitting. This radical [10] has a very large isotropic splitting (~ 60 mT); this indicates that PO_3^{2-} has a pyramidal structure with approximately sp^3 hybridization at the phosphorus atom. If PO_3^{2-} were planar, the radical would have sp^2 hybridization and should show a much smaller isotropic splitting (Section 9.3).

The natural abundance of the isotope ^{13}C ($I = \frac{1}{2}$) is 1.11% (Table H.4). The more abundant isotope ^{12}C has $I = 0$. In some cases an increased instrumental gain when taking an EPR spectrum of an organic radical reveals satellite lines arising from ^{13}C hyperfine splittings.

Consider the simple case of a molecule containing one carbon atom, for example, CO_2^-. On the average, 1.11% of these molecules are $^{13}CO_2^-$. For these molecules two "satellite" lines arise from the ^{13}C splitting. The $^{12}CO_2^-$ spectrum consists of only one line, since ^{12}C and ^{16}O have zero nuclear spin. The intensity of the $^{13}CO_2^-$ spectrum is divided between two lines; hence each line has an intensity of $(\frac{1}{2})(1.11/98.89) \times 100 = 0.561\%$ of that of the $^{12}CO_2^-$ spectrum. For molecules that contain n equivalent carbon atoms, the intensity of each satellite relative to that of the central component is $0.00561n$. In the spectrum of the benzene anion radical (Fig. 3.4), each satellite has an intensity of 3.37% of its central ^{12}C component.

3.4 HYPERFINE SPLITTINGS FROM NUCLEI WITH $I > \frac{1}{2}$

The most commonly encountered examples of nuclei with $I = 1$ are ^{2}H and ^{14}N. To say that a nucleus has a spin $I = 1$ means that in a magnetic field three orientations are allowed; these are labeled by the values of $M_I = 0, \pm 1$. These states are non-degenerate in a magnetic field, in contrast to the first-order case for two equivalent protons. Hence the spectrum should consist of three equally intense lines (Fig. 2.5).

Fremy's salt, $K_2(SO_3)_2NO$, containing the peroxylamine disulfonate (PADS) anion, offers an interesting example of an $S = \frac{1}{2}$, $I = 1$ inorganic species. In the solid phase, it occurs as a spin-paired dimer giving no ground-state EPR (but see Section 6.3.5). When dissolved in water, it gives a narrow-line EPR spectrum (Fig. 3.10) exhibiting the three-peak ^{14}N hyperfine splitting, as well as weaker lines arising from the low-abundance ^{15}N, ^{33}S and ^{17}O isotopes (Table H.4). The radical can serve as a very convenient intensity standard (for spin concentration and spectrometer sensitivity) and as a magnetic-field calibrant (Section F.1.2). It has been used to explore spin-relaxation mechanisms, using the electron-electron (ELDOR) technique (Chapter 12) [11].

For two equivalent $I = 1$ nuclei, one expects five EPR lines with an intensity distribution $1:2:3:2:1$ (Fig. 3.11). An example of a splitting from two equivalent ^{14}N nuclei is given in Fig. 3.12 for the nitronylnitroxide radical. Another species of this type, widely used as a field (g) standard (Sections F.1.2 and F.3), is the stable free radical 2,2'-diphenyl-1-picrylhydrazyl ($\mathbf{II} = $ DPPH), which, in liquid solvents, gives a five-line pattern arising from the two almost equivalent

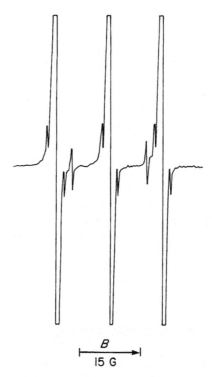

$$\overset{B}{\longmapsto\hspace{-2pt}\longrightarrow}$$
15 G

FIGURE 3.10 X-band EPR spectrum (at high gain) of the anion, $(SO_3)_2NO^{2-}$, of Fremy's salt in a 0.005 M aqueous solution at room temperature. The smaller peaks arise from ^{15}N and ^{33}S hyperfine interaction. [After J. J. Windle, A. K. Wiersema, *J. Chem. Phys.*, **39**, 1139 (1963).]

nitrogens.

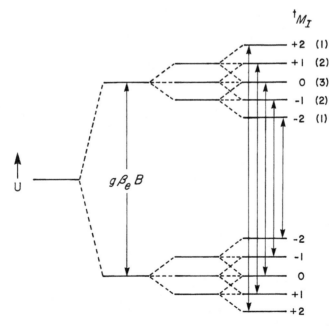

(II) 2,2′-diphenyl-1-picrylhydrazyl

In very dilute deoxygenated solution, the rich proton hyperfine structure also becomes evident.

For systems in which n equivalent nuclei of spin 1 interact with an unpaired electron, the relative intensities of the lines in a hyperfine multiplet are given by the $(n + 1)$th row of an extended Pascal triangle (Fig. 3.2b).

As mentioned before, analysis of the solution EPR spectrum does not generally identify the interacting nuclei. One method of assigning hyperfine splittings is the use of isotopic substitution. The most widely used isotope has been deuterium.

FIGURE 3.11 Energy-level diagram showing the hyperfine levels for two equivalent nuclei with $I = 1$. Each number in parentheses indicates the degeneracy of the adjacent level.

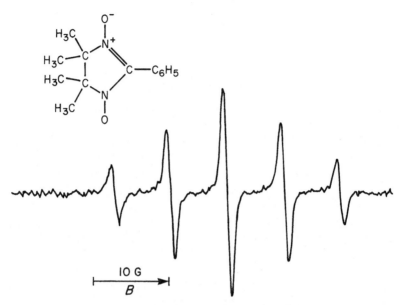

FIGURE 3.12   X-band EPR spectrum of a substituted nitronylnitroxide radical at 295 K in benzene showing splitting arising from two equivalent nitrogen atoms ($g = 2.00651$). No proton hyperfine splittings are sufficiently large to be discernible. [After J. H. Osiecki, E. F. Ullman, *J. Am. Chem. Soc.*, **90**, 1078 (1968).]

Substitution of one hydrogen atom with deuterium may permit assignment of the splittings of the remaining protons (e.g., Fig. 2.6). The two hyperfine splittings in the naphthalene anion (Fig. 3.8) were assigned by this procedure [12].

The most common nuclei with spin $I = \frac{3}{2}$ are ^{7}Li, ^{11}B, ^{23}Na, ^{35}Cl, ^{37}Cl, ^{39}K, ^{53}Cr, ^{63}Cu and ^{65}Cu. There are four nuclear-spin states, and consequently with a single such nucleus one should observe four hyperfine lines of equal intensity. Sometimes EPR spectra of radical anions exhibit small hyperfine splittings from alkali-metal cations. Such splittings indicate the presence of ion pairs in solution. Figure 3.13 shows a spectrum of the pyrazine anion radical (**III**) prepared by alkali-metal reduction of pyrazine in dimethoxyethane [13].

(III) pyrazine anion radical

In this solvent the species exists as a 1 : 1 ion pair with the alkali-metal counterion. In Fig. 3.13*b* the ^{39}K hyperfine splitting is not resolved, but the expected 25 lines from interaction with the two equivalent nitrogens and four equivalent hydrogens are all visible. In Fig. 3.13*a* each such line is split into a quartet from the ^{23}Na ($I =$

(a)

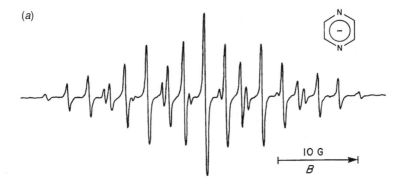

(b)

FIGURE 3.13 (*a*) X-band EPR spectrum of the pyrazine anion radical at 297 K in dimethoxyethane when Na^+ is the counterion; here ^{23}Na quartet hyperfine structure is observed. [After J. dos Santos-Veiga, A. F. Neiva-Correia, *Mol. Phys.*, **9**, 395 (1965).] (*b*) When K^+ is used as the counterion, no quartet splitting from ^{39}K is observed. [After A. Carrington, J. dos Santos-Veiga, *Mol. Phys.*, **5**, 21 (1962).]

$\frac{3}{2}$) hyperfine interaction. A reason that no ^{39}K hyperfine splitting is observed in Fig. 3.13*b* is the very small magnetic moment of the ^{39}K nucleus (Table H.4).

Because lineshape derivatives have positive and negative regions, superpositions of adjacent EPR lines exhibit regions of constructive and destructive interference. A

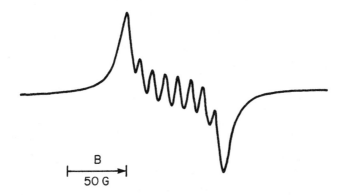

FIGURE 3.14 The X-band first-derivative spectrum of $(p\text{-}NHCOCH_3)_1TPPCo$ (N-MeIm) O_2 in toluene solution at 228 K. Here TPP designates tetraphenylporphyrin and MeIm is methylimidazole [After F. A. Walker, J. Bowen, *J. Am. Chem. Soc.*, **107**, 7632 (1985).]

good example is the spectra of complexes $LCoO_2$, in which the single unpaired electron is located mostly on the superoxide (O_2^-) ion bonded to a cobalt ion (Co^{3+}). The latter (^{59}Co, $I = \frac{7}{2}$, 100% abundance) provides the hyperfine structure: eight adjacent equally intense lines (Fig. 3.14 and Problem F.4). Here L represents the set of other ligands bonded to the cobalt ion. These species are synthetic analogs of dioxygen-carrying proteins (e.g., myoglobin and hemoglobin).

3.5 USEFUL RULES FOR THE INTERPRETATION OF EPR SPECTRA

The following are a few important rules that aid in the interpretation of isotropic EPR spectra:

1. If the applied field B is sufficiently large compared to the hyperfine splittings, the positions of lines are expected to be symmetric about a central point and are given by Eq. 3.1. If hyperfine splittings are sufficiently large, second-order interactions can cause asymmetry of the spectrum (Section 3.6). Variations in spectral linewidths may arise from a slow tumbling rate of the radical (Section 10.5.5). This may give an appearance of asymmetry. Asymmetry may also be caused by superposition of lines from radicals having different g factors.

2. A spectrum having no intense central line indicates the presence of an odd number of equivalent nuclei of half-integral spin. The observation of a central line does not exclude the presence of an odd number of nuclei.

3. For first-order spectra with hyperfine splittings from nuclei each with $I = \frac{1}{2}$, the sum $\Sigma_j n_j |a_j|$, where $j = 1 \ldots N$ for *all* nuclei, must equal the *spectral extent*. Here N is the number of sets of equivalent nuclei and n_j is the number of nuclei with the hyperfine splitting a_j (absolute values). The spectral extent is the separation (in mT) between the outermost lines, which in multi-line spectra are often very weak and may therefore be missed.

4. The stick-plot reconstruction, if it is correct, should match the experimental line positions, especially in the wings of the spectrum. If the widths of all lines are equal and there is little overlap, then the relative amplitudes should correspond to the degeneracies.

5. The separation of the pairs of adjacent outermost lines is always the smallest hyperfine splitting.

6. The total number of energy levels in the system for one value of M_S is given by $\Pi_j(2I_{(j)} + 1)^{n_j}$ (where $j = 1, \ldots, N$); n_j is the number of nuclei with spin $I_{(j)}$ in set j.

7. The maximum possible number of lines (when second-order splittings are not resolved) is given by $\Pi_j(2^tI_{(j)} + 1)$ (where $j = 1, \ldots, N$). Here the composite spin is $^tI_{(j)} = \sum_{i=1}^{n_j} I_{(j)} = n_j I_{(j)}$. If the widths are unequal or the resolution poor so that overlap is serious, it may be desirable to undertake

a computer simulation of the spectrum, based on assumed hyperfine splittings and linewidths (Section F.1.1). When several hyperfine splittings are present, or more than one radical is present, it is imperative to carry out such a simulation as a test of the analysis. These assumptions can be tested using the case of the 1,3-butadiene anion radical discussed previously.

The reader is advised to work out Problem 3.5 as an example of the application of these rules.

3.6 HIGHER-ORDER CONTRIBUTIONS TO HYPERFINE SPLITTINGS

The analysis of hyperfine splittings presented here is valid only in cases where the hyperfine coupling energy is very much smaller than the electron Zeeman energy $g\beta_e B$ (e.g., see Fig. 3.8). Where hyperfine couplings are *large* (or equivalently the applied magnetic fields are *small*), additional shifts and splittings of some lines can occur. This extra splitting is usually called 'higher-order splitting' since, to derive this effect, the energies of the levels must be calculated at least to second order [9,14,15]. This is just one example of various higher-order effects that are encountered in magnetic-resonance analysis. The reader is urged to consult Sections C.1.6, C.1.7 and C.2.2 for examples.

The higher-order splittings briefly considered here normally are observed only for relatively large hyperfine splittings and commensurately narrow linewidths. They are noted here to alert the reader to their occurrence. Note that with single nuclei, only shifts (and no such splittings) occur, and these provide no information that is not already available from the first-order spectrum.

The following is only a brief outline of second-order splittings for the case of two equivalent nuclei of spin $I = \frac{1}{2}$. The central line of the $1:2:1$ triplet arises from transitions between degenerate energy levels. When A_0 becomes a significant fraction of $g\beta_e B$, this degeneracy is lifted and one observes four equally intense lines (Fig. 3.15a). In fact, all lines except one of the central pair have been shifted downward from the 'first-order' positions. The position of the unshifted line requires no correction to provide the true g factor, since there is no nuclear magnetism contribution.

In general, for a single type of nucleus when the nuclear spins $\hat{\mathbf{I}}_i$ of a set of n equivalent nuclei ($i = 1, 2, \ldots, n$) are added vectorially to form a total spin vector ${}^t\hat{\mathbf{I}}$ with component eigenvalues ${}^t M_I$, the second-order correction to the position of each line is given (since ${}^t M_I$ is conserved in these transitions) to a good approximation by

$$\Delta B = -\frac{A_0{}^2}{2\,g\beta_e h\nu}[{}^t I({}^t I + 1) - {}^t M_I^2] \tag{3.2}$$

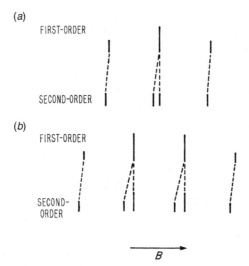

FIGURE 3.15 First-order and second-order splitting of (*a*) a $1:2:1$ triplet arising from two equivalent nuclei each with $I = \frac{1}{2}$, and (*b*) a $1:3:3:1$ quartet arising from three equivalent nuclei each with $I = \frac{1}{2}$. The second-order splittings are significant when $A_0/g\beta_e B \approx 0.01$.

valid since $S = \frac{1}{2}$. Here the projection has $2\,^t I_I + 1$ values:

$$^t M_I = -\,^t I, \; -\,^t I + 1, \ldots, +\,^t I \tag{3.3}$$

The shift is seen to be always downfield.

As a further example consider the case of three equivalent nuclei of spin $I = \frac{1}{2}$. The first- and second-order spectra are sketched in Fig. 3.15*b*. The spectrum of the CF_3 radical, shown in Fig. 3.16, displays the second-order splittings given by Eq. 3.2.

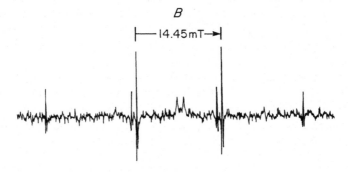

FIGURE 3.16 X-band second-derivative EPR spectrum of the CF_3 radical at 110 K in liquid C_2F_6, showing resolved second-order hyperfine splitting. The inner doublet arises from FO_2. [After R. W. Fessenden, R. H. Schuler, *J. Chem. Phys.*, **43**, 2704 (1965).]

When the energies contributed by the spin-hamiltonian term $A_0 \hat{S}_z \hat{I}_z$ are sufficiently large compared to the Zeeman terms in Eq. 2.40, then correction terms up to order 3, 4, 5, ... may be significant (Sections 2.4 and 5.3). This is the case when $A_0/g_e \beta_e \geq 10$ mT, for measurements at ~ 10 GHz. Here Breit–Rabi corrections, as discussed in Appendix C, must be made.

When several inequivalent nuclei contribute hyperfine splittings sufficiently large in magnitude that at least third-order terms are required, the *relative* signs of hyperfine splitting constants can be extracted [16].

3.7 DEVIATIONS FROM THE SIMPLE MULTINOMIAL SCHEME

On considering $S = \frac{1}{2}$ chemical systems XL_n featuring n equivalent $I = \frac{1}{2}$ nuclides, it turns out that

1. For $n > 2$, the degeneracy for the spin energy levels cannot be completely removed by any applied magnetic field.
2. For $n > 4$, certain spin states cannot occur at all in nature, consistent with the (generalized) Pauli exclusion principle.

The exact theory [17] predicts deviations from the simplest approach described above, which effects can in fact be observed when the ligand hyperfine splittings are sufficiently large (see Fig. 3.16).

3.8 OTHER PROBLEMS ENCOUNTERED IN EPR SPECTRA OF FREE RADICALS

Most of the EPR spectra encountered in this chapter refer to liquid samples in which the radicals are free to reorient rapidly. The radicals themselves generally are asymmetric species. However, the reorientation averages out any anisotropy in the g factor and in the hyperfine splittings. One should be aware that the splittings $(g_e/g)a$ may depend on both temperature and the solvent.

Free radicals are often encountered in a rigid matrix. If the host is a single crystal, one may obtain a maximum amount of information from the EPR spectra taken as a function of orientation of the crystal in the magnetic field. If the radicals are randomly oriented, one may still be able to extract a significant amount of structural information from their spectra. Analysis of such spectra requires a detailed understanding of the nature of anisotropic interactions. This subject is treated in Chapters 4 and 5, where various examples are discussed. For the spectra given in this chapter, all the lines have amplitudes that are proportional to the intensities of the lines. For other systems, this is frequently not the case. Thus radicals in media of high viscosity still undergo some reorientation or some degree of internal reorganization. Then the linewidths vary markedly throughout the spectrum. Analysis of such spectra can yield kinetic information. This subject is treated in Chapter 10. Obviously, free radicals in the gas phase are prime examples of isotropic behavior; these are discussed in Chapter 7.

3.9 SOME INTERESTING π-TYPE FREE RADICALS

Often it is convenient to label free radicals as being of the π (Section 9.2) and σ type (Section 9.3), which gives an idea as to where the unpaired electron tends to be located: in an s-type orbital or a p-type orbital. This topic will be discussed further in Chapters 5 and 9.

The hydroxyl free radical is of the π type, carrying its unpaired electron in a primarily p-type orbital on oxygen, and thus showing substantial anisotropy of the spin-hamiltonian parameters [18]. For this reason, and also because of its high mobility and its great chemical reactivity, it is very difficult to detect OH by EPR, especially in liquid solution. However, spin-trapping techniques do reveal it as having been present [19].

Another good example of a π radical ($S = \frac{1}{2}$) is $(CF_3)_2C_2S_2N$ (**IV**), which is stable and known in the solid state (dimer), liquid and even the gas phase (blue in color) [20]. It is easy to detect by EPR. When trapped dilute in solid argon at ~ 12 K, it offers a rewarding EPR spectrum [21]. The molecule (**IV**) is seen to feature a five-membered heteronuclear ring, and has symmetry C_2, down from C_{2v} because the two CF_3 groups are found via EPR to be magnetically inequivalent.

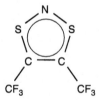

(**IV**) 4,5-bis (trifluoromethyl)-1,3,2-diathiazolyl

The unpaired electron, while delocalized somewhat, occurs primarily on the unique ring-nitrogen atom. Good agreement between the experimental and simulated EPR spectra, as well as the spin-hamiltonian parameters [g, $A(^{14}N)$, the $A(^{19}F)$ set; no $A(^{33}S)$ data] compared to these parameters obtained from quantum-mechanical modeling, was achieved.

There has been much discussion about the nature of the species present when an electron is dissolved in liquid water (e_{hydr}^-), and in many other solvents (e_{solv}^-). The first EPR paper about the former appeared in 1968 [22], dealing with a single line from an ephemeral radical obtained via pulsed electron beam irradiation. But despite myriad subsequent papers on this general topic, the actual paramagnet observed is still in doubt [23]; the species may well be hydronium, obtained via the radiation-stimulated reaction $2H_2O \Rightarrow H_3O \cdot + \cdot OH$. It too would be a π-type free radical.

REFERENCES

1. A. J. Dobbs, B. C. Gilbert, R. O. C. Norman, *J. Chem. Soc.*, 2053 (1972).
2. R. Livingston, H. Zeldes, *J. Chem. Phys.*, **44**, 1245 (1966).
3. D. H. Levy, R. J. Myers, *J. Chem. Phys.*, **41**, 1062 (1964).
4. N. M. Atherton, S. I. Weissman, *J. Am. Chem. Soc.*, **83**, 1330 (1961).
5. D. Lipkin, D. E. Paul, J. Townsend, S. I. Weissman, *Science*, **117**, 534 (1953).
6. J. R. Bolton, G. K. Fraenkel, *J. Chem. Phys.*, **40**, 3307 (1964).
7. M. Kaplan, J. R. Bolton, G. K. Fraenkel, *J. Chem. Phys.*, **42**, 955 (1965).
8. D. H. Anderson, P. J. Frank, H. S. Gutowsky, *J. Chem. Phys.*, **32**, 196 (1960).
9. R. W. Fessenden, R. H. Schuler, *J. Chem. Phys.*, **43**, 2704 (1965).
10. A. Horsfield, J. R. Morton, D. H. Whiffen, *Mol. Phys.*, **4**, 475 (1961).
11. N. M. Atherton, M. Brustolon, *Mol. Phys.*, **32**, 23 (1976).
12. T. R. Tuttle Jr., R. L. Ward, S. I. Weissman, *J. Chem. Phys.*, **25**, 189 (1956).
13. J. dos Santos-Veiga, A. F. Neiva-Correia, *Mol. Phys.*, **9**, 395 (1965).
14. R. W. Fessenden, *J. Chem. Phys.*, **37**, 747 (1962).
15. J. A. Weil, *J. Magn. Reson.*, **18**, 113 (1975).
16. R. W. Fessenden, *J. Magn. Reson.*, **1**, 277 (1969).
17. S. M. Nokhrin, J. A. Weil, D. F. Howarth, *J. Magn. Reson.* **174**(2), 209–218 (2005).
18. P. W. Atkins, M. C. R. Symons, *The Structure of Inorganic Radicals*, Elsevier, Amsterdam, Netherlands, 1967, pp. 104–107.
19. S. Laachir, M. Moussetad, R. Adhiri, A. Fahli, *Electronic J. Theor. Phys.*, **5**, 12 (2005).
20. A. G. Awere, N. Burford, C. Mailer, J. Passmore, M. J. Schriver, P. S. White, A. J. Banister, H. Oberhammer, L. H. Sutcliffe, *J. Chem. Soc., Chem. Commun.*, 66 (1987).
21. S. M. Mattar, A. D. Stephens, *J. Phys. Chem. A*, **104**, 3718 (2000).
22. E. C. Avery, J. R. Remko, B. Smaller, *J. Chem. Phys.*, **49**(2), 951 (1968).
23. A. J. Sobolewski, W. Domcke, *Phys. Chem. Chem. Phys.*, **4**(1), 4 (2002).
24. J. R. Morton, *Can. J. Phys.*, **41**, 706 (1963).
25. J. R. Morton, K. F. Preston, S. J. Strach, *J. Magn. Reson.*, **37**, 321 (1980).
26. J. R. Morton, W. E. Falconer, *J. Chem. Phys.*, **39**, 427 (1963).

NOTES

1. At this point the reader is urged to work out Problem 3.1. This should help ensure familiarity with successive splitting of hyperfine levels, as well as designation of allowed transitions and of line intensities.

2. The binomial expansion is

$$(p + q)^n = \sum_{x=0}^{n} \left[\frac{n!}{x!(n-x)} \, p^x q^{n-x} \right]$$

where n is a non-negative integer. Here $n! = 1 \times 2 \times 3 \times \ldots$, except $0! = 1$.

FURTHER READING

1. A. Carrington, A. D. McLachlan, *Introduction to Magnetic Resonance*, Harper & Row, New York, NY, U.S.A., 1967.
2. C. P. Poole Jr., H. A. Farach, *The Theory of Magnetic Resonance*, Wiley-Interscience, New York, NY, U.S.A., 1972.

PROBLEMS

3.1 (*a*) Using Fig. 3.1 as a guide, draw to scale three successive ^{19}F splittings of the spin-energy levels of the freely rotating CF_3 radical. Draw the allowed transitions, indicating intensities. Show by arrows the spin orientations for each of the M_I values. (*b*) Construct a similar plot for a radical with three protons, only two of which are equivalent. Take the splittings at constant field to be $A_1/2$ and $A_2/2$. Draw the allowed transitions, indicating intensities, and label them with the initial and final quantum numbers.

3.2 Consider Fig. 3.2. Work out the corresponding triangles for $I = \frac{3}{2}$ up to $n = 6$.

3.3 Consider the EPR spectrum of the CH_2OH radical shown in Fig. 3.6. The two proton coupling constants are cited in Section 3.2.2. How would the spectrum have appeared if the opposite assignment, $|a(CH_2)| = 0.115$ mT and $|a(OH)| = 1.738$ mT, had been made?

3.4 Use Eq. 3.1 and the liquid-solution hyperfine splittings of the 1,3-butadiene anion radical (Section 3.2.2) to specify the relative positions of all lines in the spectrum in Fig. 3.7*a*. Use the scale in that figure to measure the field value of each line relative to the center; compare with the computed values.

3.5 Complete the assignment of lines in the anthracene anion radical spectrum shown in Fig. 3.9, using the splittings given in the text and the rules given in Section 3.5 (rules 3 and 5 are particularly useful). Start with the outermost lines and move toward the center.

3.6 (*a*) When deuterium (^{2}H) is substituted for hydrogen (^{1}H) in a free radical, can one predict the value of a^D if a^H is known for the undeuterated radical? Assume that no other changes occur. (*b*) Figure 3.17 displays the spectrum

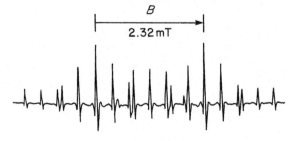

FIGURE 3.17 X-band second-derivative EPR spectrum of a mixture of CHD_2 and CH_2D in a Kr matrix at 85 K. [After R. W. Fessenden, *J. Phys. Chem.*, **71**, 74 (1967).]

of a mixture of radicals CH_2D and CHD_2. Identify the lines belonging to each spectrum and give values for each of the hyperfine splittings. Compute the ratio a^H/a^D. Compare this with the expected value.

3.7 Various dicobalt complexes $LCoO_2CoL$, closely related to the species $LCoO_2$ discussed in Section 3.4, are known [e.g., with L = 5 NH_3, 5 CN^-, 2 ethylene-diamine groups plus one (shared) NH_2 bridge, ...]. The single unpaired electron is located primarily on the bridging superoxo moiety. Draw an energy-level diagram analogous to that presented in Fig. 3.11, featuring the cobalt hyperfine splitting. Predict the EPR spectrum expected in liquid solution, that is, the number of lines and their relative intensities.

3.8 Figure 3.18 represents the spectrum obtained when a single crystal of KCl, doped with ^{33}S, is γ-irradiated. The crystal has been prepared from a sulfur-doped sample enriched in ^{33}S to an extent of 60%. The summed natural abundance of the spin-less nuclides (^{32}S and ^{34}S) is 99.3%. The spectrum has been ascribed to S_2^-.

(a) What are the various possible S_2^- species, and what are their relative abundances?

(b) How many lines are to be expected for each of these S_2^- species, and what are their relative intensities?

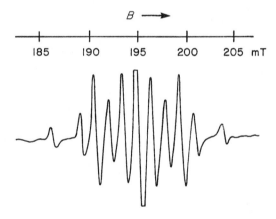

FIGURE 3.18   EPR spectrum at 9.550 GHz and 4 K of S_2^- in a γ-irradiated single crystal of KCl doped with ^{33}S (60% of S). [After J. R. Morton, *J. Phys. Chem.*, **71**, 89 (1967).]

(c) Compute the ^{33}S hyperfine splitting constant to first order.

(d) What is g at the crystal orientation used?

3.9 Consider the spectrum attributed to FPO_2^- ($S = \frac{1}{2}$) shown in Fig. 3.19 [24–26]. The strongest lines of the spectrum are those from this ion. Construct a stick-plot representation of this spectrum and extract the hyperfine splittings for ^{19}F and ^{31}P. Indicate reasons for the assignment made. The splittings are relatively

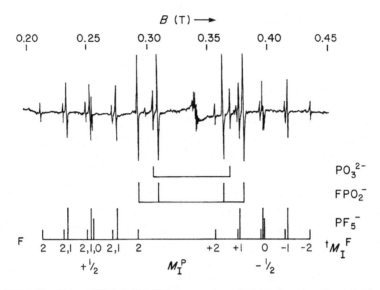

FIGURE 3.19   X-band EPR (9.510 GHz) spectrum at 295 K of a γ-irradiated single crystal of NH_4PF_6. The three radicals produced have been attributed to PF_5^-, FPO_2^- and PO_3^{2-}. The splittings of the PF_5^- lines arise from second-order interactions [16,24,25]. Here $^tI^F$ refers to the total fluorine nuclear-spin quantum number in the coupled representation (Section B.6). The central multiplet is attributed to $N_2H_4^+$. [After J. R. Morton, *Can. J. Phys.*, **41**, 706 (1963).]

large so that, for accurate work, corrections for higher-order energy terms should be made. Carry these out, using the theory given in Eq. C.25ff (see also Eqs. 3.2 and 5.10), to assess their magnitude.

3.10   The spectrum shown in Fig. 3.19 contains some lines that have now been assigned to the radical PF_5^-. The isotropic hyperfine coupling constants and g factor are as follows:

$$a^P = 135.66 \text{ mT}$$

$$a^{F(1)} = 19.78 \text{ mT}$$

$$a^{F(2)} \approx 4 \text{ mT (not resolved in Fig. 3.19)}$$

$$g = 2.00174$$

The phosphorus and major fluorine splitting patterns clearly show resolved second-order splittings.

(*a*) The unshifted line of the high-field group occurs at 384.84 mT. The separation of this line and the unshifted line of the low-field group arises from the ^{31}P hyperfine *splitting*. Why is this not quite the same as $A_0^P/g_e\beta_e$ (Section 2.4)?

(*b*) The unshifted line in the low-field fluorine splitting pattern occurs at 250.11 mT. Predict the positions of the other eight lines in the low-field group.

3.11 Consider the spectrum (Fig. 3.20—taken at high gain) arising from the 2,5-dioxy-*p*-benzosemiquinone trianion radical in basic aqueous solution. The main spectrum, consisting of a $1:2:1$ triplet from the two ring protons,

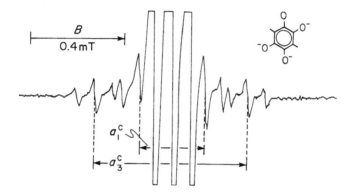

FIGURE 3.20 X-band EPR spectrum at 295 K in aqueous solution of the 2,5-dioxy-*p*-benzosemiquinone trianion radical. The off-scale triplet in the center arises from molecules having no ^{13}C nuclei. The satellite lines on the wings arise from molecules each containing one ^{13}C nucleus. [After D. C. Reitz, F. Dravnieks, J. E. Wertz, *J. Chem. Phys.*, **33**, 1880 (1960).]

is off scale in the center of the spectrum. Consider the satellite lines shown in the wings. Account for the number and relative intensities of these.

3.12 The EPR spectrum of XeF, observed in a γ-irradiated single crystal of XeF_4 [26], is given in Fig. 1.6. It is a simple example of a system in which distinct spectra are observed for different isotopic species. The relative abundances of ^{129}Xe and ^{131}Xe are 26.4% and 21.2%. The remaining 52.4% is distributed among the isotopes of mass numbers 124, 126, 128, 130, 132, 134 and 136. The XeF radicals containing these isotopes are referred to collectively as ^{even}XeF. There are 14 major lines (ignore the small doublet splitting on each line, which arises from a neighboring fluorine nucleus in the XeF_4 host). Analysis of the spectrum begins with a tabulation of expected line patterns and relative line intensities for different XeF species (Table 3.1). Use Fig. 1.6 and the following information to fill in the blanks in Table 3.1:

Radical	I	$\vert A^{Xe}\vert/h$ (MHz)	$\vert A^{F}\vert/h$ (MHz)	g
^{even}XeF	0	—	2649	1.9740
^{129}XeF	$\frac{1}{2}$	2368	2637	1.9740
^{131}XeF	$\frac{3}{2}$	701	2653	1.9740

All data quoted are for **B** parallel to the XeF axis. Mean error in A/h is ± 10 MHz.

TABLE 3.1   EPR Lines of Various XeF Isotopic Species

Species	Pattern of Lines	Line Numbers	Expected Relative Intensity of Lines	Xenon Nuclear g Factors	Mean Xenon Hyperfine Splitting ^{a} (MHz)	Mean Fluorine Hyperfine Splitting ^{a} (MHz)
^{even}XeF	One doublet	4,11	1.000	—	—	—
^{129}XeF	—	—	−1.55595	—	—	—
^{131}XeF	—	—	0.461243	—	—	—
Ratio 129/131	—	—	—	—	—	—

^{a} Mean values of the measured separation (in mT) of corresponding line components of XeF (Fig. 1.6).

3.13   Draw at least 3 of the (infinite number of) resonant forms for structure **IV**, those expected to have the most importance, showing where the unpaired electron might occur.

CHAPTER 4

ZEEMAN ENERGY (g) ANISOTROPY

4.1 INTRODUCTION

The solid state offers a broad variety of systems and phenomena observable with EPR spectroscopy. The technique is applicable to all types of solids, ranging from insulators via semiconductors to metallic conductors and superconductors. The simplest situation occurs when there is no interaction between the paramagnetic species. Much greater complexity occurs when the electron spins exist in highly correlated, magnetically concentrated systems; these may form aligned domains (examples are ferromagnetic and ferrimagnetic materials, as well as their antiferromagnetic and antiferrimagnetic counterparts, as well as superparamagnetic systems). For the most part we restrict ourselves to isolated paramagnetic centers in magnetically dilute systems.

The EPR spectra of oriented species in solids may be more complicated than for liquids; however, their analysis provides much additional useful information. One may hope to extract details of intra- and intermolecular interactions, molecular configuration, site symmetry, as well as the nature and location of neighboring atoms. Furthermore, one observes in rigid solids many paramagnetic species that are too reactive or too unstable to be detected in liquid solution. A discussion of the generation of such unpaired-electron entities is included in Appendix F.

Here we focus primarily on solids containing independent unpaired-electron species, limited to relatively few atoms in each center. When the paramagnetic center is not a normal component of the host material (and its electron is close to being localized), it is often called a '*point*' *defect*. Even with the restriction to

Electron Paramagnetic Resonance, Second Edition, by John A. Weil and James R. Bolton
Copyright © 2007 John Wiley & Sons, Inc.

magnetically dilute species, one encounters substantial differences in EPR properties, dependent on the form of the sample. One may deal with *single crystals, polycrystalline* systems (called *powders* when a sufficient number of randomly oriented small crystals are present), *amorphous,* or *glassy* systems. In single crystals the spin centers are limited to relatively few possible orientations relative to the body of the crystal itself; this results from highly organized long-range correlation of the constituent atoms. In glasses only local geometric correlation of atom positions exists. Here we refer to static disorder, there is no time dependence of any parameter.

In crystalline systems even the qualitative aspects of an EPR spectrum may be markedly dependent on the orientation of the sample, defined relative to the externally applied magnetic fields (**B** and **B**$_1$, Section 1.1). Such systems are said to be anisotropic in their behavior. Some important classes of paramagnetic systems that show such anisotropy include

1. Free radicals
2. Transition ions surrounded by ligands
3. 'Point' defects

We now undertake to discuss the important EPR aspects of such unpaired-electron centers occurring within single crystals, deferring the consideration of EPR in powders and glasses until the required background material has been presented.

In crystals the concept of symmetry is of crucial importance. The arrangements of atoms (including molecules where relevant) is classified according to a limited set of symmetry types. The latter are describable by utilizing group-theoretic considerations of symmetry operations about any given point, as well as by their long-range translational order. The latter are not essential for our purposes; in fact we need only the *eleven* so-called *proper point groups* to discuss all types of crystals encountered [1] (Table 4.1). Note that here only the rotational aspects of the repeat units of atoms

TABLE 4.1   The 11 Proper Rotation Groups of Crystals in Schoenflies Notation (International Symbol in Parentheses)

Group	$N_s{}^a$	Group	$N_s{}^a$
Cubic		Cyclic	
O (=432)	24	C_6 (=6)	6
T (=23)	12	C_4 (=4)	4
Dihedral		C_3 (=3)	3
D_6 (=622)	12	C_2 (=2)	2
D_4 (=422)	8	C_1 (=1)	1
D_3 (=32)	6		
D_2 (=222)	4		

^{a} Number of symmetry-related sites in each group. N_s gives the maximum number of magnetic-resonance symmetry-related spectra for each group (for details, see Ref. 1).

are relevant, and no attention need be paid to the presence or absence of crystal faces. Various tools (optical, diffraction by x rays, neutrons, and other particles, as well as magnetic resonance) are available to establish the proper point group of any given chemical system.

In addition to the crystal symmetry, the local symmetry at any unpaired-electron center is of prime importance. Such a center could be an impurity embedded in the crystal structure, and perhaps distorting it. We can classify this situation into one of three categories in order of decreasing local symmetry:

1. *Cubic*. Here there are three sub-systems, termed *cubal, octahedral* and *tetrahedral* (Fig. 4.1). In these, anisotropy of EPR properties is absent. In different terminology, we see that all three principal values are equal for each parameter matrix encountered.

2. *Uniaxial*. Here there is linear rotational symmetry (at least three-fold) about a unique axis contained in each paramagnetic species embedded in the crystal. Anisotropy is observable except with the field **B** in the plane perpendicular to the unique axis. Two principal values coincide but these differ from the third in each parameter matrix. This case is sometimes simply called 'axial'.

3. *Rhombic*. This is the general case, implying anisotropy for all rotations and the presence of three unequal principal values in each parameter matrix. In the literature this case is often called 'orthorhombic'.

These concepts are applicable to all magnetic properties of unpaired-electron species. For instance, S can take on any value $\frac{1}{2}, 1, \frac{3}{2}, \dots$. In this chapter we concentrate on the anisotropy of the line positions, that is, the g factors. As noted in Section 1.12 the g factor shift $g - g_e$ arises from the electromagnetic-field effects provided by the other electrons and nuclei in the magnetic species. In general these fields provide an anisotropic environment, and one should thus expect the g factor to also be anisotropic.

Specifically, the anisotropy of the g factor arises from admixture to the electron-spin angular momentum of a (generally small) amount of orbital angular

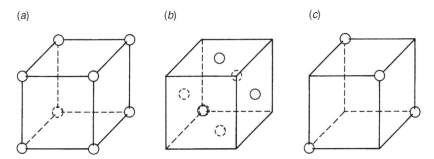

(a)   (b)   (c)

FIGURE 4.1 The subclasses of the cubic symmetry class: (*a*) cubal (eight nearest neighbors to a central point); (*b*) octahedral (six nearest neighbors); (*c*) tetrahedral (four nearest neighbors).

momentum. However, the latter need not be considered explicitly since its effect can be replaced by imputing anisotropy to *g*.

For the moment, we treat the *g* matrix as incorporating a set of parameters, without asking questions as to why their values are what they are. Quantum-mechanical models for evaluating principal *g* factors exist but are not simple, and we defer discussion of these ideas to Section 4.8.

4.2 SYSTEMS WITH HIGH LOCAL SYMMETRY

Before undertaking a general discussion of the line-position anisotropy, that is, of **g**, it is instructive to examine an unpaired-electron species located in an isotropic medium, namely, a cubic host crystal. Here, if one considers the time average over the rapid vibrational excursions, there is cubic symmetry about any normal lattice site. For an unpaired electron at such a site, *g* is strictly a scalar *constant*, and the spin hamiltonian has the form

$$\hat{\mathcal{H}} = g\beta_e(B_x\hat{S}_x + B_y\hat{S}_y + B_z\hat{S}_z) \tag{4.1}$$

For cubic *local* surroundings, the EPR line position is isotropic. The *g* factor in the simple resonance equation (Eq. 1.19) is independent of the magnetic-field direction only in isotropic systems. For example, an electron in a negative-ion vacancy (*F* center) in an alkali halide (Fig. 4.2*a*) is found to be delocalized symmetrically about the center of an octahedron of cations. Here the *g* factor is isotropic, as are

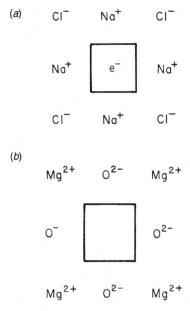

FIGURE 4.2   (*a*) Model of the *F* center in NaCl (cubic symmetry) and (*b*) model of the *V*⁻ center in MgO (tetragonal symmetry).

other properties of a system with local octahedral symmetry (i.e., proper point group O; Table 4.1).

On the other hand, the symmetry may be reduced from octahedral to tetragonal by applying an external stress along any one of the three [100]-type directions.[1] Alternatively, one may encounter (or introduce) an imperfection along one of these axes. The positive-ion vacancies (V centers) offer an example. The V^- center (earlier called V_1) in MgO or CaO (rock-salt structure; proper point group O) has one unpaired electron [3–6]. In an ideal crystal the Mg^{2+} and O^{2-} ions are all at sites of octahedral symmetry. On low-temperature x-ray irradiation, V^- is formed when an electron is removed from any one of six oxygen ions adjacent to a (preexisting) magnesium-ion vacancy; as a result there is a small displacement of the resulting O^- ion away from the vacancy. The geometric configuration of this defect center is shown in Fig. 4.2b. This ion carries an unpaired electron in a p-type orbital. The distortion leaves a fourfold axis of symmetry (i.e., uniaxial symmetry = tetragonal symmetry). It is customary to label this unique axis as the Z direction. It is taken to be horizontal in Fig. 4.2b.

If **B** is parallel to **Z** and $v = 9.0650$ GHz, an EPR line is observed at 323.31 mT. When the MgO crystal is rotated so that **B** remains in the YZ plane, the line of the V^- center shifts from 323.31 to 317.71 mT as the field direction changes from **Z** to **Y**. The variation in line position with orientation is shown in Fig. 4.3. Then we can define the parameters

$$g_\perp = \frac{hv}{\beta_e B_\perp} = \frac{6.62607 \times 10^{-34}\,\text{J s} \times 9.0650 \times 10^9\,\text{s}^{-1}}{9.27401 \times 10^{-24}\,\text{J T}^{-1} \times 0.31771\,\text{T}} = 2.0386 \quad (4.2a)$$

$$g_\parallel = \frac{hv}{\beta_e B_\parallel} = 2.0033 \quad (4.2b)$$

Here $g_\perp$ and $g_\parallel$ are the g factors appropriate to the magnitudes $B_\perp$ and $B_\parallel$ of the field when it is perpendicular and parallel to the symmetry axis (i.e., **Z**).

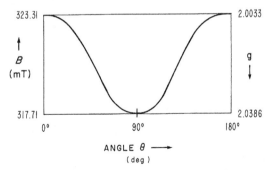

FIGURE 4.3   Angular dependence of the EPR spectrum of the V^- center in MgO, for **B** ∥ **X** (=[100]). Angle 0° indicates **B** ∥ **Z** (=[001]). Resonant-field values at g extrema are given at the left of the figure, for microwave frequency 9.0650 GHz. The corresponding g factors are shown at the right.

Such a uniaxial case is also encountered when transition ions (Section 8.2), with electron configuration nd^1 ($S = \frac{1}{2}$), are studied in tetragonal orthophosphates (group C_4; Table 4.1). The EPR parameters of Ti^{3+} ($n = 3$), Zr^{3+} ($n = 4$) and Hf^{3+} ($n = 5$) have all been measured at 77 K in single crystals of $ScPO_4$ (for Ti^{3+}), $LuPO_4$ (for Zr^{3+}) and YPO_4 (for Hf^{3+}) and display very similar g factors [7]. For instance, Ti^{3+} substituted for Sc^{3+} yields $g_\perp = 1.961$ and $g_\parallel = 1.913$.

The shape of the curve in Fig. 4.3 is found to be well represented by an effectual g factor given by the positive square root of

$$g^2 = g_\perp{}^2 \sin^2 \theta + g_\parallel{}^2 \cos^2 \theta \qquad (4.3)$$

where θ is the angle between **B** and the symmetry axis of the defect. We see that the two parameters $g_\perp$ and $g_\parallel$ allow us to find the line position at any arbitrary orientation. It is shown later that Eq. 4.3 is a special case of a more general expression (Eqs. 4.6). Equation 4.3 is applicable to all systems possessing a local symmetry axis of order 3 or higher. For such a system (i.e., one having uniaxial symmetry), the spin hamiltonian (in the absence of hyperfine interaction) is

$$\hat{\mathcal{H}} = \beta_e[g_\perp(B_X\hat{S}_X + B_Y\hat{S}_Y) + g_\parallel B_z\hat{S}_Z] \qquad (4.4a)$$

This can be as the product of a row vector, a square matrix and a column vector:

$$\hat{\mathcal{H}} = \beta_e[B_X \quad B_Y \quad B_Z] \cdot \begin{bmatrix} g_\perp & 0 & 0 \\ 0 & g_\perp & 0 \\ 0 & 0 & g_\parallel \end{bmatrix} \cdot \begin{bmatrix} \hat{S}_X \\ \hat{S}_Y \\ \hat{S}_Z \end{bmatrix} \qquad (4.4b)$$

$$= \beta_e \mathbf{B}^{\mathrm{T}} \cdot g \cdot \hat{\mathbf{S}} \qquad (4.4c)$$

The superscript T is useful in indicating transposition of a spatial column vector to the same vector expressed as a row vector (Section A.4). Thus use of the g-matrix concept allows a convenient representation of anisotropy in the energy as a function of the B-field direction. In other words, complete knowledge of the (infinite) set of g factors for any given chemical system can be encapsulated in a 3×3 'parameter' matrix.[2] The g parameters in Eqs. 4.4 are elaborated in the next section and are considered in some detail in Section 4.8.

4.3   SYSTEMS WITH RHOMBIC LOCAL SYMMETRY

The systems to be considered now are the most complex ones, those with rhombic local symmetry, which are the ones most commonly encountered (e.g., in organic media). As an example, consider the defect center shown in Fig. 4.4, which is found in those alkali halides (e.g., KBr) having the rock-salt structure. Here the defect is the superoxide ion $O_2{}^-$, a paramagnetic diatomic molecule that has a

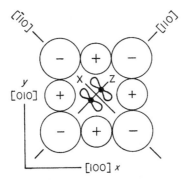

FIGURE 4.4 Projection onto plane xy of a unit cell for an alkali-halide crystal having the rock-salt structure, showing a substitutional O_2^- ion site. The molecular axis **Z** is on a crystal two-fold rotational symmetry axis, [1 1 0]. The oxygenic p lobes holding the unpaired electron are explicit, and are held 'in place' along [−1 1 0] as a result of polarization local distortion by one nearest-neighbor anion. The neighbors above and below the superoxide anion are cations.

single unpaired electron, which replaces a diamagnetic Cl^- ion [10]. It is convenient to choose the O—O interatomic direction as the axis **Z** of the local coordinate system. The axis **X** of the right-handed coordinate system is taken to lie in the plane defined by the two parallel $2p$ orbitals shown in Fig. 4.4, whereas **Y** is directed out of this plane.[3] Note that the axes **X** and **Y** are not equivalent. The symmetry of the defect is rhombic rather than uniaxial because of interaction with neighboring atoms. The spin hamiltonian (ignoring any hyperfine interactions) is

$$\hat{\mathcal{H}} = \beta_e(g_X B_X \hat{S}_X + g_Y B_Y \hat{S}_Y + g_Z B_Z \hat{S}_Z) \tag{4.5}$$

Here we encounter the rhombic case with $g_X = 1.9268$, $g_Y = 1.9314$ and $g_Z = 2.5203$ [10]. If it were possible to have all such diatomic defects present along a single crystal direction (say, indices [110] as indicated in Fig. 4.4), the spectrum according to Eq. 4.5 would consist of only a single orientation-dependent line. Consider the EPR line arising from such a set of O_2^- ions (in reality, there are five other sets; see Section 4.5). The g factors g_X, g_Y and g_Z are obtainable from the line positions measured with the field along the X, Y and Z directions. The effectual value of g for an arbitrary orientation is then given by the positive square root of

$$g^2 = g_X{}^2 \cos^2 \theta_{B,X} + g_Y{}^2 \cos^2 \theta_{B,Y} + g_Z{}^2 \cos^2 \theta_{B,Z} \tag{4.6a}$$

$$= g_X{}^2 c_X{}^2 + g_Y{}^2 c_Y{}^2 + g_Z{}^2 c_Z{}^2 \tag{4.6b}$$

Here $\theta_{B,X}$, $\theta_{B,Y}$ and $\theta_{B,Z}$ are the angles between the field **B** and the X, Y and Z axes. It is convenient to represent the cosines of these angles by the symbols c_X, c_Y and c_Z (Eq. 4.6b). These are referred to as the *direction cosines*.[4] Note that Eq. 4.6b is equivalent to the product

$$g^2 = [c_X \quad c_Y \quad c_Z] \cdot \begin{bmatrix} g_X^2 & 0 & 0 \\ 0 & g_Y^2 & 0 \\ 0 & 0 & g_Z^2 \end{bmatrix} \cdot \begin{bmatrix} c_X \\ c_Y \\ c_Y \end{bmatrix} \tag{4.7}$$

The simple form of Eqs. 4.6 and of the parameter matrix in Eq. 4.7 result from the use of the known principal-axis system (Section A.5) of the defect center. More generally, when one measures the *g* factors in ignorance of the principal axes, the off-diagonal elements of the *g* matrix are non-zero. Indeed, it would have been logical to have measured line positions as a function of rotation about the ⟨100⟩-type axes in the case of the cubic crystal. A more careful notation, as well as the technique of arriving at the values in the matrix in Eq. 4.7 from such measurements, is discussed in the next section.

4.4 CONSTRUCTION OF THE *g* MATRIX

In recognition of the fact that in general *g* is a matrix, the spin hamiltonian of Eq. 4.5 may be written as Eq. 4.4c. We note from Eq. 2.16a that Eq. 4.4c is equivalent to considering a generalized electron magnetic moment

$$\hat{\boldsymbol{\mu}} = \beta_e \mathbf{g} \cdot \hat{\mathbf{S}} \tag{4.8}$$

which is taken to interact with the field **B** (see Eqs. 1.14).

Alternatively, the product $\mathbf{B}^T \cdot \mathbf{g}$ in Eq. 4.4c may be regarded as a vector resulting from a transformation of the actual field **B** to an effective field

$$\mathbf{B}_{\text{eff}} = \mathbf{g}^T \cdot \mathbf{B}/g_e \tag{4.9a}$$

or equivalently

$$\mathbf{B}_{\text{eff}}^T \equiv \mathbf{B}^T \cdot \mathbf{g}/g_e \tag{4.9b}$$

The magnitude of the effective field is given by

$$B_{\text{eff}} = [(\mathbf{g}^T \cdot \mathbf{B})^T \cdot (\mathbf{g}^T \cdot \mathbf{B})]^{1/2}/g_e \tag{4.10a}$$

$$= [\mathbf{B}^T \cdot \mathbf{g} \cdot \mathbf{g}^T \cdot \mathbf{B})]^{1/2}/g_e \tag{4.10b}$$

$$= \{[\mathbf{n}^T \cdot (\mathbf{g} \cdot \mathbf{g}^T) \cdot \mathbf{n}]^{1/2}/g_e\}B \tag{4.10c}$$

where

$$\mathbf{n} = \mathbf{B}/B \tag{4.11a}$$

$$= \begin{bmatrix} c_x \\ c_y \\ c_z \end{bmatrix} \tag{4.11b}$$

is the unit vector along **B**. In concert with Eq. 1.22*b*, we define

$$g = [\mathbf{n}^T \cdot (\mathbf{g} \cdot \mathbf{g}^T) \cdot \mathbf{n})]^{1/2} \tag{4.12}$$

where of course *n* is a function of crystal orientation relative to **B**. The sign of *g* is positive for most systems.[5] Then the parameter *g* is seen to be a scalar not dependent on the magnitude *B*, but orientation-dependent since it is a function of **n** defining the direction of the vector **B**.

The spin angular momentum taken to be quantized along $\mathbf{B}_{eff}$. Thus $\mathbf{B}_{eff}{}^T \cdot \hat{\mathbf{S}}$ yields $B_{eff} M_S$, where M_S ranges by unit values from $-S$ to $+S$ (Section B.4). The eigenvalues of the spin hamiltonian (Eq. 4.4*c*) for $S = \frac{1}{2}$ are such that the two electron-spin energy levels are

$$U_1 = -\tfrac{1}{2} g \beta_e B \tag{4.13a}$$

and

$$U_2 = +\tfrac{1}{2} g \beta_e B \tag{4.13b}$$

The energy-level separation thus is

$$\Delta U = U_2 - U_1 \tag{4.14a}$$

$$= g \beta_e B \tag{4.14b}$$

The *g* factor for an arbitrary field orientation is unknown until the matrix $\mathbf{g} \cdot \mathbf{g}^T$ has been established, by EPR spectroscopic measurement of $\Delta U(\mathbf{n})$. We see from Eq. 4.12 that it is the matrix product $\mathbf{g} \cdot \mathbf{g}^T$ that is the measurable matrix, rather than **g** itself. Because **g** is not necessarily symmetric (across its diagonal), it is not trivial to set $\mathbf{g}^T$ equal to **g**. Thus the notation g^2 in Eqs. 4.3 and 4.6 is seen to be not entirely satisfactory.

We adopt the definition $\mathbf{g}\mathbf{g} \equiv \mathbf{g} \cdot \mathbf{g}^T$, and now explore some of the properties of this parameter matrix.[2] Even if **g** is asymmetric, **gg** is always symmetric. Thus we need write explicitly only the diagonal and upper off-diagonal elements. In any arbitrary cartesian coordinate system *x*, *y*, *z* fixed in the crystal, **gg** is not diagonal, so that

$$g^2 = [c_x \quad c_y \quad c_z] \cdot \begin{bmatrix} (\mathbf{g}\mathbf{g})_{xx} & (\mathbf{g}\mathbf{g})_{xy} & (\mathbf{g}\mathbf{g})_{xz} \\ & (\mathbf{g}\mathbf{g})_{yy} & (\mathbf{g}\mathbf{g})_{yz} \\ & & (\mathbf{g}\mathbf{g})_{zz} \end{bmatrix} \cdot \begin{bmatrix} c_x \\ c_y \\ c_z \end{bmatrix} \tag{4.15}$$

with a form as in Eq. 4.12 for *g* itself. One may interpret the double subscripts as follows. For example, component $(\mathbf{g}\mathbf{g})_{yx}$ may be considered as the contribution to **gg** along the axis **y** when the magnetic field is applied along **x**. That such contributions are to be expected for the case of the $O_2{}^-$ paramagnetic center may be seen from Fig. 4.4. Since the axis **x** is not orthogonal to the axes **X** or **Z** of the

O_2^- ion, any field B_x has components along both the X and Z directions. There are thus components of magnetization along the X and Z axes.[6] Hence even at this special orientation the off-diagonal component $(\mathbf{gg})_{yx}$ is non-zero. However, since $\mathbf{z}$ ($\equiv \mathbf{Y}$) is a principal axis, the components $(\mathbf{gg})_{xz} = (\mathbf{gg})_{zx}$ and $(\mathbf{gg})_{yz} = (\mathbf{gg})_{zy}$ vanish.

Following the procedure outlined in Section A.5.2, we now turn to the general case of the calculation of matrix $\mathbf{gg}$ from sets of measurements, for which

$$
\begin{aligned}
g^2 = {} & (\mathbf{gg})_{xx} \sin^2 \theta \cos^2 \phi + 2(\mathbf{gg})_{xy} \sin^2 \theta \cos \phi \sin \phi \\
& + (\mathbf{gg})_{yy} \sin^2 \theta \sin^2 \phi + 2(\mathbf{gg})_{xz} \cos \theta \sin \theta \cos \phi \\
& + 2(\mathbf{gg})_{yz} \cos \theta \sin \theta \sin \phi + (\mathbf{gg})_{zz} \cos^2 \theta
\end{aligned}
\tag{4.16a}
$$

The $(\mathbf{gg})_{ij}$ elements can be determined from experiment by successive rotations of the crystal with $\mathbf{n}$ fixed (or alternatively rotations of the field, i.e., of $\mathbf{n}$, with the crystal fixed) in the xz, yz, and xy planes. For the xz plane ($\phi = 0$), if θ is the angle between $\mathbf{B}$ and the z axis, $c_x = \sin \theta$, $c_y = 0$, and $c_z = \cos \theta$. Then

$$
g^2 = [\sin \theta \quad 0 \quad \cos \theta] \cdot
\begin{bmatrix}
(\mathbf{gg})_{xx} & (\mathbf{gg})_{xy} & (\mathbf{gg})_{xz} \\
& (\mathbf{gg})_{yy} & (\mathbf{gg})_{yz} \\
& & (\mathbf{gg})_{zz}
\end{bmatrix}
\cdot
\begin{bmatrix}
\sin \theta \\
0 \\
\cos \theta
\end{bmatrix}
\tag{4.16b}
$$

and

$$
g^2 = (\mathbf{gg})_{xx} \sin^2 \theta + 2(\mathbf{gg})_{xz} \sin \theta \cos \theta + (\mathbf{gg})_{zz} \cos^2 \theta
\tag{4.16c}
$$

Similarly, for rotation in the yz plane ($\phi = 90°$), $c_x = 0$, $c_y = \sin \theta$, and $c_z = \cos \theta$ so that

$$
g^2 = (\mathbf{gg})_{yy} \sin^2 \theta + 2(\mathbf{gg})_{yz} \sin \theta \cos \theta + (\mathbf{gg})_{zz} \cos^2 \theta
\tag{4.17}
$$

Likewise, for rotation in the xy plane ($\theta = 90°$), $c_x = \cos \phi$, $c_y = \sin \phi$, and $c_z = 0$ and hence

$$
g^2 = (\mathbf{gg})_{xx} \cos^2 \phi + 2(\mathbf{gg})_{xy} \sin \phi \cos \phi + (\mathbf{gg})_{yy} \sin^2 \phi
\tag{4.18}
$$

It is evident that in each plane only three measurements are necessary in principle to obtain the three parameters available therefrom, although many are made in practice to attain precision. For the xz plane, measurements with $\theta = 0$ and $90°$ give the values $(\mathbf{gg})_{zz}$ and $(\mathbf{gg})_{xx}$. The value of $(\mathbf{gg})_{xz} = (\mathbf{gg})_{zx}$ can be determined with the best precision at $45°$ and at $135°$. In fact, one only requires three 'independent' planes to determine $\mathbf{gg}$, and these need not be orthogonal.

Following the evaluation of the six independent components of the $\mathbf{gg}$ matrix, it is possible to transform it to a diagonal form. This is accomplished by finding a matrix $\mathbf{C}$ such that

$$
\underbrace{\begin{bmatrix} C_{Xx} & C_{Xy} & C_{Xz} \\ C_{Yx} & C_{Yy} & C_{Yz} \\ C_{Zx} & C_{Zy} & C_{Zz} \end{bmatrix}}_{\mathbf{C}} \cdot \underbrace{\begin{bmatrix} (gg)_{xx} & (gg)_{xy} & (gg)_{xz} \\ & (gg)_{yy} & (gg)_{yx} \\ & & (gg)_{zz} \end{bmatrix}}_{\mathbf{gg}} \cdot \underbrace{\begin{bmatrix} C_{Xx} & C_{Yx} & C_{Zx} \\ C_{Xy} & C_{Yy} & C_{Zy} \\ C_{Xz} & C_{Yz} & C_{Zz} \end{bmatrix}}_{\mathbf{C}^{\mathrm{T}}}
$$

$$
= \underbrace{\begin{bmatrix} (gg)_X & 0 & 0 \\ & (gg)_Y & 0 \\ & & (gg)_Z \end{bmatrix}}_{^{\mathrm{d}}\mathbf{gg}} \tag{4.19}
$$

We note that once the matrix has been cast into the diagonal form $^{\mathrm{d}}\mathbf{gg}$, to display the principal values, we can dispense with the double labeling of the latter. As indicated in Eq. 4.19, the components of $\mathbf{C}$ are in fact the direction cosines connecting the molecular axes X, Y, Z of the paramagnetic defect with the laboratory axes x, y, z. Matrix $\mathbf{C}^{\mathrm{T}}$ is the transpose of $\mathbf{C}$; both are real orthogonal so that $\mathbf{C}^{\mathrm{T}} = \mathbf{C}^{-1}$. The procedure for finding the matrix $\mathbf{C}$ that diagonalizes a given matrix $\mathbf{gg}$ is given in Section A.5.5.

Generally, by definition, $\mathbf{X}, \mathbf{Y}$ and $\mathbf{Z}$ are the principal axes of the matrix $\mathbf{gg}$. If the magnetic species has any proper axes of symmetry, then these axes (if there is more than one, when these are orthogonal) coincide with $\mathbf{X}, \mathbf{Y}$ or $\mathbf{Z}$; if there are planes of symmetry, these must be perpendicular to $\mathbf{X}, \mathbf{Y}$ or $\mathbf{Z}$. For molecules of low symmetry, the principal axes may be in any direction (dictated by the local fields) but are necessarily orthogonal to each other. The principal directions and hence the matrix $\mathbf{C}$ are the same for $\mathbf{gg}$ and for the symmetric matrix $\mathbf{g}$. One of the principal directions corresponds to a minimum value of g and another to a maximum. In principle, each principal-axis vector can be taken arbitrarily to point in either sense along its direction, that is, its sign has no physical meaning. However, we do conventionally choose matrix $\mathbf{C}$ to represent a proper rotation.

At first sight, it appears that in Eq. 4.19 we started with six parameters and ended up with only three. However, three of the original set have been utilized in arriving at the new coordinate system, that is, the principal-axis system inherent to the spin species studied. Thus, in general, $\mathbf{gg}$ contains three pieces of geometric information and three of physical (quantum-mechanical) import.

Once the principal values of $\mathbf{gg}$ are found, one wishes to obtain the matrix $\mathbf{g}$ itself. Here there are two types of problem: of matrix asymmetry and of signs. If $\mathbf{g}$ is an asymmetric matrix, then its principal-axis system is not an orthogonal one, thus differing from the set obtained for $\mathbf{gg}$. There seems to be no way of arriving experimentally at the 'true' matrix $\mathbf{g}$ obtainable from theory, whereas it is trivial to obtain $\mathbf{gg}$ from $\mathbf{g}$. However, one can arrive at a 'conventional' matrix $\mathbf{g}$, as is done everywhere in the literature. The method is to take the positive square root of each diagonal element of $^{\mathrm{d}}\mathbf{gg}$[7] and then to change the resulting diagonal matrix $^{\mathrm{d}}\mathbf{g}$ back to the laboratory coordinate system by using the reverse ($\mathbf{g} = \mathbf{C}^{\mathrm{T}} \cdot {}^{\mathrm{d}}\mathbf{g} \cdot \mathbf{C}$) of the similarity transformation, Eq. 4.19. The resulting symmetric matrix **g** reproduces the experimental data (line positions and intensities) but may differ from the theoretically derived *g* matrix. Problem 4.4 gives an opportunity to establish the principal values of a *g* matrix.

4.5   SYMMETRY-RELATED SITES

We have seen in Section 4.3, in the example dealing with the O_2^- ions, that chemically identical species can occur at various orientations, dictated by the proper point group symmetry of the crystal. These different symmetry-related sites for any given spin species become different and thus distinguishable in magnetic-resonance spectroscopy when the field **B** is applied.

For instance, consider the **gg** matrices for O_2^- in KBr referred to previously in this chapter. There are six different possible orientations ($\langle 110 \rangle$)[1] of these ions within the crystal, whose EPR line positions are described by six distinct matrices $^\alpha$**gg** with $\alpha = 1, \ldots, 6$. These matrices have identical sets of principal values but differ in the orientation of their principal-axis sets. The mathematical relations between them, called *similarity transformations* (Section A.5.5), are dictated by the crystal symmetry and can be written as

$$^\alpha\mathbf{gg} = {}^\alpha\mathbf{R} \cdot {}^1\mathbf{gg} \cdot {}^\alpha\mathbf{R}^{\mathrm{T}} \tag{4.20}$$

where the 3×3 matrices $^\alpha$**R** are properties of the whole crystal, that is, of its proper rotation group and *not* of the local symmetry of the spin species. There are only 11 distinct cases, covering all possible crystal systems. A listing of the eleven groups and number N_s of symmetry-related sites for each is given in Table 4.1. A listing of the matrices $^\alpha$**R** ($\alpha = 1, 2, \ldots, N_s$) is to be found in Ref. 1. Note that one matrix (^{1}**R**) in the set of N_s matrices $^\alpha$**R** is always the 3×3 identity matrix $\mathbf{1}_3$. In all but one case [i.e., the triclinic crystal (symmetry C_1)], EPR spectra from more than one site are in general visible. For the octahedral group O, appropriate to KBr, $N_s = 24$. However, because of the special orientations of the O_2^- ions along two-fold symmetry axes of KBr (Fig. 4.4), the $N_s = 24$ matrices $^\alpha$**gg** superpose in identical sets of four [1], yielding only six different matrices $^{\alpha'}$**gg** ($\alpha' = 1, 2, \ldots, 6$). At general orientations of **B** there are thus six distinct EPR lines, with equal intensities, unless an external stress is applied to spoil the crystal symmetry. In special experimental situations, such as **B** scanning the (001) plane, some of these lines superimpose (Fig. 4.5).

Similar considerations hold for the other spin-hamiltonian parameters to be discussed (e.g., symmetry-related hyperfine coupling matrices). The reader should understand that in general, for single-crystal EPR, there are N_s spectra, some of which may exactly superimpose. This site effect obviously leads to greater complexity of the observed spectrum. At times this causes trouble for analysis. Often, however, use can be made of the occurrence of the symmetry-related spectra to measure an unknown matrix (say, ^{1}**gg**), from far fewer field orientations than would be required

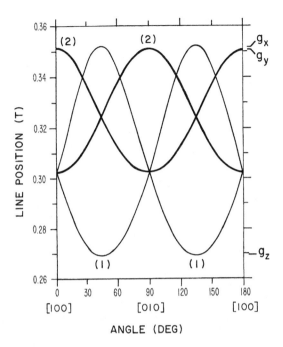

FIGURE 4.5   The EPR line positions at fixed frequency ($\nu = 9.5$ GHz) for the distinct sites of O_2^- in KBr, as a function of crystal rotation about axis [001], with the magnetic field scanning the plane (001). The number of superimposed lines is indicated within parentheses. [After H. R. Zeller, W. Känzig, *Helv. Phys. Acta*, **40**, 845 (1967).]

in the absence of distinct sites. For example, as delineated in Ref. 1, rotation of the field in a single suitably chosen crystal plane may suffice to obtain all the parameters. Note that once one matrix (e.g., $^1\mathbf{gg}$) is known, all N_s matrices $^\alpha\mathbf{gg}$ are at hand via Eq. 4.20. Furthermore, if one studies a crystal for which the proper point group is unknown, information about this group can be adduced from the observed spectra [1].

   Clearly all the preceding mathematical manipulations, as well as the analogous additional ones to follow in Chapters 5 and 6, are amenable to computer techniques. Thus, very large line-position data sets for known field orientations (and sites) can be utilized to produce the spin-hamiltonian parameters such as **g**. Automatic statistical error analysis can be incorporated [12]. An example can be found in Ref. 4 of Chapter 2.

4.6   EPR LINE INTENSITIES

The intensity (area under the absorption curve; see Section F.2) of each EPR line is dependent on various factors [8, Chapter 2; 13,14]. These include

1. The square of the transition moment (Section C.1.4), that is, of the matrix element of the amplitude (time-independent part) of the excitation spin

hamiltonian (note Eq. 2.16*a*)

$$\hat{\mathcal{H}}_1 = -\mathbf{B}_1^\mathsf{T} \cdot \hat{\boldsymbol{\mu}} = -B_1 \hat{\mu}_{B_1} \tag{4.21a}$$

$$= g\beta_e \mathbf{B}_1^\mathsf{T} \cdot \hat{\mathbf{S}} + \cdots \tag{4.21b}$$

between the initial and final states (eigenfunctions of the spin hamiltonian). This term embodies the operative magnetic-dipole selection rule, including the required orientation of $\mathbf{B}_1$ relative to $\mathbf{B}$ (Sections 1.11, C.1.4 and E.1.1).

2. The number and frequency of the photons applied to the spin system, that is, the magnitude B_1 and frequency v of $\mathbf{B}_1$.[8]

3. The population difference ΔN of the two states involved in the transition (Section 10.2.2). This is given by the Boltzmann distribution when the amplitude of $\mathbf{B}_1$ is not sufficiently high to alter ΔN, which in turn depends inversely on the absolute temperature of the ensemble and generally depends on the field $\mathbf{B}$. It is proportional to the number N of spins in the sample. In some experimental circumstances ΔN can be negative, which implies energy emission by the spin system.

4. Spectrometer characteristics (Appendixes E and F).

   In the absence of power saturation (i.e., when condition 3 holds), the transition probability is proportional to $(g\beta_e B_1)^2$. Since g is anisotropic in some systems, it follows that the line intensity can also vary under rotation of the paramagnetic species relative to fields $\mathbf{B}$ and $\mathbf{B}_1$ [13,14]. More specifically, the intensity depends on the orientations relative to the anisotropic sample of both the source ($\mathbf{B}_1$) and the direction of detection. These two directions can differ, for example, by use of crossed coils or of a bimodal microwave cavity. Generally, however, these coincide since most often a single resonator is utilized.

   One can describe the situation empirically by using transition-probability factors, Einstein coefficients A and B [15], of two types: spontaneous downward jumps (with accompanying photon emission by the spin system) and radiation-induced upward and downward jumps (with accompanying photon absorption and emission). Between any two spin states, labeled ℓ and u (of energy $U_\ell < U_u$ and populations $N_\ell > N_u$), the transition (spin flip) rates (Section 10.2.3) for isolated spins are given by

$$\frac{dN_\ell}{dt} = -\frac{dN_u}{dt} = A_{u\ell} N_u + B_{u\ell} \rho_v N_u - B_{\ell u} \rho_v N_\ell \tag{4.22}$$

Here

$$A_{u\ell} = \frac{64\pi^4 \mu_0 v_{u\ell}^3}{3hc^3} |\langle \ell | \hat{\mu}_{B_1} | u \rangle|^2 \tag{4.23a}$$

$$= \frac{8\pi h v_{u\ell}^3}{c^3} B_{u\ell} \tag{4.23b}$$

and

$$B_{u\ell} = B_{\ell u} \tag{4.24}$$

Thus there is only one independent Einstein coefficient. Here $h\nu_{u\ell} = U_u - U_\ell = g\beta_e B$ in the simplest case, and the lineshape is taken to be infinitely sharp (Dirac δ; see Section A.7), at $\nu = \nu_{u\ell}$. The relevant magnetic-dipole excitation operator is $\hat{\mu}_{B_1} = \hat{\mu}^T \cdot \mathbf{B}_1 / B_1$. The electromagnetic radiation density ρ_ν is proportional to B_1^2. In practice, the spontaneous jumps are very unlikely for EPR ($A_{u\ell} \approx 10^{-12}\,\text{s}^{-1}$ at $B = 1$ T) unless special coherence effects arising from correlated spin motions are present.

When the spectrum is taken at constant frequency by sweeping B, rather than by sweeping ν, an additional factor ($\sim g^{-1}$) arises [16,17] as a result of the conversion from frequency to field variables. It often is important to take this effect into account to obtain faithful single-crystal and powder simulations.

4.7 STATISTICALLY RANDOMLY ORIENTED SOLIDS

In this section one reaches a middle ground between the effectively isotropic systems of the first three chapters and the highly oriented solids dealt with in the earlier part of this chapter. In crystalline powders, each spin center has virtually the same properties as it would have in a large crystal. However, the principal axes of the crystallite components of the overall paramagnetic system may assume *all* possible orientations relative to the direction of the magnetic field. Even in the absence of hyperfine splitting and other zero-field splittings, one expects to have the EPR spectrum spread over the entire field range δB determined by the principal g components of the system. Fortunately, however, the lines are not uniformly distributed throughout δB, so that extrema and other features may be measurable within δB and can yield valuable information.

The first powder model considered is that of a system with $S = \frac{1}{2}$ and no interacting magnetic nuclei, and possessing uniaxial local symmetry. Subsequently, in the second model, the rhombic case will be examined.

For a single crystal one would then obtain EPR lines at positions such as those given in Figs. 4.3 and 4.5. On grinding such a crystal to a sufficiently fine powder, one expects that all orientations of the unique g axis are equally probable. Hence there are some crystallites in resonance at all fields B between $B_\perp$ (the field corresponding to $g_\perp$) and $B_\parallel$ (the field corresponding to $g_\parallel$). The field variable B (noting Eq. 4.3) is given by

$$B = \frac{h\nu}{g\beta_e} \tag{4.25a}$$

$$= [g_\perp{}^2 \sin^2\theta + g_\parallel{}^2 \cos^2\theta]^{-1/2} \frac{h\nu}{\beta_e} \tag{4.25b}$$

$$= [g_\perp{}^2 - (g_\perp{}^2 - g_\parallel{}^2)\cos^2\theta]^{-1/2} \frac{h\nu}{\beta_e} \tag{4.25c}$$

where θ is the angle between the magnetic field and the symmetry axis direction of any particular spin species in the ensemble. We need to sum over all values of θ.

Note that, in practice, the line positions and intensities are the same for a crystallite with a given orientation and those for the inverted orientation. Thus only at most half the unit sphere (Fig. 4.6) needs to be considered. Furthermore, one need not pay attention to symmetry-related species; each gives the same powder spectrum.

Since all orientations are taken to be equally probable, it is desirable to have a measure of orientation that reflects this. It is convenient to use the concept of a solid angle subtended by a bounded area $\mathcal{A}$ on the surface of a sphere of radius r. The given solid angle Ω is defined to be

$$\Omega = \frac{\mathcal{A}}{r^2} \tag{4.26}$$

that is, 4π times the ratio of the surface area $\mathcal{A}$ to the total surface area of the sphere. Consider a small powder sample at the center of a hypothetical sufficiently large sphere (Fig. 4.6). One may translate the statement that all orientations of the unique axis are equally probable into the statement that the number of crystallite axes contained in unit solid angle is equal for all regions of the sphere. Taking the coordinate axes embedded in the sphere as fixed relative to the magnetic-field direction, the orientation of each crystallite axis is measured by its angle θ relative to the direction of the applied field **B**, taken to be along the polar direction (labeled **z**).

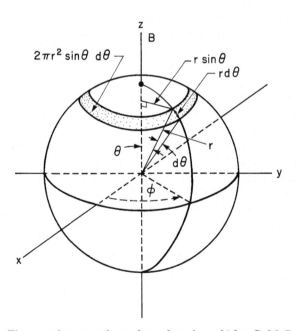

FIGURE 4.6 Element of area on the surface of a sphere. [After G. M. Barrow, *Physical Chemistry*, 2nd ed., McGraw-Hill, New York, NY, U.S.A., 1966, p. 803.]

Consider a circumpolar infinitesimal element of area (Fig. 4.6). The area of this element is $2\pi(r \sin \theta)r \, d\theta$. Hence the solid angle $d\Omega$ it subtends is given by

$$d\Omega = \frac{2\pi r^2 \sin \theta \, d\theta}{r^2} = 2\pi \sin \theta \, d\theta \tag{4.27}$$

Then $P(\theta)d\theta = d\Omega/4\pi$ is the fraction of the symmetry axes (of any sufficiently large set of crystallites) occurring between angles θ and $\theta + d\theta$. This is proportional to the probability $P(B)dB$ of a spin system experiencing a *resonant* field between B and $B + dB$, that is

$$P(\theta)d\theta = \frac{1}{2}\sin \theta \, d\theta \propto P(B)dB \tag{4.28}$$

or

$$P(B) = C\frac{1}{2}\frac{\sin \theta}{dB/d\theta} \tag{4.29}$$

where C is the normalization constant required to make the total probability unity. In the example above, one might as well consider simply a unit sphere, since r does not enter relation (4.29).

It is worthwhile to understand the significance of the numerator and the denominator in Eq. 4.29. The proportionality of $P(B)$ (and therefore of line intensity) to $\sin \theta$ reflects the very large number of systems with symmetry axes nearly perpendicular to the field direction, that is, systems with axes approximately in the equatorial plane about the field direction. By contrast, there are very few systems with the symmetry axis aligned close to the single field direction **z**. The value of $P(B)$ is large if $dB/d\theta$ is small. This implies that one has the greatest hope of seeing an EPR absorption at field values B near extrema in line positions $B(\theta)$; $B_\perp$ and $B_\parallel$ represent field extrema and therefore are such 'turning points' (see Note 6.11). On taking the derivative $dB/d\theta$ in Eq.4.25c and simplifying, one obtains

$$P(B, \theta) = \frac{C}{2}\left(\frac{h\nu}{\beta_e}\right)^2 \frac{1}{B^3|(g_\perp{}^2 - g_\parallel{}^2)\cos \theta|} \tag{4.30}$$

From this one can easily obtain $P(B)$, since B and $\cos \theta$ are linked via Eq. 4.25c. Of course, $P(B) = 0$ outside the field range given by that equation. The constant C equals 2 (Problem 4.7) For $\theta = 0$, $P(B)$ is finite when $(g_\perp{}^2 - g_\parallel{}^2)B \neq 0$. Here, since $h\nu/\beta_e = g_\parallel B_\parallel$, one finds $P(B) \propto B_\parallel{}^{-1}$. Owing to the $\cos \theta$ term in the denominator of Eq. 4.30, $P(B)$ rises monotonically to infinity as B approaches $B_\perp$, that is, as $\theta \to \pi/2$. This behavior is shown in Fig. 4.7a, where each individual line making up the EPR powder pattern has been assigned a negligible width (Dirac δ 'function' lineshape). When various (equal for each component line) amounts of broadening of the individual lines are added, the absorption line has the form shown in

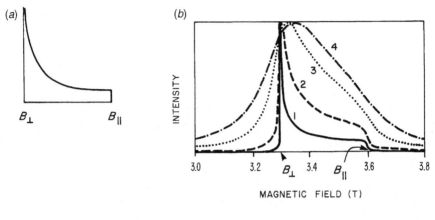

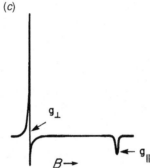

FIGURE 4.7   (*a*) Idealized absorption lineshape for a polycrystalline system containing spin centers, each having an axis of symmetry (with $g_\parallel < g_\perp$) and no hyperfine interaction. (*b*) Computed lineshapes for randomly oriented systems having uniaxial symmetry. The component (lorentzian) lines are given widths of 0.1, 1.0, 5.0 and 10.0 mT. For clarity, the displays have been normalized to equal maximum amplitudes. (*c*) First-derivative EPR powder spectrum for a system of uniaxial symmetry, with $g_\parallel < g_\perp$ [e.g., the V^- center in MgO (Section 4.2)]. [After J. A. Ibers, J. D. Swalen, *Phys. Rev.*, **127**, 1914 (1962).]

Fig. 4.7*b* (e.g., curve 1). Thus $P(B)$ must be convoluted with a suitable lineshape function to simulate an actual EPR spectrum. Figure 4.7*c* shows the first-derivative spectrum corresponding to Fig. 4.7*b* (curve 1).

Once again, we remind the reader of the need for the $1/g$ correction [16,17] referred to previously in this chapter. In the above derivation, we have ignored any anisotropy in the transition probability.

In the case of a rhombic (local-symmetry) system in powder form, the absorption pattern exhibits three primary features. Typical shapes of the absorption and of its derivative are given in Fig. 4.8. For systems of rhombic symmetry, we define the axis **Z** to be the direction that yields the g factor (g_Z) most widely separated from the other two; g_Y is the intermediate g component. In the first-derivative

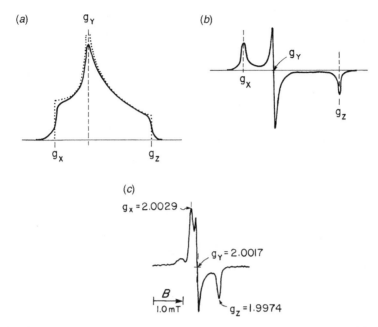

FIGURE 4.8   (a) Absorption lineshape for a randomly oriented spin system with rhombic symmetry. (b) First derivative of the curve in (a). Here $g_X > g_Y > g_Z$. (c) X-band (9.1-GHz) EPR spectrum of the CO_2^- ion on the surface of MgO powder. The extraneous peak at the left has been interpreted as belonging to a different center. [After J. H. Lunsford, J. P. Jayne, *J. Phys. Chem.*, **69**, 2182 (1965).]

presentation, the shape at each outermost field region approximates the shape of an individual component absorption line of the composite powder pattern [18], that is, summing first-derivative individual lineshapes effectively performs an integration. This is also true for the uniaxial case.

Happily, one can also obtain high-quality powder lineshape predictions, when rhombic g matrices are at hand (no zero-field effects included). Thus, following Kliava [19], for $S = \frac{1}{2}$ field-swept EPR, one has

$$L(\theta, \phi) = \kappa \frac{v_0^2 \beta_e^2 B_1^2}{8 g^3} \left[g_X^2 g_Y^2 (1 - n_Z^2) + g_X^2 g_Z^2 (1 - n_Y^2) + g_Y^2 g_Z^2 (1 - n_X^2) \right]$$

(4.31)

where n_j is the direction cosine between the jth principal axis (i.e., of $\mathbf{G} \equiv \mathbf{gg}$, Section 4.4; $j = X, Y, Z$) and $\mathbf{n}(\theta, \phi) \equiv \mathbf{B}/B$ fixed in the laboratory space.

Function L is 'several steps' ahead of function P (Eq. 4.30) in that

1. It covers rhombic functions $g(\theta, \phi)$, reducing properly to the uniaxial and isotropic cases.

2. It takes into account the anisotropy of the transition probability. The latter is proportional to $v_c^2 g_1^2 \beta_e^2 B_1^2$ [see 8, Eq. 3.10], where

$$g_1^2 \equiv \mathbf{n}_1^T \cdot \mathbf{G} \cdot \mathbf{n} - (\mathbf{n}_1^T \cdot \mathbf{G}, \cdot \mathbf{n})^2 / g^2 \qquad (4.32a)$$

and $\mathbf{n}_1 \equiv \mathbf{B}_1 / B_1$. Maryasov [20] has provided another version

$$g_1^2 = |\mathbf{g} \cdot \mathbf{n}_1 \wedge \mathbf{g} \cdot \mathbf{n}|^2 / \mathbf{g}^2 \qquad (4.32b)$$

of this relation. Because of the cross product, it shows explicitly that, for the isotropic situation, the two fields $\mathbf{B}$ and $\mathbf{B}_1$ yield zero intensity when they are parallel (at least for $|\Delta M_F| = 1$ transitions; see Section 1.13).

3. It includes the g^{-1} factor due to consideration of field sweep rather than frequency sweep conditions, as already alluded to above.

4. It can include $\kappa = \kappa(T)$, such that two spin-orientation states properly follow T^{-1} behavior (for sufficiently high temperatures).

Functions L and P both are inadequate in that they assume zero linewidths (Dirac δ) for the individual line components. Also, both assume absence of power saturation (see Section 1.5).

One approach to linewidth incorporation is to treat the distribution of resonance magnetic fields separately from the width factor [21]. One can employ a generalized function, written $Q(B, \Omega_1)$, taken as a convolution

$$Q = \int_0^{+\infty} LF \, dV \qquad (4.33)$$

of intensity function $L(B_{res}, \Omega_1)$ with a weighted lineshape function $F(B - B_{res}, \Delta_B)$, where the integration variable $V(B_{res}, \Omega_1)$ covers all resonant fields B_{res} of the spins in the system. Here Ω_1 symbolizes the set Θ, Φ, Ψ of the Euler angles describing the orientation of the macroscopic axes of the sample with respect to the static and excitation magnetic-field axes. Parameter $\Delta_B(B_{res}, \Omega_1)$ is the individual linewidth for the paramagnetic species considered. One can consider the intensity function $L = L(R, \Omega_2)$ as dependent on a random vector R summarizing the set of spin-hamiltonian parameters, with Ω_2 representing the set θ, ϕ, ψ of Euler angles describing the orientation of the local magnetic axes with respect to the macroscopic ones.

There is an abundant literature dealing with the simulation and information content of EPR powder spectra [22–26]. With the advent of efficient computers, numerical generation of the patterns, as a function of the inherent parameters (g_X, g_Y and g_Z as well as line intensity, shape and width of individual components of the packet), has become routine for the absorption or any of its derivatives [27,28]. These are plotted either as a function of field B scan (fixed-frequency experiment) or as a function of frequency v (fixed field). However, there is a subtle difference in the g dependence, according to whether field-swept or frequency-swept spectra are being considered [14]. As already mentioned above, this effect results from the dependence of the transition probability on the frequency v of the $\mathbf{B}_1$ field applied to the system. For a successful simulation, an essential aspect is to utilize a sufficiently large number of points adequately distributed on (half of) the unit sphere (Fig. 4.6).

In powders, each paramagnetic species is likely to have the same surroundings as in the single crystal. Thus the spectral parameters are expected to be the same. However, while grinding a crystal to obtain a fine powder, one tends to generate high local temperatures. In applying the theory described above, it is assumed that this causes no changes in the immediate surroundings of the spin species considered, and that no new EPR species are created (also on the surface, which is now significantly enhanced). Furthermore, in certain situations (e.g., copper complexes adsorbed on cellulose fibers [29]), there may be partial orientation of the magnetic species rather than complete randomness. In some instances, the static magnetic field $\mathbf{B}$ can cause partial ordering of crystallites [30].

With certain materials (e.g., often in glasses), one can encounter another aspect of g-factor measurement, namely, occurrence of a range of values for each of the principal values, and axis orientations, arising from differences in local surroundings [31,32]. This effect, sometimes called 'g strain', leads, in first approximation, to line broadening dependent on the magnitude of the field B used for the EPR measurement.

For EPR purposes, glassy media can be thought of as containing fixed randomly oriented spin centers. The paramagnetic species in the glass can be introduced by inclusion of suitable solutes in the original melt, or by irradiation of the glass, including ion-implantation (beam) techniques. Unlike the situation with crystalline powders, in which there is spatial correlation of the centers within each crystallite, there is virtually no spatial correlation in glasses. Nevertheless, in the absence of g strain, the g lineshape patterns tend to be the same for the two cases. On heating glasses, their viscosity decreases and the spin centers (assuming that they survive) move increasingly rapidly, thus averaging out the anisotropic parts of the g matrix. If there are no chemical changes, the ultimate g factor is the 'isotropic' one[9] given by $(|g_X| + |g_Y| + |g_Z|)/3$, which equals one-third of the trace $tr(\mathbf{g})$ in most circumstances [33].

4.8 SPIN-ORBIT COUPLING AND QUANTUM-MECHANICAL MODELING OF g

The intrinsic spin angular momentum of a free electron is associated with a g factor g_e of 2.00232. Since the ground state of most molecules (including radicals) has zero orbital angular momentum (note Section B.8), one might expect that in these cases the g factor would have precisely the free-electron value. However, as shown below, the spin-orbit interaction admixes the hypothetical 'pure-spin' ground state with certain excited states and causes a small amount of orbital angular momentum to appear in the actual ground state. The resultant circulation produces a magnetic field $\mathbf{B}_{local}$ that adds vectorially to the external field $\mathbf{B}$ (Section 1.12.1). This interaction is inversely proportional to the energy separation of the basis states. One of the results is a change in the effective *g* factor.

When we deal with electrons, their total magnetic-moment operator is the vector sum of contributions from the spin and orbital angular momenta:

$$\hat{\mu}(\mathbf{r}) = -\beta_e(\hat{\mathbf{L}} + g_e\hat{\mathbf{S}}) \qquad (4.34)$$

Here $\hat{\mathbf{L}}$ is the total electronic orbital angular-momentum operator for the ground-state configuration of the atom or ion considered. It is, of course, a spatial operator (Eqs. B.8). Note that the *g* factor for pure orbital angular momentum is unity. Then, using Eq. 2.16*a*, we can easily obtain the Zeeman hamiltonian operator $\hat{\mathcal{H}}_Z$.

In any atom, the spin and orbital (denoted by subscript 'so') angular momenta are coupled through the spin-orbit interaction term, which for the present purposes may be given as[10,11]

$$\hat{\mathcal{H}}_{so}(\mathbf{r}) = \lambda\hat{\mathbf{L}}^T \cdot \hat{\mathbf{S}} = \lambda[\hat{L}_X\hat{S}_X + \hat{L}_Y\hat{S}_Y + \hat{L}_Z\hat{S}_Z] \qquad (4.35)$$

This hamiltonian must be added to the electronic Zeeman terms, so that

$$\hat{\mathcal{H}}(\mathbf{r}) = \hat{\mathcal{H}}_Z + \hat{\mathcal{H}}_{so} = \beta_e\mathbf{B}^T \cdot (\hat{\mathbf{L}} + g_e\hat{\mathbf{S}}) + \lambda\hat{\mathbf{L}}^T \cdot \hat{\mathbf{S}} \qquad (4.36)$$

This is the energy arising from coupling of the spin magnetic moment(s) and the magnetic fields created by the orbital angular momenta.

Now consider a ground state, to be represented by $|G, M_S\rangle$, that is orbitally non-degenerate. Here G represents the spatial wavefunction and M_S, the spin state. As we shall see, the $|G, M_S\rangle$ energy levels are split by $\hat{\mathcal{H}}$ (Eq. 4.36). The energy to first order, for any S is given by the diagonal matrix element (Eq. A.90)

$$U_G^{(1)} = \langle G, M_S|g_e\beta_e B_z\hat{S}_z|G, M_S\rangle + \langle G, M_S|(\beta_e B_z + A\hat{S}_z)L_z|G, M_S\rangle \qquad (4.37)$$

since the matrix elements involving $\hat{S}_x$ and $\hat{S}_y$ vanish. The first term gives the 'spin-only' electron Zeeman energy.[12] The second term may be written as

$$\langle M_S|\beta_e B_z + \lambda\hat{S}_z|M_S\rangle\langle G|\hat{L}_z|G\rangle$$

We have shown in Section B.8 that the expectation value of $\hat{\mathbf{L}}$ for an orbitally non-degenerate state is zero in the absence of spin-orbit coupling. Hence for this case $\langle G|\hat{L}_z|G\rangle = 0$. The second-order correction to each *element* in the hamiltonian matrix (Eq. A.93) is given by

$$(\mathcal{H})_{M_S, M_S'} = -\sum_{n \neq G} \frac{\left| \langle G, M_S | (\beta_e \mathbf{B} + \lambda \hat{\mathbf{S}})^{\mathrm{T}} \cdot \hat{\mathbf{L}} + g_e \beta_e \mathbf{B}^{\mathrm{T}} \cdot \hat{\mathbf{S}} | n, M_S' \rangle \right|^2}{U_n^{(0)} - U_G^{(0)}} \quad (4.38)$$

The sum runs over all orbital states. The matrix elements of $g_e \beta_e \mathbf{B}^{\mathrm{T}} \cdot \hat{\mathbf{S}}$ vanish since $\langle G | n \rangle$ is zero. Superscripts (0) indicate the zeroth-order energies. The right-hand side of Eq. 4.38 can then be expanded to yield

$$(\mathcal{H})_{M_S, M_S'} =$$

$$-\sum_{n \neq G} \frac{\left[\langle M_S | (\beta_e \mathbf{B} + \lambda \hat{\mathbf{S}}) | M_S' \rangle \cdot \langle G | \hat{\mathbf{L}} | n \rangle][\langle n | \hat{\mathbf{L}} | G \rangle \atop \cdot \langle M_S' | (\beta_e \mathbf{B} + \lambda \hat{\mathbf{S}}) | M_S \rangle \right]}{U_n^{(0)} - U_G^{(0)}} \quad (4.39)$$

It is convenient to group together factors in of Eq. 4.39 to yield the matrix

$$-\sum_{n \neq G} \frac{\langle G | \hat{\mathbf{L}} | n \rangle \langle n | \hat{\mathbf{L}} | G \rangle}{U_n^{(0)} - U_G^{(0)}} = \begin{bmatrix} \Lambda_{xx} & \Lambda_{xy} & \Lambda_{xz} \\ \Lambda_{xy} & \Lambda_{yy} & \Lambda_{yz} \\ \Lambda_{xz} & \Lambda_{yz} & \Lambda_{zz} \end{bmatrix} = \Lambda \quad (4.40)$$

Thus the product of the two vector matrix elements, called an 'outer product' (Section A.4), yields a 3×3 matrix, Λ, symmetric in this instance. The *ij*th element of this matrix is given by

$$\Lambda_{ij} = -\sum_{n \neq G} \frac{\langle G | \hat{L}_i | n \rangle \langle n | \hat{L}_j | G \rangle}{U_n^{(0)} - U_G^{(0)}} \quad (4.41)$$

Here $\hat{L}_i$ and $\hat{L}_j$ are orbital angular-momentum operators appropriate to the x, y or z directions. Substitution of Eq. 4.40 into Eq. 4.39 yields

$$(\mathcal{H})_{M_S, M_S'} = \langle M_S' | \beta_e^2 \, \mathbf{B}^{\mathrm{T}} \cdot \Lambda \cdot \mathbf{B} + 2\lambda \beta_e \mathbf{B}^{\mathrm{T}} \cdot \Lambda \cdot \hat{\mathbf{S}} + \lambda^2 \, \hat{\mathbf{S}}^{\mathrm{T}} \cdot \Lambda \cdot \hat{\mathbf{S}} | M_S' \rangle \quad (4.42)$$

The first term on the right in Eq. 4.42 yields a constant contribution to the energy of all spin states, and represents the temperature-independent paramagnetism [35]. It causes no shifts between energy levels and hence is of no spectroscopic interest; it need not be considered further. The second and third terms in the matrix element of Eq. 4.42 constitute a hamiltonian that operates only on spin variables. When combined with the operator $g_e \beta_e \mathbf{B}^{\mathrm{T}} \cdot \hat{\mathbf{S}}$ from Eq. 4.36, it is thus called the 'spin hamiltonian' $\hat{\mathcal{H}}$. It may be written as

$$\hat{\mathcal{H}} = \beta_e^{\;2}\,\mathbf{B}^{\mathrm{T}}\cdot(g_e\mathbf{1}_3 + 2\lambda\Lambda)\cdot\hat{\mathbf{S}} + \lambda^2\,\hat{\mathbf{S}}^{\mathrm{T}}\cdot\Lambda\cdot\hat{\mathbf{S}} \tag{4.43a}$$

$$= \beta_e\mathbf{B}^{\mathrm{T}}\cdot\mathbf{g}\cdot\hat{\mathbf{S}} + \hat{\mathbf{S}}^{\mathrm{T}}\cdot\mathbf{D}\cdot\hat{\mathbf{S}} \tag{4.43b}$$

where $\mathbf{1}_3$ is the 3×3 unit matrix. Here

$$\mathbf{g} = g_e\mathbf{1}_3 + 2\lambda\Lambda \tag{4.43c}$$

and

$$\mathbf{D} = \lambda^2\Lambda \tag{4.43d}$$

In practice, $\mathbf{D}$ is made traceless by subtracting the isotropic part, $tr(\mathbf{D})/3$, since the latter has no spectroscopic context (Problem 6.3). Operator $\hat{\mathbf{S}}$ in Eq. 4.43a corresponds to the *effective* spin of the ground state. This need not be the actual spin, as is illustrated in Section 6.3. The electronic quadrupole matrix $\mathbf{D}$ (and spin-spin contributions to it) is discussed further in Chapter 6 and is of considerable EPR interest, but only if the actual or effective electronic spin is greater than $\frac{1}{2}$.

If the angular momentum of a system is due *solely* to spin angular momentum, g should be isotropic, with the value g_e. Any anisotropy or deviation from this value results from matrix Λ, and involves only contributions of the orbital angular momentum from excited states (Eqs. 4.39 and 4.40). Of course, in some cases, contributions to $\mathbf{g}$ from perturbation terms beyond the second-order ones treated herein may be appreciable, as may certain other terms needed to render the matrix invariant to the choice of coordinate system used to express the spatial wavefunctions [36].

Equation 4.43c indicates that one may immediately obtain the matrix $\mathbf{g}$ when the matrix Λ is known. As an example, we consider a P-state ion in a tetragonal electric field such that the orbital state $|L, M_L\rangle = |1, 0\rangle$ (Table B.1) lies lowest (Fig. 4.9).

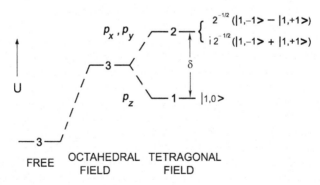

FIGURE 4.9 Displacement of the orbital energy levels of a P-state ion in an octahedral crystal field with a subsequent splitting (d) in a tetragonal electric field along $\mathbf{Z}$. The real wavefunctions p_X, p_Y and p_Z corresponding to these states are indicated.

(For the degenerate upper states $|n\rangle = |1,+1\rangle$ and $|1,-1\rangle$, one could exercise the prerogative of using the real combination forms p_x and p_y.) It is sufficient below to represent the three states as $|0\rangle$, $|+1\rangle$ and $|-1\rangle$.

Since the symmetry is tetragonal, matrix Λ is already diagonal. One principal axis is the four-fold axis $\mathbf{Z}$. The other two axes, $\mathbf{X}$ and $\mathbf{Y}$, are equivalent and are perpendicular to $\mathbf{Z}$. In this principal-axis system, the only non-zero elements of matrix Λ are the diagonal elements. The matrix elements $\langle 0|\hat{L}_z|\pm 1\rangle$ and $\langle \pm |\hat{L}_z|0\rangle$ vanish since $\hat{L}_z$ couples only states of the same M_L value (Eq. B.42e). Hence

$$\Lambda_Z = 0 \quad \text{and} \quad g_Z = g_\| = g_e \tag{4.44}$$

The value of $g_\perp$ is obtained from either Λ_X or Λ_Y (Eq. 4.41) as follows:

$$\Lambda_X = \frac{\langle 0|\hat{L}_X|+1\rangle + 1|\hat{L}_X|0\rangle + \langle 0|\hat{L}_X|-1\rangle\langle-1|\hat{L}_X|0\rangle}{\delta} \tag{4.45a}$$

$$= -\frac{1}{2\delta}\left(\langle 0|\hat{L}_-|+1\rangle\langle+1|\hat{L}_+|0\rangle + \langle 0|\hat{L}_+|-1\rangle\langle1|\hat{L}_-|0\rangle\right) \tag{4.45b}$$

$$= -\frac{1}{\delta} \tag{4.45c}$$

$$= \Lambda_Y \text{ by symmetry} \tag{4.45d}$$

Here δ is the (positive) energy splitting depicted in Fig. 4.9. Noting the relations between the matrix elements of $\hat{L}_X, \hat{L}_+$ and $\hat{L}_-$, Eqs B.42f and B.42g have been used to evaluate the matrix elements. From Eq. 4.43c one obtains

$$g_\perp = g_e - 2\lambda/\delta \tag{4.46}$$

The $V^-(O^-)$ defect center (Section 4.2, Fig. 4.2b) serves as an excellent example of a P-state ion ($S = \frac{1}{2}$), in a tetragonal local electric field. We are now in a position to interpret its g factors. According to Eq. 4.44, $g_\|$ should be very close to the free-electron value. In fact, $g_\|$ (observed) $= 2.0033$. Since λ for a positive hole on oxygen is negative (Table H.3), Eq. 4.46 predicts that $g_\perp \rangle g_e$. This is again in agreement with experiment, since $g_\perp$ (observed) $= 2.0386$ [37–39].

This procedure for calculating $\mathbf{g}$ for species with orbitally non-degenerate ground states is relatively simple; nevertheless, it demonstrates in a clear way the source of the deviations of g from the value g_e. Further examples will be found in Chapter 8.

The second term on the right side of Eq. 4.43b is effective only in systems with $S \geq 1$. One notes that this term in the spin hamiltonian is analogous to that derived for the spin-spin hamiltonian of Eq. 6.15. Experimentally, it is not possible to separate the anisotropic part of the spin-orbit coupling contribution to $\mathbf{D}$ from the spin-spin contribution.

The spin hamiltonian of Eq. 4.43b is incomplete for atoms and ions with nuclei of non-zero nuclear spin. The hyperfine and nuclear Zeeman interactions can be treated by addition of the extra terms $\hat{\mathbf{S}}^{\mathrm{T}} \cdot \mathbf{A} \cdot \hat{\mathbf{I}} - g_n \beta_n \mathbf{B}^{\mathrm{T}} \cdot \hat{\mathbf{I}}$. The hyperfine interaction is treated in Section 5.2. The spin hamiltonian then becomes

$$\hat{\mathcal{H}} = \beta_e \mathbf{B}^{\mathrm{T}} \cdot \mathbf{g} \cdot \hat{\mathbf{S}} + \hat{\mathbf{S}}^{\mathrm{T}} \cdot \mathbf{D} \cdot \hat{\mathbf{S}} + \hat{\mathbf{S}}^{\mathrm{T}} \cdot \mathbf{A} \cdot \hat{\mathbf{I}} - g_n \beta_n \mathbf{B}^{\mathrm{T}} \cdot \hat{\mathbf{I}} \qquad (4.47)$$

Clearly also, quadrupole terms for the central nucleus (if $I > \frac{1}{2}$) and hyperfine (plus quadrupole) effects from ligand nuclei can be present.

The spin hamiltonian of Eq. 4.47 is adequate for systems with $S = 1$. For $S \geq \frac{3}{2}$, still other terms must be added (Section 6.6). For instance, if $S = \frac{3}{2}$ occurs (e.g., for $3d^7$ ions), then terms linear in B and cubic in electron-spin operator components occur and hence further g-like parameters are required. For $S \geq \frac{5}{2}$, terms linear in B and fifth power in spin components are allowed. Terms non-linear in B also can occur. The various high-spin Zeeman terms make only a small contribution in the line-position analysis.

When one must begin the analysis with a lowest state that is orbitally degenerate, the application of perturbation theory, as outlined in this section, is not appropriate. Here the expectation value of the orbital angular momentum is no longer zero (Section B.8), and the spin-orbit interaction is likely to be sufficiently large that it cannot be taken on an equal footing with the Zeeman and hyperfine terms.

When $\hat{\mathcal{H}}_{\mathrm{so}}$ is added as a perturbation and dealt with separately, it generally leaves either a non-degenerate (singlet) state lowest, with no higher states populated (for even-electron systems) (hence no EPR is possible), or else it leaves a ground-state doublet (effective spin $S' = \frac{1}{2}$) that can be split by an applied magnetic field. The latter system (Kramers doublet; see Section 8.2) gives normal EPR spectra, describable by an appropriate spin hamiltonian, but often possesses very anisotropic g factors that may reach experimentally inaccessible values (e.g., $g \approx 0$). The orbitally degenerate ground state perturbed by a spin-orbit and Zeeman interaction of equal magnitude must be handled by considering both simultaneously. Examples of such odd-electron systems include the benzene anion radical (Section 9.2.1) and the Co^{2+} ion ($3d^7$) in an octahedral field as, say, in MgO (T_{ig} ground state; Section 8.3). The reader can find further details about orbitally degenerate systems, and how to deal with them in the literature [40–42].

In this chapter we have developed an understanding of g-factor anisotropy and have discussed the theoretical techniques for dealing with this phenomenon. As we shall see in the next chapter, hyperfine anisotropy can be handled in much the same way and is the source of especially valuable information about the nature of paramagnetic species. For systems with $S > \frac{1}{2}$, other important anisotropic contributions arise; these are treated in Chapter 6.

4.9 COMPARATIVE OVERVIEW

Single-crystal EPR spectra potentially give considerably more quantitative information than powder EPR spectra do, since the orientations, relative to the crystal axes, of the magnetic species in the crystal are obtainable. Also, spectral resolution is far better, since there is no overlap from myriad spectra. On the other hand, single-crystal EPR requires determination that the crystal is indeed single, and not twinned, and is adequately homogeneous. It requires determination (say, by x-ray diffraction) of the crystal-axis locations relative to the external faces or surface features, and requires careful orientation of the crystal within the EPR equipment with accurate goniometry as the crystal (or field **B**) is rotated,—maintained throughout the experiments, which are often very time-consuming. Thus long-term stability of the instrument and sample temperature is needed.

Powder EPR requires no sample orientation, and thus is experimentally quick and simple. The powder has to be adequately fine to yield superimposed spectra from sufficient crystallites (adequate number of orientations in the field **B**). There must be no reorientation effects on the individual small powder particles caused by **B**. Finally, powder EPR does require homogeneity of temperature throughout the sample, that is, adequate heat transfer between the particles.

REFERENCES

1. J. A. Weil, J. E. Clapp, T. Buch, *Adv. Magn. Reson.*, **6**, 183 (1973).

2. J. F. Nye, *Physical Properties of Crystals*, 2nd ed., Clarendon Press, Oxford, U.K., 1985. Appendix B.

3. J. E. Wertz, P. Auzins, J. H. E. Griffiths, J. W. Orton, *Faraday Discuss. Chem. Soc.*, **28**, 136 (1959).

4. W. C. O'Mara, "Trapped Hole-Centers in Magnesium Oxide", Ph.D. thesis, University of Minnesota, Minneapolis, MN, U.S.A., 1969.

5. C. J. Delbecq, E. Hutchinson, D. Schoemaker, E. L. Yasaitis, P. H. Yuster, *Phys. Rev.*, **187**, 1103 (1969).

6. F. W. Patten, F. J. Keller, *Phys. Rev.*, **187**, 1120 (1969).

7. M. M. Abraham, L. A. Boatner, M. A. Aronson, *J. Chem. Phys.*, **85**, 1 (1986).

8. A. Abragam, B. Bleaney, *Electron Paramagnetic Resonance of Transition Ions*, Clarendon Press, Oxford, U.K., 1970.

9. G. E. Pake, T. L. Estle, *The Physical Principles of Electron Paramagnetic Resonance*, 2nd ed., Benjamin, Reading, MA, U.S.A., 1973, pp. 98–99.

10. H. R. Zeller, W. Känzig, *Helv. Phys. Acta*, **40**, 845 (1967).

11. M. H. L. Pryce, *Phys. Rev. Lett.*, **3**, 375 (1959).

12. W. H. Nelson, *J. Magn. Reson.*, **38**, 71 (1980).

13. J. R. Pilbrow, *Mol. Phys.*, **16**, 307 (1969).

14. J. A. Weil, "On the Intensity of Magnetic Resonance Absorption by Anisotropic Spin Systems", in *Electronic Magnetic Resonance of the Solid State*, J. A. Weil, Ed., Canadian Society for Chemistry, Ottawa, ON, Canada, 1987, pp. 1–19.

15. P. W. Atkins, R. S. Friedman, *Molecular Quantum Mechanics*, 4th ed., Oxford University Press, Oxford, U.K., 2005, pp. 200, 538.

16. R. Aasa, T. Vänngård, *J. Magn. Reson.*, **19**, 308 (1975).

17. J. R. Pilbrow, *J. Magn. Reson.*, **58**, 186 (1984).

18. J. A. Weil, H. G. Hecht, *J. Chem. Phys.*, **38**, 281 (1963).

19. J. Kliava, *EPR Spectroscopy of Disordered Solids*, Zinatne, Riga, Latvia 1988 (in Russian).

20. A. G. Maryasov, Proc. 26th Congress (Colloque) Ampére, Athens, Greece, 1992, p. 136.

21. R. Berger, J.-C. Bissey, J. Kliava, *J. Phys. Condens. Matter*, **12**, 9347 (2000).

22. P. C. Taylor, J. F. Baugher, H. M. Kriz, *Chem. Rev.*, **75**, 203 (1975).

23. G. Van Veen, *J. Magn. Reson.*, **30**, 91 (1978).

24. Y. Siderer, Z. Luz, *J. Magn. Reson.*, **37**, 449 (1980).

25. J. A. DeGray, P. H. Rieger, *Bull. Magn. Reson.*, **8**, 95 (1986).

26. W. A. Bernhard, G. W. Fouse, *J. Magn. Reson.*, **82**, 156 (1989).

27. J. A. DeGray, P. H. Rieger, *Bull. Magn. Reson.*, **8**, 95 (1986).

28. M. She, X. Chen, X. Yu, *Can. J. Chem.*, **67**, 88 (1989).

29. S. M. Mattar, *J. Phys. Chem.*, **93**, 791 (1989).

30. J. Hulliger, L. Zoller, J. H. Ammeter, *J. Magn. Reson.*, **48**, 512 (1982).

31. W. R. Hagen, D. O. Hearshen, R. H. Sands, W. R. Dunham, *J. Magn. Reson.*, **61**, 220, 233 (1985).

32. W. R. Hagen, "*g*-Strain: Inhomogeneous Broadening in Metalloprotein EPR", in *Advanced EPR: Applications in Biology and Biochemistry*, A. J. Hoff, Ed., Elsevier, Amsterdam, Netherlands, 1989, Chapter 22.

33. M. E. Foglio, *Nuovo Cimento (Series 10)*, **B50**, 158 (1967).

34. F. K. Kneubühl, *Phys. Kondens. Materie*, **1**, 410 (1963).

35. M. Gerloch, *Magnetism and Ligand-Field Analysis*, Cambridge University Press, Cambridge, U.K., 1983, Section 7.6.2.

36. A. J. Stone, *Proc. Roy. Soc. (London)*, **A271**, 424 (1963).

37. J. E. Wertz, P. Auzins, J. H. E. Griffiths, J. W. Orton, *Discuss. Faraday Soc.*, **28**, 136 (1959).

38. W. C. O'Mara, J. J. Davies, J. E. Wertz, *Phys. Rev.*, **179**, 816 (1969).

39. M. M. Abraham, J. L. Boldú O., Y. Chen, "EPR and ENDOR of Trapped-Hole Centers in Alkaline-Earth Oxides", in *Electronic Magnetic Resonance of the Solid State*, J. A. Weil, Ed., Canadian Society for Chemistry, Ottawa, ON, Canada, 1987, p. 427. (This review cites numerous references to defects in the alkaline-earth oxides.)

40. J. E. Wertz, J. R. Bolton, *Electron Spin Resonance*, McGraw-Hill, New York, NY, U.S.A., 1972, Chapter 12.

41. S. A. Al'tshuler, B. M. Kozyrev, *Electron Paramagnetic Resonance*, Academic press, New York, NY, U.S.A., 1964. Section 3.4.

42. A. Abragam, M. H. L. Pryce, *Proc. Roy. Soc. (London)*, **A205**, 135 (1951).

NOTES

1. Miller indices enclosed in parentheses refer to individual planes, for example, (001); in square brackets to individual directions, for example, [011]; in braces to sets of equivalent planes, for example, {100}; and in angular brackets to sets of equivalent axes, for example, $\langle 111 \rangle$ and $\langle 1\bar{1}1 \rangle$ [2]. Here $\bar{1} = -1$.

2. In the literature, g has often been referred to as a 'tensor'. However, from the mathematical standpoint, $\mathbf{g}$ is a 3×3 matrix and not a tensor, whereas $\mathbf{gg}$ is a true tensor (more specifically, a tensor of rank 2). For discussion of this problem, the reader should consult Ref. 8, pp. 651 ff; also Ref. 9.

3. The O_2^- ion can be assumed to have two sets of π orbitals, pointing respectively along $\mathbf{X}$ and $\mathbf{Y}$, each formed by overlapping of collinear p orbitals pointing perpendicular to the molecular axis $\mathbf{Z}$. The two π orbitals contain three electrons. We can consider that, in Fig. 4.4, that one π orbital, namely $\mathbf{Y}$, contains two of the electrons, whereas the other ($\mathbf{X}$) contains the unpaired electron. Then it seems natural (lower energy) that $\mathbf{Y}$ should point toward two neighboring cations.

4. The three direction cosines are related by the trigonometric identity $c_X^2 + c_Y^2 + c_Z^2 = 1$. However, the relative signs of these direction cosines must be adequately specified, for example, by giving the sign of $c_X c_Y c_Z$. Thus three pieces of information are indeed required to specify the direction of $\mathbf{B}$.

5. If the matrix g arises from a spin system exhibiting only small departures from the free-spin g factor of 2.0023, then, on physical grounds, all three square roots may reasonably be taken as positive. One uses the sign convention that g for a free electron is treated as a positive quantity; the actual negative magnetogyric ratio of the free electron is allowed for by writing the spin hamiltonian in the form $\hat{\mathcal{H}} = +g\beta_e\mathbf{B}^T \cdot \hat{\mathbf{S}}$. By contrast, for nuclei one must write $\mathcal{H} = -g_n\beta_n\mathbf{g}^T \cdot \hat{\mathbf{I}}$, where g_n contains the actual sign of the magnetogyric ratio of the nucleus. For transition ions, especially heavier ones, for which g may depart greatly from the free-spin value, the correct square-root sign must usually be attained from theory by consideration of the wavefunction. If the resonance experiment can be done with circularly polarized microwaves, sign information is available since the product $det(\mathbf{g}) = g_X g_Y g_Z$ can be determined experimentally [11].

6. Consider the following crude analogy. Assume that a small cube with highly polished surfaces is at the origin of a cartesian coordinate system, with axes perpendicular to the cube faces. If a small pencil of light is directed at the cube, exactly along one of the axes, it is reflected only along that axis. For an arbitrary orientation of the cube there are components of reflected light along each of the three axes.

7. Taking this square root involves an uncertainty of sign, since a 3×3 matrix has eight possible square-root combinations of its principal values. However, see Note 4.

8. We note here that magnetic-resonance (spin-reorientation) transitions can also be induced by the bulk solid or liquid surroundings (lattice) via phonon absorption (Chapter 10).

9. We do not herein append a subindex 'o' on the isotropic part of g, as we do with the hyperfine constant (Eq. 2.38).

10. For a spherically symmetric system, this coupling is given by $\lambda\hat{\mathbf{L}}^T \cdot \hat{\mathbf{S}}$ ($=\lambda\hat{\mathbf{S}}^T \cdot \hat{\mathbf{L}}$), where λ is the spin-orbit coupling parameter (Table H.3). Since the nuclear charge increases with the atomic number (and therefore usually with the nuclear mass), the nuclear magnetic field seen by the electron(s) also increases with the atomic number. Hence λ also increases. (This parameter is not to be confused with the expansion parameter λ in Section A.6.)

11. This generally assumed expression for spin-orbit interaction is to be regarded only as a first approximation; it applies strictly only for spherical symmetry. Furthermore, the use of the terms $\beta_e\mathbf{B}^T \cdot (\hat{\mathbf{L}} + g_e\hat{\mathbf{S}})$ and $\lambda\hat{\mathbf{L}}^T \cdot \hat{\mathbf{S}}$. to proceed from the isotropic factor g_e to the symmetric matrix g implicitly requires the unpaired electrons to be in

an electric field of central symmetry. When this condition is not fulfilled, the matrix g may be asymmetric [34].

12. The first-order corrected wavefunction $|G, M_S\rangle^{(1)}$ is available from Eq. A.92 but is not needed here.

FURTHER READING

1. G. E. Pake, T. L. Estle, *The Physical Principles of Electron Paramagnetic Resonance*, 2nd ed., Benjamin, Reading, MA, U.S.A., 1973.

2. C. P. Poole Jr, H. A. Farach, *Theory of Magnetic Resonance*, Wiley, New York, NY, U.S.A., 1987.

3. M. H. L. Pryce," Paramagnetism in Crystals" (Lecture 1, International School of Physics), *Nuovo Cimento (Suppl.)*, **6**, 817 (1957).

PROBLEMS

4.1 Given that any two principal values coincide, say, $g_X = g_Y$, show that Eq. 4.6b reduces to Eq. 4.3. What is the situation when all three principal values are identical? [*Hint*: $c_X^2 + c_Y^2 + c_Z^2 = 1$.]

4.2 Calculate the spacing between lines at $g = 1.999$ and 2.000, measured at X band and W band (Table E.1). Which gives better resolution, and thus potentially higher accuracy?

4.3 Show that Eq. 4.18 can be written in the equivalent form

$$g^2 = \alpha + \beta \cos 2\phi + \gamma \sin 2\phi \qquad (4.48a)$$

where

$$\alpha = [(\mathbf{g}\mathbf{g})_{xx} + (\mathbf{g}\mathbf{g})_{yy}]/2 \qquad (4.48b)$$
$$\beta = [(\mathbf{g}\mathbf{g})_{xx} - (\mathbf{g}\mathbf{g})_{yy}]/2 \qquad (4.48c)$$

and

$$\gamma = (\mathbf{g}\mathbf{g})_{xy} \qquad (4.48d)$$

4.4 Figure 4.10 illustrates the variation of a single-line EPR spectrum of a rhombic paramagnetic defect as the magnetic field **B** scans planes ab, ac and bc of an orthorhombic crystal. Here $\nu = 9.5200$ GHz.

(*a*) By estimating values from the plots, construct the tensor $\mathbf{g}\mathbf{g}$.

(*b*) Diagonalize this tensor, and hence obtain the principal values of **g** (taken to be all positive) and the direction-cosine matrix **C**. This can be done using a suitable computer program. How are the principal directions of **g** obtainable from **C** (Section A.5.5)?

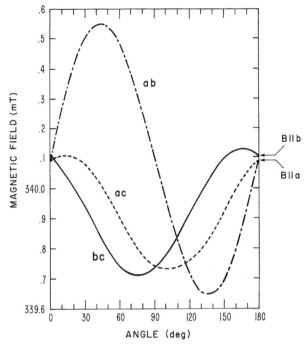

FIGURE 4.10 Variation of the resonance field B as a function of rotation in the ab $(- \cdot - \cdot - \cdot)$, ac $(\cdots)$, and bc (——) planes of a crystal. Angles are measured with respect to the a axis for (ab) and (ac) planes, and with respect to the b axis for the (bc) plane.

    (**c**) In view of the uncertainties in the input parameters, what are the error ranges on the principal g factors obtained (e.g., see Ref. 12).

4.5 Explain the degeneracies in the line positions (parenthetical numbers in Fig. 4.5), in terms of the orientations of field **B** (angles in Fig. 4.5) relative to the various positions [e.g., of the principal-axis sets ($g_x = 1.9268$, $g_y = 1.9314$, $g_z = 2.5203$)] of the O_2^- ions in KBr. [*Hint*: Draw the six equivalent positions of the ions relative to the crystal-axis (x, y, z) system. Then consider the principal axes X, Y, Z fixed on each of these molecules and how **B** rotating in the xy plane scans these.]

4.6 Consider the hypothetical asymmetric matrix

$$\mathbf{g} = \begin{bmatrix} g_1 & -g_2 & 0 \\ g_2 & g_1 & 0 \\ 0 & 0 & (g_1{}^2 + g_2{}^2)^{1/2} \end{bmatrix} \tag{4.49}$$

Calculate the matrix $\mathbf{g} \cdot \mathbf{g}^T$ and interpret the observable meaning of this result.

4.7 Derive the expression for the probability function $P(B)$ of the uniaxial g powder pattern, using Eqs. 4.25 and 4.30. Integrate the expression over the complete field range to obtain the total probability, setting this integral (Problem 4.8) equal to unity, to evaluate the normalization constant C in Eq. 4.30.

4.8 In a computer simulation of a field-swept EPR powder spectrum, the B-field range is often divided up into a number (500–5000) of segments (bins) into which are placed the intensities of all spectral lines having field positions lying within the field range of the particular bin. All bins have the same width. The overall spectrum is then plotted using the total intensities (heights and widths) of each bin. A graph of $P(B)$ versus B, derived from Eqs. 4.25c and 4.30, is given in Fig. 4.11. Integrating yields

$$\int P(B)dB = \frac{g_\perp^2 - g_\parallel^2}{\left|g_\perp^2 - g_\parallel^2\right|}\left(\frac{g_\perp^2 - (h\nu/\beta_e B)^2}{g_\perp^2 - g_\parallel^2}\right)^{1/2} + C' \qquad (4.50)$$

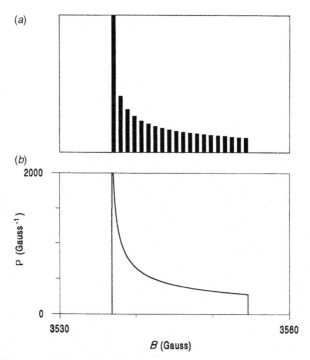

FIGURE 4.11 Plots of $P(B)$ versus B for a system with $g_\perp = 2.00$ and $g_\parallel = 1.99$ and $\nu = 9.80$ GHz. (*a*) A 20-segment histogram; (*b*) the intensity profile with a very large (>1000) number of segments.

where C' is a constant. Use Eq. 4.50 to calculate the area for each of a series of segments (limits: B_i to B_{i+1}) in the range $B_\perp$ to $B_\parallel$ and plot each of these areas in histogram form to approximate the powder spectrum. Use $g_\perp = 2.00$ and $g_\parallel = 1.99$. Calculate two histograms, one with 5 segments and one with 10 segments. Compare with the 20-segment example shown. What can you say about the number of segments (bins) necessary to generate a high-quality simulation of a powder spectrum?

4.9 Using Eqs. 2.16 and 4.8, show that the elements of the matrix **g** can generally be represented by the expression

$$g_{ij} = -g_e \frac{\partial^2 U}{\partial B_i \partial \mu_j^{(e)}}\bigg|_{\mu^{(e)}, B=0} \tag{4.51}$$

where $i, j = x, y, z$. Here U is the expectation value $\langle \mathcal{H} \rangle$ of the spin hamiltonian for a dilute spin system, and $\mu_j^{(e)}$ is $-g_e \beta_e \langle S_j \rangle$. The evaluation at the zero-spin and zero-field limits becomes important when there are high-spin and high-field terms in $\hat{\mathcal{H}}$ (see Section 6.6). Note the similarity between Eq. 4.51 and the relation (Eq. 1.13)

$$\mu_j = -\frac{\partial U}{\partial B_j}\bigg|_{B=0} \tag{4.52}$$

CHAPTER 5

HYPERFINE (A) ANISOTROPY

5.1 INTRODUCTION

In many oriented systems there may be an anisotropy in the hyperfine splittings A as well as in g. Thus, not only does each hyperfine multiplet move as a unit when the orientation is changed, but simultaneously the spacing between its component lines changes. When the hyperfine anisotropy is sufficiently great, then the qualitative appearance of the spectrum is drastically changed by rotation of a single crystal through even a relatively small angle. We temporarily ignore simultaneous changes in A and in g until we reach Section 5.4. We also restrict ourselves to electron spin $S = \frac{1}{2}$ and, for the most part, to consideration of hyperfine effects arising from a single nucleus.

A very simple example of a strongly anisotropic hyperfine interaction is that of the V_{OH} center [1,2] shown in Fig. 5.1, for which g is almost isotropic. This center in MgO consists of a linear defect $^-O\square HO^-$ in which a cation vacancy $\square$ separates a paramagnetic O^- ion and the proton of a hydroxide impurity ion (by ~ 0.32 nm). If the crystal is rotated in a (100) plane, taking θ as the angle between the defect axis and the field $\mathbf{B}$, the hydrogen hyperfine coupling $A(\theta)$ is given by an expression of the form

$$A = A_0 + (3\cos^2\theta - 1)\delta A \tag{5.1}$$

Electron Paramagnetic Resonance, Second Edition, by John A. Weil and James R. Bolton
Copyright © 2007 John Wiley & Sons, Inc.

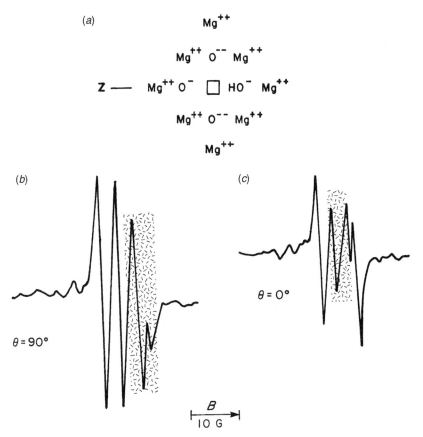

FIGURE 5.1 X-band EPR spectra of the V_{OH} center in MgO. These spectra show almost purely anisotropic hyperfine splitting. Lines arising from other (related) defects have been masked. (*a*) Structure of the defect. The symmetry axis of the defect (tetragonal crystal axis) is labeled **Z**. (*b*) Line components for **B** perpendicular to **Z**. (*c*) Line components for **B** parallel to **Z**.

Specifically, consistent with Eq. 2.2, it was found experimentally that

$$A/g_e\beta_e = 0.0016 + 0.08475(3\cos^2\theta - 1) \text{ mT} \tag{5.2}$$

which ranges from 0.1711 mT for $\theta = 0°$, becoming zero when $\cos^2\theta = (1 - 0.0016/0.08475)/3$, to -0.0831_5 mT for $\theta = 90°$. The doublet splitting (Fig. 5.1) is sufficiently small that it equals the magnitude of $A/g_e\beta_e$, with no higher-order terms needed (at 9–10 GHz). We see from Eq. 5.2 that, for this center, the proton hyperfine splitting happens to be almost purely anisotropic. In most systems, the isotropic contribution A_0 is in fact of the same order of magnitude as

δA. Then Eq. 5.1 is not applicable, and more complicated expressions are required (Section 5.3.2).

To analyze anisotropic hyperfine effects properly, one must embark on detailed consideration of the 3×3 hyperfine coupling matrix $\mathbf{A}$, which describes the physical aspects phenomenologically. We shall see that this is not a trivial matter. However, eventual attainment of parameter matrix $\mathbf{A}$ from a set of EPR measurements yields a rich harvest, revealing much detail about the local geometric configuration of a paramagnetic center and about the distribution of the nuclei and unpaired electron(s) in it. In fact, it is primarily these hyperfine effects that cause EPR spectroscopy to be such a rewarding structural tool.

5.2 ORIGIN OF THE ANISOTROPIC PART OF THE HYPERFINE INTERACTION

The origin of the *isotropic* hyperfine interaction was discussed in Chapter 2. Interaction between an electron and a nuclear dipole some distance away was rejected there as a source of the splittings observed in a liquid of low viscosity, since this interaction is time-averaged to zero. However, in more rigid systems, it is precisely this dipolar interaction that gives rise to the observed anisotropic component of hyperfine coupling. The classical expression for the dipolar interaction energy between an electron and nucleus separated by a distance r can be shown [3–5] to be

$$U_{\text{dipolar}}(\mathbf{r}) = \frac{\mu_0}{4\pi}\left[\frac{\boldsymbol{\mu}_e{}^{\mathrm{T}}\cdot\boldsymbol{\mu}_n}{r^3} - \frac{3(\boldsymbol{\mu}_e{}^{\mathrm{T}}\cdot\mathbf{r})(\boldsymbol{\mu}_n{}^{\mathrm{T}}\cdot\mathbf{r})}{r^5}\right] \tag{5.3}$$

Here $\mathbf{r}$ represents the vector joining the unpaired electron and a nucleus (Fig. 2.2). Vectors $\boldsymbol{\mu}_e$ and $\boldsymbol{\mu}_n$ are the classical electron- and nuclear-magnetic moments. For both, $\boldsymbol{\mu}^{\mathrm{T}}\cdot\mathbf{r} = \mathbf{r}^{\mathrm{T}}\cdot\boldsymbol{\mu}$. Superscript 'T' indicates the transpose (Section A.4). We see that the energy of magnetic interaction between the spins varies as r^{-3}, and is independent of the sign of $\mathbf{r}$. Note that the dipolar interaction exists whether or not there is an externally applied field.

For a quantum-mechanical system, the magnetic moments in Eq. 5.3 must be replaced by their corresponding operators. For the sake of simplicity, we shall here ignore the g anisotropy in the magnetic moment (Eq. 4.8). Thus both g and g_n are taken to be isotropic. The hamiltonian (using Eq. 1.9 for $\boldsymbol{\mu}$ in operator form) thus is

$$\hat{\mathcal{H}}_{\text{dipolar}}(\mathbf{r}) = -\frac{\mu_0}{4\pi}g\beta_e g_n\beta_n\left[\frac{\hat{\mathbf{S}}^{\mathrm{T}}\cdot\hat{\mathbf{I}}}{r^3} - \frac{3(\hat{\mathbf{S}}^{\mathrm{T}}\cdot\mathbf{r})(\hat{\mathbf{I}}^{\mathrm{T}}\cdot\mathbf{r})}{r^5}\right] \tag{5.4}$$

That $\hat{\mathcal{H}}_{\text{dipolar}}(\mathbf{r})$ describes an anisotropic interaction can be seen by expanding the vectors in Eq. 5.4, yielding

$$\hat{\mathcal{H}}_{\text{dipolar}}(\mathbf{r}) = -\frac{\mu_0}{4\pi} g\beta_e g_n\beta_n \left[\frac{r^2 - 3x^2}{r^5}\hat{S}_x\hat{I}_x + \frac{r^2 - 3y^2}{r^5}\hat{S}_y\hat{I}_y + \right.$$

$$\frac{r^2 - 3z^2}{r^5}\hat{S}_z\hat{I}_z - \frac{3xy}{r^5}(\hat{S}_x\hat{I}_y + \hat{S}_y\hat{I}_x) -$$

$$\left. \frac{3xz}{r^5}(\hat{S}_x\hat{I}_z + \hat{S}_z\hat{I}_x) - \frac{3yz}{r^5}(\hat{S}_y\hat{I}_z + \hat{S}_z\hat{I}_y) \right] \tag{5.5}$$

Coordinates x, y, z of the electron are taken with respect to axes fixed in the sample (e.g., a crystal). The point nucleus is placed at the origin. Note that, as is discussed in Section 5.3.2.3, the nucleus may be at the center of the electron distribution, or removed from it.

On averaging the hamiltonian (Eq. 5.5) over the electron distribution (i.e., integrating over the spatial variables), this becomes a spin hamiltonian, having the form

$$\hat{\mathcal{H}}_{\text{dipolar}}(\mathbf{r}) = -\frac{\mu_0}{4\pi} g\beta_e g_n\beta_n \times$$

$$[\hat{S}_x \ \hat{S}_y \ \hat{S}_z] \cdot \begin{bmatrix} \left\langle \dfrac{r^2 - 3x^2}{r^5} \right\rangle & \left\langle -\dfrac{3xy}{r^5} \right\rangle & \left\langle -\dfrac{3xz}{r^5} \right\rangle \\ & \left\langle \dfrac{r^2 - 3y^2}{r^5} \right\rangle & \left\langle -\dfrac{3yz}{r^5} \right\rangle \\ & & \left\langle \dfrac{r^2 - 3z^2}{r^5} \right\rangle \end{bmatrix} \cdot \begin{bmatrix} \hat{I}_x \\ \hat{I}_y \\ \hat{I}_z \end{bmatrix} \tag{5.6a}$$

$$= \hat{\mathbf{S}}^{\mathrm{T}} \cdot \mathbf{T} \cdot \hat{\mathbf{I}} \tag{5.6b}$$

The angular brackets imply that the average over the electron spatial distribution has been performed. Note that the dependence on electron-nuclear distance is r^{-3} in all elements, and that the average depends on which orbital the unpaired electron is in. Note also that matrix $\mathbf{T}$ is symmetric about its main diagonal and is traceless.

The full spin hamiltonian requires the addition of the isotropic hyperfine term $A_0\hat{\mathbf{S}}^{\mathrm{T}} \cdot \hat{\mathbf{I}}$, that is, the contact interaction (Eq. 2.39b) as well as the electron Zeeman and nuclear Zeeman terms. Thus[1]

$$\hat{\mathcal{H}} = g\beta_e\mathbf{B}^{\mathrm{T}} \cdot \hat{\mathbf{S}} + \hat{\mathbf{S}}^{\mathrm{T}} \cdot \mathbf{A} \cdot \hat{\mathbf{I}} - g_n\beta_n\mathbf{B}^{\mathrm{T}} \cdot \hat{\mathbf{I}} \tag{5.7}$$

where the hyperfine parameter 3×3 matrix is

$$\mathbf{A} = A_0\mathbf{1}_3 + \mathbf{T} \tag{5.8}$$

Here A_0 is the isotropic hyperfine coupling, and $\mathbf{1}_3$ is the 3×3 unit matrix. It is useful to note that A_0 is just $tr(\mathbf{A})/3$.

When the crystal is rotated, that is, the unit vector $\mathbf{n}$ along $\mathbf{B}$ is changed, the value of $\mathbf{n}^T \cdot \mathbf{A} \cdot \mathbf{n}$ changes. In fact, from a set of such numbers [using the same procedure as for the determination of matrix $\mathbf{gg}$ given in Section 4.4 (see also Eq. A.52b)], one can arrive at the 3×3 symmetric hyperfine matrix $\mathbf{A}_{\text{sym}} \equiv (\mathbf{A} + \mathbf{A}^T)/2$, to within a factor of ± 1.[2] This matrix (together with the matrix $\mathbf{g}$ and g_n) contains all the information needed to reproduce the EPR spectral positions and relative peak intensities at all frequencies and crystal orientations.

Because magnetic-field units are often convenient, we have already defined (in Eq. 2.48) the symbol $a_0 = A_0/g_e\beta_e$ for the isotropic part of $\mathbf{A}$. We now define two analogous parameters useful when hyperfine anisotropy occurs, and which are derivable from matrix $\mathbf{T}$. Thus we have

$$a_0 = (A_1 + A_2 + A_3)/3g_e\beta_e \tag{5.9a}$$

$$b_0 = [A_1 - (A_2 + A_3)/2]/3g_e\beta_e \tag{5.9b}$$

$$c_0 = (|A_2| - |A_3|)/2g_e\beta_e \tag{5.9c}$$

Here A_i ($i = 1, 2, 3$) denotes the principal values of $\mathbf{A}$, ordered such that $|A_1| - |A_2|$ and $|A_1| - |A_3|$ are larger than or equal to $|A_2| - |A_3|$, thus selecting which parameter is A_1, and (arbitrarily) taking $|A_2| - |A_3|$ to be non-negative. These new hyperfine parameters, called the *uniaxiality* parameter b_0 and the *asymmetry* (*rhombicity*) parameter c_0, are independent of a_0 and vanish if there is no anisotropy. Note that, because of the invariance of $tr(\mathbf{A})$ to change of coordinate system, a_0 (but not b_0 and c_0) is available from $\mathbf{A}$ without diagonalizing it. In many instances, as we shall see, these parameters exhibit an intimate relationship to the fundamental quantum-mechanical properties of individual atoms.

It is useful to realize that measurement of matrix $\mathbf{A}$ can yield the *relative* signs of parameters a_0 and b_0. Often, when the relatively simple dipole-dipole model yields a value of $(b_0)_{\text{theor}}$ that is close in magnitude to that of $(b_0)_{\text{expt}}$, the actual sign of a_0 can be derived by assuming that the sign of $(b_0)_{\text{expt}}$ is given by theory (Problem 5.11). The sign of a_0 may not be the one predicted by Eq. 2.38, that is, by the sign of g_n (using Table H.4), due to the core-polarization effect [9]. This features unpairing of inner-shell electrons, often with inner s-type electron spins with a net polarization in the direction opposite to that of the total spin population on the atom (Sections 9.2.4 and 9.2.5).

5.3 DETERMINATION AND INTERPRETATION OF THE HYPERFINE MATRIX

5.3.1 The Anisotropic Breit-Rabi Case

In some instances, the hyperfine energy is not small compared to the electron Zeeman energy, so that neither term in the spin hamiltonian (Eq. 5.7) can be treated approximately. The result is the appearance of higher-order energy terms (Section 3.6), leading to unequal spacings between the hyperfine components observed in the field-swept EPR spectra (e.g., see Fig. 5.2). Here, then, the general Breit-Rabi type of approach (Appendix C) must be applied.

In practice, analytic mathematical solutions for the anisotropic Breit-Rabi problem are not available. However, accurate numerical solutions (by computer) are not difficult and yield all magnetic-resonance line positions as well as the relative intensities. Let us now briefly consider another approach, in which anisotropy is brought in as a perturbation on the isotropic hyperfine situation.

It can be shown that Relation C.26 for an isotropic $S = \frac{1}{2}$ situation can be modified [10,11] to become

$$\frac{B_{-|M_I|} - B_{|M_I|}}{2|M_I|} = \frac{\dfrac{\mathbf{n}^T \cdot \mathbf{A}_{sym} \cdot \mathbf{n}}{g\beta_e}}{1 - \left(\dfrac{\mathbf{n}^T \cdot \mathbf{A}_{sym} \cdot \mathbf{n}}{2h\nu}\right)^2} \tag{5.10}$$

where, as usual, $g = (\mathbf{n}^T \cdot \mathbf{g} \cdot \mathbf{g}^T \cdot \mathbf{n})^{1/2}$ (Eq. 4.12). The set of these relations yields the elements of $\mathbf{A}_{sym} = (\mathbf{A} + \mathbf{A}^T)/2$ directly [when g is known, say, from an even-even isotope ($I = 0$) central spectrum]. We note that at each field orientation, I (if I is an integer) or $(2I + 1)/2$ (if I is half of an odd integer) such quantities are measurable, all (nominally) giving the same value. Equation 5.10 is valid when the isotropic component $|A_0|$ is large compared to the hyperfine anisotropy.

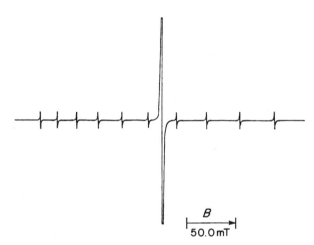

FIGURE 5.2 Computer simulation of the 10.0000 GHz field-swept EPR spectrum of a Ge^{3+} ($S = \frac{1}{2}$) center (denoted by $[GeO_4/Na]_A^0$) in crystalline α-quartz, obtained for spin-hamiltonian parameters determined at 77 K. The spectrum extends from 205.0 to 505.0 mT. The central line arises from even-isotope species ($I = 0$) of germanium, whereas the 10-line hyperfine multiplet arises from ^{73}Ge ($I = \frac{9}{2}$). The spectrum was calculated with $\mathbf{B} \| \mathbf{z}$ (=optic axis $\mathbf{c}$) and $\mathbf{B}_1 \| \mathbf{x}$ (=electrical axis a) (simulated by M. J. Mombourquette and J. A. Weil). The four-line ^{23}Na superhyperfine structure is too small to be seen at the field scale used.

For example, consider the analysis of the anisotropic splittings caused by the low-abundance isotope ^{73}Ge $(I = 9/2)$ in a Ge^{3+} center $(S = 1/2)$ found in crystalline SiO_2. The 10 hyperfine lines (Fig. 5.2) can be grouped in pairs $(M_I, -M_I)$, yielding 5 values for $\mathbf{n}^T \cdot \mathbf{A}_{sym} \cdot \mathbf{n}$. These can be averaged. Equation 5.10 thus gives this single number for a given $\mathbf{n}$, despite the unequal hyperfine spacings.

Similarly, for hydrogen atoms trapped at low temperatures within cavities in quartz crystals, the local electric fields cause anisotropy in A (i.e., admixture of p, d, ... orbitals into the nominal ground state) and in g [12]. The large isotropic part of the hyperfine splitting constant makes it important (say, for 10 GHz EPR) to use the Breit-Rabi formalism described above.

As implied, use of Eq. 5.10, together with fields and g factors measured at various orientations of the crystal with respect to $\mathbf{B}$, cannot yield $\mathbf{A}$ or $\mathbf{A}^T$. Rather, the relation is valid in the approximation

$$\mathbf{A} \cdot \mathbf{A}^T \approx A_0 \begin{bmatrix} 2A_{11} - A_0 & A_{12} + A_{21} & A_{13} + A_{31} \\ & 2A_{22} - A_0 & A_{23} + A_{32} \\ & & 2A_{33} - A_0 \end{bmatrix} \tag{5.11a}$$

$$= A_0(2\mathbf{A}_{sym} - A_0 \mathbf{1}_3) \tag{5.11b}$$

$$= A_0(\mathbf{A} + \mathbf{A}^T - A_0 \mathbf{1}_3) \tag{5.11c}$$

to the 'square' of $\mathbf{A}$. Here A_0 is $tr(\mathbf{A} + \mathbf{A}^T)/6$; that is, it is the isotropic component of $\mathbf{A}$ (and of $\mathbf{A}^T$). The magnitude of A_0 must be large compared to the anisotropic part for Eqs. 5.11 to hold. This analysis on its own does not yield the sign of A_0.

When the magnetic field B used is high enough that the higher-order effects referred to above are negligible (this is usually the case), then we can turn to the less general theory described in the following section.

5.3.2   The Case of Dominant Electron Zeeman Energy

Often, as pertains at sufficiently high magnetic fields, the electron Zeeman term in Eq. 5.7 can be assumed to be the dominant energy term; that is, the electron magnetic-moment alignment is much less affected by the nuclear magnetic moment than by $\mathbf{B}$. This allows one conveniently to quantize $\hat{\mathbf{S}}$ along $\mathbf{B}$; that is, M_S describes the eigenvalues of $\hat{\mathbf{S}}$ projected along $\hat{\mathbf{B}}$. Furthermore, higher-order hyperfine contributions (Sections 2.1, 3.6 and 5.3.1) can be taken to be negligible. By inserting the electron-spin eigenvalue vector $M_S\mathbf{n}$ for $\hat{\mathbf{S}}$ in Eq. 5.7, we obtain

$$\hat{\mathcal{H}} = g\beta_e B M_S \hat{\mathbf{1}}_3 + (M_S\mathbf{n}^T \cdot \mathbf{A} - g_n\beta_n B\mathbf{n}^T) \cdot \hat{\mathbf{I}} \tag{5.12a}$$

$$\equiv g\beta_e B M_S \hat{\mathbf{1}}_3 - g_n\beta_n \mathbf{B}_{eff}^T \cdot \hat{\mathbf{I}} \tag{5.12b}$$

As before, $\mathbf{n}$ is the unit (column) vector in the direction of $\mathbf{B}$. Here we have defined an effective magnetic field

$$\mathbf{B}_{\mathrm{eff}} = \mathbf{B} + \mathbf{B}_{\mathrm{hf}} \tag{5.13}$$

acting on the nuclear magnetic moment, where

$$\mathbf{B}_{\mathrm{hf}} = -\frac{M_S}{g_n \beta_n} \mathbf{A}^{\mathrm{T}} \cdot \mathbf{n} \tag{5.14a}$$

and thus

$$\mathbf{B}_{\mathrm{hf}}{}^{\mathrm{T}} = -\frac{M_S}{g_n \beta_n} \mathbf{n}^{\mathrm{T}} \cdot \mathbf{A} \tag{5.14b}$$

Vector $\mathbf{B}_{\mathrm{hf}}$ represents the contribution to the magnetic field *at the nucleus* arising from the electron-spin magnetic moment. We note that $\mathbf{B}_{\mathrm{eff}}$ is not necessarily parallel to $\mathbf{B}$ (Fig. 5.3), and depends on M_S. Thus the axis of quantization changes during an EPR transition so that M_I changes its meaning [13]. This is a generalization and correction to the erroneous view, generally held, that M_I is unchanged during a 'pure' electron spin flip. The magnitude of the effective field is given by

$$B_{\mathrm{eff}} = |\mathbf{B}_{\mathrm{eff}}| = [(\mathbf{B} + \mathbf{B}_{\mathrm{hf}})^{\mathrm{T}} \cdot (\mathbf{B} + \mathbf{B}_{\mathrm{hf}})]^{1/2} \tag{5.15a}$$

$$= \left[B^2 - 2\frac{M_S}{g_n \beta_n}(\mathbf{n}^{\mathrm{T}} \cdot \mathbf{A} \cdot \mathbf{n})B + \frac{1}{(2\,g_n \beta_n)^2}\mathbf{n}^{\mathrm{T}} \cdot \mathbf{A} \cdot \mathbf{A}^{\mathrm{T}} \cdot \mathbf{n} \right]^{1/2} \tag{5.15b}$$

We note from Eqs. 5.14 that the projection of the hyperfine field along $\mathbf{B}$ is proportional to $\mathbf{n}^{\mathrm{T}} \cdot \mathbf{A} \cdot \mathbf{n}$ and the magnitude of the hyperfine field, to

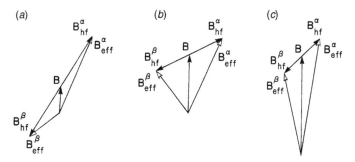

FIGURE 5.3 Vector addition of the external field $\mathbf{B}$ and of the hyperfine field $\mathbf{B}_{\mathrm{hf}}$ for $S = I = \frac{1}{2}$. The superscripts α and β refer to $M_S = +\frac{1}{2}$ and $M_S = -\frac{1}{2}$, respectively. We note that $\mathbf{B}_{\mathrm{hf}}{}^{\alpha} = -\mathbf{B}_{\mathrm{hf}}{}^{\beta}$. The 3 cases depicted above are: (a) $B \ll B_{\mathrm{hf}}$; (b) $B \approx B_{\mathrm{hf}}$; (c) $B \gg B_{\mathrm{hf}}$.

$[\mathbf{n}^T \cdot \mathbf{A} \cdot \mathbf{A}^T \cdot \mathbf{n}]^{1/2}$ (Eqs. 5.15). The field magnitude B_{hf} can be very large; for example, if the proton hyperfine coupling is ~ 3 mT (a typical value), then $B_{hf} \cong 1$ T. Remember that $\mathbf{B}_{hf}$ is the hyperfine field *at the nucleus* and *not at the electron*. The latter would be only ~ 2 mT in this case.

As is evident from Eq. 5.12b, it is most natural to quantize $\hat{\mathbf{I}}$ along $\mathbf{B}_{eff}$. However, this often is inconvenient, and hence various types of approximations are made, depending on the physical circumstances. Several cases (Fig. 5.3) are considered herein.

5.3.2.1 *General Case*

In the general case [13,14], one finds the occurrence of satellite lines. As an example, we deal with the $S = \frac{1}{2}$ system but leave I unspecified. Referring to Fig. 5.3b, we consider the total resultant field $\mathbf{B}_{eff}$ at the nucleus. Vector $\hat{\mathbf{I}}$ is quantized along $\mathbf{B}_{eff}{}^{\alpha}$ for $M_S = +\frac{1}{2}$ and along $\mathbf{B}_{eff}{}^{\beta}$ for $M_S = -\frac{1}{2}$. The energies resulting from Eq. 5.12b, for $I = \frac{1}{2}$, are given by

$$U_{\alpha(e)\alpha^{\alpha}(n)} = +\tfrac{1}{2} g \, \beta_e B \; - \tfrac{1}{2} g_n \beta_n B_{eff}{}^{\alpha} \tag{5.16a}$$

$$U_{\alpha(e)\beta^{\alpha}(n)} = +\tfrac{1}{2} g \, \beta_e B \; + \tfrac{1}{2} g_n \beta_n B_{eff}{}^{\alpha} \tag{5.16b}$$

$$U_{\beta(e)\alpha^{\beta}(n)} = -\tfrac{1}{2} g \, \beta_e B \; - \tfrac{1}{2} g_n \beta_n B_{eff}{}^{\beta} \tag{5.16c}$$

$$U_{\beta(e)\beta^{\beta}(n)} = -\tfrac{1}{2} g \, \beta_e B \; + \tfrac{1}{2} g_n \beta_n B_{eff}{}^{\beta} \tag{5.16d}$$

The nuclear-spin eigenfunctions are not the same for $M_S = +\frac{1}{2}$ and $-\frac{1}{2}$, since the axis of quantization for $\hat{\mathbf{I}}$ is different in the two cases; here, as with B_{eff}, the superscripts indicate the electron-spin state. By expressing $|\alpha^{\alpha}(n)\rangle$ and $|\beta^{\alpha}(n)\rangle$ as linear combinations of $|\alpha^{\beta}(n)\rangle$ and $|\beta^{\beta}(n)\rangle$, we can show that the relation between the nuclear-spin states is

$$|\alpha^{\alpha}(n)\rangle = \cos\frac{\omega}{2}|\alpha^{\beta}(n)\rangle - \sin\frac{\omega}{2}|\beta^{\beta}(n)\rangle \tag{5.17a}$$

$$|\beta^{\alpha}(n)\rangle = \sin\frac{\omega}{2}|\alpha^{\beta}(n)\rangle + \cos\frac{\omega}{2}|\beta^{\beta}(n)\rangle \tag{5.17b}$$

where ω is the angle between $\mathbf{B}_{eff}{}^{\alpha}$ and $\mathbf{B}_{eff}{}^{\beta}$ (Sections A.5.2, A.5.5 and C.1.4).

The energy levels are given in Fig. 5.4 (also see Fig. C.2). The four possible EPR transition energies are

$$\Delta U_a = U_{\alpha(e)\beta^{\alpha}(n)} - U_{\beta(e)\alpha^{\beta}(n)} = g\beta_e B + \tfrac{1}{2} g_n \beta_n \{B_{eff}{}^{\alpha} + B_{eff}{}^{\beta}\} \tag{5.18a}$$

$$\Delta U_b = U_{\alpha(e)\beta^{\alpha}(n)} - U_{\beta(e)\beta^{\beta}(n)} = g\beta_e B + \tfrac{1}{2} g_n \beta_n \{B_{eff}{}^{\alpha} - B_{eff}{}^{\beta}\} \tag{5.18b}$$

$$\Delta U_c = U_{\alpha(e)\alpha^{\alpha}(n)} - U_{\beta(e)\alpha^{\beta}(n)} = g\beta_e B - \tfrac{1}{2} g_n \beta_n \{B_{eff}{}^{\alpha} - B_{eff}{}^{\beta}\} \tag{5.18c}$$

$$\Delta U_d = U_{\alpha(e)\alpha^{\alpha}(n)} - U_{\beta(e)\beta^{\beta}(n)} = g\beta_e B - \tfrac{1}{2} g_n \beta_n \{B_{eff}{}^{\alpha} + B_{eff}{}^{\beta}\} \tag{5.18d}$$

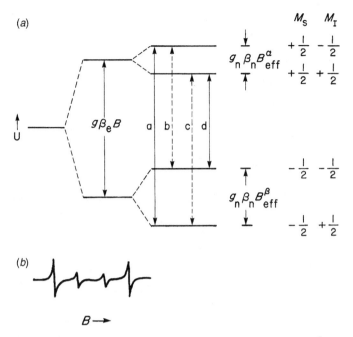

FIGURE 5.4   (*a*) Energy levels at constant field for a system with $S = I = \frac{1}{2}$ ($g_n > 0$) when B is close to B_{hf} (Fig. 5.3*b*), but with $\mathbf{B}_{eff}{}^{\alpha} < \mathbf{B}_{eff}{}^{\beta}$. Here a and d are the normally allowed transitions; b and c are usually of much lower intensity. (*b*) Observed EPR lines at constant (X-band) frequency, with relative intensities derived from Eqs. 5.19.

Since the intensities of the lines are proportional (Section C.1.4) to $|\langle M_S', M_I'|\mathbf{B_1}^T \cdot (g\beta_e\hat{\mathbf{S}} - g_n\beta_n\hat{\mathbf{I}})|M_S, M_I\rangle|^2$, the relative intensities of the lines are given by

$$\mathscr{I}_a = \mathscr{I}_d \propto \sin^2 \frac{\omega}{2} \tag{5.19a}$$

$$\mathscr{I}_b = \mathscr{I}_c \propto \cos^2 \frac{\omega}{2} \tag{5.19b}$$

Thus all four transitions can be of comparable intensity (Fig. 5.4; here $\omega \approx 70°$). Failure to recognize this has led to incorrect assignments of hyperfine splittings. The treatment shown above is still rather general, although neglecting higher-order terms (Section 5.3.1). It is instructive now to examine the preceding results for two limiting cases:

The Case of $B \ll B_{hf}$: Here $\omega \approx 180°$, and hence transitions a and d are the strong ones. We see that $g_e\beta_e$ times the separation of these two lines is very nearly the hyperfine energy, that is, $|\Delta U_a - \Delta U_d| \approx 2|g_n|\beta_n B_{hf}$, where field $B_{hf} = [\mathbf{n}^T \cdot \mathbf{A} \cdot \mathbf{A}^T \cdot \mathbf{n}/4g_n^2 \beta_n^2]^{1/2}$.

The Case of $B \gg B_{hf}$: Here $\omega \approx 0$ and hence transitions b and c of Fig. 5.4*b* are strong. Now, $g_e\beta_e$ times the separation of these two lines is given by

$|\Delta U_c - \Delta U_b| \approx 2|g_n|\beta_n B_{hf}'$, where $B_{hf}' = |\mathbf{n}^T \cdot \mathbf{A} \cdot \mathbf{n}/(2g_n\beta_n)|$. Note that this result is consistent with Eq. 5.10 (see also Eq. 2.1), since $h\nu \approx g\beta_e B$ for sufficiently large B.

We now turn to analysis of the anisotropy effects in these two limiting cases.

5.3.2.2   The Case of $B \approx B_{hf}$

This case is the one most commonly encountered and thus is analyzed in detail.[3] As before, in general $\mathbf{B}$ (taken along $\mathbf{z}$) and $\mathbf{B}_{hf}$ ($\approx \mathbf{B}_{eff}$) are not in the same direction (Fig. 5.3a). Thus $\hat{\mathbf{S}}$ and $\hat{\mathbf{I}}$ again are best quantized along different directions (along $\mathbf{B}$ for $\hat{\mathbf{S}}$ and along $\mathbf{B}_{hf}$ for $\hat{\mathbf{I}}$). The contribution to $\mathbf{B}_{hf}$ can be resolved into two components that are parallel and perpendicular to $\mathbf{B}$. The latter defines axis $\mathbf{x}$. Using Eq. 5.12b, the spin hamiltonian becomes

$$\hat{\mathcal{H}} \approx g\beta_e B M_S \hat{1}_3 - g_n\beta_n \mathbf{B}_{hf}^T \cdot \hat{\mathbf{I}} \tag{5.20a}$$

$$= g\beta_e B M_S \hat{1}_3 - g_n\beta_n[B_\perp \hat{I}_x + B_\parallel \hat{I}_z] \tag{5.20b}$$

Note that $B_{hf} = [B_\perp^2 + B_\parallel^2]^{1/2}$. For purposes of illustration, we consider the case of $I = \frac{1}{2}$, but allow S to be arbitrary.

If the spin functions for $\hat{\mathbf{I}}$ quantized along $\mathbf{z}$ are denoted by $|\alpha(n)\rangle$ and $|\beta(n)\rangle$, corresponding to $M_I = +\frac{1}{2}$ and $-\frac{1}{2}$, then the nuclear-spin hamiltonian matrix in terms of these is[4] (Sections A.5, B.5 and C.1.2)

$$
\bar{\mathcal{H}} = \begin{array}{c} \langle\alpha(n)| \\ \\ \langle\beta(n)| \end{array}
\begin{array}{cc} |\alpha(n)\rangle & \qquad\qquad |\beta(n)\rangle \end{array}
\left[\begin{array}{cc} g\beta_e B M_S - \dfrac{g_n\beta_n B_\parallel}{2} & -\dfrac{g_n\beta_n B_\parallel}{2} \\ \\ & g\beta_e B M_S + \dfrac{g_n\beta_n B_\parallel}{2} \end{array} \right] \tag{5.21}
$$

On diagonalizing this matrix, the energies for this system are found to be

$$U_{M_S,M_I} = g\beta_e B M_S + |g_n\beta_n B_{hf}|M_I \tag{5.22a}$$

$$= g\beta_e B M_S + (\mathbf{n}^T \cdot \mathbf{A} \cdot \mathbf{A}^T \cdot \mathbf{n})^{1/2}|M_S|M_I \tag{5.22b}$$

and the energy eigenfunctions are admixtures of $|\alpha(n)\rangle$ and $|\beta(n)\rangle$. Here, as discussed in Section 5.3.2.1, M_S and M_I represent spin components taken along two different directions, and M_I changes sign during the electron spin flip. However, it is convenient and conventional (although incorrect) to write Eq. 5.22a as

$$U_{M_S,M_I} = g\beta_e B M_S + A M_S M_I \tag{5.22c}$$

where now M_I is taken as constant when M_S changes sign, and $A = (\mathbf{n}^T \cdot \mathbf{A} \cdot \mathbf{A}^T \cdot \mathbf{n})^{1/2}$.

We now discuss the determination of matrix $\mathbf{A}$ from a set of EPR spectra taken at a suitable set of crystal orientations.

First, we present an illuminating example. Consider a single electron in a hybrid orbital

$$|sp\rangle = c_s|s\rangle + c_p|p\rangle \tag{5.23}$$

centered on the interacting nucleus located at the origin. Here $|c_s|^2 + |c_p|^2 = 1$. We take state $|sp\rangle$ to be an s orbital admixed with a p orbital (Table B.1), whose axis (z') is taken to lie in the xz plane at angle θ_p from z (Fig. 5.5), the latter chosen to be along **B**. Thus **A** is symmetric and uniaxial about this axis. Note that the direction **x** is *defined* by the relative directions of axis z' and **B** (and is arbitrary if θ_p is 0 or 180°). We take g to be isotropic and neglect the nuclear Zeeman term in Eq. 5.12a. Since $\hat{\mathbf{S}}$ is quantized along **z**, terms in $\hat{S}_x$ and $\hat{S}_y$ in Eq. 5.5 may be neglected. In the present case, the analogous situation does not hold for $\hat{\mathbf{I}}$, so that the term in $\hat{S}_z\hat{I}_x$ contributes. Using polar angle θ and azimuthal angle ϕ for **r**, one can substitute $r\cos\theta$ for z, $r\sin\theta\cos\phi$ for x, and $r\sin\theta\sin\phi$ for y in Eq. 5.5. The relevant effective hyperfine magnetic field components (along **x** and **z**; see Fig. 5.5) are then given by

$$B_\perp = +\frac{M_S}{g_n\beta_n}3\delta A\sin\theta_p\cos\theta_p \tag{5.24a}$$

$$B_\parallel = -\frac{M_S}{g_n\beta_n}[A_0 + \delta A(3\cos^2\theta_p - 1)] \tag{5.24b}$$

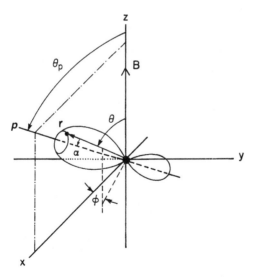

FIGURE 5.5 The hybrid orbital $|sp\rangle$ in a magnetic field **B** showing the vector **r** from the nucleus to the unpaired electron, as well as relevant angles.

so that

$$B_{hf} = \{9(\delta A)^2 \sin^2 \theta_p \cos^2 \theta_p + [A_0 + \delta A(3\cos^2 \theta_p - 1)]^2\}^{1/2}/2g_n\beta_n \quad (5.25a)$$

$$= \{(A_0 - \delta A)^2 + 3(2A_0 + \delta A)\delta A \cos^2 \theta_p\}^{1/2}/2g_n\beta_n \quad (5.25b)$$

In the preceding equations

$$\delta A = \frac{\mu_0}{4\pi} g\beta_e g_n\beta_n \left(\frac{3\cos^2 \alpha_p - 1}{2r^3}\right) \quad (5.26)$$

Here α is the angle between $\mathbf{r}$ and the axis $\mathbf{z}'$ of the p orbital. The angular brackets in Eq. 5.26 (Eq. A.57) as before indicate an average over the electronic wavefunction, that is, over $\mathbf{r}$. The part of the hyperfine field (Eq. 5.14) arising from the *isotropic* hyperfine interaction (s orbital) is in fact oriented along $\mathbf{z}$, since it is a scalar inter-action (Eq. 5.8). This contribution to A_0 in Eq. 5.24b is proportional to $|c_s|^2$ (Eq. 2.38). For an s orbital, the bracketed quantity vanishes, while for a p_z orbital it is simply $(2/5)\langle r^{-3}\rangle_p$.[5] Note that the bracket contains the factor $|c_p|^2$. The expression for δA represents a first approximation, since, if excited states (say, $p_{x'}$ and $p_{y'}$) are sufficiently close to the ground state, other terms must be added to the right side of Eq. 5.26.

From Eqs. 5.22 and 5.25, we have

$$U_{M_S,M_I} = g\beta_e B M_S + [(A_0 - \delta A)^2 + 3(2A_0 + \delta A)\delta A \cos^2 \theta_p]^{1/2}M_S M_I \quad (5.27a)$$

$$= g\beta_e B M_S + A M_S M_I \quad (5.27b)$$

The general form for A (Eqs. 5.27) was made available in 1960 [16]. Clearly, unless $B_\perp$ vanishes, the correct nuclear-spin eigenfunctions for the spin hamiltonian (Eq. 5.21) are admixtures of $|\alpha(n)\rangle$ and $|\beta(n)\rangle$. Note that it is the sign of $\delta A/A_0$ that is important in Eq. 5.27a.

At constant microwave frequency, EPR transitions occur at the resonant fields

$$B = \frac{h\nu}{g\beta_e} - \frac{g_e}{g}[(a_0 - b_0)^2 + 3(2a_0 + b_0)b_0 \cos^2 \theta_p]^{1/2}M_I \quad (5.28a)$$

$$= \frac{h\nu}{g\beta_e} - \left(\frac{g_e}{g}\right)a M_I \quad (5.28b)$$

where $a(\theta_p) = A(\theta_p)/g_e\beta_e$ (as in Eq. 2.48), $a_0 = A_0/g_e\beta_e$ and $b_0 = \delta A/g_e\beta_e$ (Eqs. 5.9a,b). The sign of a can be taken as positive, since we are dealing only with first-order hyperfine effects here.

It is of interest to consider two limiting cases:

1. $|A_0| \ll |\delta A|$. Here a is given [15] by

$$a = |b_0(1 + 3\cos^2\theta_p)^{1/2}| \qquad (5.29)$$

2. $|A_0| \gg |\delta A|$. The square root in Eqs. 5.28a may then be expanded to give

$$a \cong |a_0 + b_0(3\cos^2\theta_p - 1)| \qquad (5.30)$$

For intermediate cases, the general relation (Eqs. 5.28) must be used.

It would appear at first glance that A (θ_p dependence in Eq. 5.27) does not average to A_0 for a molecule tumbling in a liquid. However, one must realize that it is the hyperfine magnetic field at the nucleus, and not the energy, that is averaged over all orientations. It is clear from Eq. 5.24a that $B_\perp$ averages to zero, whereas $B_\parallel$ averages to $-M_S A_0 / g_n \beta_n$, as required. The energy for the tumbling system *is not* obtained by averaging U_{M_S, M_I} (Eqs. 5.27).

We now return to the general problem of obtaining $\mathbf{A}$ in the case $B \ll B_{\text{hf}}$. As in Eqs. 5.12–5.14, the hyperfine interaction is considered in terms of the hyperfine field $\mathbf{B}_{\text{hf}}$ at the nucleus. From Eq. 5.22, it is clear that the hyperfine part of the transition energy ΔU is proportional to B_{hf}. The difference of transition energies ΔU is given by $g_n \beta_n B_{\text{hf}} = A$, and is proportional to the magnitude $[\mathbf{n}^T \cdot \mathbf{A} \cdot \mathbf{A}^T \cdot \mathbf{n}]^{1/2}$ of vector $\mathbf{A}^T \cdot \mathbf{n}$ (Eq. 5.14a). With reference to the allowed (fixed-field) transitions k and m of Fig. 2.4a, which occur at frequencies ν_k and ν_m, one has $h(\nu_k - \nu_m) = A$. For fixed-frequency spectra (Fig. 2.4b), the spacing between lines is $B_m - B_k = A/g\beta_e$ at sufficiently high fields.

The procedure for evaluating the elements of the hyperfine matrix is analogous to that for evaluating the g matrix in Section 4.4, since $\mathbf{A}^T \cdot \mathbf{n}$ is a vector akin to $\mathbf{g}^T \cdot \mathbf{n}$. In the present case,

$$A^2 = (\mathbf{A}^T \cdot \mathbf{n})^T \cdot (\mathbf{A}^T \cdot \mathbf{n}) = \mathbf{n}^T \cdot (\mathbf{A} \cdot \mathbf{A}^T) \cdot \mathbf{n} = \mathbf{n}^T \cdot \mathbf{A}\mathbf{A} \cdot \mathbf{n} \qquad (5.31a)$$

where $\mathbf{A}\mathbf{A}$ by definition is $\mathbf{A} \cdot \mathbf{A}^T$. Thus (Eq. 4.11$b$) one has

$$A^2 = [c_x \quad c_y \quad c_z] \cdot \begin{bmatrix} (\mathbf{A}\mathbf{A})_{xx} & (\mathbf{A}\mathbf{A})_{xy} & (\mathbf{A}\mathbf{A})_{xz} \\ & (\mathbf{A}\mathbf{A})_{yy} & (\mathbf{A}\mathbf{A})_{yz} \\ & & (\mathbf{A}\mathbf{A})_{zz} \end{bmatrix} \cdot \begin{bmatrix} c_x \\ c_y \\ c_z \end{bmatrix} \qquad (5.31b)$$

The task at hand (compare with Eq. 4.15) is thus the evaluation of the elements of the matrix $\mathbf{A}\mathbf{A}$, which is symmetric and hence contains only six independent components.[6] From Eq. 5.31b one obtains (Eqs. A.52)

$$A^2 = (\mathbf{AA})_{xx} \sin^2 \theta \cos^2 \phi + 2(\mathbf{AA})_{xy} \sin^2 \theta \cos \phi \sin \phi +$$
$$(\mathbf{AA})_{yy} \sin^2 \theta \sin^2 \phi + 2(\mathbf{AA})_{xz} \cos \theta \sin \theta \cos \phi +$$
$$2(\mathbf{AA})_{yz} \cos \theta \sin \theta \sin \phi + (\mathbf{AA})_{zz} \cos^2 \theta \qquad (5.32)$$

We note that

$$A^2 = (g_e \beta_e a)^2 \qquad (5.33)$$

where $(g_e/g)a$ is the experimental (first-order) splitting, which must be measured at suitable orientations.

Once matrix $\mathbf{AA}$ has been obtained from the EPR spectra, the next task is to diagonalize it. Note that all three of its principal values are non-negative. If we take their square roots, we can obtain the magnitudes that are usually reported in the literature. These are not necessarily those of the principal values of the symmetrized hyperfine matrix $(\mathbf{A} + \mathbf{A}^T)/2$. As already mentioned,[1] the true matrix $\mathbf{A}$ in general is asymmetric; that is, $\mathbf{A} \neq \mathbf{A}^T$. In most of the literature, it is at this point in the analysis that the hyperfine matrix is *assumed* to be symmetric, and it is that matrix $(\mathbf{A})$ that is reported. This equals the true matrix $\mathbf{A}$ reported only if in fact $\mathbf{A} = \mathbf{A}^T$ for the latter. Luckily, knowledge of the reported matrix usually suffices to fully characterize the EPR spectra of the spin species studied, but this does not necessarily suffice when exact quantum-mechanical modeling of the molecule is the objective.

As stated above, the magnitudes of the elements of the diagonal form are obtained from the square roots of the principal values of $\mathbf{AA}$. The relative signs of the principal values become available when the fields B used are sufficiently large that the nuclear Zeeman term in Eq. 5.7 affects the spectra. In some instances, signs and likely asymmetry become available from quantum-mechanical modeling of the molecular species of interest.

Consider the especially simple system when we encounter uniaxial symmetry. In this case, Eq. 5.32 becomes

$$A^2 = A_\perp^2 \sin^2 \theta + A_\parallel^2 \cos^2 \theta \qquad (5.34)$$

where θ is the angle between the unique axis and $\mathbf{B}$.

Returning to the more general anisotropic case, we now apply the expressions presented above to actual experimental hyperfine coupling data to obtain a matrix $\mathbf{A}$ for the a-fluorine atom of the $^-OOC{-}CF{-}CF_2{-}COO^-$ radical di-anion. This species is obtained by irradiation of hydrated sodium perfluorosuccinate [17]. This π-type radical has its unpaired electron primarily in a non-bonding $2p$ orbital on the trigonal carbon atom but, as we shall see, with appreciable spin population also on the α-fluorine atom. Thus the $s + p$ example just presented (Eq. 5.23) is relevant but is not quite general enough. The crystal structure is monoclinic, with

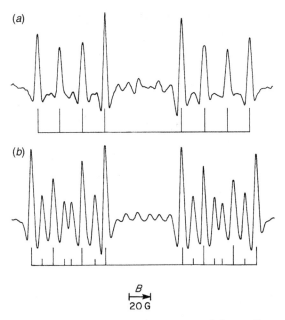

FIGURE 5.6 (*a*) Second-derivative X-band spectrum of the perfluorosuccinate radical dianion at 300 K for **B** ∥ **b** at 9.000 GHz; (*b*) similar spectrum at 35.000 GHz showing the greatly increased intensity of the forbidden transitions. [After M. T. Rogers, D. H. Whiffen, *J. Chem. Phys.*, **40**, 2662 (1964).]

unit-cell parameters $a = 1.14$, $b = 1.10$, $c = 1.03$ nm and $b = 106°$. Here β is the angle between the a and the c axes. An orthogonal $a'bc$ axis system is chosen, taking $\mathbf{a}'$ to be perpendicular to the bc plane. Figure 5.6*a* exhibits a typical X-band EPR spectrum taken at 300 K, displaying the substantial α-fluorine splitting as well as the smaller ones from the β-fluorine atoms. The g factors range from 2.0036 to 2.0060 but are herein treated as isotropic. In Fig. 5.7 the hyperfine splittings from the α atom are plotted as the magnetic field explores the $a'b$, bc and $a'c$ planes of the single crystal for both the allowed and the 'forbidden' lines.

With the crystal point group symmetry C_2 at hand here, the radicals occur in two different orientations (Section 4.5) related by two-fold axis **b**, as well as translation (and possibly inversion). Thus, generally, spectra from both are present and care must be taken in the analysis not to mix these up. However, these do superimpose exactly for **B** in the $a'c$ plane or for **B** parallel to **b** [18].

The elements of the α-^{19}F **AA** matrices for the two sites can be obtained by interpolation from these plots, using values at a set of special angles. Such data are listed in Table 5.1 (one can average the duplicate measurements). However, for better precision, least-squares fits should be made (using plots of A^2 versus rotation angle) to all the experimental data, including the forbidden lines.

TABLE 5.1   Selected Hyperfine Splittings ^{a} and the Components of AA for the α-Fluorine Atom of the $^-$OOC—CF—CF$_2$—COO$^-$ Radical Ion

Plane	Angle (deg)	$(A/h)^2$ (MHz^2)	Tensor Element
$a'b$	0	$1.61 \times 10^4 =$	$(\mathbf{AA})_{a'a'}/h^2$
	90	$16.24 \times 10^4 =$	$(\mathbf{AA})_{bb}/h^2$
	45	13.84×10^4 ⎫	Difference $=$
	135	4.29×10^4 ⎭	$2(\mathbf{AA})_{a'b}/h^2$
bc	0	$16.48 \times 10^4 =$	$(\mathbf{AA})_{bb}/h^2$
	90	$2.72 \times 10^4 =$	$(\mathbf{AA})_{cc}/h^2$
	45	9.67×10^4 ⎫	Difference $=$
	135	9.99×10^4 ⎭	$2(\mathbf{AA})_{bc}/h^2$
ca'	0	$2.69 \times 10^4 =$	$(\mathbf{AA})_{cc}/h^2$
	90	$1.59 \times 10^4 =$	$(\mathbf{AA})_{a'a'}/h^2$
	45	2.69×10^4 ⎫	Difference $=$
	135	1.42×10^4 ⎭	$2(\mathbf{AA})_{a'c}/h^2$

^{a} Measured at 300 K with $\nu = 9.000$ GHz. Only the data for one radical site are displayed.
Source: Data from M. T. Rogers, D. H. Whiffen, *J. Chem. Phys.*, **40**, 2662 (1964).

The matrix obtained from the limited data in Table 5.1 is

$$\mathbf{AA}/h^2 = \begin{bmatrix} 1.60 & \pm 4.78 & 0.64 \\ & 16.36 & \mp 0.16 \\ & & 2.71 \end{bmatrix} \times 10^4 \ (\text{MHz})^2 \qquad (5.35)$$

The choices in sign for two of the matrix elements are associated with the presence of the two symmetry-related types of radical sites (Problem 5.6). Both matrices have the same set of principal values.[7]

Note that the qualitative appearance (Fig. 5.7) of the plots of hyperfine splittings versus orientation indicates the relative importance of off-diagonal elements of **AA**. For example, if the relevant off-diagonal element is comparable in magnitude with the diagonal elements it spans, then the plot of the splitting in the given plane is very asymmetric about its center. However, if the off-diagonal element is relatively small, then the plot is close to symmetric. Figure 5.7a is a good example of the former case [i.e., $(\mathbf{AA})_{xy}$ is relatively large], whereas Fig. 5.7b represents the latter case [i.e., $(\mathbf{AA})_{yz}$ is small].

Either of the two matrices 5.35 may now be diagonalized by subtracting a parameter 1 from each diagonal element and setting the resulting determinant equal to zero (Section A.5.5). Expansion of the determinant yields the following cubic equation:

$$\lambda^3 - 20.67\lambda^2 + 51.56\lambda - 1.30 = 0 \qquad (5.36)$$

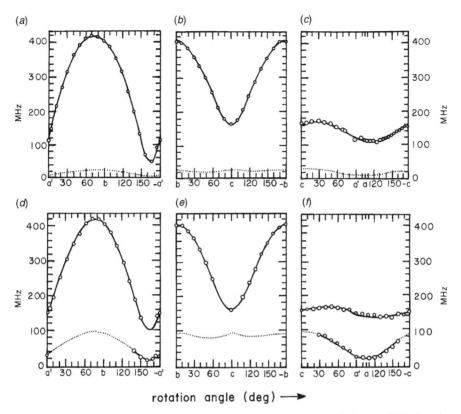

FIGURE 5.7 Angular dependence of the hyperfine line splitting (MHz) in the $^-$OOC—CF—CF$_2$—COO$^-$ radical at 300 K, yielding the data in Table 5.1. The uncertainty of data represented by large circles is greater than that for the small circles. Curves are drawn for only one of the two symmetry-related sites (the upper signs of the direction-cosine matrix of Table 5.1). Dotted lines correspond to spectral lines with relative intensity less than 20% of the total absorption intensity. (a)–(c) The microwave frequency is 9.000 GHz. **B** is in the $a'b$, bc and $a'c$ planes in (a)–(c); (d)–(f) spectra analogous to (a)–(c) but for a frequency of 35.000 GHz. [After M. T. Rogers, D. H. Whiffen, *J. Chem. Phys.*, **40**, 2662 (1964).]

The roots of this equation are 17.77, 2.87 and 0.025 (all $\times 10^4$ MHz^2); hence

$$^{\mathrm{d}}\mathbf{AA}/h^2 = \begin{bmatrix} 17.77 & 0 & 0 \\ & 2.87 & 0 \\ & & 0.025 \end{bmatrix} \times 10^4 \ (\mathrm{MHz})^2 \qquad (5.37)$$

The smallest principal value is not accurately determined from the present data. Other orientations are required to obtain a more accurate value. By taking square roots, we obtain $(\mathbf{A} + \mathbf{A}^T)/2$; this equals $\mathbf{A}$ if as usual the latter is *assumed* to be symmetric. Thus we have

$$^d\mathbf{A}/h = \begin{bmatrix} 421.5 & 0 & 0 \\ & 169.4 & 0 \\ & & 16 \end{bmatrix} \text{MHz} \tag{5.38}$$

where there is an ambiguity as to the sign of each principal element of $^d\mathbf{A}$. It is possible to obtain the correct signs for the diagonal elements of $^d\mathbf{A}$, if the nuclear Zeeman term (the final term in Eq. 5.12a) is significant. This term accounts for the difference between the 9-GHz separations in Figs. 5.7a–c, for which the nuclear Zeeman term is negligible, and the 35-GHz separations of Figs. 5.7d–f, for which the full theory must be used. All three signs turn out to be positive (see below). The matrix $\mathbf{A}$ in the crystal coordinate system, obtained from the complete data set (Fig. 5.7), is presented in Table 5.2, as are its principal values and directions. Here small corrections (Eqs. 5.15 and 5.16) were made to account for the effect of the nuclear Zeeman term. In other words, the approximation $B \ll B_{hf}$ is not quite adequate. All three principal values were chosen to be positive.

As is now evident, the analysis of a complex spectrum, which may contain 'forbidden' transitions (e.g., lines for which $\Delta M_S = \pm 1$, $\Delta M_I = \pm 1$), is often aided by using two different microwave frequencies. Figures 5.6a and 5.6b illustrate the spectrum of the $^-$OOC—CF—CF$_2$—COO$^-$ radical at 9 and at 35 GHz, that is, cases $B \ll B_{hf}$ and $B \approx B_{hf}$. The latter spectrum clearly shows the 'forbidden' transitions.

We now demonstrate the use of a high microwave frequency in determining the relative signs of hyperfine matrix elements. The measurements to be considered are the ones made at 35.000 GHz. Thus here we revisit the case $B \approx B_{hf}$. The α-^{19}F hyperfine splittings are computed in the following manner.

The main-line splitting in the [100] direction is used as an example. We obtain $\mathbf{B}_{hf}^T$ (Eq. 5.14b) from

$$-M_S\mathbf{n}^T \cdot \mathbf{A}/h = 2M_S \, [-23.5, \; \mp 51.7, \; -16.2] \text{ MHz} \tag{5.39}$$

TABLE 5.2 The α-^{19}F Hyperfine Matrix $\mathbf{A}_{sym}$ of the Perfluorosuccinate Radical Dianion and Its Principal Values and Direction [a]

Matrix $\mathbf{A}/h$ (MHz)			Principal Values (MHz)	Direction Cosines relative to Axes $a'bc$		
46.9	± 103.3	32.5	421	0.267	± 9.964	0.011
—	392.7	∓ 5.9	165	0.208	∓ 0.068	0.976
—	—	157.7	11	0.941	∓ 0.258	-0.219

[a] The upper and lower signs refer consistently to the two sets of radical sites. These data were obtained at 300 K with $\nu = 9.000$ GHz.

Sources: Data from M. T. Rogers, D. H. Whiffen, *J. Chem. Phys.*, **40**, 2662 (1964); also see L. D. Kispert, M. T. Rogers, *J. Chem. Phys.*, **54**, 3326 (1971).

by use of matrix $\mathbf{A}/h$ in Table 5.2. With $B = 1.2475$ T, we obtain (using Table H.4) the value $g_n\beta_n B/h = 50.0$ MHz, so that we have

$$g_n\beta_n \mathbf{B}_{\text{eff}}{}^{\alpha} T/h = [50.0 - 23.5, \ \mp 51.7, \ -16.2] \text{ MHz} \qquad (5.40)$$

Hence, for both sites, we have

$$g_n\beta_n B_{\text{eff}}{}^{\alpha} T/h = 60.3 \text{ MHz} \qquad (5.41a)$$

$$g_n\beta_n B_{\text{eff}}{}^{\beta} T/h = 91.3 \text{ MHz} \qquad (5.41b)$$

Thus the two hyperfine splittings (Eqs. 5.18) are

$$|\Delta U_a - \Delta U_d|/h = 151.6 \text{ MHz} \qquad (5.42a)$$

$$|\Delta U_c - \Delta U_b|/h = 31.0 \text{ MHz} \qquad (5.42b)$$

and are entered in Table 5.3, choice 1. They agree, as they should, with the observed points for $\mathbf{B}\|\mathbf{a}'$ in Figs. 5.7d and 5.7f. Since $\omega = 101°$, the relative intensities (Eqs. 5.19) are 0.60 and 0.40, for the a,d and b,c transitions.

The other choice 1 entries in Table 5.3 were calculated in a similar manner. The calculations were repeated with the other sign choices. It is clear, from an appropriate statistical analysis, that the sign choice that gives the best agreement with experiment is the one for which all principal values have the same sign. A positive sign is chosen, since the maximum hyperfine coupling is expected on theoretical grounds to be positive ($g_n > 0$) for an unpaired electron in a $2p_z$ orbital on the α-fluorine atom (note Eq. 5.43, below).

We turn now to interpretation of the hyperfine anisotropies of the perfluorosuccinate ion. Studies at various temperatures (77–300 K) of the EPR characteristics of the $^-$OOC—CF—CF$_2$—COO$^-$ radical disclose that the spectra, and hence the parameter matrices, change markedly as the crystal is cooled from room temperature [19,20]. This indicates that in fact the molecules are oscillating rapidly at 300 K, so that the matrix $\mathbf{A}$ given in Table 5.2 represents dipolar interactions time-averaged over these distortions and vibrations. Thus $\mathbf{A}$ must not be interpreted in terms of bond directions and angles of a static molecule.[8] There are, in fact, two crystallographically non-equivalent positions bearing radicals I and II, as shown in Fig. 5.8. The mean lifetime τ (Chapter 10) describing the interconversion between these states is given by $\tau^{-1} = 9.9 \times 10^{12} \exp(-\Delta U^{\ddagger}/RT)$ in units of s^{-1}, with activation energy $\Delta U^{\ddagger} = 15.26$ kJ mol^{-1} [19].

Table 5.4 presents the fluorine hyperfine principal values and directions for radicals I and II measured at 77 K [20]. In both species, the α-fluorine matrix yields the maximum splitting when $\mathbf{B}$ is along the Z axis (i.e., perpendicular to the plane of the trigonal carbon atom).[9] This type of matrix is characteristic of a nucleus interacting primarily with electron-spin density in a p orbital on the same

TABLE 5.3 Observed and Calculated Splittings (MHz), Obtained with Various Sign Choices for the Principal Values of A/h, of the α-Fluorine Nucleus in the Perfluorosuccinate Radical Dianion

Direction Cosines of Field **B**	Observed Splittings	Calculated Splittings (and Relative Intensities)			
		Sign Choices			
		(1)	(2)	(3)	(4)
[1,0,0]	153	152 (0.59)	154 (0.58)	153 (0.58)	154 (0.58)
	29	31 (0.41)	18 (0.42)	21 (0.42)	85 (0.42)
[0,1,0]	407	407 (0.41)	407 (1.00)	407 (1.00)	407 (0.99)
	—	96 (0.00)	96 (0.00)	96 (0.00)	96 (0.01)
[0,0,1]	162	163 (0.96)	164 (0.95)	164 (0.95)	163 (0.96)
	—	97 (0.04)	96 (0.05)	96 (0.05)	97 (0.04)
[cos 30°,0, cos 60°]	170	169 (0.75)	171 (0.73)	177 (0.70)	176 (0.70)
	65	61 (0.25)	54 (0.27)	25 (0.30)	32 (0.30)
[cos 50°,0, −cos 40°]	148	149 (0.63)	152 (0.62)	156 (0.61)	154 (0.61)
	48	54 (0.37)	44 (0.38)	28 (0.39)	37 (0.39)
[−cos 20°, cos 70°,0]	—	110 (0.18)	112 (0.20)	112 (0.20)	111 (0.19)
	17	19 (0.82)	1 (0.80)	46 (0.80)	14 (0.81)
[0,cos 60°, cos 30°]	252	252 (0.95)	253 (0.94)	266 (0.86)	266 (0.86)
	—	84 (0.05)	83 (0.06)	22 (0.14)	30 (0.14)

$$
\begin{array}{llllll}
\text{Sign choices:} & (1) & +421 & +165 & +11 & \text{(MHz)} \\
& (2) & +421 & +165 & -11 & \\
& (3) & +421 & -165 & +11 & \\
& (4) & +421 & -165 & -11 &
\end{array}
$$

Sources: Taken from M. T. Rogers, D. H. Whiffen, *J. Chem. Phys.*, **40**, 2662 (1964); see also L. D. Kispert, M. T. Rogers, *J. Chem. Phys.*, **54**, 3326 (1971).

atom (π-type radical; see Chapter 9). For unit p-orbital spin population (i.e., $|c_p|^2 = 1$; see Section 9.2), the anisotropic part of the hyperfine matrix would be expected to be (using Table H.4)

$$
\mathbf{T}^{\mathrm{F}}/g_e\beta_e =
\begin{bmatrix}
-62.8 & 0 & 0 \\
& -62.8 & 0 \\
& & +125.6
\end{bmatrix} \text{mT}
\tag{5.43}
$$

The maximum value ($2b_0$) is linked to the direction of the p orbital. From the numerical magnitude (14.2 mT for radical I) of the Z principal component of $\mathbf{T}^{\alpha\text{-F}}$ (Table 5.4), and use of Table H.4, one may deduce that the actual spin population is $\rho^{\alpha\text{-F}} \approx T_z/T_z^{\mathrm{F}} = 14.2/125.6 = 0.113$. This result may be interpreted as

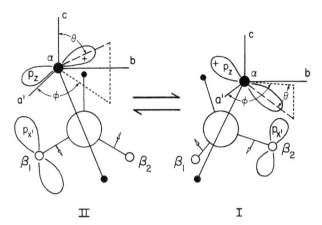

FIGURE 5.8 Newman projections along the C_α—C_β bond showing the two configurations of the I and II perfluorosuccinate radicals, which exist below 130 K. Rapid exchange between I and II lead to the room-temperature configuration reported in Ref. 17. For reference, the a', b and c axes are drawn relative to the α-fluorine p_z orbital. [After C. M. Bogan, L. D. Kispert, *J. Phys. Chem.*, **77**, 1491 (1973).]

evidence for a partial donation of unpaired-electron population to the $2p_z$ orbital of the α-fluorine from the $2p_z$ orbital of the α-carbon atom. In this analysis, we have taken the at-a-distance dipolar interaction between this fluorine nucleus and the unpaired-electron population on the carbon atom to be negligible.

The β-fluorine interaction, too, is very anisotropic, in contrast to β-proton hyperfine interactions, which are almost isotropic. The observed large anisotropy can arise only if there is a net spin population in a p orbital on the fluorine atom. Spin population in an s orbital would produce only an isotropic hyperfine interaction. The orientation of the principal axes of the β-fluorine hyperfine matrices strongly suggests that the interaction that leads to spin population in the β-fluorine p orbitals arises from a direct overlap of these orbitals with the α-carbon $2p_z$ orbital. There is some evidence from NMR and EPR work in solution that such p-π interactions are important [21,22].

In closing this section, we note that in some systems one may observe additional weak lines not accounted for by considering 'forbidden' transitions of the primary paramagnetic species (e.g., see Fig. 5.6). An example is the case of hydrogen atoms trapped in irradiated frozen acids such as H_2SO_4. In the EPR spectrum, weak sets of lines are separated from the corresponding allowed lines by $g_n\beta_nB/g_e\beta_e$. That is, they are proportional to the nuclear resonance frequency of the proton at the field B used for the EPR experiment [23,24]. The weak lines arise from 'matrix' protons that undergo a 'spin flip' when the electron spins of nearby trapped hydrogen atoms are reoriented. The coupling is dipolar and the intensity of the weak lines varies approximately as B^{-2}. In principle, such spin-flip lines from protons of the hydration water molecule should be observable in the perfluorosuccinate radical system.

TABLE 5.4 Principal Values and Direction Cosines of the ^{19}F Hyperfine Matrices ^{a} for the $^{-}$OOCCFCF$_2$COO^{-} Radicals I and II

Designation	Principal Values ^{b} (mT)	Spherical Coordinates ^{c} θ (deg)	ϕ	Direction Cosines ^{d} ($a'bc$)		
		Radical I				
$A_{\alpha I}$	21.7	61.26	$\mp$95.62	-0.08059	∓ 0.87256	0.48090
	(+)0.7	—	—	0.28079	± 0.44191	0.85198
	(+)0.2	—	—	0.95592	∓ 0.20822	-0.20704
$A_{\beta_1 I}$	12.2	52.3	$\mp$49.2	0.51634	∓ 0.59896	0.61207
	+4.1	—	—	0.52889	± 0.78516	0.32217
	+4.1	—	—	-0.67354	± 0.15737	0.72220
$A_{\beta_2 I}$	0.9	32.1	$\mp$37.2	0.42339	∓ 0.32094	0.84719
	-0.3	—	—	-0.67714	∓ 0.73336	0.06059
	-0.2	—	—	0.60185	∓ 0.59930	-0.52782
		Radical II				
$A_{\alpha II}$	22.4	55.78	$\pm$62.83	0.37758	± 0.73565	0.56236
	(+)0.5	—	—	-0.15214	∓ 0.54978	0.82134
	(+)0.4	—	—	0.91339	∓ 0.39568	-0.09566
$A_{\beta_1 II}$	12.5	45.6	$\pm$10.6	0.70222	± 0.13162	0.69969
	+4.4	—	—	-0.68619	∓ 0.13689	0.71442
	+4.1	—	—	-0.18981	± 0.98180	0.00581
$A_{\beta_1 II}$	1.4	20.7	$\mp$9.5	0.34903	∓ 0.05819	0.93630
	-0.1	—	—	0.09252	± 0.99533	0.02739
	-0.2	—	—	0.93253	∓ 0.07697	-0.35278

^{a} Measured at 77 K with $\nu = 9$ GHz (from Ref. 19).

^{b} The small matrix elements may be in error by $+0.3$ mT in certain cases. Where the relative signs are somewhat uncertain, they have been placed in parentheses.

^{c} The angle between the given principal direction and the c axis is q; the angle between a and the projection of the principal direction onto the $a'b$ plane is ϕ.

^{d} The upper and lower signs refer consistently to the two sets of radical sites related by the twofold crystal axes. The upper signs correspond to the two conformations correlated by the radical motion, in the cases of radicals I and II, and the 300 K radical.

5.3.2.3   The Case of B ≫ B_hf   When the external magnetic field **B** is sufficiently large, $\hat{\mathbf{I}}$ may be taken effectively to be quantized along **B** (Fig. 5.3c); Eq. 5.7 may then be written

$$\hat{\mathcal{H}} = g_e\beta_e B\hat{S}_B + (\mathbf{n}^T \cdot \mathbf{A} \cdot \mathbf{n})\hat{S}_B\hat{I}_B - g_n\beta_n B\hat{I}_B \tag{5.44}$$

where $\hat{S}_B = \hat{\mathbf{S}}^T \cdot \mathbf{n}$ and $\hat{I}_B = \hat{\mathbf{I}}^T \cdot \mathbf{n}$. In other words, we ignore $B_\perp$ in Eq. 5.21. We see that $\hat{\mathcal{H}}$ is diagonal as is; that is, replacing $\hat{S}_B$ by M_S and $\hat{I}_B$ by M_I yields the energy eigenvalues.

If we wish, we can call $\mathbf{n}^T \cdot \mathbf{A} \cdot \mathbf{n}$ a diagonal element of **A** (in an appropriate coordinate system), that is, A_{BB}. With this notation, the energy eigenvalues of $\hat{\mathcal{H}}$ in Eq. 5.44 are

$$U_{M_S, M_I} = g\beta_e B M_S + A_{BB}M_S M_I - g_n\beta_n B M_I \tag{5.45}$$

for any fixed $\mathbf{n} = \mathbf{B}/B$. The allowed transition energies are

$$U_{M_S+1, M_I} - U_{M_S, M_I} = g\beta_e B + A_{BB}M_I \tag{5.46}$$

For given M_S, transitions occur for both M_I and $-M_I$. We cannot obtain the sign of A_{BB} since we cannot know which observed transition is which. Thus only $|A_{BB}|$ is measurable. Equivalently, as the analysis in Section 5.3.1 reveals, we can say that only $\mathbf{A} \cdot \mathbf{A}^T$ is obtainable from a set of measurements at various orientations of **B**.

As a specific example, again consider **B** pointed along **z**, that is, $\mathbf{n} = \mathbf{z}$. Since both $\hat{\mathbf{S}}$ and $\hat{\mathbf{I}}$ are now quantized along **B**, one may neglect all terms involving x and y components of $\hat{\mathbf{S}}$ and $\hat{\mathbf{I}}$ in Eq. 5.5. Consider the special case of the unpaired electron located in a pure p orbital (whose axis is in the xz plane as shown in Fig. 5.5) centered on the interacting nucleus, that is, a uniaxial situation. With the substitution $z = r\cos\theta$, Eq. 5.6 effectively reduces (using Eq. 5.26) to

$$\hat{\mathcal{H}}_{\text{dipolar}} = \frac{\mu_0}{4\pi}g\beta_e g_n\beta_n \left\langle \frac{3\cos^2\theta_p - 1}{r^3} \right\rangle \hat{S}_z\hat{I}_z \tag{5.47a}$$

$$= \delta A(3\cos^2\theta_p - 1)\hat{S}_z\hat{I}_z \tag{5.47b}$$

$$= A_{zz}\hat{S}_z\hat{I}_z \tag{5.47c}$$

Here again θ_p is the angle between **z** and the axis **p** of the p orbital (Fig. 5.5). Equations 5.1 and 5.2 are consistent with this result (taking $\theta = \theta_p$). The reader may wish to consider Problem 5.4 when considering derivation of Eq. 5.47b from Eq. 5.47a. If the electron is interacting with a nucleus not at the center of the p orbital, Eq. 5.47b still holds; however, an appropriate average must be taken over the electronic wavefunction. We now explore this case.

When a nucleus giving rise to hyperfine effects is situated away from, rather than at, the center of the unpaired-electron distribution, then the observed hyperfine splitting tends to be small. The distribution quantity $|\psi(\text{at external nucleus})|^2$ gives rise only to a small isotropic hyperfine term of the contact type given by Eq. 2.38. The at-a-distance magnetic dipole interaction drops off rapidly as R increases; here R is the distance between the nucleus considered and the center of the unpaired-electron distribution (say, some other nucleus, with charge number Z). As a simplest example, let us consider the latter to be a $1s$ orbital of a one-electron atom and take the external nucleus to have no electrons of its own (Fig. 5.9). The dipolar part (Eq. 5.6) of the hyperfine matrix **A** for this one-electron uniaxial system is given [5,25] by

$$\mathbf{T} = \frac{\mu_0}{4\pi} g\beta_e g_n\beta_n R^{-3}f \begin{bmatrix} 2 & 0 & 0 \\ & -1 & 0 \\ & & -1 \end{bmatrix} \tag{5.48}$$

where

$$f(R) = 1 - e^{-\rho R}[1 + \rho R + \tfrac{1}{2}\rho^2 R^2 + \tfrac{1}{6}\rho^3 R^3] \tag{5.49}$$

with $\rho = 2Z/r_b$; here r_b is the Bohr radius. The product $R^{-3}f$ can be shown to go to zero with an $\sim R^{-3}$ dependence as $R \to 0$, and also to go to zero as $R \to \infty$. Table 5.5 presents some values of both the isotropic and anisotropic hyperfine parameters as a function of distance R between a proton and a hydrogenic electron ($1s$) cloud. Relations similar to those of Eqs. 5.48 and 5.49 have been derived [26] for p electrons and applied to free-radical systems.

The V_{OH} center measured at X band, considered in Section 5.1, is a good example of the case of an external nucleus (the proton), with $B \gg B_{\text{hf}}$. For instance, $B \approx 320$ mT as compared to $B_{\text{hf}} = 53.8$ mT for $\theta = 0$.

In closing this discussion, we note that in most EPR situations involving an external nucleus, one deals with the case $B \gg B_{\text{hf}}$. Here the nuclear Zeeman term cannot

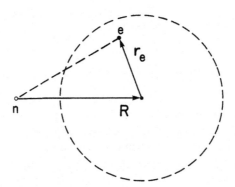

FIGURE 5.9 Model of the hyperfine anisotropic interaction between an unpaired electron, distributed according to a spherically symmetric function $\psi(\mathbf{r}_e)$, and an 'external' and 'uncharged' nucleus n.

TABLE 5.5   The Isotropic and Anisotropic Hyperfine Parameters for a Bare Proton at Distance R from the Center of a $1s$ Unpaired Electron Distribution[a]

R (m)	f	a_0 (mT)	$T_{max}/g_e\beta_e$ (mT)
0	0	50.77	0
1.0×10^{-11}	0.0006296	34.79	3.552
r_b	0.1428765	6.871	5.440
1.0×10^{-10}	0.5223026	1.160	2.947
2.0×10^{-10}	0.9431053	0.026	0.6652
5.0×10^{-10}	0.9999918	0.000	0.0451
∞	1	0	0

[a] Here $a_0 = [2\mu_0/3]g_n\beta_n|\psi(R)|^2 = [2\mu_0/3\pi r_b^3]g_n\beta_n \exp(-2R/r_b)$ (Eq. 2.5), where r_b is the Bohr radius (Table H.1).

generally be ignored. We also wish to emphasize that the satellite peaks discussed in this chapter depend on anisotropy, and hence are not observable in liquids (because of tumbling averaging) or in gases.

5.4   COMBINED g AND HYPERFINE ANISOTROPY

Generally, simultaneous anisotropy of g and A occurs, and the principal-axis systems of $\mathbf{g}$ and $\mathbf{A}$ do not coincide except in instances of high local symmetry for the species dealt with; for an example of this latter situation, see Ref. 27. This leads to additional complexity. Thus, for instance, in Eqs. 5.6 for the dipole-dipole interaction, one must replace g in $\mathbf{T}$ by the matrix multiplicant $\mathbf{g}$ [15]. In general, one obtains energy expressions (e.g., Eq. 6.55c) involving combination matrices $\mathbf{g} \cdot \mathbf{A} \cdot \mathbf{A}^T \cdot \mathbf{g}^T$, which must be deconvoluted using explicit knowledge of $\mathbf{g}$.

The most favorable case occurs when there exist several isotopes of the nucleus of interest (e.g., C, O, Mg, Si), at least one having a nuclear spin of zero. For those molecules with zero-spin nuclei, $\mathbf{g}$ can be obtained by the method discussed in Chapter 4; these results can then be utilized when analyzing the hyperfine effects arising from the spin-bearing nuclei. When this is not possible, then special techniques can yield relevant energy expressions [9, Section 3.8; 10,28], such as Eq. 6.54, or generalized numerical (computer) techniques can be applied. Often in the literature $\mathbf{g}$ is taken to be isotropic or is taken (in a first approximation) to have principal axes coinciding with those of $\mathbf{A}$. In particular, the latter assumption is a dangerous practice when dealing with a low-symmetry species.

Clearly it is not possible for a powder to yield information about the orientation of the principal axes of, say, a g matrix relative to the laboratory frame, as one can for a single crystal. However, the relative orientations of such axes, when more than one parameter matrix (say, $\mathbf{g}$ and $\mathbf{A}$) is involved, can be derived from EPR powder-pattern analysis [29].

5.5 MULTIPLE HYPERFINE MATRICES

When more than one nucleus with non-zero spin is part of the paramagnetic center being considered, some new features can be encountered:

1. When neither nuclear hyperfine interaction in such a pair is large compared to the electron Zeeman interaction, then first-order theory, as developed in Chapters 2 and 3, remains valid; contributions of the two nuclei to the EPR line positions are then additive, a feature that was implicitly assumed up to this point.

2. When second-order contributions (Eqs. 3.2 and C.29) need to be considered, then cross-terms involving parameters of both members of all pairs of nuclei enter the energy (and hence line-position) equations (Section 6.7). This is true *even though* no interaction terms between nuclear magnetic moments are included in the spin hamiltonian [11]. A cross-term occurs only when both nuclei of a pair exhibit hyperfine anisotropy. Energy terms containing simultaneous contributions from more than two nuclei are absent in this approximation, so that sets of nuclei can be considered pair-wise. In addition, the direct nuclear magnetic dipolar interactions should, in principle, be included in the spin hamiltonians; in practice these are found to be negligible.

5.6 SYSTEMS WITH $I > \frac{1}{2}$

The nuclear-spin angular-momentum direction is linked to the actual shape of the nucleus, that is, to the axis of symmetry of its (time-averaged) electrical charge distribution. When a nucleus has a spin I greater than $\frac{1}{2}$, any electric-field gradient acting on that nucleus can orient its charge ellipsoid and hence its spin direction. Such a gradient is caused primarily, of course, by the electron distribution in the immediate neighborhood. Thus this tendency to align the nucleus affords a means of examining the relative shapes and potency of the atomic orbitals centered at the nucleus in question. As the reader may guess, electrons populating s orbitals, due to their sphericity, cannot act in this fashion. Of the non-s orbitals, p orbitals are more effective than, say, d orbitals with the same principal quantum number.

The energy of alignment, called the *nuclear quadrupole energy*, is derivable from a suitable spin-hamiltonian term [30,31]

$$\hat{\mathcal{H}}_Q = \hat{\mathbf{I}}^{\mathrm{T}} \cdot \mathbf{P} \cdot \hat{\mathbf{I}} \tag{5.50}$$

valid for $I > \frac{1}{2}$, where

$$\mathbf{P} = P \begin{bmatrix} \eta - 1 & 0 & 0 \\ & -\eta - 1 & 0 \\ & & 2 \end{bmatrix} \tag{5.51a}$$

is the nuclear quadrupole coupling matrix in its principal-axis system, with

$$P = \frac{e^2 q_{efg} Q}{4I(2I - 1)} \qquad (5.51b)$$

and **P** is symmetric and traceless. Here $-|e|q_{efg}$, given in units of J C^{-1} m^{-2}, is just the electric-field gradient of largest magnitude seen by the nucleus (by definition, along the primary principal axis **Z** of **P**). Parameter $|e|Q$ describes the electrical shape of the nucleus and is a fixed number ($+$ or $-$) for each isotopic species; it is obtainable from tables (Q is in units of m^2; see Table H.4 footnote regarding tabulations of Q). The asymmetry parameter η describes the deviation of the field gradient from uniaxial symmetry about **Z**; it is dimensionless and is zero when there is local uniaxial symmetry.

Analysis of energy expressions [30,31] derivable from Eq. 5.50 reveals that the local electric-field gradient splits the nuclear-spin state energies already at zero magnetic field B.[10] Thus spectroscopy between such levels is possible and is called *nuclear quadrupole resonance* (NQR). When there is one or more unpaired electrons, that is, in EPR work, such energy contributions are present, in addition to the now familiar hyperfine effects. One can say that there is a competition to align the nuclear spin by several agents, namely, the local electric-field gradient, the local magnetic field originating from the unpaired electron(s), and the externally applied field. These complications must be dealt with when analyzing EPR spectra of solids containing nuclei with $I > \frac{1}{2}$ (Problem C.5).

It should be emphasized that adding the term $\hat{\mathcal{H}}_Q$ (Eq. 5.50) to the spin hamiltonian of an unpaired-electron system does not affect the EPR transitions *to first order*, that is, all energy levels are shifted equally to this approximation. It is the second-order energy contribution (as sketched in Section 6.7), as well as higher ones, which affects the spectrum. ENDOR effects (Chapter 12) are more sensitive to $\hat{\mathcal{H}}_Q$.

We content ourselves herein by stating that the nuclear quadrupole matrix **P** is obtainable via EPR, and the parameters q_{efg} and η therein can give very detailed and valuable information about the electron distributions near such nuclei.

5.7 HYPERFINE POWDER LINESHAPES

The calculation of the expected lineshape for hyperfine splitting in a powder is considered for the case of an isotropic g factor, and $S = I = \frac{1}{2}$. Here the dipolar interaction of the unpaired electron with the single nucleus is to be considered for all possible orientations of the electron-nucleus vector **r** of Fig. 2.2. The angle θ between this vector and the applied field can vary from 0 to π. We adopt the hyperfine parameter $A(\theta) = g_e \beta_e a(\theta)$ as our variable. As before, we assume that $B \gg |a|$, so that there is linearity between B and a (Eq. 5.28).

Let us now, for tutorial purposes, adopt the simple example of the single electron in a hybrid sp orbital previously treated (Section 5.3.2.2). Careful consideration should convince the reader that ranging over the angle θ_p between the p-orbital direction and the field **B** is exactly equivalent to ranging over θ, the angle between the nucleus-electron vector and **B** (Fig. 5.5). Differentiating with respect to the latter angle, while taking g to be isotropic and assuming frequency v to be fixed, one obtains

$$\frac{\sin\theta}{da/d\theta} = -\frac{[(a_0 - b_0)^2 + 3(2a_0 + b_0)b_0 \cos^2\theta]^{1/2}}{3(2a_0 + b_0)b_0 \cos\theta} \tag{5.52}$$

and hence, via Eq. 5.28a, we obtain

$$\frac{\sin\theta}{da/d\theta} = -\frac{g[(1-\xi)^2 + 3(2+\xi)\xi\cos^2\theta]^{1/2}}{3g_e M_I(2+\xi)\xi\cos\theta} \tag{5.53}$$

where

$$\xi \equiv b_0/a_0 = \delta A/A_0 \tag{5.54}$$

The magnitude of the lineshape given by Eq. 5.53 is just $P_{|M_I|}(B)$, giving the field-swept spectrum for either $M_I = \pm\frac{1}{2}$.[11] For $M_I = +\frac{1}{2}$, the relevant field range is $B \leq B_r = hv/g\beta_e$, while for $M_I = -\frac{1}{2}$, one has $B \geq B_r$. Note that, unlike the analogous g-matrix powder pattern (Eq. 4.30 and Fig. 4.7), the envelope extent does not depend on B. The absorption of course consists of two separate envelopes for $P(B)$, since $M_I = \pm\frac{1}{2}$ (Eq. 5.28). This pair of envelopes is the powder extension of the ordinary isotropic $I = \frac{1}{2}$ hyperfine doublet and is centered at B_r under the present assumptions. The overall 'mean' envelope separation is $|A_0|/g\beta_e$.

Figure 5.10 illustrates the total lineshape $P(B)$ plotted versus B for a number of values of ξ. The individual lineshape here is taken to have negligible width. Note that the outer edges of each envelope correspond to the angles $\theta = 0°$ (180°) and 90°. In all cases, $P_{|M_I|}(B)$ has a finite value at $\theta = 0°$ and increases monotonically toward infinity at $\theta = 90°$. The value $\xi = 0$ represents the pure isotropic case. The value $\xi = -2$ produces a *pseudo*-isotropic case.[12] At $\xi = +1$, $P(B)$ is non-zero and independent of B over a finite range, except at $B = B_r$, at which a singularity exists.

The curves in Fig. 5.10 have been drawn assuming a non-zero width inherent in the basic lineshape, yielding a broadening similar to that given in Fig. 4.7b is found. The first-derivative lineshape is very similar to that shown in Fig. 4.7c except that there is a duplication, with *opposite* phases, since there are two hyperfine components. As was the case for g anisotropy in Section 4.7, the outer lines appear as absorption lines in the first-derivative presentation.

Figure 5.11 illustrates the EPR spectrum of the FCO radical, which is randomly oriented in a CO matrix at 4.2 K. For this radical, g is essentially isotropic and is close to g_e. Although the symmetry is not quite uniaxial, it is considered to be so for purposes of illustration. The separation of the outermost lines is given by $|A_0 + 2\delta A|/g_e\beta_e \cong 51.4$ mT, whereas the separation of the inner lines is given by

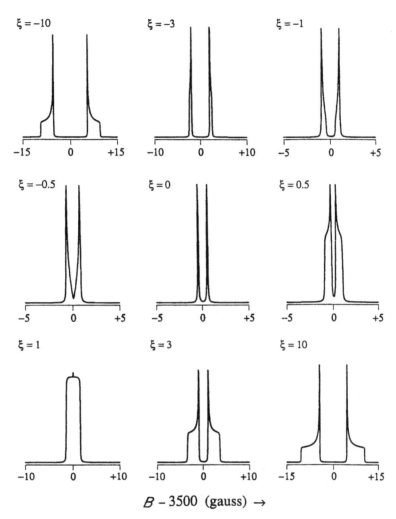

B – 3500 (gauss) →

FIGURE 5.10 Hyperfine absorption lineshapes (Eq. 5.53) for a randomly oriented paramagnetic species with $S = I = \frac{1}{2}$ and having an isotropic g factor (g_e), for nine selected values of $\xi = dA/A_0$. These are plots of the envelopes $P(\theta)$ versus B. The individual lines are simulated using lorentzian lineshapes with linewidth $\Delta B_{1/2}$ of 0.05 G. The total areas under the curves are equal; the P and B scales vary.

$|A_0 - \delta A|/g_e \beta_e \cong 24.6$ mT. From this one may deduce that *either* $A_0/h \cong \pm 940$ MHz, $\delta A/h \cong \pm 250$ MHz, or $A_0/h = \pm 20$ MHz, $\delta A/h \cong \pm 710$ MHz. The former assignment is the correct one, but one requires additional information to resolve the ambiguity [32].

As a second example, we cite briefly the use of powder/glass patterns to extract liquid-solution parameters. Thus, for the naphthalene cation radical, which is very

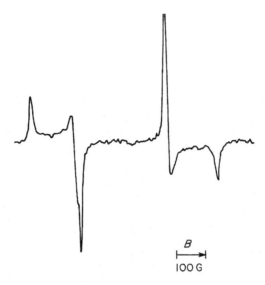

B

$\longmapsto$ 100 G

FIGURE 5.11   EPR spectrum of FCO in a CO matrix at 4.2 K. The microwave frequency is 9123.97 MHz. The hyperfine interaction is not quite uniaxial, as is seen by the incipient splitting of the second peak from the left. [After F. J. Adrian, E. L. Cochran, V. A. Bowers, *J. Chem. Phys.*, **43**, 462 (1965).]

unstable in liquid solvents, simulation of the EPR pattern for the radical studied in boric acid glass at 300 K yields the principal values of the proton hyperfine matrices, from which the traces yield the two $|A_0{}^H|$ values (Table 9.3) [33].

Naturally, deviations from the powder-pattern model described above arise in general, including cases when

1. Approximate Eq. 5.28 is not adequate.
2. There is anisotropy in g, transition probability, and/or linewidth.
3. The nucleus has spin greater than $\frac{1}{2}$.
4. There are several spin-bearing nuclei.
5. The total electron spin is greater than $\frac{1}{2}$.

It is then necessary to use other relations,[13] and possibly to employ computer simulation [36]. It *may* be possible in simple cases to determine some or all of the components of **g** and **A**. However, the reader is warned that there are strong possibilities for misassignments. Figure 5.12 illustrates the idealized first-derivative lineshapes for some simple cases. Note the difference in phase occurring for A anisotropy as compared to g anisotropy. The problems associated with small hyperfine splittings and satellite lines (such as those discussed in Section 5.3.2.2) can be very considerable [37].

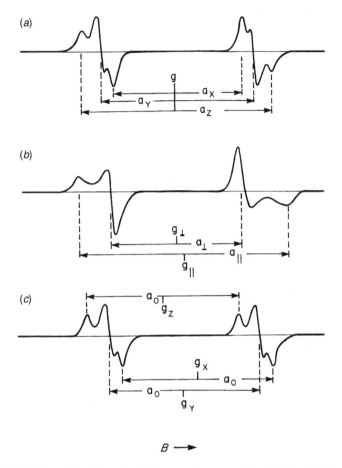

FIGURE 5.12 Examples of first-derivative powder spectra of radicals exhibiting hyperfine splitting from one nucleus with $I = \frac{1}{2}$: (*a*) isotropic *g* factors and $a_Z > a_Y > a_X > 0$; (*b*) uniaxial symmetry, $g_{\parallel} < g_{\perp}$; $a_{\parallel} > a_{\perp} > 0$; (*c*) isotropic hyperfine splitting $a_0 > 0$ and $g_X < g_Y < g_Z$. [After P. W. Atkins, M. C. R. Symons, *The Structure of Inorganic Radicals*, Elsevier, Amsterdam, Netherlands, 1967, p. 270.]

We cannot go into details of EPR spectra derived from partially aligned molecules here. An example is the anisotropic X-band spectrum of the ion $C_2F_4^-$, generated by γ irradiation of tetrafluoroethylene in a crystalline methylcyclohexane-d_{14} matrix [38]. An excellent summary of preferential orienting of paramagnetic species is available in the book by Weltner [39].

Finally, we turn to an example in which a species acts crystalline in one dimension and powder-like in the other two. This is the situation for NO_2 molecules ($S = \frac{1}{2}$) adsorbed at 20 K as monolayers on a Kr/Ag substrate [40]. Computer simulation reproduces the experimental EPR spectrum nicely and indicates that the planes of the NO_2 molecules are coplanar with the surface.

REFERENCES

1. P. W. Kirklin, P. Auzins, J. E. Wertz, *J. Phys. Chem. Solids*, **26**, 1067 (1965).

2. B. Henderson, J. E. Wertz, *Adv. Phys.*, **17**, 749 (1968).

3. W. Cheston, *Elementary Theory of Electric and Magnetic Fields*, Wiley, New York, NY, U.S.A., 1964, p. 151.

4. P. W. Atkins, R. S. Friedman, *Molecular Quantum Mechanics*, 4th ed., Oxford University Press, Oxford, U.K., 2005, Section 13.16.

5. R. Skinner, J. A. Weil, *Am. J. Phys.*, **57**, 777 (1989).

6. F. K. Kneubühl, *Phys. Kondens. Materie*, **1**, 410 (1963).

7. H. M. McConnell, *Proc. Natl. Acad. Sci. U.S.A.*, **44**, 766 (1958).

8. M. Rudin, A. Schweiger, Hs. H. Günthard, *Mol. Phys.*, **46**, 1027 (1982).

9. A. Abragam, B. Bleaney, *Electron Paramagnetic Resonance of Transition Ions*, Clarendon, Oxford, U.K., 1970, Section 17.6.

10. J. A. Weil, *J. Magn. Reson.*, **4**, 394 (1971).

11. J. A. Weil, *J. Magn. Reson.*, **18**, 113 (1975).

12. J. A. Weil, *Can. J. Phys.*, **59**, 841 (1981).

13. J. A. Weil, J. H. Anderson, *J. Chem. Phys.*, **35**, 1410 (1961).

14. G. T. Trammell, H. Zeldes, R. Livingston, *Phys. Rev.*, **110**, 630 (1958).

15. H. Zeldes, G. T. Trammell, R. Livingston, R. W. Holmberg, *J. Chem. Phys.*, **32**, 618 (1960).

16. S. M. Blinder, *J. Chem. Phys.*, **33**, 748 (1960).

17. M. T. Rogers, D. H. Whiffen, *J. Chem. Phys.*, **40**, 2662 (1964).

18. J. A. Weil, J. E. Clapp, T. Buch, *Adv. Magn. Reson.*, **6**, 183 (1973).

19. L. D. Kispert, M. T. Rogers, *J. Chem. Phys.*, **54**, 3326 (1971).

20. C. M. Bogan, L. D. Kispert, *J. Phys. Chem.*, **77**, 1491 (1973).

21. P. Scheidler, J. R. Bolton, *J. Am. Chem. Soc.*, **88**, 371 (1966).

22. W. A. Sheppard, *J. Am. Chem. Soc.*, **87**, 2410 (1965).

23. H. Zeldes, R. Livingston, *Phys. Rev.*, **96**, 1702 (1954).

24. G. T. Trammell, H. Zeldes, R. Livingston, *Phys. Rev.*, **110**, 630 (1958).

25. J. Isoya, J. A. Weil, P. H. Davis, *J. Phys. Chem. Solids*, **44**, 335 (1983).

26. H. M. McConnell, J. Strathdee, *Mol. Phys.*, **2**, 129 (1959); For an erratum, see R. M. Pitzer, C. W. Kern, W. N. Lipscomb, *J. Chem. Phys.*, **37**, 267 (1962).

27. H. Zeldes, R. Livingston, *J. Chem. Phys.*, **35**, 563 (1961).

28. H. A. Farach, C. P. Poole Jr., *Adv. Magn. Reson.*, **5**, 229 (1971).

29. B. M. Peake, P. H. Rieger, B. H. Robinson, J. Simpson, *J. Am. Chem. Soc.*, **102**, 156 (1980).

30. R. S. Drago, *Physical Methods in Chemistry*, Saunders, Philadelphia, PA, U.S.A., 1977, Chapter 14.

31. C. P. Slichter, *Principles of Magnetic Resonance*, 3rd ed., Springer, New York, NY, U.S.A., 1990, Chapter 10.

32. F. J. Adrian, E. L. Cochran, V. A. Bowers, *J. Chem. Phys.*, **43**, 462 (1965).

33. G. S. Owen, G. Vincow, *J. Chem. Phys.*, **54**, 368 (1971).

34. E. L. Cochran, F. J. Adrian, V. A. Bowers, *J. Chem. Phys.*, **33**, 156 (1961).

35. R. Neiman, D. Kivelson *J. Chem. Phys.*, **35**, 156 (1961).

36. R. Aasa, T. Vanngaård, *J. Magn. Reson.*, **19**, 308 (1975).

37. R. Lefebvre, J. Maruani, *J. Chem. Phys.*, **42**, 1480 (1965).

38. J. R. Morton, K. F. Preston, J. T. Wang, F. Williams, *Chem. Phys. Lett.*, **64**, 71 (1979).

39. W. Weltner Jr., *Magnetic Atoms and Molecules*, Scientific and Academic Editions, Van Nostrand Reinhold, New York, NY, U.S.A., 1983, pp. 95–102.

40. M. Zomack, K. Baberschke, *Surf. Sci.*, **178**, 618 (1986).

41. D. Pooley, D. H. Whiffen, *Trans. Faraday Soc.*, **57**, 1445 (1961).

NOTES

1. Strictly speaking, the nuclear Zeeman spin-hamiltonian operator should be written in terms of a parameter matrix, that is, as $-\beta_n \mathbf{B}^{\mathrm{T}} \cdot \mathbf{g}_n \cdot \hat{\mathbf{I}}$, in complete analogy with the electronic Zeeman operator discussed in detail in this chapter. Taking $\mathbf{g}_n$ as a matrix allows inclusion of the well-known chemical shift and other phenomena. In practice, in EPR, these anisotropy effects generally are negligibly small. Thus we take g_n as a scalar in this book. Note also that the alternate choice of the hyperfine term $\hat{\mathbf{I}}^{\mathrm{T}} \cdot \mathbf{A} \cdot \hat{\mathbf{S}}$ would imply that $\mathbf{A}$ here is the transpose of $\mathbf{A}$ occurring in Eq. 5.7. Matrix $\mathbf{A}$ need not be symmetric when g is anisotropic; indeed, there are various known examples for which $\mathbf{A}$ is asymmetric [6–8].

2. That sign becomes measurable under conditions when the nuclear Zeeman term appreciably affects the observed spectrum. Note also that $\mathbf{n}^{\mathrm{T}} \cdot \mathbf{A} \cdot \mathbf{n} = \frac{1}{2}\mathbf{n}^{\mathrm{T}} \cdot (\mathbf{A} + \mathbf{A}^{\mathrm{T}}) \cdot \mathbf{n}$ (Problem 5.2). From such values, one can extract $A_{ij} + A_{ji}$, but not A_{ij} or A_{ji}, when $i \neq j$.

3. This problem was first considered by Zeldes et al. [15] (see also Blinder [16]).

4. The spin-hamiltonian matrix is not, in fact, diagonal since $\hat{\mathbf{I}}$ is *not* quantized along $\mathbf{z}$.

5. $\langle 3 \cos^2\alpha - 1 \rangle = 4/7$ for a d_{z^2} orbital and 8/15 for an f_{z^3} orbital.

6. As is the case for $\mathbf{gg}$, $\mathbf{AA}$ is a true tensor; however, $\mathbf{A}$ is not.

7. As we see, site splitting here leads to a pair of 'equal and opposite' off-diagonal elements, since the cartesian crystal-axis system was appropriately chosen with respect to the crystal symmetry. Each site has only one appropriate sign for each off-diagonal element, but care is required to assign the correct pairing. Thus in Eq. 5.35 the value $+4.78$ is to be associated with -0.16 and -4.78 with $+0.16$. This correlation cannot be determined experimentally from the original three rotations, but it can be obtained from other crystal positions, especially that with the field in the directions $(3^{-1/2}, 3^{-1/2}, \pm 3^{-1/2})$.

8. This is a general caveat. Interpretation of spin-hamiltonian parameters should be done with awareness of their temperature dependence, that is, in light of possible dynamic effects in the paramagnetic entity being investigated.

9. This is in contrast to the typical α-proton anisotropic hyperfine matrix in other radicals.

10. This, then, is a contribution to the zero-field splitting in addition to the hyperfine splitting.

11. A relation similar to Eq. 5.53 was developed by Blinder [16].

12. Such a case could easily be mistaken for a purely isotropic hyperfine interaction. The only way to tell would be to examine (if possible) the system in a liquid of low viscosity where the true isotropic hyperfine splitting would be obtained.

13. The case of rhombic symmetry and an isotropic g factor is considered by Blinder [16]; see also Cochran et al. [34]. The case of uniaxial symmetry and comparable hyperfine and g anisotropy is considered by Neiman and Kivelson [35].

FURTHER READING

1. B. Bleaney, "Hyperfine Structure and Electron Paramagnetic Resonance", in *Hyperfine Interactions*, A. J. Freeman, R. B. Frankel, Eds., Academic Press, New York, NY, U.S.A., 1967.

2. C. P. Poole Jr., H. A. Farach, *The Theory of Magnetic Resonance*, Wiley-Interscience, New York, NY, U.S.A., 1972.

PROBLEMS

5.1 From the table below obtained from the 10.0000 GHz EPR spectrum of $[GeO_4/Na]_A^0$ in irradiated α-quartz (Fig. 5.2):

   (*a*) Label the dectet peaks with M_I values.

   (*b*) Obtain g_{zz} from the even-isotope peak (here by definition **z** is parallel to the direction of **B**).

   (*c*) Evaluate the matrix element $|A_{sym}/g_e\beta_e|_{zz}$ using Eq. 5.10.

   (*d*) Calculate the natural abundance of ^{73}Ge.

^{73}Ge Isotopic Species		Even-Ge Isotopic Species (^{70,72,74,76}Ge)	
B (mT)	Relative Intensity	B (mT)	Relative Intensity
231.114	0.245	358.507	27.85
249.184	0.239	—	—
269.376	0.234	—	—
291.823	0.230	—	—
316.621	0.229	—	—
343.821	0.229	—	—
373.421	0.231	—	—
405.371	0.234	—	—
439.573	0.239	—	—
475.892	0.246	—	—

5.2 Prove that $\mathbf{n}^T \cdot \mathbf{A} \cdot \mathbf{n} = \mathbf{n}^T \cdot \mathbf{A}^T \cdot \mathbf{n}$ even when **A** is an asymmetric matrix. Here **n** can be taken to be the unit vector along **B** (Section A.5.2).

5.3 Show that the mean value of $3 \cos^2\alpha - 1$ for a $2p_z$ orbital (Table B.1) is $\frac{4}{5}$ (Eq. 5.26 and Fig. 5.5).

5.4 Consider a pure p orbital $\psi_p = (3/4\pi)^{1/2} \cos\alpha$ pointed along a unit vector $\mathbf{p}$ lying in the xz plane (Fig. 5.5). Prove the relation

$$\langle 3 \cos^2\theta - 1 \rangle = \tfrac{1}{2}(3 \cos^2\alpha - 1)(3 \cos^2\theta_p - 1) \qquad (5.55)$$

between the polar angles

   α: between $\mathbf{p}$ and $\mathbf{r}$
   θ: between $\mathbf{r}$ and $\mathbf{z}$
   θ_p: between $\mathbf{p}$ and $\mathbf{z}$

Vector $\mathbf{r}$ gives the position of the electron relative to the nucleus at the origin. Note that ψ_p is independent of the azimuthal angle β measured relative to axis $\mathbf{p}$. You may wish to use the integrals

$$\langle \sin^2\alpha \sin^2\beta \rangle_\Omega = \tfrac{1}{5} \qquad (5.56a)$$

$$\langle \sin^2\alpha \sin\beta \cos\beta \rangle_\Omega = 0 \qquad (5.56b)$$

$$\langle \cos^2\alpha \rangle_\Omega = \tfrac{3}{5} \qquad (5.56c)$$

where subscript Ω indicates integration over the whole range of α and β.

5.5 The hyperfine matrix for an unpaired electron in a pure p orbital at angle θ_p from $\mathbf{z}$ with its axis in the xz plane (Section 5.3.2.3 and Fig. 5.5) is

$$\mathbf{A} = \delta A \begin{bmatrix} 3\sin^2\theta_p - 1 & 0 & 3\sin\theta_p \cos\theta_p \\ & -1 & 0 \\ & & 3\cos^2\theta_p - 1 \end{bmatrix} \qquad (5.57)$$

The nucleus is at the origin.

(*a*) Diagonalize $\mathbf{A}$.

(*b*) Find and discuss its principal values and directions.

(*c*) Find expressions for a_0, b_0 and c_0 (Eqs. 5.9).

5.6 Consider a 3×3 real symmetric matrix $\mathbf{Y}_1$ taken in a cartesian coordinate system with axes x, y, z labeling both the rows and the columns. Show that a similarity transformation $\mathbf{R} \cdot \mathbf{Y}_1 \cdot \mathbf{R}^{-1} = \mathbf{Y}_2$ with a rotation matrix

$$\mathbf{R} = \begin{bmatrix} \cos \Omega & 0 & -\sin \Omega \\ 0 & 1 & 0 \\ \sin \Omega & 0 & \cos \Omega \end{bmatrix} \tag{5.58}$$

describing a rotation about axis $\mathbf{y}$ yields a matrix $\mathbf{Y}_2$ differing only from $\mathbf{Y}_1$ in the signs of the elements Y_{xy} and Y_{yz}, if $\Omega = 180°$. Prove that $\mathbf{Y}_1$ and $\mathbf{Y}_2$ have the same principal values, for any Ω.

5.7 The hyperfine matrices for the C—H proton of the two sites of the $HC(OH)CO_2^-$ radical in irradiated anhydrous $H_2C(OH)CO_2Li$ have been measured in relation to the axes of the crystal with the following results [41]:

$$\begin{array}{c} \mathbf{A}/h \\ (\text{MHz}) \end{array} = \begin{bmatrix} -51 & \mp13 & +14 \\ & -77 & \pm6 \\ & & -35 \end{bmatrix} \quad \mathbf{C} = \begin{bmatrix} 0.43 & 0.74 & 0.51 \\ \pm0.88 & \mp0.48 & \mp0.03 \\ -0.23 & -0.46 & 0.86 \end{bmatrix}$$

where $\mathbf{C} \cdot \mathbf{A} \cdot \mathbf{C}^{-1}$ is diagonal. The accuracy is limited by the linewidth, which is ~ 15 MHz.

(*a*) Verify that the direction-cosine matrix $\mathbf{C}$ diagonalizes $\mathbf{A}$.

(*b*) In matrix $\mathbf{C}$, which are the principal vectors, the rows or columns?

(*c*) Specify the direction of each principal axis in terms of the angles it makes with the axes $\mathbf{x}$, $\mathbf{y}$ and $\mathbf{z}$.

5.8 The di-*t*-butyl nitroxide radical may be introduced as a substitutional impurity into 2,2,4,4-tetramethylcyclobutane-1,3-dione. In terms of the Cartesian axis system abc' of the monoclinic crystal, the elements of the matrices $\mathbf{AA}$ (in mT^2) and $\mathbf{gg}$ are as follows:

$(\mathbf{AA})_{xx} = 0.5486$	$(\mathbf{gg})_{xx} = 4.03081$
$(\mathbf{AA})_{xy} = 0.2666$	$(\mathbf{gg})_{xy} = -0.00057$
$(\mathbf{AA})_{xz} = 0.1110$	$(\mathbf{gg})_{xz} = +0.00501$
$(\mathbf{AA})_{yy} = 10.189$	$(\mathbf{gg})_{yy} = 4.00955$
$(\mathbf{AA})_{yz} = -0.7042$	$(\mathbf{gg})_{yz} = +0.00092$
$(\mathbf{AA})_{zz} = 0.4386$	$(\mathbf{gg})_{zz} = 4.02834$

(*a*) Show that the principal values of the parameters are

$(\mathbf{AA})_X = 0.6162$	$(\mathbf{gg})_X = 4.03314$
$(\mathbf{AA})_Y = 0.3136$	$(\mathbf{gg})_Y = 4.02688$
$(\mathbf{AA})_Z = 10.246$	$(\mathbf{gg})_Z = 4.00868$

(b) Find the direction cosines (principal axes) of the **A** and the **g** matrices (which are taken to be symmetric), using the technique described in Section A.5.5. Are the two matrices coaxial?

(c) The crystal structure indicates that the plane of

is in the ac' plane of the crystal. The N—O bond forms an angle of about $34°$ with the a axis. One usually assumes that the largest principal value of the **A** matrix corresponds to the direction of the $2p_z$ orbital. Use the direction cosines derived in part (b) to verify the validity of this assumption. [Data taken from W. R. Knolle, 'An Electron Spin Resonance Study of Fluorinated Radicals', Ph.D. dissertation, University of Minnesota, 1970. This system was first studied by O. H. Griffith, D. W. Cornell, H. M. McConnell, *J. Chem. Phys.*, **43**, 2909 (1965).]

5.9 Find the principal axes of matrices 5.37 and 5.38 (note Section A.5.5). The elements of the required direction-cosine matrix **C** are obtained from Eq. 5.35 by substituting in turn the three values of l into the equations

$$(1.60 - \lambda_i)C_{i1} \pm 4.78C_{i2} + 0.64C_{i3} = 0$$

$$\pm 4.78C_{i1} + (16.36 - \lambda_i)C_{i2} \mp 0.16C_{i3} = 0$$

$$0.64C_{i1} \mp 0.16C_{i2} + (2.71 - \lambda_i)C_{i3} = 0$$

where $i = 1, 2, 3$ labels the principal value dealt with. The second subindex refers to the spatial coordinate (x, y, z).

5.10 Derive the expression for the limit, as a_0 becomes negligible compared to b_0, of the hyperfine splitting expression (Eq. 5.28b)

$$a = [(a_0 - b_0)^2 + 3b_0(2a_0 + b_0)\cos^2 \theta_p]^{1/2} \qquad (5.59)$$

(which is valid for the usual case in which the hyperfine field is much larger than the externally applied field). Show also (by binomial expansion of the square root) that, if b_0^2 can be neglected in comparison with a_0^2, the general expression reduces to the familiar form

$$a = |a_0 + b_0(3\cos^2 \theta_p - 1)| \qquad (5.30)$$

where a is taken to be positive and $g = g_e$ is assumed.

5.11   Consider a single unpaired-electron system in which one atom X has an unpaired-electron population $|c_p|^2 = 0.25$ in a pure p orbital. Ignore the rest of the unpaired-electron density as being remote from the hyperfine nucleus, ^{7}Li, of interest. Take $g = g_e$. The measured ^{7}Li hyperfine matrix is

$$\frac{\mathbf{A}}{g_e\beta_e} = \begin{bmatrix} 0.1470 & 0 & 0 \\ & 0.0981 & 0.0002 \\ & & 0.0889 \end{bmatrix} \text{mT}$$

but its absolute sign is unknown.

(*a*) Determine $|a_0|$, $|b_0|$ and c_0.

(*b*) Show that the distance R between X and the ^{7}Li nucleus is ~ 0.26 nm, adopting Eq. 5.48 with $f = 1$. What, then, is the sign of b_0?

(*c*) What is the sign of a_0?

5.12   Often the parameter $\langle r^{-3} \rangle$ in dipolar formulas (e.g., Eqs. 5.6a, 5.26, 6.15a, 6.41 and 9.18a, b, as well as Problem 6.9) is used to estimate an average inter-spin distance $\langle r \rangle$. Consider a set (1,2,3) of values of r, say, in nanometer (nm) units. Calculate $\langle r \rangle$, $\langle r^{-1} \rangle^{-1}$, $\langle r^3 \rangle^{1/3}$ and $\langle r^{-3} \rangle^{-1/3}$, and compare these values. What do you conclude?

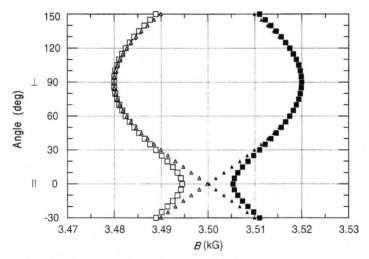

FIGURE 5.13   A graph of rotation angle versus EPR line position, created for an $S = I = \frac{1}{2}$ case. The g factor was taken to be isotropic at g_e, and the frequency was 9.8087 GHz. There are four EPR transitions possible. The two plotted are for jumps $1 \leftrightarrow 4$ (occurring at the lower magnetic field B) and $2 \leftrightarrow 3$, among the four energy levels (note Fig. 5.4). We are dealing with only a single type of crystal site here. The hyperfine matrix $\mathbf{A}$ was taken to be uniaxial with values of 0.20 and 40.00 G for $A_{\parallel}/g_e\beta_e$ and $A_{\perp}/g_e\beta_e$. The nuclear Zeeman term (isotropic for ^{1}H) was included in the curves with boxed points and absent in the curves with triangles. (Prepared by M. J. Mombourquette and J. A. Weil; artificial data.)

5.13 Using the information in Fig. 5.13 and its caption, sketch the powder spectra (with and without inclusion of the nuclear Zeeman energy) that you would expect to observe in a field-swept EPR experiment, assuming $\mathbf{B}_1 \perp \mathbf{B}$. Assume zero linewidth for this purpose. You could generate this spectrum using an appropriate computer program, using (close to) zero linewidths.

(*a*) Explain the details of your derived spectra.

(*b*) In this special hyperfine situation, how many of the four EPR transitions are required to adequately represent the primary spectrum?

(*c*) Calculate $\xi \equiv \delta A / A_0$. Compare the derived powder spectra with appropriate one depicted in Fig. 5.10. Under what conditions does the latter give incorrect results?

CHAPTER 6

SYSTEMS WITH MORE THAN ONE UNPAIRED ELECTRON

6.1 INTRODUCTION

Nearly all the species considered in previous chapters had only one unpaired electron (viz., $S = \frac{1}{2}$). In principle, an EPR spectrum is obtainable for any system with an odd number of electrons ($S = \frac{1}{2}, \frac{3}{2}, \frac{5}{2}, \ldots$). For systems with an *even* number of electrons ($S = 1, 2, 3, \ldots$), there is no such guarantee; however, as we shall see, many systems of the latter type are accessible to EPR spectroscopy.

We begin by considering the theory of systems with two unpaired electrons, initially ignoring all nuclear spins.[1] Such systems include (1) atoms or ions in the gas phase (e.g., oxygen atoms), (2) small molecules in the gas phase (e.g., O_2), (3) organic molecules containing two or more unpaired electrons (e.g., naphthalene excited to its metastable triplet state) in solid-state solutions and crystals, (4) inorganic molecules (e.g., CCO in rare-gas matrices), (5) 'point' defects in crystals containing more than one unpaired electron (e.g., the F_t center in MgO), (6) biradicals in fluid solution and the solid state and (7) certain transition-group (e.g., V^{3+} and Ni^{2+}) and rare-earth ions. Systems 1 and 2 are dealt with in Chapter 7.

In all chemical species, if the highest occupied electronic level is orbitally nondegenerate and is doubly occupied by electrons, the ground state must be a spin singlet (Fig. 6.1a). If one of these electrons is excited to an unoccupied orbital by absorption of a quantum of the appropriate energy (Fig. 6.1b), the system is still in a singlet state, since allowed transitions occur without change of multiplicity. However, the molecule may then undergo intersystem crossing to a metastable triplet state, with a change of spin (Fig. 6.1c). This process is attributed to the

Electron Paramagnetic Resonance, Second Edition, by John A. Weil and James R. Bolton
Copyright © 2007 John Wiley & Sons, Inc.

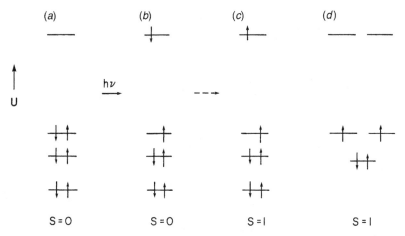

FIGURE 6.1 Energy levels and spin configuration of a six-electron system in: (a) its singlet ground state; (b) after electronic excitation (the spins remain paired and hence the state is a singlet); (c) after the molecule goes to a metastable triplet state via a radiationless process (the triplet state is usually lower in energy than the singlet state [in (b)] because of decreased interelectronic repulsion); and (d) configuration of a four-electron system with a triplet ground state. The lowest level is symbolic of filled orbitals below the degenerate pair; it is irrelevant whether the filled orbitals are degenerate or non-degenerate.

presence of spin-orbit coupling, molecular rotation and/or hyperfine interaction in the presence of an external field **B**. It is highly selective in populating the three sublevels; that is, the process leads to spin polarization.

A triplet ground state (Fig. 6.1d) requires, in view of the Pauli exclusion principle [1], that this state have at least a two-fold orbital degeneracy (or near degeneracy). Low-lying orbitals filled with electrons have been shown in Fig. 6.1 to emphasize that it is the set of highest *occupied* orbitals that is important in determining the multiplicity of the state.

6.2 SPIN HAMILTONIAN FOR TWO INTERACTING ELECTRONS

For the case of two electrons there are four spin states. One way of representing these states is to construct simple product spin states (the *uncoupled representation*; Section B.6)

$$\alpha(1)\alpha(2) \quad \alpha(1)\beta(2) \quad \beta(1)\alpha(2) \quad \beta(1)\beta(2) \qquad \text{Set (6.1)}$$

In a paramagnetic center of moderate size, such that the two electrons interact appreciably, it is more advantageous to combine these configurations into combination states (the *coupled representation*) because the system separates in energy into a triplet state and a singlet state.[2] These coupled functions are either *symmetric* or *antisymmetric* with respect to exchange of the electrons. The combination functions are (Section B.6)

Symmetric	**Antisymmetric**	
$\alpha(1)\alpha(2)$		
$\dfrac{1}{\sqrt{2}}[\alpha(1)\beta(2) + \beta(1)\alpha(2)]$	$\dfrac{1}{\sqrt{2}}[\alpha(1)\beta(2) - \beta(1)\alpha(2)]$	Set (6.2)
$\beta(1)\beta(2)$		
Triplet state, $S = 1$	**Singlet state, $S = 0$**	

The multiplicity of the state with total spin $S = 1$ is $2S + 1 = 3$; hence it is called a *triplet state*. The state with $S = 0$ analogously is called a *singlet state*. If the two electrons occupy the same spatial orbital, then only the antisymmetric or singlet state is possible because of the restrictions imposed by the Pauli exclusion principle [1]. However, if each electron occupies a different orbital, then both the singlet and triplet states exist.[3]

6.2.1 Electron-Exchange Interaction

The singlet and triplet states are split apart in energy by the electron-exchange interaction, represented by the spin hamiltonian

$$\hat{\mathcal{H}}_{\text{exch}} = \sum_{ij} J_{ij}\hat{S}_{1i}\hat{S}_{2j} \qquad (i, j = 1, 2, 3) \qquad (6.3a)$$

$$= \tfrac{1}{2}(\hat{\mathbf{S}}_1{}^{\text{T}} \cdot \mathbf{J} \cdot \hat{\mathbf{S}}_2 + \hat{\mathbf{S}}_2{}^{\text{T}} \cdot \mathbf{J}^{\text{T}} \cdot \hat{\mathbf{S}}_1) \qquad (6.3b)$$

where $\hat{\mathbf{S}}_1$ and $\hat{\mathbf{S}}_2$ are the electron-spin operators for electrons 1 and 2 (Section B.10 and Problem B.11). Indices i and j label spatial coordinates.

Here **J** is a 3×3 matrix that takes into account the electric (coulombic) interaction between the two unpaired electrons, but not the important magnetic interaction that is introduced in Section 6.2.2. Details concerning the theory of exchange-coupled systems are available in the literature [2–4].

For our purposes, we consider only the most important part of the exchange-energy operator (Eq. 6.3), that is, the isotropic part[4]

$$(\hat{\mathcal{H}}_{\text{exch}})_{\text{iso}} = J_0 \, \hat{\mathbf{S}}_1{}^{\text{T}} \cdot \hat{\mathbf{S}}_2 \qquad (6.4)$$

where $J_0 = tr(\mathbf{J})/3$ is the isotropic electron-exchange coupling constant, which to a first approximation is given [5] by the exchange integral

$$J_0 = -2\langle \phi_a(1)\phi_b(2)| \frac{e^2}{4\pi\varepsilon_0 r} |\phi_a(2)\phi_b(1)\rangle \qquad (6.5)$$

Here ϕ_a and ϕ_b are different normalized spatial molecular-orbital wavefunctions, evaluated while considering the electrons to be non-interacting, ε_0 is the permittivity of the vacuum, and r is the inter-electron distance.[5] Whether the singlet or the triplet state lies lower depends on the sign of J_0. The interaction between two hydrogen atoms is a textbook example; here $J_0 > 0$ and the singlet (bonding) state lies lowest. In the molecular-orbital description of H_2, J_0 is a major contribution to the total binding energy. If $J_0 < 0$, as is the case in some species, the triplet state has the lower energy (Fig. 6.2).

In analogy with the isotropic electron-nuclear hyperfine interaction (Eq. C.2b), Eq. 6.3 can be written as

$$\left(\hat{\mathcal{H}}_{\text{exch}}\right)_{\text{iso}} = J_0 \left[\hat{S}_{1z}\hat{S}_{2z} + \tfrac{1}{2}(\hat{S}_{1+}\hat{S}_{2-} + \hat{S}_{1-}\hat{S}_{2+})\right] \tag{6.6}$$

Application of this operator to the spin wavefunctions shows that the singlet and triplet states are separated by the energy $|J_0|$ (Fig. 6.2), with eigenfunctions as listed in set 6.2. Experimentally, one can observe which state (the singlet or triplet) is lower in energy by studying the EPR signal intensity (i.e., the spin-level populations) as the temperature approaches 0 K.

Note that J_0 is the analog of the isotropic hyperfine coupling parameter introduced in Eq. 2.39b. As pointed out, the four energy levels of the hydrogen atom are similarly split at zero field into a triplet and a singlet (Appendix C). There is also considerable analogy to the NMR isotropic spin-spin coupling constant J. The magnitude of J_0 decreases with increasing r (Eq. 6.5), so that $|J_0|$ is very small if the two electrons are on the average sufficiently far apart. In this case, if the exchange energy is not large compared to the magnetic dipolar interaction energy, then the two-electron system is called a *biradical*.

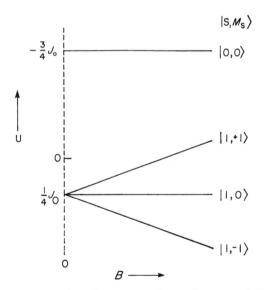

FIGURE 6.2 The state energies of a system of two electrons exhibiting an exchange interaction, for $J_0 < 0$. When J_0 is positive, the singlet state lies lower in energy.

6.2.2 Electron-Electron Dipole Interaction

In addition to electron exchange, which splits the states into a singlet and a triplet, there exists another important interaction, also quadratic in the electron spin, namely, the anisotropic magnetic dipole-dipole interaction.[6] This interaction causes the three-fold degeneracy of the triplet state to be removed even in zero magnetic field; the latter effect often is called *zero-field splitting*.[7]

The dipole-dipole interaction for the coupling of two unpaired electrons is analogous to the corresponding interaction (Eq. 5.3) between electronic and nuclear magnetic dipoles, which gives rise to the anisotropic hyperfine interaction (Fig. 2.2); that is, the electron-spin electron-spin dipolar interaction is given by the hamiltonian

$$\hat{\mathcal{H}}_{ss}(\mathbf{r}) = \frac{\mu_0}{4\pi}\left[\frac{\hat{\boldsymbol{\mu}}_1^T \cdot \hat{\boldsymbol{\mu}}_2}{r^3} - \frac{3(\hat{\boldsymbol{\mu}}_1^T \cdot \mathbf{r})(\hat{\boldsymbol{\mu}}_2^T \cdot \mathbf{r})}{r^5}\right] \tag{6.7}$$

Here the inter-electron vector $\mathbf{r}$ is defined as in Fig. 2.2 with $\boldsymbol{\mu}_n$ replaced by $\boldsymbol{\mu}_e$. The magnetic-moment operators may be replaced by the corresponding spin operators, to yield

$$\hat{\mathcal{H}}_{ss}(\mathbf{r}) = \frac{\mu_0}{4\pi}g_1 g_2\beta_e^2\left[\frac{\hat{\mathbf{S}}_1^T \cdot \hat{\mathbf{S}}_2}{r^3} - \frac{3(\hat{\mathbf{S}}_1^T \cdot \mathbf{r})(\hat{\mathbf{S}}_2^T \cdot \mathbf{r})}{r^5}\right] \tag{6.8}$$

where g_1 and g_2 are the g factors for electrons 1 and 2, taken here to be isotropic. Henceforth, for simplicity, we assume that $g_1 = g_2 = g$. Expansion of the scalar products in Eq. 6.8 yields

$$\hat{\mathcal{H}}_{ss}(\mathbf{r}) = \frac{\mu_0}{4\pi}\frac{(g\beta_e)^2}{r^5}\Big[(r^2 - 3x^2)\hat{S}_{1x}\hat{S}_{2x} + (r^2 - 3y^2)\hat{S}_{1y}\hat{S}_{2y} +$$
$$(r^2 - 3z^2)\hat{S}_{1z}\hat{S}_{2z} - 3xy(\hat{S}_{1x}\hat{S}_{2y} + \hat{S}_{1y}\hat{S}_{2x}) -$$
$$3xz(\hat{S}_{1x}\hat{S}_{2z} + \hat{S}_{1z}\hat{S}_{2x}) - 3yz(\hat{S}_{1y}\hat{S}_{2z} + \hat{S}_{1z}\hat{S}_{2y})\Big] \tag{6.9}$$

Because the two electrons are coupled, it is more convenient to express $\hat{\mathcal{H}}_{ss}$ in terms of the total spin operator $\hat{\mathbf{S}}$, defined by

$$\hat{\mathbf{S}} = \hat{\mathbf{S}}_1 + \hat{\mathbf{S}}_2 \tag{6.10}$$

This is accomplished by expanding the appropriate operators, for example

$$\hat{S}_x^2 = (\hat{S}_{1x} + \hat{S}_{2x})^2 = \hat{S}_{1x}^2 + \hat{S}_{2x}^2 + \hat{S}_{1x}\hat{S}_{2x} + \hat{S}_{2x}\hat{S}_{1x} \tag{6.11}$$

Note that $\hat{\mathbf{S}}_1$ and $\hat{\mathbf{S}}_2$ commute since they are operators for different particles. Hence

$$\hat{S}_{1x}\hat{S}_{2x} = \tfrac{1}{2}\hat{S}_x^2 - \tfrac{1}{4}\hat{\mathbf{1}}_3 \tag{6.12}$$

since the eigenvalues of $\hat{S}_{1x}^2$ and $\hat{S}_{2x}^2$ are both $\frac{1}{4}$ (Section B.6). Similar expressions are obtained for the y and z terms. An expression for $\hat{\mathbf{S}}_1^T \cdot \hat{\mathbf{S}}_2$ then follows (note

Eq. B.49). The circumflex about the unit matrix symbol signals that we are dealing with a three-dimensional quantum-mechanical space (i.e., the spin triplet state), rather than with ordinary 3-space.

Using the angular-momentum commutation relations (Eqs. B.12), the following transformation can be derived

$$\hat{S}_{1x}\hat{S}_{2y} + \hat{S}_{2x}\hat{S}_{1y} = \tfrac{1}{2}(\hat{S}_x\hat{S}_y + \hat{S}_y\hat{S}_x) \tag{6.13}$$

with similar expressions for the xz and yz components.

Substitution of these expressions into Eq. 6.9, together with the identity $r^2 = x^2 + y^2 + z^2$, yields

$$\hat{\mathcal{H}}_{ss}(\mathbf{r}) = \frac{\mu_0}{4\pi}\frac{(g\beta_e)^2}{r^5}\frac{1}{2}\Big[(r^2 - 3x^2)\hat{S}_x^{\,2} + (r^2 - 3y^2)\hat{S}_y^{\,2} + (r^2 - 3z^2)\hat{S}_z^{\,2} -$$
$$3xy(\hat{S}_x\hat{S}_y + \hat{S}_y\hat{S}_x) - 3xz(\hat{S}_x\hat{S}_z + \hat{S}_z\hat{S}_x) - 3yz(\hat{S}_y\hat{S}_z + \hat{S}_z\hat{S}_y)\Big] \tag{6.14}$$

The factor of $\tfrac{1}{2}$ arises from the conversion from the $\mathbf{S}_1$, $\mathbf{S}_2$ basis to the $\mathbf{S}$ basis. Because all matrix elements connecting the triplet and singlet manifolds are zero in $\hat{\mathcal{H}}_{ss}(\mathbf{r})$ as well as in $\hat{\mathbf{S}}$, one can switch from consideration of the full manifold to working separately with the triplet and singlet parts. The corresponding dipolar energy for the latter ($S = 0$) is, of course, zero.

Equation 6.14 can be converted into a spin-hamiltonian form by suitable integration; it can then be written more conveniently in matrix form as

$$\hat{\mathcal{H}}_{ss} = \frac{\mu_0}{8\pi}(g\beta_e)^2\begin{bmatrix}\hat{S}_x & \hat{S}_y & \hat{S}_z\end{bmatrix} \cdot \begin{bmatrix} \left\langle\dfrac{r^2 - 3x^2}{r^5}\right\rangle & \left\langle\dfrac{-3xy}{r^5}\right\rangle & \left\langle\dfrac{-3xz}{r^5}\right\rangle \\[2mm] & \left\langle\dfrac{r^2 - 3y^2}{r^5}\right\rangle & \left\langle\dfrac{-3yz}{r^5}\right\rangle \\[2mm] & & \left\langle\dfrac{r^2 - 3z^2}{r^5}\right\rangle \end{bmatrix} \cdot \begin{bmatrix}\hat{S}_x \\ \hat{S}_y \\ \hat{S}_z\end{bmatrix} \tag{6.15a}$$

$$= \hat{\mathbf{S}}^T \cdot \mathbf{D} \cdot \hat{\mathbf{S}} \quad \text{for } S = 0, 1 \tag{6.15b}$$

$$= \hat{\mathbf{S}}_1^{\,T} \cdot \mathbf{D} \cdot \hat{\mathbf{S}}_2 + \hat{\mathbf{S}}_2^{\,T} \cdot \mathbf{D} \cdot \hat{\mathbf{S}}_1 \quad (= 2\hat{\mathbf{S}}_1^{\,T} \cdot \mathbf{D} \cdot \hat{\mathbf{S}}_2) \tag{6.15c}$$

Note that $\hat{\mathbf{S}}_1^{\,T} \cdot \mathbf{D} \cdot \hat{\mathbf{S}}_1 = \hat{\mathbf{S}}_2^{\,T} \cdot \mathbf{D} \cdot \hat{\mathbf{S}}_2 = 0$. The last form results from the interchangeability of the individual spins S_1, S_2. Operator $\hat{\mathcal{H}}_{ss}$ is sometimes called the 'electronic quadrupole spin hamiltonian'. As before, the angular brackets indicate that the elements of the parameter matrix $\mathbf{D}$ are averages over the electronic spatial wavefunction. As with the matrices encountered in Chapters 4 and 5, $\mathbf{D}$ can be diagonalized, to

^{d}**D**. The diagonal elements of ^{d}**D** are D_X, D_Y and D_Z. By convention, D_Z is taken to be the principal value with the largest absolute magnitude and D_Y has the smallest absolute magnitude when $D_X \neq D_Y$, producing a set ordered in energy.

We see from the sum of the diagonal elements of Eq. 6.15a that **D** is a matrix with a trace of zero:

$$tr(\mathbf{D}) = D_X + D_Y + D_Z = 0 \tag{6.16}$$

In the principal-axis system of **D**, Eq. 6.15b becomes

$$\hat{\mathcal{H}}_{ss} = D_X \hat{S}_X^2 + D_Y \hat{S}_Y^2 + D_Z \hat{S}_Z^2 \tag{6.17}$$

As always, the principal-axis system is defined by the details of the interaction giving rise to the 3×3 matrix. The axes lie on symmetry elements, for example, along molecular symmetry axes when such are present.

The dipole-dipole coupling between the two unpaired electrons is not the only interaction that can lead to a spin-hamiltonian term of the form of Eq. 6.15b. Coupling between the electron spin and the electronic orbital angular momentum (Section 4.8) gives rise to a term of the same form (Eq. 6.15b), as does the generalized anisotropic exchange interaction [8].

The effective spin hamiltonian for two interacting electrons, obtained by adding Eqs. 6.4 and 6.15b to the electron Zeeman term, is

$$\hat{\mathcal{H}} = g\beta_e \mathbf{B}^T \cdot \hat{\mathbf{S}} + \hat{\mathbf{S}}^T \cdot \mathbf{D} \cdot \hat{\mathbf{S}} + \tfrac{1}{2} J_0 \left[\hat{\mathbf{S}}^2 - \tfrac{3}{2} \hat{\mathbf{1}}_3 \right] \tag{6.18}$$

where the form of the last term arises from the vector cosine sum rule (Eq. B.49). If $J_0 < 0$ and $|J_0| \gg k_b T$, only the lower ($S = 1$) state is populated. Conversely, if $J_0 \gg k_b T$, only the diamagnetic ($S = 0$) state is thermally populated. Furthermore, since the exchange term of Eq. 6.18 contributes only a common constant to the energy of each of the three triplet states, it is neglected in the next section.

6.3 SYSTEMS WITH $S = 1$ (TRIPLET STATES)

From our previous discussion we see that the spin hamiltonian for an $S = 1$ state is

$$\hat{\mathcal{H}} = g\beta_e \mathbf{B}^T \cdot \hat{\mathbf{S}} + \hat{\mathbf{S}}^T \cdot \mathbf{D} \cdot \hat{\mathbf{S}} \tag{6.19a}$$

If **D** is expressed in its principal-axis system, Eq. 6.19a may be written as

$$\hat{\mathcal{H}} = g\beta_e \mathbf{B}^T \cdot \hat{\mathbf{S}} + D_X \hat{S}_X^2 + D_Y \hat{S}_Y^2 + D_Z \hat{S}_Z^2 \tag{6.19b}$$

Note that, although there are actually two unpaired electrons (each $S = \frac{1}{2}$) in the triplet molecule, we use an effective spin $S' = 1$ to describe its magnetic properties, with the singlet ignored. Thus, while there really are four spin states, only three are *active*. The use of an effective *spin*, one that generates the multiplicity needed for the states considered, is common and convenient in magnetic resonance. For one, it allows a simple formulation of the spin hamiltonian describing the system.

6.3.1 Spin Energies and Eigenfunctions

It is often convenient to use the eigenfunctions $|M_S\rangle = |+1\rangle$, $|0\rangle$ and $|-1\rangle$ of $\hat{S}_Z$ as a basis set (Fig. 6.2); these are the eigenfunctions of $\hat{\mathcal{H}}$ (Eq. 6.19*b*) in the limit as $B \to 0$. However, they are *not* eigenfunctions of $\hat{\mathcal{H}}_{ss}$ (Eqs. 6.15*b* and 6.17). Hence it is necessary to set up the spin-hamiltonian matrix for $\hat{\mathcal{H}}$ and find its energy eigenvalues and eigenstates. For present purposes, it is not necessary to know how the principal-axis system is oriented but only that it exists. Equation 6.19*b* can be written as

$$\hat{\mathcal{H}} = g\beta_e(B_X\hat{S}_X + B_Y\hat{S}_Y + B_Z\hat{S}_Z) + D_X\hat{S}_X^2 + D_Y\hat{S}_Y^2 + D_Z\hat{S}_Z^2 \qquad (6.19c)$$

If we take the quantization axis for $\hat{\mathbf{S}}$ along principal axis **Z**, then the required spin matrices are those given in Eq. B.77. Substitution of these into Eq. 6.19*c* with subsequent matrix addition and multiplication yields spin-hamiltonian matrix

$$
\mathcal{H} =
\begin{array}{c c}
 & \begin{array}{c c c} |+1\rangle \quad\quad\quad & |0\rangle \quad\quad\quad & |-1\rangle \end{array} \\
\begin{array}{c} \langle+1| \\[18pt] \langle 0| \\[18pt] \langle-1| \end{array} &
\left[
\begin{array}{c c c}
g\beta_e B_Z + \dfrac{1}{3}D & \dfrac{1}{\sqrt{2}}g\beta_e(B_X - iB_Y) & \dfrac{1}{2}(D_X - D_Y) \\[12pt]
\dfrac{1}{\sqrt{2}}g\beta_e(B_X + iB_Y) & D_X + D_Y & \dfrac{1}{\sqrt{2}}g\beta_e(B_X - iB_Y) \\[12pt]
\dfrac{1}{2}(D_X - D_Y) & \dfrac{1}{\sqrt{2}}g\beta_e(B_X + iB_Y) & -g\beta_e B_Z + \dfrac{1}{3}D
\end{array}
\right]
\end{array}
$$

$$(6.20)$$

where $D/3 = (D_X + D_Y)/2 + D_Z$. The secular determinant (Eq. A.69*a*) is obtained from $\mathcal{H}$ by subtracting energy U from all diagonal elements. Setting the corresponding determinant equal to zero, one obtains

$$
\begin{vmatrix}
g\beta_e B_Z + \dfrac{1}{2}D_Z - U & \dfrac{1}{\sqrt{2}}g\beta_e(B_X - iB_Y) & \dfrac{1}{2}(D_X - D_Y) \\[12pt]
\dfrac{1}{\sqrt{2}}g\beta_e(B_X + iB_Y) & -D_Z - U & \dfrac{1}{\sqrt{2}}g\beta_e(B_X - iB_Y) \\[12pt]
\dfrac{1}{2}(D_X - D_Y) & \dfrac{1}{\sqrt{2}}g\beta_e(B_X + iB_Y) & -g\beta_e B_Z + \dfrac{1}{2}D_Z - U
\end{vmatrix} = 0 \qquad (6.21)
$$

Here Eq. 6.16 has been used to simplify terms. The situation is especially simple when $\mathbf{B} \parallel \mathbf{Z}$. Then $B_X = B_Y = 0$, and Eq. 6.21 becomes

$$
\begin{vmatrix}
g\beta_e B_Z + \dfrac{1}{2}D_Z - U & 0 & \dfrac{1}{2}(D_X - D_Y) \\[2mm]
0 & -D_Z - U & 0 \\[2mm]
\dfrac{1}{2}(D_X - D_Y) & 0 & -g\beta_e B_Z + \dfrac{1}{2}D_Z - U
\end{vmatrix} = 0 \qquad (6.22)
$$

The solution $U = -D_Z$ is obtained by inspection. Expansion of the remaining 2×2 determinant gives the other two energies as

$$
U_{X,Y} = \tfrac{1}{2}\{D_Z \pm [4g^2\beta_e^2 B_Z^2 + (D_X - D_Y)^2]^{1/2}\} \qquad (6.23)
$$

In zero magnetic field, the energies are

$$
U_X = \tfrac{1}{2}[D_Z - (D_X - D_Y)] = -D_X \qquad (6.24a)
$$

$$
U_Y = \tfrac{1}{2}[D_Z + (D_X - D_Y)] = -D_Y \qquad (6.24b)
$$

$$
U_Z = -D_Z \qquad (6.24c)
$$

Thus $|U_Z| > |U_X| \geq |U_Y|$ at $B = 0$, in accordance with our convention (Section 6.2.2). We note that the zero in energy lies between the smallest and the largest principal D values, and that all the degeneracy is removed except in the uniaxial case. In the literature, the notation $\chi = -D_X$, $\mathcal{Y} = -D_Y$ and $Z = -D_Z$ has sometimes been used.

Since the trace of $\mathbf{D}$ is zero, only two independent energy parameters are required. It is common to designate these as

$$
D = \tfrac{3}{2}D_Z \qquad (6.25a)
$$

$$
E = \tfrac{1}{2}(D_X - D_Y) \qquad (6.25b)
$$

Useful expressions for D and E are obtainable from the matrix $\mathbf{D}$ in Eq. 6.15a taken in its principal-axis system and will be given later, by Eqs. 6.41 and 9.18a, b. Note that D and E are analogous to the hyperfine parameters b_0 and c_0 of Eqs. 5.9b, c. Equations 6.24 can now be written as

$$
U_X = \tfrac{1}{3}D - E \qquad (6.26a)
$$

$$
U_Y = \tfrac{1}{3}D + E \qquad (6.26b)
$$

$$
U_Z = -\tfrac{2}{3}D \qquad (6.26c)
$$

Thus, by our convention, if $D > 0$, then $E < 0$. In terms of D and E, the spin-hamiltonian operator (Eq. 6.19*b*) becomes

$$\hat{\mathcal{H}}_n = g\beta_e \mathbf{B}^{\mathrm{T}} \cdot \hat{\mathbf{S}} + D(\hat{S}_Z^2 - \tfrac{1}{3}\hat{S}^2) + E(\hat{S}_X^2 - \hat{S}_Y^2) \tag{6.27}$$

It is important to note that the values of D and E are not unique. They depend on which axis is chosen as **Z**. The convention [9] already stated (see text after Eq. 6.15) ensures that $|D/3| \geq |E|$. One often is ignorant of the absolute signs of D and E since the EPR line positions depend only on their *relative* signs. Thus the values quoted for D and E often are absolute magnitudes. The sign of D can be determined from relative intensity measurements of EPR lines at low temperatures [10], by optically detected EPR [11], from static magnetic susceptibility data [12], or possibly by comparison with other spin-hamiltonian parameters (e.g., hyperfine and quadrupolar). The sign of E depends on the specific assignment of the axes **X** and **Y** and thus has no physical meaning except in terms of the convention that we have chosen. It is sometimes convenient to express D and E in magnetic-field units, that is, $D' = D/g_e\beta_e$ and $E' = E/g_e\beta_e$. It is not uncommon to express these parameters in cm^{-1}, that is, by defining $\overline{D} \equiv D/hc$ and $\overline{E} \equiv E/hc$.

The energies of the three states as a function of magnetic field are plotted in Fig. 6.3 for **B** parallel to **Z**. Here it is assumed that $D > 0$. When $D < 0$, the states at zero field are reversed in their energy order. For systems with uniaxial

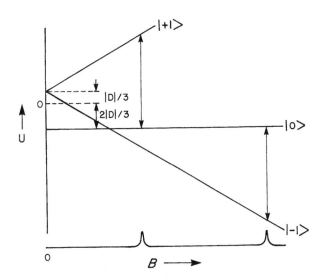

FIGURE 6.3 The state energies and corresponding eigenfunctions (high-field labels) as a function of applied magnetic field B for a system of spin $S = 1$ and **B** $\parallel$ **Z**, shown for $D > 0$ and $E = 0$. The two primary transitions, of type $\Delta M_S = \pm 1$, are indicated for a constant-frequency spectrum. When $E \neq 0$, then the degeneracy at $B = 0$ is lifted, and the corresponding energies vary non-linearly with B (Eqs. 6.30).

symmetry one has $D_X = D_Y$ and hence $E = 0$ (Fig. 6.3). When $E \neq 0$, for systems with rhombic symmetry, all three states are non-degenerate at zero field. For naphthalene in its lowest triplet state (Section 6.3.4), the energies and transitions are shown in Figs. 6.4a–c for **B** parallel to **X**, **Y** and **Z**. The lowest-field transition in each figure is of the '$\Delta M_S = \pm 2$' type, as explained in Section 6.3.2. Note that this nomenclature is tainted, since M_S is not a strictly valid quantum number at low magnetic fields.

The eigenfunctions (kets) of $\hat{\mathcal{H}}$ (Eq. 6.19b) are linear combinations of the kets $|M_S\rangle = |+1\rangle$, $|0\rangle$ and $|-1\rangle$. The coefficients are obtained by substitution of the eigenvalues of Eq. 6.24 into the determinant 6.21 and solving the corresponding secular equations (e.g., as in Section A.5.5). The coefficients depend on the magnitude of **B**. It is convenient to define an auxiliary 'mixing' angle $\gamma = \frac{1}{2}\tan^{-1}(E/g\beta_e B)$.

Consider the situation **B** parallel to the principal axis **Z** with $D > 0$. The upper state $[-\sin\gamma\,|-1\rangle - \cos\gamma\,|+1\rangle]$ becomes T_X (e.g., Eq. 6.28a) at

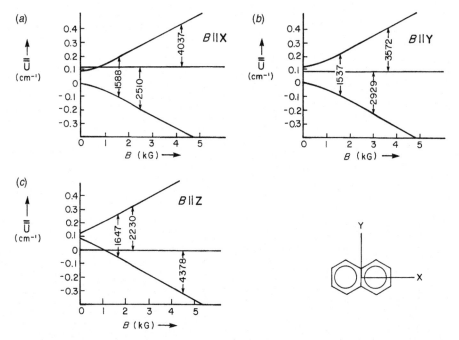

FIGURE 6.4 Spin system energies $(U + 2D/3)/hc$ as a function of applied magnetic field B for naphthalene in its lowest triplet state (which lies $\sim 21{,}000\text{ cm}^{-1}$ above the singlet ground state), measured at $T = 77$ K. Clearly, $D > 0$. The transition at lowest field in each case is allowed only for the microwave magnetic field $\mathbf{B}_1$ parallel to **B**. We note that it is relatively isotropic as compared to the usual EPR transitions. The latter yielded $\overline{D} = 0.1003(6)$ and $\overline{E} = -0.0137(2)\text{ cm}^{-1}$. (a) **B** $\parallel$ **X**; (b) **B** $\parallel$ **Y**; (c) **B** $\parallel$ **Z**. The resonance magnetic fields at $\nu = 9.272$ GHz are indicated vertically, in gauss. The proton hyperfine energies are ignored. [After C. A. Hutchison Jr., B. W. Mangum, *J. Chem. Phys.*, **34**, 908 (1961).]

$B = 0$ ($\gamma = -\pi/4$, if $E < 0$), whereas the other mixed state i[cos γ $|-1\rangle$ − sin γ $|+1\rangle$] is T_Y (e.g., Eq. 6.28*b*) at $B = 0$. Note that the upper two levels merge at $B = 0$ for uniaxial symmetry ($E = 0$), and that the high-field states ($\gamma = 0$) are usually renormalized to be $|+1\rangle$ and $|-1\rangle$ (highest and lowest energies).

In the limit as $B \to 0$ with **B** parallel to the principal axis **Z**, the zero-field triplet eigenfunctions [13] are[8]

$$|T_X\rangle = \frac{1}{\sqrt{2}}(|-1\rangle - |+1\rangle) \tag{6.28a}$$

$$|T_Y\rangle = \frac{i}{\sqrt{2}}(|-1\rangle + |+1\rangle) \tag{6.28b}$$

$$|T_Z\rangle = |0\rangle \tag{6.28c}$$

Note that the zero-field functions $|T_X\rangle$, $|T_Y\rangle$ and $|T_Z\rangle$ are the same linear combinations of angular-momentum eigenfunctions as for the orbital functions for $\ell = 1$ (Fig. 4.9); that is, they transform like p orbitals. Here each of the three eigenstates corresponds to a situation in which the spin angular momentum vector lies in one of the three principal planes (e.g., XY) of **D**.

It is sometimes convenient to choose the functions in Eqs. 6.28 as the basis set, since they are the eigenfunctions of $\hat{\mathcal{H}}_{ss}$ at zero field. In the presence of a magnetic field, the spin-hamiltonian matrix then becomes (note Problem 6.2)

$$\mathcal{H} = \begin{array}{c} \\ \langle T_X| \\ \langle T_Y| \\ \langle T_Z| \end{array} \overset{\displaystyle \begin{array}{ccc} |T_X\rangle & |T_Y\rangle & |T_Z\rangle \end{array}}{\begin{bmatrix} -D_X & -ig\beta_e B_Z & +ig\beta_e B_Y \\ +ig\beta_e B_Z & -D_Y & -ig\beta_e B_X \\ -ig\beta_e B_Y & +ig\beta_e B_X & -D_Z \end{bmatrix}} \tag{6.29}$$

From this it is clear that when **B** is parallel to **X**, or **Y**, or **Z**, the energy of the *corresponding* state has its zero-field value, and is independent of B.

For **B** along **Z**, one has $U_Z = -2D/3$ and, using Eqs. 6.16 and 6.25, one again finds Eq. 6.23, now written in the form

$$U_{X,Y} = \tfrac{1}{3}D \pm [E^2 + (g\beta_e B)^2]^{1/2} \tag{6.30}$$

Clearly, when **B** ∥ **Z**, the eigenfunctions of the spin hamiltonian are of the form $b_1|T_X\rangle + ib_2|T_Y\rangle$, $ib_2|T_X\rangle + b_1|T_Y\rangle$ and $|T_Z\rangle$. The coefficients b_1 and b_2 are real, depend on B [10] and are not difficult to obtain (e.g., Problem 6.4). Alternatively, the eigenfunctions can, of course, be written in terms of the $|M_S\rangle$ set.

When **B** is sufficiently large in magnitude compared to D' and is parallel to **Z**, the off-diagonal elements in the spin-hamiltonian matrix 6.20 can be neglected. Then the eigenstates $|M_S\rangle$ of $\hat{S}_Z$ are eigenstates of the spin hamiltonian of Eq. 6.19b. Hence the spin energies can be labeled by the value of M_S.

As stated above, in the special case in which **B** lies along any one of the three principal axes, there exists an exact analytic solution, valid for all field positions of the primary ($\Delta M_S = +1$) EPR lines. For example, if **B** $\parallel$ **Z**, Eqs. 6.26c and 6.30 give the energy separations corresponding to the two allowed $\Delta M_S = +1$ lines as

$$U_X - U_Z = h\nu = +D + [E^2 + (g\beta_e B_{aZ})^2]^{1/2} \tag{6.31a}$$

$$U_Z - U_Y = h\nu = -D + [E^2 + (g\beta_e B_{bZ})^2]^{1/2} \tag{6.31b}$$

where B_{aZ} corresponds to the transition at lower field and B_{bZ} to that at higher field when $D > 0$ (and vice versa for $D < 0$). These are valid above the level-crossing region visible in Fig. 6.3; that is, the square-root term dominates over $|D|$. One can then derive from Eqs. 6.31 the exact and general expression

$$|D| = \frac{(g\beta_e)^2}{4h\nu}(B_{hZ}{}^2 - B_{\ell Z}{}^2) \tag{6.32a}$$

valid for either sign of D. Here the subindices h and ℓ denote the higher and lower resonant fields. Similarly, for **B** $\parallel$ **X**

$$|D - 3E| = \frac{(g\beta_e)^2}{2h\nu}(B_{hX}{}^2 - B_{\ell X}{}^2) \tag{6.32b}$$

and for **B** $\parallel$ **Y**

$$|D + 3E| = \frac{(g\beta_e)^2}{2h\nu}(B_{hY}{}^2 - B_{\ell Y}{}^2) \tag{6.32c}$$

Equations 6.32 are very useful for obtaining the zero-field parameters $|D|$ and $|E|$, especially for statistically randomly oriented samples (Section 6.3.3). Note the simplification when $E = 0$.

One possible problem with the previous considerations is that the **D** principal-axis directions in the single crystal may not yet be known. Another is that the assumption that g is isotropic, made above, is not appropriate.

From the secular equation for the $S = 1$ spin-hamiltonian matrix (Eq. 6.29) generalized for anisotropic g, one can obtain in general a cubic equation in energy eigenvalues U, valid for any direction of **B**. It is possible to derive an exact equation from this giving the magnetic fields of the magnetic-resonance transitions, all observed at each crystal orientation always at the same constant frequency [14,15] to yield the useful anisotropic parameter $g^2 d_1$ (see Eq. 6.55b) as

$$\mathbf{n}^T \cdot \mathbf{g} \cdot \mathbf{D} \cdot \mathbf{g}^T \cdot \mathbf{n} = \frac{det(\mathbf{D})}{(\beta_e B)^2} +$$

$$\pm \frac{1}{3^{3/2}} \left[\frac{(h\nu)^2 - tr(\mathbf{D}^2) - (g\beta_e B)^2}{(\beta_e B)^2} \right] \times$$

$$[(2g\beta_e B)^2 - (h\nu)^2 + 2tr(\mathbf{D}^2)]^{1/2} \qquad (6.33)$$

where $\mathbf{n}$ is the unit vector along $\mathbf{B}$, and $g = (\mathbf{n}^T \cdot \mathbf{g} \cdot \mathbf{g}^T \cdot \mathbf{n})^{1/2}$ (Eq. 4.12). Here field B is the resonant field for both $|\Delta M_S| = 1$ transitions as well as the $|\Delta M_S| = 2$ transition. Only the left-hand side of Eq. (6.33) depends on the field orientation. Having measured frequency ν and the sets of magnetic fields B (i.e., at each of various orientations), it is then possible to arrive at the unknown matrices $\mathbf{D}$ and $\mathbf{g}$, using numerical fitting techniques.

An easier to visualize but approximate technique for arriving at $\mathbf{D}$ from experimental data is available from perturbation theory, valid when the electron Zeeman energy $g\beta_e B$ is sufficiently large compared to $|D|$. We can utilize such an expression (Eq. 6.54) for the spin-hamiltonian energies, to obtain the transition energies in the approximate forms

$$U(0) - U(-1) = g\beta_e B_h - \tfrac{3}{2} d_1 +$$

$$\frac{1}{4g\beta_e B_h} \left[d_2 - \tfrac{3}{2} d_1{}^2 + \tfrac{1}{2} [tr(\mathbf{D}^2) - 2d_{-1} det(\mathbf{D})] \right] + \cdots \qquad (6.34a)$$

$$U(+1) - U(0) = g\beta_e B_\ell + \tfrac{3}{2} d_1 +$$

$$\frac{1}{4g\beta_e B_h} \left[d_2 - \tfrac{3}{2} d_1{}^2 + \tfrac{1}{2} [tr(\mathbf{D}^2) - 2d_{-1} det(\mathbf{D})] \right] + \cdots \qquad (6.34b)$$

where $d_n = \mathbf{n}^T \cdot \mathbf{g} \cdot \mathbf{D}^n \cdot \mathbf{g}^T \cdot \mathbf{n}/g^2$ and B_h and B_ℓ are the magnetic fields of the higher- and lower-field transitions.[9] By simple manipulation, one may obtain the expression

$$\Delta B = B_h - B_\ell \qquad (6.35a)$$

$$= \frac{3d_1}{g\beta_e} - \frac{1}{4g^2 \beta_e{}^2} \left(\frac{1}{B_h} - \frac{1}{B_\ell} \right) \times$$

$$[d_2 - \tfrac{3}{2} d_1{}^2 + \tfrac{1}{2} [tr(\mathbf{D}^2) - 2d_{-1} det(\mathbf{D})]] + \cdots \qquad (6.35b)$$

for the field separation between the $\Delta M_S = \pm 1$ transitions, valid at any field orientation.[10] We see that in the first approximation, for isotropic g, one has

$B_h - B_\ell \approx 3\mathbf{n}^T \cdot \mathbf{D} \cdot \mathbf{n}/g\beta_e$; that is, one obtains the magnitude of $\mathbf{D}$ projected along $\mathbf{n}$. It is thus possible, by measuring the field separations at various directions $\mathbf{n}$, to arrive directly at a first approximation to $\mathbf{D}$. This matrix may then be refined by using Eq. 6.33 or Eq. 6.35. In practice, matrix $\mathbf{D}$ (and simultaneously all other spin-hamiltonian parameters: $\mathbf{g}$, sets of $\mathbf{A}_i$ and $\mathbf{P}_i$, etc.) is obtained numerically by computer fitting of the observed line positions.

Note that, while Eqs. 6.33–6.35 in this section assume that $B > 0$, it is quite feasible to do EPR studies at $B = 0$. This is possible whenever zero-field spin energy-level splittings exist and can be connected by matching photon energies $h\nu$ of an applied excitation field $\mathbf{B}_1$ (Appendix E).

In certain systems, the literature routinely contains citations of effective g values, as defined in Chapter 1. A prominent example is the high-spin Fe^{3+} $3d^5$ EPR peak found in many circumstances in various glasses, which occurs at '$g = 4.3$' (e.g., see Refs. 16–20), when measured at X band. Its presence is a useful indicator of the presence of this ion, and indicates some aspects of its surroundings. However, the reality here is that this 'g' takes on this value because of the presence of sizable zero-field electronic quadrupole energy (parameters D and E: see Eqs. 6.25) in addition to the electron-spin Zeeman term, and is frequency-dependent. The actual g value is very close to being isotropic, nearly at g_e. This situation is in sharp contrast with the occurrence of a true g value of 4.13 ($\sim 30/7$ in theory) for low-spin (effective $S' = \frac{1}{2}$) Fe^{1+} $3d^7$ in octahedral sites (see Fig. 1.11).

6.3.2 '$\Delta M_S = \pm 2$' Transitions

At high fields, where the quantum numbers $M_S = +1$, 0 and -1 are meaningful in that they correspond to the eigenfunctions of the spin hamiltonian, a '$\Delta M_S = \pm 2$' transition is not allowed. However, at low fields, the eigenfunctions become linear combinations of the high-field states (Eqs. 6.28) and quantum numbers M_S are no longer strictly applicable. Thus the usual $\Delta M_S = \pm 1$ selection rule does not apply. The '$\Delta M_S = \pm 2$' transition is permitted for the microwave field $\mathbf{B}_1$ *parallel* to the static field $\mathbf{B}$. This can be shown by taking the $\hat{S}_z$ matrix element for the states

$$c_2|-1\rangle + c_1|+1\rangle \qquad \text{and} \qquad -\mathrm{i}[c_1{}^*|-1\rangle - c_2{}^*|+1\rangle]$$

between which the '$\Delta M_S = \pm 2$' transition occurs (Fig. 6.4 and Problem 6.4). As we saw, the coefficients are functions of angle γ. One also finds that, when $\mathbf{B}$ is at an arbitrary orientation relative to the principal axes of $\mathbf{D}$, the $|+1\rangle$, $|0\rangle$ and $|-1\rangle$ states are all mixed by the spin-spin interaction. Hence '$\Delta M_S = \pm 2$' transitions can be seen in a normal EPR cavity [21] (i.e., with the microwave field perpendicular to the static field). These are single-photon transitions. Since the non-zero parts of the intensity arise from the same states (e.g., $\langle +1|$ and $|+1\rangle$) on both sides of the transition matrix element, it follows that no net angular momentum change in the spin system is involved (π transition: see Appendix D). Note that the state $|0\rangle$ does not enter into the mechanism.

The position of the low-field side of the '$\Delta M_S = \pm 2$' transition in randomly oriented solids does not correspond to that of the low-field X, Y or Z components from Fig. 6.4 but rather occurs at a turning point B_{min} [21,22].

As we have seen, the angular dependence of all the (single-photon) lines in the triplet spectrum for any fixed frequency ν is given [14, 23] by all the non-negative real solutions for B of Eq. 6.33. This holds for the $\Delta M_S = \pm 2$ transition. Then, for isotropic g, the minimum possible value

$$B_{min} = \frac{1}{2g\beta_e}[(h\nu)^2 - 2(D_X^2 + D_Y^2 + D_Z^2)]^{1/2} \tag{6.36a}$$

$$= \frac{1}{g\beta_e}\left[\frac{(h\nu)^2}{4} - \frac{D^2 + 3E^2}{3}\right]^{1/2} \tag{6.36b}$$

of the resonant field occurs when the square-root factor in Eq. 6.33 becomes zero. The orientation of the direction **B** at which B_{min} is achieved is not generally a principal axis of **D**. Note that the low-field edge of the derivative line for a randomly oriented triplet system can be used to estimate $D^* \equiv (D^2 + 3E^2)^{1/2}$, which is a measure of the root-mean-square zero-field splitting. In some cases, D and E can be approximately determined if the shape of the '$\Delta M_S = \pm 2$' line is analyzed [24]. However, if the zero-field splitting parameters are sufficiently large compared with the microwave photon energy $h\nu$, no '$\Delta M_S = 2$' transition can occur. In any case, the preceding equations make no prediction about the intensity of such a transition.

Finally, it should be mentioned that for significantly high power levels (large B_1), double-quantum (two-photon) transitions are observable [25,26]. These are between states $|\pm 1\rangle$, and occur near $g = 2$.

6.3.3 Randomly Oriented Triplet Systems

Triplet molecules in liquid solution are difficult to detect. While the rotational motions do tend to remove the zero-field splittings (D), very rapid tumbling is required to do so, and the associated spin-lattice relaxation (τ_1 much shorter than for $S = \frac{1}{2}$ radicals) broadens the lines [27].

Few triplet systems have been investigated in the oriented solid state. This arises largely from the difficulties of preparing single crystals of adequate size with well-defined orientation of guest molecules at an appropriate concentration. The observation of a '$\Delta M_S = \pm 2$' line in the region of $g \approx 4$ was the stimulus for the detection of triplet states in numerous non-oriented systems [24]. The relatively large amplitude of the '$\Delta M_S = \pm 2$' lines is associated with their small anisotropy. Subsequently, it was recognized [23] that even for non-oriented systems one can detect the ordinary $\Delta M_S = \pm 1$ transitions at 'turning points'.[11] General conditions for the occurrence of off-axis extra lines in the EPR powder (and glass phase) patterns have been derived, using third-order perturbation theory applied to $S > \frac{1}{2}$ systems [30].

It is instructive to mention that high-quality triplet-state EPR lines can be obtained from aromatic molecules (e.g., anthracene-d_{10}) dissolved in low-density stretched polyethylene films [31]. These molecules then occur oriented within the film. The photo-excited spectra are strongly anisotropic, as becomes evident by placing **B** along various different directions relative to the stretch axis. The parameters obtained are consistent with those derived from single-crystal measurements.

For simplicity, we now consider an ensemble of triplet-state molecules randomly oriented in a solid matrix. From an evaluation of d_1 in Eqs. 6.35, the field separation ΔB of the two allowed $\Delta M_S = \pm 1$ transitions to a first approximation (with isotropic g) is seen to be given by

$$\Delta B = B_h - B_\ell = \frac{3}{g\beta_e} \left[D_X \sin^2 \theta \cos^2 \phi + D_Y \sin^2 \theta \sin^2 \phi + D_Z \cos^2 \theta \right]^{1/2} \quad (6.37a)$$

$$= \frac{1}{g\beta_e} [D(3\cos^2 \theta - 1) + 3E \sin^2 \theta \cos 2\phi]^{1/2} \quad (6.37b)$$

where θ is the polar angle (between **B** and axis **Z** of a given molecule) and ϕ is the azimuthal angle. If $\overline{B}$ is the average field $(B_h + B_\ell)/2$, then the orientation dependence of each line is given by

$$B_h - \overline{B} = \overline{B} - B_\ell = \frac{1}{2g\beta_e} [D(3\cos^2 \theta - 1) + 3E \sin^2 \theta \cos^2 \phi]^{1/2} \quad (6.38)$$

as can be derived from Eqs. 6.34.

For the uniaxial case, as the field changes its orientation from $\theta = 0°$ to $\theta = 90°$, the line positions relative to $\overline{B}$ change from $|D|/g\beta_e$ to $-|D|/2g\beta_e$. By applying an analysis similar to that given in Sections 4.7 and 5.7, we can express the probability distribution for a given upper-field transition as follows:

$$P(B_h) \propto \frac{g\beta_e}{6|D \cos \theta|} \quad (6.39)$$

The calculated shapes for the $\Delta M_S = \pm 1$ lines are given in Fig. 6.5. The separation between the outer vertical lines in Fig. 6.5a (which represents the theoretical lineshape) is approximately $2|D|/g\beta_e$, while that between the two inner lines is $|D|/g\beta_e$. The high-field portion of the triphenylbenzene di-anion (**I**) spectrum in Fig. 6.6 shows a satisfying correspondence with the derivative spectrum in Fig. 6.5b. Note that the triplet powder pattern (for $\Delta M_S = \pm 1$) tends to consist of equal, but oppositely signed, contributions, just as was the case for the hyperfine-dominated spectra (see Figs. 5.10 and 5.11), for analogous reasons.

(I) triphenylbenzene dianion

The analysis can readily be extended to a randomly oriented triplet system with $E \neq 0$. The theoretical lineshape is given in Fig. 6.7a, and the derivative spectrum is given in Fig. 6.7b.[12] The separation of outermost lines is again $2|D|/g\beta_e$, whereas that of the intermediate and inner pairs is $(|D + 3E|)/g\beta_e$ and $(|D - 3E|)/g\beta_e$. There is a close correspondence between Figs 6.7b and 6.8, which gives the spectrum of the first excited triplet state of naphthalene in a rigid, non-oriented ('glassy') matrix at 77 K. The compound used was actually $C_{10}D_8$ instead of $C_{10}H_8$ so as to minimize linewidth contributions from unresolved hyperfine splittings. The pairs of lines correspond with those given in Fig. 6.4 for a single crystal. In the $g = 2$ region an additional line is seen at high microwave power. This line has been identified as a double-quantum transition [25]. For observations of the $\Delta M_S = \pm 1$ lines in the random non-oriented sample, one requires a far greater EPR spectrometer sensitivity than for an equivalent concentration in a single crystal. In the former case, only a small fraction of all molecules in the triplet state contribute to any of the observable derivative lines. The $\Delta M_S = \pm 1$ lines are seen to be weak compared with the '$\Delta M_S = \pm 2$' line.

For rigid media in which the geometric configurations of host and guest molecules are markedly dissimilar, the linewidths in the triplet spectrum may be many times broader than in cases where host and guest are very similar {specifically, diphenyl-methylene $(C_6H_5{-}C{-}C_6H_5)$ in diphenyldiazomethane $[C_6H_5{-}C(N_2){-}C_6H_5]$ shows a linewidth of 1.7 mT; in n-pentane $[CH_2(CH_2)_3CH_3]$ the linewidth is 9.4 mT [32]}. Thus in a dissimilar host-matrix system, it appears likely that a range of solute-solvent configurations is tolerated; the various configurations display a distribution of D and E values.

In nonrigid media, EPR absorption for triplet-state systems is not observed unless $|D|$ and E are sufficiently small. If intramolecular spin-spin interactions are modulated at a rapid rate because of molecular reorientations, one expects

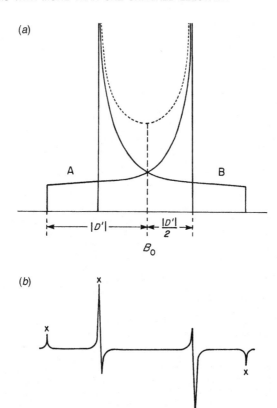

FIGURE 6.5 (a) Theoretical EPR absorption spectrum for a randomly oriented triplet system (with $E = 0$) for a given value of D and ν (taking $g = g_e$). A zero linewidth is assumed. The solid curve B corresponds to the curve of Fig. 4.7a; the solid curve A represents a reflection of B about the central field B_0. The central (small dash) trough is the sum of A and B. Compare with Fig. 5.10 ($\xi = 10$). (b) First-derivative curve computed from (a) after assuming a non-zero linewidth. Only the field region corresponding to $\Delta M_S = \pm 1$ is shown. The points marked x correspond to the resonant field values when the magnetic field is oriented along $\mathbf{Z}$ (cusp-shaped lines) or perpendicular to $\mathbf{Z}$. [After E. Wasserman, L. C. Snyder, W. A. Yager, *J. Chem. Phys.*, **41**, 1763 (1964).]. Note and tentatively explain the difference between the idealized first-derivative spectrum shown here, and the one depicted in Fig. 5.11.

a spread in the components of $\mathbf{D}$. Since the trace of $\mathbf{D}$ is zero, the contribution of the term $\hat{\mathbf{S}}^T \cdot \mathbf{D} \cdot \hat{\mathbf{S}}$ may become negligible. Two limiting cases may be considered:

1. When $|D|$ and $|E|$ are large, the modulations of the spin-spin interaction lead to so short a spin lifetime that the averaged spectrum has undetectably broad lines.
2. When $|D|$ and $|E|$ are very small, the line-broadening effects in non-rigid media are also small. In the absence of hyperfine splitting, one sees a single line, as if the spectrum were due to a system with $S = \frac{1}{2}$.

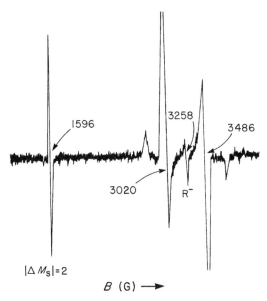

FIGURE 6.6 Triplet-state EPR spectrum of a rigid solution of the di-anion of 1,3,5-triphenylbenzene in methyltetrahydrofuran at 77 K; $\nu = 9.150\ \mathrm{GHz}$. The line R^- arises from the mono-negative ion (see discussion in Section 6.3.6.2). [After R. E. Jesse, P. Biloen, R. Prins, J. D. W. van Voorst, G. J. Hoijtink, *Mol. Phys.*, **6**, 633 (1963).]

A system that may be an example of case 2 is the set of four ions shown below [structure **II**; R could be $C(CH_3)_3$]. Here two ketyl radical anions (formed by reaction of carbonyl compounds and an alkali metal) are bound by two alkali ions to form a quartet cluster [33].

$$
\begin{array}{ccc}
\mathrm{R} & \mathrm{Na^+} & \mathrm{R} \\
\diagdown & & \diagup \\
\mathrm{C}\!-\!\mathrm{O^-} & & {}^-\mathrm{O}\!-\!\mathrm{C} \\
\diagup & & \diagdown \\
\mathrm{R} & \mathrm{Na^+} & \mathrm{R}
\end{array}
$$

(II)

Such systems have very small $\bar{D}$ values ($0.007{-}0.015\ \mathrm{cm^{-1}}$), in solid CH_3DMF at 77 K. At room temperature, in liquid DMF, they show a seven-component composite spectrum (intensity ratios: $1:2:3:4:3:2:1$) arising from two equivalent alkali (^{23}Na, $I = \frac{3}{2}$) nuclei, with each proton-split component looking just *as if* the second ketyl unit were not present [34].

6.3.4 Photo-excited Triplet-State Entities

We have now developed the theory necessary to interpret the EPR spectra of triplet ($S = 1$) systems and are thus in a position to examine specific examples of the application of EPR to these systems.

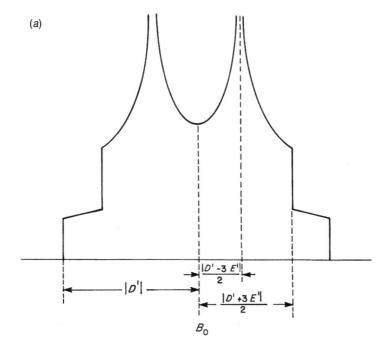

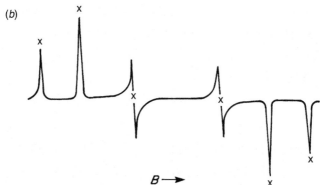

FIGURE 6.7 (*a*) Theoretical EPR absorption spectrum (centered at a field B_0) of a randomly oriented triplet system for given values of D', E' and v (and $g = g_e$). A zero linewidth is assumed. (*b*) First-derivative curve computed from (*a*) after assuming a non-zero linewidth. Only the transitions corresponding to $\Delta M_S = \pm 1$ are shown. The points marked x correspond to resonant-field values at which the magnetic field is oriented along one of the principal axes of **D** of the system. [After E. Wasserman, L. C. Snyder, W. A. Yager, *J. Chem. Phys.*, **41**, 1763 (1964).]

There is a very wide range of possibilities for triplet systems. We begin in this section with the most important category, namely, the large number of systems that are diamagnetic ($S = 0$) in the ground state but have relatively long-lived

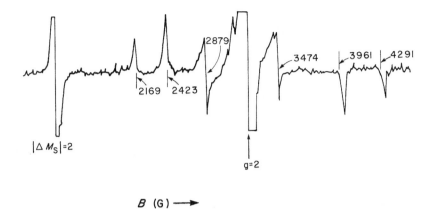

FIGURE 6.8 EPR spectrum at 9.08 GHz of photo-excited triplet perdeuteronaphthalene ($C_{10}D_8$) in a glassy mixture ('EPA') of hydrocarbon solvents at 77 K. Lines in the region of $g = 2$ arise from free radicals ($S = \frac{1}{2}$) and from double-quantum transitions. [After W. A. Yager, E. Wasserman, R. M. R. Cramer, *J. Chem. Phys.*, **37**, 1148 (1962).]

excited triplet states generated by steady-state or flash irradiation. Thereafter we consider thermally excited triplet entities and, finally, ground-state triplet species.

After irradiation with visible or ultraviolet light, many aromatic hydrocarbons in rigid solutions at low temperature exhibit excited states of unusually long lifetime—some of the order of minutes, as manifested by the long-lived glow (phosphorescence) remaining after turning off the incident light. This behavior is the result of the existence of a metastable state, which is populated via other excited states. G. N. Lewis et al. [35] postulated in 1941 that this long-lived state is a spin triplet state and that direct excitation to, or emission from, this state is spin-forbidden (to first approximation). Following Lewis' prediction, magnetic-susceptibility experiments on excited aromatic molecules in rigid media yielded results in qualitative accord with the triplet nature of the state. That is, on irradiation there is an increase in paramagnetism; this decays on cessation of irradiation, with the same decay rate constant as that of phosphorescence.

Aromatic hydrocarbons have been the focus of much of the early EPR triplet work, partly because of their availability and stability, their well-defined π-electron systems, and their long triplet lifetimes (no heavy atoms). It is relatively easy to prepare magnetically dilute systems containing small amounts of the molecules of interest in a diamagnetic optically inert medium. Thus specific photo-excitation of the molecules from their diamagnetic (singlet) ground states to populate their (lowest) triplets should allow study by EPR. However, a number of early experiments failed to detect such triplets. One reason for the initial failures is the marked anisotropy of the EPR line positions, arising from the dipolar interaction between the two electrons coupled to give $S = 1$. A second reason is the low sensitivity of the spectrometers at the time (1950–1955) when these first attempts were made.

Once the cause of the earlier failures was recognized, a successful observation of the lowest excited triplet state of naphthalene was achieved by irradiating single crystals of durene (1,2,4,5-tetramethylbenzene) containing a small fraction of naphthalene.[13] Since the two molecules are similar in shape, the naphthalene directly replaces durene in the lattice.

Optical studies indicate that the singlet-triplet splitting in naphthalene is $\sim20,000$ cm^{-1}, and hence that J_0 is large enough to ensure that the singlet excited state does not affect the magnetic properties of the system. Note that the EPR spectrum of a triplet system yields no explicit information about the exchange parameter(s). An electronic energy diagram, patterned after the one originated by A. Jablonski in 1933, is shown in Fig. 6.9.

The EPR spectra observed for naphthalene are precisely in accord with the expectations for a system with $S = 1$. The positions of EPR lines for the three principal-axis orientations of the field are given in Figs. 6.4a–c. It was found [10,37] that $\bar{D} = 0.1003$ cm^{-1}, $\bar{E} = -0.0137$ cm^{-1} ($D' = 107.3$ mT and $E' = -14.7$ mT), and g (isotropic) $= 2.0030$. The principal-axis system for **D**, as related to the molecular frame, is shown in Figs. 6.4 and 6.9. The lines in the vicinity of $h\nu/2g\beta_e$ were considered in Section 6.3.2. The zero-field splitting parameters shown above are relatively small. This is consistent with the Pauli exclusion

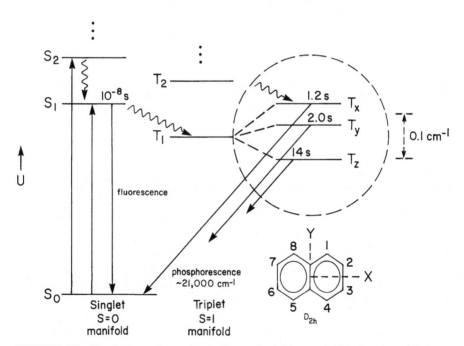

FIGURE 6.9 The lowest electronic singlet and triplet energy levels of naphthalene, showing photon absorption, fluorescence and phosphorescence transitions and their mean lifetimes, as well as radiation-less transitions (wavy lines). The zero-field splittings of the lowest triplet are indicated at the right.

principle and the coulombic repulsion between the two mobile unpaired electrons, which causes them to stay apart, decreasing the dipolar interaction energy and hence $|D|$.

The line positions as a function of orientation for triplet naphthalene in a single crystal of durene, for the magnetic field oriented in the xy, xz and yz planes of the crystal, are shown in Fig. 6.10. The spectra include contributions from the two types of sites in the unit cell, one of which (site 1) is scanned in its **D** principal planes. The reader is urged to interpret the angular-dependence

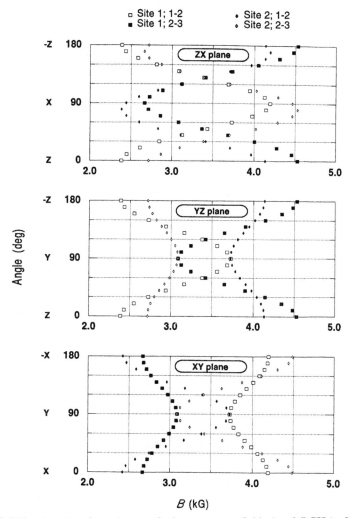

FIGURE 6.10 Angular dependence of the resonant field (at 9.7 GHz) for triplet naphthalene in durene, as a function of rotation with **B** in several planes for the two symmetry-related molecules (the planes are defined in Fig. 6.4). Hyperfine effects are ignored. [After C. A. Hutchison Jr., B. W. Mangum, *J. Chem. Phys.*, **34**, 908 (1961).]

curves and to extract the zero-field splitting parameters D and E (Problems 6.5 and 6.6).

For the naphthalene triplet state in a durene single crystal, with $\mathbf{B}$ along the X or Y principal axes, a $1:4:6:4:1$ quintet can be resolved at 77 K [38]. By employing variously deuterated samples, it was determined that the hyperfine splitting $a = 0.561$ mT arises from the 1,4,5,8 protons and $a = 0.229$ mT from the 2,3,6,7 protons (these values refer to $\mathbf{B} \parallel \mathbf{Z}$). These hyperfine splittings are very similar to those of the naphthalene anion considered in Chapter 3 (Fig. 3.8) (see Section 9.2.2 for further discussion).

Benzene itself has been studied by EPR [39, 40]. The data indicate that the molecule in its lowest triplet state no longer has D_{6h} symmetry; that is, it is distorted. There is interconversion among the (three) energy-equivalent configurations, as is evident from the linewidth behavior. Such transfer of the triplet excitation, here intramolecular, is a general phenomenon. Thus diffusion of such triplet excitons [41] can populate triplet states in molecules that were not originally excited by the ultraviolet irradiation. For example, the EPR signal of phenanthrene can decrease while that of naphthalene increases after irradiation in biphenyl crystals doped with both [42].

There exist inorganic systems that display photo-excited metastable triplet states with optical and magnetic properties closely analogous to those of the aromatic π systems. We consider the d^0 transition ions (e.g., V^{5+}, Cr^{6+}, Mn^{7+}, Mo^{6+}) in oxides [43]; for example, the $VO_4{}^{3-}$ ion in YVO_4 or in $Ba_3(VO_4)_2$, which, while diamagnetic in its singlet ground state, exhibits an EPR spectrum when illuminated. Presumably the optical excitation shifts electrons into the previously empty d shell with accompanying distortion of the already elongated oxygen tetrahedron (Fig. 6.11). Studies of EPR at various frequencies (4–23 GHz) and magnetic fields (including $B = 0$) have yielded electronic quadrupole ($\mathbf{D}$) splittings (and mean lifetimes) of the lowest triplet levels (Fig. 6.12) for the vanadate ion dilute (4%) in YPO_4 at 1.2 K [44]. Note the magnitude of the zero-field splittings as compared to those found for the delocalized p electrons in aromatic systems. These sensitive measurements, yielding the spin-hamiltonian parameters $\mathbf{g}$, $\mathbf{D}$ and an estimate of $\mathbf{A}(^{51}V)$ (Fig. 6.11), were carried out by means of optical detection of the EPR signals. Here square-wave modulation at 300 Hz of excitation field $\mathbf{B}_1$ results in modulation of the phosphorescence detected synchronously at the same frequency (Chapter 12).

6.3.5 Thermally Accessible Triplet Entities

Sections 6.3.4 and 6.3.6 of this chapter deal with systems in which the triplet state of interest is a photo-excited state or the ground state. In either case we tacitly assume that the separation of the triplet state from a nearby singlet state is large enough so that one need not consider mixing of the two states. An additional interesting case is that in which the singlet-triplet separation is small enough to make the triplet state thermally accessible but still not so small as to cause serious mixing of states. The singlet-triplet separation is approximately $|J_0|$, where the exchange

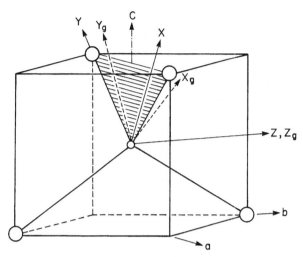

FIGURE 6.11 The structure of the YPO$_4$ unit cell, showing a V^{5+} having replaced a P^{5+} ion at the center. For the ground-state singlet, the local symmetry is D_{2d} (as in pure YPO$_4$), featuring two reflection planes (ac and bc) intersecting at two-fold axis c. It is believed that distortion removes one such plane in the lowest triplet state. Axes **X**, **Y**, **Z** indicate the principal axes of **D**, with axis **Z** normal to the remaining reflection plane and axis **Y** along a V–O direction. Axes **X**$_g$, **Y**$_g$, **Z**$_g$ denote the principal axes of **g**. [After W. Barendswaard, R. T. Weber, J. H. van der Waals, *J. Chem. Phys.*, **87**, 3731 (1987).]

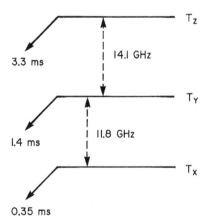

FIGURE 6.12 The zero-field splitting of the lowest $S = 1$ state of V^{5+} in YPO$_4$ at 1.2 K, also giving the mean lifetime of each level. The sign of D here was taken to be positive; if it were negative, then the order of the levels would be reversed. The **X**, **Y**, **Z** labeling follows this book's convention [W. Barenwaard, R. T. Weber, J. H. van der Waals, *J. Chem. Phys.*, **87**, 3731 (1987)].

interaction between two electrons is given by $J_0 \hat{\mathbf{S}}_1^{\mathrm{T}} \cdot \hat{\mathbf{S}}_2$ (Eqs. 6.3–6.5). When the triplet lies higher than the singlet ($J_0 > 0$), the relative population of the triplet state is governed by the Boltzmann factor $3 \exp[-J_0/k_bT]$. In general, for a given population of a paramagnetic state, the intensity $\mathscr{I}$ of EPR absorption is given by a Curie-law (Eq. 1.16 and Section 10.3.4) dependence, that is, $\mathscr{I} \propto 1/T$. The integrated intensity $\mathscr{I}$ of EPR absorption arising from a thermally excited triplet state should depend on temperature as

$$\mathscr{I} \propto T^{-1}[3 + \exp(J_0/k_bT)]^{-1} \tag{6.40}$$

Thus a study of the temperature dependence of the EPR intensity (i.e., area $\mathcal{A}$; see Section E.1) permits a determination of the value of J_0.

A clear-cut example of a thermally accessible triplet state is provided by the F_t 'point' defect in MgO [45]. This center is thought to be a neutral trivacancy, that is, a missing linear $(\mathrm{O{-}Mg{-}O})^{2-}$ fragment replaced by two electrons. It gives no EPR spectrum at very low temperatures, unless ultraviolet (uv)-irradiated. Alternatively, a spectrum is generated by warming above 4 K. This indicates that the triplet state for the two electrons lies above the singlet so that $J_0 > 0$. An analysis of the temperature dependence yields $\bar{J}_0 = 56(7)$ cm^{-1}.[14] At arbitrary orientations of the magnetic field, the EPR spectrum consists of six lines (pairs of $\Delta M_S = \pm 1$ transitions, one for each of the three distinct orientations of the $\mathrm{O}^{2-}{-}\mathrm{Mg}^{2+}{-}\mathrm{O}^{2-}$ axes in the cubic crystal). The 300 K line positions as a function of rotation in the (001) and (110) planes are shown in Fig. 6.13. An analysis of these data reveals that $D' = 30.7$ mT and $E = 0$, with $g = 2.0030(5)$. From Eqs. 6.41 or 9.18a, the average interelectronic distance is found to be 4.5 Å; this compares well with the relevant oxygen-oxygen distance of 4.2 Å in MgO [45].

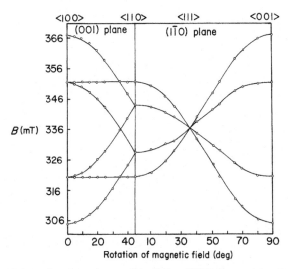

FIGURE 6.13   Orientation dependence of the X-band EPR lines arising from the F_t center in MgO at 300 K. [After B. Henderson, *Br. J. Appl. Phys.*, **17**, 851 (1966).]

Another interesting example is the observation of a triplet EPR spectrum in powdered samples of Fremy's salt, represented here by $K_4[(SO_3)_2NO]_2$ to emphasize its spin-paired dimeric structure [46]. Here $\overline{D} = +0.076$ cm^{-1} and $\overline{E} = +0.0044$ cm^{-1}. The area of the half-field peak was found to increase exponentially (Eq. 6.40) with temperature (250–350 K), yielding the singlet-triplet gap energy $\overline{J}_0 = 2180$ cm^{-1}. Since no hyperfine (^{14}N) splitting is observed, this spectrum has been assigned to a *triplet exciton*, which is an excited state that migrates rapidly through the crystal lattice.

6.3.6 Ground-State Triplet Entities

A triplet species need be no larger than an atom if it has an appropriate set of degenerate orbitals. The list of ground-state triplet atoms includes C, O, Si, S, Ti and Ni. The two most prominent ground-state triplet diatomic molecules are O_2 and S_2 (Chapter 7).

6.3.6.1 *Carbenes and Nitrenes*

The H—C—H fragment (methylene or 'carbene') is one of the simplest molecular systems, and the triplet nature of its ground state has been established spectroscopically. The EPR spectrum of methylene has been reported (Table 6.1) for both CH_2 and CD_2. For the former, $\overline{D} = 0.69$ cm^{-1} and $\overline{E} = 0.003$ cm^{-1}; for the latter, $\overline{D} = 0.75$ cm^{-1} and $\overline{E} = 0.011$ cm^{-1}. The difference presumably arises from the effect of zero-point vibration. Many substituted methylenes have also been studied (Table 6.1). The non-zero value of E for some of these molecules indicates that for them there is

TABLE 6.1 **Zero-Field Splitting Parameters for Triplet Ground-State Molecules**

Molecule	$|\overline{D}|$ (cm^{-1})	$|\overline{E}|$ (cm^{-1})	Reference
H—C—H	0.69	0.003	a
D—C—D	0.75	0.011	a
H—C—C≡N	0.8629	0	b
H—C—CF₃	0.712	0.021	c
H—C—C₆H₅	0.5150	0.0251	d
H—C—C≡C—H	0.6256	0	b
H—C—C≡C—CH₃	0.6263	0	b
H—C—C≡C—C₆H₅	0.5413	0.0035	b
C₆H₅—C—C₆H₅	0.4055	0.0194	d
N≡C—C—C≡N	1.002	<0.002	e
N—C≡N	1.52	<0.002	e

[a] R. A. Bernheim, H. W. Bernard, P. S. Wang, L. S. Wood, P. S. Skell, *J. Chem. Phys.*, **53**, 1280 (1970); **54**, 3223 (1971); R. A. Bernheim, R. J. Kempf, E. F. Reichenbecher, *J. Magn. Reson.*, **3**, 5 (1970); E. Wasserman, V. J. Kuck, R. S. Hutton, E. D. Anderson, W. A. Yager, *J. Chem. Phys.*, **54**, 4120 (1971).
[b] R. A. Bernheim, R. J. Kempf, J. V. Gramas, P. S. Skell, *J. Chem.Phys.*, **43**, 196 (1965).
[c] E. Wasserman, L. Barash, W. A. Yager, *J. Am. Chem. Soc.*, **87**, 4974 (1965).
[d] E. Wasserman, A. M. Trozzolo, W. A. Yager, R. W. Murray, *J. Chem. Phys.*, **40**, 2408 (1964).
[e] E. Wasserman, L. Barash, W. A. Yager, *J. Am. Chem. Soc.*, **87**, 2075 (1965).

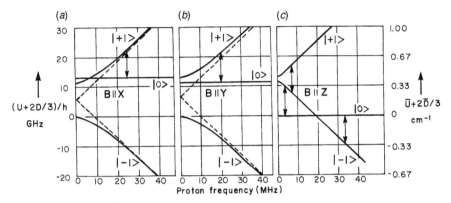

FIGURE 6.14   Energy levels of the fluorenylidene molecule (**III**) in its triplet ground state as a function of applied magnetic field (measured in proton NMR fluxmeter frequency units). The EPR transitions indicated are at $\sim$9.7 GHz. (*a*) **B** $\parallel$ **X**; (*b*) **B** $\parallel$ **Y**; (*c*) **B** $\parallel$ **Z**. Here $|\overline{D}| = 0.4078$ and $|\overline{E}| = 0.0283$ cm^{-1}. [After C. A. Hutchison Jr., G. A. Pearson, *J. Chem. Phys.*, **47**, 520 (1967).]

no axis of symmetry of order 3 or greater; this indicates that the molecules are non-linear. For such systems, the maximum number of peaks (six $\Delta M_S = \pm 1$ transitions) is expected in the glass-phase EPR spectrum, just as for naphthalene in the excited triplet state (Figs. 6.7 and 6.8). When the system is nearly uniaxial, the parameter E may be so small that one may only be able to set an upper limit for its value. An increase in the extent of the conjugated system attached to the methylene carbon atom may lead to a decrease in the parameter D as is evident from Table 6.1.

Figure 6.14 shows the energy-level diagrams for the fluorenylidene molecule (**III**) [47]. The molecule is generated in its ground triplet state by irradiation of diazofluorene at 77 K. It is thus to be regarded as a derivative of methylene. If the zero-field splitting D is large compared with hn for a microwave quantum, only certain of the lines allowed by the selection rules are observed. When **B** is parallel to axis **X** or **Y**, only one transition is observed for $\nu \approx 9.7$ GHz. Since the ordinate in Fig. 6.14 is expressed in gigahertz, the frequency required to cause a transition between adjacent levels is immediately apparent from it. For **B** $\parallel$ **Z**, three transitions are expected and are observed. Two of these are between the levels designated by $|0\rangle$ and $|-1\rangle$. Note (Fig. 6.14c) that the '$\Delta M_S = \pm 2$' transition occurs at an intermediate value of the magnetic field. The parameters measured, $|\overline{D}| = 0.4078$ and $|\overline{E}| = 0.0283$ cm^{-1}, are appreciably larger than those of naphthalene, as expected (Section 6.3.4). The reader should compare the resulting transitions allowed at X band (Figs. 6.4 and 6.14).

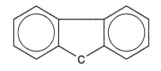

(III) fluorenylidene

Another organic triplet species of considerable experimental and theoretical interest is trimethylenemethane (**IV**), which can be represented as

(**IV**) trimethylenemethane

This free radical, TMM,[15] can be prepared by γ irradiation of methylenecyclopropane. EPR studies [49] reveal that it is a triplet ground-state species, rather than a biradical; that is, $|J_0|$ is relatively large compared to $|D|$, and parameter J_0 is negative (see Fig. 6.2). The radical has four π electrons, and is close to uniaxial (planar with symmetry D_{3h}) as inferred from the parameters $\overline{D} = 0.0248$ cm^{-1} and $\overline{E} < 0.003$ cm^{-1}, at 77 K. The temperature dependence of these parameters, and of the proton hyperfine matrix (six equivalent protons, principal values A_i/h of -14, -38 and -26 MHz) suggests that anisotropic rotational effects occur. Because the proton nuclear Zeeman term is not small compared to the hyperfine values in this anisotropic system, all four hyperfine transitions per proton are observed (Section 5.3.2.1). The relative sign of **D** and **A**, obtainable from the EPR data, discloses that D is positive.

From Eq. 6.15a (**D** diagonal) and Eq. 6.25a, it is easily shown that

$$D = \frac{3\mu_0}{16\pi} g^2 \beta_e^2 \langle r^{-3} \rangle \langle 1 - 3\cos^2\theta \rangle \tag{6.41}$$

Here angle θ is the angle between the inter-electron (spin) vector **r** and the principal axis **Z** of **D**. It is evident that the sign of D is determined solely by $\langle 1 - 3\cos^2\theta \rangle$. If the dipoles were fixed at two points (in which case $\mathbf{r}/r = \mathbf{Z}$ and $\frac{1}{2}D_Z = -D_Y = -D_X$), Eq. 6.41 would yield $D = -(3\,\mu_0/8\pi)g^2\beta_e^2\langle r^{-3}\rangle$, that is, D would be negative. For a triplet-state atom not in an electric field, the spherical symmetry dictates that $D = 0$. In non-spherical systems, D can be negative or positive. The latter sign occurs in binuclear triplet species when the internuclear axis is perpendicular to **Z**, as dictated by the unpaired-electron distribution. In trimethylenemethane (**IV**), the two unpaired electrons can be considered as being equally distributed at three points (carbons) in a plane normal to the symmetry axis, so that $D > 0$.

6.3.6.2 *Dianions of Symmetric Aromatic Hydrocarbons*

A molecule may have a ground triplet state in its neutral, cationic or anionic form. Here there is one unpaired electron in each of a pair of degenerate orbitals. Thus, as is shown in Fig. 6.1d, the lowest-energy state (ground state) is that corresponding to single occupation with parallel spins of the highest two occupied levels.

Degenerate orbital energy levels are found in molecules with an n-fold ($n \geq 3$) axis of symmetry. Molecules of this type do not *necessarily* have a triplet ground state. The situation depends on the sign of the electron-exchange integral (Eq. 6.5). If J_0 is positive, then the singlet state lies lower. This is the case, for instance, in the coronene di-anion with alkali counterions (**V**) [49,50].

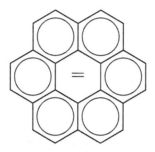

(V) coronene di-anion

Occurrence of one electron in each of two degenerate orbitals of symmetrically substituted benzenes may be achieved if the di-anion can be formed. Triplet ground states have been demonstrated for symmetric molecules such as the 1,3,5-triphenylbenzene (**I**) and triphenylene di-anions (**VI**).

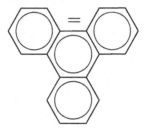

(VI) triphenylene di-anion

These ions possess degenerate antibonding orbitals, analogous to those shown in Fig. 6.15 for the hypothetical benzene di-anion (note Table 9A.2). For the triphenyl-benzene di-anion (ground triplet state, Fig. 6.6) D is less ($\overline{D} = 0.042$ cm^{-1}) than that of the neutral excited triplet-state molecule ($\overline{D} = 0.111$ cm^{-1}) [51]. The orbital occupation is very different for these two cases; calculations show that in the excited triplet molecule, there is a greater interaction (leading to a larger D value) between two electrons in the 'paired' bonding and antibonding orbitals than between two electrons in the antibonding orbitals of the ground-state di-anion.

6.3.6.3 Inorganic Triplet Species

Other than O_2 and S_2 (considered in Chapter 7) and some transition-ion complexes, there are not many stable inorganic molecules that exist in a triplet ground state. Some unstable species can be trapped in low-temperature matrices. Excellent examples are the isoelectronic molecules CCO

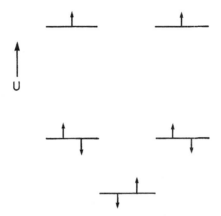

FIGURE 6.15 Triplet π-electron configuration of the hypothetical benzene di-anion.

and CNN, prepared by reaction of C atoms with CO or N_2 with subsequent trapping in a frozen rare-gas matrix at 4 K [52]. Both of these molecules have large values of D ($\overline{D} = 0.7392$ cm^{-1} for CCO and $\overline{D} = 1.1590$ cm^{-1} for CNN in solid neon) such that $D > h\nu$ at X band. Thus only one transition is seen in each case. The ^{13}C and ^{14}N hyperfine parameters are also reported. The fact that $E = 0$ indicates that these molecules are linear.

Another quite different example of a ground-state triplet is the quasi-tetrahedral $[AlO_4]^+$ 'point' defect in a quartz single crystal [53]. This center is believed to contain two electron holes forming a triplet spin system in which the two unpaired electrons are $\sim 0.26_5$ nm apart on adjacent (and symmetry-related) oxygens in the AlO_4 entity. For this center at ~ 35 K, $D' = -69.8$ mT and $E' = 6.3$ mT. These values are in very good agreement with those calculated on the basis of a model placing an unpaired-electron population of 0.76 in two oxygen $2p$ orbitals perpendicular to the Al—O—Si plane. As expected [3], the hyperfine matrix from the central ^{27}Al ion is accurately approximated by $\mathbf{A} = (\mathbf{A}_1 + \mathbf{A}_2)/2$, where matrices $\mathbf{A}_i$ ($i = 1, 2$) stem from the corresponding radicals ($S = \frac{1}{2}$) centered on the two oxygens. Similarly, $\mathbf{g} = (\mathbf{g}_1 + \mathbf{g}_2)/2$. The hyperfine splittings arising from the low-abundance isotope ^{29}Si, that is, from outer silicon atoms bonded to the oxygens, are half as large as the corresponding ones from the $S = \frac{1}{2}$ species.

Various transition ions provide examples of $S = 1$ systems, for example, V^{3+} and Ni^{2+} in the $3d$ series (Chapter 8). The latter ion, dilute in K_2MgF_4 where it is in a Mg^{2+} site surrounded by a slightly distorted octahedron of F^- ions, yielded $\overline{D} = -0.425$ cm^{-1} and $\overline{E} = -0.065$ cm^{-1} (with isotropic $g = 2.275$) at 1.6 K [54].

6.4 INTERACTING RADICAL PAIRS

The earliest EPR work in this specialty area is that done by Bleaney and Bowers [55] and also by Kumagai et al. [56], on cupric acetate monohydrate, in which Cu^{2+} $3d^1$ ions, each influenced by the local electric field, also interact pairwise and reveal an effective electron spin of $S = 1$ ($g_\perp = 2.08$ and $g_{||} = 2.42$) at 90 K. The electronic quadrupolar parameters are $D/hc = 0.34$ cm^{-1} and $E/hc = 0.01$ cm^{-1}. There are 5×5 orbital states, with a singlet lowest, each such state with spin degeneracy of 2×2 (neglecting nuclear Cu spins; these do, however, give rise to observed hyperfine structure). The resulting ground state is a singlet (diamagnetic), while the lowest excited state is a triplet. The latter becomes appreciably populated even at quite low temperatures (>50 K) and gives rise to the observed EPR spectrum. The splitting J_0/hc (about $+300$ cm^{-1}; see Eq. 6.4) between the singlet and triplet states, as determined from the temperature dependence of the EPR intensity, is caused primarily by the electronic exchange interaction. Thus, here, $|J_0| \gg g\beta_e B$. The anisotropy of the g value indicates presence of appreciable spin-orbit interaction.

In the case of strongly coupled spin $= \frac{1}{2}$ identical pairs with nuclear spins, such as described above, the hyperfine coupling parameters have magnitudes half as large as the corresponding values for the single entities.

The case of weakly coupled radical pairs ($|J_0| \ll g\beta_e B$) has been successfully treated by Itoh et al. [57]. Here the singlet and triplet are mixed by the Zeeman and hyperfine interactions, and J_0 can be evaluated from the EPR line positions. This has been done for pairs of RR′C=NO free radicals created by irradiation of single crystals of glyoximes [57]; for R $=$ R′ $=$ CH$_3$, $J_0/hc = +0.200$ cm^{-1}.

Obviously, far more detail on the spin-pair systems will be sought. Thus, the exact distances between the pair units, in both the singlet and triplet states, will be welcomed, as will the orientations of these axes within the crystals. Advanced techniques, such as ELDOR in ESE studies (see Chapters 11 and 12), are expected to be helpful in such efforts [58].

6.5 BIRADICALS

As indicated above, a biradical is a molecule containing two unpaired electrons that, on the average, are so far apart that interactions between them are sufficiently weak and energy classification into a singlet and a triplet manifold is not useful.[2] More generally, this category includes radical pairs, in which the spins are located on close but separate species.

The border region between *triplet-state species* and *biradicals* is, of course, nebulous, as it is set by the magnitudes of D and E compared to that of the singlet-triplet energy separation. We note that as D and $E \to 0$, the half-field transition remains pretty well in place (see Eq. 6.36b), but declines rapidly in intensity. For triplet states, $|D/hc|$ typically is in the range 0.1–2.5 cm^{-1} (Section 6.3), but it is this magnitude compared to that of exchange energy parameter J_0 which is the crucial factor. Thus TMM previously discussed has a triplet ground state with $D/hc \sim +0.025$ cm^{-1}, while J_0/hc is estimated to be -5000 cm^{-1} [59].

Consider a biradical composed of two identical molecular fragments, each containing one unpaired electron as well as one magnetic nucleus giving rise to hyperfine splitting. We consider the isotropic case. The spin hamiltonian appropriate to this system is [60,61][16]

$$\hat{\mathcal{H}} = g\beta_e B(\hat{S}_{1z} + \hat{S}_{2z}) + A_0(\hat{\mathbf{S}}_1{}^T \cdot \hat{\mathbf{I}}_1 + \hat{\mathbf{S}}_2{}^T \cdot \hat{\mathbf{I}}_2) + J_0\hat{\mathbf{S}}_1{}^T \cdot \hat{\mathbf{S}}_2 \qquad (6.42)$$

Here we consider the biradical to exist in liquid solution so that the anisotropies arising from **g**, **D** and **T** are averaged to zero. We neglect the effect of the nuclear Zeeman terms, the cross-hyperfine interactions, and the nucleus-nucleus spin coup-lings. If $|A_0| \ll g\beta_e B$, then the hyperfine terms may be taken to first order only, and Eq. 6.42 can be approximated as

$$\hat{\mathcal{H}} = g\beta_e B(\hat{S}_{1z} + \hat{S}_{2z}) + A_0(\hat{S}_{1z}\hat{I}_{1z} + \hat{S}_{2z}\hat{I}_{2z}) + J_0\hat{\mathbf{S}}_1{}^T \cdot \hat{\mathbf{S}}_2 \qquad (6.43)$$

where **B** ∥ **z**.

First, consider the limiting case where $|J_0| \gg |A_0|$. The zero-order spin hamiltonian is then

$$\hat{\mathcal{H}}_0 = g\beta_e B(\hat{S}_{1z} + \hat{S}_{2z}) + J_0\hat{\mathbf{S}}_1{}^T \cdot \hat{\mathbf{S}}_2 \qquad (6.44a)$$

which can be expanded (Section C.1) to yield

$$\hat{\mathcal{H}}_0 = g\beta_e B(\hat{S}_{1z} + \hat{S}_{2z}) + J_0\left[\hat{S}_{1z}\hat{S}_{2z} + \tfrac{1}{2}(\hat{S}_{1+}\hat{S}_{2-} + \hat{S}_{1-}\hat{S}_{2+})\right] \qquad (6.44b)$$

The appropriate eigenfunctions are

$$|1,+1\rangle \equiv |+\tfrac{1}{2}, +\tfrac{1}{2}\rangle \qquad (6.45a)$$

$$|1,0\rangle \equiv \frac{1}{\sqrt{2}}\left(|+\tfrac{1}{2}, -\tfrac{1}{2}\rangle + |-\tfrac{1}{2}, +\tfrac{1}{2}\rangle\right) \qquad (6.45b)$$

$$|1,-1\rangle \equiv |-\tfrac{1}{2}, -\tfrac{1}{2}\rangle \qquad (6.45c)$$

$$|0,0\rangle \equiv \frac{1}{\sqrt{2}}\left(|+\tfrac{1}{2}, -\tfrac{1}{2}\rangle - |-\tfrac{1}{2}, +\tfrac{1}{2}\rangle\right) \qquad (6.45d)$$

The quantum numbers in the right-hand set of kets (uncoupled representation) refer to the eigenvalues of $\hat{S}_{1z}$ and $\hat{S}_{2z}$, whereas those on the left (coupled represen-tation) refer to the quantum numbers S and M_S arising from $\hat{\mathbf{S}}^2$ and $\hat{S}_z$, where $\hat{S}_z = \hat{S}_{1z} + \hat{S}_{2z}$.

The energies to first order (in A_0 and J_0; note and compare Figs. 6.2 and 6.15, in which A_0 is neglected) are

$$U_{1,\,+1}{}^{(1)} = +g\beta_e B + J_0/4 + A_0{}^t M_I/2 \qquad (6.46a)$$

$$U_{1,0}{}^{(1)} = +J_0/4 \qquad (6.46b)$$

$$U_{1,\,-1}{}^{(1)} = -g\beta_e B + J_0/4 - A_0{}^t M_I/2 \qquad (6.46c)$$

$$U_{0,0}{}^{(1)} = -3J_0/4 \qquad (6.46d)$$

Here ${}^t M_I = M_{I_1} + M_{I_2}$. The only allowed EPR transitions are $|1, \pm 1\rangle \leftrightarrow |1,0\rangle$, which are degenerate at a transition energy of $g\beta_e B + A_0 M_I/2$ and are independent of J_0. In the case of $I_1 = I_2 = 1$, the first-order spectrum would consist of five lines separated by $|a_0|/2$ (where $a_0 = A_0/g_e\beta_e$) with intensity ratios 1–2–3–2–1, since ${}^t M_I = -2, -1, 0, +1, +2$. An example of this case is the EPR spectrum of the biradical tetramethyl-2,2,5,5-pyrrolidoneazine-3 dioxyl1,1' (**VII**), shown in Fig. 6.16a, where the line separation is only 0.740 mT so that $a_0({}^{14}\text{N}) = 1.480$ mT [62].

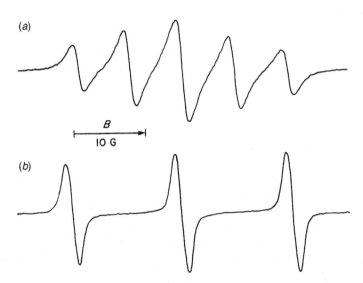

FIGURE 6.16   (a) X-band EPR spectrum of biradical **VII** showing interaction of both unpaired electrons with nitroxide ${}^{14}\text{N}$ nuclei. The spacing is $|a_0|/2$, where $|a_0| = 1.480$ mT is the spacing for the corresponding monoradical (one NO group replaced by NH). [After R. M. Dupeyre, H. Lemaire, A. Rassat, *J. Am. Chem. Soc.*, **87**, 3771 (1965).] (b) X-band EPR spectrum of the biradical **VIII**, in which the nitroxide groups are isolated from each other. The hyperfine splitting (1.56 mT) is just the same as that of the corresponding monoradical. This is an illustration of the case $|J_0| \ll |A_0|$. [After R. Briere, R. M. Dupeyre, H. Lemaire, C. Morat, A. Rassat, P. Rey, *Bull. Soc. Chim. France*, No. 11, 3290 (1965).]

(VII) tetramethyl-2,2,5,5-pyrrolidoneazine-3 dioxyl-1,1′

Now consider the second limiting case, of $|J_0| \ll |A_0|$. The zero-order spin hamiltonian is

$$\hat{\mathcal{H}}_0 = g\beta_e B(\hat{S}_{1z} + \hat{S}_{2z}) + A_0[\hat{S}_{1z}\hat{I}_{1z} + \hat{S}_{2z}\hat{I}_{2z}) \tag{6.47}$$

This operator is separable into two parts

$$\hat{\mathcal{H}}_0(1) = g\beta_e B\hat{S}_{1z} + A_0\hat{S}_{1z}\hat{I}_{1z} \tag{6.48a}$$

$$\hat{\mathcal{H}}_0(2) = g\beta_e B\hat{S}_{2z} + A_0\hat{S}_{2z}\hat{I}_{2z} \tag{6.48b}$$

The eigenfunctions of $\hat{\mathcal{H}}_0(1)$ are

$$|+\tfrac{1}{2}, M_I(1)\rangle \quad \text{and} \quad |-\tfrac{1}{2}, M_I(1)\rangle$$

with eigenvalues

$$U_{+1/2}{}^{(0)} = +g\beta_e B/2 + A_0 M_I(1)/2 \tag{6.49a}$$

$$U_{-1/2}{}^{(0)} = -g\beta_e B/2 - A_0 M_I(1)/2 \tag{6.49b}$$

One obtains analogous results for $\hat{\mathcal{H}}_0(2)$. Thus this case may be considered as two non-interacting systems with $S = \tfrac{1}{2}$. When $I = 1$, the first-order EPR spectrum consists of three lines separated by $|a_0|$. An example of this case is shown in Fig. 6.16b, where the biradical is di(tetramethyl-2,2,6,6-piperidinyl-4 oxyl-1)terephthalate (**VIII**).

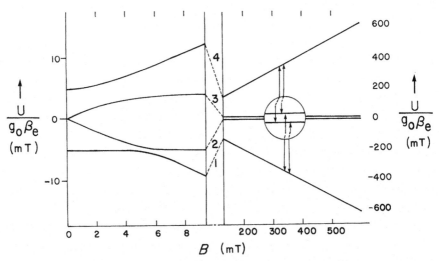

(**VIII**) di(tetramethyl-2,2,6,6-piperidinyl-4 oxyl-1)terphthalate

The intermediate case of $|J_0| \approx |A_0|$ gives rise to a complex group of lines. The intensity and position of these lines are a strong function of $|J_0/A_0|$ [63–65]. Thus J_0 can be extracted from the solution EPR spectrum.

The general case, when one encounters anisotropic parameters and where field **B** is of arbitrary magnitude, can be dealt with using the spin hamiltonian

$$\hat{\mathcal{H}} = \beta_e \mathbf{B}^{\mathrm{T}} \cdot \mathbf{g}_1 \cdot \hat{\mathbf{S}}_1 + \beta_e \mathbf{B}^{\mathrm{T}} \cdot \mathbf{g}_2 \cdot \hat{\mathbf{S}}_2 + \tfrac{1}{2}(\hat{\mathbf{S}}_1{}^{\mathrm{T}} \cdot \mathbf{J} \cdot \hat{\mathbf{S}}_2 + \hat{\mathbf{S}}_2{}^{\mathrm{T}} \cdot \mathbf{J}^{\mathrm{T}} \cdot \hat{\mathbf{S}}_1)$$
$$+ \text{nuclear terms as needed} \tag{6.50}$$

The EPR line positions and relative intensities are obtainable from Eq. 6.50 by numerical (computer) solution. An example of the energy levels for such a case is depicted in Fig. 6.17.

FIGURE 6.17   The energy levels $U(M_{S_1}, M_{S_2})$ as a function of applied magnetic field B for a biradical system not containing any nuclear spin. The primary EPR transitions ($\Delta M_{S_i} = \pm 1$), involving the two $M_{S_i} = 0$ levels, are shown. No nuclear spins occur here. The spin-hamiltonian parameters (Eq. 6.50) used to generate the figure are taken from J. Isoya, W. C. Tennant, Y. Uchida, J. A. Weil, *J. Magn. Reson.*, **49**, 489 (1982).

Note that the three matrices in Eq. 6.50 are attainable by computer analysis of single-crystal EPR rotational data, where **J** includes the isotropic exchange parameter J_0 as well as the spin-spin dipolar interaction and anisotropic exchange.

6.6 SYSTEMS WITH $S > 1$

A few organic high-spin radicals are known. For instance, the assembly of the three diphenylhydrazyl groups (Section E.1.2) mounted meta to each other on a central 1,3,5-tricyanobenzene yields a stable triradical ($S = \frac{3}{2}$), exhibiting $\Delta M_S = \pm 1, \pm 2$ and ± 3 transitions [66]. Perhaps the highest spin multiplicity known for organic molecules at this time is an undecet ground state, arising from the five sets of unpaired electrons formally located at five methylene carbons held between phenyl groups, in the *meta*-linked molecule C_6H_5—C—$(C_6H_4$—C—C_6H_4—C—$)_2C_6H_5$ (having $S = 5$) prepared and studied by EPR [67]. More accurately, five of the unpaired electrons are delocalized in a π orbital; the other five occur in σ non-bonding orbitals.

Monatomic high-spin species can be created (e.g., by irradiation procedures), and have been studied since the mid-1950s or 50. A good example is the nitrogen atom, whose electronic ground state ($S = \frac{3}{2}$) is ${}^4S_{3/2}$ (see Section B.7). Such atoms can be trapped and stabilized in solid matrices (e.g., in molecular nitrogen at low temperatures [68], in azides [69], and in fullerene cages [70]), in liquids (e.g., superfluid helium [71]), and in gas phase (see Section 7.2). For N^0, typical parameters are $g = 2.002$, $\overline{D} = 0.010 \text{ cm}^{-1}$, $\overline{E} = -0.002 \text{ cm}^{-1}$ (the latter two measured, of course, in solids) and $a_0({}^{14}N) = 0.5 \text{ mT}$. Because of the high mobility of the atom, D and E are quite temperature-sensitive.

Numerous other inorganic systems offer high-spin ($S > 1$) species. These include clusters of adjacent unpaired electrons (i.e., of F centers) in alkali halides. For instance, three F centers forming an equilateral triangle in the (1,1,1) plane of a KCl crystal constitute an $S = \frac{3}{2}$ center, which has been studied by both EPR and ENDOR [72].

Clusters of transition atoms also lead to high-spin systems. Thus Mn_2 and Mn_5, in rare-gas matrices, are amenable to study by EPR [73]. For Mn_2, since the exchange parameter J_0 is negative ($J_0' = -11.0 \text{ T}$, i.e., antiferromagnetic coupling), this molecule features a diamagnetic ground state. However, higher spin states ($S = 1,2,3$) can be thermally populated and yield EPR spectra with characteristic ${}^{55}Mn$ ($I = \frac{5}{2}$) hyperfine structure. The pentamer cluster Mn_5 has a total spin $S = \frac{25}{2}$.

More recently, single-molecule magnets (SMMs) have been studied at very high frequencies (40–200 GHz). EPR spectroscopy has yielded information about the energy levels and power saturation behavior of molecular nanomagnet crystals, including ferric complexes (abbreviated 'Fe$_8$') and manganese complexes (abbreviated 'Mn$_{12}$' [74,75]). Both have effective spin: $S' = 10$.

Clearly, the primary domain of high-spin species is that of transition ions (Chapter 8), which have been a fruitful source of EPR progress since the beginnings of this field.

6.7 HIGH-SPIN AND HIGH-FIELD ENERGY TERMS

For magnetic species with $S > 1$, additional terms (e.g., Eqs. 8.17) should be added to the spin hamiltonian (Eq. 6.18), at least in principle. These usually make only relatively small contributions to the total energy of the system but show up as corrections to the EPR line positions and intensities. They are not as simple as the terms in Eq. 6.18, since tensorial entities greater than rank 2 (not expressible as matrices) enter, and each term brings in a multitude of parameters to be obtained from the EPR data. Table 6.2 indicates what types of terms are in principle allowed. This table is a listing of what terms can be present for each value of S, ignoring nuclear-spin terms. The symbol $\hat{S}$ implies the possible presence of all three operators $\hat{S}_x$, $\hat{S}_y$ and $\hat{S}_z$. Operator $\hat{S}^2$ indicates the term $\hat{\mathbf{S}}^T \cdot \mathbf{D} \cdot \hat{\mathbf{S}}$ discussed in this chapter. Similarly, the symbol B here implies the possible presence of B_x, B_y and B_z. Integers n, n' and n'' are positive odd integers; thus $n = 1$ indicates the usual electron Zeeman term, and all other integers n (and all n', n'') are usually safely ignored (except in highly accurate measurements).

Column 2 in Table 6.2 includes the high-field situation (for any S) in which one considers that all spin-hamiltonian parameters are field-dependent. For instance, the Zeeman splitting factor can be written as the series

$$g = g^{(0)} + g^{(2)}B^2 + g^{(4)}B^4 + \cdots \tag{6.51}$$

Here $g^{(0)}$ is the usual g factor, and the $g^{(2)}$ term describes an energy term having the form of components of $\mathbf{B}^3\hat{\mathbf{S}}$ (e.g., $B_x^3\hat{S}_y$), which may become appreciable in the analysis of ultrahigh-field EPR spectra.

TABLE 6.2 Possible Electron-Spin Magnetic-Field Terms in the Spin Hamiltonian[a]

S		$\hat{\mathcal{H}}$ Terms			
$\frac{1}{2}$	$\hat{S}B^n$				
1	$\hat{S}B^n$	$\hat{S}^2$			
$\frac{3}{2}$	$\hat{S}B^n$	$\hat{S}^2$	$\hat{S}^3B^{n'}$		
2	$\hat{S}B^n$	$\hat{S}^2$	$\hat{S}^3B^{n'}$	$\hat{S}^4$	
$\frac{5}{2}$	$\hat{S}B^n$	$\hat{S}^2$	$\hat{S}^3B^{n'}$	$\hat{S}^4$	$\hat{S}^5B^{n''}$
.					
.					
.					

[a] The meaning of the symbols $\hat{S}$ and B in this table are defined in the text (Sections 6.6 and 8.3). Exponents n, n' and n'' generally are positive integers.

In accordance with Table 6.2, a term of the form $\mathbf{B}\hat{S}^3$ can occur for $S > 1$ [76–78]. With smaller S, for example, $S = \frac{1}{2}$, $\hat{S}^3$ can always be written as a linear combination of terms at most linear in $\hat{S}$. Terms of this type can arise from the product in perturbation theory of the normal Zeeman term $g_e\beta_e\mathbf{B}^{\mathrm{T}} \cdot \hat{\mathbf{S}}$ and the square $(\lambda\hat{\mathbf{L}}^{\mathrm{T}} \cdot \hat{\mathbf{S}})^2$ of the spin-orbit term [79]. For octahedral or tetrahedral symmetries, the additional terms have the form

$$g'\beta_e[B_X\hat{S}_X{}^3 + B_Y\hat{S}_Y{}^3 + B_Z\hat{S}_Z{}^3 - \mathbf{B}^{\mathrm{T}} \cdot \hat{\mathbf{S}}(3\,\hat{\mathbf{S}}^{\mathrm{T}} \cdot \hat{\mathbf{S}} - \hat{1})/5]$$

Here the operator form of the last term is a notational formality: $\hat{\mathbf{S}}^{\mathrm{T}} \cdot \hat{\mathbf{S}} = S(S + 1)$ and $\hat{1} = 1$. With lower symmetry, the number of BS^3-type terms increases, and there is proliferation of the parameters describing the effect [78]. Conversely, if description of an experimental spectrum requires such a term, this confirms the identification of an $S > \frac{3}{2}$ state. If there is a nucleus contributing splitting, an additional term of the form $\hat{\mathbf{S}}^3\,\hat{\mathbf{I}}$ may also be required.

The derivation and treatment of the terms in Table 6.2 are outside the scope of this book. The reader is referred to a variety of sources for this type of treatment [78, 80–86]. Note the special diagrammatic methods in Ref. 83.

6.8 THE SPIN HAMILTONIAN: A SUMMING UP

In the previous chapters, we focused on quantitative description of the Zeeman split-tings (g), nuclear hyperfine and quadrupolar effects (A, P), and the electronic quad-rupolar and high-spin interactions $(D, \ldots)$. In general, these parameters can all occur simultaneously for any given paramagnetic species. Thus, to describe the EPR spectra (as well as other spectra, e.g., ENDOR), it is necessary to add together all possible terms[17] into a single spin hamiltonian

$$\hat{\mathcal{H}} = g\beta_e\mathbf{B}^{\mathrm{T}} \cdot \hat{\mathbf{S}} + \hat{\mathbf{S}}^{\mathrm{T}} \cdot \mathbf{D} \cdot \hat{\mathbf{S}} + \sum_{i=1}^{N}(\hat{\mathbf{S}}^{\mathrm{T}} \cdot \mathbf{A}_i \cdot \hat{\mathbf{I}}_i - \beta_n\mathbf{B}^{\mathrm{T}} \cdot \mathbf{g}_{ni} \cdot \hat{\mathbf{I}}_i + \hat{\mathbf{I}}_i{}^{\mathrm{T}} \cdot \mathbf{P}_i \cdot \hat{\mathbf{I}}_i + \cdots)$$

$$(6.52)$$

encountered for any electron spin (composite or not) and N nuclear spins, all contributing to the spectrum. Which terms must be included to analyze any given spectrum is a matter of judgment and experience, added to an understand-ing of the chemical system being investigated. The correct spin hamiltonian yields the observed positions and relative intensities of the lines, in the absence of dynamic effects (Chapter 10). However, exact energies and transition probabilities obtained by parameter fitting (i.e., repeated diagonalization of $\hat{\mathcal{H}}$ by computer) are always possible, generally with some loss of intuitive under-standing. One important mental exercise regarding Eq. 6.52 is to note which terms depend on B and which do not (zero-field terms), since the choice of spectral region (*B* region) selected to work in is significantly affected by this aspect.

Let us now discuss the *signs* associated with any parameter matrix **Y**. The *relative* signs of the matrix elements Y_{ij} are all obtainable from the positions of the EPR lines. However, determination of the *absolute* sign (i.e., which of $\pm\mathbf{Y}$ is correct) may not always be possible. If several matrices (e.g., **g**, **D**, **A**, **P**) are important in the spin hamiltonian, the relative signs of some pairs may be available from the data, even though the absolute sign of the set is not. In some cases the relative signs of **A** and **P** for a particular nucleus may be fixed by the data, but not necessarily with respect to **A** and **P** of some other nucleus also exhibiting line splittings. Special auxiliary measurements may be needed (see Note 4 in Chapter 4 regarding matrix **g** and Section 6.3.1 for matrix **D**) to arrive at the most complete sign information. In summary, determination of the signs of spin-hamiltonian parameters is a non-trivial task.

An approach to visualization of the energy terms arising from a given $\hat{\mathcal{H}}$ is to resort to a perturbation treatment. We now undertake a brief discussion of the results of this approximation technique.

Consider a simplified version of Eq. 6.52, namely

$$\hat{\mathcal{H}} = g\beta_e \mathbf{B}^{\mathrm{T}} \cdot \hat{\mathbf{S}} + \hat{\mathbf{S}}^{\mathrm{T}} \cdot \mathbf{D} \cdot \hat{\mathbf{S}} + \hat{\mathbf{S}}^{\mathrm{T}} \cdot \mathbf{A} \cdot \hat{\mathbf{I}} \tag{6.53}$$

of Eq. 6.52, incorporating the now familiar terms from Chapters 4 and 5. For sufficiently large field *B*, the first term dominates (if $M_S \neq 0$) and the other two can be treated as perturbations [87, 88]. The resulting single-nucleus second-order energy expressions, valid for any *S* and *I*, and for any coordinate system are

$$U(M_S, M_I) = g\beta_e B M_S + \tfrac{1}{2} d_1 \left[3M_S^2 - S(S+1)\right] +$$

$$\frac{1}{2\,g\beta_e B}(d_2 - d_1{}^2)\left[8M_S^2 + 1 - 4S(S+1)\right]M_S +$$

$$\frac{1}{8\,g\beta_e B}\left[tr(\mathbf{D})^2 - 2d_2 + d_1{}^2 - 2d_{-1}\,det(\mathbf{D})\right]\left[2S(S+1) - 2M_S^2 - 1\right]M_S +$$

$$KM_S M_I + \frac{1}{2\,g\beta_e B}\left\{\tfrac{1}{2}\left[tr(\mathbf{A}\cdot\mathbf{A}^{\mathrm{T}}) - k_1{}^2\right]M_S\left[I(I+1) - M_I{}^2\right] - \cdot\right.$$

$$\frac{det(\mathbf{A})}{K}\left[S(S+1) - M_S^2\right]M_I + (k_1{}^2 - K^2)M_S\,M_I{}^2 +$$

$$\left. 2(e - d_1)K\left[3M_S^2 - S(S+1)\right]M_I\right\} \tag{6.54}$$

Here

$$g^2 = \mathbf{n}^T \cdot \mathbf{g} \cdot \mathbf{g}^T \cdot \mathbf{n} \tag{6.55a}$$

$$d_n = \mathbf{n}^T \cdot \mathbf{g} \cdot \mathbf{D}^n \cdot \mathbf{g}^T \cdot \mathbf{n}/g^2 \tag{6.55b}$$

$$K^2 = \mathbf{n}^T \cdot \mathbf{g} \cdot \mathbf{A} \cdot \mathbf{A}^T \cdot \mathbf{g}^T \cdot \mathbf{n}/g^2 \tag{6.55c}$$

$$k_1 = \mathbf{n}^T \cdot \mathbf{g} \cdot \mathbf{A} \cdot \mathbf{A}^T \cdot \mathbf{A} \cdot \mathbf{A}^T \cdot \mathbf{g}^T \cdot \mathbf{n}/g^2 K^2 \tag{6.55d}$$

$$e = \mathbf{n}^T \cdot \mathbf{g} \cdot \tfrac{1}{2}(\mathbf{D} \cdot \mathbf{A} \cdot \mathbf{A}^T + \mathbf{A}^T \cdot \mathbf{A} \cdot \mathbf{D}) \cdot \mathbf{g}^T \cdot \mathbf{n}/g^2 K^2 \tag{6.55e}$$

and, as before, $\mathbf{n} = \mathbf{B}/B$. Equation 6.55a is the same as Eq. 4.12. We note the presence in Eq. 6.54 of terms in B^{-1}; higher-order perturbation treatment would give further terms, in B^{-2}, and so on. The terms in d_{-1}, d_1, d_2, K, k_1 and e demonstrate the intermingling of the three spin-hamiltonian terms in Eq. 6.53 in arriving at the energy expression.

Obviously, from the energy expression 6.54, it is now possible to derive anisotropic spectroscopic frequencies $[U(M_S', M_I') - U(M_S, M_I)]/h$ to be compared with experiment (e.g., see Eqs. 6.34). Similar perturbation techniques are available to derive transition moments, that is, relative intensities [89].

When more than one spin-bearing nucleus is present, there are naturally hyperfine terms in Eq. 6.53, for each nucleus, with resulting energy terms in Eq. 6.54 [90]. Even though no terms coupling the nuclear spins directly have been added to the spin hamiltonian (since such interaction energies are negligibly small), pair-wise cross-terms that depend on M_{I_i} and M_{I_j} will nevertheless occur in Eq. 6.54; these arise from electron-mediated dipolar interactions. As usual, for each nucleus with $I > \frac{1}{2}$, quadrupole terms (Eq. 5.50) should be added to Eq. 6.53.

As more and more terms are included (e.g., nuclear-quadrupole energies [90]), the perturbation energy expressions become increasingly complicated and unwieldy. Clearly, also, the limited applicability of perturbation theory must be kept in mind. Nevertheless, Eq. 6.54 (or its variants) have proved very useful and yield important insights.

Finally, it should be pointed out that the general spin hamiltonian (Eq. 6.52), including all high-spin terms, can be formulated in a far more compact and mathematically elegant form, involving spherical-tensor operators [76,80–85,90]. Thus all the matrices ($\mathbf{g}, \mathbf{D}, \mathbf{A}_i, \mathbf{P}_i$) can readily be expressed in terms of the relevant expansion coefficients. The higher-order terms (not formulatable in terms of 3×3 matrices) are also easily presented. However, all too many different versions and notations regarding this operator structure appear in the literature.

6.9 MODELING THE SPIN-HAMILTONIAN PARAMETERS

Since the mid-1990s with advent of ever more capable rapid computing systems, the accurate modeling of all the magnetic resonance parameters (e.g., $\mathbf{g}, \mathbf{A}, \mathbf{P}, \mathbf{D}, \ldots$) has become feasible, and thus there is now quite a burgeoning literature on this topic.

We can cite the text *Calculation of NMR and EPR Parameters* [91], which presents a nice overview. As specific examples, we can indicate the density functional theoretical (DFT) calculation of g matrices, from first principles, say, for paramagnetic diatomic molecules and defect centers in crystalline quartz [92]. For the latter medium, the detailed structural configuration and the ^{17}O, ^{27}Al and ^{29}Si hyperfine parameter matrices for the famed center $[AlO_4]^0$ have been very successfully calculated [93,94]. Years of such work by Ziegler and his group [95–97], and others, has borne much fruit; for instance, their inclusion of spin orbit coupling has indicated the general occurrence not only of spin singlet-triplet mixing but also the resulting appearance of non-zero spin density on the atoms even of diamagnetic molecules. Simultaneous inclusion of higher-order spin-orbit and spin-polarization effects in a relativistic calculation of electronic g matrics were reported in 2005 [98].

REFERENCES

1. P. W. Atkins, R. S. Friedman, *Molecular Quantum Mechanics*, 4th ed., Oxford University Press, Oxford, U.K., 2005, pp. 225–227.

2. A. Bencini, D. Gatteschi, *EPR of Exchange Coupled Systems*, Springer, Berlin, Germany, 1990.

3. J. Owen, E. A. Harris, "Pair Spectra, Exchange Interactions", in *Electron Paramagnetic Resonance of Transition Ions*, S. Geschwind, Ed., Plenum Press, New York, NY, U.S.A.,1972, Chapter 6.

4. A. Abragam, B. Bleaney, *Electron Paramagnetic Resonance of Transition Ions*, Clarendon, Oxford, U.K., 1970, Section 9.3.

5. G. E. Pake, T. L. Estle, *The Principles of Electron Paramagnetic Resonance*, 2nd ed., Benjamin, Reading, MA, U.S.A., 1973, pp. 157–161.

6. N. W. Ashcroft, N. D. Mermin, *Solid State Physics*, Saunders, Philadelphia, PA, U.S.A., 1976, p. 680.

7. J. E. Harriman, *Theoretical Foundations of Electron Spin Resonance*, Academic Press, New York, NY, U.S.A., 1978, p. 191.

8. R. E. Coffman, A. Pezeshk, *J. Magn. Reson.*, **70**, 21 (1986).

9. P. L. Hall, B. R. Angel, J. P. E. Jones, *J. Magn. Reson.*, **15**, 64 (1974).

10. A. W. Hornig, J. S. Hyde, *Mol. Phys.*, **6**, 33 (1963).

11. H. Schuch, F. Seiff, R. Furrer, K. Möbius, K. P. Dinse, *Z. Naturforsch.*, **29a**, 1543 (1974).

12. Y. Yamaguchi, N. Sakamoto, *J. Phys. Soc. Jpn.*, **27**, 1444 (1969).

13. H. F. Hameka, L. J. Oosterhoff, *Mol. Phys.*, **1**, 358 (1958).

14. P. Kottis, R. Lefebvre, *J. Chem. Phys.*, **39**, 393 (1963), Eq. 6.

15. O. Claesson, A. Lund, *J. Magn. Reson.*, **41**, 106 (1980), Eq. 2.

16. T. Castner Jr., G. S. Newell, W. C. Holton, C. P. Slichter, *J. Chem. Phys.*, **32**(3), 668 (1960).

17. J. S. Griffith, *Mol. Phys.*, **8**, 213 (1964).

18. R. Aasa, T. Vänngård, *Ark. Kemi*, **24**(18), 331 (1965).

19. D. G. McGavin, W. C. Tennant, *J. Magn. Reson.*, **62**, 357 (1985).

20. R. Berger, J. Kliava, E. Yahiaoui, J.-C. Bissey, P. K. Zinsou, P. Béziade, *J. Non-Cryst. Solids*, **189**, 151 (1995).

21. M. S. de Groot, J. H. van der Waals, *Mol. Phys.*, **3**, 190 (1960).

22. J. H. van der Waals, M. S. de Groot, *Mol. Phys.*, **2**, 333 (1959).

23. P. Kottis, R. Lefebvre, *J. Chem. Phys.*, **41**, 379 (1964).

24. I. M. Brown, P. J. Phillips, D. Parikh, *Chem. Phys. Lett.*, **132**(3), 273 (1986).

25. W. A. Yager, E. Wasserman, R. M. R. Cramer, *J. Chem. Phys.*, **37**, 1148 (1962).

26. S. I. Weissman, *J. Chem. Phys.*, **29**, 1189 (1958).

27. M. S. de Groot, J. H. van der Waals, *Physica*, **29**, 1128 (1963).

28. J.-Ph. Grivet, J. Mispelter, *Mol. Phys.*, **27**, 15 (1974).

29. W. V. Sweeney, D. Coucouvanis, R. E. Coffman, *J. Chem. Phys.*, **59**, 369 (1973).

30. I. V. Ovchinnikov, V. N. Konstantinov, *J. Magn. Reson.*, **32**, 179 (1978).

31. Y. Teki, T. Takui, K. Itoh, *J. Chem. Phys.*, **88**, 6134 (1988).

32. A. M. Trozzolo, E. Wasserman, W. A. Yager, *J. Chim. Phys.*, **61**, 1663 (1964).

33. N. Hirota, S. I. Weissman, *J. Am. Chem. Soc.*, **86**, 2538 (1964).

34. N. Hirota, S. I. Weissman, *J. Am. Chem. Soc.*, **82**, 4424 (1960).

35. G. N. Lewis, D. Lipkin, T. T. Magel, *J. Am. Chem. Soc.*, **63**, 3005 (1941).

36. D. S. McClure, *J. Chem. Phys.*, **22**, 1668 (1954); **24**, 1 (1956).

37. C. A. Hutchison Jr., B. W. Mangum, *J. Chem. Phys.*, **34**, 908 (1961).

38. N. Hirota, C. A. Hutchison Jr., P. Palmer, *J. Chem. Phys.*, **40**, 3717 (1964).

39. M. S. de Groot, J. H. van der Waals, *Mol. Phys.*, **6**, 545 (1963).

40. M. S. de Groot, I. A. M. Hesselmann, J. H. van der Waals, *Mol. Phys.*, **16**, 45 (1969).

41. F. Gutmann, H. Keyzer, L. E. Lyons, *Organic Semiconductors*, Part B, Krieger, Malabar, FL, U.S.A., 1983, Chapters 4, 5 and 13.

42. N. Hirota, C. A. Hutchison Jr., *J. Chem. Phys.*, **42**, 2869 (1965).

43. J. H. van der Waals, *Int. Rev. Phys. Chem.*, **5**, 219 (1986).

44. W. Barendswaard, R. T. Weber, J. H. van der Waals, *J. Chem. Phys.*, **87**, 3731 (1987).

45. B. Henderson, *Br. J. Appl. Phys.*, **17**, 851 (1966).

46. B. D. Perlson, D. B. Russell, *J. Chem. Soc. Chem. Commun.*, 69 (1972); *Inorg. Chem.*, **14**, 2907 (1975).

47. C. A. Hutchison Jr., G. A. Pearson, *J. Chem. Phys.*, **47**, 520 (1967).

48. K. Komaguchi, M. Shiotani, A. Lund, *Chem. Phys. Lett.*, **265**, 217 (1997).

49. O. Claesson, A. Lund, T. Gillbro, T. Ichikawa, O. Edlund, H. Yoshida, *J. Chem. Phys.*, **72**, 1463 (1980).

50. M. Glasbeek, A. J. W. Visser, G. A. Maas, J. D. W. van Voorst, G. J. Hoytink, *Chem. Phys. Lett.*, **2**, 312 (1968).

51. R. E. Jesse, P. Biloen, R. Prins, J. D. W. van Voorst, G. J. Hoijtink, *Mol. Phys.*, **6**, 633 (1963).

52. G. R. Smith, W. Weltner Jr., *J. Chem. Phys.*, **62**, 4592 (1975).

53. R. H. D. Nuttall, J. A. Weil, *Can. J. Phys.*, **59**, 1886 (1981).

54. Y. Yamaguchi, *J. Phys. Soc. Jpn.*, **29**, 1163 (1970).

55. B. Bleaney, K. Bowers, *Proc. R. Soc. (London)*, **214**, 451–465 (1952).

56. H. Kumagai, H. Abe, J. Shimada, *Phys. Rev.* **87**, 385–387 (1952).

57. K. Itoh, H. Hayashi, S. Nagakura, *Mol. Phys.* **17**(6), 561–577 (1969).

58. A. G. Maryasov, Y. D. Tvsetkov, J. Raap, *Appl. Magn. Reson.* **14**, 101–113 (1998).

59. W. T. Borden, E. R. Davidson, *Ann. Rev. Phys. Chem.*, **30**, 125 (1979).

60. D. C. Reitz, S. I. Weissman, *J. Chem. Phys.*, **33**, 700 (1960).

61. G. R. Luckhurst, *Mol. Phys.*, **10**, 543 (1966).

62. R. M. Dupeyre, H. Lemaire, A. Rassat, *J. Am. Chem. Soc.*, **87**, 3771 (1965).

63. S. H. Glarum, J. H. Marshall, *J. Chem. Phys.*, **47**, 1374 (1967).

64. R. Brière, R. M. Dupeyre, H. Lemaire, C. Morat, A. Rassat, P. Rey, *Bull. Soc. Chim. France*, No. **11**, 3290 (1965).

65. A. Nakajima, H. Ohya-Nishiguishi, Y. Deguchi, *Bull. Chem. Soc. Jpn.*, **45**, 713 (1972).

66. G. Kothe, K.-H. Wassmer, A. Naujok, E. Ohmes, J. Rieser, K. Wallenfels, *J. Magn. Reson.*, **36**, 425 (1979).

67. I. Fujita, Y. Teki, T. Takui, T. Kinoshita, K. Itoh, *J. Am. Chem. Soc.*, **112**, 4074 (1990).

68. D. M. Lindsay, *J. Chem. Phys.*, **81**(7), 3356 (1984).

69. F. F. Carlson, *J. Chem. Phys.*, **39**(5), 1206 (1963).

70. N. Weiden, H. Käss, K.-P. Dinse, *J. Phys. Chem. B*, **103**, 9826 (1999).

71. E. B. Gordon, A. A. Pel'menev, O. F. Pugachev, V. V. Khmelenko, *Sov. J. Low Temp. Phys.*, **8**, 299 (1982).

72. H. Seidel, M. Schwoerer, D. Schmid, *Z. Phys.*, **182**, 398 (1965).

73. C. A. Baumann, R. J. Van Zee, S. V. Bhat, W. Weltner Jr., *J. Chem. Phys.*, **78**, 190 (1983).

74. B. Cage, S. E. Russek, S. Zipse, J. M. North, N. S. Dalal, *Appl. Phys. Lett.*, **87**, 82501 (2005).

75. S. Hill, S. Maccagnano, K. Park, E. M. Achey, J. M. North, N. S. Dalal, *Phys. Rev. B*, **65**, 224410 (2002).

76. B. Bleaney, *Proc. Phys. Soc. (London)*, **A73**, 939 (1959).

77. G. F. Koster, H. Statz, *Phys. Rev.*, **113**, 445 (1959).

78. D. G. McGavin, W. C. Tennant, J. A. Weil, *J. Magn. Reson.*, **87**, 92 (1990).

79. M. Sato, A. S. Rispin, H. Kon, *Chem. Phys.*, **18**, 211 (1976).

80. A. Abragam, M. H. L. Pryce, *Proc. Roy. Soc. (London)*, **A205**, 135 (1951).

81. H. A. Buckmaster, R. Chatterjee, Y. H. Shing, *Phys. Stat. Solidi*, **A13**, 9 (1972).

82. H. A. Buckmaster, R. Chatterjee, V. M. Malhotra, *J. Magn. Reson.*, **43**, 417 (1981).

83. R. Skinner, J. A. Weil, *J. Magn. Reson.*, **21**, 271 (1976); **29**, 223 (1978).

84. R. Calvo, *J. Magn. Reson.*, **26**, 445 (1977).

85. A. B. Roitsin, *Phys. Stat. Sol.*, **B104**, 11 (1981).

86. V. G. Grachëv, *Sov. Phys. JETP*, **65**, 1029 (1987).

87. J. A. Weil, *J. Magn. Reson.*, **18**, 113 (1975).

88. M. Iwasaki, *J. Magn. Reson.*, **16**, 417 (1974).

89. D. B. Fulton, R. Skinner, J. A. Weil, *J. Magn. Reson.*, **69**, 41 (1986); 82, 12 (1989).

90. M. J. Mombourquette, W. C. Tennant, J. A. Weil, *J. Chem. Phys.*, **85**, 68 (1986).

91. M. Kaupp, M. Buhl, V. G. Malkin, Eds., *Calculation of NMR, EPR Parameters*, Wiley-VCH, Darmstadt, Germany, 2004. Also see: I. Malkin, O. L. Malkina, V. G. Malkin, M. Kaupp, *J. Chem. Phys.*, **123**, 244103 (2005).

92. C. J. Pickard, F. Mauri, *Phys. Rev. Lett.*, **88**(8), 086403 (2002).

93. G. Pacchioni, F. Frigoli, D. Ricci, J. A. Weil, Phys. Rev. B, **63**, 054102 (2001).

94. J. To, A., A. Sokol, S. A. French, N. Kaltsoyannis, C. R. A. Catlow, *J. Chem. Phys.*, **122**, 144704 (2005).

95. S. Patchkovskii, T. Ziegler, *J. Chem. Phys.*, **111**(13), 5730 (1999).

96. S. Patchkovskii, T. Ziegler, *J. Phys. Chem.*, **105**, 5490 (2001).

97. S. K. Wolff, T. Ziegler, *J. Chem. Phys.*, **109**(3), 895 (1998).

98. I. Malkin, O. L. Malkina, V. G. Malkin, M. Kaupp, *J. Chem. Phys.*, **123**, 244103 (2005).

99. M. M. Abraham, J. L. Boldú, O., Y. Chen, "EPR and ENDOR of Trapped-Hole Centers in Alkaline-Earth Oxides", in *Electronic Magnetic Resonance of the Solid State*, J. A. Weil, Ed., Canadian Society for Chemistry, Ottawa, ON, Canada, 1987, pp. 427ff.

100. V. C. Sinclair, J. M. Robertson, A. McL. Mathieson, *Acta Crystallogr.*, **3**, 251 (1950).

101. D. Haarer, D. Schmid, H. C. Wolf, *Phys. Stat. Sol.*, **23**, 633 (1967).

102. D. Feller, W. T. Borden, E. R. Davidson, *J. Chem. Phys.*, **74**, 2256 (1981).

103. D. W. Wylie, A. J. Shuskus, C. G. Young, O. R. Gilliam, P. W. Levy, *Phys. Rev.*, **125**, 451 (1962).

104. J. Ullrich, A. Angerhofer, J. V. von Schutz, H. C. Wolf, *Chem. Phys. Lett.*, **140**, 416 (1987).

NOTES

1. Most species containing two unpaired electrons have many other electrons. It is assumed that the spin states of the two electrons are not affected by the other electrons that occupy molecular orbitals, two paired electrons in each.

2. Note the analogous behavior of the $S = I = \frac{1}{2}$ system (Section C.1).

3. The very words 'singlet' and 'triplet' imply approximate (but most useful) labels, since all two-electron states in reality are mixtures of both.

4. Some authors [5] use the notation $-2J\hat{\mathbf{S}}_1^{\mathrm{T}} \cdot \hat{\mathbf{S}}_2$, and others [6] use $-J\hat{\mathbf{S}}_1^{\mathrm{T}} \cdot \hat{\mathbf{S}}_2$. Thus care must be taken to maintain self-consistency, correct numerical values.

5. We note that exchange is just one manifestation of the coulomb interaction between the electrons and is related to their capability to form a chemical bond.

6. There also exists an electron-spin electron-spin contact interaction, analogous to the Fermi contact interaction that is the mechanism of isotropic hyperfine interaction (Section 2.3.3). However, the magnitude of this term is very small [7]. To the extent that it is present, it contributes to J_0.

7. Of course, there is a zero-field splitting arising from the exchange interaction, which splits the singlet and triplet states. Also in the presence of hyperfine interaction there is a zero-field splitting (Section C.1.3), but this is far smaller than that of the dipolar zero-field splittings found in most systems with $S \geq 1$.

8. For arbitrary orientations of **B**, these equations do not apply, since then none of the principal axes of **D** correspond to the axis of quantization.

9. We consider $d_1 > 0$. When it is negative, B_h and B_l must be interchanged in Eqs. 6.34a,b.

10. Here we have assumed that $D > 0$; for $D < 0$, the field expression must be reversed.

11. We saw earlier that, in simple cases (Figs. 4.7, 4.8, 5.11, and 5.12), the first-derivative peaks in powder patterns occur at field locations for which the line positions are at extrema, which occur with the field **B** along principal directions of matrix **g** or **A**. In more complex cases, the peaks in powder spectra show up at field orientations where the line positions $B(\theta, \phi)$ are nearly constant with respect to field orientation [29–31]. These zero-slope positions of B, called 'turning points', are not necessarily linked to principal-axis locations of individual spin hamiltonian parameter matrices.

12. When matrices **g** and **D** are *not* coaxial, for example, for low-symmetry triplet species, the powder spectra can be considerably more complicated with not easily recognized patterns.

13. Use of dilute single crystals instead of pure naphthalene greatly lengthens the lifetime of the excited triplet state in a particular molecule. In the pure crystal, rapid migration of the triplet excitation usually leads to effective quenching. This technique has been used in some fundamental studies of the excitation of oriented molecules by polarized light [36].

14. The number in parentheses indicates the error in the last place(s).

15. The closely related cation free radical TMM$^+$ also has been studied in detail by EPR [48].

16. Note that, in analogy with the hyperfine interaction term (Eq. C.1), the dot product must be retained for the exchange term.

17. What types and forms of terms are actually allowed to occur in a general $\hat{\mathcal{H}}$ is dictated by various physical principles. For a discussion of this, see Refs. 2, 78, and 80–86.

FURTHER READING

1. C. A. Hutchison Jr., "Magnetic Resonance Spectra of Organic Molecules in Triplet States in Single Crystals", in *The Triplet State*, A. B. Zahlan, Ed., Cambridge University Press, Cambridge, U.K., 1967, pp. 63–100.

2. S. P. McGlynn, T. Azumi, M. Kinoshita, *Molecular Spectroscopy of the Triplet State*, Prentice-Hall, Englewood Cliffs, NJ, U.S.A., 1969.

3. Yu. N. Molin, K. M. Salikhov, K. I. Zamaraev, *Spin Exchange–Principles and Applications in Chemistry and Biology*, Springer, Berlin, Germany, 1980.

4. J. Owen, E. A. Harris, "Pair Spectra, Exchange Interactions", in *Electron Paramagnetic Resonance*, S. Geschwind, Ed., Plenum Press, New York, NY, U.S.A., 1972, Chapter 6.

5. D. W. Pratt, "Magnetic Properties of Triplet States", in *Excited States*, Vol. 4, E. C. Lim, Ed., Academic Press, New York, NY, U.S.A., 1979, pp. 137–236.

6. M. Weissbluth, "Electron Spin Resonance in Molecular Triplet States", in *Molecular Biophysics*, B. Pullman, M. Weissbluth, Eds., Academic Press, New York, NY, U.S.A., 1965, pp. 205–238.

7. W. Weltner Jr., *Magnetic Atoms and Molecules*, Van Nostrand Reinhold, New York, NY, U.S.A., 1983, Chapters 3 and 5.

8. M. Baumgarten, in *EPR of Free Radicals in Solids*, A. Lund, M. Shiotane, Eds., Kluwer, Dordrecht, Netherlands, 2003, Chapter 12.

PROBLEMS

6.1 By substitution of the appropriate spin matrices, derive the spin hamiltonian matrix of Eq. 6.20. Express this in terms of D and E.

6.2 (*a*) Obtain the spin matrices $\mathbf{S}_x$, $\mathbf{S}_y$, $\mathbf{S}_z$, $\mathbf{S}_x^2$, $\mathbf{S}_y^2$ and $\mathbf{S}_z^2$ using the triplet-state eigenfunctions given in Eqs. 6.28 as a basis set. (*b*) Use these 3×3 spin matrices to obtain the spin-hamiltonian matrix 6.29.

6.3 Show that the isotropic part of an electronic quadrupole matrix $\mathbf{D}$ affects all spin levels (states $|M_S\rangle$) equally, so that it generally can be ignored, since it cannot be measured spectroscopically.

6.4 Show that the $\hat{S}_z$ operator causes a transition between the spin states $c_2|-1\rangle - c_1|+1\rangle$ and $-i[c_1^*|-1\rangle + c_2^*|+1\rangle]$. Compare the intensity (at any field B) of this transition with that obtained using the $\hat{S}_x$ operator in the basis in which the spin hamiltonian matrix is diagonal (*Hint*: See Section C.1.4.). Thus justify the statements made in Section 6.3.2 concerning the relative orientation of the static and excitation magnetic fields required for observation of the '$\Delta M_S = \pm 2$' transition.

6.5 Use the procedure outlined in Section 6.3.1 to extract the matrix $\mathbf{D}$ for the naphthalene triplet in a single crystal of durene from the angular-dependence curves given in Fig. 6.10. Diagonalize $\mathbf{D}$ and obtain the values of D and E.

6.6 (*a*) Derive Eqs. 6.32 from Eqs. 6.31, and their analogs for $\mathbf{B} \parallel \mathbf{X}$ and for $\mathbf{B} \parallel \mathbf{Y}$.

  (*b*) Use the magnetic-field positions and microwave frequency from Fig. 6.4 to obtain D and E for the lowest triplet state of naphthalene, using Eqs. 6.32.

6.7 Show that on crystal rotation, in the case of uniaxial symmetry and at sufficiently high frequency v, the maximum field spacing between the $\Delta M_S = \pm 1$ transitions (for $S = 1$, $g = g_e$) is given by $\Delta B = 2|D|/g_e\beta_e = (3\,\mu_0/4\pi)g_e\beta_e \langle r^{-3}\rangle = 5570.85\,\langle r^{-3}\rangle$ with B in mT and r in Å.

6.8 Derive the matrix $\mathbf{D}$ for the F_t center in MgO, using the information given in the text and in Fig. 6.13.

6.9 The V^0 center produced by x irradiation of MgO is an example of an $S = 1$ system in a local electric field of tetragonal symmetry. This defect consists of two positive holes (missing electrons) on opposite sides of a positive-ion vacancy, that is, the array $O^-\Box\;O^-$ instead of $O^{2-}\Box\;O^{2-}$. The spin hamiltonian for this system is Eq. 6.19*b*. For the case of $\mathbf{B} \parallel \mathbf{Z}$, where $\mathbf{Z}$ is the tetragonal axis $\langle 001 \rangle$ of the defect, the energy-level scheme and the allowed transitions for a system of this type are depicted in Fig. 6.3. The NMR *proton* resonance frequencies at the magnetic fields corresponding to the two transitions shown occur at 13.3345 and 15.2680 MHz; $\nu = 9.4174$ GHz.

(a) Calculate the energies of the states in zero field and from these obtain the zero-field splitting.

(b) Write expressions for the energies of the two transitions.

(c) From these, find the value of $\overline{D}$.

(d) Obtain the value of g (here $g_\parallel$).

(e) What feature of the spectrum could prove that in zero field the $M_S = +1$ states lie below the $M_S = 0$ state, rather than above it?

(f) The separation of the pair of lines is given approximately by

$$\Delta B = (\mu_0/4\pi)(3\mu^2/2g_e\beta_e)\langle r^{-3}\rangle |3\cos^2\theta - 1|$$

as one expects for interacting dipoles aligned by field **B**. From this, calculate $\langle r^{-3}\rangle$ and hence estimate the separation of the two dipoles of spin $\frac{1}{2}$ [99]. (The magnetic moment of a hole has the same absolute magnitude as that of the electron.)

6.10 The zero-field splitting parameters for the triplet exciton in anthracene are given as $\overline{D} = -0.0058$ and $\overline{E} = 0.0327 \text{cm}^{-1}$. The low value of $\overline{D}$ is deceptive, since (for this crystal-axis system) $\overline{E} > |\overline{D}|$. After ascertaining the direction cosines of the axes of the anthracene molecules relative to the crystal axes [100], show that the parameters ascribable to the individual molecules are $\overline{D} = 0.0688$ and $\overline{E} = -0.0081 \text{cm}^{-1}$ [101].

6.11 Consider equations 6.36. In that context, can you justify the statement 'There will be exactly four **B**-field orientations at which the lowest possible half-field line position $B_{\min}$ of a given triplet-state species occurs, if it occurs at all'? If so, do so and provide a critique citing at least five conditions that must be met for the statement to be valid.

6.12 Consider the proton hyperfine structure in the X-band EPR spectrum of triplet trimethylenemethane.

(a) Justify that 4096 hyperfine lines are to be expected for each of the two primary ($\Delta M_S = \pm 1$) transitions.

(b) The isotropic proton splitting (in mT) is given (Chapter 9) by

$$a_\text{H} \approx -2.46\rho_\pi{}^a - 0.19\rho_\pi{}^c \tag{6.56}$$

at least approximately. Here the superindices a and c denote the adjacent and central carbon atoms. Using the relation

$$3\rho_\pi{}^a + \rho_\pi{}^c \approx 1 \tag{6.57}$$

estimate the unpaired-electron populations ρ_π in the molecule [102].

6.13 Nitrogen atoms trapped at 77 K in irradiated potassium azide exhibit an EPR spectrum reproduced by the spin hamiltonian

$$\hat{\mathcal{H}} = \beta_e \mathbf{B}^{\mathrm{T}} \cdot \mathbf{g} \cdot \hat{\mathbf{S}} + D\hat{S}_Z^2 + E(\hat{S}_X^2 - \hat{S}_Y^2) + \hat{\mathbf{S}}^{\mathrm{T}} \cdot \mathbf{A} \cdot \hat{\mathbf{I}} \qquad (6.58)$$

with $g = 2.001$, $\overline{D} = +0.0143 \text{ cm}^{-1}$, $\overline{E} = -0.00199 \text{ cm}^{-1}$ and $\overline{A} = 0.00051 \text{ cm}^{-1}$ [103]. Parameters g and $\overline{A}$ are isotropic. What is the spin S, and why? Draw semi-quantitatively the EPR spectrum expected at 9.2 GHz for the field orientation yielding the largest fine-structure splitting, including a field scale (mT).

6.14 The application of a scannable radiofrequency field $\mathbf{B}_1$ to a sample containing molecules in a photo-excited triplet state can yield zero-field transitions detected directly in the transmitted light intensity. Such absorption-detected magnetic resonance in a photosynthetic reaction center at 12 K has yielded relatively strong absorptions in the 890 nm light, at frequencies $\nu = 466$ and 658 MHz, and a weaker one at 191 MHz [104]. Using Fig. 6.3, estimate values of $|D|/h$ and $|E|/h$ from these data.

CHAPTER 7

PARAMAGNETIC SPECIES IN THE GAS PHASE

7.1 INTRODUCTION

By definition, atoms and molecules observed in the gas phase differ from those in condensed phases in that they are almost perfectly free to perform translational motion. Since there is no observable energy splitting caused by such motion (i.e., of the center of mass of the species), there are no direct spectroscopic consequences. On the other hand, literally free molecular rotation does allow observation of the quantized rotational energy levels; their splittings are often of the same magnitude as those of the Zeeman spin states. The ensuing rotation-magnetic interactions have major effects on the EPR spectra of diatomic and polyatomic molecules.

Angular momenta remain in the forefront for gas-phase systems when understanding of the EPR transitions is to be attained. There are four sources for these: electronic (total) orbital and spin, rotation of the nuclear framework and nuclear spin(s). Here the set of total angular-momentum vectors $\hat{\mathbf{F}}$ of the atoms or molecules can be assumed to be randomly oriented, but each is fixed in its direction until disturbed by a collision (a relatively rare event on the EPR time scale in most studies). More accurately, the quantum number M_F remains constant but, in fact, cannot be measured until an external magnetic field is applied, that is, until a quantization direction is specified.

Collisions with the walls of the container or between atoms or molecules may cause appreciable effects on relaxation times and linewidths. Literally no electrically charged species have been observed in the gas phase by EPR,[1] mostly because large concentrations of ions, sufficient to allow standard EPR experiments, have not been attainable.

Electron Paramagnetic Resonance, Second Edition, by John A. Weil and James R. Bolton
Copyright © 2007 John Wiley & Sons, Inc.

208

From the multiplicity and locations of the lines observed for a paramagnetic gas-phase sample in the EPR spectrometer, one can obtain detailed information about many interesting molecular parameters. Furthermore, both qualitative and quantitative chemical analyses are relatively easy and can be carried out as a function of time to enable studies of reaction kinetics.

For molecules with electric dipoles, the spectrum generally contains lines of the electric-dipole transition type, rather than only the relatively weak magnetic-dipole absorptions. Indeed, in most of the literature studies, the observed spectrum consists of electric-dipole lines. However, specially designed magnetic-resonance cavities can be used so as to confine the molecules to regions where essentially there are only excitation fields of the magnetic type ($\mathbf{B}_1$) and none of the electric type ($\mathbf{E}_1$). We note that the term 'EPR' does not necessarily imply the presence of magnetic-dipole transitions, but rather applies to the resonance spectroscopy of paramagnetic species.

We consider primarily species with one or more unpaired electrons, for example, doublet and triplet states; however, species without net electron spin, but that exhibit electronic orbital magnetism, also give EPR spectra. We discuss monatomic species first, followed by diatomics and then by simple polyatomics. Virtually no work has been done on more complicated gas-phase molecules, due in part to the complexity of the rotational-magnetic patterns.

7.2 MONATOMIC GAS-PHASE SPECIES

Open-shell atoms, since they are generally unstable toward dimerization, are usually created in steady-state concentrations from suitable molecules by means of an electric discharge, thermal dissociation, electron bombardment, or photolysis; these atoms usually are created externally and subsequently diffused into the EPR cavity. Obviously, the prime example of this type of species is the hydrogen atom, in any of its three isotopic varieties. Details of the relevant spin hamiltonian and the consequent spectroscopic implications have been discussed in Sections 2.4 and 2.5, as well as in Appendix C. The atomic parameters (g, A, ...) are now well known, at least in the ground and lower excited electronic states. Much of the more recent EPR work involves chemical reactions of atomic hydrogen. Many other hydrogenic atoms (e.g., the alkali-atom species) are also well known. However, most studies of these species have been made by means of atomic beams, detecting the arrival of the atoms at some location, rather than by standard EPR techniques. Two-electron species, such as helium atoms excited into triplet states, have also been studied by such techniques [1].

Of special chemical interest are the various EPR investigations of atomic species, including O, S, Se, Te, N, P, As, Sb, F, Cl, Br, I and Ar. For some of these entities, several electronic states (e.g., $^4S_{3/2}$, $^2D_{5/2}$, $^2D_{3/2}$ of the ground-state configuration of nitrogen [2], including direct measurement of the $^2D_{5/2}$-$^2D_{3/2}$ fine-structure interval by use of far-infrared laser electronic paramagnetic resonance (LEPR) spectroscopy [3]) have been measured.

For the halogen atoms, the single hole in the p electronic shell leads to spin-$\frac{1}{2}$ species; the ground state is labeled $^2P_{3/2}$.[2] Since $J = \frac{3}{2}$, the allowed high-field (B) transitions occur between M_J values $-\frac{3}{2} \leftrightarrow -\frac{1}{2}$, $-\frac{1}{2} \leftrightarrow +\frac{1}{2}$ and $+\frac{1}{2} \leftrightarrow +\frac{3}{2}$. Each such absorption exhibits $2I + 1$ hyperfine peaks. Experiments dealing with transition-group species appear to be lacking.

Note that we are dealing with spherically symmetric systems here; that is, the unpaired electron responds to the central coulomb field of the nucleus and the other electrons. Thus the electronic orbital angular momentum about the nucleus is important. The orbital operator $\hat{\mathbf{L}}$ is added to the spin operator $\hat{\mathbf{S}}$ to give the total electronic angular-momentum operator $\hat{\mathbf{J}}$; hence the associated quantum numbers L, S, J and M_J are defined and serve to label the energy states (Section B.7), when nuclear hyperfine and quadrupole splittings are ignored.

Atomic fluorine can be produced by means of an electric discharge [4], and consists of a single isotope (^{19}F) having a nuclear spin of $I = \frac{1}{2}$. Thus here (and above) the true total angular momentum $\hat{\mathbf{F}} = \hat{\mathbf{J}} + \hat{\mathbf{I}}$ (Section B.7) must be invoked. Its quantum number F is integral and, for given J, takes on $2I + 1$ values, ranging from $|J - I|$ to $J + I$. The corresponding Zeeman levels for the ground-state term are shown in Fig. 7.1, as are the 'allowed' transitions observed with $\mathbf{B}_1 \perp \mathbf{B}$ in a

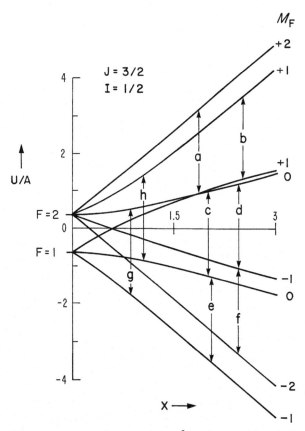

FIGURE 7.1 Zeeman pattern of the ground-state $^2P_{3/2}$ levels of atomic fluorine, showing possible EPR transitions at fixed frequency. Here the field variable is $x = (g_J - g_n)\beta_e B / A$, where A is the zero-field (hyperfine) splitting energy. [After H. E. Radford, V. W. Hughes, V. Beltrán-López, *Phys. Rev.*, **123**, 153 (1961).]

fixed-frequency EPR experiment. The electronic g factor is, to first approximation, given by the Landé formula

$$g = 1 + [J(J + 1) - L(L + 1) + S(S + 1)]/[2J(J + 1)] \qquad (7.1)$$

generally applicable to non-relativistic atoms [5, p. 245]. For instance, g ($J = \frac{1}{2}$) was found to be 0.66561(3). Two-photon EPR lines, from transitions involving three spin-energy levels, have been observed in this system [6]. Transitions of type $\Delta M_J = \pm 1$, $\Delta M_I = \pm 1$, as well as the primary EPR jumps of type $\Delta M_J = \pm 1$, $\Delta M_I = 0$, have been observed for all the usual halogen atoms in their ground states, $^2P_{3/2}$ [7].

Atomic iodine has been studied in its first excited state, $^2P_{1/2}$ [8]. The iodine atoms were produced from dissociation of I_2 by the reaction

$$I_2(^1\Sigma_g{}^+) + O_2(^1\Delta) \rightarrow 2I(^2P_{3/2}) + O_2(^3\Sigma) \qquad (7.2)$$

The excited state is reached via the rapid equilibrium

$$I(^2P_{3/2}) + O_2(^1\Delta) \rightleftharpoons I(^2P_{1/2}) + O_2(^3\Sigma) \qquad (7.3)$$

The singlet dioxygen was generated chemically. All except the diamagnetic species (ground-state I_2) are observable by EPR; their relative concentrations are obtainable by double integration of the first-derivative lines. Thus the equilibrium constant $K = 2.9$ for the latter reaction at 295 K can be derived from EPR measurements. The measured $g = 0.6664$ for the excited iodine atom is within 0.2% of that given by the Landé formula (Eq. 7.1).

It is of some interest to discuss EPR intensities and relaxation times of the gas-phase species. Clearly, to attain adequate signal sizes in such rarefied samples, one must maximize the effective sample volume. This implies a need for particularly good magnetic-field homogeneity. For atomic hydrogen, various mechanisms affecting the line breadth (i.e., lifetime broadening, Doppler plus collision broadening and narrowing, and electron exchange between atoms) have been discussed in considerable detail [9]. The relatively long relaxation times τ_1 for S-state atoms tend to lead to narrow EPR lines; the non-zero orbital angular momentum in all other atoms allows much more efficient relaxation via collisions and hence to serious line broadening.

7.3 DIATOMIC GAS-PHASE SPECIES

Perhaps the most important paramagnetic diatomic molecule is dioxygen, due in part to the crucial part it plays in the biosphere. Its microwave absorption spectrum was of early and special interest during the development of radar usage in the 1940s.

The electronic ground state of O_2 is conventionally symbolized by $^3\Sigma_g{}^-$ [3]. For our purposes, it is important that there is (nominally) no electronic orbital angular momentum (similar to the ground state of H). The two unpaired electrons form a triplet state (consistent with Hund's rules [5, p. 240]). We also note that no permanent electric-dipole moment is present, since there is a center of symmetry. The EPR transitions are all of the magnetic-dipole type and thus are of much lower intensity than those in molecules where electric-dipole transitions are possible. There are three isotopes of interest: ^{16}O (99.759%, $I = 0$), ^{17}O (0.037%, $I = \frac{5}{2}$) and ^{18}O (0.204%, $I = 0$).

The gas-phase O_2 molecules undergo end-over-end tumbling, with quantized rotational angular momentum described by a suitable spatial operator $\hat{N}$. Thus the total angular-momentum operator is $\hat{J} = \hat{N} + \hat{S}$ (excluding nuclear spin that enters only when ^{17}O is present). Quantum-statistical considerations [5, p. 295] dictate that for the most abundant entity, $^{16}O^{16}O$, the rotational quantum number N can have only odd values. Hence N is never zero, and, since the energy is proportional to $N(N + 1)$, some rotation is always present.

If the spin behavior were independent of the rotational status, the triplet-state EPR features discussed in Chapter 6 would be present.[4] However, in practice, the rotation of the component charges of the molecule gives rise to magnetic fields that interact with the spin magnetic moment. Thus the spin-energy triplets and the infinite ladder of rotational levels are inextricably interwoven. As a result, the EPR spectrum consists of a virtually infinite (but countable) set of lines [11,12]. A portion of an X-band EPR spectrum of O_2 is shown in Fig. 7.2. The corresponding line positions are indicated in Table 7.1. The relative intensities depend, among other factors, on the population of the states, that is, on the temperature of the gas [13].

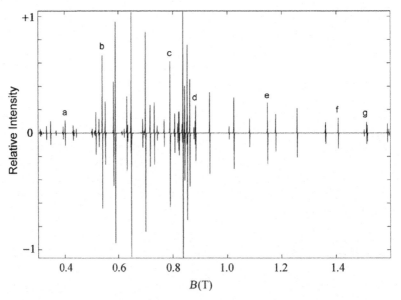

FIGURE 7.2   Portion (0.3–1.6) of a simulated field-swept EPR spectrum (9.14456 GHz) of gaseous dioxygen $^{16}O^{16}O$ at 100 K (∼0.1 torr), with limit $N < 15$. Selected line positions and relevant quantum numbers are given in Table 7.1. [Spectrum produced by Dr. S. Stoll, ETH, Zürich; unpublished data. The parameters were taken from Ref. 15.]

TABLE 7.1 Selected EPR Lines of $^{16}O^{16}O$ at $\nu = 9.14456$ GHz, T = 100 K (see Fig. 7.2)

Line	B (T)	N	J_1, M_{J_1}	$\longleftrightarrow$	J_2, M_{J_2}
a	0.401812	5	6, −3		4, −2
b	0.540939	1	1, −1		1, 0
c	0.790107	5	4, 0		6, +1
d	0.862578	7	6, 0		6, +1
e	1.148942	3	4, +2		4, +3
f	1.407802	7	7, −7		7, −6
g	1.512384	5	6, +4		6, +5

Since the spin and rotational energies are not independent, spin relaxation can take place via changes in the tumbling; this in turn is sensitive to the collisions experienced by the molecule. Such behavior should be compared with the situation for atomic species described in Section 7.2. As a result, the linewidths depend on pressure, but the lines are sharp (Fig. 7.2) under attainable conditions (e.g., 0.01 mT for pressures of ∼0.2 torr).

Thanks to its stability, dioxygen is very useful as a comparison standard for measurement of concentrations of other gas-phase free radicals [14].

The symmetry of the O_2 molecule, of course, gives rise to uniaxiality in the parameter matrices needed to describe its EPR spectrum. It is important to realize that these parameters are not averaged by the molecular tumbling, unlike the situation in liquids. Thus, for example, both $g_\perp$ and $g_\parallel$ are measurable. When intermolecular and wall collisions are negligible, gas-phase molecules do not tumble incoherently; the orientation of each molecule's *total* angular-momentum vector is random but does not vary with time. When a constant external magnetic field **B** is introduced, each $\hat{\mathbf{J}}$ is quantized along a specific direction. This is along an effective field differing slightly from that of **B** because of the g anisotropy, which arises from spin-orbit and spin-rotation coupling.

The Zeeman part of the relevant hamiltonian may be written

$$\hat{\mathcal{H}}_Z = \beta_e[g_\perp \hat{\mathbf{S}}^T \cdot \mathbf{B} + (g_z - g_\perp)\hat{S}_z B_z + g_{rot}\hat{\mathbf{N}}^T \cdot \mathbf{B}] \tag{7.4}$$

where g_{rot} is a g-type parameter associated with the rotational magnetic moment, while $g_\perp$ and $g_z = g_\parallel$ are associated with the electron spin. The z direction is along the internuclear axis. The electronic g factors are given (Eqs. 4.38 and 4.41) approximately by $g_z = g_e$ and

$$g_\perp = g_e - 2\lambda \sum_{n \neq G} \frac{\langle G | \hat{L}_x | n \rangle \langle n | \hat{L}_x | G \rangle}{U_n^{(0)} - U_G^{(0)}} \tag{7.5}$$

Here λ is the molecular spin-orbit coupling parameter, n labels the electronic states and G denotes the orbitally non-degenerate (spatial) ground state.

Experimentally, it is observed that g_z is slightly less than g_e (i.e., $g_z - g_e \approx 10^{-4}$), due to relativistic effects. The spin-orbit term in $g_\perp$ brings in the dependence on the internuclear distance caused by mixing in of excited states (e.g., Π states) other than of the Σ type. The effect is small: $g_\perp - g_e \approx 3 \times 10^{-3}$, presumably due to the small magnitude of λ. The rotational g factor is very small ($g_{\text{rot}} \approx 10^{-4}$); that is, the magnetic moment associated with $\mathbf{N}$ is almost negligible for low rotational states.

The primary zero-field hamiltonian describing the rotational energies has the form

$$\mathcal{H}_0 = B_{\text{rot}} \hat{\mathbf{N}}^{\text{T}} \cdot \hat{\mathbf{N}} + D(\hat{S}_Z^2 - \tfrac{1}{3}\hat{S}^2) + \mu_{\text{rot}} \hat{\mathbf{N}} \cdot \hat{\mathbf{S}} \tag{7.6}$$

The first term on the right describes the rigid-rotor model, the second term has the form of the familiar electron spin-spin interaction (but also includes spin-orbit effects), and the third covers the rotational-spin magnetic coupling. For greater accuracy, other (small) terms involving $\hat{\mathbf{N}}^{\text{T}} \cdot \hat{\mathbf{N}}$ should be included in Eq. 7.6. Since $\hat{\mathbf{N}}$ is a spatial operator (containing first derivatives with respect to angular coordinates describing the molecular orientation), operator $\hat{\mathcal{H}}_0$ should not be regarded as a spin hamiltonian. Solutions of the matrices $\hat{\mathbf{H}}_0 + \hat{\mathbf{H}}_Z$, attainable via computer diagonalization, lead to the energy levels as a function of applied field B, as typified by Fig. 7.3. These energies can be analyzed in terms of rotational spin triplets labeled $J = N - 1, N, N + 1$. The EPR transitions are primarily of type $|\Delta M_J| = 1$. When required, terms describing nuclear-hyperfine and quadrupole energies must be added to Eq. 7.6.

Studies of O_2 enriched with ^{17}O ($I = \tfrac{5}{2}$) have led to detailed understanding of the interaction between the magnetic moments of the molecule and the nuclear-spin

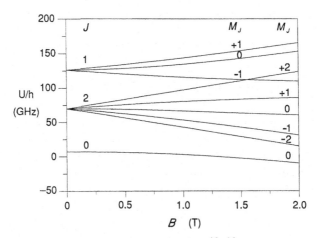

FIGURE 7.3 Zeeman splitting of the energy levels of $^{16}O^{16}O$ in its lowest rotational state ($N = 1$). One of the EPR transitions within this manifold is indicated (labeled b) in Fig. 7.2 and Table 7.1 for $\nu = 9.14456$ GHz. (Spectrum taken by S. M. Manley, J. A. Weil; unpublished data.)

moment; also, the local electric-field gradient is obtained via measurement by EPR of the nuclear-quadrupole parameter [15]. Small differences in hamiltonian parameters arise from the mass differences among the isotopes ^{16}O, ^{17}O and ^{18}O.

To this point, we have ignored the possibility of vibrational excitation, considering all molecules to be in their zero-point vibrational states labeled by quantum number $v = 0$. By utilization of the microwave discharge technique, O_2 in its $v = 1$ state has in fact been prepared and measured by EPR [16]. It should be noted that the parameters in $\hat{\mathcal{H}}_0$ and $\hat{\mathcal{H}}_Z$ are in fact functions of v, and can be assumed to be series expansions in the quantity $(v + \frac{1}{2})^n$ with $v = 0,1,2,\ldots$.

As was already indicated, molecular oxygen can also be observed by EPR when it is in its electronic excited state $^1\Delta_g$.[3] This species is metastable relative to the ground state since the conversion between a singlet and a triplet spin state is slow because it is spin-forbidden. Here the paramagnetism arises entirely from the electron orbital motions, with no spin component. Thus the total angular-momentum operator is the sum of the projection of $\hat{L}$ onto figure axis Z and of the rotation vector (which is normal to Z). EPR spectra have been measured and analyzed for states $J = 2$ and 3, including ^{17}O hyperfine effects [17,18]. Note that here, and in various other non-Σ free radicals, the orbital angular momentum exerts its maximal magnetic effects. This is unlike the situation with such species in condensed phase where the local electric fields tend to quench the angular momenta more or less completely.

EPR studies of S_2, OS, OSe and FN, molecules that are valence-isoelectronic with O_2, have been reported. In S_2, since the atoms are heavier than those in O_2, the spin-orbit coupling effects are appreciably more significant. Parameter D in Eq. 7.6 is larger, as is the difference $g_Z - g_\perp$ in Eq. 7.4. For the heteronuclear molecules, the electric-dipole transitions, spanning different rotational levels (e.g., $\Delta N \neq 0$, $|\Delta M_J| = 1$), predominate.

As another example, consider the NO molecule (which is isoelectronic with O_2^-). The unpaired electron resides in a π molecular orbital;[5] hence $\Lambda = \pm 1$ and $\Sigma = \pm\frac{1}{2}$.[3] Thus $\Omega = \Lambda + \Sigma = \frac{1}{2}$ and $\frac{3}{2}$. These two states ($^{2|\Sigma|+1}\Pi_\Omega$), non-degenerate because of spin-orbit coupling ($\sim\lambda|\Lambda\Sigma|$), are characterized as $^2\Pi_{1/2}$ and $^2\Pi_{3/2}$. In NO, they are separated by $\Delta U/hc = 124\ \mathrm{cm}^{-1}$, with $^2\Pi_{1/2}$ as the ground state. The rotational spacing is $\sim 5\ \mathrm{cm}^{-1}$.

The application of a magnetic field causes M_J splitting $g\beta_e B$ of the levels obtainable from the appropriate hamiltonian (Zeeman plus electron orbital and spin, rotational, and interaction terms between these) [19]. The molecular g factor is given [20] to good approximation by the relation

$$g = \frac{(\Lambda + 2\Sigma)(\Lambda + \Sigma)}{J(J + 1)} \tag{7.7}$$

analogous to that for the Landé g factor, Eq. 7.1. For the $^2\Pi_{1/2}$ state ($\Lambda = +1$, $\Sigma = -\frac{1}{2}$), we see that the spin and orbital magnetic moments virtually cancel each other, so that $g \approx 0$, resulting in a 'non-magnetic' ground state [5, p. 393], even though there is an unpaired electron. Note that this is in contradistinction to $^2P_{1/2}$ states in atoms (see Eq. 7.1). The $^2\Pi_{3/2}$ state ($\Lambda = +1$, $\Sigma = +\frac{1}{2}$) has a strong magnetic moment and gives rise to intense EPR signals at temperatures $T \leq \Delta U/k_b = 177$ K. Here several of the lowest rotational levels are appreciably populated, but only vibrational ground-state molecules are present. Molecular rotation admixes the two electronic states.

The total angular momentum (excluding nuclear spin) is defined to be $\hat{\mathbf{J}} = \hat{\mathbf{\Lambda}} + \hat{\mathbf{R}} + \hat{\mathbf{S}}$, where $\hat{\mathbf{\Lambda}}$ points along the molecular axis. Here $\hat{\mathbf{R}}$ is the operator of molecular rotation (if present) perpendicular to $\hat{\mathbf{\Lambda}}$. Quantum number N ($= 1, 2,$ $3, \ldots$) corresponds to the allowed values of $\hat{\mathbf{\Lambda}} + \hat{\mathbf{R}}$. The total angular-momentum quantum number J runs from $|\Sigma - N|$ to $\Sigma + N$ for given N. For example, consider the $J = \frac{3}{2}$ manifold of $^2\Pi_{3/2}$, which consists of a quartet of states labeled by M_J (the component of $\hat{\mathbf{J}}$ along $\mathbf{B}$, a *space-fixed* axis) $= +\frac{3}{2}, +\frac{1}{2}, -\frac{1}{2}, -\frac{3}{2}$. From Eq. 7.7, we see that $g \approx \frac{4}{5}$. Similarly, there are six states for $J = \frac{5}{2}$, and $g \approx \frac{12}{35}$.

The energy splittings of two rotational states of NO ($^2\Pi_{3/2}$) as a function of magnetic field are illustrated in Fig. 7.4. The three (five) allowed transitions for $J = \frac{3}{2}(\frac{5}{2})$ correspond to $\Delta M_J = \pm 1$. In fixed-frequency EPR they occur at somewhat different magnetic fields as a result of a second-order Zeeman interaction term connecting the $J = \frac{3}{2}$ and $\frac{5}{2}$ states. As J increases, g becomes progressively smaller (Eq. 7.7), and hence transitions within these states are observed at increasingly high magnetic fields (or lower microwave frequencies).

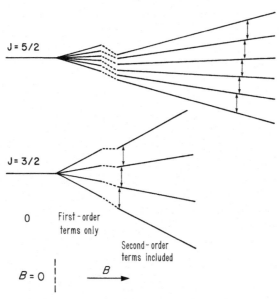

FIGURE 7.4   Splitting of the $J = \frac{3}{2}$ and $J = \frac{5}{2}$ states of the $^2\Pi_{3/2}$ manifold of NO as a function of the applied magnetic field. The displacements of the first-order Zeeman levels by a second-order interaction are shown by the dashed lines. The Λ-type doubling and hyperfine splittings are not included.

Since $\Lambda \neq 0$ in NO, there is a latent residual orbital degeneracy that has not been shown in Fig. 7.4. In fact, the magnetic interaction associated with the orbital and the rotational angular momenta leads to a removal of the Λ degeneracy [16]. The splitting ('Λ-type doubling') of the states labeled ' $+$ ' and ' $-$ ', increases with increasing J. The lifting of the degeneracy makes possible four transitions for each line. Of these, two ($\pm \leftrightarrow \mp$) are induced by the *electric* component of the microwaves and two ($\pm \leftrightarrow \mp$, degenerate) by the *magnetic* component. Since electric-dipole transitions are approximately 100–1000 times more intense than magnetic-dipole transitions, the former are the more easily observed. Thus the experimental arrangement must be such that the gas molecule can be exposed to regions of the cavity where the excitation field E_1 is large. For this purpose a cylindrical TE_{011} cavity (Appendix E) can be used [21].

The EPR spectra of NO also exhibit hyperfine splittings. Figure 7.5 exhibits separate EPR spectra of $^{14}N^{16}O$ and $^{15}N^{16}O$ for both the $J = \frac{3}{2}$ and the $J = \frac{5}{2}$ states of $^2\Pi_{3/2}$. Since $I = 1$ for ^{14}N and $I = \frac{1}{2}$ for ^{15}N, and the hyperfine splittings are large compared to the Λ splitting, one observes two and three sets of the three transitions shown for the $J = \frac{3}{2}$ states. Each is doubled by the removal of the Λ degeneracy. The isotropic and anisotropic parts of the hyperfine-splitting matrix are both obtainable from the gas-phase EPR measurements. Interpretation of the $J = \frac{5}{2}$ spectrum is left as Problem 7.6.

Many other heteronuclear diatomic molecules have been studied by EPR methods. The list includes the haloxides ClO, BrO and IO, as well as CF, FN (^1D), FS, FSe, GeH, HO, HS, HSe, HTe, NS, OS and OSe. Many of these species are of special interest to radioastronomers, since their spectra show them to exist extra-terrestrially. In our atmosphere, the reaction

$$H + O_3 \rightarrow HO + O_2 \tag{7.8}$$

has been observed in the night-air glow. Hydroxyl radicals produced in the laboratory have been detected by gas-phase EPR in various rotational levels of both spin components ($^2\Pi_{3/2}$ and $^2\Pi_{1/2}$ when rotating) of the electronic ground state [22], including some vibrationally excited states.

Most of the free radicals listed above have been studied in their ground states ($^2\Pi$) and thus contain one unpaired electron per molecule. The spin orientation here is strongly linked to the molecular orientation (i.e., internuclear axis) through spin-orbit coupling; hence the observed microwave transitions, while tunable by the applied magnetic field B, are of the electric-dipole type.

7.4 TRIATOMIC AND POLYATOMIC GAS-PHASE MOLECULES

The list of triatomic molecules so far studied by EPR is not extensive; it includes the linear species NCO and NCS ($^2\Pi_{3/2}$ ground state) and non-linear species HCO, NF$_2$ and NO$_2$.

Very interesting vibronic effects have been explored in NCO and NCS by means of EPR [23]. In the ground state ($^2\Pi_{3/2}$), the spectrum resembles that of analogous

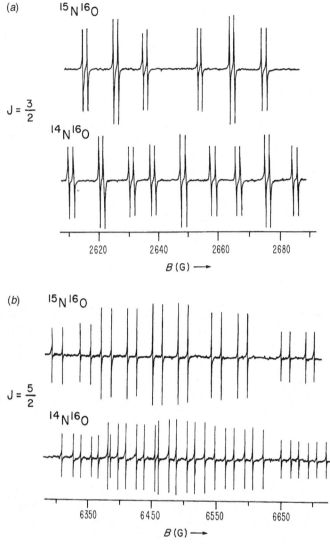

FIGURE 7.5 EPR spectra of gas-phase $^{15}N^{16}O$ and $^{14}N^{16}O$ in the $^2\Pi_{3/2}$ state, at $\nu = 2.8799$ GHz (S band). The transitions shown are of the electric-dipole type and correspond to $\Delta J = 0$, $\Delta M_J = \pm 1$, $\Delta M_I = 0$ and $\pm \leftrightarrow \mp$. Spectra are shown for molecules: (*a*) refers to the $J = \frac{3}{2}$ state, while (*b*) refers to the $J = \frac{5}{2}$ state. [After R. L. Brown, H. E. Radford, *Phys. Rev.*, **147**, 6 (1966).]

diatomic species. However, in thermally accessible states, bending vibrational motion becomes active and destroys the linearity; they then exhibit changed molecular parameters explainable by intermixing of vibrational and electronic states (Renner effect).

With non-linear molecules, except those of very high symmetry, there is no first-order spin-orbit coupling; hence the spin is only weakly coupled to the molecular framework. It follows that any electron-resonance electric-dipole transition can be found by varying B only if the spectrometer frequency is close to that of the corresponding zero-field line in the first place. Magnetic-dipole transitions tend to be weaker than these by factors of 10^3-10^4. Thus high concentrations and utmost sensitivity are called for to detect any of the thousands of EPR lines.

Difluoroamine radicals (NF_2) dimerize appreciably to form tetrafluorohydrazine; from the EPR intensity measured over the temperature range 340–435 K, the enthalpy change of dissociation of N_2F_4 was found to be 81(4) kJ mol^{-1} [24].

Almost no gas-phase free radicals containing more than three atoms have been observed by EPR. One exception is $(CF_3)_2NO$, which has been investigated carefully as a function of both radical concentration and total pressure P_t, using various diamagnetic inert diluent gases [25]. The increasing collisional relaxation effects on the spin-rotational coupling as P_t is raised remove these splittings, narrow the line and permit resolution of the ^{14}N and ^{19}F hyperfine splittings.

With sufficiently small species, such as NH_2 and CH_3 trapped in inert-gas matrices, EPR shows that virtually free rotation can occur and that nuclear-spin statistics must be applied to understand the relative hyperfine peak intensities at the lowest temperatures [26].

7.5 LASER ELECTRON PARAMAGNETIC RESONANCE

Since 1968, when *laser electron paramagnetic resonance* (LEPR) [3,27,28] first made its appearance, this type of spectroscopy has yielded much information about gas-phase free radicals. The idea is to observe resonant absorption by optical detection via magnetic-field (B) scanning of far-infrared lines of such radicals excited by suitable lasers. Here the frequency ν of the laser must be sufficiently close to the zero-field (rotational) frequencies of the free radical investigated. The limited frequency range 100–3000 GHz in which lasers have been operative has allowed only molecules of low mass to be investigated. Since the sensitivity of magnetic resonance goes up sharply with increasing frequency, LEPR offers an advantage of an increase in sensitivity by as much as 10^6 over ordinary EPR. The first LEPR absorption detected was at $\nu = 891$ GHz (HCN laser) for the transition between the levels $N = 3$, $J = 4$, $M_J = -4$ and $N = 5$, $J = 5$, $M_J = -4$ of ground-state O_2, occurring at $B = 1.6418$ T. Since then, many free radicals (e.g., HC, HN, HS and ^2HS, HF$^+$, HCl$^+$, HBr$^+$, HSi, CH_2, HO_2, HSO) have been investigated by LEPR; accordingly, a large number of molecular parameters including hyperfine couplings have become available.

7.6 OTHER TECHNIQUES

A very useful procedure, deserving an acronym of its own, is *magnetic resonance induced by electrons* (MRIE). Electron-impact excitation of a gas in or near the resonance cavity is used to create interesting free radicals in excited states of pre-existing species.

For instance, ground-state N_2 ($^1\Sigma_g$) can be excited to a paramagnetic metastable state ($^3\Sigma_u$) and then studied. Similarly, ionization to create first N_2^+ ($^2\Sigma_u$) and eventually the ground state ($^2\Sigma_g$) of this cation can be accomplished.

In a 1977 review [29], the techniques used for MRIE are organized into four categories:

1. MOMRIE = microwave optical magnetic resonance induced by electrons
2. ACS = anticrossing spectroscopy
3. LCS = level-crossing spectroscopy
4. MBMR = molecular beam magnetic resonance

All these methods yield information of the same type as do ordinary EPR methods. They are obtained with much improved spectral sensitivity, since it is not the low-energy photon associated with a transition between magnetic sublevels that is detected directly. Thus, in techniques 1–3, it is an emitted optical photon that is observed.

As indicated above, far-infrared laser magnetic resonance (LMR) also has played a role in determining energy parameters.

7.7   REACTION KINETICS

EPR spectroscopy certainly also has played an important role in establishing atom concentrations and characterizing reaction kinetics for paramagnetic species in the gas phase [30,31]. For instance, the recombination of nitrogen atoms (via three-body collisions) as a function of total pressure has been studied by this means [32].

7.8   ASTRO-EPR

One can easily define a burgeoning category that can be called 'astro-EPR' spectroscopy. It is well known that many of the species found in space are free radicals [33], and thus can and do give EPR signals. The most prominent is the famous 1420.4 MHz (21.105 cm) line (see Section C.1.6) arising from spin transitions of $^1H^0$ in its ground electronic state, which can be used to measure local magnetic fields present at various locations in the Universe [34], and to measure the H-atom concentrations there [35], including motions relative to the Earth. Spin exchange processes occurring during collisions between such atoms enter importantly for the realization of 21 cm tomography of the Dark-Age Universe [36]. As another example, space-probe EPR measurements of O_2 in the outer atmosphere ('limb') of the Earth have yielded special information about the solar radiance there [37]. Various 'coronium' ions, found in the solar corona, for instance Fe^{13+} (Fe XIV), which has a $3s^2 3p^1$ ground state ($^2P_{1/2}$), are expected to show EPR.

REFERENCES

1. V. Hughes, G. Tucker, E. Rhoderick, G. Weinreich, *Phys. Rev.*, **91**, 828 (1953).

2. V. Beltrán-López, J. Rangel G., A. González-Nucamendi, J. Jiménez-Mier, A. Fuentes-Maya, *Phys. Rev.*, **A39**, 58 (1989).

3. U. Bley, P. B. Davies, I. H. Davis, M. Koch, F. Temps, *J. Chem. Phys.*, **90**(2), 628 (1989).

4. H. E. Radford, V. W. Hughes, V. Beltrán-López, *Phys. Rev.*, **123**, 153 (1961).

5. P. W. Atkins, *Molecular Quantum Mechanics*, 2nd ed., Oxford University Press, Oxford, U.K., 1983.

6. P. Friedmann, R. N. Schindler, *Z. Naturforsch.*, **26a**, 1090 (1971).

7. M. S. de Groot, C. A. de Lange, A. A. Monster, *J. Magn. Reson.*, **10**, 51 (1973).

8. H. V. Lilenfeld, R. J. Richardson, F. E. Hovis, *J. Chem. Phys.*, **74**, 2129 (1981).

9. J. P. Wittke, R. H. Dicke, *Phys. Rev.*, **103**, 620 (1956).

10. S. Foner, H. Meyer, W. H. Kleiner, *J. Phys. Chem. Solids*, **18**, 273 (1961).

11. R. Beringer, J. G. Castle Jr., *Phys. Rev.*, **75**, 1963 (1945); **81**, 82 (1951).

12. M. Tinkham, M. W. P. Strandberg, *Phys. Rev.*, **97**, 937, 951 (1955).

13. I. B. Goldberg, H. O. Laeger, *J. Phys. Chem.*, **84**, 3040 (1980).

14. I. B. Goldberg, A. J. Bard, "Electron Spin Resonance Spectroscopy," in *Treatise on Analytical Chemistry*, 2nd ed., P. J. Elving, Ed., Vol. 10, Wiley, New York, NY, U.S.A., 1983, Chapter 3.

15. P. Gerber, *Helv. Phys. Acta*, **45**, 655 (1972).

16. T. J. Cook, B. R. Zegarski, W. H. Breckenridge, T. A. Miller, *J. Chem. Phys.*, **58**, 1548 (1973).

17. C. A. Arrington Jr., A. M. Falick, R. J. Myers, *J. Chem. Phys.*, **55**, 909 (1971).

18. T. A. Miller, *J. Chem. Phys.*, **54**, 330 (1971).

19. J. Graybeal, *Molecular Spectroscopy*, McGraw-Hill, New York, NY, U.S.A., 1988, Sections 14.7 and 14.9.

20. G. Herzberg, *Molecular Spectra and Molecular Structure. I. Spectra of Diatomic Molecules*, 2nd ed., Vol. 1, Van Nostrand, Princeton, NJ, U.S.A., 1950, p. 301.

21. R. L. Brown, H. E. Radford, *Phys. Rev.*, **147**, 6 (1966).

22. J. M. Brown, M. Kaise, C. M. L. Kerr, D. J. Milton, *Mol. Phys.*, **36**, 553 (1978).

23. A. Carrington, A. R. Fabris, B. J. Howard, N. J. D. Lucas, *Mol. Phys.*, **20**, 961 (1971).

24. L. H. Piette, F. A. Johnson, K. A. Booman, C. B. Colburn, *J. Chem. Phys.*, **35**, 1481 (1961).

25. T. J. Schaafsma, D. Kivelson, *J. Chem. Phys.*, **49**, 5235 (1968).

26. W. Weltner Jr., *Magnetic Atoms and Molecules*, Van Nostrand Reinhold, New York, NY, U.S.A., 1983, pp. 120–121.

27. K. M. Evenson, H. P. Broida, J. S. Wells, R. J. Mahler, M. Mizushima, *Phys. Rev. Lett.*, **21**, 1038 (1968).

28. M. Mizushima, K. M. Evenson, J. A. Mucha, D. A. Jennings, J. M. Brown, *J. Mol. Spectrosc.*, **100**, 303 (1983).

29. T. A. Miller, R. S. Freund, *Adv. Magn. Reson.*, **9**, 49 (1977).

30. A. A. Westenberg, *Prog. React. Kinet.*, **7**, 23 (1973).

31. K. H. Hoyermann, H. Gg. Wagner, J. Wolfrum, *Ber. Bunsen Ges.*, **71**, 599, 603 (1967).

32. T. C. Marshall, *Phys. Fluids*, **5**(7), 743 (1962).

33. G. Herzberg, *The Spectra and Structures of Simple Free Radicals*, Dover, New York, NY, U.S.A., 1988.

34. A. Cooray, S. R. Furlanetto, *Mon. Not. Roy. Astron. Soc.*, **359**, L47 (2005).

35. J. M. Dickey, N. M. McClure-Griffiths, B. M. Gaensler, A. J. Green, *Astrophys. J.*, **585**, 801 (2003).

36. B. Zygelman, *Astrophys. J.*, **622**, 1356 (2005).

37. M. J. Schwartz, W. R. Read, W. Van Snyder, *IEEE Trans. Geosci. Remote Sens.*, **44**(5), 1182 (2006).

NOTES

1. Thus we need not concern ourselves with the circular motions of such species about the applied magnetic field **B**. The *cyclotron resonance* occurs at a frequency that is proportional to the net charge.

2. For an explanation of atomic-state notation, see Section B.7.

3. The electronic-state notation for diatomic molecules is explained in Section B.7.

4. It is possible to trap individual diatomic molecules within sufficiently roomy cavities in condensed media. For instance, EPR studies of molecular oxygen confined in a single-crystal clathrate reveal [10] that some hindered rotational motion is present but that here the spectrum is dominated by the triplet-state electronic quadrupole effect ($\hat{\mathbf{S}}^T \cdot \mathbf{D} \cdot \hat{\mathbf{S}}$ term in $\hat{\mathcal{H}}_{ss}$ as discussed in Chapter 6).

5. In linear molecules, individual orbitals, for which the electron distribution is cylindrically symmetric about the internuclear axis, are referred to as s orbitals. Orbitals for which the electron distribution has a single nodal plane containing this axis are called π orbitals.

FURTHER READING

1. J. M. Brown, "Electron Resonance of Gaseous Free Radicals", in *Magnetic Resonance*, MTP International Review of Science, Butterworths, London, U.K. 1972, Vol. 4, C. A. McDowell, Ed., Chapter 7.

2. J. M. Brown and A. Carrington, *Rotational Spectroscopy of Diatomic Molecules*, Cambridge University Press, Cambridge, UK, 2003. While this book does not cover EPR spectroscopy, it does give a very detailed coverage of "Effective Hamiltonians" (Ch. 7), of which the "Spin Hamiltonian", so important in EPR, is a prime example.

3. A. Carrington, *Microwave Spectroscopy of Free Radicals*, Academic Press, London, U.K., 1974.

4. G. R. Eaton, S. S. Eaton, K. M. Salikhov, in *Foundations of Modern EPR*, World Scientific, Singapore, 1998, Chapters D1 and D2.

5. G. W. Hills, "Laser Magnetic Resonance Spectroscopy", *Magn. Reson. Rev.*, **9**(1–3), 15 (1984).

6. A. Hudson, K. D. J. Root, "Halogen Hyperfine Interactions", *Adv. Magn. Reson.*, **5**, 6–12 (1971).

7. D. H. Levy, "Gas-Phase Magnetic Resonance of Electronically Excited Molecules", *Adv. Magn. Reson.*, **6**, 1 (1973).

8. T. A. Miller, "The Spectroscopy of Simple Free Radicals", *Annu. Rev. Phys. Chem.*, **27**, 127 (1976).

9. W. Weltner Jr., *Magnetic Atoms and Molecules*, Van Nostrand Reinhold, New York, NY, U.S.A., 1983, Chapter 4.

10. A. Westenberg, "Use of ESR for the Quantitative Determination of Gas-Phase Atom and Radical Concentrations", *Prog. React. Kinet.*, **7**, 23 (1973).

PROBLEMS

7.1 Predict the ground-state EPR spectrum of gaseous atomic nitrogen ^{14}N, and also of ^{15}N.

7.2 Discuss and contrast the electronic ground states of the three gas-phase diatomic molecules: N_2, NO and O_2. Include suitable labeling (Section B.7) for these states.

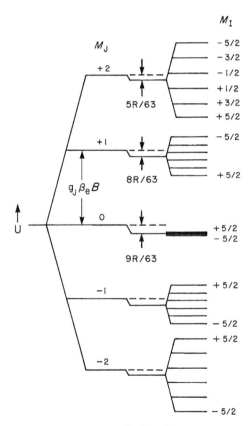

FIGURE 7.6 The magnetic splittings for $^{16}O^{17}O$ ($^1\Delta_g$) with $J = 2$ at $B > B_{rot}/g_e\beta_e$ (X-band). The splittings are shown accurate to second-order perturbations. For a Δ term, $\Lambda = 2$. The double arrow indicates one of the four EPR transitions (which are superimposed, to first order). Here $g_J \approx 2/3$. Energy parameter R equals $(\Lambda\beta_e B)^2/6B_{rot}$. [After C. A. Arrington Jr., A. M. Falick, R. J. Myers, *J. Chem. Phys.*, **55**, 909 (1971).]

7.3 The hamiltonian $\hat{\mathcal{H}}_0 + \hat{\mathcal{H}}_z$ describing $^{16}O^{16}O$ and $^{16}O^{18}O$ in the electronic ground state $^3\Sigma_g^-$ is given by Eqs. 7.4 and 7.6. What restrictions are there on the rotational quantum numbers N, and why? What factors determine the EPR lines to be expected at room temperature for each species?

7.4 Consider the molecule $^{16}O^{17}O$ in its excited state $^1\Delta_g$. For a D term, $|\Lambda| = 2$. While it has no unpaired electron, it nevertheless gives an EPR spectrum. For the angular-momentum state $J = 2$ ($\hat{\mathbf{J}} = \hat{\mathbf{L}} + \hat{\mathbf{N}}$), the Zeeman energy levels are shown in Fig. 7.6, including also the hyperfine splittings from the ^{17}O nucleus ($I = \frac{5}{2}$). Formulate the selection rules for ΔM_J and ΔM_I, and predict the number of lines allowed by these selection rules [13].

7.5 Derive the Landé formula (Eq. 7.1) for g.

7.6 Interpret the EPR spectra of the $J = \frac{5}{2}$ state of NO ($^2\Pi_{3/2}$) shown in Fig. 7.5.

CHAPTER 8

TRANSITION-GROUP IONS

8.1 INTRODUCTION

The transition-group, rare-earth and actinide ions, that is, the members of the $3d$, $4d$, $5d$, $4f$ and $5f$ groups, have been the subject of a host of EPR investigations. Of the approximately 116 'known' elements, 55 belong to these series.

Transition-ion complexes and salts have played a seminal role in many aspects of EPR, including the development of the spin-hamiltonian concept. Their importance was based on

1. Availability of various numbers of unpaired electrons per species (total spin $S = 0 \rightarrow \frac{7}{2}$)
2. Availability of species with simple local symmetries (e.g., cubic) and well-characterized neighbors to the central ion
3. Ease of preparation and stability, and yet with a variety of possible oxidation states
4. Availability of reasonably applicable and adequate electronic theory, for example, the crystal-field model

As time progressed, a trend to investigate lower-symmetry transition-ion species developed, especially since these have considerable importance in chemical catalysis and biomedical applications. Numerous excellent treatises give summaries and

Electron Paramagnetic Resonance, *Second Edition*, by John A. Weil and James R. Bolton
Copyright © 2007 John Wiley & Sons, Inc.

furnish details of these accomplishments—most based on EPR investigations [1; 2, pp. 89–201; 3; 4, Chapter 7; 5, Chapter 3; 6,7].

The skill in interpreting transition-ion properties lies partly in proper ordering of the relevant energy terms: interelectron repulsion, spin-orbit interaction and Zeeman energies. The order in which various perturbation treatments are carried out and their success depend on this. Generally, the Zeeman term is small in comparison to the others (but it is crucial, of course, for EPR). The spin-orbit energy tends to be small compared to the repulsion term for ions in the upper half of the periodic table; however, it is of major importance in f-electron systems, largely neglected in this book. The last two energy types can be of equal magnitude, and thus handling them depends on the system considered and can be problematic.

Transition-group positive ions are usually complexed with simple negative ions, neutral molecules, or with bulky polyatomic entities; any of these may be referred to as a *ligand*. Observation of the EPR spectra of transition-group complexes at low concentrations in the solid state is generally no more difficult than EPR of free radicals. However, the widths of transition-group EPR lines tend to be larger because of the short relaxation times; low temperatures may then be required to lengthen the relaxation time sufficiently for an EPR spectrum to be observable. We treat only the case in which the paramagnetic atoms or ions are sufficiently far apart that they act independently. In that case no linewidth effects from interaction between them are observed.

One aspect that makes transition elements interesting subjects for study by EPR or other techniques is their variable valence (Table 8.1). This feature is a characteristic of their unfilled electron shells. As an example, the readiness of iron to change between the +2 and +3 states provides sites for electron transfer in biological oxidation-reduction systems. The observation of hyperfine splitting may serve to identify the central nucleus in such a species. The nucleus need not be that of the

TABLE 8.1   Ground-State Properties of Free d^n Ions

Number n of d Electrons	S	L	J	Orbital Degeneracy	Term Symbol[a]	Examples $(3d^n)$
	(of the Ground State)					
1	$\frac{1}{2}$	2	$\frac{3}{2}$	5	$^2D_{3/2}$	Sc^{2+}, Ti^{3+}, V^{4+} [b], Cr^{5+}
2	1	3	2	7	$^3F_2$	Ti^{2+}, V^{3+}, Cr^{4+}
3	$\frac{3}{2}$	3	$\frac{3}{2}$	7	$^4F_{3/2}$	Ti^+, V^{2+}, Cr^{3+}, Mn^{4+}
4	2	2	0	5	$^5D_0$	V^+, Cr^{2+}, Mn^{3+}, Fe^{4+}
5	$\frac{5}{2}$	0	$\frac{5}{2}$	1	$^6S_{5/2}$	Cr^+, Mn^{2+}, Fe^{3+}, Co^{4+}
6	2	2	4	5	$^5D_4$	Mn^+, Fe^{2+}, Co^{3+}
7	$\frac{3}{2}$	3	$\frac{9}{2}$	7	$^4F_{9/2}$	Fe^+, Co^{2+}, Ni^{3+}
8	1	3	4	7	$^3F_4$	Fe^0, Co^+, Ni^{2+}, Cu^{3+}
9	$\frac{1}{2}$	2	$\frac{5}{2}$	5	$^2D_{5/2}$	Ni^+, Cu^{2+}

[a] See Section B.8.

[b] Includes the ubiquitous vanadyl ion, VO^{2+}. See: P. Chand, V. K. Jain, G. C. Upreti, *Magn. Reson. Rev.*, **14**, 49 (1988).

host ion; it may instead be a foreign nucleus that is present naturally or is introduced by doping.

8.2 THE ELECTRONIC GROUND STATES OF d-ELECTRON SPECIES

We begin by concentrating on d-electron systems and note that only S-, D- and F-type ions are expected (Table 8.1). The possible values of J for each case range from $|L - S|$ to $L + S$ in integral steps. For example, the $3d^1$ ion Ti^{3+}, which has a ground-state configuration of ^2D, has two possible values of J: $2 - \frac{1}{2} = \frac{3}{2}$ and $2 + \frac{1}{2} = \frac{5}{2}$. The J value is indicated on the term symbol by a right subscript (Section B.7). In the present example, the two states are designated by $^2D_{3/2}$ and $^2D_{5/2}$. They are separated in energy because of the 'spin-orbit coupling' (Section 4.8). In Ti^{3+}, the $^2D_{3/2}$ state lies lower in energy. In general, if the d shell is less than half full, the state with the minimum value of J lies lowest. The reverse is true if the d shell is more than half full.

The energy of a free ion changes from the value U_0 in the absence of an external magnetic field to U in the presence of the field. Here

$$U = U_0 + g\beta_e BM_J \tag{8.1}$$

where M_J is the quantum number for the component of $\hat{\mathbf{J}}$ along field $\mathbf{B}$. The factor g here is called the *Landé factor* and is given by Eq. 7.1. Note that this factor is different for each value of J. The primary free-ion EPR transitions are those for which $\Delta M_j = \pm 1$.

In treating the d^n configurations, use of the Russell-Saunders term symbols (Table 8.1) to classify the free ions implies that the Coulomb interactions between their d electrons have been taken into account at the start. For $n > 1$, the effect of the ligands is usually assumed to be relatively small compared to this effect, but not compared to the spin-orbit energy.

As is evident from Table 8.1, the values of L and S are the same for electronic configurations d^n and d^{10-n}. This is consistent with the fact that, in quantum mechanics, treatments of electrons and holes (electrons missing from a shell) are mathematically equivalent. Physically, certain parameters (e.g., spin-orbit parameter) describing these do differ in sign (Table H.3).

The key to understanding the EPR characteristics of a given transition-ion species is the nature of its electronic ground state,[1] since generally all other states are too high in energy to be populated. It is useful to imagine the energy-level modification in which a free transition ion is taken into the local bonded situation of interest. Generally, the free-ion energy is (close to) orbitally degenerate ($2L + 1$, see Table 8.1), and this energy degeneracy is removed by the electric fields arising from ligands. The spin degeneracies ($2S + 1$ for the free ion) are lifted by spin-orbit coupling (Section 4.8), to yield degeneracies as listed. The final degeneracy is determined by the local symmetry at the ion (Table 8.2). In this book we cannot discuss

TABLE 8.2   Degeneracy of Energy Levels for d^n Transition Ions in Various Local Electric Fields[a]

Configuration	d^1	d^2	d^3	d^4	d^5	d^6	d^7	d^8	d^9
Orbital Degeneracy in Electric Fields of Various Symmetries[b]									
Free ion	5	7	7	5	1	5	7	7	5
Octahedral[c]	$2,3^d$	$1,2\cdot3$	$1^d,2\cdot3$	$2^d,3$	1	$2,3^d$	$1,2\cdot3^d$	$1^d,2\cdot3$	$2^d,3$
Tetrahedral	$2^d,3$	$1^d,2\cdot3$	$1,2\cdot3$	$2,3^d$	1	$2^d,3$	$1^d,2\cdot3$	$1,2\cdot3$	$2,3^d$
Trigonal	$1,2\cdot2$	$3\cdot1,2\cdot2$	$3\cdot1,2\cdot2$	$3\cdot1,2$	1	$3\cdot1,2$	$3\cdot1,2\cdot2$	$3\cdot1,2\cdot2$	$3\cdot1,2$
Tetragonal	$3\cdot1,2$	$3\cdot1,2\cdot2$	$3\cdot1,2\cdot2$	$3\cdot1,2$	1	$3\cdot1,2$	$3\cdot1,2\cdot2$	$3\cdot1,2\cdot2$	$3\cdot1,2$
Rhombic	$5\cdot1$	$7\cdot1$	$7\cdot1$	$5\cdot1$	1	$5\cdot1$	$7\cdot1$	$7\cdot1$	$5\cdot1$
Spin Degeneracy in Electric Fields of Various Symmetries for a Single Orbital Level[b]									
Free ion	2	3	4	5	6	5	4	3	2
Octahedral	2	3	4	2,3	2,4	2,3	4	3	2
Trigonal	2	1,2	$2\cdot2$	$1,2\cdot2$	$3\cdot2$	$1,2\cdot2$	$2\cdot2$	1,2	2
Tetragonal	2	1,2	$2\cdot2$	$3\cdot1,2$	$3\cdot2$	$3\cdot1,2$	$2\cdot2$	1,2	2
Rhombic	2	$3\cdot1$	$2\cdot2$	$5\cdot1$	$3\cdot2$	$5\cdot1$	$2\cdot2$	$3\cdot1$	2

[a] An important limiting case, square planar, is not covered herein.
[b] $m\cdot n$ means that there are m sets of states of n-fold degeneracy.
[c] Fields of tetrahedral symmetry invert the order of these states.
[d] Lower or lowest state.
Source: After W. Gordy, W. V. Smith R. F. Trambarulo, *Microwave Spectroscopy*, Wiley, New York, NY, U.S.A., 1953, p. 225.

details of the theory leading to these ground states. In addition to the works already cited [1; 2, pp. 89–201; 3; 4, Chapter 7; 5, Chapter 3; 6,7], excellent texts on the topic of transition ions in various local surroundings exist [8–11].

We can, however, arrive at a qualitative understanding of the basic aspects. Consider the imposition of negatively charged ligands around the free ion, placed (initially) in a cubic array (Fig. 4.1). In reality, the ligands are favorably received, and via a combination of coulombic attraction and bonding interactions, the total energy is lowered. However, for the non-bonding open-shell electrons, a repulsive situation prevails (e.g., see Fig. 4.9, valid for a P-state ion). For the octahedral case (Fig. 4.1b), the negative charges are placed at distances $+R$ on the coordinate axes (x, y, z), as shown in Fig. 8.1, whereas for the tetrahedral case (Fig. 4.1c), the four charges are placed at points $[a(R/\sqrt{3}), b(R/\sqrt{3}), c(R/\sqrt{3})]$, with sign sets $(a, b, c) = (+, +, +), (+, -, -), (-, +, -)$ and $(-, -, +)$.

Consider now the five d functions (Fig. 8.2), which are eigenfunctions of the cubic crystal-field hamiltonian [12, Sections 8.10–8.11].[2] The square of these functions (or a linear combination thereof) gives the angular distribution of the open-shell electrons. It is clear that in the octahedral situation, orbitals d_{xy}, d_{xz} and d_{yz} are further removed from the negative charges than are d_{z^2} and $d_{x^2-y^2}$. By symmetry, the first three are an energetically equivalent set; the same is true of the latter two. The splitting is denoted by an energy parameter Δ.

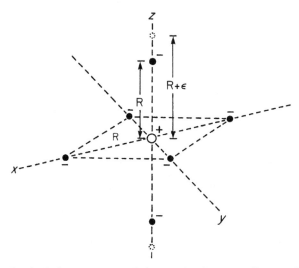

FIGURE 8.1 Octahedral arrangement of six negative ions at a distance d from a central positive ion (solid circles). A tetragonal distortion resulting from an increased separation $(R + \varepsilon)$ along the z axis is shown by open circles.

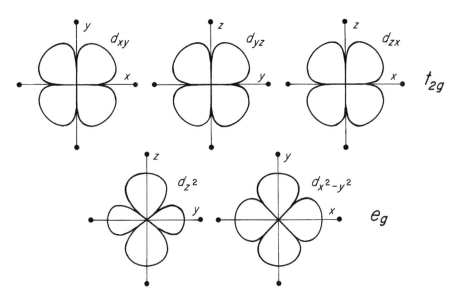

FIGURE 8.2 Representation of d orbitals showing the relation of the orbital lobes to the x, y and z axes. Each of the subscripts should be multiplied by r^{-2}; for example, the orbital usually referred to as d_{z^2} is given more fully as $(3z^2 - r^2)r^{-2}$. The orbitals indicated in the figure are representations of real wavefunctions, obtained by taking linear combinations of the imaginary wavefunctions that are eigenfunctions of $\hat{L}_z$. The symbols t_{2g} and e_g denote spatial triplet and doublet states for a d electron in an octahedral electric field (also see Note 4).

Let us now consider D-state ions. The same quantum-mechanical calculation for energy-level splittings in a crystal field is valid for all such ions. The splitting parameter D depends on the type of cubic electric field and the electron configuration d^n treated.[3] The usual group-theoretic notation, T_2 for the energy-degenerate triplet of states and E for the doublet set, is appropriate for these cubic cases.[4] Note the subindex g (gerade = even) applicable in the octahedral and cubal cases (for which the inversion through the origin is a symmetry operation). Figure 8.3a depicts the energetics for the D-state ions d^1 and d^6 in the octahedral case. The opposite energy-level order is correct in the tetrahedral (and cubal)[5] case (Fig. 8.3b). For d^4 and d^9 ions, the figures given above are correct if the octahedral and tetrahedral labels are interchanged. Thus the assumption of equivalence of electron and hole states continues to have validity.

Next suppose that the symmetry is lowered (Fig. 8.1) to tetragonal by either elongating or compressing the cube (charge positions) along one cartesian axis, say, $\mathbf{z}$ (i.e., by setting the sign of the distortional splitting parameter δ to be $+$ or $-$). This is indicated in Fig. 8.3 for both the octahedral and tetrahedral

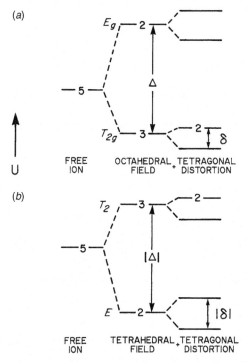

FIGURE 8.3   Splittings and degeneracies of orbital levels of d^1 or d^6 ions in two types of electric field caused by negative charges: (a) octahedral field ($\Delta > 0$) plus tetragonal distortion; (b) tetrahedral field ($\Delta < 0$) plus tetragonal distortion. For d^4 and d^9 ions, (a) applies to tetrahedral fields and (b) to octahedral fields. Shifting of the center of gravity of the set of levels is ignored.

cases. The energy eigenvalues and corresponding eigenfunctions are

$$U_1 = \frac{3}{5}\Delta + \frac{2}{3}\delta \qquad d_{x^2-y^2} \tag{8.2a}$$

$$U_2 = \frac{3}{5}\Delta - \frac{2}{3}\delta \qquad d_{z^2} \tag{8.2b}$$

$$U_3 = -\frac{2}{5}\Delta + \frac{2}{3}\delta \qquad d_{xy} \tag{8.2c}$$

$$U_{4,5} = -\frac{2}{5}\Delta - \frac{1}{3}\delta \qquad d_{xz}, d_{yz} \tag{8.2d}$$

The sign choices for the D-state ions are indicated in Table 8.3.

We have neglected the spin-orbit effects above. These can now be brought in by perturbation theory, leading to formulation of matrices **g** and **D** (Section 4.8). Examples follow.

A considerable body of data relating orbital and spin degeneracies in electric fields of various symmetries has been collected in Table 8.2, which illustrates the progressive reduction in orbital degeneracy of the lowest state as one goes from the free ion through the uniaxial symmetries (trigonal or tetragonal) to rhombic symmetry.

We recall (Chapter 4) that, when there is orbital degeneracy, a modified technique must be used to derive the spin hamiltonian since the standard one fails. Note also that for odd numbers of unpaired electrons no crystal field can remove all degeneracy. A theorem by Kramers, based on consideration of time-reversal symmetry, guarantees that no electric field can accomplish this [13]. However, externally applied magnetic fields can do so [4, Section 15.4; 5, Section 3.8].

The conclusions about ultimate degeneracy presented above are misleading in that it can be shown, by the Jahn-Teller theorem, that orbital degeneracy in non-linear molecular systems never persists [4, Chapter 2; 5, Section 3.9; 12, Section 10.11; 5, Sections 3.9 and 10.4; 6, Section 3.9]. The lowest total energy is achieved by a spontaneous distortion to a nuclear configuration of reduced symmetry, leaving a state with only Kramers degeneracy (effective $S = \frac{1}{2}$). When there are several configurations with equal (or almost equal) energies, there may be jumps between these, depending on the size of k_bT. Such a 'dynamic' Jahn-Teller effect is manifested (Section 10.5.3.4) in the EPR spectra [14]. The understanding of this effect requires

TABLE 8.3   Crystal-Field Parameter Sign Choices for D-State Ions

d^1, d^6		d^4, d^9
$\Delta > 0, \delta > 0$	oct, elong	$\Delta > 0, \delta > 0$
$\Delta > 0, \delta < 0$	oct, comp	$\Delta > 0, \delta < 0$
$\Delta < 0, \delta > 0$	tth, comp	$\Delta < 0, \delta > 0$
$\Delta < 0, \delta < 0$	otth, elong	$\Delta < 0, \delta < 0$

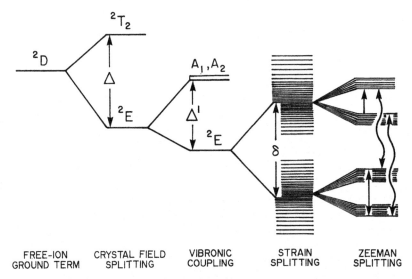

FREE-ION   CRYSTAL FIELD   VIBRONIC   STRAIN   ZEEMAN
GROUND TERM   SPLITTING   COUPLING   SPLITTING   SPLITTING

FIGURE 8.4 Schematic energy-level diagram implied by the interpretation of the EPR parameters. The ^2D free-ion term is split by the cubic crystal field into ^2E and $^2T_2$ states with the ^2E state lowest. A weak-to-moderate vibronic interaction results in a series of admixed electronic/vibrational states; the resulting ground vibronic state is also a ^2E state. The first excited vibronic state is sufficiently far removed that the ground state can be treated as isolated. This state is split by random internal strains into two Kramers doublets, the degeneracy of which is removed by an applied magnetic field. EPR transitions (represented by straight arrows) can be induced by the microwave field. The wavy arrows represent vibronic relaxation processes, which can produce an averaging of a portion of the anisotropic EPR pattern. This type of relaxation should not be confused with spin-lattice relaxation. [After J. R. Herrington, L. A. Boatner, T. J. Aton, T. L. Estle, *Phys. Rev.*, **B10**, 833 (1974).]

a detailed examination of the interaction between the crystal vibrations and the electronic distribution (vibronic effect), as sketched in Fig. 8.4. As shown there, electric-field inhomogeneities (i.e., strain) within the crystal also are important, splitting each resulting vibronic state into two Kramers doublets.

The preceding theoretical approach is oversimplified; luckily it works quite well. More and more, just as with organic radicals, the energy-level manifold, electron distribution and vibronic factors are being analyzed using large-scale computer-based ab initio molecular-orbital theory.

8.3 THE EPR PARAMETERS OF d-ELECTRON SPECIES

At this point it is appropriate to consider specific ions in a variety of crystal fields where the ground state is orbitally non-degenerate. In such cases, the theory described in Section 4.8 is applicable. From Table 8.4 we note that there are relatively few D-state ion types that have a non-degenerate ground state. The d^1(tth)

TABLE 8.4   Of d^n Ions with Orbitally Non-Degenerate Ground States in Cubic Local Crystal Fields^a

| State | Octahedral | | Tetrahedral | |
	Low-Field	High-Field	Low-Field	High-Field
^1A	—	d^6	—	d^4
^3A	d^8	d^8	d^2	d^2
^4A	d^3	d^3	d^7	d^7
^6A	d^5	—	d^5	—

^{a} Presence of lower-symmetry crystal fields of proper sign and direction in most cases produces orbitally non-degenerate ground states.

and d^1 (cubal) ions have the E state lowest, but a further tetragonal distortion removes the orbital degeneracy, as shown in Fig. 8.3b. For a compression along $\mathbf{z}$, the $|0\rangle = d_{z^2}$ state lies lowest; conversely, for any elongation along $\mathbf{z}$, the hybrid state $[|+2\rangle + |-2\rangle]/\sqrt{2} = d_{x^2-y^2}$ lies lowest (Eqs. 8.2). The former case is considered first, since it is relatively simple to treat. We note the resemblance to the p^1 case discussed in Section 4.8.

The g components obtained by using Eqs. 4.38 and 4.41 are

$$g_z = g_{||} = g_e + 2\lambda\Lambda_z = g_e \tag{8.3}$$

Here $\Lambda_z = 0$ since $\hat{L}_x$ couples only states with the same M_L components. Furthermore, we obtain

$$g_x = g_\perp = g_e + 2\lambda\Lambda_x = g_e - 2\lambda \sum_{M_L \neq G} \frac{\langle G|\hat{L}_x|M_L\rangle \langle M_L|\hat{L}_x|G\rangle}{U_{M_L}^{(0)} - U_G^{(0)}} \tag{8.4}$$

The only states coupled to $|0\rangle$ by $\hat{L}_x$ are $|+1\rangle$ and $|-1\rangle$ (Eqs. B.42f, g), and hence

$$g_\perp = g_e - \frac{2\lambda}{\Delta}\left(\langle 0|\tfrac{1}{2}\hat{L}_-|+1\rangle \langle +1|\tfrac{1}{2}\hat{L}_+|0\rangle + \langle 0|\tfrac{1}{2}\hat{L}_+|-1\rangle\langle -1|\tfrac{1}{2}\hat{L}_-|0\rangle\right) \tag{8.5a}$$

The matrix elements in Eq. 8.5a are analogous to those in Eq. 4.44, but now $L = 2$. Application of Eqs. B.42f and B.42g to evaluate these elements gives

$$g_\perp = g_e - 6\lambda/\Delta \tag{8.5b}$$

The central-ion hyperfine parameters for this case are approximated (e.g., see Ref. 4, p. 456) by

$$A_\perp = A_0 - \left(\frac{2}{7} + \frac{17}{7}\frac{\lambda}{\Delta}\right)P_d \tag{8.6a}$$

$$A_{\parallel} = A_0 + \left(\frac{4}{7} + \frac{34}{7}\frac{\lambda}{\Delta}\right)P_d \tag{8.6b}$$

where (Eq. 5.26)

$$P_d = \frac{\mu_0}{4\pi}g_e g_n \beta_e \beta_n \langle r^{-3}\rangle_d \tag{8.7}$$

Thus the uniaxiality parameter (Eq. 5.9b) is

$$b_0 = \frac{1}{g_e\beta_e}\left(\frac{2}{7} + \frac{17}{7}\frac{\lambda}{\Delta}\right)P_d \tag{8.8}$$

A good example of a $3d^1$ (tth + ttg) ion is Cr^{5+} in CrO_4^{3-} doped in Ca_2PO_4Cl single crystals [15,16], where $g_{\parallel} = 1.994$ and $g_{\perp} = 1.950$. Here one deals (to a first approximation) with a compressed tetrahedron, with the unpaired electron residing primarily in the Cr $3d_{z^2}$ orbital.

In the case of an elongated tetrahedron, the g factors are (Problem 8.2)

$$g_{\parallel} \approx g_e - \frac{8\lambda}{\Delta} \tag{8.9a}$$

$$g_{\perp} \approx g_e - \frac{2\lambda}{\Delta} \tag{8.9b}$$

For d^n ions with $S > \frac{1}{2}$, the zero-field terms $\hat{\mathbf{S}}^T \cdot \mathbf{D} \cdot \hat{\mathbf{S}}$ enter and may dominate the EPR characteristics. The theory presented in Chapter 4 (e.g., Eq. 4.42) and Chapter 6 is applicable. Matrix $\mathbf{D}$ arises predominantly because of spin-orbit coupling, rather than spin-spin interaction, and often leads to large anisotropy of line positions.

The case of F-state ($\ell = 3$) ions yields similar results. For ions d^3 or d^8 in an octahedral electric field, the non-degenerate A_{2g} state lies lowest, whereas for d^2 or d^7 ions the T_{1g} states lie lowest (Figs. 8.5a and 8.5b). As in a D-state ion, the order of levels is inverted in a tetrahedral field.

We now turn to a specific example of an F-state transition ion, namely, the ^3F ground state of the free d^8 ion. In an octahedral field, the sevenfold orbitally degenerate states split into two triply degenerate states ($^3T_{1g}$ and $^3T_{2g}$) and one non-degenerate state ($^3A_{2g}$). This is depicted in Fig. 8.5. From the 7×7 hamiltonian matrix for the octahedral crystal-field energy [9, p. 69], one finds the eigenvalues

$$U(T_{1g}) = +\tfrac{3}{5}\Delta \qquad \text{triply degenerate} \tag{8.10a}$$

$$U(T_{2g}) = -\tfrac{1}{5}\Delta \qquad \text{triply degenerate} \tag{8.10b}$$

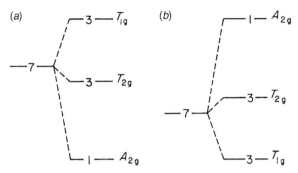

FIGURE 8.5 Energy splitting of the states of an F-state ion in a local electric field: (a) d^3 or d^8 ions in an octahedral field; (b) d^2 or d^7 ions in an octahedral field. In a tetrahedral field, the splittings for d^3 or d^8 ions are obtained by level inversion of (a) (with removal of subindex g, since now there is no center of symmetry). For d^2 or d^7 ions they are obtained by level inversion of (b).

$$U(A_{2g}) = -\tfrac{6}{5}\Delta \qquad \text{non-degenerate} \qquad (8.10c)$$

where we take $\Delta > 0$ for d^8. The corresponding eigenstates (for $L = 3$) are

$$T_{1g}\begin{cases} |t'_{-1}\rangle \equiv \left(\tfrac{3}{8}\right)^{1/2}|-1\rangle + \left(\tfrac{5}{8}\right)^{1/2}|+3\rangle \\ |t'_{+1}\rangle \equiv \left(\tfrac{3}{8}\right)^{1/2}|+1\rangle + \left(\tfrac{5}{8}\right)^{1/2}|-3\rangle \\ |t'_0\rangle \equiv |0\rangle \end{cases} \qquad (8.11a)$$

$$T_{2g}\begin{cases} |t''_{+1}\rangle \equiv \left(\tfrac{5}{8}\right)^{1/2}|-1\rangle - \left(\tfrac{3}{8}\right)^{1/2}|+3\rangle \\ |t''_{-1}\rangle \equiv \left(\tfrac{5}{8}\right)^{1/2}|+1\rangle - \left(\tfrac{3}{8}\right)^{1/2}|-3\rangle \\ |t''_0\rangle \equiv \left(\tfrac{1}{2}\right)^{1/2}|+2\rangle + \left(\tfrac{1}{2}\right)^{1/2}|-2\rangle \end{cases} \qquad (8.11b)$$

$$A_{2g}|a\rangle \equiv \left(\tfrac{1}{2}\right)^{1/2}|+2\rangle - \left(\tfrac{1}{2}\right)^{1/2}|-2\rangle \qquad (8.11c)$$

(The subscript on the t designates the expectation value of the fictitious angular-momentum operator $\hat{L}_x{}'$, taking $L' = 1$ for each triplet manifold.)

One can begin the analysis of g, to zero order, by ignoring the effects of spin-orbit coupling on the ground-state wavefunction. The energies in a magnetic field $\mathbf{B} \parallel \mathbf{z}$ are obtained from the Zeeman hamiltonian (note Eq. 4.33)

$$\hat{\mathcal{H}}_Z = \beta_e B_z(\hat{L}_z + g_e\hat{S}_z) \qquad (8.12)$$

Since the ground state is orbitally non-degenerate, the contribution of $\hat{L}_z$ is zero (Section B.8). The ground-state wavefunction (Eq. 8.11c) including spin may be written

$$|G\rangle = \frac{1}{\sqrt{2}}(|+2, M_S\rangle - |-2, M_S\rangle) \qquad (8.13)$$

where $M_S = -1, 0, +1$. Thus the energies in a magnetic field are (Fig. 8.6)

$$U_{\pm 1} = \pm 2\beta_e B_z \qquad (8.14a)$$

$$U_0 = 0 \qquad (8.14b)$$

Hence in this approximation $g_Z = g_e$. Furthermore, $g_X = g_Y = g_e$, as is required by the octahedral symmetry.

The spin-orbit coupling hamiltonian operator $\hat{\mathcal{H}}_{so}$ (Eq. 4.32) causes an admixture of excited states into the ground state. (In crystal fields of lower than octahedral symmetry, since $S = 1$, there will also be a zero-field splitting of the spin degeneracy.) The calculation of these effects is most conveniently approached

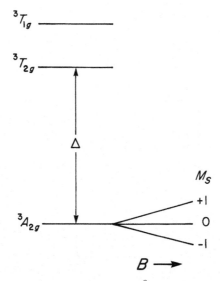

FIGURE 8.6 Energy splitting of the states of a d^8 ion in an octahedral field and, for the ground state, in an added magnetic field. Here, zero-field splittings are ignored. The same diagram applies to a d^2 ion in a tetrahedral field (with removal of subindex g, since now there is no center of symmetry).

through the matrix Λ of Eqs. 4.37 and 4.38. Since the field is octahedral, we expect this matrix to be not only diagonal but also scalar ($\lambda \mathbf{1}_3$), and only one component of Λ needs to be calculated, for example, Λ_z. The only excited state that contributes to Λ_z is the $|t_0''\rangle$ function of the T_{2g} state (Eq. 8.11b). Recall that Δ is the energy separation between the T_{2g} states and the A_{2g} state (Fig. 8.6). Thus

$$\Lambda_z = -\frac{\left|\frac{1}{\sqrt{2}}\left(\langle+2| - \langle-2|\right)|\hat{L}_z|\frac{1}{\sqrt{2}}\left(|+2\rangle + |-2\rangle\right)\right|^2}{\Delta} \qquad (8.15a)$$

$$= -\frac{4}{\Delta} \qquad (8.15b)$$

Substitution of Eq. 8.15b into Eq. 4.41 gives

$$g_z = g_x = g_y = g_e - \frac{8\lambda}{\Delta} \qquad (8.16)$$

Thus, as expected here, g is isotropic and matrix $\mathbf{D}$ is zero. Since there are no excited states close to the ground state to cause relaxation broadening, EPR spectra of $3d^8$ ions can often be seen at room temperature or at 77 K.

The Ni^{2+} ion is a most important example of a $3d^8$(oct) ion. Here $S = 1$. The g factor may be estimated from Eq. 8.16 if optical absorption data are available. For example, in the $Ni(NH_3)_6^{2+}$ ion an optical band (assigned to the $^3T_2 \leftarrow {}^3A_2$ transition) is observed at $10700 \, cm^{-1}$. If one takes the free-ion value of λ ($-325 \, cm^{-1}$), g is calculated to be 2.245. The experimental value is 2.162 [17]. The discrepancy is probably due to the presence of some covalent bonding. Alternatively, one can use g and Δ to compute an effective value of λ, that is, $-211 \, cm^{-1}$. In a wide range of octahedral environments, g varies from about 2.10 to 2.33; similar behavior is found for the Cu^{3+} ion.

It is observed that Ni^{2+} often gives rise to quite broad EPR lines, even in presumably octahedral environments. For example, in MgO, where other substitutional ions may give lines $\sim$0.05 mT wide, the Ni^{2+} linewidth may be as much as 4 mT (Fig. 8.7). Since Ni^{2+} has an *even* number of electrons, Kramers theorem does not apply. Hence residual lattice strains (local electric fields) may cause the $|0\rangle$ state to be shifted by varying amounts relative to the $|+1\rangle$ and $|-1\rangle$ states (Fig. 8.6). In other words, deviations from octahedral symmetry can cause a zero-field splitting D (Section 6.3.1), which may be positive or negative.

Consider the case of $^{63}Cu^{3+}$ ion in Al_2O_3 [18]. The spin hamiltonian for this $3d^8$ ion in the uniaxial local electric field present may be written (compare with

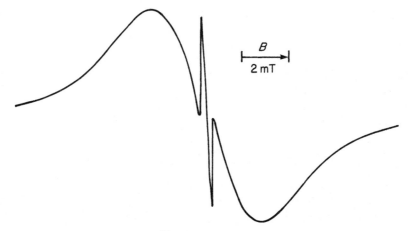

FIGURE 8.7 EPR spectrum of Ni^{2+} ($S = 1$) in MgO at 115 K with $\nu = 9.155$ GHz. The broad line is the superposition of the transitions $|-1\rangle \leftrightarrow |0\rangle$ and $|0\rangle \leftrightarrow |+1\rangle$. The narrow central line is the transition $|-1\rangle \leftrightarrow |+1\rangle$ effected by the absorption of two quanta, observable at sufficiently high microwave power (Section 6.3.2). [After S. R. P. Smith, F. Dravnieks, J. E. Wertz, *Phys. Rev.*, **178**, 471 (1969).]

Eq. 6.27) in the form

$$\hat{\mathcal{H}}_n = g_\parallel \beta_e B \hat{S}_z \cos \theta + \tfrac{1}{2} g_\perp \beta_e B (\hat{S}_+ + \hat{S}_-) \sin \theta +$$

$$D\left[\hat{S}_2^2 - \tfrac{1}{3} S(S + 1)\hat{\mathbf{1}}_3\right] + A_\parallel \hat{S}_z \hat{I}_z + \tfrac{1}{2} A_\perp (\hat{S}_+ \hat{I}_- + \hat{S}_- \hat{I}_+) \qquad (8.17)$$

where θ is the angle between **B** and the symmetry axis **z**. Here $S = 1$, and the matrices **g**, **D**, and **A** are taken to be co-uniaxial; nuclear quadrupole effects are neglected (see Problem 8.6). These parameters have been accurately measured [18] by both EPR spectroscopy and (more accurately) ENDOR (see Chapter 12)

For an S-state ion there is no orbital degeneracy to be removed. However, the local electric field, together with spin-orbit and electron spin-spin interactions, can cause some removal of spin-state degeneracy; the details are not simple [19]. The ions Mn^{2+} and Fe^{3+} are both most prominent in EPR spectroscopy. We choose the latter as an example of an S-state ion, partly because (unlike Mn^{2+}) it does not feature the extra complication of copious hyperfine structure.[6]

In the pure cubic (e.g., octahedral) situation, one might expect isotropic behavior for an S-state ion. Nevertheless, although D and E are zero and g is isotropic, there now exists (because of the high spin, $S = \tfrac{5}{2}$) a new zero-field energy term, as outlined above (Section 6.6). This term produces EPR anisotropy. For this case, one can write [20]

$$\hat{\mathcal{H}}^{(4)} = \frac{1}{8} U^{(4)} \left[\hat{S}_x^{\ 4} + \hat{S}_y^{\ 4} + \hat{S}_z^{\ 4} - \frac{707}{16} \hat{\mathbf{1}}_6 \right] \tag{8.18a}$$

$$= \frac{1}{8} U^{(4)} \left[112 \hat{S}_z^{\ 4} - 760 \hat{S}_z^{\ 2} + 567 \hat{\mathbf{1}}_6 + 8(\hat{S}_+^{\ 4} + \hat{S}_-^{\ 4}) \right] \tag{8.18b}$$

where **x**, **y** and **z** are the fourfold axes of the octahedron and $\hat{\mathbf{1}}_6$ is the 6×6 unit matrix. The resulting zero-field splitting and EPR spectrum for $\mathbf{B} \parallel \mathbf{z}$ are shown in Fig. 8.8. For Fe^{3+} in $SrTiO_3$, the energy parameter $U^{(4)}/g_e\beta_e$ is 0.45 mT [21]. The high-spin (electronic hexadecapole $\sim S^4$) terms, such as those in Eq. 8.18, are more sensitive to the details of this ion's local surroundings than are the lower (electronic quadrupole $\sim S^2$) terms. With decrease in symmetry away from octahedral, the latter terms (i.e., $\hat{S}^T \cdot \mathbf{D} \cdot \hat{S}$) enter and in fact dominate. The energy-level

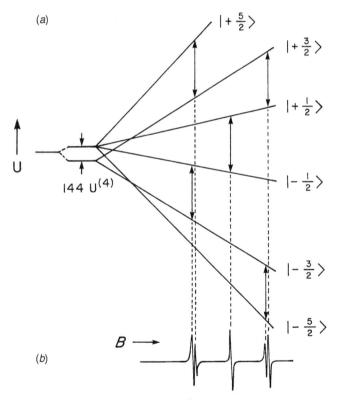

FIGURE 8.8 (*a*) Energy-level diagram for a d^5 ion in an octahedral crystal field. (*b*) The allowed EPR spectrum is shown for $h\nu \gg 144U^{(4)}$. The diagram applies only for **B** parallel to a principal axis of the octahedron.

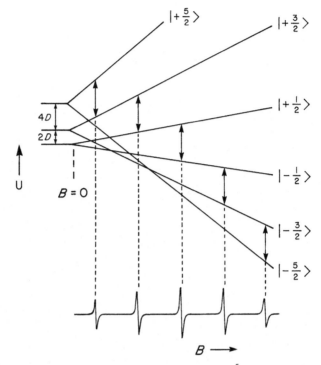

FIGURE 8.9 Energy levels and allowed transitions for a d^5 ion in a weak tetragonal field with **B** parallel to the tetragonal axis.

scheme for tetragonal symmetry ($|D| > E = 0$) is shown in Fig. 8.9, for **B** parallel to the tetragonal axis. We note that at $B = 0$, there are three Kramers doublets. The five-line EPR spectrum is highly anisotropic, despite the S-state nature of the ion. Such anisotropy, arising primarily from **D**, is shown in Fig. 8.10 for a high-spin Fe^{3+} species in crystalline quartz. Here the local symmetry at each site is only C_2. Note the separate spectra arising from the three symmetry-related sites (Section 4.5).

Transition-group ions, of course, often occur in glasses, and can be advantageously studied therein by means of EPR. As an example, we can cite the multiple oxidation states of Cr ions in various borate-aluminate glasses, studied in detail by EPR and optical means [22]. For other relevant literature, see the 1990 review by Griscom [23].

8.4 TANABE-SUGANO DIAGRAMS AND ENERGY-LEVEL CROSSINGS

It is important for various reasons to know how the relative energies of the lowest levels of a given transition ion vary as a function of the local electric field. In

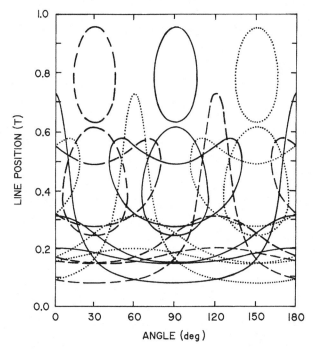

FIGURE 8.10 Anisotropy in the EPR spectrum of a high-spin Fe^{3+} ion in crystalline silicon dioxide, produced mostly by the spin-hamiltonian term **S**T.**D**.**S** [with $D/h = 9.5225(1)$ and $|E|/h = 1.7744(1)$ GHz]. The line positions at 9.915 GHz and $T = 20$ K are shown as a function of crystal rotation about the crystal threefold screw symmetry axis **c**. All symmetry-related sites are included; solid curves, site 1; dashed curves, site 2; dotted curves, site 3. Angle $0°$ is at **B** ∥ $\mathbf{a}_1$, which is a crystal two-fold axis. Note the trigonal symmetry evident in the figure. [After L. E. Halliburton, M. R. Hantehzadeh, J. Minge, M. J. Mombourquette, J. A. Weil, *Phys. Rev.*, **B40**, 2076 (1989).]

general, it is possible to calculate these energies, but the change of energy with local-field strength is different for each state, and the dependence is usually non-linear. A diagram of the excited-state energies, relative to that of the ground state, as a function of the local-field strength is called a Tanabe-Sugano diagram [24]. Figure 8.11 (appropriate to d^4 ions, i.e., to V^+, Cr^{2+}, Mn^{3+} and Fe^{4+}) has been taken from a set of such diagrams of relative ion energy versus relative octahedral field strength. Each parameter is divided by a reference energy U_o appropriate to the particular ion-host system. (The reference energies referred to here are the *Racah parameters*.)

The diagram for a particular d^n case is applicable to isoelectronic ions, for example, Fe^0, Co^+, Ni^{2+} and Cu^{3+} (Table 8.1). The Tanabe-Sugano diagrams are valuable guides to the interpretation of optical absorption or emission spectra. They are of interest for our purposes because they clarify the occurrence (for some ions) of two different spin states in different hosts. We see that in a relatively strong local field, some excited states of a given d^n ion may approach the ground

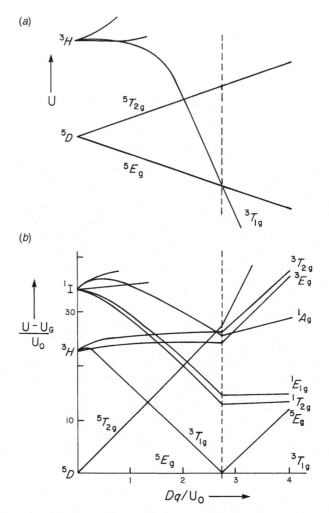

FIGURE 8.11 Splitting of the energy levels of a d^4 ion as a function of octahedral field strength Dq (note that $Dq = \Delta/10$). The vertical dashed line indicates the cross-over between the two different ground states. The ground-state energy symbol is U_G. A number of levels, which increase monotonically in energy with increasing Dq, have not been included. [Representation (b) is due to Tanabe and Sugano [24]; it is applicable to all $3d^4$ ions, since both $U - U_G$ and Dq are scaled by an energy characteristic (U_0) of the particular ion.]

state closely enough to make significant contributions both to g factors and to the zero-field splitting parameters D and E. Furthermore, for some d^n cases (d^4, d^5, d^6, d^7) the decrease in energy with increasing octahedral crystal field of at least one of the excited states is much greater than that of the initial ground state (Fig. 8.11). Hence, for magnitudes of the crystal field beyond some critical value, there is a switch to a new ground state. This is represented symbolically in Fig. 8.11 for some of the states of a d^4 ion.

The regions at the left- and right-hand sides of Fig. 8.11 correspond to high-spin and low-spin behavior, referring to ground-state electron distributions that lead to maximum and minimum total spin.[7] High- and low-spin cases are in principle found for most d^n ions. Thus the ground states indicated in Table 8.1 are observed for most, but by no means all, ligands.

Various methods exist for investigating high-spin/low-spin transitions. As an example, consider the investigation [25] of the d^6 ion Fe^{2+} in crystalline $Fe(1\text{-propyltetrazole})_6(BF_4)_2$, where it is thought to be in a trigonally distorted octahedron. The ions Mn^{2+} and Cu^{2+} were used as spin probes to check on their ferrous neighbors, and their EPR spectra were taken over a wide temperature range, including the high-spin ($S = 2$)/low-spin ($S = 0$) transition (to ~ 128 K). It was concluded that their spin change is associated with a crystallographic phase change. The T_0 value observed may depend on the B field applied [26].

8.5 COVALENCY EFFECTS

In previous sections of this chapter, we have dealt primarily with ligand-field theory, examining the ground-state unpaired-electron configurations and predicted g factors. It is incumbent on us at this point to convey that the latter are only first approximations to reality in that the very important covalency effects have not been dealt with [4, Chapter 20]. These shift the g factors, sometimes even overturning the sign of $g - g_e$ predicted from the electron/hole concept. Thus the literature is full of more sophisticated (and quite successful) calculations of g matrices for specific systems [27–30].

Covalence, of course, affects the size of the transition-ion hyperfine splittings, and even more dramatically the observed hyperfine structure from ligand nuclei [31,32]. We note the approximate proportionality between hyperfine splittings and spin density on adjacent and more remote atoms (Section 9.2). It should be clear that reliable estimation of covalence effects require large-scale quantum-mechanical calculations, say, of the self-consistent-field molecular-orbital type. The effects manifest themselves as additional important terms in the equations for **g**, **D** and **A** (i.e., in Eqs. 4.41, 4.42 and 5.8). For instance, the effects of spin-orbit coupling at ligand atoms become important. The theory of the anisotropic hyperfine interaction with ligand nuclei is available [33,34].

Many EPR studies have been carried out on complexes deviating from the stable 18-valence-electron (18-v.e.) closed-shell configuration, due to one-electron oxidation or reduction. While such species often are said to have the unpaired electron in an orbital on the central metal ion, this idea is not exact. Thus the ^{13}C hyperfine structure in the EPR spectrum of the 17-v.e. complex $V(CO)_6$ suggests that it is a π radical with considerable unpaired-electron population on the ligands. Similarly, the 19-v.e. sandwich complex cobaltocene (Cp_2Co, where Cp is cyclopentadiene), yields EPR parameters [e.g., $A(^{59}Co)$] whose interpretation requires considerable

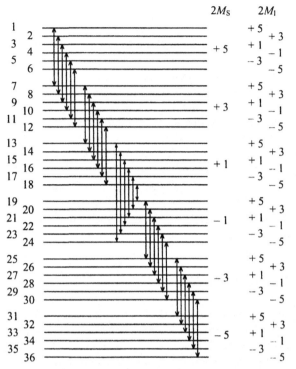

FIGURE 8.12 Stylized energy-level diagram for $B \gg 0$, not to scale. The most important EPR spin flips are displayed (with appropriate energy-level labels), albeit not with identical transition energies such as would be operative for field-swept (fixed-frequency) spectra.

(42%) ligand unpaired-electron populations. Reviews of 17-v.e. species [35] and 19-v.e. species [36] are available.

The spectrum of Mn^{2+} is quite complicated, but can be analyzed quantitatively, featuring g anisotropy, electronic quadrupole high-spin ($S = \frac{5}{2}$) effects (D and E), as well as hyperfine splittings from the 100% isotope ^{55}Mn ($I = \frac{5}{2}$). In the random powder (actually putty) phase, Mn^{2+} in plasticine (modeling clay) is stable, cheap and convenient to use, say, as a standard [37]. A stylized energy-level diagram and a breakdown of an actual field-swept X-band spectrum (48 transitions shown) are depicted in Figs. 8.12 and 8.13.

8.6 A FERROELECTRIC SYSTEM

Certain materials can be thought of as containing electric dipoles, which behave much like their ferromagnetic and paramagnetic analogs. One such substance is potassium ferrocyanide trihydrate, which is diamagnetic, with a ferroelectric transition ('Curie') temperature of $T_c = -26°C$. Crystals doped with V^{2+} ($S = \frac{3}{2}$, 99.8% ^{51}V

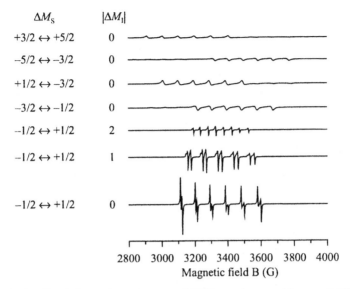

FIGURE 8.13 Simulations of seven types of EPR powder transitions at 9.40000 MHz: (A) $M_S: -\frac{1}{2} \leftrightarrow +\frac{1}{2}$, $|\Delta M_I| = 0$; (B) $M_S: -\frac{1}{2} \leftrightarrow \frac{1}{2}$, $|\Delta M_I| = 1$; (C) $M_S: -\frac{1}{2} \leftrightarrow +\frac{1}{2}$, $|\Delta M_I| = 2$; (D) $M_S: -\frac{3}{2} \leftrightarrow -\frac{1}{2}$, $\Delta M_I = 0$; (E) $M_S: +\frac{1}{2} \leftrightarrow \frac{3}{2}$, $\Delta M_I = 0$; (F) $M_S: -\frac{3}{2} \leftrightarrow -\frac{3}{2}$, $\Delta M_I = 0$; (G) $M_S: +\frac{3}{2} \leftrightarrow \frac{5}{2}$, $\Delta M_I = 0$. The first-derivative linewidths ΔB_{pp} for the individual lorentzian lines were 2.8 G for (A)–(C) and 14.0 G for (D)–(G).

with $I = \frac{7}{2}$) in small concentrations yield very nice EPR spectra with well-resolved hyperfine octets [38]. Thermal studies ($-180 \leftrightarrow 20°C$) with single crystals yield values of parameters D and E depending linearly on the local spontaneous electric polarization. Above T_c, the spin hamiltonian becomes close to temperature-independent. Quite dramatic linewidth changes (for transitions $\Delta M_s \pm \frac{3}{2} \leftrightarrow \pm \frac{1}{2}$) occur in the region near T_c. We thus have here a good example of how transition ions can be used as probes to explore phase transitions and other solid-state characteristics.

8.7 SOME *f*-ELECTRON SYSTEMS

So far in this book, we have almost completely neglected the topic of *f*-electron system EPR. We cannot cover this subject in any depth but have to content ourselves with discussing a few such systems.

Our primary choice is the U^{5+} ion, which has the configuration $5f^1$ in its ground state. Its properties, and those of other uranium cations, are discussed in a 1984 review [39]. Whereas $4f$ electrons occur relatively concentrated near the nucleus, $5f$ electrons are closer to the periphery and hence are more affected by the electric field of the surroundings. The relativistic effects, including the spin-orbit interaction, are also more prominent for the latter.

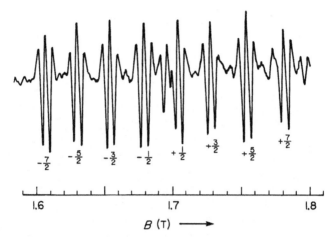

FIGURE 8.14 The second-derivative EPR spectrum (9.309 GHz) of a U^{5+} oxide center (at 77 K) embedded in crystalline LiF, taken with **B** ∥ (100). The primary spectrum arises from ^{235}U ($I = \frac{7}{2}$) in the enriched sample and shows a doublet splitting attributed to a $^{19}F^-$ nearest-neighbor ion. The center of the spectrum shows remnant ^{238}U lines. [After V. Lupei, A. Lupei, S. Georgescu I. Ursu, *J. Phys. C: Solid State Phys.*, **9**, 2619 (1976).]

When uranium oxide is doped into LiF, the U^{6+} ion (which is diamagnetic, $5f^2$) enters along with oxide ions (i.e., as $UO_5{}^{4-}$) and fits into the structure, replacing Li^+ plus five F^- ions. Irradiation produces the U^{5+} ion, with five oxide ions and one fluoride ion as neighbors, which yields quite sharp EPR lines and reveals a hyperfine doublet arising from the nearest-neighbor fluoride ion. Consistent with theory developed for the isoelectronic ion Np^{6+} [40], the g factor is *negative* and of small magnitude [$g_{||} = -0.3935(5)$ and $g_\perp = -0.5912(5)$] [41]. Figure 8.14 shows an EPR spectrum for the species obtained with uranium highly enriched in ^{235}U ($I = \frac{7}{2}$). In this instance, one has the unusual case where the nuclear quadrupole effects are at least as large as the hyperfine coupling, as was discerned from the line-position anisotropy.

A quite different view of actinides is to use them as radiation sources, embedded in materials and monitored by EPR. Thus poly(vinyl alcohol) (PVA) films have been doped with $^{238}U(VI)$ and also with $^{239}Pu(IV)$ to provide in situ α irradiaction, with and without accompanying external γ irradiation, which revealed formation of various free radicals and some view of their reactions [42].

REFERENCES

1. W. Low, *Paramagnetic Resonance in Solids*, Solid State Physics, Supplement 2, Academic Press, New York, NY, U.S.A., 1960.

2. B. R. McGarvey, "Electron Spin Resonance", in *Transition-Metal Chemistry*, R. L. Carlin, Ed., Vol. 3, Marcel Dekker, New York, NY, U.S.A., 1966.

3. J. W. Orton, *Electron Paramagnetic Resonance*, Iliffe, London, U.K., 1968.

4. A. Abragam, B. Bleaney, *Electron Paramagnetic Resonance of Transition Metal Ions*, Oxford University Press, London, U.K., 1970.

5. G. E. Pake, T. L. Estle, "Ligand or Crystal Fields", in *The Physical Principles of Electron Paramagnetic Resonance*, 2nd ed., Benjamin, Reading, MA, U.S.A., 1973.

6. S. A. Al'tshuler, B. M. Kozyrev, *Electron Paramagnetic Resonance in Compounds of Transition Elements*, 2nd ed. (revised) (Engl. transl.), Wiley, New York, NY, U.S.A., 1974.

7. J. R. Pilbrow, *Transition Ion Electron Paramagnetic Resonance*, Oxford University Press, Oxford, U.K., 1990.

8. C. Ballhausen, *Introduction to Ligand Field Theory*, McGraw-Hill, New York, NY, U.S.A., 1962.

9. B. N. Figgis, *Introduction to Ligand Fields*, Interscience, New York, NY, U.S.A., 1966 (1986 reprint).

10. M. Gerloch, *Magnetism and Ligand Field Analysis*, Cambridge University Press, Cambridge, U.K., 1983.

11. J. S. Griffith, *The Theory of Transition Metal Ions*, Cambridge University Press, Cambridge, U.K., 1961.

12. P. W. Atkins, R. S. Friedman, *Molecular Quantum Mechanics*, 4th ed., Oxford University Press, Oxford, U.K., 2005.

13. P. H. E. Meijer, *Physica*, **26**, 61 (1960).

14. H. Bill, "Observation of the Jahn-Teller Effect with EPR", in *The Dynamic Jahn-Teller Effect in Localized Systems*, Yu. E. Perlin and M. Wager, Eds., North Holland, Amsterdam, Netherlands, 1984, Chapter 13.

15. E. Banks, M. Greenblatt, B. R. McGarvey, *J. Chem. Phys.*, **47**, 3772 (1967).

16. B. R. McGarvey, "Charge Transfer in the Metal-Ligand Bond as Determined by Electron Spin Resonance", in *Electron Spin Resonance of Metal Complexes*, T. F. Yen, Ed., Plenum Press, New York, NY, U.S.A., 1969, pp. 1 ff.

17. T. Garofano, M. B. Palma-Vittorelli, M. U. Palma, F. Persico, "Motional and Exchange Effects in the ESR Behavior of Ni^{2+} and Mn^{2+} Hexammine Cubic Crystals", in *Paramagnetic Resonance*, Vol. 2, W. Low, Ed., Academic Press, New York, NY, U.S.A., 1963, pp. 582 ff.

18. W. E. Blumberg, J. Eisinger, S. Geschwind, *Phys. Rev.*, **130**, 900 (1963).

19. A. S. Chakravarty, *J. Chem. Phys.*, **39**, 1004 (1963).

20. J. E. Wertz, J. R. Bolton, *Electron Spin Resonance*, McGraw-Hill, New York, NY, U.S.A., 1972, p. 268.

21. K. A. Müller, *Helv. Phys. Acta*, **31**, 173 (1958).

22. S. Ram, K. Ram, B. S. Shukla, *J. Mater. Sci.*, **27**, 511 (1992).

23. D. L. Griscom, *Glass: Sci. Technol.*, **48**, 151 (1990).

24. Y. Tanabe, S. Sugano, *J. Phys. Soc. Jpn.*, **9**, 753, 766 (1954).

25. A. Ozarowski, B. R. McGarvey, *Inorg. Chem.*, **28**, 2262 (1989).

26. S. Sasaki, T. Kambara, *J. Phys. C: Solid State Phys.*, **15**, 1035 (1982).

27. W. C. Lin, *J. Magn. Reson.*, **68**, 146 (1986).

28. E. Buluggiu, A. Vera, *J. Magn. Reson.*, **41**, 195 (1980).

29. M. Moreno, *Chem. Phys. Lett.*, **76**, 597 (1980).

30. E. Dalgaard, J. Linderberg, *J. Chem. Phys.*, **65**, 692 (1976).

31. J. Owen, J. H. M. Thornley, *Rep. Prog. Phys.*, **29**, 675 (1966).

32. E. Šimánek, Z. Šroubek, "Covalent Effects in EPR Spectra—Hyperfine Interactions", in *Electron Paramagnetic Resonance*, S. Geschwind, Ed., Plenum Press, New York, NY, U.S.A., 1972, Chapter 8.

33. B. R. McGarvey, "Superhyperfine Interaction for Ions with Non-quenched Orbital Angular Momentum", in *Electronic Magnetic Resonance of the Solid State*, J. A. Weil, Ed., Canadian Society for Chemistry, Ottawa, ON, Canada, 1987, pp. 83–97.

34. N. M. Atherton, A. J. Horsewill, *J. Chem. Soc. Faraday Trans. II*, **76**, 660 (1980).

35. M. C. Baird, *Chem. Rev.*, **88**, 1217 (1988).

36. D. Astruc, *Chem. Rev.*, **88**, 1189 (1988).

37. R. Rahimi-Moghaddam, Y. Upadrashta, M. J. Nilges, J. A. Weil, *Appl. Magn. Reson.*, **24**, 113 (2003).

38. D. E. O'Reilly, G. E. Schacher, *J. Chem. Phys.*, **43**(12), 4222 (1965).

39. I. Ursu, V. Lupei, *Bull. Magn. Reson.*, **6**, 162 (1984).

40. C. A. Hutchison Jr., B. Weinstock, *J. Chem. Phys.*, **32**, 56 (1960).

41. V. Lupei, A. Lupei, S. Georgescu, I. Ursu, *J. Phys. C: Solid State Phys.*, **9**, 2619 (1976).

42. M. Kumar, R. M. Kadam, A. R. Dhobale, M. S. Sastry, *J. Nucl. Radiochem. Sci.*, **1**(2), 77 (2000).

43. S. A. Marshall, T. T. Kikuchi, A. R. Reinberg, *Phys. Rev.*, **125**, 453 (1962).

44. V. A. Atsarkin, É. A. Gerasimova, I. G. Matveeva, A. V. Frantsesson, *Sov. Phys. JETP*, **16**, 903 (1963).

45. B. Henderson, J. E. Wertz, *Defects in the Alkaline Earth Oxides*, Halstead-Wiley, New York, NY, U.S.A., 1977, p. 7.

NOTES

1. The same is, of course, true for *any* EPR species.

2. These continue to be eigenfunctions when tetragonal distortions along the quantization axes are included.

3. The literature contains various notations for the splitting parameter: $\Delta = 10Dq$ and $\Delta = 120B_4$ [4, Fig. 7.3].

4. Other representations (e.g., A_{2g} for a non-degenerate state) also are employed later. Lower-case letters ($a_2, e, t_2, \ldots$) refer to one-electron orbitals.

5. Consistent with the relations $\Delta_{tth} = -(4/9)\Delta_{oct}$ and $\Delta_{cubal} = -(8/9)\Delta_{oct}$ (holding for equal R values).

6. Since there is a low-abundance magnetic isotope (^{57}Fe, $I = \frac{1}{2}$), there is access to hyperfine parameters here, too. Despite the predominance of d orbitals, which vanish at the nucleus, the S-state ions have sizable isotropic hyperfine components a_0 arising from core polarization (Section 5.2) caused by the d^5 electrons.

7. In an octahedral field, the d levels split so as to leave the triply degenerate t_{2g} orbitals lower in energy than the e_g orbitals. For example, in d^5 ions two situations arise, depending on the magnitude of the local electric field. In the high-spin case, Hund's rule applies; that is, the state $t_{2g}^3 e_g^2$ with maximum spin multiplicity has the lowest energy. In the low-spin case the local-field splitting is so large that electrons occupy only the lower group of levels, that is, t_{2g}^5.

FURTHER READING

1. A. Carrington, A. D. McLachlan, *Introduction to Magnetic Resonance*, Harper & Row, New York, NY, U.S.A., 1967, Chapter 10.

2. F. A. Cotton, *Chemical Applications of Group Theory*, 3rd ed., Wiley, New York, NY, U.S.A., 1990, Chapter 9.

3. B. E. Douglas, D. H. McDaniel, J. J. Alexander, *Concepts and Models of Inorganic Chemistry*, 3rd ed., Wiley, New York, NY, U.S.A., 1994.

4. B. N. Figgis, M. A. Hitchman, *Ligand Field Theory and Its Applications*, Wiley-VCH, New York, NY, U.S.A., 2000.

5. F. S. Ham, "Jahn-Teller Effects in EPR Spectra", in *Electron Paramagnetic Resonance*, S. Geschwind, Ed., Plenum Press, New York, NY, U.S.A., 1972, Chapter 1.

6. D. S. McClure, *Electronic Spectra of Molecules and Ions in Crystals*, Academic, New York, NY, U.S.A., 1959.

7. L. A. Sorin, M. V. Vlasova, *Electron Spin Resonance of Paramagnetic Crystals* (Engl. transl.), Plenum Press, New York, NY, U.S.A., 1973.

8. S. Sugano, Y. Tanabe, H. Kamimura, *Multiplets of Transition-Metal Ions in Crystals*, Academic Press, New York, NY, U.S.A., 1970.

9. J. E. Wertz, J. R. Bolton, *Electron Spin Resonance*, McGraw-Hill, New York, NY, U.S.A., 1972, Chapters 11 and 12.

PROBLEMS

8.1 Complete the table given below, giving the d-electron configurations for octahedral and tetrahedral symmetry. Here one neglects the d-electron repulsion but takes into account both the Pauli exclusion principle and Hund's rules. (P. W. Atkins, R. S. Friedman, *Molecular Quantum Mechanics*, 4th ed., Oxford University Press, Oxford, U.K., 2005, pp. 274 ff.)

n	\multicolumn{10}{c}{d^n}
	1	2	3	4	5	6	7	8	9	10
oct	t_{2g}^1						$t_{2g}^6 e_g$			
S	$\frac{1}{2}$						$\frac{1}{2}$			
tth	e^1				$e^4 t_2^2$					
S	$\frac{1}{2}$				1					

8.2 Consider the $3d^1$ cation surrounded by four anions such that it is exposed to a tetrahedral electric field with a tetragonal distortion along $\mathbf{z}$, causing the $(1/\sqrt{2})[|+2\rangle - |-2\rangle]$ state to be lowest. Use the Λ-matrix formalism to show that

$$g_\parallel \approx 2 - 8\lambda/\Delta \qquad (8.19a)$$

$$g_\perp \approx 2 - 2\lambda/\Delta \qquad (8.19b)$$

in this case. What approximations have been made?

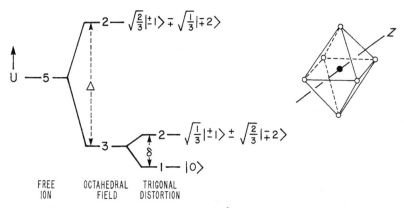

FIGURE 8.15   Splitting of state energies of a $d^1(\text{oct} + \text{trg})$ ion. The trigonal axis ($\mathbf{z}$) is a body diagonal of the circumscribing cube.

8.3   The wavefunctions (kets) and energies for a $3d^1$ ion in an octahedral field with trigonal distortion (along axis $\mathbf{z}$) are given in Fig. 8.15. Derive the expressions for $g_\parallel$ and $g_\perp$ in this case and thus show that these expressions are the same as for the tetragonal distortion case, that is

$$g_\parallel \approx g_e \tag{8.20a}$$

$$g_\perp \approx g_e - 2\lambda/\delta \tag{8.20b}$$

where terms involving Δ^{-1} have been neglected.

8.4   Derive the following equations from the dependence of $\mathbf{g}$ and $\mathbf{D}$ on Λ:

$$D = \tfrac{1}{2}\lambda\big[g_z - \tfrac{1}{2}(g_X + g_Y)\big] \tag{8.21a}$$

$$E = \tfrac{1}{4}\lambda(g_X - g_Y) \tag{8.21b}$$

8.5   For Ni^{2+} in Al_2O_3, $g_\parallel = 2.1957$ and $g_\perp = 2.1859$. Compute $\overline{D} = D/hc$ using the equations of Problem 8.4 and compare with the experimental value of $-1.375\ \text{cm}^{-1}$ [43]. For Cr^{3+} in $MgWO_4$, $g_\parallel = 1.966$, $g_\perp = 1.960$ and $\overline{D} = +0.795\ \text{cm}^{-1}$ [44]. Again calculate $\overline{D}$. Comment on any differences between the calculated and experimental $\overline{D}$ values.

8.6   As we have seen, the spin hamiltonian for a $3d^8$ ion in a uniaxial local electric field may be written in the form Eq. (8.17).

   (*a*) Use second-order perturbation theory (Sections 6.7, A.6 and C.1.7) for the electronic quadrupole plus hyperfine energy added to the electron Zeeman energy to show that primary EPR transitions when $\theta = 0$, occur at (near) magnetic fields given by

$$B_{-1 \to 0} = B' + \frac{D}{g_\| \beta_e} - \frac{A_\| M_I}{g_\| \beta_e} - \frac{A_\perp^2}{2 g_\|^2 \beta_e^2 B'} [I(I+1) - M_I^2 - M_I] \quad (8.22a)$$

$$B_{0 \to +1} = B' - \frac{D}{g_\| \beta_e} - \frac{A_\| M_I}{g_\| \beta_e} - \frac{A_\perp^2}{2 g_\|^2 \beta_e^2 B'} [I(I+1) - M_I^2 + M_I] \quad (8.22b)$$

where $B' = h\nu / g_\| \beta_e$ is sufficiently large that perturbation theory may be applied, and M_S and M_I are 'good' quantum numbers. The second-order contributions from matrix $\mathbf{D}$ (see Eqs. 6.34) cancel out in the present case. Note that the nuclear Zeeman energy is taken to be negligible.

(*b*) Given that $D < 0$, $S = 1$ and $I = \frac{3}{2}$ (for $\mathbf{B} \parallel \mathbf{z}$), use the resonance field positions, obtained from Eq. 8.22 and listed below for the hyperfine quartets, to confirm the sign and magnitude of the hyperfine parameters given below for the $^{63}Cu^{3+}$ ion in Al_2O_3 [18]; the frequency used was 9.0420 GHz:

Transition	Fields B (mT)	Transition	Fields B (mT)
$\lvert-1\rangle \to \lvert 0\rangle$	115.7	$\lvert 0\rangle \to \lvert+1\rangle$	504.1
$\lvert-1\rangle \to \lvert 0\rangle$	113.4	$\lvert 0\rangle \to \lvert+1\rangle$	501.6
$\lvert-1\rangle \to \lvert 0\rangle$	120.0	$\lvert 0\rangle \to \lvert+1\rangle$	508.2
$\lvert-1\rangle \to \lvert 0\rangle$	126.9	$\lvert 0\rangle \to \lvert+1\rangle$	505.9

Parameters: $\Delta(^3T_2)/hc = 21000$ cm^{-1} (see Fig. 8.6);
$g_\| = 2.0788$; $g_\perp = 2.0772$; $D/hc = -0.1884$ cm^{-1};
$E = 0$; $^{63}A_\|/hc = -0.00644$ cm^{-1}; $^{63}A_\perp/hc = -0.00601$ cm^{-1}.

These line positions are off from field values obtained via exact numerical solutions by an average value of 12 mT. This improves somewhat at higher frequencies. The relevant energy-level diagram is depicted in Fig. 8.16.

8.7 The $3d^3$ ion Cr^{3+} has been observed in association with a cation vacancy in a nearest-cation site along a $\langle 100 \rangle$ direction in CaO [45]. The local symmetry is tetragonal, with $g_\| = 1.9697$, $g_\perp = 1.9751$, and $|D|/hc = 0.13606$ cm^{-1}; the value of D, given in Ref. 45, presumably should be negative.
(*a*) Predict the value of S, and present the reasoning behind your result.
(*b*) Use $\lambda = 91$ cm^{-1} for the Cr^{3+} free ion to estimate $\Delta_\|$ and $\Delta_\perp$.
(*c*) Is the measured value of D/hc compatible with the observed g factors?
(*d*) Sketch the expected EPR spectra for $\mathbf{B}$ parallel to the axes of the cubic crystal.

8.8 For a $3d^5$ (*high-spin*) ion in an octahedral crystal field, show that the relative intensities of the five fine-structure EPR lines are $5:8:9:8:5$. (*Hint:* Compute the matrix elements of $\hat{S}_+$.)

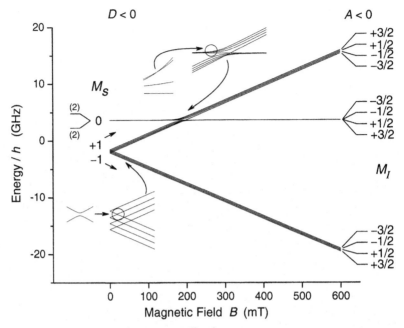

FIGURE 8.16 Energy-level diagram for ^{63}Cu^{3+} in Al$_2$O$_3$ at ~1.5 K. The EPR parameters used [18] in the computer calculation are $g_{\parallel} = 2.0788$, $g_{\perp} = 2.0772$, $D_{\parallel} = 64.71$ mT; $D_{\perp} = -129.42$ mT; $A_{\parallel} = -6.19$ mT; $A_{\perp} = -6.64$ mT. Note the two-fold (Kramers) degeneracy of all levels at zero magnetic field. The avoided crossing of two levels at $B \cong 3$ mT is indicated, as are the multiple level crossings at B near to 200 mT.

8.9 The expression

$$\left[\frac{g_{\parallel} A_{\parallel}}{g} \cos^2 \theta + \frac{g_{\perp} A_{\perp}}{g} \sin^2 \theta \right] M_S M_I$$

for the energy of a state $|M_S, M_I\rangle$ may be used for a system with uniaxial symmetry if $|A_{\parallel}| \gg |A_{\perp}|$. Show that it may be transformed into the more usual expression

$$\left[a_0 + b_0(3 \cos^2 \theta - 1) \right] M_S M_I$$

for angular variation of hyperfine splitting, if the anisotropy of the g factor is neglected.

8.10 Consider the loops depicted in the Fig. 8.10 plot of line positions versus angle between the crystal and **B**. Interpret this behavior. (*Hint:* Consider the change with angle of the energy-level spacings, which need not be linear, and remember that we are dealing with a fixed-frequency experiment.)

8.11 Estimate the splittings of the two doublets indicated in Fig. 8.8, using Eq. 8.18*b*. Compare your answer with the results published by Müller [21].

CHAPTER 9

THE INTERPRETATION OF EPR PARAMETERS

9.1 INTRODUCTION

An adequate interpretation of the parameters of the spin hamiltonian usually requires the application of molecular quantum mechanics. We already noted the explicit expressions for matrices **g** and **D** (Eqs. 4.41 and 4.42), as well as the relation of the isotropic and anisotropic parts of the hyperfine matrix to $|\psi(0)|^2$ (Eq. 2.38) and $\langle r^{-3} \rangle$ (Eq. 5.6a). In general, for quantitative interpretation, a large-scale analysis by computer is required. Happily, however, relatively simple and successful analytical quantum-mechanical models exist that serve well for tutorial purposes. This chapter introduces some of these techniques. One key aspect for EPR analysis is to understand which, among the infinite number of electronic states, is the ground state, and to realize its properties. Part of the power of EPR spectroscopy is that (unlike optical spectroscopy) it deals only with spin transitions within a single electronic state.

Before going further, we need to tighten up our knowledge of spin densities and unpaired-electron populations.[1] Like other densities, spin densities ρ_s can be evaluated in selected local regions and thus depend on location within a paramagnetic atom or molecule. Thus they can be integrated over part or the total volume, yielding dimensionless physically useful parameters ρ called the 'unpaired-electron populations' on the species considered.

There can be a simple proportionality between spin density $\langle \rho_s \rangle_n$ at a nucleus n and unpaired-electron population ρ (see Section 5.6). For example,

Electron Paramagnetic Resonance, *Second Edition*, by John A. Weil and James R. Bolton
Copyright © 2007 John Wiley & Sons, Inc.

$\langle \rho_s \rangle_p = \rho |\psi|_p^2$ for atomic hydrogen; both functions are evaluated at the proton p [1,2]. In general, however, both are multi-electron functions.

Consider, as an example, a Gd(3+) ion ($S = \frac{7}{2}$) embedded in some complex. The spin density is very high at the cation, with only small magnitudes occurring on the nearest-neighbor ligands. The unpaired-electron population here is 7. In most free radicals, $\rho = 1$. More examples follow.

In situations where the anisotropic hyperfine splitting on some nucleus n can be considered to be purely dipolar and is uniaxial (see Eqs. 5.48 and 5.49 near the limit $f = 1$), the distance r between the electron-spin species, requiring unpaired-electron population correction factors $k(e)$, and external nucleus n carrying a point magnetic population [no corrections; $k(n) = 1$ for all our purposes] can be approximated by

$$ r = \left(\frac{\mu_0}{4\pi} \right) \frac{g\beta_e g_n \beta_n k(e)k(n)}{T_\perp} \tag{9.1} $$

and thus this distance can be estimated [3]. Matrix **T** is defined in Eq. 5.8, and is expressed in energy units. We note that $T_\perp$ varies as approximately r^{-3} and may contain geometry factors. The distance r is, of course, set by the electromagnetic quantum-mechanical interactions between all atoms present, and is hardly affected by the magnetic dipole-dipole effect. The above Equation 9.1 may be compared to its isotropic equivalent, Eq. 2.51.

Since we have considered the transition ions in Chapter 8, we focus in this chapter on the interpretation of EPR parameters from free radicals and triplet states. Free radicals are classified into organic radicals, inorganic radicals and point defects in crystalline solids. A short discussion of EPR in metals and semiconductors is also included.

It is useful to distinguish between σ-type and π-type free radicals. The former type features one unpaired electron in an orbital having no nodal plane, whereas the latter has one unpaired electron in a molecular orbital that has such a symmetry element. Often the nodal plane in π-type radicals extends over several atoms; this arises from overlap between the p orbitals on each atom and implies that the unpaired electron is delocalized over the system. By contrast, σ-type radicals tend to have unpaired electrons primarily localized on one atom.

9.2 π-TYPE ORGANIC RADICALS

Among the various molecular-orbital theoretical approaches [4], Hückel molecular-orbital (HMO) theory is the simplest. We shall apply this theory to some relatively simple paramagnetic species, with a view to understanding the isotropic hyperfine splittings exhibited by these π-type radicals. Some details of HMO theory are to be found in Appendix 9A at the end of this chapter.

Most of the radicals examined in Chapter 3 are conjugated molecules, containing paired electrons in low-lying σ orbitals and the remainder in π orbitals. The distinguishing characteristic in modeling such compounds, diamagnetic or paramagnetic, is the overlap of p orbitals on adjacent atoms. Such overlapping permits the electrons in these orbitals to be delocalized as a π system over the molecular skeleton. One may, to a good approximation, describe the energy states of these electrons separately from those of the others, in terms of molecular orbitals generated from linear combinations of the atomic $2p_z$ orbitals. For example, each of the six $2p_z$ orbitals in benzene has a node in the molecular plane, defined to be the xy plane. Hence the molecular orbitals arising from combinations of $2p_z$ orbitals (Tables 9A.1 and 9A.2) are referred to as π *orbitals*. The ground state of benzene consists of three π orbitals containing six electrons, with the other three π orbitals unoccupied.

Each unpaired electron of a π-type radical is expected to be distributed over the molecular framework. For example, in the benzene monoanion the average relative probability of finding its unpaired electron in the vicinity of any one carbon atom is $\frac{1}{6}$, as required by symmetry. For other monocyclic radicals a similar uniform distribution should be found. The equivalence of each position in a given monocyclic radical leads to the ^{1}H hyperfine splitting patterns shown in Figs. 3.3*a*–*h*.

For radicals with lower symmetry, there is no such obvious guide to the unpaired-electron distribution. The HMO approach provides valuable guidance toward determining this distribution. The information of interest is contained in the expression for the particular spatial molecular orbital

$$\psi_i = \sum_{j=1}^{n} c_{ij}\phi_j \tag{9.2}$$

occupied by the unpaired electron. Here n is the number of atomic orbitals ϕ_j, which are orthonormal. Since ψ_i is normalized, one has

$$\sum_{j=1}^{n} |c_{ij}|^2 = 1 \tag{9.3}$$

The magnitude squared of the coefficient c_{ij} is the relative probability that the electron in molecular orbital ψ_i is in atomic orbital ϕ_j. Thus $|c_{ij}|^2$ measures the unitless unpaired π-electron population ρ_j 'on' atom j when this atom bears only a single orbital occurring in ψ_i:

$$\rho_j = |c_{ij}|^2 \tag{9.4}$$

It follows that

$$\sum_{j=1}^{n} \rho_j = 1 \tag{9.5}$$

As an example, consider the radical anion of 1,3-butadiene [5]. The EPR spectrum displayed in Fig. 3.7*a* was analyzed on the basis of a quintet of lines of relative intensities $1:4:6:4:1$ with a proton hyperfine splitting of 0.762 mT;

each line of the quintet is split further into a $1:2:1$ triplet with a proton hyperfine splitting of 0.279 mT. The considerable difference between the two splitting 'constants' suggests a highly non-uniform unpaired-electron distribution.

The butadiene anion has five π electrons. Reference to the molecular orbitals of Table 9A.1 shows that, consistent with the Pauli exclusion principle, the unpaired electron must reside in ψ_3. From Eq. 9.4 the unpaired-electron populations are found to be $\rho_1 = \rho_4 = 0.36$ and $\rho_2 = \rho_3 = 0.14$. Thus HMO theory predicts that the end carbon atoms should have the higher unpaired-electron densities. We note that these are indeed the positions at which the larger proton hyperfine splittings are observed and that the ratio of the hyperfine splittings, $a_1/a_2 = 2.7$,[2] agrees satisfactorily with the ratio of the unpaired-electron populations, $\rho_1/\rho_2 = 2.6$.

This correspondence seems to point to some sort of linear relation between the (isotropic) proton hyperfine splitting parameters a_k and the unpaired π-electron populations of the carbon atoms in π-type organic radicals. Indeed, such a relation has been proposed [6–9]; it may be written for proton k as

$$a_k = Q_{\rho k} \tag{9.6}$$

where ρ_k is the unpaired π-electron population at the adjacent carbon atom k and Q is a proportionality constant expressed in magnetic-field units. The origin of Eq. 9.6 is considered in Section 9.2.4. An examination of Fig. 9.1 shows that for most π-type

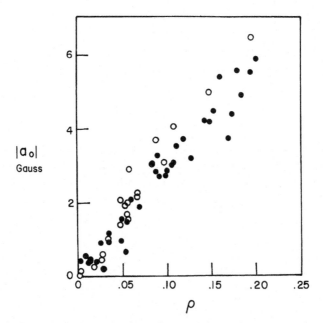

FIGURE 9.1 Experimental proton hyperfine splitting parameters $|a_0|$ versus HMO unpaired-electron populations ρ for a group of aromatic hydrocarbon radical ions. Open circles refer to positive ions and filled circles to negative ions. [After I. C. Lewis, L. S. Singer, *J. Chem. Phys.*, **43**, 2712 (1965).]

organic radicals the correlation (Eq. 9.6) is good. A similar and even more extensive data set, featuring a good linear plot of the hydrogen $1s$-orbital unpaired-electron population against the adjacent carbon $2p_z$-orbital unpaired-electron population, has been presented by Pople et al. [2]. Here, rather than the crude HMO method, the much more advanced INDO molecular-orbital technique was utilized.[3]

Theoretical estimates of Q place it in the range from -2 to -3 mT. The significance of the negative sign is explained in Section 9.2.4. The first confirmation of the negative sign of Q in Eq. 9.6 was obtained by an analysis of the splittings in the malonic acid radical (HOOC—CH—COOH), created by irradiation of the acid [10]. The simplest π radical is, of course, methyl (CH_3). It has $S = \frac{1}{2}$ and exhibits a proton hyperfine splitting constant of -2.304 mT [11]. This agrees nicely with Eq. 9.6 with $\rho = 1$, providing strong evidence that CH_3 indeed is planar (cf. CF_3, Section 9.3).

In certain other molecules, it is possible to establish a value of Q semi-empirically from the experimental hyperfine splittings. For instance, in the cyclic polyene radicals C_5H_5 (cyclopentadienyl) (**I**), $C_6H_6^+$, C_7H_7 (cycloheptatrienyl) (**II**) and $C_8H_8^-$, which are planar, the unpaired-electron population is known from the molecular symmetry. Thus an experimental determination of the hyperfine splitting constant a in these molecules provides an estimate of Q. Table 9.1 gives the experimental values of a and the corresponding values of Q for these monocyclic radicals.

(I) cyclopentadienyl radical   **(II) cycloheptatrienyl radical**

There is an appreciable variation in Q for these monocyclic radicals. If one compares the values for the two neutral radicals or for the two negatively charged

TABLE 9.1   Proton Hyperfine Splitting Parameters for Monocyclic Radicals

Radical	Temperature[a] (K)	a^H (mT)	Q (mT)	Reference
C_5H_5	~200	0.600	3.00	b
$C_6H_6^+$	298	0.428	2.57	c
$C_6H_6^-$	173	0.375	2.25	d
C_7H_7	298	0.395	2.77	e, f
$C_8H_8^-$	~298	0.321	2.57	g

[a] Some of the hyperfine splittings have been found to be temperature-dependent.[b, c, f]
[b] R. W. Fessenden, S. Ogawa, *J. Am. Chem. Soc.*, **86**, 3591 (1964). See also: M. Iwasaki, K. Toriyama, K. Nunome, *J. Chem. Soc., Chem. Commun.*, 320 (1983).
[c] M. K. Carter, G. Vincow, *J. Chem. Phys.*, **47**, 292 (1967).
[d] J. R. Bolton, *Mol. Phys.*, **6**, 219 (1963).
[e] A. Carrington, I. C. P. Smith, *Mol. Phys.*, **7**, 99 (1963).
[f] G. Vincow, M. L. Morrell, W. V. Volland, H. J. Dauben Jr., F. R. Hunter, *J. Am. Chem. Soc.*, **87**, 3527 (1965).
[g] T. J. Katz, H. L. Strauss, *J. Chem. Phys.*, **32**, 1873 (1960).

radicals, the variation is much smaller, correctly suggesting that the charge on the radical has some effect on Q.

An understanding of the hyperfine properties of protons in conjugated hydrocarbon radicals is aided by classifying these as alternant or non-alternant. A molecule (and its ions) is defined as alternant if one may label alternate positions of the carbon skeleton with an asterisk and have no two adjacent positions both 'starred' or both 'unstarred'. All linear systems are alternant, as are also those cyclic systems that have no rings made up of an odd number of atoms, for example, anthracene (Fig. 3.9). In contrast, the C_5H_5 and C_7H_7 radicals are non-alternant, as is the azulene anion (**III**). When there is more than one way of starring atoms, by convention one adopts that designation that gives the largest number of starred atoms. If the numbers of starred and of unstarred positions are equal, the hydrocarbon is called *even-alternant*; if not, it is called *odd-alternant*.

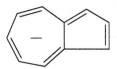

(**III**) azulene anion

Odd-alternant hydrocarbon radicals have a very useful property that permits rapid calculation of the unpaired-electron populations without actually determining molecular-orbital coefficients. As an example, consider the non-bonding semi-occupied orbital ψ_4 of the benzyl radical ($C_6H_5CH_2$) (**IV**)

(**IV**) benzyl radical

$$\psi_4 = 0\phi_1 - 0.378\phi_2 + 0\phi_3 + 0.378\phi_4 + 0\phi_5 - 0.378\phi_6 + 0.756\phi_7 \qquad (9.7)$$

Having starred this odd-alternant radical appropriately, one assigns equal and opposite coefficients about unstarred positions having two neighbors. One begins by assigning the coefficient $-c$ to atom 2, $+c$ to 4, $-c$ to 6, and finally $+2c$ to 7 to cancel the contributions from atoms 2 and 6. The squares of the coefficients must sum to unity; hence $c = 1/\sqrt{7} = 0.378$. The unpaired-electron population is then $\frac{1}{7}$ at atoms 2, 4 and 6, and $\frac{4}{7}$ at atom 7. The simple procedure employed here for determining ρ values saves much effort, as compared with the direct HMO calculation (Problem 9A.4). This procedure may also be applied to even-alternant hydrocarbons if non-bonding orbitals are present (e.g., cyclooctatetraene).

TABLE 9.2   Benzyl Radical^a Hyperfine Parameters a_i (in mT)

Protons on Carbon Atoms	^{1}H Splitting Parameters a_i	
	Experimental^{b,c}	HMO Calculated
2,6	−0.49	−0.40
4	−0.61	−0.40
7 (CH$_3$)	−1.59	(−1.59)
3,5	+0.15	0

^{a}See structure **IV**.
^{b}W. T. Dixon, R. O. C. Norman, *J. Chem. Soc.*, 4857 (1964).
^{c}A. Carrington, I. C. P. Smith, *Mol. Phys.*, **9**, 137 (1965).

The experimental hyperfine splittings for the benzyl radical (**IV**) are given in Table 9.2. Using the splitting for position 7 to fix Q, the hyperfine splittings for positions 2, 4 and 6 are calculated to be −0.40 mT. No hyperfine splitting would be expected for protons at positions 3 and 5 because the atomic orbital coefficients are zero. The significance of the small positive hyperfine splitting observed for protons at these positions is discussed later in this chapter. Although there are significant deviations from predictions, one can regard the calculated values as being in remarkable agreement with experiment, considering the crudity of the approach.

9.2.1   Anions and Cations of Benzene and Some of Its Derivatives

Benzene represents a classic hydrocarbon for study of the effects of substituents in removing the degeneracies of energy levels and for modifying the unpaired-electron distribution in its ±1 ions. In common with numerous other monocyclic systems, the π HMO molecular orbital of lowest energy in benzene is non-degenerate; the next four higher orbitals form degenerate pairs (Table 9A.2). It is customary to use the group-theoretic labeling (group D_{6h}). Here it is sufficient to note that e always refers to degenerate pairs of orbitals, whereas a and b refer to non-degenerate orbitals. The set of six molecular orbitals for benzene is given in Table 9A.2 in order of *increasing* energy (bottom to top). The bracketed orbitals are degenerate. Note that in the a orbital there is no change of sign and hence no vertical nodal plane. This is the orbital of lowest energy. In increasing order of energy, the e_1 orbitals have two oppositely signed (+/−) regions and a nodal plane, the e_2 orbitals two nodal planes, whereas the b orbital, that of highest energy, has three nodal planes.

In the benzene anion, the extra (7th) π electron is in the e_2 set of orbitals, whereas in the cation an electron is missing from the e_1 set. Hyperfine splitting data for the anion allow us to show how occupancy of the e_2 orbitals changes with substitution of the benzene ring.

The liquid-solution EPR spectrum of the benzene anion at −100°C is shown in Fig. 3.4. The spectrum consists of seven lines with intensities characteristic of hyperfine interaction from six equivalent protons. This result is expected from the symmetry of the molecule, but it is instructive to see how it arises from the Hückel molecular orbitals given in Table 9A.2. The six π electrons of the neutral benzene go into the three bonding molecular orbitals, but the addition of an extra electron to make the benzene anion creates a new problem. The lowest unoccupied molecular orbital in benzene is doubly degenerate. Hence the unpaired electron is expected to occupy equally the two e_2 antibonding molecular orbitals.[4] The coefficients at each of the atoms for these orbitals are given at the right of Fig. 9.2. It is evident that the wavefunction of one e_2 state, A, is antisymmetric with respect to reflection in a vertical plane passing through carbon atoms labeled 1 and 4; S, the other, is symmetric with respect to reflection in the same plane. A is termed the 'antisymmetric' orbital and S the 'symmetric' orbital.

The total unpaired-electron population at a given position is obtained by taking one-half the sum of the electron populations (squares of coefficients) at that position for each of the two orbitals. For example, at positions 1 and 2, $\rho_1 = \frac{1}{2}\left(0 + \frac{1}{3}\right) = \frac{1}{6}$ and $\rho_2 = \frac{1}{2}\left(\frac{1}{4} + \frac{1}{12}\right) = \frac{1}{6}$. Thus all positions are equivalent.

Although in the benzene anion the orbitals A and S are, in the first approximation, equally occupied, the population balance is extremely delicate. The introduction of substituents serves to remove the degeneracy, that is, makes one orbital more energetic than the other. Thus the effect of substituents on the EPR spectrum of the benzene anion is best understood by considering the limiting spectra anticipated when the unpaired-electron distribution approximates that of the A or of the S orbitals.

The EPR spectrum in Fig. 9.2 is that of the p-xylene anion [12]. It is significant that the splitting from the CH_3 protons is too small to be resolved. This phenomenon is to be expected when the unpaired electron resides predominantly in the A orbital.

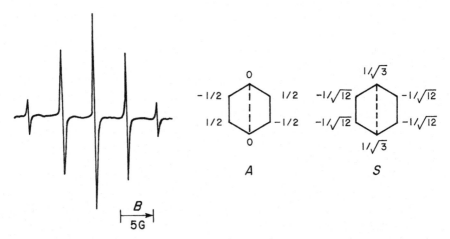

FIGURE 9.2   EPR spectrum of the p-xylene anion, with the atomic orbital coefficients of the antisymmetric (A) and symmetric (S) molecular orbitals of benzene at the right. The symmetry is defined with respect to the perpendicular plane (dashed) passing through the center of the molecule. Solvent is dimethoxyethane, and temperature is $-70°C$. [After J. R. Bolton, A. Carrington, *Mol. Phys.*, **4**, 497 (1961).]

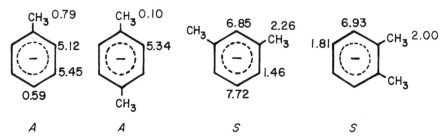

FIGURE 9.3 Proton hyperfine splittings *a* (in gauss) for various methyl-substituted benzene anions. Symbols *A* and *S* indicate whether the antisymmetric or the symmetric π orbitals lie lowest for these molecules. [After J. R. Bolton, A. Carrington, *Mol. Phys.*, **4**, 497 (1961); J. R. Bolton, *J. Chem. Phys.*, **41**, 2455 (1964).]

The hyperfine splittings that have been observed for various methyl-substituted benzenes are given in Fig. 9.3. The data show that the introduction into benzene of even one CH_3 group removes the degeneracy of *A* and *S*. The methyl groups are rapidly rotating, that is, effectively linear. The electronic properties of the substituent determine whether the *A* or the *S* orbital has the lower energy. The methyl group is considered to be electron-releasing in conjugated systems. For the toluene anion, the *A* orbital has a vertical nodal plane through the 1 and 4 positions, whereas the *S* orbital has a large unpaired-electron population $\left(\frac{1}{3}\right)$ at these positions. Repulsion between the electrons on the methyl group and the large negative charge at positions 1, 2, 5 and 6 in the *S* orbital causes the latter to be destabilized relative to *A*.

The *Q* value of -2.25 mT for the benzene anion may be used to estimate the toluene anion hyperfine splittings. Because of the node through the 1 and 4 positions, one should expect little or no hyperfine splitting from the methyl protons or the proton *para* to the methyl group. An unpaired-electron population of $\frac{1}{4}$ should give rise to a hyperfine splitting of ~0.56 mT. The measured hyperfine splittings (Fig. 9.3) show that the unpaired-electron distribution does approximate that of the *A* orbital.

Even the substitution of a deuterium nucleus for a proton in the benzene anion is sufficient to bring about a measurable split of the energies of the *S* and *A* orbitals as indicated by small departures of the proton hyperfine splittings from those in the benzene anion [13].

(V) benzene-d_1 anion

9.2.2 Anions and Cations of Polyacenes

For some of the EPR spectra analyzed in Chapter 3, it was not possible to assign the observed hyperfine splittings on the basis of the spectrum alone. Thus, for the mono-anion of naphthalene (**VI**) (Fig. 3.8), it is not obvious which set of four equivalent protons should be assigned as yielding the larger hyperfine splitting. The same uncertainty is found for the two quintet splittings in the spectrum (Fig. 3.9) of the anion of anthracene (**VII**). It thus is very desirable to have a simple and rational basis for the assignment of these hyperfine splittings, as presented here. In addition, it is helpful that the relative magnitudes of hyperfine splittings can be predicted without making detailed calculations.

In the HMO approximation, alternant hydrocarbons have orbital energies symmetrically disposed about the central energy α. *Odd*-alternant hydrocarbons have a non-bonding orbital at this energy (e.g., Fig. 9A.1). Orbitals with energies symmetrically disposed about α involve the same atomic orbitals, with coefficients that have the same *absolute* magnitudes. Therefore the squares of the coefficients of the highest bonding orbital and of the lowest antibonding orbital of an even-alternant hydrocarbon are identical. Hence *the unpaired-electron distribution is predicted to be identical in the cation and anion radicals corresponding to a given diamagnetic parent.* This statement is often referred to as the *pairing theorem* and is found to apply to a high degree of approximation [14,15].

The EPR spectra of the anions and cations of some of the polyacenes [naphthalene (**VI**), anthracene (**VII**), tetracene (**VIII**) and pentacene (**IX**)] have been studied. The proton hyperfine splittings for these molecules are listed in Table 9.3. It is evident that the hyperfine splittings are similar for protons in corresponding positions in the anion and the cation of a given molecule. These results are in reasonable accord with the pairing theorem. The agreement in reality is even better than is apparent, since Q depends somewhat on the excess charge density [16,17].

9.2.3 *g* Factors of π Radicals

The g factors of π radicals have been the focus of considerable theoretical attention, basically using the theory outlined in Section 4.8. Typically, $g - g_e = (1 \text{ to } 4) \times 10^{-4}$. For aromatic radicals in the liquid phase, Stone [18,19] showed that

$$g - g_e = g_{(0)} + g_{(1)}\lambda + g_{(2)}\lambda^2 \tag{9.8}$$

where the $g_{(i)}$ are (semi-empirical) parameters and λ is the coefficient of the resonance integral β in the HMO of the unpaired electron. One can classify the excited states required (Eqs. 4.38 and 4.41) to calculate **g** into different types; for example, excitation of the odd π electron into an antibonding σ orbital, and excitation of any σ-bonding electron into the semi-occupied π orbital. The theory fits well except when there is a degenerate (or almost so) ground state, for example, for the benzene radical anion, in which case complex corrections (for vibronic coupling and ion pairing) must be made [20–22].

TABLE 9.3 Proton Hyperfine Splitting Parameters in Polyacene Anions and Cations

Molecule	Position	$|a_+^H|$ (mT)	$|a_-^H|$ (mT)
Naphthalene[a] (**VI**)	1	0.540	0.495
	2	0.160	0.187
Anthracene[b] (**VII**)	9	0.6533	0.5337
	1	0.3061	0.2740
	2	0.1379	0.1509
Tetracene[c] (**VIII**)	5	0.5061	0.4226
	1	0.1694	0.1541
	2	0.1030	0.1162
Pentacene[d] (**IX**)	6	0.5083	0.4263
	5	0.3554	0.3032
	1	0.0975	0.0915
	2	0.0757	0.0870

[a](+) Estimated from simulating the X-band EPR spectrum taken in boric acid glass at ~300 K; G. Vincow, P. M. Johnson, *J. Chem. Phys.*, **39**, 1143 (1963) and G. S. Owen, G. Vincow, *J. Chem. Phys.*, **54**, 368 (1971); (−) N. M. Atherton, S. I. Weissman, *J. Am. Chem. Soc.*, **83**, 1330 (1961).
[b](+) and (−) J. R. Bolton, G. K. Fraenkel, *J. Chem. Phys.*, **40**, 3307 (1964).
[c](+) J. S. Hyde, H. W. Brown, *J. Chem. Phys.*, **37**, 368 (1962); (+) and (−) J. R. Bolton, unpublished work [see J. R. Bolton, *J. Chem. Phys.*, **71**, 3702 (1967)].
[d](+) and (−) J. R. Bolton, *J. Chem. Phys.*, **46**, 408 (1967).

9.2.4 Origin of Proton Hyperfine Splittings

As discussed, it has been found for planar conjugated organic radicals that proton hyperfine splittings are proportional to the unpaired π-electron population on the carbon adjacent to the proton (Eq. 9.6). Isotropic proton hyperfine splittings were shown in Chapter 2 to arise when a net unpaired-electron density exists at the proton. In π radicals, the unpaired electron can be considered to reside in a π-molecular orbital constructed from a linear combination of $2p_z$ carbon atomic orbitals. However, each such $2p_z$ orbital has a node in the plane of the molecule and, since this plane also contains the adjacent proton, there should be no unpaired-electron density at the proton and hence no hyperfine splitting. In spite of this node, the numerous spectra in Chapter 3 demonstrate that isotropic proton hyperfine splittings *do* occur in π radicals. Out-of-plane proton vibrations are found to give a negligible effect. Rather, the concept of unpaired-electron density must be reexamined in order to resolve this paradox. It was assumed that an electron in a conjugated molecule does not influence the other electrons in the radical. However, in reality the other electrons are affected. Thus, in some regions of the molecule, 'paired' electrons become slightly unpaired. (This is one of the several effects that go under the name of 'electron correlation'). Thus the actual *spin density* at the proton (Eq. 2.51) is not simply related to the nominal unpaired-electron population on the adjacent carbon atom. Factor Q in Eq. 9.6 brings in this effect, which we now discuss.

Consider a C—H fragment of a conjugated system. If spin α is assigned to the one electron in the $2p_z$ orbital on the carbon atom, there are two possibilities for assigning the spins in the C—H σ bond; these are shown in Fig. 9.4. Here it is assumed that the carbon atom has its $2p_z$ orbital perpendicular to the C—H bond; the $2p_x$ and $2p_y$ orbitals plus the $2s$ orbital of the carbon atom form trigonal sp^2 hybrids. The hydrogen atom bonds to one of these three coplanar hybrids.

If there were no electron in the $2p_z$ orbital, the electron configurations (a) and (b) of Fig. 9.4 would be equally probable; hence the spin density at the proton would be zero. However, when a $2p_z$ electron is present, say, with spin α, configurations (a) and (b) are no longer equally probable. This effect is often called *spin polarization*. It has been demonstrated from atomic spectroscopy that when two different, but equivalent, orbitals on the same atom are singly occupied by electrons, the more stable arrangement is the one with the electron spin components M_S equal (one of Hund's rules). Thus configuration (a), in which the two electrons shown on the

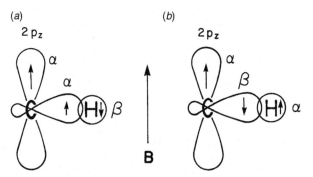

FIGURE 9.4 Possible electron-spin configurations in the σ-orbital bonding the carbon atom to the hydrogen atom in a C—H fragment, for a spin in the $2p_z$ orbital of that carbon: (a) spins parallel in the σ bonding orbital and the $2p_z$ orbital of carbon; (b) Corresponding spins antiparallel.

carbon atom have parallel spins, is more stable and hence more probable than (*b*), for which the spins are antiparallel; that is, as a consequence of the *positive* spin density at the carbon nucleus, there is a net *negative* electron-spin density (i.e., excess of β spin over α spin) at the proton. Conversely, if the spin state of the electron in the carbon p_z orbital is β, then α spin predominates at the proton. Detailed treatment of the effect demonstrates that Eq. 9.6 is close to quantitative [2,6–9], with Q negative. Of course, in a conjugated radical the unpaired-electron population ρ_i at a given carbon atom is less than unity. Note that for the ensemble of molecules, spin states β $\left(M_S = \frac{1}{2}\right)$ are the more populated (Fig. 1.2).

The concepts discussed above can be expressed elegantly and quantitatively in terms of a suitable mathematical formalism. We saw (Eq. 2.51) that the sign and magnitude of a_0 can be obtained quantum-mechanically by introduction of the spin-density operator $\hat{\rho}_s$. Here, for each nucleus,

$$\hat{\rho}_s(\mathbf{r}_s) = \langle \hat{S}_z \rangle^{-1} \sum_k \hat{S}_{kz} \, \delta(\mathbf{r}_k - \mathbf{r}_n) \tag{9.9}$$

is the spin-density operator, where the sum is over all electrons; $\hat{\mathbf{S}} = \sum_k \hat{\mathbf{S}}_k$ is the total electron-spin operator and $\langle \hat{S}_z \rangle$ is the expectation value $M_S = \sum_k M_{S_k}$ of $\sum_k \hat{S}_{kz}$ for the state $\psi(\mathbf{r}_k)$ considered (we assume $M_S \neq 0$). Clearly, ψ contains both spatial and electron-spin variables. The factor $\delta(\mathbf{r}_k - \mathbf{r}_n)$ is the famous Dirac delta 'function' (Section A.7) [4,23,24], here three-dimensional, with dimension of volume^{-1}. It has meaning only within a definite integral $\int_V F \, \delta \, dV$ of some spatial function $F(\mathbf{r}_k)$ (e.g., $\psi^*\psi$), which it sets to zero except at the single point $\mathbf{r}_k = \mathbf{r}_n$. Thus the integral becomes $F|_{\mathbf{r}_k=\mathbf{r}_n}$ when the volume V includes nucleus n.

9.2.5 Sign of the Proton Hyperfine Splitting Constant

The negative sign (Eq. 9.6) of Q implies that the hyperfine parameter a_k for the proton of a C—H fragment is negative, and that the spin density there is opposite in sign to that in the adjacent carbon $2p_z$ orbital.

Although this sign information can be obtained from comparison of the isotropic and anisotropic hyperfine couplings (Section 5.2 and Problem 5.11), another method, involving a verification of the signs by a proton magnetic resonance, is now examined.

This procedure involves the measurement of proton magnetic resonance line-shifts for paramagnetic molecules [25]. The NMR lines must be narrow enough relative to the magnitudes of the lineshifts to permit the measurement of the latter. Thus these lines must not be broadened too much by the relaxation of the proton spins in the presence of the nearby electron spin. This implies that the latter must relax relatively much more rapidly. The proton NMR spectrum of the biphenyl anion at room temperature is shown in Fig. 9.5. The chemical shifts in this spectrum are huge, compared to those found for protons in diamagnetic molecules, and arise from the local magnetic fields generated by each hyperfine interaction. Referring to Fig. 9.5, one notes that there are two lines shifted to the high-field side of the resonance position for diamagnetic molecules. These correspond to a negative value of a_i for two sets of

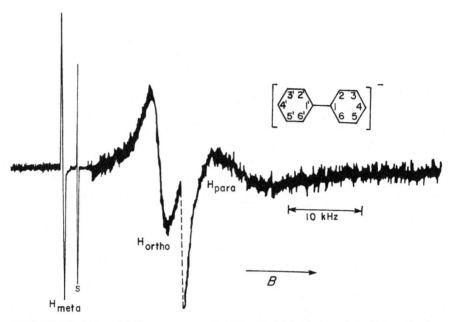

FIGURE 9.5 Proton NMR spectrum at 60 MHz of a 1 M solution of the biphenyl anion, structure (**X**) in diglyme CH_3—O—$(CH_2$—CH_2—O—$)_2CH_3$ at room temperature. The concentration of neutral biphenyl is negligible. The line S arises from the solvent. Various peaks have been measured with different radiofrequency power, gain and modulation. [After G. W. Canters, E. de Boer, *Mol. Phys.*, **13**, 495 (1967).]

protons in the radical. This result is expected from Eq. 9.6 with a negative value of Q and positive unpaired-electron population at the carbon atom. However, one line is shifted downfield; it must therefore correspond to a *positive* a_i for one set of protons. This result is understandable in terms of the new concept, negative spin density, introduced in the previous section. If proton hyperfine splittings are less than $\sim$0.6 mT, it may be possible to observe paramagnetic chemical shifts ΔB for free radicals in liquid solution [26,27].

The chemical shift ΔB is given by

$$\Delta B = B_i - B_0 = -\frac{g g_e \beta_e^2 B_0}{4 g_p \beta_n k_b T} a_i \qquad (9.10)$$

for the ith proton, where B_i is the resonance field for the shifted line and B_0 is the field corresponding to the unshifted proton resonance line [28,29]. The derivation of Eq. 9.10 is left to the reader, as Problem 9.2. It follows from Eq. 9.10 that there is a negative (downfield) chemical shift if a_i is positive, and vice versa.

The proton NMR spectrum of the biphenyl anion at room temperature is shown in Fig. 9.5. The chemical shifts in this spectrum are huge, compared to those found for protons in diamagnetic molecules, and arise from the local magnetic fields generated by each hyperfine interaction. Referring to Fig. 9.5, one notes that there are two lines shifted to the high-field side of the resonance position for diamagnetic molecules. These correspond to a *negative* value of a_i for two sets of protons in the radical. This result is expected from Eq. 9.6 with a negative value of Q and positive unpaired-electron population at the carbon atom. However, one line is shifted down-field; it must therefore correspond to a *positive* a_i for one set of protons. This result is understandable in terms of the new concept, *negative spin density*, introduced in the previous section.

If only one proton is attached to each carbon atom of a conjugated radical and if all π spin densities are positive, the extent (Section 3.5) of the EPR spectrum cannot exceed the value of $|Q|$. For most radicals, the spectral extent does not exceed ~ 2.7 mT (i.e., $|Q|$). However, for some radicals [e.g., the biphenyl anion (**X**) and the perinaphthenyl radical (**XI**)] the spectral extent is considerably in excess of this value. An extra-large spectral extent can be understood if negative π unpaired-electron populations occur. The normalization condition for unpaired-electron population requires that the *algebraic* sum of all such populations be unity for free radicals. If some populations are negative, then others must be correspondingly more positive. Consequently, the sum of the *absolute values* of the unpaired-electron populations can be greater than unity. Since the spectral extent depends only on the absolute magnitude of the hyperfine splittings, negative spin densities result in a (seemingly) unusually large spectral extent.

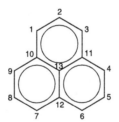

(**XI**) perinaphthenyl radical

In the biphenyl anion spectrum in Fig. 9.5, the line that is shifted downfield must be assigned to positions at which the π spin density is *negative*. One would not have inferred this fact from the spectral extent. However, there are appreciable spin densities at positions that have no attached protons. The magnitude of the shift for the low-field line indicates that this line arises from protons having the *smallest* magnitude of hyperfine splitting. From the solution EPR spectrum, the smallest splitting arises from a set of four equivalent protons. These can be either the protons at positions 2,6,2′,6′ or 3,5,3′,5′ (Fig. 9.5). Molecular-orbital studies indicate that the latter assignment should be made [30].

The HMO theory is too crude to yield negative π-electron spin densities at carbon atoms. With a generalized definition [2] of spin density, distinguishing between spin-α and -β states, more advanced MO schemes do yield these with both signs, and place non-zero spin density on the protons. Thus, for the latter nuclei, one obtains the relation

$$a_p = \frac{2\mu_0}{3} g\beta_e g_p \beta_n \frac{|\psi|_p^2 \rho_p}{\langle \hat{S}_z \rangle} \tag{9.11}$$

to be compared with Eqs. 2.38 and 2.51. Here the wavefunction evaluated at the site of the proton is, of course, s-like.

One of the more advanced MO theories, which allows for negative π unpaired-electron populations, is due to McLachlan [30,31]. This theory uses Hückel orbitals as unperturbed functions, and brings in electron interaction and correlation. When there are N carbon atoms in the conjugated system, the expression obtained for the unpaired-electron population at carbon atom t is

$$\rho_t = |c_{mt}|^2 + \lambda \sum_{r=1}^{N} \pi_{rt} |c_{mr}|^2 \tag{9.12}$$

Here c_{mt} is the coefficient of atom t in the mth molecular orbital that contains the unpaired electron. λ is a dimensionless parameter that may be varied to provide a best fit to the spectral extent. It is usually given a value between 1.0 and 1.2. Symbol π_{rt} is the dimensionless mutual atom-atom polarizability defined by

$$\pi_{rt} = -4\beta \sum_{j}^{\text{bonding}} \sum_{k}^{\text{antibonding}} \frac{(c_{jr}c_{kr}^{*})(c_{jt}c_{kt})}{U_k - U_j} \tag{9.13}$$

The Hückel coefficients c are for atoms r and t in molecular orbitals j and k. U_k and U_j are the Hückel energies of the k and j levels. The summations need not include non-bonding levels, since their effects cancel out in the summations.

9.2.6 Methyl Proton Hyperfine Splittings and Hyperconjugation

Examination of Fig. 9.3 reveals that splittings from some methyl protons exceed those caused by some ring protons. Hence there must be some mechanism that couples the methyl protons to the π system.

An effective model for the coupling mechanism is that of *hyperconjugation* (defined below), which provides a direct link of the methyl hydrogen atoms with electrons in the π system. It is well known that two fragments of a molecule may interact if there is a compatibility in the symmetry properties (and energies) of their wavefunctions. A single $2p_z$ orbital or a π orbital is antisymmetric with respect to the plane of the molecule; that is, it changes sign on reflection in the plane. The atomic orbitals of the three hydrogen atoms may be combined to give

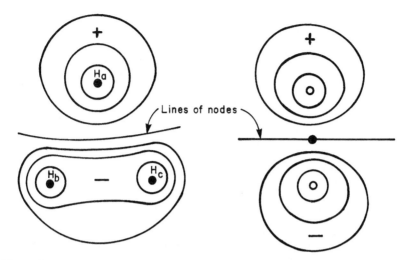

FIGURE 9.6   Schematic representation of a three-hydrogen (3H)-atom molecular orbital of the same symmetry as the p atomic orbital in a conjugated radical. [After C. A. Coulson, *Valence*, Oxford University Press, London, U.K., 1961, p. 362.]

a molecular orbital with the same symmetry as a π orbital. Such a combination is

$$\psi = c_1\phi_1 - c_2(\phi_2 + \phi_3) \tag{9.14}$$

This symmetry is shown schematically in Fig. 9.6. ψ can be considered as a *pseudo-π* orbital. Hence it may be regarded as part of the π system. Because the methyl protons form a part of the π system, the spin density at the protons has the same sign as that on the carbon bonded to the methyl group. It is to be recalled that the hyperfine splitting a_i of a proton i is proportional (Eq. 9.11) to the *square* $|\psi|_p^2$ of the wavefunction at the proton. On CH_3 the protons have identical spin densities, and hence the hyperfine splitting constants from H_a, H_b and H_c of Fig. 9.6 all have the same sign and magnitude.

That the spin density on β protons

is opposite in sign to that on α protons

was established by observing an opposite shift of the lines from the two types of protons in a nuclear magnetic resonance experiment [32,33].

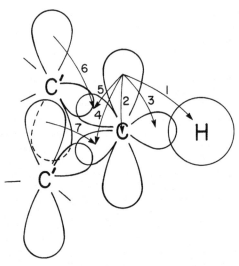

FIGURE 9.7   Spin polarization contributions to the ^{13}C and to the proton hyperfine splittings in a $(C')_2CH$ fragment. The numbered interactions are (1) Q_{CH}^H; (2) Q_s^C; (3) Q_{CH}^C; (4,5) $Q_{CC'}^C$; (6,7) $Q_{C'C}^C$.

9.2.7   Hyperfine Splitting from Nuclei Other than Protons

When isotropic proton hyperfine splittings were considered in Section 9.2.4, it was necessary to consider only the interaction of the π unpaired electron with the σ electrons in one bond (i.e., the C—H bond). However, in the case of nuclei that form part of the framework of a conjugated molecule, the interactions with several bonds must be considered. The hyperfine splittings by ^{13}C are considered first, but the model should generally be applicable to other nuclei, such as ^{14}N, ^{17}O, ^{19}F and ^{33}S [34]. This model is essentially a generalization of the treatment given for the C—H fragment. It has been observed that experimental ^{13}C hyperfine splittings are not simply proportional to the π unpaired-electron population on the same carbon atom. Rather, it is also necessary to include contributions from the populations on neighboring carbon atoms. Figure 9.7 illustrates the several interactions that are present, characterized by appropriate Q parameters. The notation used is as follows. In each Q parameter, the superscript designates the atom giving rise to the hyperfine splitting. In symbol Q_s^C, the subscript indicates the polarization of the carbon $1s$ electrons by the local π unpaired-electron population. In analogous symbols, the first subscript designates the atom on which the population is contributing to the spin polarization; the two subscripts together indicate the bond that is being polarized.

Consider the $(C')_2CH$ fragment shown in Fig. 9.7. By analogy with the C—H fragment, Q_{CH}^C and $Q_{CC'}^C$ are expected to be *positive*, whereas $Q_{C'C}^C$ and Q_{CH}^H should be *negative*. A consideration [34] of the combined contributions leads to the following relation

$$a_i^C = \left(Q_s^C + \sum_{j=1}^{3} Q_{CX_j}^C \right) \rho_i + \sum_{j=1}^{3} Q_{X_jC}^C \rho_j \tag{9.15}$$

where atoms X_j are those bonded to carbon atom $i(\equiv C)$. Quantitative calculations [34] of the spin-polarization constants in Eq. 9.15 yield the following results (in mT):

$$Q_s^C = -1.27, \quad Q_{CH}^H = +1.95, \quad Q_{CC'}^C = +1.44 \quad Q_{C'C}^C = -1.39$$

Inserting these values into Eq. 9.15, for $(C')_2CH$, one obtains (in mT)

$$a_i^C = 3.56\rho_i - 1.39 \sum_j \rho_j \tag{9.16a}$$

where the hydrogen unpaired-electron population has been deemed negligible. Similarly, for a $(C')_3C$ fragment, one obtains

$$a_i^C = 3.05\rho_i - 1.39 \sum_j \rho_j \tag{9.16b}$$

These relations are equally applicable to neutral radicals and $+1$ and -1 radical ions. The results of such estimates are displayed in Table 9.4 for the anthracene cation and anion [35]; the sets of experimental and calculated hyperfine splitting constants agree nicely. In this case it was possible to obtain an independent estimate of the π unpaired-electron populations from proton hyperfine splittings, with the aid of Eq. 9.6 and the normalization condition $\sum_i \rho_i = 1$: these are included in Table 9.4. The agreement is very satisfactory, considering that the parameters were calculated

TABLE 9.4 Calculated and Experimental ^{13}C Hyperfine Splitting Parameters and Unpaired-Electron Populations in the Anthracene Cation and Anion [a,b]

| | ^{13}C Hyperfine Splitting Parameters $|a_i^C|$ (mT) | | | |
Position i	Cation	Anion	Calculated	ρ Experimental [c]
9	+0.848	+0.876	+0.842	0.220
11	−0.450	−0.459	−0.490	−0.021
1	—	+0.357	+0.337	0.107
2	±0.037	−0.025	−0.033	0.054

[a] See structure **VII** in Table 9.3.
[b] From J. R. Bolton, G. K. Fraenkel, *J. Chem. Phys.*, **40**, 3307 (1964).
[c] Calculated from averaged hyperfine splitting constants (Table 9.3) using $Q_{CH}^H = -2.70$ mT and the normalization condition for unpaired-electron populations (Eq. 9.5).

from such an approximate theory. Similar comparison for other radicals show that Eq. 9.15 is widely applicable for ^{13}C splittings in aromatic hydrocarbons.

In nitrogen heterocyclic aromatic molecules, ^{14}N substitutes for carbon atoms; hence one might expect that Eq. 9.15 would also apply to ^{14}N. This is probably correct; however, experience has shown that here the effect of π unpaired-electron populations on neighboring atoms is small. This implies that the factor $Q_{C'N}^N$ must be small; certain estimates place it in the range from -0.4 to $+0.4$ mT [36–40]. In view of the small contribution from neighbors, many workers have used a simpler equation similar to Eq. 9.6 for ^{14}N hyperfine splittings.

Hyperfine splittings from ^{17}O [41] and ^{33}S [42] have also been interpreted in terms of an equation similar to Eq. 9.15.

It might be expected that since fluorine substitutes for hydrogen in aromatic molecules, an equation such as Eq. 9.6 would also hold for fluorine hyperfine splittings; that is, if ρ_C is *positive*, one expects that a^F would be *negative*. However, it has been shown conclusively that fluorine hyperfine splittings are *positive* in such molecules [43]. The non-bonding p electrons on the fluorine apparently participate in partial double bonding with the conjugated system to which the fluorine atom is attached; that is, some of the electron density in fluorine p orbitals is delocalized into the π system of the molecule. This electron transfer results in a net π spin density on the fluorine atom, having the same sign as that on the adjacent carbon atom. One expects that the local contribution to a^F (i.e., π unpaired-electron population on F) predominates; this would result in a *positive* fluorine hyperfine splitting constant (see Section 5.3.2.2 for some discussion of this topic).

9.2.8 One-Dimensional Chain Paramagnets

Almost all the systems considered so far have been ones where unpaired electrons are located on isolated relatively small molecules or defects. One-dimensional chain paramagnetic systems represent a class in which unpaired electrons are delocalized over a system of macroscopic dimension. One example is the π system polyacetylene $(CH)_n$ (**XII**), consisting of very long conjugated chains of two types: *cis* and *trans*.

C1   C2

T1   T2

(**XII**) Polyacetylene

In principle, these species should be diamagnetic with double bonds in fixed positions and π electrons delocalized over the chains. In practice, even in highly purified materials, there are defects that give rise to paramagnetism, and EPR signals are observed [44,45]. These signals tend to be single lines near $g = g_e$; they are thus relatively uninformative.

There is considerable interest, for both *cis* and *trans* materials, in the formation of regions ('domain walls') at which type-1 switches to type-2 bond distribution, with unpaired spins present. Much effort [^{2}H and ^{13}C doping, advanced ENDOR and ESEEM techniques (Chapter 12)] is being brought to bear on this complex system [45,46], especially to discern whether the domain walls are fixed or mobile (solitons). Chemical doping experiments show that polyacetylenes are semiconductors capable of being transformed into excellent electrical conductors that yield EPR signals with dysonian lineshapes (Section 9.6) [44].

A second example of macroscopic π systems involves certain organic molecules that are strong electron donors or acceptors, and can exhibit strong EPR signals under appropriate conditions.

The *p*-phenylenediamines are strong donors. For example, the species forms readily and is called 'Wurster's blue cation' (**XIII**). It exhibits a complex multi-line ^{1}H and ^{14}N hyperfine pattern [47] in aqueous solution, and is known to dimerize to some extent in non-aqueous solutions [48]. In the solid state (e.g., the perchlorate salt), it crystallizes in long parallel one-dimensional chains[5] and undergoes antiferromagnetic spin pairing at low temperatures.

(**XIII**) Wurster's blue cation

Among the strong acceptors, tetracyanoethylene (**XIV**) and tetracyanoquinodimethane (**XV**) have been of considerable interest, since they readily form mono-anion radicals. For instance, pairwise clustering of such species in the crystal form leads to thermally accessible singlet and triplet species, in which the triplet excitation (exciton) is mobile [49]. Many of these materials are semiconductors.

(**XIV**) tetracyanoethylene

(**XV**) tetracyanoquinodimethane

As an inorganic example, we can cite chains of platinum atoms, bonded directly to each other, but each liganded with various oxygen anions. Thus we deal with one-dimensional (1D) chains of nascent metallic character; these materials are called 'platinum blues'. In one example, a paramagnetic one, there is a mixed chain $Pt(II)_3Pt(III)$ having spin $S = \frac{1}{2}$. EPR studies have yielded principal values $g_\perp = 2.509$ and $g_\parallel = 1.978$, and nicely resolved ^{195}Pt (33.8% natural abundance) hyperfine structure showing that the unpaired electron is highly delocalized along the chain, with all four Pt atoms close to equivalent, with the parallel direction along the mean chain direction z [50]. The Pt—Pt bonding can be described in terms of d_{z^2} orbital overlap, in first approximation.

In summary, EPR has played a prominent role in the study of electron spin-spin interactions in these systems. A review [51], up to 1966, is available.

9.3 σ-TYPE ORGANIC RADICALS

For the radicals considered thus far, the unpaired electron is located primarily in a carbon $2p_z$ (or π) orbital. Small isotropic hyperfine couplings ($|a_0| < 100\,\text{MHz}$) are observed, but these arise primarily from the indirect mechanism described in Section 9.2.3. The nuclei are usually located at or near the nodal plane of the $2p_z$ orbital.

There are a number of known radicals that exhibit proton hyperfine couplings with isotropic components of the order of 150–400 MHz. These couplings are far too large to be explained by the indirect mechanism, and one is forced to conclude that the wave-function of the unpaired electron has considerable density at the nucleus considered. Thus, the unpaired electron is located primarily in the σ orbital that would normally form a σ bond between that nucleus and some atom (such as hydrogen) absent in the radical. Most σ orbitals have a considerable s-orbital component.

In the ethynyl radical C≡C—H, the unpaired electron occupies primarily an orbital pointing outward, that is, one that would be directed toward a second proton in the acetylene molecule. Likewise in the vinyl radical HC=CH$_2$, the unpaired electron is primarily in an orbital that would attach a hydrogen atom to form the ethylene molecule. Yet another example is the formyl radical HCO, derived from formaldehyde H_2CO, in which the bond angle is thought to be 120° [52]. A closely related radical is FCO.

In each of these cases, the sign of the ^1H (or ^{19}F) hyperfine coupling constant is believed to be positive, arising from a considerable s component at the hydrogen (or fluorine) atom. For instance, in the HCO radical, the unpaired-electron population in the $1s$ orbital of hydrogen is approximately 0.27, since the proton hyperfine coupling constant is $0.27 \times 1420 = 384$ MHz [53]. This is an unusually large proton coupling. In terms of resonance structures, it can be assumed to imply considerable presence of H + CO in the ground state.

The magnitude of isotropic ^{13}C hyperfine splittings provides a direct indication of whether there is a significant s-orbital contribution on carbon. A pure s orbital would yield a ^{13}C hyperfine splitting of ~ 135 mT (Table H.4). An sp^3 hybrid for a tetrahedral configuration would give 25% of this value. For example, in the π radical

CH_3, the a^C hyperfine splitting constant is only 3.85 mT [54], showing this radical to be close to planar. In the CF_3 radical, $a^C = 27.1$ mT [54]. This large increase in the ^{13}C hyperfine splitting can be explained in terms of a large pyramidal distortion in CF_3. Thus the latter is a σ radical.

It is interesting to compare the isotropic parts of the proton couplings in the formyl (HC=O) and the vinyl radicals. For the latter in a rigid medium, the couplings (in mT) are 1.57 for $H_{(1)}$, 3.43 for $H_{(2)}$ (*cis*), and 6.85 for $H_{(3)}$ (*trans*) [55]. (Problem 10.7 explores the apparent changes in couplings when this radical is observed in liquid solution.) Even the largest value is considerably less than that (13.7 mT) for HCO, which likely has a bond angle of $\sim 125°$. The difference arises from the large variation of coupling constant with bond angle. From the value for $H_{(1)}$ in the vinyl radical, the HCC bond angle is estimated to be $140-150°$.

Hyperfine couplings in σ-type radicals may also exhibit large anisotropy. For example, in FCO, the principal hyperfine matrix components $A(^{19}F)/h$ are 1437.5, 708.2 and 662.0 MHz [55]. Presumably a large spin polarization of the CF bond occurs, arising from configuration interaction between the ground state and a low-lying excited state describable in terms of atomic F and CO.

It is possible to estimate the spin distribution in s radicals by using a molecular-orbital theory, such as the INDO method [2] (which includes all valence-shell atomic orbitals).

9.4 TRIPLET STATES AND BIRADICALS

The triplet state of naphthalene, too, can be discussed in terms of the HMO model. Thus one unpaired electron is in the highest bonding orbital, whereas the other was transferred (Section 6.3.4) from there to the lowest antibonding orbital. In accordance with the pairing theorem, the orbital coefficients of these are equal in magnitude. The unpaired-electron populations obtained experimentally and from various theoretical approaches are listed in Table 9.5. The π-electron populations (Problem 9A.3) sum to 1 (and not 2), consistent with the operation of the Pauli exclusion principle [1, Section 8.5]. These parameters yield a good approximation to the set of proton hyperfine couplings (Section 9.3.4),[6] which are seen to be much the same as those of the naphthalene anion (Section 3.2.2) and cation (Table 9.3), despite the presence here of two unpaired electrons.

It is of interest to calculate the value of D when the two coupled electrons are on the same carbon atom, namely, for CH_2. This hydrocarbon is one of the smallest molecular species with a low-lying triplet state, that is, its $^3B_1$ ground state. It is thus a favorite molecule for theoretical calculations (see Ref. 56 for a summary and also Section 6.3.6.1). Experiment and ab initio calculations agree that CH_2 is non-linear. For a bond angle of $135°$, the latter yield $\bar{D} = 0.81$ cm^{-1}, $\bar{E} = 0.05$ cm^{-1} [57]. EPR spectroscopy yields $\bar{D} = 0.76(2)$ cm^{-1}, when a correction for motional effects is made [58]. In these small molecules, one must be concerned about a possible contribution to D from spin-orbit coupling. In O_2, this contribution

TABLE 9.5 Unpaired-Electron Populations for Naphthalene in Its Lowest Triplet State[a,b]

Source of Data	Spin Population[c]			
	ρ_1	ρ_2	ρ_9	ρ_1/ρ_2
From the anisotropic part of the proton hyperfine splitting	0.219	0.062	−0.063	3.5
From the isotropic part of the proton hyperfine splitting[d]	0.220	—	—	—
From HMO calculations (Problem 9A.3)	0.181	0.069	0	2.6
From advanced MO calculations:				
Amos[e]	0.235	0.048	−0.066	4.89
Pariser[f]	0.168	0.074	0.015	2.27
Goodman and Hoyland[g]	0.198	0.052	0	3.81
Atherton and Weissman[h]	0.220	0.083	−0.106	2.65

[a] See structure **VI** in Table 9.3.
[b] N. Hirota, C. A. Hutchison Jr., P. Palmer, *J. Chem. Phys.*, **40**, 3717 (1964).
[c] For position labeling, see structure **VI**.
[d] Using $Q_{CH}^H/h = -66.50 \times 10^6 \text{ s}^{-1}$.
[e] A. T. Amos, *Mol. Phys.*, **5**, 91 (1962).
[f] R. Pariser, *J. Chem. Phys.*, **24**, 250 (1962).
[g] L. Goodman, J. R. Hoyland, *J. Chem. Phys.*, **39**, 1068 (1963).
[h] N. M. Atherton, S. I. Weissman, *J. Am. Chem. Soc.*, **83**, 1330 (1961).

is appreciable. However, calculations for CH_2 disclose that the spin-orbit contribution to D and E is small as compared to the spin-spin interaction [57].[7]

The large value of D in $H-C-C\equiv N$, notwithstanding the possibility of delocalization in the $C\equiv N$ group, is probably due to the existence of a negative unpaired-electron population on the central atom of the $C-C\equiv N$ group. This π system is akin to that of the allyl radical $H_2C(CH)CH_2$ (Problem 9.4 and Appendix 9A). The expected negative populations on the central atom would thus lead to an increased positive population on each outer carbon atom and hence to an increased value of D. Such an effect cannot occur with the $H-C-CF_3$ molecule listed in Table 6.1. The effect is probably operative in the molecule $H-C-C\equiv C-H$ also, and assuredly also in $N\equiv C-C-C\equiv N$, where there are five π-electron centers.

Table 6.1 includes the parameters for one nitrene. These species, $N-R$, are isoelectronic with the carbenes. The parent compound for the nitrenes is $N-H$. It has been estimated that $\bar{D} = 1.86 \text{ cm}^{-1}$ for this fragment [59]. For $N-C\equiv N$, the reduction in the D value by delocalization is probably somewhat offset by the enhancement of the positive population on the nitrogen atoms, due to a negative spin density on the carbon.

9.5 INORGANIC RADICALS

The assignment and interpretation of the EPR spectra of inorganic radicals have been a very active field of investigation. It is not possible to give a complete coverage; however, we shall attempt to outline the major features with some examples.

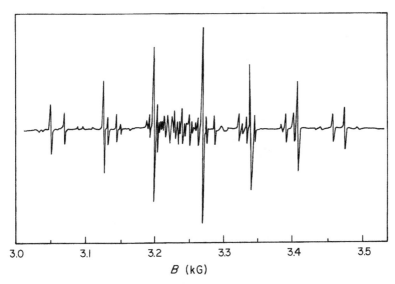

FIGURE 9.8 EPR spectrum of the V_k center (Cl_2^-) in x-irradiated KCl at 77 K with the magnetic field parallel to the [100] direction in the (100) plane, with $v = 9.263$ GHz. [After T. G. Castner, Jr. W. Känzig, *J. Phys. Chem. Solids*, **3**, 178 (1957).]

Identification of Radical Species. As in the case of organic radicals, the values of principal components of hyperfine matrices can provide the major clues in the identification of species resulting from the irradiation of inorganic materials. For example, x irradiation of LiF at 77 K produces (among others) a species that exhibits a 1 2 1 triplet EPR spectrum for **B** ‖ [100]. Such a pattern implies hyperfine interaction with two nuclei of spin $\frac{1}{2}$. The principal values of the g matrix are $g_X = 2.0234$, $g_Y = 2.0227$ and $g_Z = 2.0031$, indicative of nearly uniaxial symmetry. The hyperfine splitting shows uniaxial behavior, with $a_{\parallel} = 88.7$ mT and $a_{\perp} = 5.9$ mT [60]. The species responsible is undoubtedly the F_2^- ion (V_K center). If the experiment is done with KCl, the spectra (Fig. 9.8) from the molecular ions $(^{35}Cl—^{35}Cl)^-$, $(^{35}Cl—^{37}Cl)^-$ and $(^{37}Cl—^{37}Cl)^-$ provide redundant and incontrovertible (and redundant) identification that the center here is Cl_2^-. Interpretation of Fig. 9.8 is left as a problem for the reader.

In other cases the appearance of hyperfine structures is not sufficient to provide a positive identification. For example, γ-irradiated KNO_3 exhibits the EPR spectrum shown in Fig. 9.9. There are at least three radical species, each of which contains a nitrogen atom, as evidenced by the triplet hyperfine splittings. However, the assignment to specific radicals requires further information. The reasonable possibilities can be listed as NO_2, NO_2^{2-}, NO_3, and NO_3^{2-}. The experimental results for the hyperfine and g matrix principal values are listed in Table 9.6.

The identification requires a knowledge of the theoretical predictions of the structure and orbital sequence in each radical; in addition, one requires information from

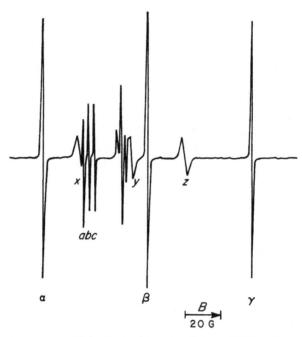

FIGURE 9.9 Spectra of radicals obtained on γ irradiation of KNO_3. Species 1 (lines α, β and γ) has been assigned as the NO_2 radical. Species 2 (lines a, b and c) has been assigned as the NO_3^{2-} radical. [After R. Livingston, H. Zeldes, *J. Chem. Phys.*, **41**, 4011 (1964).]

studies of these radicals in other host matrices. In various hosts, NO_2 exhibits a ^{14}N hyperfine coupling with little anisotropy and an isotropic hyperfine coupling of about 150 MHz [61]. The small anisotropy arises from the fact that NO_2 is usually rotating about its two-fold axis, even in a solid. *Fixed* NO_2 exhibits

TABLE 9.6 Hyperfine and g Matrices for Radical Species Found in γ-Irradiated KNO_3

Species	g Components	^{14}N Hyperfine Components (MHz)
1	$g_\parallel = 2.006^a$	$A_\parallel/h = 176^a$
	$g_\perp = 1.996$	$A_\perp/h = 139$
2	$g_\parallel = 2.0031^b$	$A_\parallel/h = 12.08^b$
	$g_\perp = 2.0232$	$A_\perp/h = 9.80$
3	$g_\parallel = 2.0015^a$	$A_\parallel/h = 177.6^a$
	$g_\perp = 2.0057$	$A_\perp/h = 89.0$

^{a}H. Zeldes, "Paramagnetic Species in Irradiated KNO_3", in *Paramagnetic Resonance*, Vol. 2, W. Low, Ed., Academic Press, New York, NY, U.S.A., 1963, p. 764.
^{b}R. Livingston, H. Zeldes, *J. Chem. Phys.*, **41**, 4011 (1964).

considerable anisotropy. The large hyperfine coupling arises from the fact that the unpaired electron is located primarily in a non-bonding orbital on nitrogen. The g matrix is virtually isotropic, with $g_{iso} \approx 2.000$. Comparison with Table 9.6 indicates that species 1 is probably the NO_2 radical.

In NO_3 (the symmetric isomer; D_{3h}) the unpaired electron is located in an orbital composed largely of non-bonding oxygen p orbitals lying in the plane of the molecule. Thus the nitrogen hyperfine coupling is expected to be very small. Examination of the results in Table 9.6 suggests that species 2 may be this NO_3 radical.

Species 3 exhibits considerable isotropic and anisotropic hyperfine interaction. NO_3^{2-} is a reasonable possibility, since this ion is expected to be not quite planar [62], that is, a slightly distorted π-type radical. The distortion would introduce some s character into the orbital of the unpaired electron and thus account for the large isotropic hyperfine coupling (~ 120 MHz).

Structural Information. When a radical species has been identified, the g and hyperfine matrices can provide considerable information about the detailed geometric and electronic structure of the radical. The NO_2 radical (observed in $NaNO_2$ [63]) is an excellent example. From Table H.4 one notes that a single electron in a $2s$ orbital on a free nitrogen atom would give rise to an isotropic hyperfine coupling of 1540 MHz. From the observed value of $A_0/h = 151$ MHz, the unpaired-electron population in the nitrogen $2s$ orbital is computed to be $\rho_s = \frac{151}{1540} = 0.10$. Similarly, from the maximum value in the anisotropic hyperfine matrix, the population in the nitrogen $2p_x$ orbital is computed to be $\rho_p = \frac{12}{48} = 0.25$. Hence the $2p/2s$ ratio is 2.5. A simple consideration of orbital hybridization suggests that the bond angle is between $130°$ and $140°$. This is in good agreement with gas-phase vibrational analysis [64] and microwave results ($134°$) [65]. Presumably, the unpaired-electron populations for the nitrogen $2p$ and $2s$ orbitals do not add up to unity because there is some population in $2p$ orbitals on the oxygen atoms.

When isotropic hyperfine couplings are small, as for species B in Table 9.6, one must beware of interpreting these in terms of a percentage of s character in the orbital of the unpaired electron. The indirect mechanism leading to isotropic hyperfine coupling (Section 9.2.3) may give the major contribution. Generally, if $|\rho_s| < 0.05$ as computed above, then an interpretation in terms of a bond angle is dubious.

It is interesting to compare the EPR results for isoelectronic series of radicals. Table 9.7 contains the data for the ClO_3, SO_3^-, and PO_3^{2-} radicals, as well as for the NO_2 and CO_2^- radicals. It is clear that as the atomic number of the central atom decreases, the tetratomic radicals become more pyramidal (as evidenced by the decreased ratio ρ_p/ρ_s); the triatomic radicals become more bent.

As a final example of inorganic radicals, we cite the EPR of adsorbed oxygenic species [66]. The $S = \frac{1}{2}$ ions O^-, O_2^- and O_3^- on the surfaces of various materials all show characteristic spectra, corroborated with the help of ^{17}O enrichment, and undergo chemical interconversions of catalytic interest.

TABLE 9.7 Comparison of the EPR Data for Central Atoms of Some Isoelectronic ($S = \frac{1}{2}$) Radicals

Radical	Host	g			Hyperfine (MHz)				Unpaired-Electron Populations				Reference
		g_{XX}	g_{YY}	g_{ZZ}	T_{XX}/h	T_{YY}/h	T_{ZZ}/h	A_0/h	ρ_s	ρ_p	ρ_p/ρ_s	$\rho_s + \rho_p$	
$^{35}\text{ClO}_3$	KClO_4	2.0132	2.0132	2.0066	−40.5	−40.5	81	342	0.076	0.34	4.5	0.42	a
$^{33}\text{SO}_3^-$	$\text{K}_2\text{CH}_2(\text{SO}_3)_2$	—	—	—	−37	−39	75	353	0.13	0.49	3.8	0.62	b
$^{31}\text{PO}_3^{2-}$	$\text{Na}_2\text{HPO}_3 \cdot 5\text{H}_2\text{O}$	2.001	2.001	1.999	−148	−148	297	1660	0.16	0.53	3.3	0.69	c
$^{14}\text{NO}_2$	NaNO_2	2.0057	2.0015	1.9910	−22.3	37.0	−14.8	153	0.099	0.44	4.4	0.54	d
$^{13}\text{CO}_2^-$	NaO_2CH	2.0032	2.0014	1.9975	−32.0	78.0	−46.0	468	0.14	0.66	4.7	0.80	e

[a]P. W. Atkins, J. A. Brivati, N. Keen, M. C. R. Symons, P. A. Trevalion, *J. Chem. Soc.*, 4785 (1962).
[b]G. W. Chantry, A. Horsfield, J. R. Morton, J. R. Rowlands, D. H. Whiffen, *Mol. Phys.*, **5**, 233 (1962).
[c]A. Horsfield, J. R. Morton, D. H. Whiffen, *Mol. Phys.*, **4**, 475 (1961).
[d]H. Zeldes, R. Livingston, *J. Chem. Phys.*, **35**, 563 (1961).
[e]D. W. Ovenall, D. H. Whiffen, *Mol. Phys.*, **4**, 135 (1961). See also T. A. Vestad, H. Gustafsson, A. Lund, E. O. Hole, E. Sagstuen, *Phys. Chem. Chem. Phys.*, **6**, 3017 (2004).

9.6 ELECTRICALLY CONDUCTING SYSTEMS

Electrically conducting systems represent another example (see Section 9.2.8) of inter-acting electrons; in this case the cooperativity extends over the entire macroscopic sample. We consider metals, metal ammonia and amine solutions, semiconductor, and graphitic materials. The analysis of EPR lineshapes and linewidths can in principle yield information about the electrical conductivity, conduction-electron g factor and spin relaxation time, the electron state density on the Fermi energy surface and carrier diffu-sion parameters. Frequently, especially in solids, mobile electrons are called 'itinerant'.

9.6.1 Metals

Metals may be visualized as a matrix of fixed cations in a sea of highly delocalized (conduction) electrons; as they are highly mobile, they are able to interact with each

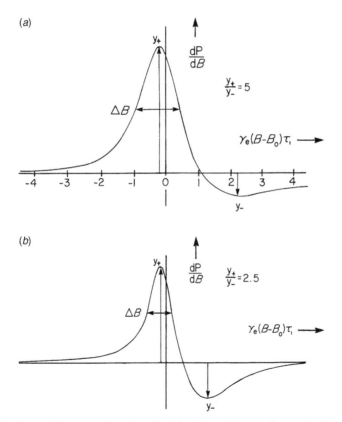

FIGURE 9.10   (a) First derivative of the ideal dysonian absorption line in the X-band region; (b) typical first-derivative EPR spectrum observed in colloidal samples of Na(s), with mean particle diameter small compared to the skin depth. Horizontal scale is not the same as in (a). [After F. Vescial, N. S. VanderVen, R. T. Schumacher, *Phys. Rev.*, **134**, A1286 (1964).]

other. EPR signals are observed [67]; however, only the layers near the surface contribute, since the excitation field $\mathbf{B}_1$ penetrates only a short distance (~ 1 μm) into the metal (skin effect).

The magnetic susceptibility in metals has a diamagnetic component due to the circulation of electrons in the field $\mathbf{B}$. This is opposed by the normal paramagnetic component due to the unpaired electrons. The g factors of the observed EPR spectra are close to g_e. For example, in sodium metal, $g - g_e = 9.7(3) \times 10^{-4}$ in both the liquid and solid phases [68]. The EPR lineshape typically is asymmetric (Fig. 9.10), arising from a mixture of absorption and dispersion effects. This admixture arises because the electron diffusion relative to the surface occurs in times that typically are long compared to the spin-relaxation times, as explained by Dyson and others [69,70].

It has been possible to study S-state ions in metals [71], and thus to learn details of the interaction between the conduction electrons and the inserted spin probes (gadolinium ions). The observed g shifts (2.01 – 1.88) correlate nicely with certain properties of the pure alloys used as solvents.

9.6.2 Metals Dissolved in Ammonia and Amine Solutions

When alkali or alkaline-earth metals (M) are dissolved in liquid ammonia or amines, ionization takes place to produce metal cations and solvated electrons. The latter (blue color when dilute in liquid NH_3) exhibit very sharp EPR lines (width 0.002 mT!) with $g = 2.0012(2)$, independent of concentration (<1 M) and of cation M^+ [72–74].

In concentrated solutions (bronze in color), the electronic conductivity becomes metallic, rather than electrolytic, and the EPR line broadens, becoming dysonian in shape [74]. Furthermore, solid cubic complexes $M(NH_3)_x$ can be isolated [e.g., $Li(NH_3)_4$ [75]] that exhibit EPR lines with dysonian shapes characteristic of normal metallic behavior [76].

Dilute solutions of metals in amines exhibit EPR spectra with resolved ^{14}N hyperfine splittings, which give some insight into the structure and dynamics of the interaction of the electron with the surrounding solvent molecules [77].

With crown ethers, such as 18C6 (**XVI**) (inert ligands capable of encapsulating alkali cations), it is possible to isolate stable electrides [e.g., $Cs^+(18C6)_2e^-$] containing

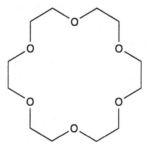

(XVI) 18C6 crown ether

close to a unity stoichiometric ratio of electron anions to metal cations. The crystals exhibit a single dysonian line at $g \approx g_e$ (linewidth 0.05 mT) down to 3 K [77,78]. The temperature dependence of the electrical conductivity suggests that the material is a semiconductor with a band-gap energy of 0.9(1) eV.

9.6.3 Semiconductors

Semiconductors, like insulators, have virtually continuous electronic energy bands, derived from orbitals based on all atoms in the crystal. The highest occupied band (valence band) is virtually filled with electrons and is separated from the next virtually unoccupied band (conduction band) by an energy gap (band-gap) that has few or no energy levels. In insulators the band gap is very large (>4 eV), so that thermal excitation of electrons from the valence band to the conduction band is rare. In semiconductors the band gap is smaller ($1-3$ eV), so that electron (and hole) conductivity, arising from promotion of electrons between these two bands, is possible at moderate temperatures. This conductivity may be enhanced greatly by doping with appropriate donors (n-type) or acceptors (p-type), which leads to formation of paramagnetic species.[8]

EPR has proved to be an important tool in the study of semiconductors, particularly in identifying and elucidating the structure of point defects and impurity ions. For example, the tetrahedral structure of solid Si can be damaged by electron irradiation, generating defects (V^+, V^0 and V^-) at which electrons are trapped next to Si atoms with 'dangling' bonds [81–83]. The neutral vacancy (V^0) has four interacting dangling bonds that interact to produce spin pairing and thus is diamagnetic. The V^- and V^+ species have $S = \frac{1}{2}$ and exhibit EPR spectra, often with resolvable ^{29}Si hyperfine splittings.

Center V^+ exhibits an EPR spectrum featuring three equally intense prominent peaks, each flanked by weaker ^{29}Si hyperfine doublets [74,84]. On applying uniaxial stress to the crystal, one can alter the relative intensities of the three peaks (Fig. 9.11). The explanation for the triplet is that any one of three energy-equivalent distortions occurs at each vacancy site, differing in the location of the one-electron bond and the two-electron bond formed between the four tetrahedral silicon neighbors. External stress redistributes these bond configurations among each other.

The mixed semiconductors (III–V or II–VI) have also been widely studied by EPR/ENDOR. The anion anti-site center in p-type GaP [85] is an example of a center in which a group-V atom occupies a group-III atom site, forming a "double donor". For example, the center $P^{4+}(P^{3-})_4$ exhibits an EPR spectrum (Fig. 9.12) with $g = 2.007(3)$, consisting of an isotropic hyperfine doublet ($a_0 = 103$ mT) arising from the central P ion; each of these lines is split into an (anisotropic) $1:4:6:4:1$ quintet (~ 9 mT) from interaction with the four tetrahedrally disposed P neighbors [85]. A superior technique for investigating such defects, and others in semiconductors, features optical detection of EPR and ENDOR (Chapter 12).

For example, a 1992 study [86] reports detection of the microwave-modulated luminescence (at 0.8 eV) from the first-neighbor ^{31}P shell of the phosphorus antisite

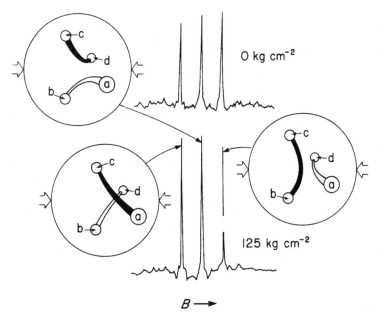

FIGURE 9.11   Changes in the 20-GHz EPR spectrum of the silicon vacancy center V^+ (at 4 K) under compressional stress. The insets sketch the defect bonding pattern corresponding to each line. Here **B** ∥ [100]. The stress was applied along [100]. [After G. D. Watkins, *J. Phys. Soc. Jpn.*, 18, Suppl. **2**, 22 (1963).]

in zinc-doped InP, yielding $|A_\parallel|/h = 368.0(5)$ and $|A_\perp|/h = 247.8(5)$ MHz for each of the four nuclei.

Numerous other magnetic defects—for example, clusters of vacancies, interstitials, and transition ions—also occur in semiconductors, but these cannot be discussed here.

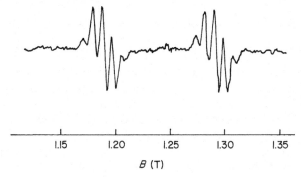

FIGURE 9.12   EPR spectrum of the $^{31}P^{4+}$ ion in the P_{Ga}^+ anti-site center $[P^{4+}(P^{3-})_4]$ in the II–V semiconductor GaP (34.8 GHz, **B** ∥ [100], 20 K). [After U. Kaufmann, J. Schneider, *Festkörperprobleme*, **20**, 87 (1980).]

Clusters of molecules also show semiconductor properties. An interesting example is lithium phthalocyanine (LiPc) (**XVII**), in which the π-radical rings

(**XVII**) lithium phthalocyanine

stack linearly with the Li atoms superimposed. The EPR spectrum of the solid consists of an exchange-narrowed sharp line ($\Delta B_{pp} \approx 0.005$ mT) at $g = 2.0015$ [87,88]. This broadens dramatically as a result of exchange interactions when O_2 (Section 10.5.3.1), diffusing rapidly through the channels in the crystal, is admitted (and does so reversibly). The substance is chemically stable and offers a sensitive and rapid means of measuring O_2 concentrations in solutions by means of EPR.

9.6.4 Graphitic Compounds

Graphitic intercalation compounds are distinct in that they are highly anisotropic. A comprehensive review of the status of the conduction EPR field of these conducting 'metallic' materials became available in 1997 [89].

9.7 TECHNIQUES FOR STRUCTURAL ESTIMATES FROM EPR DATA

Despite the ultimate need for complex large-scale numerical modeling, various more or less empirical but relatively simple techniques have been developed to attain structural information from the electron-spin electron-spin interaction parameters (D, J, set of B_4^m discussed in Section 9.7.2).

9.7.1 The Newman Superposition Model

This empirical technique [90–92], applied mostly to transition ions embedded in a symmetric crystal structure (e.g., in a mineral), can describe the electronic quadrupole matrix **D** in terms of additive uniaxial crystal-field contributions from the nearest-neighbor ions. It can give information about the coordination number, ligands and local symmetry, and has most often been applied to S-state ions (Mn^{2+}, Fe^{3+} and Gd^{3+}) in oxides and halides.

The Newman model postulates that

$$D = \frac{1}{2}D_0 \sum_i (3\cos^2\theta_i - 1)\left(\frac{R_D}{R_i}\right)^{t_D} \tag{9.17}$$

for metal ion M and ligand type X. The sum is over the nearest-neighbor ligands (all of the same type), polar angle θ_i gives the direction between ligand X_i and axis z of a cartesian set fixed at M, and the distance between M and X_i is R_i. Parameters $D_0(M, X, R_0)$, R_D and t_D are evaluated empirically; typically, the reference distance is $0.19 \le R_D \le 0.21$ nm and $t_D = 8 \pm 1$.[9] For parameter E, an equation differing from Eq. 9.17 only in the form of the angular factor is appropriate.

For systems MnX_6^{4-} with X = Cl, Br and I, studies reveal that parameter D_0 increases monotonically with increasing covalence of the Mn—X bonds [94]. Its complex behavior depends, for instance, on local distortions.

As a second example, we cite the good success of the Newman model in the interpretation of the S^2 and S^4 parameters for Fe^{3+} in a cation site of Li_2O, where two neighbor sites appear to be Li^+ vacancies [95]. However, the model is none too successful in some same cases [96].

9.7.2 The Pseudo-cube Method

The fourth-order terms, that is, measured coefficients of spin-hamiltonian terms quartic in the components of (Section 6.6 and Eqs. 8.17), are even more sensitive than those (i.e., **D**) quadratic in $\hat{S}$. They are found to be useful, despite the fact that they are seldom available with the same accuracy as **D**, in learning about the location of S-state ions and their local environment.

The method of analysis, developed by Michoulier and Gaite [97,98], depends on transforming the fourth-order measured parameters to various rotated coordinate frames (other than the lab crystal frame) until one is found exhibiting the highest local symmetry around the paramagnetic ion being investigated. Various criteria for this have been developed. For example, the sites of Fe^{3+} ions in $KTiOPO_4$ can be identified uniquely as being type Ti(1) rather than Ti(2), by means of the pseudo-cube method [99].

9.7.3 Distances from Parameter D

In triplet-state systems, some rough estimates of interelectron distances are available from the principal values of **D**, that is, from D and E that depend on the mean distance (i.e., on r^{-3}) between the two electrons with parallel spins. In particular, from Eqs. 6.15 and 6.25, one has (see also Eq. 6.41)

$$D = \frac{3\mu_0}{16\pi}(g\beta_e)^2\left\langle\frac{r^2 - 3Z^2}{r^5}\right\rangle \tag{9.18a}$$

$$E = \frac{3\mu_0}{16\pi}(g\beta_e)^2\left\langle\frac{Y^2 - X^2}{r^5}\right\rangle \tag{9.18b}$$

where X, Y, Z are the components of the interelectron vector expressed in the principal-axis system. Thus experimental values (e.g., obtained from Eqs. 6.32) can provide information about the spatial disposition of the two electrons, if the averages over the electron positions in Eqs. 9.18 can be modeled. This analysis is valid only if the interaction is predominantly dipolar in nature, that is, if there is no significant contribution from spin-orbit coupling to **D**.

9.7.4 Eatons' Interspin-Distance Formula

It has proved possible to extract mean distances r between spin-$\frac{1}{2}$ centers via the simple formula

$$\frac{\mathcal{A}(\Delta M_S = \pm 2)}{\mathcal{A}(\Delta M_S = \pm 1)} = k_r r^{-6} \tag{9.19}$$

where the left side contains the ratio of the integrated areas (under the absorption curve) for the two types of transitions possible (Section 6.3), corrected for any hyperfine effects present. The proportionality factor k_r is obtainable by a suitable procedure [100,101]. The recommended value is $k_r = 1.95 \times 10^{-3}\,\text{nm}^6$. This method is valid when the dipole-dipole interaction dominates over anisotropic exchange, typically for $r > 0.4\,\text{nm} = 4\,\text{Å}$. The method has been applied, for example, to obtain r for an interacting Cu^{2+} ($3d^9$)—nitroxyl spin-labeled species [100].

9.7.5 Summary

EPR is rapidly becoming an excellent tool for discerning atom positions, as well as bond lengths and directions, in paramagnetic species. Because of its sensitivity, this technique can furnish such information where non-spectroscopic methods (e.g., x-ray diffraction) fail. A recent journal issue is devoted to this topic [102].

REFERENCES

1. A. Carrington, A. D. McLachlan, *Introduction to Magnetic Resonance*, Harper & Row, New York, NY, U.S.A., 1967, p. 81.

2. J. A. Pople, D. L. Beveridge, P. A. Dobosh, *J. Am. Chem. Soc.*, **90**, 4201 (1968).

3. D. L. Tierney, H. Huang, P. Martásek, L. J. Roman, R. B. Silverman, B. M. Hoffman, *J. Am. Chem. Soc.*, **122**, 7869 (2000).

4. M. Weissbluth, *Atoms and Molecules*, Academic press, New York, NY, U.S.A., 1978.

5. D. H. Levy, R. J. Myers, *J. Chem. Phys.*, **41**, 1062 (1964).

6. H. M. McConnell, *J. Chem. Phys.*, **24**, 764 (1956).

7. H. M. McConnell, D. B. Chesnut, *J. Chem. Phys.*, **28**, 107 (1958).

8. S. I. Weissman, *J. Chem. Phys.*, **25**, 890 (1956).

9. R. Bersohn, *J. Chem. Phys.*, **24**, 1066 (1956).

10. T. Cole, C. Heller, H. M. McConnell, *Proc. Natl. Acad. Sci. U.S.A.*, **45**, 525 (1959).

11. R. W. Fessenden, R. H. Schuler, *J. Chem. Phys.*, **39**, 2147 (1963).

12. J. R. Bolton, A. Carrington, *Mol. Phys.*, **4**, 497 (1961).

13. R. G. Lawler, J. R. Bolton, G. K. Fraenkel, T. H. Brown, *J. Am. Chem. Soc.*, **86**, 520 (1964).

14. A. D. McLachlan, *Mol. Phys.*, **2**, 271 (1959); **5**, 51 (1962).

15. J. Koutecky, *J. Chem. Phys.*, **44**, 3702 (1966).

16. J. P. Colpa, J. R. Bolton, *Mol. Phys.*, **6**, 273 (1963).

17. J. R. Bolton, *J. Chem. Phys.*, **43**, 309 (1965).

18. A. J. Stone, *Proc. Roy. Soc. (London)*, **A271**, 424 (1963).

19. A. J. Stone, *Mol. Phys.*, **6**, 509 (1963); **7**, 311 (1964).

20. R. A. Rouse, M. T. Jones, *J. Magn. Reson.*, **19**, 294 (1975).

21. R. E. Moss, A. J. Perry, *Mol. Phys.*, **22**, 789 (1971).

22. M. T. Jones, T. C. Kuechler, S. Metz, *J. Magn. Reson.*, **10**, 149 (1973).

23. M. Weissbluth, "The Triplet State in Molecular Biophysics", in *Molecular Biophysics*, B. Pullman, M. Weissbluth, Eds., Academic Press, New York, NY, U.S.A., 1965, pp. 694–696.

24. R. Skinner, J. A. Weil, *Am. J. Phys.*, **57**, 777 (1989).

25. R. S. Drago, *Physical Methods of Chemistry*, Saunders, Philadelphia, PA, U.S.A., 1977, Chapter 12.

26. E. de Boer, C. MacLean, *Mol. Phys.*, **9**, 191 (1965).

27. K. H. Hausser, H. Brunner, J. C. Jochims, *Mol. Phys.*, **10**, 253 (1966).

28. H. M. McConnell, C. H. Holm, *J. Chem. Phys.*, **27**, 314 (1957).

29. D. R. Eaton, W. D. Phillips, "Nuclear Magnetic Resonance of Paramagnetic Molecules", in J. S. Waugh, Ed., *Advances in Magnetic Resonance*, Vol. 1, Academic Press, New York, NY, U.S.A., 1965.

30. A. D. McLachlan, *Mol. Phys.*, **3**, 233 (1960).

31. See also L. Salem, *The Molecular Orbital Theory of Conjugated Systems*, Benjamin, New York, NY, U.S.A., 1966, Chapter 5.

32. A. Forman, J. N. Murell, L. E. Orgel, *J. Chem. Phys.*, **31**, 1129 (1959).

33. D. Lazdins, M. Karplus, *J. Am. Chem. Soc.*, **87**, 920 (1965).

34. M. Karplus, G. K. Fraenkel, *J. Chem. Phys.*, **35**, 1312 (1961).

35. J. R. Bolton, G. K. Fraenkel, *J. Chem. Phys.*, **40**, 3307 (1964).

36. R. L. Ward, *J. Am. Chem. Soc.*, **84**, 332 (1962).

37. J. C. M. Henning, C. de Waard, *Phys. Lett.*, **3**, 139 (1962).

38. D. H. Geske, G. R. Padmanabhan, *J. Am. Chem. Soc.*, **87**, 1651 (1965).

39. J. C. M. Henning, *J. Chem. Phys.*, **44**, 2139 (1966).

40. C. L. Talcott, R. J. Myers, *Mol. Phys.*, **12**, 549 (1967).

41. M. Broze, Z. Luz, B. L. Silver, *J. Chem. Phys.*, **46**, 4891 (1967).

42. P. D. Sullivan, *J. Am. Chem. Soc.*, **90**, 3618 (1968).

43. D. R. Eaton, A. D. Josey, W. D. Phillips, R. E. Benson, *Mol. Phys.*, **5**, 407 (1962).

44. I. B. Goldberg, H. R. Crowe, P. R. Newman, A. J. Heeger, A. G. MacDiarmid, *J. Chem. Phys.*, **70**, 1132 (1979).

45. S. Roth, H. Bleier, *Adv. Phys.*, **36**, 385 (1987).

46. E. J. Hustedt, H. Thomann, B. H. Robinson, *J. Chem. Phys.*, **92**, 978 (1990).

47. J. R. Bolton, A. Carrington, J. dos Santos-Veiga, *Mol. Phys.*, **5**, 615 (1962).

48. A. Kawamori, A. Honda, N. Joo, K. Suzuki, Y. Ooshika, *J. Chem. Phys.*, **44**(11), 4364 (1966).

49. M. T. Jones, D. B. Chesnut, *J. Chem. Phys.*, **38**, 1311 (1963).

50. P. Arrizabalaga, P. Castan, M. Geoffroy, J.-P. Laurent, *Inorg. Chem.*, **24**, 3656 (1985).

51. P. L. Nordio, Z. G. Zoos, H. M. McConnell, *Annu. Rev. Phys. Chem.*, **17**, 237 (1966).

52. F. J. Adrian, E. L. Cochran, V. A. Bowers, *J. Chem. Phys.*, **36**, 1661 (1962).

53. F. J. Adrian, E. L. Cochran, V. A. Bowers, *J. Chem. Phys.*, **43**, 462 (1965).

54. R. W. Fessenden, R. H. Schuler, *J. Chem. Phys.*, **43**, 2704 (1965).

55. E. L. Cochran, F. J. Adrian, V. A. Bowers, *J. Chem. Phys.*, **40**, 213 (1964).

56. J. F. Harrison, *Acc. Chem. Res.*, **7**, 378 (1974).

57. S. R. Langhoff, *J. Chem. Phys.*, **61**, 3881 (1974).

58. E. Wasserman, R. S. Hutton, V. J. Kuck, W. A. Yager, *J. Chem. Phys.*, **55**, 2593 (1971).

59. J. A. R. Coope, J. B. Farmer, C. L. Gardner, C. A. McDowell, *J. Chem. Phys.*, **42**, 54 (1965).

60. T. G. Castner Jr., W. Känzig, *J. Phys. Chem. Solids*, **3**, 178 (1957).

61. P. W. Atkins, M. C. R. Symons, *J. Chem. Soc.*, 4794 (1962).

62. A. D. Walsh, *J. Chem. Soc.*, 2296 (1953).

63. H. Zeldes, R. Livingston, *J. Chem. Phys.*, **35**, 563 (1961).

64. G. E. Moore, *J. Opt. Soc. Am.*, **43**, 1045 (1953).

65. G. R. Bird, *J. Chem. Phys.*, **25**, 1040 (1956).

66. J. H. Lunsford, *Catal. Rev.*, **8**, 135 (1973).

67. R. N. Edmonds, M. R. Harrison, P. P. Edwards, *Annu. Rep. Prog. Chem.*, **C82**, 265 (1985).

68. R. A. B. Devine, R. Dupree, *Philos. Mag.*, **21**, 787 (1970).

69. F. J. Dyson, *Phys. Rev.*, **98**, 349 (1955).

70. G. Feher, A. F. Kip, *Phys. Rev.*, **158**, 225 (1967).

71. M. Peter, D. Shaltiel, J. H. Wernick, H. J. Williams, J. B. Mock, R. C. Sherwood, *Phys. Rev.*, **126**(4), 1395 (1962).

72. C. A. Hutchison Jr., R. C. Pastor, *J. Chem. Phys.*, **21**, 1959 (1953).

73. T. P. Das, *Adv. Chem. Phys.*, **4**, 303 (1962).

74. R. S. Alger, *Electron Paramagnetic Resonance Techniques and Applications*, Wiley-Interscience, New York, NY, U.S.A., 1968, Section 6.3.

75. W. S. Glaunsinger, M. J. Sienko, *J. Chem. Phys.*, **62**, 1873, 1883 (1975).

76. J. L. Dye, Prog. *Inorg. Chem.*, **32**, 327 (1984).

77. P. P. Edwards, *J. Solution Chem.*, **14**, 187 (1985); *J. Phys. Chem.*, **88**, 3772 (1984).

78. D. Issa, A. Ellaboudy, R. Janakiraman, J. L. Dye, *J. Phys. Chem.*, **88**, 3847 (1984).

79. F. C. Rong, W. R. Buchwald, E. H. Poindexter, W. L. Warren, D. J. Keeble, *Solid-State Electron.*, **34**, 835 (1991).

80. E. H. Poindexter, *Semicond. Sci. Technol.*, **4**, 961 (1989).

81. M. Stutzmann, *Z. Phys. Chem. N.F.*, **151**, 211 (1989) (in English).

82. B. Henderson, *Defects in Crystalline Solids*, E. Arnold, London, U.K., 1972, Section 5.3.

83. G. D. Watkins, "EPR Studies of Lattice Defects in Semiconductors", in *Defects and Their Structure in Non-metallic Solids*, B. Henderson and A. E. Hughes, Eds., Plenum Press, New York, NY, U.S.A., 1976, p. 203.

84. G. D. Watkins, "EPR and Optical Absorption Studies in Irradiated Semiconductors", in *Radiation Damage in Semiconductors*, F. L. Vook, Ed., Plenum Press, New York, NY, U.S.A., 1968, pp. 67–81.

85. U. Kaufmann, J. Schneider, A. Räuber, *Appl. Phys. Lett.*, **29**, 312 (1976); U. Kaufmann, J. Schneider, *Festkörperprobleme*, **20**, 87 (1980).

86. H. C. Crookham, T. A. Kennedy, D. J. Treacy, *Phys. Rev.*, **B46**, 1377 (1992).

87. P. Turek, J.-J. André, A. Giraudeau, J. Simon, *Chem. Phys. Lett.*, **134**, 471 (1987).

88. X.-S. Tang, M. Moussavi, G. C. Dismukes, *J. Am. Chem. Soc.*, **113**, 5914 (1991).

89. A. M. Ziatdinov, *Mol. Phys. Rep.*, **18/19**, 149–157 (1997).

90. D. J. Newman, *Adv. Phys.*, **20**, 197 (1971).

91. D. J. Newman, W. Urban, *Adv. Phys.*, **24**, 793 (1975).

92. M. Moreno, *J. Phys. Chem. Solids*, **51**, 835 (1990).

93. Y. Y. Yeung, *J. Phys. C. Solid State Phys.*, **21**, 2453 (1988).

94. M. Heming, G. Lehmann, *Chem. Phys. Lett.*, **80**, 235 (1981); see also M. Heming, G. Lehmann, "Superposition Model for the Zero-Field Splittings of 3d-Ion EPR: Experimental Tests, Theoretical Calculations and Applications", in *Electronic Magnetic Resonance of the Solid State*, J. A. Weil, Ed., Canadian Society for Chemistry, Ottawa, ON, Canada, 1987, pp. 163–174.

95. J. M. Baker, A. A. Jenkins, R. C. C. Ward, *J. Phys. Condens. Matter*, **3**, 8467 (1991).

96. M. J. Mombourquette, J. A. Weil, *J. Chem. Phys.*, **87**, 3385 (1987).

97. J. Michoulier, J. M. Gaite, *J. Chem. Phys.*, **56**, 5205 (1972).

98. J. M. Gaite, "Study of the Structural Distortion around S-State Ions in Crystals, Using the Fourth-Order Spin-Hamiltonian Term of the EPR Spectral Analysis", in *Electronic Magnetic Resonance of the Solid State*, J. A. Weil, Ed., Canadian Society for Chemistry, Ottawa, ON, Canada, 1987, pp 151–174.

99. N. M. Nizamutdinov, N. M. Khasanson, G. R. Bulka, V. M. Vinokurov, I. S. Rez, V. M. Garmash, N. I. Ravlova, *Sov. Phys. Crystallogr.*, **32**, 408 (1987).

100. S. S. Eaton, K. M. More, B. M. Sawant, G. R. Eaton, *J. Am. Chem. Soc.*, **105**, 6560 (1983).

101. R. E. Coffman, A. Pezeshk, *J. Magn. Reson.*, **70**, 21 (1986).

102. Various authors, *Appl. Magn. Reson.*, **3**(2), 1–381 (1992).

103. R. W. Fessenden, R. H. Schuler, *J. Chem. Phys.*, **39**, 2147 (1963).

104. A. Hinchliffe, N. M. Atherton, *Mol. Phys.*, **13**, 89 (1967).

105. S. H. Glarum, J. H. Marshall, *J. Chem. Phys.*, **44**, 2884 (1966).

NOTES

1. There is confusion in the literature as to the usage of terms such as 'spin density' and 'spin population'. We prefer to use 'density' in the sense that dimensions of volume^{-1} are implied. Thus electron probability density has the units m^{-3}, and charge density has units C m^{-3}, and spin density has units m^{-3}. The term "spin population" is not recommended, since it can also suggest the Boltzmann distribution among the spin states. Rather, 'unpaired-electron population' is used herein to denote the unit-less quantity equaling the square of (unitless) wavefunction coefficients, or algebraic sums thereof (which can be negative).

2. We deal in this chapter with isotropic hyperfine splitting constants. For convenience, we drop the subscript 0 that indicates this.

3. Reference 2 applies and discusses the unrestricted self-consistent-field molecular-orbital scheme, based on the Hartree-Fock-Roothaan equations, which resorts to intermediate neglect of differential overlap (INDO).

4. However, various other effects enter. The Jahn-Teller distortion (Section 8.2), including vibronic coupling, and the nearby cation (e.g., K$^+$) affect the degeneracy.

5. See the series of papers by J. Kommandeur and co-workers, *J. Chem. Phys.*, **47**, 391–413 (1967).

6. Using relations such as Eq. 9.6 for adjacent as well as more distant carbon atoms.

7. In Chapter 4 we deal with the opposite extreme: the case in which the zero-field splittings arise entirely from spin-orbit coupling.

8. The silicon metal-oxide-semiconductor field-effect transistor (MOSFET) is a dominant device in the electronics industry. The whole unit can be mounted in a magnet, and the recombination of electrons and holes can be observed by monitoring its electrical characteristics: electrically detected magnetic resonance (EDMR) [79,80].

9. A higher value, $t_d = 16$, has more recently been recommended [93].

10. It is unusual to have two different hyperfine splittings for two hydrogen atoms bonded to the same carbon atom. This implies that Q is not the same for the two hydrogen atoms. An explanation for this effect has been proposed [104].

FURTHER READING

Relations Between Hyperfine Splittings and Spin Densities

N. M. Atherton, *Principles of Electron Spin Resonance*, Prentice-Hall, New York, NY, U.S.A., 1993. (Chapter 3 contains a quite detailed discussion of the relationship between spin density and unpaired-electron population.)

E. T. Kaiser, L. Kevan, Eds., *Radical Ions*, Wiley-Interscience, New York, NY, U.S.A., 1968. (Chapters 1, 4, 5 and 6 deal with spin densities, radical cations, orbital degeneracy in substituted benzenes, and anion radicals.)

J. D. Memory, *Quantum Theory of Magnetic Resonance Parameters*, McGraw-Hill, New York, NY, U.S.A., 1968, Chapters 7 and 8. Relations between hyperfine splittings and spin densities are treated in terms of valence-bond and molecular-orbital theories.

Organic Radicals

F. Gerson, W. Huber, *Electron Spin Resonance of Organic Radicals*, Wiley-VCH, Weinheim, Germany, 2003. (Chapter 3 covers spin densities and unpaired-electron populations.)

E. T. Kaiser, L. Kevan, Eds., *Radical Ions*, Wiley-Interscience, New York, NY, U.S.A., 1968.

Inorganic Radicals

P. W. Atkins, M. C. R. Symons, *The Structure of Inorganic Radicals*, Elsevier, Amsterdam, Netherlands, 1967.

J. R. Morton, "Electron Spin Resonance Spectra of Oriented Radicals", *Chem. Rev.*, **64**, 453 (1964).

PROBLEMS

9.1 The proton hyperfine splittings for the naphthalene anion are 0.495 and 0.187 mT (Section 3.2.2). Based on the molecular orbitals of naphthalene (Problem 9A.3), how should these hyperfine splittings be assigned? How does the ratio of hyperfine splittings compare with the ratio of the squares of the atomic-orbital coefficients for the molecular orbital containing the odd electron?

9.2 Given that the proton NMR transition energies in a free radical containing a proton with hyperfine splitting a_i are

$$h\nu = |g_p\beta_n B_i - g_e\beta_e a_i M_S| \qquad (9.20)$$

where B_i is the NMR resonance field, derive Eq. 9.10 assuming that the energy-level populations are given by the Boltzmann distribution.

9.3 Proton NMR spectra of ethylbenzene at 56.4 MHz are shown in Fig. 9.13*a* without and in Fig. 9.13*b* with the corresponding monoanion as solute. From the shifts seen in the latter, confirm that the hyperfine splittings for the CH_2 and the *para* protons of the group are +0.080 and −0.087 mT, respectively. In this system, electron transfer is so rapid that all ethylbenzene molecules participate; the shifts are proportional to the mole fraction of the reduced form.

9.4 Calculate the unpaired-electron populations in the allyl radical, H_2CCHCH_2, from the Hückel molecular orbitals and energies given in Fig. 9A.1, taking $\lambda = 1.1$. Compare the results with the populations derived from the experimental hyperfine splittings [103] given below, taking $Q = -2.70$ mT. Assume that the smaller hyperfine splitting is positive, corresponding to a negative unpaired-electron population on the middle carbon atom. The two primary resonance structures of the allyl radical, with hyperfine

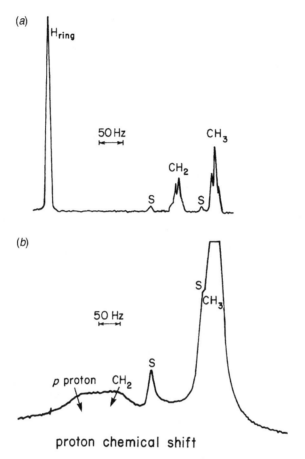

FIGURE 9.13 Proton magnetic resonance spectra at 56.4 MHz of (*a*) 1.93 M ethylbenzene, and (*b*) 1.93 M ethylbenzene plus 4.5×10^{-2} M ethylbenzene anion. The solvent is liquid d^8-tetrahydrofuran at $-75°C$. Peaks marked *S* are due to an impurity. [After E. de Boer, J. P. Colpa, *J. Phys. Chem.*, **71**, 21 (1967).]

splittings, are[10]

0.406 mT

$$\left. \begin{array}{l} 1.393 \text{ mT} \\ 1.483 \text{ mT} \end{array} \right\}$$
$$\text{H}_2\text{C}=\overset{\overset{\text{H}}{|}}{\text{C}}-\text{CH}_2 \quad \longleftrightarrow \quad \text{H}_2\text{C}-\overset{\overset{\text{H}}{|}}{\text{C}}=\text{CH}_2$$

9.5 The proton hyperfine splittings for the 1,3-butadiene anion are -0.762 and -0.279 mT.

(*a*) What is the average value of Q?

(*b*) Explain why $|Q|$ is so low. (Usually Q ranges from -2.5 to -3.0 mT.)

9.6 The ^{1}H and ^{13}C hyperfine splittings (including the signs) have been measured [105] for the radical

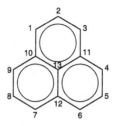

(XI) perinaphthenyl

$$a_1{}^H = -0.6270 \text{ mT} \quad a_2{}^H = 0.1833 \text{ mT} \quad a_1{}^C = 0.979 \text{ mT}$$

$$a_2{}^C = -0.792 \text{ mT} \quad a_{10}{}^C = -0.792 \text{ mT} \quad a_{13}{}^C = 0.332 \text{ mT}$$

The wavefunction for the non-bonded orbital is as follows:

$$\psi_{NB} = \frac{1}{\sqrt{6}}(\phi_1 - \phi_3 + \phi_4 - \phi_6 + \phi_7 - \phi_9)$$

(**a**) Assume $Q_{CH}{}^H = -2.7$ mT, and calculate ρ_1 and ρ_2.

(**b**) ρ_{10} and ρ_{13} have been computed from theory and are given as $\rho_{10} = -0.054$ and $\rho_{13} = +0.044$. Use this spin distribution to calculate the ^{13}C splitting constants. (Remember that positions 10 and 13 have three carbon atoms bonded to the central carbon, whereas positions 1 and 2 have two carbons and a proton.) How do these compare with the experimental ^{13}C splittings?

9.7 The statement has been made that the value of Q determines the total extent (Section 9.2.5) of the π-radical EPR spectrum. For the benzene anion the spectral extent is $\sim$2.25 mT, for CH$_3$ $\sim$6.9 mT, and for perinaphthenyl, $\sim$4.3 mT. Comment on the magnitudes of these values.

9.8 Interpret the spectrum shown in Fig. 9.8, which arises from the Cl$_2{}^-$ ion in KCl.

APPENDIX 9A HÜCKEL MOLECULAR-ORBITAL CALCULATIONS

A brief summary of the HMO approach to the calculation of orbital energies and unpaired-electron distributions in π-electron systems is given here. Because of the crude assumption of non-interaction among the electrons, we can treat all anions, neutral molecules, and cations using the same theory. Thus the σ system of H$_2{}^+$,

H_2 and H_2^- can serve as one basic example; these calculations yield equations equally applicable to the π-electron states of the molecules $C_2H_4^+$, C_2H_4 and $C_2H_4^-$. This approach has been widely described in textbooks and in intermediate-level chemistry courses. Hence here we shall only map out successive steps and summarize intermediate and working-level expressions. Detailed molecular-orbital calculation procedures and tabulations of the results for many molecules are given in the references at the end of this appendix.

1. Define the molecular orbitals to be linear combinations

$$|\psi_i\rangle = c_{i1}|\phi_1\rangle + c_{i2}|\phi_2\rangle + \cdots + c_{in}|\phi_n\rangle \tag{9A.1}$$

of n normalized atomic orbitals. The total energy expectation value for the ith molecular orbital ($i = 1, \ldots, n$) is given by $\langle\psi_i|\hat{\mathcal{H}}|\psi_i\rangle$. We shall not need to establish the form of the hamiltonian $\hat{\mathcal{H}}$ explicitly. For the present we set $n = 2$; that is, we consider systems such as H_2 or the C_2H_4 π system for which one has two molecular orbitals

$$|\psi_1\rangle = c_{11}|\phi_1\rangle + c_{12}|\phi_2\rangle \tag{9A.2a}$$
$$|\psi_1\rangle = c_{21}|\phi_1\rangle + c_{22}|\phi_2\rangle \tag{9A.2b}$$

of interest. It is useful to define two parameters

$$H_{i,j} \equiv \langle\phi_i|\hat{\mathcal{H}}|\phi_j\rangle = H_{ji} \tag{9A.3a}$$
$$S_{i,j} \equiv \langle\phi_i|\phi_j\rangle = S_{ji} \tag{9A.3b}$$

in terms of the atomic orbitals ($i, j = 1, 2, \ldots, n$).

2. Determine the ratio of the coefficients c_i in each state (we suppress the first index here) by setting the derivatives $\partial U/\partial c_1$ and $\partial U/\partial c_2$ equal to zero. Then rewrite the two resulting equations in terms of the parameters H_{ij} and S_{ij} ($i, j = 1, 2$)

$$c_1(H_{11} - US_{11}) + c_2(H_{12} - US_{12}) = 0 \tag{9A.4a}$$
$$c_1(H_{12} - US_{12}) + c_2(H_{22} - US_{22}) = 0 \tag{9A.4b}$$

3. Set $H_{11} = H_{22}$, $S_{11} = S_{22} = 1$, and $S_{12} = 0$

4. Write determinantal equations

$$\begin{vmatrix} H_{11} - U & H_{12} \\ H_{12} & H_{11} - U \end{vmatrix} \tag{9A.5}$$

noting that c_1 and c_2 are the variables. Solution of the resulting determinant yields the energies

$$U_1 = H_{11} + H_{12} \qquad\qquad (9A.6a)$$
$$U_2 = H_{11} - H_{12} \qquad\qquad (9A.6b)$$

of the two levels. The ratio c_1/c_2 is found to be $+1$ for the orbital with energy U_1 and -1 for the orbital with energy U_2. The coefficient c_1 is determined by the normalization condition $\langle \psi | \psi \rangle = 1$. The final result is that the wavefunctions are

$$|\psi_1\rangle = \frac{1}{\sqrt{2}}(|\phi_1\rangle + |\phi_2\rangle) \quad \text{(energy } U_1\text{)} \qquad\qquad (9A.7a)$$

$$|\psi_2\rangle = \frac{1}{\sqrt{2}}(|\phi_1\rangle - |\phi_2\rangle) \quad \text{(energy } U_2\text{)} \qquad\qquad (9A.7b)$$

for the lower and the upper states, since H_{11} and H_{12} are both negative.

In the HMO description of ground-state $H_2{}^+$ and $C_2H_4{}^+$, the single electron occupies the lower level. For ethylene in its ground state, the two π electrons occupy the lower level of this diamagnetic molecule.

It is important to be able to establish the energy levels for linear conjugated systems of n atoms. Each of the n molecular orbitals is taken to be a linear combination of n atomic orbitals (Eq. 9A.1). The secular determinant is set equal to zero. The integrals H_{ij} and S_{ij} are the numerical parameters already encountered. Thus, generalizing Eq. 9A.5, one has

$$\begin{vmatrix} H_{1n} - US_{1n} & \cdots & \cdots & \cdots & H_{nn} - US_{nn} \\ \cdots & \cdots & \cdots & \cdots & \cdots \\ H_{13} - US_{13} & H_{23} - US_{23} & H_{33} - US_{33} & \cdots & H_{3n} - US_{3n} \\ H_{12} - US_{12} & H_{22} - US_{22} & H_{23} - US_{23} & \cdots & H_{2n} - US_{2n} \\ H_{11} - US_{11} & H_{12} - US_{12} & H_{13} - US_{13} & \cdots & H_{1n} - US_{1n} \end{vmatrix} = 0$$

$$\qquad\qquad (9A.8)$$

where the rows are arranged in increasing order of the energy H_{ii}. The following simplifying assumptions are made:

1. $S_{ii} = 1$, $S_{ij} = 0$ if $i \neq j$.
2. All H_{ij} $(i \neq j) = \beta$ if atoms are bonded and zero otherwise. The numerical parameter β is called the *resonance integral* (a negative quantity).
3. All $H_{ii} = \alpha$. The numerical parameter α is called the *Coulomb integral* (a negative quantity).

These symbols, used as matrix elements, should not be confused with the spin functions α and β used elsewhere in this book.

Application of Eq. 9A.7 to the allyl molecule leads to the determinantal equation

$$
\begin{array}{cc}
\text{H} & \text{H} \\
\diagdown & \diagup \\
\text{C} =\!\!= \text{CH} =\!\!= \text{C} \\
\diagup & \diagdown \\
\text{H} & \text{H}
\end{array}
$$

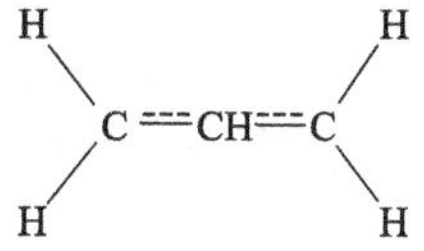

$$
\begin{vmatrix}
\alpha - U & \beta & 0 \\
\beta & \alpha - U & \beta \\
0 & \beta & \alpha - U
\end{vmatrix} = 0
\qquad (9A.9a)
$$

On dividing all terms by β and making the substitution $x = (\alpha - U)/\beta$, one obtains the determinantal equation

$$
\begin{vmatrix}
x & 1 & 0 \\
1 & x & 1 \\
0 & 1 & x
\end{vmatrix} = 0
\qquad (9A.9b)
$$

The three eigenvalues, obtained by expansion of the determinant, and the corresponding wavefunctions are

$$
U_3 = \alpha - \sqrt{2}\beta \quad \psi_3 = \frac{1}{2}\phi_1 - \frac{1}{\sqrt{2}}\phi_2 + \frac{1}{2}\phi_3
\qquad (9A.10a)
$$

$$
U_2 = \alpha \qquad\qquad \psi_2 = \frac{1}{\sqrt{2}}\phi_1 + 0\,\phi_2 - \frac{1}{\sqrt{2}}\phi_3
\qquad (9A.10b)
$$

$$
U_1 = \alpha + \sqrt{2}\beta \quad \psi_1 = \frac{1}{2}\phi_1 + \frac{1}{\sqrt{2}}\phi_2 + \frac{1}{2}\phi_3
\qquad (9A.10c)
$$

The orbital energy levels for and spin configurations of the allyl radical, cation and anion are shown in Fig. 9A.1.

The coefficients for the set of corresponding molecular orbitals can be obtained from the secular determinant (Eq. 9A.9b) by writing each line as an equation and substituting each eigenvalue ($x = -\sqrt{2}$, 0 or $\sqrt{2}$) in turn, and by applying the normalization condition (Eq. 9.3)

Calculation of the four Hückle molecular orbitals and energies of 1,3-butadiene is given as a problem at the end of this appendix; the results are quoted in Table 9A.1.

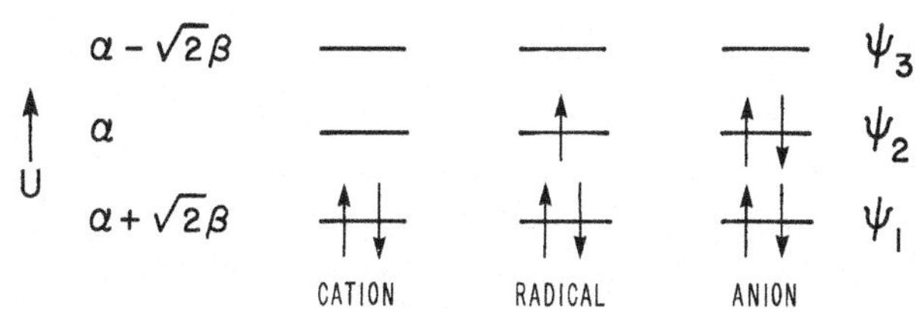

FIGURE 9A.1 The orbital energy levels of the allyl cation, radical and anion. Here ψ_i ($i = 1,2,3$) is the $2p_z$ atomic orbital on carbon atom i.

TABLE 9A.1   Molecular Orbitals and Energies of 1,3-Butadiene

Molecular Orbitals	Orbital Energies
$\psi_4 = 0.371\phi_1 - 0.600\phi_2 + 0.600\phi_3 - 0.371\phi_4$	$U_4 = \alpha - \frac{1}{2}(\sqrt{5}+1)\beta$
$\psi_3 = 0.600\phi_1 - 0.371\phi_2 - 0.371\phi_3 + 0.600\phi_4$	$U_3 = \alpha - \frac{1}{2}(\sqrt{5}-1)\beta$
$\psi_2 = 0.600\phi_1 + 0.371\phi_2 - 0.371\phi_3 - 0.600\phi_4$	$U_2 = \alpha + \frac{1}{2}(\sqrt{5}-1)\beta$
$\psi_1 = 0.371\phi_1 + 0.600\phi_2 + 0.600\phi_3 + 0.371\phi_4$	$U_1 = \alpha + \frac{1}{2}(\sqrt{5}+1)\beta$

TABLE 9A.2   Molecular Orbitals and Energies of Benzene

Molecular Orbitals	Orbital Energies
$\psi(b) = \frac{1}{\sqrt{6}}(\phi_1 - \phi_2 + \phi_3 - \phi_4 + \phi_5 - \phi_6)$	$U(b) = \alpha - 2\beta$
$\begin{bmatrix} \psi(e_2) = \frac{1}{2}(\phi_2 - \phi_3 + \phi_5 - \phi_6) \\ \psi(e_2) = \frac{1}{\sqrt{12}}(2\phi_1 - \phi_2 - \phi_3 + 2\phi_4 - \phi_5 - \phi_6) \end{bmatrix}$	$U(e_2) = \alpha - \beta$ $U(e_2) = \alpha - \beta$
$\begin{bmatrix} \psi(e_1) = \frac{1}{2}(\phi_2 + \phi_3 - \phi_5 - \phi_6) \\ \psi(e_1) = \frac{1}{\sqrt{12}}(2\phi_1 + \phi_2 - \phi_3 - 2\phi_4 - \phi_5 + \phi_6) \end{bmatrix}$	$U(e_1) = \alpha + \beta$ $U(e_1) = \alpha + \beta$
$\psi(a) = \frac{1}{\sqrt{6}}(\phi_1 + \phi_2 + \phi_3 + \phi_4 + \phi_5 + \phi_6)$	$U(b) = \alpha + 2\beta$

The neutral molecule has four π electrons. Following the rules, these must be assigned to the molecular orbitals of lowest energy (i.e., two to ψ_1 and two to ψ_2, since β is negative) to describe the ground state.

For other conjugated systems one may proceed in an analogous fashion. The secular determinant for linear conjugated systems contains the values $\alpha - U$ on the diagonal, with β one position off the diagonal, and zero elsewhere. For cyclic systems there are other non-zero off-diagonal terms. The resulting $n \times n$ determinant may easily be solved by computers; however, the task is simplified if the determinant is factorable; this can often be accomplished if of the symmetry properties of the molecule are employed using straightforward methods of group theory [A1,A2]. The π molecular orbitals of benzene (Table 9A.2) are entirely determined by symmetry. For further information regarding HMO theory, see Refs. A3–A8.

HMO References

A1. A. Streitwieser Jr., *Molecular Orbital Theory*, Wiley, New York, NY, U.S.A., 1961. [Chapters 2 and 3 describe in detail the procedures for calculations of orbital energies and wavefunctions of hydrocarbons. Chapter 4 describes refinements of the method and Chapter 5 deals with applications to molecules having hetero (N, O, S or halogen] atoms.

A2. F. A. Cotton, *Chemical Application of Group Theory*, 3rd ed., Wiley-Interscience, New York, NY, U.S.A., 1990. (The treatment of monocyclic systems in Chapter 7 is of special interest, as is Chapter 8, dealing with inorganic complexes).

A3. J. N. Murrell, S. F. A. Kettle, J. M. Tedder, *Valence Theory*, Wiley, New York, NY, U.S.A., 1965. (Chapter 15 deals with the π-electron theory of organic molecules. Section 15.8, "A Critique of Hückel Theory," gives some insight into the successes of the HMO approach.)

A4. L. Salem, *The Molecular Orbital Theory of Conjugated Systems*, Benjamin, New York, NY, U.S.A., 1966, Chapter 1.

A5. M. J. S. Dewar, *The Molecular Orbital Theory of Organic Chemistry*, McGraw-Hill, New York, NY, U.S.A., 1969, Chapter 5.

A6. C. A. Coulson, A. Streitwieser, *Dictionary of π-Electron Calculations*, Freeman, San Francisco, CA, U.S.A., 1965.

A7. E. Heilbronner, Jr. P. A. Straub, *Hückel Molecular Orbitals*, Springer, New York, NY, U.S.A., 1966.

A8. P. W. Atkins, *Molecular Quantum Mechanics*, 2nd ed., Oxford University Press, Oxford, U.K., 1983, Section 10.9.

HMO Problems

9A.1   Set up the secular equation for the cyclopropenyl (C_3H_3) radical and solve for the orbital energies. Draw an orbital energy diagram and show the distribution of electrons among the π orbitals.

9A.2   Set up the secular equation for the 1,3-butadiene and solve for the energies. Substitute the energies into the secular equations and determine the coefficients in the four π molecular orbitals (Table 9A.1).

9A.3   The seven lowest-lying Hückel molecular orbitals

$$\psi_n = c_{n1}\phi_1 + c_{n2}\phi_2 + c_{n3}\phi_3 + \cdots + c_{n10}\phi_{10}$$

are shown below for naphthalene, in order of increasing energy; a structure showing the atom numbering is given in Table 9.3.

ψ_n	c_{n1}	c_{n2}	c_{n3}	c_{n4}	c_{n5}	c_{n6}	c_{n7}	c_{n8}	c_{n9}	c_{n10}
ψ_7	0	−0.408	0.408	0	0	0.408	−0.408	0	0.408	−0.408
ψ_6	0.425	−0.263	−0.263	0.425	−0.425	0.263	0.263	−0.425	0	0
ψ_5	0.425	0.263	−0.263	−0.425	0.425	0.263	−0.263	−0.425	0	0
ψ_4	0	0.408	0.408	0	0	0.408	0.408	0	−0.408	−0.408
ψ_3	0.400	0.174	−0.174	−0.400	−0.400	−0.174	0.174	0.400	0.347	−0.347
ψ_2	0.263	0.425	0.425	0.263	−0.263	−0.425	−0.425	−0.263	0	0
ψ_1	0.301	0.231	0.231	0.301	0.301	0.231	0.231	0.301	0.461	0.461

(*a*) Without doing any calculations, sketch approximately the set of HMO energies for naphthalene, and show the orbital occupation by electrons for the −1, 0 and +1 charged species.

(*b*) Compare ψ_5 with ψ_6, and ψ_4 with ψ_7. What identities may be written for corresponding c_{ni} values of the related pairs of molecular orbitals?

(*c*) What is the significance of a zero value of c_{ni}?

(*d*) Sketch the locations of nodal planes for all these orbitals.

9A.4 Calculate the unpaired-electron populations at each of the carbon atoms in the benzyl radical ($C_6H_5CH_2$), taking into account the following attributes of odd-alternant hydrocarbons.

1. There are two possible numbers of starred atoms, depending on the starting point. Choose the configuration with the larger number of starred atoms.

2. The unpaired electron resides in a non-bonding orbital, for which one notes that (a) the molecular-orbital coefficients of unstarred atoms are zero, and (b) the sum of the molecular-orbital coefficients of atoms about a starred position is zero.

Starting at one of the starred atoms in the benzene ring, assign the relative values of coefficients at each atom. From the requirement that the sum of the squares of the coefficients is equal to 1, ascertain the unpaired-electron population at each position.

Equations 10.19 are not correct at small fields B (i.e., at $B \leq B_1$) since the susceptibilities in reality do not vanish at $\omega_B = 0$, but rather must be modified in light of the condition expressed in Note 10.

The dynamic susceptibilities χ' and χ'' have definite and important meanings, representing the dispersion and power absorption of the magnetic-resonance transition (see Section F.3.5). The latter is especially important and is the radiative part of $d(-\mathbf{M}^T \cdot \mathbf{B}_{total})/dt$ (Eqs. 1.8 and 1.14a). It follows [9, Chapter 2; 20] that the power $P_a(\omega)$ absorbed by the magnetic system from the linearly polarized excitation field (Eq. 10.25) is

$$P_a(\omega) = 1 \frac{\omega \chi'' B_1{}^2}{\mu_0 V} \tag{10.30}$$

per unit sample volume.[13] Hence, with use of Eq. 10.29b

$$P_a(\omega) = \frac{\pi}{\mu_0} \frac{B_1{}^2}{(1 + \gamma_e{}^2 B_1{}^2 \tau_1 \tau_2)^{1/2}} \omega \; \omega_B \chi^0 Y \tag{10.31}$$

Note that $Y(\omega - \omega_B)$ is a normalized lorentzian function (Table F.1a) and depends on the experimental conditions [i.e., on B_1 via the linewidth $\Gamma = (1 + \gamma_e{}^2 B_1{}^2 \tau_1 \tau_2)^{1/2}/\tau_2$]. The increase in linewidth as saturation sets in can be discussed in terms of 'lifetime broadening'. Increasing the microwave power produces spin transitions at a faster rate and hence decreases the mean spin-orientation lifetime. As B_1 becomes very large, τ_1 becomes proportional to $B_1{}^{-2}$ (Eq. 10.13) and hence Γ becomes insensitive to B_1 (when τ_2 is non-zero).

As long as $\gamma_e{}^2 B_1{}^2 \tau_1 \tau_2 \ll 1$, this saturation term can be neglected, and both P and dP/dB are proportional to $B_1{}^2$. When the absorption line is strongly saturated ($\gamma_e{}^2 B_1{}^2 \tau_1 \tau_2 \gg 1$), according to the Bloch theory, both χ' and χ'' decrease with increasing power P_0 (e.g., Fig. F.8a), and P_a becomes constant. However, note that the theory fails when $B_1 > B$ and $\tau_1 \neq \tau_2$, as in solids (see Note 11.6 and Chapter 8 of Ref. 11).

The various equations in this section have been written in terms of ω and $\omega_B = -\gamma_e B$. It has been implied that ω is the continuous variable and ω_B is a constant for the spin species at hand, with resonance centered at a particular value $\omega_r = \omega_B$. In EPR, of course, usually ω is held constant and B is scanned. It is relatively easy to switch to B as the variable,[14] with $\omega = \omega_B$ held fixed; now ω_B is the angular frequency at a particular value B_r of B as given by the resonance condition.

When we switch to field-sweep conditions, the lineshape function, now $Y(B - B_r)$, is a lorentzian with half width at half-height given by

$$\Gamma = \frac{1}{|\gamma_e| \tau_2} (1 + \gamma_e{}^2 B_1{}^2 \tau_1 \tau_2)^{1/2} \tag{10.32}$$

In more recent work, the effect of field modulation on the magnetic field $\mathbf{B}$ appearing in the Bloch equations has been treated successfully, via perturbation theory [27]. Expression for both the absorption and dispersion phenomena are given.

10.4 LINEWIDTHS

We classify spectral lines into those that are *homogeneously broadened* and those that are *inhomogeneously broadened*.

10.4.1 Homogeneous Broadening

Homogeneous line broadening for a set of spins occurs when all these see the same net magnetic field and have the same spin-hamiltonian parameters. (More correctly, the local fields need not be identical at any one instant, but need only give the same time-averaged field over sufficiently short intervals.) This means that the lineshape (i.e., the transition probability as a function of magnetic field) is the same for each dipole. The resulting line usually has a lorentzian shape (Fig. F.1). This is in accord with the predicted Bloch absorption lineshape (Eq. 10.31), which is lorentzian with a linewidth (half-width at half-height) in frequency (ω) units of τ_2^{-1} under non-saturating conditions. In general, one often can define an effective τ_2 by equating it to $|\kappa\gamma_e\Gamma|^{-1}$, where Γ (mT) now is half the linewidth at half-height in the absence of microwave power saturation (Eq. 10.32 and Section F.3.4) and κ is a factor that depends on the lineshape. For lorentzian lines $\kappa = 1$, whereas for gaussian lines $\kappa = (\pi \ln 2)^{1/2}$.

We turn now to models for the transverse relaxation time τ_2, a topic that lies outside the realm of the Bloch equations. To visualize one possible contribution to τ_2, one can return to consideration of the individual spins. It has been established that τ_2 often is a measure of the interaction between spins. In this case, if $\tau_2 = \infty$, the spins are completely isolated from each other, whereas $\tau_2 = 0$ implies very strong coupling, such that there are no local variations in the spin temperature. However, the latter limit is of no relevance here since we do not deal with strongly coupled (ferromagnetic or antiferromagnetic) systems. The spins can interact via magnetic dipolar coupling. Note that mutual spin flips of paired spins cause no change in energy of the spin system but do affect the lifetime (τ_1) of each spin. The propagation of magnetization through the lattice via such flips, called *spin diffusion*, causes equilibration to the same spin temperature throughout the system of equivalent spins; τ_2 is a measure of this rate.

Another model invoking random sudden fast events considers collisions (e.g., in gas-phase radicals) that reorient the spin magnetic moments. This leads to Bloch-type equations where now $\tau_1 = \tau_2$ is the mean time period between collisions [28]. The actual linewidth is determined by $(2\tau_1)^{-1} + (1 + \gamma_e^2 B_1^2 \tau_1\tau_2)^{1/2}\tau_2^{-1}$ rather than by the second term alone (Eq. 10.32). The lifetime broadening produced by the first term is missing in the Bloch formulation (which predicts a Dirac δ-function absorption in the limit $\tau_2 \to \infty$). The factor 2 in the first term arises since $Z \approx 2\tau_1$, as discussed earlier. Some techniques for measuring τ_1 and τ_2 are described in Section F.6.

10.4.2 Inhomogeneous Broadening

We briefly consider the inhomogeneous case. Here the line-broadening mechanism distributes the resonance frequencies over an unresolved band, without broadening the lines arising from individual equivalent spins. Generally the unpaired electrons in a sample are not all subjected to exactly the same B values. Thus, at any given time, only a small fraction of the spins is in resonance as the external magnetic field is swept through the 'line'. The observed line is then a superposition of a large number of individual components (referred to as 'spin packets'), each slightly shifted from the others. The resultant envelope often has approximately a gaussian shape (Fig. F.2).[15] It thus is possible to choose B_1 so as to power saturate some selected portion of the EPR line, decreasing its intensity there (this is known as 'hole burning'). The following are some causes of inhomogeneous broadening for a given spin (chemical) species:

1. An inhomogeneous external magnetic field.
2. Unresolved hyperfine structure (e.g., for F centers in KCl), occurring when the number of hyperfine components from nearby nuclei is so great that no structure is observed. Hence one detects the envelope of a multitude of lines. [These may be resolved by the technique of electron-nuclear double resonance (ENDOR; Chapter 12).]
3. Anisotropic interactions in randomly oriented systems in the solid state. Here the distribution of local magnetic fields resulting from the anisotropic g and hyperfine interactions gives rise to the inhomogeneity. In this case the line-shape may be highly unsymmetric (Chapters 4 and 5).
4. Dipolar interactions with other fixed paramagnetic centers. These may impose a random local field at a given unpaired electron, arising from dipolar fields from other electron spins (Chapter 6).

The values of τ_1 and of τ_2 may be the same for all packets, or they may differ. Experimental techniques (e.g., double-field modulation) exist to detect homogeneous spin-packet lines within an inhomogeneously broadened EPR line [29].

The lines making up an inhomogeneous packet may have different widths, and this can cause some strange effects. For instance, when the central lines are broad compared to the outer ones, the absorption may show a minimum there, leading (say) to a set of inverse-phase 'lines' to the sides of the central region of the first-derivative spectrum [30].

In some of the above cases, the local magnetic fields giving rise to inhomogeneous broadening can be averaged out via sufficiently rapid dynamic effects (e.g., tumbling, collisions, exchange), yielding homogeneously broadened lines, as discussed in the next sections.

10.5 DYNAMIC LINESHAPE EFFECTS

We now open a range of topics centering on dynamic processes that lead to homogeneous line broadening, ignoring the static mechanisms considered in Section 10.4.2 that lead to inhomogeneous broadening.

Any dynamic process in and around the paramagnetic center can cause lineshape effects. Some such processes are: hindered rotation, tumbling of the molecule in a viscous liquid, interactions with other paramagnetic species and chemical reactions (e.g., acid-base equilibria and electron-transfer reactions). This broadening arises from dynamic fluctuations in the local field at the unpaired electron(s). If the changes occur sufficiently slowly, one observes lines assignable to distinct species (e.g., conformers). However, as the rate of fluctuations increases, the EPR lines broaden and finally coalesce into a single line (or set of lines), the position of which is the weighted average of the original line positions.

10.5.1 Generalized Bloch Equations

There are many theoretical models that can be used to simulate the effects of dynamic magnetic-field fluctuations on EPR lines. Some of these are summarized in Section 10.5.2. However, we have chosen to start with the *generalized Bloch equation model*, since it is easy to understand conceptually and since computer calculations can be carried out readily for it.[16]

Consider a radical that can exist in either of two distinct forms or environments, a and b (i.e., each has a distinctive EPR spectrum). For the sake of simplicity, assume that the probabilities for these forms are f_a and f_b (where $f_a + f_b = 1$) and that each form gives rise to a single EPR line of lorentzian shape, one at resonance field B_a and the other at a higher field B_b (Fig. 10.7a). The line separation is $\Delta B_0 = B_b - B_a$ and often depends on **B**. In other words, the two species generally have different g factors. When we use the word *slow* or *fast*, we mean interconversion rates (e.g., local magnetic-field fluctuations) that are slow or fast compared to the characteristic parameter $|\gamma_e|\Delta B_0$. The actual time taken for a molecule to react to such an event is assumed to be very short compared to the inverse of this.

We start with the complex Bloch equation (Eq. 10.22) in the rotating frame. Here the magnetization M_z has been replaced by $M_z{}^0$, since we assume that the microwave power has been set low enough so as to avoid saturation. It also follows that the inherent linewidth is $\tau_2{}^{-1}$ (Eq. 10.32), not necessarily the same for chemical forms a and b. We now use the abbreviated notation $G = M_{+\phi}$. Equation 10.23 yields

$$\alpha_a = \tau_{2a}{}^{-1} - i(\omega_{Ba} - \omega) \tag{10.33a}$$

$$\alpha_b = \tau_{2b}{}^{-1} - i(\omega_{Bb} - \omega) \tag{10.33b}$$

Relaxation times τ_{2a} and τ_{2b} represent the inverse linewidths for forms a and b in the absence of dynamic processes (and of power saturation). These are taken to be independent of temperature. We note that $\gamma_a \neq \gamma_b$ implies $g_a \neq g_b$.

The functions G_a and G_b can be considered in the same sense as concentrations in chemical kinetics. Thus for the reaction

$$a \underset{k_b}{\overset{k_a}{\rightleftharpoons}} b \tag{10.34}$$

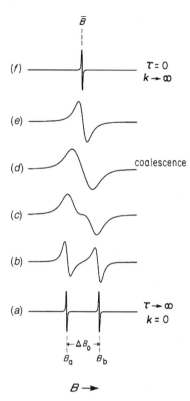

FIGURE 10.7 Synthetic first-derivative spectra showing the effect of increasing rate of interconversion between species a and b of an unpaired-electron species: (a) slow-rate limit ($\tau \to \infty$); (b) moderately slow rate ($\tau \gg |\bar{\gamma}\Delta B_0|$); ($c$) faster rate showing spectral lineshifts; (d) coalescence point; (e) fast-rate limit ($\tau \to 0$), where it was assumed that $\Gamma_{0a} = \Gamma_{0b}$ ($=\Gamma_0$) and that $f_a = f_b$. Note from Eq. 10.47 that $\Delta B = [(\Delta B_0)^2 - 2(\bar{\gamma}\tau)^{-2}]^{1/2}$ for the two-line spectra. The values $\Delta B_0 = 1.8$ mT and $\Gamma_0 = 0.03$ mT were used. Quantity $\bar{\gamma}$ is defined in Eq. 10.37.

we can introduce chemical or physical dynamics into the Bloch equations by adding first-order kinetic terms to Eq. 10.22, yielding

$$\frac{dG_a}{dt} + \alpha_a G_a = i\gamma_a B_1 M_{za} + k_b G_b - k_a G_a \qquad (10.35a)$$

$$\frac{dG_b}{dt} + \alpha_b G_b = i\gamma_b B_1 M_{zb} + k_a G_a - k_b G_b \qquad (10.35b)$$

These linearly coupled equations can be solved at steady state ($dG_a/dt = dG_b/dt = 0$) for G_a and G_b. It is assumed that relaxation times τ_{1a} and τ_{1b} are sufficiently short that the thermal equilibrium between the spins is maintained. Thus for

$dM_{za}/dt = dM_{zb}/dt = 0$ (Eqs. 10.20c and 10.24c), we can utilize

$$M_{za} = f_a \gamma_a M_z^0 / \bar{\gamma} \qquad (10.36a)$$

and

$$M_{zb} = f_b \gamma_b M_z^0 / \bar{\gamma} \qquad (10.36b)$$

where

$$\bar{\gamma} = f_a \gamma_a + f_b \gamma_b \qquad (10.37)$$

The total complex transverse magnetization G is then given by

$$G = G_a + G_b \qquad (10.38a)$$

$$= iB_1 M_z^0 \frac{f_a \gamma_a (\alpha_b + k_a + k_b) + f_b \gamma_b (\alpha_a + k_a + k_b)}{(\alpha_a + k_a)(\alpha_b + k_b) - k_a k_b} \qquad (10.38b)$$

where $\alpha_{a,b}$ are complex numbers. Consistent with the chemical balance condition $f_a k_a = f_b k_b$, the population fractions obey the relations $f_a = \tau_a/(\tau_a + \tau_b)$ and $f_b = \tau_b/(\tau_a + \tau_b)$. By using $\tau_a^{-1} = k_a$ as well as $\tau_b^{-1} = k_b$, and by defining an inverse lifetime $\tau^{-1} = \tau_a^{-1} + \tau_b^{-1}$, Eq. 10.38a can be written in the alternate form

$$G = iB_1 M_z^0 \frac{\bar{\gamma} + \tau(f_a \gamma_a \alpha_b + f_b \gamma_b \alpha_a)}{\tau \alpha_a \alpha_b + f_a \alpha_a + f_b \alpha_b} \qquad (10.38c)$$

The intensity of absorption is proportional to the imaginary part of G (Section 10.3.4). The lineshapes as a function of τ are shown in Fig. 10.7 (first-derivative presentation). Before considering the general lineshape function, we consider two limits:

1. *Slow Dynamics.* Here lifetimes τ_a and τ_b are long compared to $|\gamma_e \Delta B_0|^{-1}$. We expect two separate lines. For instance, when B is near $B_a = -\omega/\gamma_a$ then, assuming $G_b \approx 0$, Eq. 10.35a for steady-state conditions yields

$$G_a = if_a \gamma_a B_1 M_z^0 \frac{1}{\alpha_a + k_a} \qquad (10.39)$$

and on taking the imaginary part (Section A.1), one obtains

$$M_{y\phi\, a} = -f_a B_1 M_z^0 \frac{\Gamma_{0a} + k_a/|\gamma_a|}{(\Gamma_{0a} + k_a/|\gamma_a|)^2 + (B_a - B)^2} \qquad (10.40)$$

which represents $\chi''(B)$, that is, the power absorbed per unit volume (Eq. 10.30). This is a lorentzian line with an absorption half-width at half-height of

$$\Gamma_a = \Gamma_{0\alpha} + |\gamma_a \tau_a|^{-1} \qquad (10.41)$$

where $\tau_a (= k_a^{-1})$ is the average lifetime of the form a. There is an exactly analogous lineshape expression for form b.

Thus we see that each line is broadened (but not shifted) by the onset of the dynamic process (Fig. 10.7b). By measuring the increase in linewidth, one can determine the rate constants for the dynamic process.

2. *Fast Dynamics.* When the two forms are interchanging very rapidly, such that τ_a and τ_b are very short, then the terms in τ can be neglected in Eq. 10.38b, and hence

$$G \approx i\,\bar{\gamma}B_1 M_z^0 \frac{1}{f_a\alpha_a + f_b\alpha_b} \tag{10.42}$$

On taking the imaginary part, one obtains

$$M_{y\phi} = -B_1 M_z^0 \frac{\bar{\Gamma}}{(\bar{\Gamma})^2 + (\bar{B} - B)^2} \tag{10.43}$$

where the weighted averages

$$\bar{\Gamma} = f_a\Gamma_{0a} + f_b\Gamma_{0b} \tag{10.44}$$
$$\bar{B} = f_aB_a + f_bB_b \tag{10.45}$$

have been used. Equation 10.43 clearly represents a lorentzian line of width $\bar{\Gamma}$ (halfwidth at half-height; Table F.1a) centered at the field $\bar{B}$ (Fig. 10.7e). A more detailed analysis shows that, as the system approaches the fast limit, the lineshape is centered at $\bar{B}$ with a lorentzian lineshape but that the linewidth is given by (Fig. 10.7d)

$$\Gamma = \bar{\Gamma} + f_a f_b \tau\,|\bar{\gamma}|\,(\Delta B_0)^2 \tag{10.46}$$

Thus again kinetic rate constants can be obtained from the changes in the linewidth of the single line.

3. *Intermediate Dynamics.* It is possible to derive [33,34] a general expression for the lineshape by taking the imaginary part of Eq. 10.38. As the system progresses from the slow-rate region into the intermediate region, the two lines are seen not only to broaden but also to shift inward (Fig. 10.7c). By determining the fields at which the denominator of the imaginary part of Eq. 10.38 has minima, one can derive that the separation of the two lines is given [33] by

$$\Delta B = [(\Delta B_0)^2 - 2(\bar{\gamma}\tau)^{-2}]^{1/2} \tag{10.47}$$

valid when the first right-hand term dominates. Eventually, the two lines coalesce into a single broad line centered at $\bar{B}$ (Fig. 10.7d). The coalescence point (defined as the point at which the second derivative of the absorption changes sign at

$B = \overline{B}$) is found to occur at a τ value

$$\tau = \frac{2\sqrt{2}}{|\overline{\gamma}|\Delta B_0} \tag{10.48}$$

Note that this value (which is in s rad^{-1}, since γB is an angular frequency) generally depends on the measurement frequency used, since ΔB_0 does. The coalescence phenomenon is a manifestation of the lifetime-broadening relation expressed in Eq. B.74. If one writes this as $\Delta t \, \Delta \omega \approx 1$, where $\Delta \omega$ is the separation of the two lines in angular frequency units, then Δt represents the smallest average time period during which the states a and b may be distinguished. If the lifetime τ is less than Δt, then only one central line is observed, since the two states cannot be distinguished.

We see, then, that EPR spectroscopy can yield rate data even for a chemical system in a steady-state condition. Thus via lineshape simulations (usually produced by computer) of spectra taken over an appropriate range of temperatures T, and using the exact formula for $im[G(\tau)]$, one can assemble an Arrhenius plot of $\ln(\tau^{-1})$ versus T^{-1}, the slope of which yields the activation energy for the chemical process at hand (Eq. 10.34). Linewidths Γ_{0a} and Γ_{0b} must be known and must not be too temperature-sensitive; the same is true for the equilibrium constant $K = f_a/f_b$, obtainable from the relative areas of the absorption curves available until they merge. Examples are presented below.

If there are more than two sites (or other than a $1:1$ stoichiometry in reaction 10.34), then the lineshapes can be more complicated. Various modifications of the Bloch formalism have been discussed [17, pp. 224–225; 35].

10.5.2   Other Theoretical Models

When there is observable zero-field splitting (e.g., hyperfine effects at either site), the Bloch formalism for the EPR lineshape of chemically or physically dynamic species is not adequate. Happily the more advanced density-matrix approach (see Note 11.3), first developed for the analogous NMR situations, does yield formulas useful in these cases. Summaries of the theory and applications of the whole dynamic field are available [35–37].

We now turn to some specific mechanisms that can cause lineshape effects. They all share a common feature, namely, that the spin hamiltonian of the species under-goes sudden and random changes, either in its parameters ($g, A, D, \ldots$) or possibly even in its form.

10.5.3   Examples of Line-Broadening Mechanisms

We now turn to various examples of thermal effects on the EPR spectral lineshapes. We remind ourselves (Section F.2) that the peak-to-peak amplitude of each derivative line is proportional to the relative intensity of the corresponding transition. However, under certain conditions, linewidths can vary with temperature, and from one line to another in a given spectrum. The result is a departure of the proportionality between the first-derivative amplitude and the line intensity, since the derivative amplitude is inversely proportional to the *square* of the linewidth (Tables F.1). Thus small changes in linewidths can cause large changes in the relative amplitudes of various lines in the spectrum. We shall see examples of this effect in the following sections.

10.5.3.1 Electron-Spin Exchange The term *electron-spin exchange* is here reserved for a bimolecular reaction in which the unpaired electrons of two free radicals exchange their spin orientations.[17]

Electron-spin exchange was first observed in the EPR spectra of the $(SO_3)_2NO^{2-}$ radical [38]. Here we consider such exchange for the analogous case of di-*t*-butyl nitroxide radicals in liquid solution [39,40]. Figure 10.8*a* displays the hyperfine (^{14}N) triplet spectrum observed at a very low radical concentration. At a higher concentration (Fig. 10.8*b*), the lines clearly are broadened. The exchange of electron-spin states between two radicals with the same nuclear-spin state does

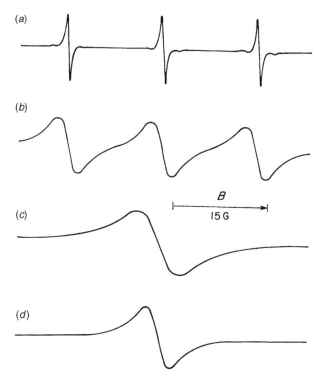

FIGURE 10.8 First-derivative spectra of the di-*t*-butyl nitroxide radical in ethanol at room temperature at various radical concentrations: (*a*) 10^{-4} M; (*b*) 10^{-2} M; (*c*) 10^{-1} M; (*d*) pure liquid nitroxide. (Spectra taken by J. R. Bolton.)

not change the resonant field, that is, the width. From the additional linewidth one can calculate τ, using

$$\Gamma = \Gamma_0 + |2\gamma_e\tau|^{-1} \qquad (10.49)$$

(compare with Eq. 10.41) valid when all encounters are effective. However, in the present case, the second right-hand term must be multiplied by the statistical factor of $\frac{2}{3}$ since one-third of the encounters between radicals result in no lineshape effect. It is important to note that it is $\frac{1}{2}\tau$, which is the electron-spin exchange rate per *molecule* and thus τ should be proportional to the inverse concentration $[R]^{-1}$ of radicals. The second-order rate constant is given by

$$k_{(2)} = \frac{1}{2\tau[R]} \qquad (10.50)$$

and is independent of $[R]$. For $(t\text{-butyl})_2NO$ in dimethylformamide, $k_{(2)} = 7.5 \times 10^9$ $M^{-1}\ s^{-1}$ [39]. This large value of $k_{(2)}$ indicates that spin exchange must occur with a high probability, virtually on each encounter, since this rate constant approximates that of a diffusion-controlled reaction.

As the concentration of $(t\text{-butyl})_2NO$ continues to increase, the lines coalesce to a single line (Fig. 10.8c) that becomes narrower at even higher concentrations (Fig. 10.8d). The latter type of spectrum is often said to be *exchange-narrowed* since the electron spins are exchanging so rapidly that the time-averaged hyperfine field is close to zero.[18] Generally, electron-spin exchange is to be avoided if resolved hyperfine structure with narrow lines is desired.

For instance, dissolved molecular oxygen causes line broadening, linear with temperature but which appears to be independent of solvent viscosity [41]. Since 1989 or so, this phenomenon has proved useful in the quantitative determination of O_2 concentrations, that is, oximetry in liquids, which is of special use in biomedical applications [42].

The electron-spin exchange effect on linewidths is not the same as the intermolecular magnetic dipole-dipole effect. Both are effective only during collisions in liquids. Electron-spin exchange is a quantum-mechanical effect (Section 6.2.1), which in liquids produces a much greater broadening than does the dipole-dipole effect. This can be shown by the following example. Suppose that the radical concentration is 10^{-3} M and the electron-spin exchange rate constant $k_{(2)}$ is $10^{10}\ M^{-1}\ s^{-1}$. From Eqs. 10.47 and 10.48 one calculates $\Gamma - \Gamma_0 = 0.06$ mT. However, at the same concentration, the dipolar broadening would contribute only ~ 0.001 mT to the linewidth.

10.5.3.2 Electron Transfer   Electron transfer between a radical and a diamagnetic species is very similar to electron-spin exchange in its effect on the spectrum. The first such reaction studied was that between naphthalene anions and neutral naphthalene molecules [43]

$$\text{naph(1)}^- + \text{naph(2)} \rightleftharpoons \text{naph(1)} + \text{naph(2)}^- \qquad (10.51)$$

However, the role of the cation (e.g., Na^+ or K^+) is not entirely negligible.

It should be noted that naphthalene molecules are distinguishable by virtue of the many (2^8) different arrangements of the proton spins. In fact, the solution EPR spectrum of the naphthalene anion shows 25 distinct resonant field positions, most of which are degenerate. Thus when an electron-transfer reaction occurs, the resonance field for the electron is usually shifted. If the transfer rate is small compared to the separation between resonance lines (slow-transfer-rate region), the effect is to cause a broadening of each resonance line in the spectrum. The broadening is, in general, not the same for each hyperfine component. For example, there are 36 times as many molecules with a resonance field corresponding to the central line as for those molecules with a resonance field corresponding to one of the outermost lines. Since the probability of a jump between molecules with the same resonant field is much greater for molecules contributing to the central line, one might expect that this line would be narrower than lines toward the outside of the spectrum. This phenomenon has been observed [44] and indicates that the spin-orientation (α or β) lifetime of an electron (characterized by τ_1) is much longer than the average residence lifetime 2τ of the electron on a given naphthalene molecule; hence this mechanism causes line broadening of the τ_2 type.

The broadening is given by Eqs. 10.41 and 10.50 (note that the concentration [naph] of neutral naphthalene must be used in Eq. 10.50). Measurement of the linewidth as a function of this concentration enables one to obtain the second-order electron-transfer rate constant. For the naphthalene anion in tetrahydrofuran [40], $k_{(2)} = 5.7 \times 10^7 \, \text{M}^{-1} \, \text{s}^{-1}$. This value is almost 100 times smaller than the diffusion-controlled rate constant. Thus one concludes that electron transfer occurs in only a small fraction of the collisions of a radical ion with neutral naphthalene molecules. This low efficiency may be due to a transfer mechanism involving the alkali cation positioned near the anion.

In a similar system, namely, the electron transfer between benzophenone and its anion, the spectrum coalesces to a quartet of equally intense lines at high concentrations of benzophenone [45,46]. This observation indicates that in this case the transfer process involves a sodium *atom*, instead of a single electron. Similarly, EPR study of the transfer of H atoms between 2,4,6-tri-*t*-butylphenol (**I**) and the corresponding phenoxy radical yields the second-order rate constant $k_{(2)} = 500 \, \text{M}^{-1} \, \text{s}^{-1}$ at 30°C [47].

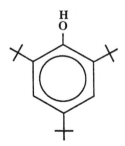

(**I**) 2,4,6-tri-*t*-butylphenol

10.5.3.3 *Proton Transfer* The previous two examples illustrate changes in the magnetic environment that arise from exchange of electron-spin states or from transfer of an unpaired electron from one molecule to another. However, environmental changes can also occur if a chemical reaction exchanges one or more nuclei in the molecule with nuclei in the solvent. Usually such reactions are too slow to have an effect on an EPR spectrum, although effects on NMR spectra can be very pronounced. In the case of proton exchange, reaction rates are sometimes large enough to produce detectable effects.

A good example of proton transfer is that of the CH_2OH radical considered in Chapter 3. Figure 10.9 displays its spectrum in aqueous methanol solution at two pH values. The OH doublet is resolved at the higher pH, but as the pH is lowered, the doublet spacing decreases; the lines broaden somewhat and finally collapse into a single line. In this case the OH proton is rapidly exchanging with H^+ ions. The proton-exchange rate may be estimated from the line separation given by Eq. 10.47. The second-order rate constant is 1.76×10^8 M^{-1} s^{-1} [48].

10.5.3.4 *Fluxional Motion* The internal motions of unpaired-electron species can give rise to striking EPR effects. We cite a relevant example involving an organic radical cation in Section 10.5.4.2, and one dealing with an inorganic complex at the end of Section 10.5.5.2.

Here we limit ourselves to a 'simple' inorganic solid-state example: $Cu(H_2O)_6^{2+}$ ($S = \frac{1}{2}$), dilute in zinc fluosilicate crystals. These ions occur in any of three configurations corresponding to equivalent tetragonal distortions of the ligand 'octahedron', caused by the Jahn-Teller effect (Section 8.2). Reorientational jumps between these configurations cause major lineshape distortions away from the expected regular (superimposed) ^{63,65}Cu hyperfine quartets. The lineshapes are depicted in Fig. 10.10 and exhibit major dynamic broadening, strongly dependent

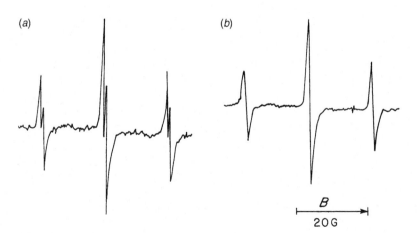

FIGURE 10.9 X-band first-derivative spectra of the CH_2OH radical in aqueous solution at room temperature: (*a*) pH = 1.40 (here $a_{OH} = 0.96$ G and $\Gamma_0 = 0.30$ G). (*b*) pH = 1.03. [After H. Fischer, *Mol. Phys.*, **9**, 149 (1965).]

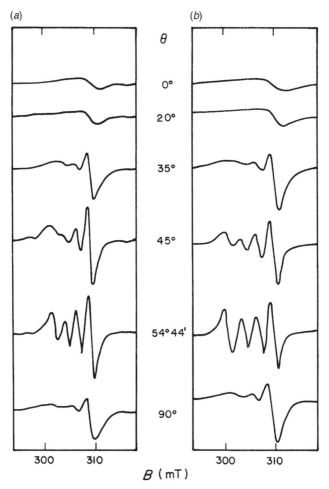

FIGURE 10.10 Angular dependence of the EPR spectrum of the $Cu(H_2O)_6{}^{2+}$ complex in zinc fluosilicate at 45 K: (*a*) experimental and (*b*) theoretical. The magnetic field lies in the (110) plane and makes an angle θ with the [001] axis. The spectra in (*a*) were measured at 9.5 GHz. The lineshapes in (*b*) were computed by convoluting each of the four lorentzians having half-widths Γ defined by $|\gamma_e \tau_2|^{-1}$ with gaussians having half-widths $\Gamma = (2 \ln 2)^{1/2} \times 0.6$ mT attributed to unresolved proton hyperfine splittings. Here τ_2 is a function of M_I (see Note 20). The jump frequency used was 5×10^9 s^{-1}. [After Z. Zimpel, *J. Magn. Reson.*, **85**, 314 (1989).]

on M_I and highly anisotropic. These effects have been successfully modeled [49] using a discrete-jump density-matrix approach (see Note 11.3).

10.5.4 Linewidth Variation: Dynamic Hyperfine Contributions

Variation in linewidths with M_I can be caused by sufficiently rapid changes in EPR hyperfine splittings arising from chemical processes or from internal rearrangements within a molecule. Such variations become evident on cooling, which produces a decrease in the tumbling rate of radicals in solution. This section and the next qualitatively survey the types of effects that can be encountered and present several examples. For a survey of the detailed theory, the reader is referred to a comprehensive review [50].

10.5.4.1 *Single Nucleus*

Consider a radical that gives rise to hyperfine splitting from a single nucleus with spin I. Suppose that the radical can exist in two forms, a and b, which can interconvert. Let a_a and a_b be the hyperfine splittings for these. At slow interconversion rates, two spectra (usually superimposed) should be observed. Each spectrum consists of $2I+1$ lines corresponding to the possible values of M_I. In the limit of fast interconversion, a single spectrum is observed; it consists of $2I+1$ lines, with a mean hyperfine splitting given by

$$\bar{a} = f_a a_a + f_b a_b \qquad (10.52)$$

where f_a and f_b are the mole fractions of a and b.

Consider a specific example in which $I = \frac{3}{2}$, $a_a = 1.00$ mT, $a_b = 0.10$ mT, $f_a = 0.75$, and $f_b = 0.25$. Figure 10.11a displays a stick diagram of the spectrum that would be observed in the region of slow interconversion between a and b. Two distinct four-line spectra are apparent. Figure 10.11b displays the average spectrum in the region of rapid interconversion. It is important to note that in going from species a to species b, *the value of M_I does not change*. Thus there is a one-to-one correspondence between lines in Fig. 10.11a and lines in Fig. 10.11b.

In Section 10.5.1 it was pointed out that in the region of rapid interconversion, the linewidth is given by Eq. 10.46. Referring again to Fig. 10.11, it is clear that the $M_I = \pm\frac{1}{2}$ lines exhibit small shifts on conversion from species a to species b, whereas the $M_I = \pm\frac{3}{2}$ lines exhibit much larger shifts. Thus the latter can be expected to be broader than the former (Fig. 10.11c). In general, for a nuclear spin I [50], the width is given by

$$\Gamma = \bar{\Gamma} + f_a f_b \tau |\bar{\gamma}| (a_a - a_b)^2 M_I^2 \qquad (10.53)$$

A good example of this effect is shown by the sodium naphthalenide spectrum [51] (Fig. 10.12) in a tetrahydrofuran/diethyl ether (1:3) solvent mixture. Ion pairing may be inferred, since a marked hyperfine splitting from ^{23}Na ($I = \frac{3}{2}$) is observed. At $-60°$C, all four lines of a given ^{23}Na multiplet have roughly the same amplitude and hence nearly the same width (Fig. 10.12a). However, as the temperature is lowered, the $M_I = \pm\frac{3}{2}$ lines broaden relative to the $M_I = \pm\frac{1}{2}$ lines (Figs. 10.12b,c). These spectra have been interpreted in terms of a rapidly established equilibrium between two ion pairs, one having a large and the other a small ^{23}Na hyperfine splitting. The differential broadening of the $M_I = \pm\frac{3}{2}$ and the $M_I = \pm\frac{1}{2}$ lines may be used to obtain a value for the rate constant for interconversion, using Eq. 10.53 and neglecting any anisotropy conditions (Section 10.8). As the temperature is lowered, the ^{23}Na

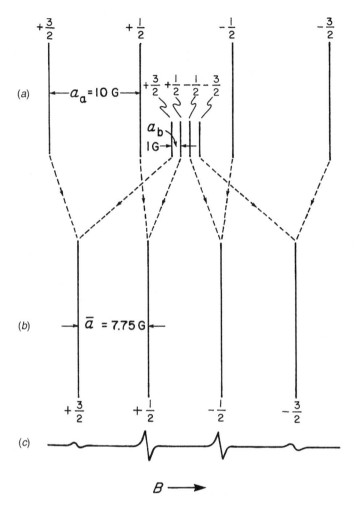

FIGURE 10.11 Stick-diagram representation of a spectrum in the limits of (*a*) slow and (*b*) fast exchange for two forms of a radical exhibiting a four-line spectrum from a nucleus of spin $\frac{3}{2}$ and $g_n > 0$; (*c*) simulated EPR spectrum.

hyperfine splitting decreases. The relative amounts of the two ion pairs and hence the equilibrium constant at each temperature can be obtained by using Eq. 10.53. Thus thermodynamic as well as kinetic information can be obtained from a study of these effects (see Problem 10.7 for a quantitative analysis of this system).

10.5.4.2 *Multiple Nuclei* Let us consider dynamic effects in a radical with two equivalent nuclei, each having a non-zero nuclear spin. As an example, consider the hypothetical *cis*-1,2-dichloroethylene anions forming ion pairs with Na$^+$ (Fig. 10.13). Two cases are distinguished:

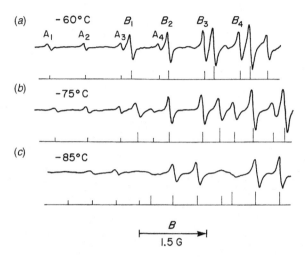

FIGURE 10.12 X-band first-derivative spectra of the low-field portion of the sodium naphthalenide spectrum at temperatures -60, -75 and $-85°$C. The solvent is a mixture of tetrahydrofuran (25%) and diethyl ether (75%). The lines marked A_i and B_i ($i = 1 - 4$) are the ^{23}Na quartets for the two outermost ^{1}H hyperfine line components. The stick spectra indicate the relative peak-to-peak heights in the absence of broadening effects. Compare with the spectrum in Fig. 3.8. [After N. Hirota, *J. Phys. Chem.*, **71**, 127 (1967).]

1. The dynamic modulation is in-phase, that is, the hyperfine splittings increase or decrease in unison during the jump. This case is illustrated in Fig. 10.13*a*. In both structures the two protons are equivalent at any instant. Such nuclei are said to be *completely equivalent*.

2. The modulation of the two hyperfine splittings is exactly out-of-phase, that is, when one increases, the other decreases. This case is illustrated in Fig. 10.13*b*. The proton closer to the Na$^+$ yields a hyperfine splitting that is different from that of the other proton. When the Na$^+$ ion jumps to the other end of the molecule, these hyperfine splittings are interchanged. On the *average* the hyperfine splittings for the two protons have the same value. Such nuclei are then *dynamically equivalent* (as contrasted with *instantaneously equivalent*).

First consider the general case of in-phase modulation of hyperfine splittings (case 1 above). Since the nuclei are completely equivalent, the positions of the spectral lines may at all times be described by the total nuclear-spin quantum number $^tM_I = \Sigma_i M_{I(i)}$. This means that the widths of the lines can be treated as if there were one interacting nucleus with a total nuclear spin of $^tI = \Sigma_i I_i$. Consequently, this represents a case analogous to that given in Section 10.5.4.1. That is, the linewidths vary as $^tM_I^2$ (Eq. 10.53).

Next consider the case of out-of-phase modulation of hyperfine splittings (case 2 above). The fact that the two nuclei are not *instantaneously* equivalent leads to an interesting phenomenon in the EPR spectrum. This is commonly

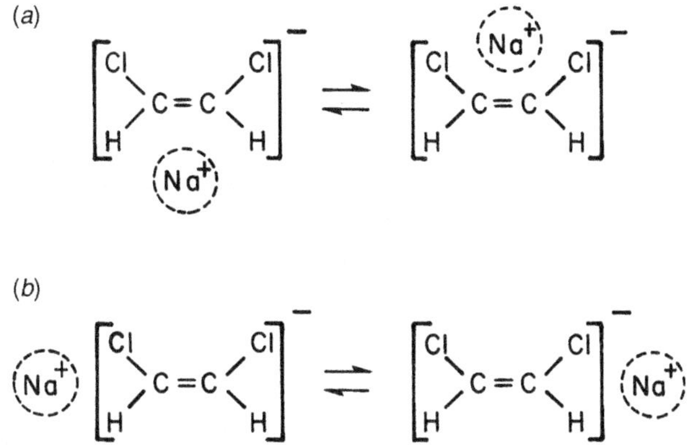

(a)

(b)

FIGURE 10.13 Ion-pair structural equilibria for the hypothetical *cis*-1,2-dichloroethylene anion, with the sodium cation located (*a*) below or above the horizontal plane and (*b*) at either end of the anion.

referred to as the *alternating linewidth effect*, which was first observed in the spectrum of the monocation of 1,4-dihydroxytetramethylbenzene (dihydroxydurene) [52] (Fig. 10.14) and also the related dinitrodurene monoanion [53] (Fig. 10.15).

The interpretation of this striking effect is aided by the consideration of two nuclei with $I = 1$. The model assumed is one in which alternately one hyperfine splitting is relatively large and the other relatively small. It is further assumed that the radical can exist in two thermodynamically equivalent states, a and b, that is, states of the same energy.

In this model for the dinitrodurene anion, the variation occurs in the ^{14}N hyperfine splitting from the two nitro groups, each having either of two rotated positions relative to the plane of the molecule (Fig. 10.16). If one group has a given orientation, then the other has the other orientation. Thus $f_a = f_b = \frac{1}{2}$ and $\tau_a = \tau_b = 2\tau$. The hyperfine splittings are assumed values, but they are probably not far from the actual values. In the thermal region of infrequent interconversion between a and b, one would observe the spectrum shown in Fig. 10.17a. As the rate of interconversion increases, the lines coalesce as shown in Fig. 10.17b. The widths are given by Eq. 10.46. Three line components do not shift in going from a to b, and hence these lines remain sharp. These are the two outside lines and one component of the central line. The $^tM_I = \pm 1$ lines appear broad because of the sizable field shifts involved. Two components of the $^tM_I = 0$ line undergo a large magnetic-field shift (Fig. 10.17b), and hence these two components are usually not detected; instead, the single sharp and unshifted central component is seen. The appearance of the spectrum in Fig. 10.15 can now be understood [46]. Knowledge of the two nitrogen hyperfine splittings (in the limit of slow exchange) and the use of Eq. 10.46 permit the interconversion rate $\frac{1}{2}\tau$ to be obtained from the width of the $^tM_I = \pm 1$ components. In Fig. 10.17c the hyperfine splittings are completely averaged, and the

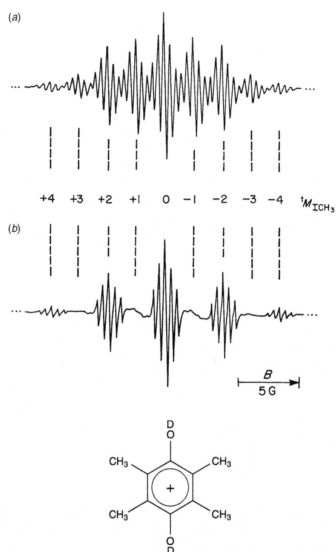

FIGURE 10.14 EPR spectra of the cation radical of p-dideuteroxydurene in $D_2SO_4/$ CH_3NO_2 at (a) $+60°C$ and (b) $-10°C$. The hyperfine splittings are $a_{CH_3}^{H} = 0.205$ mT and $a_{OD}^{D} = 0.042$ mT. Only 9 of the 13 methyl-proton multiplets are shown. [After P. D. Sullivan, J. R. Bolton, *Adv. Magn. Reson.*, **4**, 39 (1970).]

five-line spectrum characteristic of two equivalent nuclei of spin 1 is obtained (Fig. 3.12).

The whole field, describing the relevant EPR techniques and results shedding light on the dynamics of degenerate tautomerism in free radicals, has been amply reviewed [54].

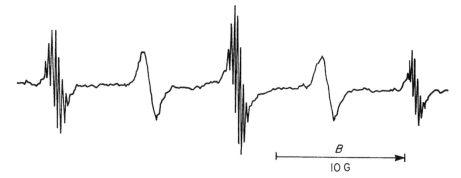

FIGURE 10.15 First-derivative spectrum of the dinitrodurene anion in dimethyl formamide at room temperature. The major groups are due to hyperfine splitting from the two nitrogen nuclei, and the minor splittings arise from the protons. [After J. H. Freed, G. K. Fraenkel, *J. Chem. Phys.*, **37**, 1156 (1962).]

10.5.5 Molecular Tumbling Effects

Let us now return to the case of a single unpaired-electron species and consider first the effects arising from its anisotropic spin-hamiltonian parameters. These can appear in addition to the phenomena described above.

Tumbling rates depend on the shape and size of molecules, on their interactions with their surroundings (e.g., solvent), and on the available thermal energy (temperature). The EPR effects of tumbling depend on the magnitudes of the anisotropies of the spin-hamiltonian parameters.[19] It is useful to consider two categories:

1. *Isotropic Tumbling.* Here the probabilities P_Ω of rotations about different axes by the same angle are all equal.
2. *Anisotropic Tumbling.* Here P_Ω depends on which molecular axis of rotation is active.

In theoretical models, rotations are treated either as a Brownian-type process or as a discontinuous-jump process.

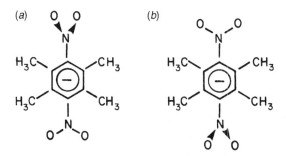

FIGURE 10.16 Two structures for the dinitrodurene anion radical. The nitro-group ^{14}N hyperfine splittings (in mT) are (*a*) 1.4 (coplanar) and (*b*) 0.05 (perpendicular).

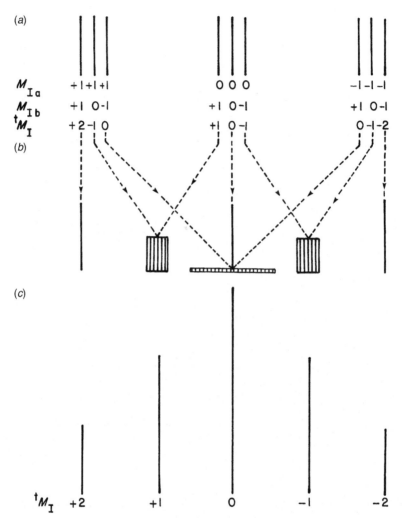

FIGURE 10.17   Representation of the spectra of a radical containing two inequivalent ^{14}N nuclei subjected to an out-of-phase modulation, $^{t}M_I = M_{Ia} + M_{Ib}$: (*a*) slow-interconversion limit; (*b*) intermediate rate of interconversion: blocks represent linewidths and amplitudes; (*c*) fast-interconversion limit.

In the Brownian (rotational diffusion) model, each molecule in the ensemble is assumed to rotate continuously and freely about some axis, with an arbitrary angular speed. At arbitrary intervals, the rotation axis and speed change instantaneously and randomly (say) because of a collision with another molecule.

In the jump model, each molecule is taken to be stationary at some arbitrary orientation for some random period of time. It then jumps instantaneously to some other fixed orientation.

In both cases, the average interval between perturbations is called the *mean residence time* or the 'lifetime' of the orientation.

10.5.5.1 Dipolar Effects

In Chapters 4 and 5 it was shown that in solids the *g* factors and the hyperfine couplings can be very orientation-dependent. It has also been indicated that for free radicals in solutions of low viscosity, anisotropic interactions are averaged to zero. Such averaging does not necessarily occur if the solvent has a moderate viscosity. One can think of the tumbling as a series of events, in which each initial situation changes to the subsequent one differing by a resonant field shift ΔB_0.

To illustrate what may happen to a solution spectrum when the molecular tumbling rate is decreased, consider the spectra of the *p*-dinitrobenzene anion in dimethylformamide, as shown in Fig. 10.18. At 12°C one observes a 'normal' spectrum in which the relative first-derivative amplitudes are proportional to the degeneracies of the corresponding transition energies. However, at −55°C the appearance of the spectrum has changed drastically, although the line positions are unaltered. The change in the spectrum results from variation in the widths of the several lines. Note that this variation is not symmetric about the central line.

To understand the origin of these effects, consider once again the di-*t*-butyl nitroxide radical. This is a fairly simple case, since the *g* matrix and the nitrogen hyperfine matrix (^{14}N, $I = 1$) have the same principal axes, and each matrix is approximately uniaxial [55] (see also Problem 5.8). Figure 10.19*a* illustrates the spectrum obtained for the randomly oriented radicals in the solid phase. The parallel and perpendicular features of such spectra were considered in Chapter 5. The linewidth of each hyperfine component has [56] the form

$$\Gamma = \alpha + \beta M_I + \gamma M_I^2 + \cdots \tag{10.54a}$$

The coefficients depend on the anisotropies of *g* and of the hyperfine splittings and on the mean tumbling rate (set by the solvent viscosity).[20] Parameters β and γ go to zero as this rate increases. The sign of the unpaired-electron population at the atom giving rise to the hyperfine splitting enters, as discussed below.

Figure 10.19*b* illustrates the nitroxide spectrum obtained from a dilute solution of moderate viscosity. The tumbling rate here is sufficiently rapid that the line positions (but not the linewidths) correspond to those of the completely averaged spectrum in Fig. 10.19*c*. Noting from Eq. 10.54*a* that the linewidth depends on the sign and magnitude of M_I, it is clear why the three lines have different widths. Since the highest-field line ($M_I = -1$) is broader than the lowest-field line ($M_I = +1$), it follows that $\beta < 0$. In interpreting this result, the following measured parameters [55] for di-*t*-butyl nitroxide were used: $a_\parallel = +3.18$ mT, $a_\perp = +0.68$ mT, so that $a_0 = +1.51$ mT (for ^{14}N), $g_\perp = 2.007$ and $g_\parallel = 2.003$. The theory [59], using these parameters, yields $\beta < 0$. Had a_0 and $a_\parallel/a_\perp$ each been negative, then the high-field line would have been narrower than the low-field line (Problem 10.10). These conclusions are reversed when $g_\parallel > g_\perp$. This linewidth phenomenon is the basis for one method [60] of measuring the signs of isotropic hyperfine splittings (Section 2.6).

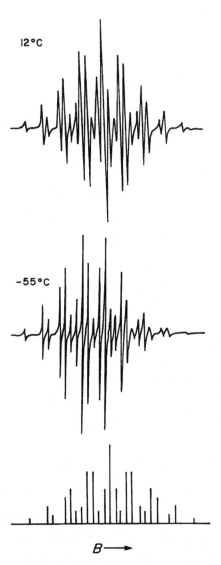

FIGURE 10.18   Electron spin resonance spectra of the negative ion of *p*-dinitrobenzene in dimethylformamide. The stick plot is based on the hyperfine splittings: $|a^N| = 0.151$ mT and $|a^H| = 0.112$ mT. [After J. H. Freed, G. K. Fraenkel, *J. Chem. Phys.*, **40**, 1815 (1964).]

In general, the linewidths in spectra exhibiting the effects of finite tumbling rates can be approximated [61] by the following relation

$$\Gamma = \alpha + \sum_i \beta_i {}^t M_{I(i)} + \sum_i \gamma_i {}^t M_{I(i)}{}^2 + \sum_{i<j} \gamma_{i,j} {}^t M_{I(i)} {}^t M_{I(j)} \qquad (10.54b)$$

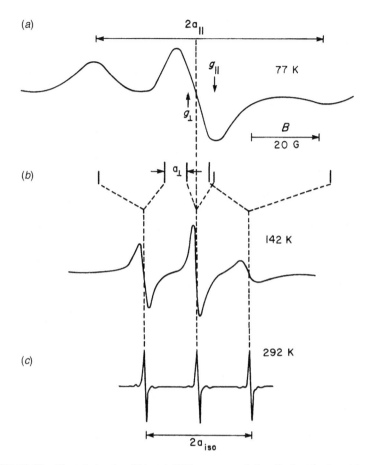

FIGURE 10.19 First-derivative X-band EPR spectra of the di-t-butyl nitroxide radical: (*a*) at 77 K (solid); (*b*) at 142 K (viscous ethanol solution); (*c*) at 292 K (low-viscosity ethanol). Single-crystal data are from Ref. 56. Note that the linewidths in spectrum *b* correspond to the spacings between the coalescing lines (see stick diagram). (Spectra taken by J. R. Bolton.)

valid near the fast limit. Here $^tM_{I(i)}$ and $^tM_{I(j)}$ refer to the total z component of the nuclear-spin quantum number for sets i and j of *completely* equivalent nuclei. If we assume that all isotropic hyperfine splittings are negative, then the high-field lines have $^tM_{I(i)} > 0$. This is an arbitrary assignment, since the signs of the hyperfine splittings are seldom known.

The parameters in Eqs. 10.54 are now considered in more detail. Coefficient α is a constant term including all line-broadening effects that are the same for all hyperfine components. The coefficients β_i depend on anisotropy and arise from cross-product terms of the g and hyperfine matrices [specifically, of $(\mathbf{g} - g_0\mathbf{1}_3) \cdot \mathbf{T}$, where $g_0 = tr(\mathbf{g})/3$]. They cause the spectrum to appear asymmetric. In certain cases the parameters β_i can be calculated [50]. A special-interest case occurs when the nucleus in question has a p orbital that is part of a π-electron system containing an unpaired electron. For such nuclei the high-field components are broader (i.e., lower first-derivative amplitude) if the ^{13}C isotropic hyperfine splitting constant is positive. This is the case if the π-electron unpaired-electron population ρ_i is positive. The opposite is true if ρ_i is negative. It also assumes that $g_\parallel < g_\perp$, which is true for most π-electron radicals.

As an example, consider the highly resolved EPR spectrum of the 2,5-dioxy-p-benzosemiquinone trianion [62] shown in Fig. 10.20 (cf. Fig. 3.20). The two ^{13}C splittings of $|a_1{}^C| = 0.263$ mT and $|a_3{}^C| = 0.666$ mT are indicated on the figure. Note that for each of the ^{13}C lines, the amplitudes are such that the high-field line is somewhat broader than the corresponding low-field line. This implies that the product $a^C \rho^C > 0$. At position 1, it is reasonably certain that

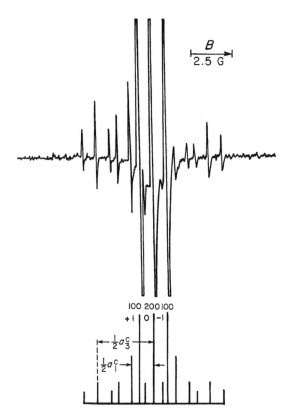

FIGURE 10.20 First derivative of the X-band EPR spectrum of 2,5-dioxy-1,4-benzosemiquinone in KOH solution at room temperature. The lines arising from proton splittings are off scale in the spectrum. The quantum numbers $(+1, 0, -1)$ and the relative intensities $(100, 200, 100)$ of the proton lines are indicated to approximate scale relative to the smallest ^{13}C lines. [After M. R. Das, G. K. Fraenkel, *J. Chem. Phys.*, **42**, 1350 (1965).]

$\rho_1{}^C > 0$; hence $a_1{}^C > 0$. At position 3 the very small proton hyperfine splitting (0.079 mT) implies that $\rho_3{}^C$ is very small. The large magnitude of the ^{13}C splitting must then arise from unpaired-electron population on neighboring carbon atoms. From Eq. 9.15 this contribution to $a_3{}^C$ is seen to be negative. Since this represents the largest contribution, $a_3{}^C < 0$ and hence $\rho_3{}^C < 0$. It should be emphasized that this type of argument does not apply to proton hyperfine splittings.

The coefficients γ_i in Eqs. 10.54 are a function only of the hyperfine anisotropy. Where they can be calculated, they provide information on the rotational correlation time for tumbling of the radical in the liquid. This correlation time τ_c can be regarded as roughly the average time required for rotation through ~ 1 radian about a principal axis. The assumption is made that τ_c is isotropic. In the special cases where the nucleus in question is part of the skeleton of a π-electron system, the coefficients can be used to make assignments of hyperfine splittings. In these cases it has been shown that the coefficients γ_i are proportional to the square of the π-electron unpaired-electron population on the interacting atom [63]; thus relative broadening of hyperfine components can indicate which of a pair of splittings is to be assigned to the position having the higher π unpaired-electron population.

The γ_{ij} are coefficients that arise from the products of the hyperfine matrices of nuclei from different equivalent sets. It has been shown [64] that these coefficients can yield information about the relative signs of different hyperfine splittings.

Modeling of the coefficients in Eqs. 10.54 is mathematically complicated but can be quite successful. As an example, we cite the study by Campbell and Freed [65] of the slow-motional EPR spectra of vanadyl (VO^{2+}, $S = \frac{1}{2}$) complexes.

Slow tumbling dynamics of species (e.g., nitroxides) in solids also gives rise to EPR lineshape effects. They have been carefully investigated both experimentally and theoretically [66].

10.5.5.2 Spin-Rotation Interaction In the gas phase, molecules are free to rotate. This rotational motion is quantized, and transitions between the rotational energy levels may be detected in a microwave spectrum if the molecule has a permanent electric dipole moment. In such molecules this rotational motion also generates a magnetic moment because the electrons do not rigidly follow the movement of the nuclear framework. If the molecule has a net electron *spin* magnetic moment, this is coupled to the rotational magnetic moment by a dipole-dipole interaction. The effect of this coupling is analogous to magnetic electron-dipole/electron-dipole couplings. However, the interaction is *not* averaged to zero in the gas phase since the rotational angular momentum and magnetic moment vectors are collinear and fixed in space. Gas-phase EPR spectra are very complex as a result of this 'spin-rotation' interaction (Section 7.3).

For liquids of low viscosity (i.e., at sufficiently high temperatures), molecules in the liquid state may have an opportunity to undergo a few cycles of rotation before a collision occurs. Hence the rotational magnetic moment generated can couple with the electron-spin magnetic moment [67,68]. It has been shown that this effect broadens all lines equally [68]. The linewidths generally vary as T/η, where η is the

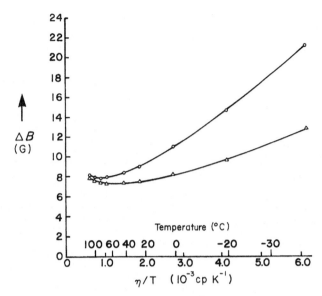

FIGURE 10.21   Peak-to-peak widths of the $M_I = -\frac{7}{2}$ line (o) and the $M_I = -\frac{1}{2}$ line (Δ) of 5×10^{-4} M VO^{2+} in deoxygenated toluene, as a function of solution viscosity [1 centipoise (cP) $= 10^{-3}$ kg m^{-1} s^{-1}). [After R. Wilson, D. Kivelson, *J. Chem. Phys.*, **44**, 154 (1966).]

coefficient of viscosity and T is the absolute temperature. As discussed above, broadening due to anisotropic hyperfine effects generally varies as η/T (and depends on M_I). Hence one can expect to find an optimum temperature for best resolution of an EPR spectrum. Figure 10.21 shows the linewidth variation with temperature for the $M_I = -\frac{7}{2}$ and $-\frac{1}{2}$ lines in the EPR spectrum of the VO^{2+} ion (vanadyl acetylacetonate) in toluene. The former line shows a much greater linewidth variation with η/T than does the latter. The temperature corresponding to the minimum linewidth is also different for the two lines.

10.5.6   General Example

We close this section by presenting an example of an EPR system in which there are three simultaneous effects: (1) linewidth alternation, (2) anisotropy lineshape distortion and (3) chemical concentration variation. These have been observed [69] in the $4d^9$ $(S = \frac{1}{2})$ neutral complex *bis*(1,2-*bis*(diphenylphosphino)ethane)rhodium(0), denoted by [Rh(dppe)$_2$]^0. Here, the rhodium atom is at the center of a set of four phosphorus (^{31}P, 100%, $I = \frac{1}{2}$) atoms that form a distortable ligand system. The EPR spectra (Fig. 10.22) reveal a symmetric five-line pattern at 270 K, arising from four equivalent ^{31}P hyperfine interactions. As the temperature T is lowered, line splitting and the alternating linewidth effect become prominent. Analysis, in terms of two sets of two *completely equivalent* nuclei, yields rate constants as depicted in Fig. 10.23a, yielding an enthalpy of activation $\Delta H^{\ddagger}$ of 14.7 kJ mol^{-1}

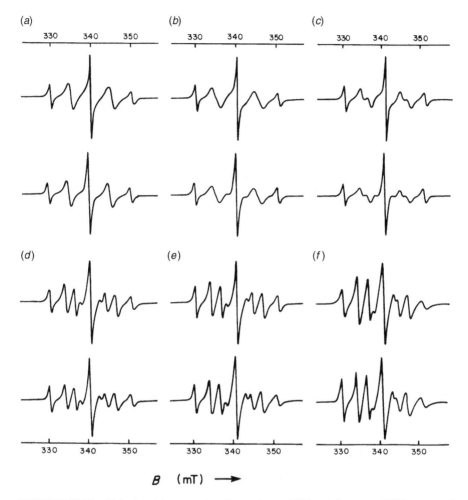

FIGURE 10.22 Calculated (upper set) and experimental X-band (lower set) EPR spectra of [Rh(dppe)$_2$]^0 in toluene at six temperatures: (a) 259 K, (b) 249 K, (c) 237 K, (d) 219 K, (e) 209 K, (f) 199 K. [After K. T. Mueller, A. J. Kunin, S. Greiner, T. Henderson, R. W. Kreilick, R. Eisenberg, *J. Am. Chem. Soc.*, **109**, 6313 (1987).]

and an entropy of activation $\Delta S^{\ddagger}$ of -19 J mol^{-1} K^{-1} for the interchange of two ligand configurations. Furthermore, analysis of simultaneous 'anisotropic' broadening as T is lowered yields the coefficients of Eq. 10.54, while line-area measurements disclose a decrease in concentration of the radicals describable in terms of the equilibrium constant $K = [(1 - \alpha)\alpha^2]/2c_t$ (Fig. 10.23b), where α is the fraction of Rh(dppe)$_2$ species existing in the solution as the paramagnetic monomer, and c_t is the total concentration of such species. A charge-transfer reaction

$$2[\text{Rh(dppe)}_2]^0 \rightleftharpoons [\text{Rh(dppe)}_2]^+[\text{Rh(dppe)}_2]^- \tag{10.55}$$

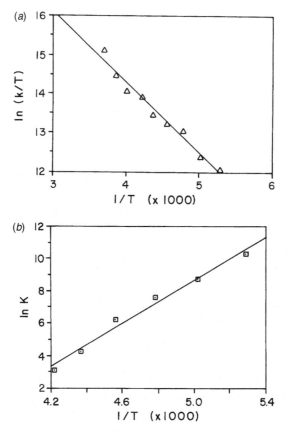

FIGURE 10.23 (*a*) Eyring plot for the fluxional process (*k* has units s^{-1}; T in K) and (*b*) Van't Hoff plot for the bimolecular equilibrium, both involving $[Rh(dppe)_2]^0$. [After K. T. Mueller, A. J. Kunin, S. Greiner, T. Henderson, R. W. Kreilick, R. Eisenberg, *J. Am. Chem. Soc.*, **109**, 6313 (1987).]

is believed to occur, with $\Delta H = -55.6\,\mathrm{kJ\,mol^{-1}}$ and $\Delta S = -207\,\mathrm{J\,mol^{-1}\,K^{-1}}$ derived from the $K(T)$ data.

10.6 LONGITUDINAL DETECTION

It is instructive to distinguish between longitudinal (i.e., M_z) and transverse ($M_{x,y}$) magnetization, that is, EPR intensity detection in different directions relative to the Zeeman field **B** ($\|\mathbf{z}$). The latter is the usual case. Longitudinal detection of EPR (LODEPR), denoting longitudinal detection of the macroscopic magnetization M_z, has been presented in the literature [70,71], featuring an $S = \frac{1}{2}$ system (DPPH) irradiated by two nearly resonant waves (both $\perp$ **B**) having two somewhat different frequencies. Applications include direct evaluation of B_1 mean values and of relaxation-time variations.

10.7 SATURATION-TRANSFER EPR

With inhomogeneously broadened lines, power saturation in one narrow region of the EPR spectrum can be observed, as can its recovery and spreading of saturation via spin diffusion to the rest of the line. Monitoring the recovery of the spin populations after giving the system extra energy (i.e., saturation recovery measurements) offers a potent method of measuring relaxation times.

The term *saturation transfer* refers to the diffusion of the z component of the magnetization. Its efficiency depends sensitively on the motional dynamics of the unpaired-electron species present. With application of sufficiently great amplitudes of B_1 and use of special techniques [e.g., observation of dispersion first derivatives, special field-modulation conditions or use of two B_1 sources at off-set frequencies (ELDOR; see Chapter 12)], it is possible to obtain valuable information about relatively slow molecular motions (correlation times τ_c in the range of 10^{-3} to 10^{-11} s). This is especially useful in biomedical systems, as reviewed by Hyde [72].

Theoretical and experimental aspects of saturation-transfer EPR are discussed in various literature references [73–75]. For example, spin diffusion between DPPH molecules dissolved in polystyrene was found [73] to have the characteristic time constant $\tau_d = 20$ ms, probably set by proton spin flips on neighbor solvent molecules. Various instruments have been designed especially to perform saturation-recovery work (see Appendix E), and comparison of the effectiveness of cw EPR relative to pulse EPR is found in the more recent literature [76].

10.8 TIME DEPENDENCE OF THE EPR SIGNAL AMPLITUDE

The EPR spectrometer (like any spectrometer) can be used to follow the rates of chemical reactions. In addition there are situations where an ensemble of radicals is generated (e.g., photochemically) in a non-equilibrium set of spin-state populations. This is called *chemically induced dynamic electron polarization* (CIDEP). The subsequent 'thermalization' of the spin populations leads to a time dependence that in some cases is superimposed on the decay of radical concentrations. We now briefly discuss these topics, deferring pulsed EPR aspects to the next chapter.

10.8.1 Concentration Changes

EPR line intensities are often very useful for the study of chemical reactions. We consider three cases:

1. The reaction is so fast that one can only detect (paramagnetic) products by EPR. One common technique for dealing with this is *spin trapping* (see Chapter 13 for references). Here a reactive free radical is identified indirectly, by allowing

it to add to an appropriate chemical species to produce a more stable radical detectable by EPR and identifiable by its hyperfine pattern. A good example is the hydroxyl free radical trapped in liquid solution by use of the diamagnetic *spin trap* N-*t*-butyl-α-phenylnitrone (PBN), via the reaction

$$(CH_3)_3C\overset{+}{-}N\overset{H}{=}\overset{|}{C}-C_6H_5 + OH \longrightarrow (CH_3)_3C-N-\overset{H}{\underset{|}{C}}-C_6H_5 \qquad (10.56)$$
$$\underset{O^-}{|} \qquad\qquad\qquad\qquad \underset{O \ \ OH}{}$$

For the nitroxide radical formed, the primary triplet with $a(^{14}N) = 1.53$ mT, occurring at $g = 2.0057$, is split into doublets with $a(^1H_a) = 0.275$ mT; there is no resolved splitting from the OH group [77]. Numerous spin traps have been synthesized, and the EPR solution characteristics for the nitroxides formed by reactions with various free radicals (e.g., OH, O_2^-, CH_3) have been tabulated [78] so as to allow ready identification of R.

2. The reaction can be followed by ordinary cw EPR, using intensities of lines. There is a large literature describing such work, ranging from reactions that take weeks to go to completion (e.g., diffusion-controlled reactions in solids) to reactions ended in seconds. An example is cited in Section 10.5.6. Most of the work has dealt with organic species. We must content ourselves by merely citing references [79–81] to a few appropriate review articles. We can discuss some general aspects. For the moment let us assume that the populations of the electron-spin energy levels are at thermal equilibrium. Each EPR signal amplitude should be proportional to the concentration of some paramagnetic species in the sample, provided that the linewidth is independent of concentration, an assumption that is not always valid. The time course of the signal amplitude may be followed by fixing the magnetic field at one of the extrema of a first-derivative line (usually the most intense) in the spectrum.[21] If more than one paramagnetic species contributes to the EPR spectrum, the time course of each can be followed independently if the magnetic field is alternately positioned on specific lines arising from given species, assuming that they are sufficiently resolved in the spectrum.

3. The reaction is too fast to be followed by ordinary techniques and spectrometers, where limits are set by the inductance of the magnetic-field sweep coils, as well as the time constants and bandwidth of the amplifiers. With some systems, one can create unpaired-electron species very rapidly by use of flash lamps or pulsed electron beams. The chemical situation can then be sampled by EPR for time periods immediately after each pulse, at selected fixed B fields. Such time-resolved techniques have been reviewed fairly recently [82–84]. They extend the available kinetic information into the nanosecond region. As an example, we cite the flash photo-excitation of certain cycloalkanones into their lowest triplet states, from which they rapidly progress via ring opening to a series

of biradicals [85]. The relative occupancies of the singlet and triplet sublevels of the biradicals can be followed in detail, as a function of time (μs).

We now turn to one informative phenomenon, chemically induced dynamic electron polarization, which is readily observed and can be studied in this fashion.

10.8.2 Chemically Induced Dynamic Electron Polarization

In some situations the paramagnetic entities are created such that the populations of the electron-spin energy levels are not at thermal equilibrium. There may be an excess population in the upper state (causing net photon emission) or in the lower state (enhanced photon absorption). The excess population then decays toward thermal equilibrium at a characteristic rate (Section 10.2.2). Since the EPR signal amplitude is directly proportional to the population difference of spin levels (Sections 4.6 and F.2), this decay results in a time dependence of this amplitude.

The CIDEP effect was first observed with H atoms, freshly generated via 2.8 MeV electron pulse radiolysis in liquid methane at ~ 100 K [86]. The phenomenon manifests itself by the unusual appearance of the two hyperfine lines: in first-derivative presentation they occur with *opposite* phase; that is, one is emissive (E) and the other absorptive (A), in contrast to Fig. 1.4. Here the population differences of the two M_I states (Fig. 2.4) have opposite signs, as the result of complex polarization effects ensuing from the initial creation of the radical pair H and CH_3, and subsequent recombinations.

Much more recent CIDEP studies of atomic ^{1}H and ^{2}H in ice have furnished information about geminate recombination of hydrogen atoms and hydroxyl radicals [87].

Steady-state photolysis of liquid benzene solutions of the fullerene C_{60} yielded the radical HC_{60}, where the 3.3 mT hydrogen doublet shows striking E–A polarization in the cw EPR spectrum [88].

The CIDEP phenomenon has been found with numerous other radicals, and has been studied as a function of time elapsed after the creation pulse. As an example, we cite the $(CH_3)_2COH$ radical [82], for which time-integration spectroscopy gives the EPR spectra depicted in Figs. 10.24 and 10.25.

The details of the mechanisms describing the kinetics of the CIDEP effect have been rewardingly investigated and reviewed [89–92]. Suffice it here to mention only that there is a complex contribution from two chemical mechanisms: creation of free-radical anion/cation pairs and of triplet systems, with interconversion between these, often accompanied by electron- and proton-transfer effects and creation of secondary radicals [93–95]. These are of special importance in the energetics of the photochemistry occurring in natural and synthetic photosynthesis.

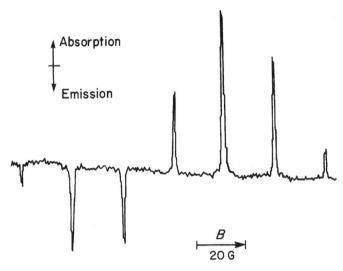

FIGURE 10.24 Time-integration EPR spectra of the $(CH_3)_2COH$ radical created by digital summation of the signals between 1.00 and 2.00 ms of the 20 ns photolysis flash from a 308-nm excimer laser (repetition rate ~ 20 Hz). The sample contained 1–4 (V:V) propanone in propan-2-ol. [After K. A. McLauchlan, D. G. Stevens, *Mol. Phys.*, **57**, 223 (1986).]

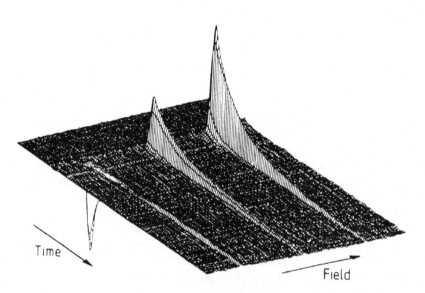

FIGURE 10.25 A two-dimensional spectrum showing the field and time variation of the central three of the seven EPR lines (Fig. 10.24) of the radical-pair mechanism spin-polarized radical $(CH_3)_2COH$. [After K. A. McLauchlan, D. G. Stevens, *Mol. Phys.*, **57**, 223 (1986).]

10.9 DYNAMIC NUCLEAR POLARIZATION

A technique closely associated with EPR spectroscopy is dynamic nuclear polarization (DNP), which leads to enhancement of NMR signals by power-saturating EPR lines in the same sample, thereby changing the spin populations in the nuclear manifold. The mechanisms involved are not simple. A good summary of DNP can be found in the literature [96–98]. For our present purposes, it suffices to note that EPR here is an essential tool for other purposes.

10.10 BIO-OXYGEN

The ground state of dioxygen is of course, a spin triplet (see Section 7.3). Because of the considerable importance of dissolved O_2 in animals and plants, major efforts have been expanded using EPR to measure its concentrations and kinetics in such systems, and thus there exists a considerable literature. For example, one can cite 'The measurement of oxygen in vivo using EPR techniques' [99,100], referring mostly to work carried out at low frequencies (100–4000 MHz) to counteract the dielectric-loss effects of liquid water. Such oximetry allowed non-invasive measurement of the local equilibrium pressure of oxygen down to 0.5 torr, for example, in mammalian tissue (including tumors).

By contrast, EPR imaging studies of dioxygen deep in tissues are rapidly being developed [101]. Here a contrast agent (triarylmethyl probes), which interacts with O_2 to broaden its own EPR line, providing O_2-concentration images of normal and tumor tissues in living mice.

It seems appropriate to point out that the relaxation effects of MRI contrast agents: paramagnetic species [Gd(III) and Mn(II) complexes, Fe(III) oxides and porphyrins, free radicals] acting on nearby nuclei (usually protons) have become of prime importance, in that magnetic resonance imaging now uses this phenomenon extensively to contrast various sites in mammalian tissues. Clearly the average distance r between, say, the Gd^{3+} ion with its seven unpaired electrons and the nearest proton of a solvent water molecule is a key aspect, since the dipolar relaxation mechanism has a r^{-6} dependence. This distance has been measured (0.31 ± 0.01 nm) using from the anisotropic ^1H hyperfine splitting, via pulsed ENDOR spectroscopy [102]. The correction factor $k(e)$ for the Gd^{3+} ion was taken to be 1: see Eq. 9.2.

10.11 SUMMARY

It should now be apparent to the reader that a large number of different mechanisms can contribute to lineshape effects. In this chapter it has been possible to survey the origins of some of these only briefly. Although these effects contribute to the complexity of EPR spectra, analysis can yield valuable structural and kinetic information. For example, use of nitroxide radicals (R_1R_2NO) incorporated as 'spin labels' into substances of biomedical importance has yielded much valuable information regarding the structure, dynamics and chemical behavior of such species (for references, see Chapter 13). In some cases, ignorance of the lineshape effects makes it most difficult to interpret an EPR spectrum (e.g., the vinyl radical in liquid solution; see Problem 10.7). For a quantitative interpretation of these lineshape phenomena, an understanding of the theory of relaxation processes is required.

REFERENCES

1. P. M. Morse, *Thermal Physics*, 2nd ed., Benjamin, New York, NY, U.S.A., 1969, Chapter 24.

2. O. Svelto, *Principles of Lasers*, 3rd ed., Plenum Press, New York, NY, U.S.A., 1989, Chapter 2.

3. B. Schwarzschild, *Phys. Today*, **43**, 17 (1990).

4. I. Waller, *Z. Phys.*, **79**, 370 (1932).

5. J. H. Van Vleck, *J. Chem. Phys.*, **7**, 72 (1939).

6. N. Bloembergen, E. M. Purcell, R. V. Pound, *Phys. Rev.*, **73**, 679 (1948).

7. A. G. Redfield, *Adv. Magn. Reson.*, **1**, 1 (1965).

8. A. A. Manenkov, R. Orbach, Eds., *Spin-Lattice Relaxation in Ionic Solids*, Harper & Row, New York, NY, U.S.A., 1966.

9. G. E. Pake, T. L. Estle, *The Physical Principles of Electron Paramagnetic Resonance*, 2nd ed., Benjamin, New York, NY, U.S.A., 1973.

10. C. P. Poole Jr., H. A. Farach, *Relaxation in Magnetic Resonance, Dielectric and Mössbauer Applications*, Academic Press, New York, NY, U.S.A., 1971.

11. L. A. Sorin, M. V. Vlasova, *Electron Spin Resonance of Paramagnetic Crystals*, Plenum Press, New York, NY, U.S.A., 1973, Chapter 5.

12. R. Orbach, "Spin Relaxation in Solids", in D. ter Haar, Ed., *Fluctuation, Relaxation and Resonance in Magnetic Systems*, Oliver and Boyd, London, U.K., 1961, pp. 219–229.

13. R. de L. Kronig, *Physica*, **6**, 33 (1939).

14. C. B. P. Finn, R. Orbach, W. P. Wolf, *Proc. Phys. Soc. (London)*, **77**, 261 (1961).

15. R. Orbach, *Proc. Roy. Soc. (London)*, **A264**, 458 (1961).

16. F. Bloch, *Phys. Rev.*, **70**, 460 (1946).

17. J. A. Pople, W. G. Schneider, H. J. Bernstein, *High-Resolution Nuclear Magnetic Resonance*, McGraw-Hill, New York, NY, U.S.A., 1959.

18. A. Carrington, A. D. McLachlan, *Introduction to Magnetic Resonance*, Harper & Row, New York, NY, U.S.A., 1967.

19. F. A. Bovey, L. Jelinski, P. A. Mirau, *Nuclear Magnetic Resonance Spectroscopy*, Academic Press, San Diego, CA, U.S.A., 1988, Section 1.7.

20. C. P. Slichter, *Principles of Magnetic Resonance*, 3rd ed., Springer, New York, NY, U.S.A., 1990, Chapter 2.

21. T. M. Barbara, *J. Magn. Reson.*, **98**, 608 (1992).

22. G. Whitfield, A. G. Redfield, *Phys. Rev.*, **106**, 918 (1957).

23. B. Goss-Levi, *Phys. Today*, **59**(9), 20 (2006).

24. A. Abragam, *The Principles of Nuclear Magnetism*, Oxford University Press, Oxford, U.K., 1961, Chapter 12.

25. B. Yu-Kuang Hu, *Am. J. Phys.*, **57**, 821 (1989).

26. G. Feher, *Bell Syst. Tech. J.*, **36**, 449 (1957).

27. R. D. Nielsen, E. J. Hustedt, A. H. Beth, B. H. Robinson, *J. Magn. Reson.*, **170**, 345 (2004).

28. M. A. Garstens, *Phys. Rev.*, **93**, 1228 (1954).

29. M. Perić, B. Rakvin, A. Dulčić, *J. Magn. Reson.*, **63**, 88 (1985).

30. S. R. P. Smith, F. Dravnieks, J. E. Wertz, *Phys. Rev.*, **178**, 471 (1969).

31. H. S. Gutowsky, D. W. McCall, C. P. Slichter, *J. Chem. Phys.*, **21**, 279 (1953).

32. P. D. Sullivan, J. R. Bolton, *Adv. Magn. Reson.*, **4**, 39 (1970).

33. H. S. Gutowsky, C. H. Holm, *J. Chem. Phys.*, **25**, 1228 (1956).

34. M. T. Rogers, J. C. Woodbrey, *J. Phys. Chem.*, **66**, 540 (1962).

35. C. S. Johnson Jr., *Adv. Magn. Reson.*, **1**, 33 (1965).

36. R. M. Lynden-Bell, "The Calculation of Lineshapes by Density Matrix Methods", in *Progress in NMR Spectroscopy*, J. W. Emsley, J. Feeney, L. H. Sutcliffe, Eds., Pergamon Press, London, U.K., 1967, Chapter 2.

37. G. Binsch, "The Study of Intramolecular Rate Processes by Dynamic Nuclear Magnetic Resonance," in *Topics in Stereochemistry*, Vol. 3, E. L. Eliel, N. L. Allinger, Eds., Wiley-Interscience, New York, NY, U.S.A., 1968, pp. 97–192.

38. J. P. Lloyd, G. E. Pake, *Phys. Rev.*, **94**, 579 (1954).

39. W. Plachy, D. Kivelson, *J. Chem. Phys.*, **47**, 3312 (1967).

40. T. A. Miller, R. N. Adams, *J. Am. Chem. Soc.*, **88**, 5713 (1966).

41. M. J. Povich, *J. Phys. Chem.*, **79**, 1106 (1975); also *Anal. Chem.*, **47**, 346 (1975).

42. J. S. Hyde, W. K. Subczynski, "Spin-label Oximetry", in *Biological Magnetic Resonance*, Vol. 8, *Spin Labeling Theory and Applications*, L. J. Berliner, J. Reuben, Eds., Plenum Press, New York, NY, U.S.A., 1989, pp. 399–425.

43. R. L. Ward, S. I. Weissman, *J. Am. Chem. Soc.*, **79**, 2086 (1957).

44. P. J. Zandstra, S. I. Weissman, *J. Chem. Phys.*, **35**, 757 (1961).

45. S. I. Weissman, *Z. Elektrochem.*, **64**, 47 (1964).

46. F. C. Adam, S. I. Weissman, *J. Am. Chem. Soc.*, **80**, 1518 (1958).

47. R. W. Kreilick, S. I. Weissman, *J. Am. Chem. Soc.*, **84**, 306 (1962).

48. H. Fischer, *Mol. Phys.*, **9**, 149 (1965).

49. Z. Zimpel, *J. Magn. Reson.*, **85**, 314 (1989).

50. G. K. Fraenkel, *J. Phys. Chem.*, **71**, 139 (1967).

51. N. Hirota, *J. Phys. Chem.*, **71**, 127 (1967).

52. J. R. Bolton, A. Carrington, *Mol. Phys.*, **5**, 161 (1962).

53. J. H. Freed, G. K. Fraenkel, *J. Chem. Phys.*, **37**, 1156 (1962).

54. N. N. Bubnov, S. P. Solodovnikov, A. I. Prokofev, M. I. Kabachnik, *Russ. Chem. Rev.*, **47**, 549 (1978).

55. L. J. Libertini, O. H. Griffith, *J. Chem. Phys.*, **53**, 1359 (1970); see also O. H. Griffith, D. W. Cornell, H. M. McConnell, *J. Chem. Phys.*, **43**, 2909 (1965).

56. A. Hudson, G. R. Luckhurst, *Chem. Rev.*, **69**, 191 (1969).

57. A. Rockenbauer, P. Simon, *J. Magn. Reson.*, **18**, 320 (1975).

58. S. K. Hoffmann, Z. Zimpel, M. Augustyniak, W. Hilczer, M. Szpakowska, *J. Magn. Reson.*, **98**, 1 (1992).

59. A. Carrington, A. Hudson, G. R. Luckhurst, *Proc. Roy. Soc. (London)*, **A284**, 582 (1965).

60. E. de Boer, E. L. Mackor, *J. Chem. Phys.*, **38**, 1450 (1963).

61. J. H. Freed, G. K. Fraenkel, *J. Chem. Phys.*, **40**, 1815 (1964).

62. M. R. Das, G. K. Fraenkel, *J. Chem. Phys.*, **42**, 1350 (1965).

63. J. R. Bolton, G. K. Fraenkel, *J. Chem. Phys.*, **41**, 944 (1964).

64. J. H. Freed, G. K. Fraenkel, *J. Chem. Phys.*, **41**, 699 (1964).

65. R. F. Campbell, J. H. Freed, *J. Phys. Chem.*, **84**, 2668 (1980).

66. S. Lee, S.-Z. Tang, *Phys. Rev.* B, **32**, 7116 (1985).

67. G. Nyberg, *Mol. Phys.*, **12**, 69 (1967).

68. R. Wilson, D. Kivelson, *J. Chem. Phys.*, **44**, 154 (1966); P. W. Atkins and D. Kivelson, *J. Chem. Phys.*, **44**, 169 (1966).

69. K. T. Mueller, A. J. Kunin, S. Greiner, T. Henderson, R. W. Kreilick, R. Eisenberg, *J. Am. Chem. Soc.*, **109**, 6313 (1987).

70. F. Chiarini, M. Martinelli, F. A. Rolla, S. Santucci, *Phys. Lett.*, **44A**(2), 91 (1973).

71. M. Martinelli, L. Pardi, C. Pinzino, S. Santucci, *Phys. Rev. B*, **16**(1), 164 (1977).

72. J. S. Hyde, "Saturation Transfer Spectroscopy", in *Methods in Enzymology*, Vol. 49, Part G, C. H. W. Hirs, S. N. Timasheff, Eds., Academic Press, New York, NY, U.S.A., 1978, Chapter 19.

73. R. Boscaino, F. M. Gelardi, R. N. Mantegna, *J. Magn. Reson.*, **70**, 251 (1986).

74. R. Boscaino, F. M. Gelardi, R. N. Mantegna, *J. Magn. Reson.*, **70**, 262 (1986).

75. N. B. Galloway, L. R. Dalton, *Chem. Phys.*, **30**, 445 (1978); **32**, 189 (1978); **41**, 61 (1979).

76. R. D. Nielsen, S. Canaan, J. A. Gladden, M. H. Gelb, C. Mailer, B. H. Robinson, *J. Magn. Reson.*, **169**, 129 (2004).

77. J. R. Harbour, V. Chow, J. R. Bolton, *Can. J. Chem.*, **52**, 3549 (1974).

78. G. R. Buettner, *Free Radical Biol. Med.*, **3**, 259 (1987).

79. R. O. C. Norman, *Chem. Soc. Rev.*, **8**, 1 (1979); *Pure Appl. Chem.*, **51**, 1009 (1979).

80. H. Fischer, H. Paul, *Acc. Chem. Res.*, **20**, 200 (1987).

81. Ya. S. Lebedev, *Prog. React. Kinet.*, **17**, 281 (1992).

82. K. A. McLauchlan, D. G. Stevens, *Mol. Phys.*, **57**, 223 (1986).

83. K. A. McLauchlan, D. G. Stevens, *Acc. Chem. Res.*, **21**, 54 (1988).

84. A. D. Trifunac, R. G. Lawler, D. M. Bartels, M. C. Thurnauer, *Prog. React. Kinet.*, **14**, 43 (1986).

85. G. L. Closs, M. D. E. Forbes, *J. Am. Chem. Soc.*, **109**, 6185 (1987).

86. R. W. Fessenden, R. H. Schuler, *J. Chem. Phys.*, **39**, 2147 (1963).

87. D. M. Bartels, P. Han, P. W. Percival, *Chem Phys.*, **164**, 421 (1992).

88. J. R. Morton, K. F. Preston, P. J. Krusic, L. B. Knight Jr., *Chem. Phys. Lett.*, **204**, 481 (1993).

89. J. K. S. Wan, S.-K. Wong, D. A. Hutchinson, *Acc. Chem. Res.*, **7**, 58 (1974).

90. F. J. Adrian, *Rev. Chem. Interm.*, **3**, 3 (1979).

91. C. D. Buckley, K. A. McLauchlan, *Mol. Phys.*, **54**, 1 (1985).

92. M. C. Depew, J. K. S. Wan, *Magn. Reson. Rev.*, **8**, 85 (1983).

93. P. J. Hore, K. A. McLauchlan, *Mol. Phys.*, **42**, 533 (1981).

94. O. E. Yakimchenko, Ya. S. Lebedev, *Russ. Chem. Rev.*, **47**, 531 (1978).

95. Z. Wang, J. Tang, J. R. Norris, *J. Magn. Reson.*, **97**, 322 (1992).

96. B. van den Brandt, H. Glättli, I. Grillo, P. Hautle, H. Jouve, J. Kohlbrecher, J. A. Konter, E. Leymarie, S. Mango, R. May, A. Michels, H. B. Stuhrmann, O. Zimmer, *Nucl. Instrum. Meth. Phys. Res. A*, **526**, 81 (2004).

97. T. Guiberteau, D. Grucker, *J. Magn. Reson. B*, **110**, 47 (1996).

98. L. R. Becerra, G. J. Gerfen, B. F. Bellew, J. A. Bryant, D. A. Hall, S. J. Inati, R. T. Weber, S. Un, T. F. Prisner, A. E. McDermott, K. W. Fishbein, K. E. Kreischer, R. J. Temkin, D. J. Singel, R. G. Griffin, *J. Magn. Reson. A*, **117**, 28 (1995).

99. H. M. Swartz, R. B. Clarkson, *Phys. Med. Biol.*, **43**, 1957 (1998).

100. H. J. Halpern, C. Yu, M. Peric, D. Barth, D. J. Grdina, B. A. Teicher, *Proc. Natl. Acad. Sci. U.S.A.*, **91**, 13047 (1994).

101. H. J. Halpern, M. Elas, C. Mailer, B. B. Williams, A. Parasca, E. D. Barth, C. A. Pelizzari, V. S. Subramanian, K. Ichikawa, G. A. Rinard, R. W. Quine, G. R. Eaton, S. S. Eaton, *Acad. Radiol.*, **10(8)**, 947 (2003).

102. P. Caravan, A. V. Astashkin, A. M. Raitsimring, *Inorg. Chem.*, **42**, 3972 (2003).

103. G. Feher, J. P. Gordon, E. Buehler, E. A. Gere, C. D. Thurmond, *Phys. Rev.*, **109**, 221–222 (1958).

104. E. L. Cochran, F. J. Adrian, V. A. Bowers, *J. Chem. Phys.*, **40**, 213 (1964).

NOTES

1. There are at least as many definitions of the word 'dispersion' as there are letters in the name. The reader is invited to make a list! As a general definition, we may try 'transformation of a more orderly and more homogeneous system to a less orderly and less homogeneous system, without loss of energy by it. This evolution may be a temporary and non-random process'.

2. Electrons and protons obey Fermi-Dirac statistics, but it can be shown that Maxwell–Boltzmann statistics apply, to an adequate approximation, when the interactions between the spins are sufficiently weak; that is, the spin moments act independently [1]. There is then no limit for the number of spins having the same energy. However, spin-spin interaction must be adequate to maintain thermal equilibrium *within* the spin system in order for T_s to be meaningful.

3. Spin systems with negative spin temperatures cool by passing through $-\infty$ to $+\infty$ and then to finite positive T_s values.

4. The quantity $\rho_\nu \, d\nu$ is the total energy of the set of photons with energy between $h\nu$ and $h(\nu + d\nu)$ per unit volume per unit frequency interval. Density ρ_ν has SI units of J m^{-3} Hz^{-1}. For a resonator, producing a linearly polarized magnetic field (Eq. 10.23a), $\int_0^\infty \rho_\nu \, d\nu = 2B_1{}^2/\mu_0$. Here B_1 may be a function of position.

5. Everywhere in space there is a background blackbody radiation at 3 K [3].

6. The concept of a spin-lattice relaxation time goes back to the work of Waller in 1932 [4] and of Van Vleck in 1939 [5]. However, the modern developments of the concept largely derive from the work of Bloembergen et al. [6] in 1948, and work by Redfield [7] in 1957.

7. A compilation of papers on spin-lattice relaxation and its mechanisms has been published [8]. Also the books by Pake and Estle [9], Poole and Farach [10] and Sorin and Vlasova [11], and a review by Orbach [12] contain good discussions of this topic.

8. Vector **M** is in motion at system equilibrium, which can be defined as the situation when the time-*average* values of the components of **M** over $n > 1$ cycles are independent of n.

9. Note that γ_e has a negative sign (Section 1.7), leading to a clockwise sense (while looking along vector **B**) of rotation of **M** relative to the direction of **B**; that is, ω_B is positive.

10. Equation 10.19c is valid only when $B \gg B_1$, since **M** relaxes with respect to the *instantaneous* magnetic field $\mathbf{B} + \mathbf{B}_1$, not the static field **B** [22] (Problem 11.5).

11. Note that $M_z \to M_z^{\ 0}$ as $\tau_1 \to 0$. Spin polarization is seen to be predicted by the Bloch equations (see Eq. 10.24c) for $B_1 > 0$ even when $B = 0$. This effect has been observed (i.e., $M_z \neq 0$, $\mathbf{z} \perp \mathbf{B}_1$) for electrons and recently [23] even for nuclei.

12. Often the concept of a complex magnetic susceptibility $\chi = \chi' - i\chi''$ is convenient. The connection between $\chi'(\omega)$ and $\chi''(\omega)$ is an example of the Kramers-Kronig relation [24,25]. Strictly speaking, in solids these susceptibilities are second-rank tensors rather than scalars (see Note 1.16).

13. Often EPR spectrometers measure signals proportional to $P^{1/2}$ rather than to P (Section E.1.8). For a discussion, see the detailed article by Feher [26].

14. However, ω_B is rarely a simple function of B (e.g., Eq. 6.54), so the formalism described above does not in fact apply. In the simplest case, $\omega_B = -(g/g_e)\gamma_e B$.

15. It must be noted carefully that τ_2 has been related to the inverse width of a *homogeneously* broadened line. The inverse width of an *inhomogeneously* broadened line has not been linked to any actual relaxation time; however, the inverse width of each component spin packet is a measure of τ_2 for that packet.

16. The modified Bloch equation model was introduced in 1953 by Gutowsky et al. [31] to explain dynamic processes in NMR; however, this approach is easily adapted for EPR [32]. Since EPR deals with larger energy splittings, it can probe kinetic phenomena involving a shorter time scale, that is, more frequent events.

17. Electron exchange is detectable only if the colliding radicals have *different* electron-spin states (M_S). There is no way of detecting exchanges if their initial spin states are identical.

18. A similar exchange-narrowed spectrum is observed for most pure solid free radicals. Here the strong exchange arises from the permanent overlap of molecular wavefunctions.

19. If the spin-hamiltonian parameters are all isotropic, rotation cannot be detected by magnetic measurements.

20. In single crystals, the individual linewidths may depend on M_I (in fact, as the square root of the function in Eq. 10.54a) and may be anisotropic. This is a static effect, arising from either a distribution of hyperfine parameters (A strain) and/or from the presence of exchange coupling ($|J| \approx |A/(g_e\beta_e)|$) [57,58]. Rotational effects due to dipolar motions may, of course, also occur.

21. In some cases it is preferable to turn off the modulation amplitude and detect in the absorption mode. In this case the field should be placed at the center of the strongest line.

FURTHER READING

1. A. Abragam, *The Principles of Nuclear Magnetism*, Oxford University Press, Oxford, U.K., 1961.

2. L. Allen, J. H. Eberly, *Optical Resonance and Two-level Atoms*, Dover, New York, NY, U.S.A., 1987.

3. N. Bloembergen, "The Concept of Temperature in Magnetism", *Am. J. Phys.*, **41**, 325 (1973).

4. J. H. Freed, "Molecular Rotational Dynamics in Isotropic and Oriented Fluids Studied by ESR", in *Rotational Dynamics of Small and Macromolecules*, Th. Dorfmüller, R. Pecora, Eds., Springer, Berlin, Germany, 1987, pp. 89–142.

5. N. Hirota, H. Ohya-Nishigushi, "Electron Paramagnetic Resonance", in *Techniques of Chemistry*, 4th ed., Vol. 4, *Investigations of Rates and Mechanisms of Reactions*, C. F. Bernasconi, Ed., Wiley, New York, NY, U.S.A., 1986, Chapter 11.

6. Yu. N. Molin, K. M. Salikhov, K. I. Zamaraev, *Spin Exchange—Principles and Applications for Chemistry and Biology*, Springer, Berlin, Germany, 1980.

7. L. T. Muus, P. W. Atkins, K. A. McLauchlan, J. B. Pedersen, "Chemically Induced Magnetic Polarization", NATO ASI Series, Vol. C34, Reidel, Dordrecht, Netherlands, 1977. See also P. J. Hore, C. G. Joslin, K. A. McLauchlan, "Chemically Induced Dynamic Electron Polarization", *J. Chem. Soc.* (Specialist Periodical Report on Electron Spin Resonance), **5**, 1 (1979).

8. R. Orbach, H. J. Stapleton, "Electron Spin-Lattice Relaxation", in *Electron Paramagnetic Resonance*, S. Geschwind, Ed., Plenum Press, New York, NY, U.S.A., 1972, Chapter 2.

9. K. J. Standley, R. A. Vaughan, *Electron Spin Relaxation Phenomena in Solids*, Adam Hilger, London, U.K., 1969.

10. D. ter Haar, "Simple Derivation of the Bloch Equation", *Am. J. Phys.*, **34**, 1164 (1966).

11. I. I. Rabi, N. F. Ramsey, J. Schwinger, "Use of Rotating Coordinates in Magnetic Resonance Problems", *Rev. Mod. Phys.*, **26**, 167 (1954).

12. M. Weissbluth, *Photon-Atom Interactions*, Academic, Boston, MA, U.S.A., 1989, Chapter 3.

PROBLEMS

10.1 (*a*) Derive the expression $\Delta N/N = (1 - e^{-x})/(1 + e^{-x})$, where $x = \Delta U/k_b T$, and (*b*) show by series expansion that $\Delta N/N \approx x/2$ when $|x| \ll 1$.

10.2 When a spin system is placed into a homogeneous constant magnetic field (of magnitude B), it gives up energy to the field, and reaches a steady-state population. Denote its total energy W(B) by W, which is negative

(*a*) Show that the ultimate population difference (Eq. 10.3) for a spin $\frac{1}{2}$ system at a sufficiently high temperature is given by

$$\Delta N = -NW/(2k_b T_s) \qquad (10.57)$$

where N is the total (large) number of spins, and T_s is the spin-system temperature defined via the Boltzmann distribution. Give a simple expression for $W(B)$.

(*b*) Consider a sudden feeding in of energy $|2W|$ to the spin system, causing an exact population inversion. Predict what will happen as the spin system then returns to steady state (see Ref. 103).

10.3 Linewidths in EPR spectra have been reported to range from 1.5 to ≥ 100 mT.

  (*a*) Compute the *minimum possible* value of τ_1 for lines of width 1.5, 0.1 and 100 mT.

  (*b*) Discuss possible methods that might be used to measure the spin-lattice relaxation time τ_1 (Section F.6).

  (*c*) What are the distinguishing characteristics of the relaxation times τ_1 and τ_2?

10.4 Given that

$$\frac{A_{u\ell}}{B_{u\ell}} = \frac{8\pi h \nu^3}{c^3} \tag{10.58}$$

where c is the speed of light, and the radiation density ρ_ν is given by the Planck blackbody law

  (*a*) Show that a two-level spin system comes to equilibrium such that the spin temperature equals the temperature of the blackbody radiation.

  (*b*) Given that $A_{u\ell}^{-1} = 10^4$ years for uncorrelated spins, calculate τ_1 for this case (assume that $B = 1$ T).

10.5 Derive the limiting expressions for the magnetization components (Eqs. 10.24) as τ_2 goes to infinity. (*Hint:* See Section A.7.) Plot the corresponding susceptibilities (Eqs. 10.27–10.29) and compare with Fig. 10.6. What is the meaning attributable to parameter τ_2?

10.6 From the linewidths in Fig. 10.8, at concentrations of 10^{-1} and 10^{-2} M, determine the second-order rate constant for electron-spin exchange.

10.7 Consider the spectrum of sodium naphthalenide shown in Fig. 10.12. The isotropic sodium hyperfine splittings, in the limit of slow conversion between two distinct ion pairs (*a* and *b*), are $a_a = 0.22$ mT and $a_b = 0.08$ mT.

  (*a*) From the magnitude of the hyperfine splitting (computed using the scale in the figure) and using Eq. 10.52, compute the equilibrium constant for the ion-pair interconversion reactions (Eq. 10.34).

  (*b*) Plot $\ln K$ versus $1/T$ to obtain the enthalpy change ΔH° for this interconversion.

  (*c*) Using the scale in the figure and the relative amplitudes, compute approximate linewidths for lines B_1, B_2, B_3 and B_4 in Figs. 10.12*a–c*.

  (*d*) Using Eq. 10.53 and noting that $\Gamma = (\sqrt{3}/2)\Delta B_{pp}$ for lorentzian lines, where ΔB_{pp} is the peak-to-peak width shown in Fig. F.1, compute the mean lifetime τ at each of the three temperatures. Note that $\tau_i = k_i^{-1}$, for $i = a,b$.

  (*e*) From the relation $\tau = \tau_a \tau_b / (\tau_a + \tau_b)$ and the equilibrium constant, compute τ_a and τ_b at the three temperatures.

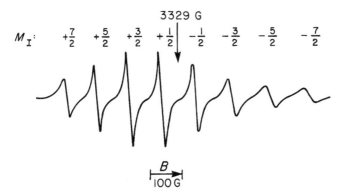

$$M_I: \quad +\tfrac{7}{2} \quad +\tfrac{5}{2} \quad +\tfrac{3}{2} \quad +\tfrac{1}{2} \quad -\tfrac{1}{2} \quad -\tfrac{3}{2} \quad -\tfrac{5}{2} \quad -\tfrac{7}{2}$$

FIGURE 10.26 First-derivative spectrum of the VO^{2+} ion in deoxygenated toluene solution at 236 K. [After R. Wilson, D. Kivelson, *J. Chem. Phys.*, **44**, 154 (1966).]

(*f*) Plot $\ln 1/\tau_a$ and $\ln 1/\tau_b$ versus $1/T$ to obtain the activation energies for the forward and reverse reactions.

10.8 Consider the vinyl radical $HC{=}CH_2$. Draw the expected eight-line (first-order) stick spectrum, labeled with M_I values, taking $a_1 = 1.57$ mT, $a_2(trans) = 3.43$ mT, and $a_3(cis) = 6.85$ mT, as is observed in rigid media [104]. In liquid solution, due to rapid interconversion between the two tautomers, only two hyperfine splittings (1.57 and 10.28 mT) are found [86]. Explain this observation, drawing a suitable correlation diagram between the stick spectra.

10.9 The spectrum of the VO^{2+} ion is given in Fig. 10.26. The ^{51}V nucleus, in 99.75% natural abundance, has $I = \tfrac{7}{2}$.
   (*a*) Determine relative linewidths for each component, using the fourth line as a reference.
   (*b*) Assuming that $g\| < g_\perp$, determine the sign of the hyperfine splitting.
   (*c*) Determine the coefficients α, β and γ by application of Eq. 10.54*a* to this spectrum.

10.10 (*a*) From the spectrum of the *p*-dinitrobenzene anion at $-55°C$ in Fig. 10.18, determine the relative linewidths of each component assuming that the central line has a unit linewidth.
   (*b*) Assign a value of $'M_H$ and $'M_N$ to each line component.
   (*c*) Determine the sign of β_N and β_H (Eq. 10.54*b*). (*Hint*: β_N should be determined from the line components for which $'M_H$ is zero.)
   (*d*) Determine the sign of γ_{NH} (Eq. 10.54*b*). (*Hint*: Compare the relative widths of the proton hyperfine components for which $'M_N = -1$ and $'M_N = +1$.)

10.11   Referring to Fig. 10.19*b*, construct a first-order stick diagram for the di-*t*-butyl nitroxide radical under the incorrect assumption (see text) that a_0 has the opposite sign, using $a_\parallel = +0.21$ mT, $a_\perp = -2.37$ mT, and the correct *g* factors. From this diagram, predict the relative widths of the three hyperfine lines in a solution of moderate viscosity. What is actually observed?

10.12   Rotation of molecules containing pairs of coupled identical magnetic dipoles $g\beta\mathbf{J}$, aligned in an external magnetic field **B**, causes spin-lattice relaxation and line broadening, modeled statistically [18, p. 192] by the expression

$$\frac{1}{\tau_1} = \frac{2\mu_0^2}{5}\left(\frac{g^4\beta^4 J(J+1)}{\hbar^2}\right)\left\{\frac{\tau_c}{1+\omega_B^2\tau_c^2} + \frac{4\tau_c}{1+4\omega_B^2\tau_c^2}\right\}\langle r^{-6}\rangle \qquad (10.59a)$$

$$\frac{1}{\tau_2} = \frac{2\mu_0^2}{10}\left(\frac{g^4\beta^4 J(J+1)}{\hbar^2}\right)\left\{3\tau_c + \frac{5\tau_c}{1+\omega_B^2\tau_c^2} + \frac{2\tau_c}{1+4\omega_B^2\tau_c^2}\right\}\langle r^{-6}\rangle$$

$$(10.59b)$$

for the relaxation times. Here *r* is the average distance between the dipoles and τ_c is the rotational correlation time. The latter is a measure of the frequency of random rotations of the molecules. Plot τ_1 and τ_2 versus τ_c for electrons, using reasonable values of *g*, ω_B and *r*. Discuss in physical terms the functional behavior obtained.

10.13   The term 'adiabatic' occurs throughout the magnetic resonance literature. In classical thermodynamics, wherein time dependence never appears for any process, diabatic and adiabatic signify heat transfer or none across the boundary between the system considered and its surroundings. Elsewhere, time is of the essence in this nomenclature, and the denotations 'slow' and 'fast' enter. Write a succinct discussion of the term 'adiabatic', describing its meaning and usage within EPR spectroscopy.

CHAPTER 11

NON-CONTINUOUS EXCITATION OF SPINS

11.1 INTRODUCTION

Until now we have considered continuous-wave (cw) situations in which $\mathbf{B}_1$ is sinusoidal but has a maximum amplitude constant with time. We now turn to the important ideas relevant when the excitation amplitude is time-dependent, say, when B_1 is pulsed. The use of such excitations in EPR has been developing slowly but steadily since the early 1960s. Much of the pioneering work in pulsed EPR was carried out by W. B. Mims, beginning in 1961.

We continue in this chapter to make use of the concept of frequencies, namely, those associated with applied magnetic excitation fields ($\mathbf{B}_1$) and with transition energies. The Larmor frequency corresponding to 'precession' of each particle's spin magnetic moment can be defined as $\nu_B = g\beta B/h$ (Problem 1.5). We remember that $\nu_B = \nu_{\alpha\beta} = (U_\alpha - U_\beta)/h$ for the $S = \frac{1}{2}$ spin system. In quantum mechanics, the concept of precession is linked to the Heisenberg principle, as applied to angular momentum; that is, the spin components perpendicular to $\mathbf{B}$ are depicted by a 'cone of uncertainty' (Fig. 11.1) and only the cone axis direction if selected is exactly measurable.

In practice, the form of $\mathbf{B}_1(t)$ can be chosen to fit the requirements of the experimentalist, at least in principle. The direction of $\mathbf{B}_1$ (or several such fields, applied consecutively) must be specified relative to that of $\mathbf{B}$, and to the orientation of the sample if the latter is an anisotropic single crystal. The frequency (or several such) must be selected, as well as the phase relation(s) between the relevant sinusoids. The amplitudes $B_1(t)$ are crucial in determining the observed effects. Of course, for every change in B_1, the excitation spin-hamiltonian term $-\hat{\boldsymbol{\mu}}^T \cdot \mathbf{B}_1$ changes concurrently.

Electron Paramagnetic Resonance, *Second Edition*, by John A. Weil and James R. Bolton
Copyright © 2007 John Wiley & Sons, Inc.

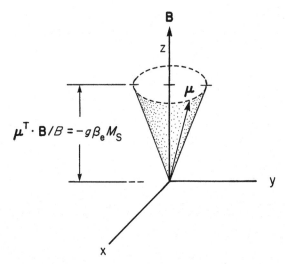

FIGURE 11.1 The precession model for the behavior of a spin magnetic moment $\langle \boldsymbol{\mu} \rangle$ in a static magnetic field **B**. The situation is depicted for a negative value of $\alpha g \beta M_J$ (Eq. 1.9).

For our present purposes, we take $\mathbf{B}_1(t)$ as a monochromatic sinusoid, constant in amplitude[1] over desired periods of time, instantaneously turned on from zero or turned off after desired intervals. Such square-wave behavior approximates very well the achievable actual pulses of excitation magnetic fields. It is important to note that such a rapid step change in B_1 is equivalent, in the region of the step, to the presence of field components having a range of frequencies superimposed on that of the basic frequency ν of field $\mathbf{B}_1$; that is, it is a generalized Fourier series of various sinusoids [1,2].

The field $\mathbf{B}_1$ is taken as linearly polarized, in the region of the sample, and generally oriented at 90° to field **B**. As we saw (Eq. 10.23), such a field can profitably be regarded as the superposition of two equal-amplitude rotating components (frequency ν), one moving clockwise and the other counterclockwise (Fig. 10.5, and Eqs. 10.23 and 10.24), both vectors circulating about **B**. Only one component is effective (see Appendix D) in causing magnetic resonance (σ-type) transitions; that is, the effective excitation field has magnitude B_1.

Initially, we once again consider single spins, and thereafter we shall graduate to ensembles of these. We temporarily ignore hyperfine effects and other inter-spin interactions.

11.2 THE IDEALIZED B_1 SWITCH-ON

Consider now the familiar (cw) situation of a spin magnetic moment $\hat{\boldsymbol{\mu}} = \alpha g \beta \hat{\mathbf{J}}$ (Eq. 1.9) in a sinusoidal (frequency ν) excitation field $\mathbf{B}_1$ that is linearly polarized, oriented at 90° to static Zeeman field **B** ∥ **z**. Both fields are assumed to be homogeneous. As indicated, we can adopt a vector model to visualize the situation (Fig. 11.1). In the absence of $\mathbf{B}_1$, the spin $\hat{\mathbf{J}}$ acts much as if it were precessing about $\mathbf{B}$ at its natural frequency ν_B, with the same rotational sense for both the $M_J = +\frac{1}{2}$ and $-\frac{1}{2}$ states. For non-zero B_1, there are two simultaneous precessions, one about $\mathbf{B}$ and the other about $\mathbf{B}_1$ generally at a frequency lower than ν_B (usually $B \gg B_1$). When $\mathbf{B}_1$ has basic frequency $\nu = \nu_B$, that is, at magnetic resonance, the frequency and sense of rotation of one component of $\mathbf{B}_1$ just match those of the spin moment. One may think of that $\mathbf{B}_1$ component as applying a torque on the expectation value $\langle \boldsymbol{\mu} \rangle$ of the magnetic moment, which torque is maximal since $\mathbf{B}_1$ and $\mathbf{B}$ are at right angles. Then the spin-moment vector (i.e., the axis of the cone of uncertainty) is driven by $\mathbf{B}_1$ back and forth between its J_z eigenstates, oscillating at frequency $\nu_1 = g\beta B_1/h$ with suitable energy exchange between spin and effective B_1 field.[2]

This situation is seen more clearly from the solutions of the so-called Rabi problem, that is, of the quantum-mechanical dynamic equations that furnish the probability amplitudes for the above-mentioned phenomena [3,4]. Thus, for example, if an isolated electron spin $\frac{1}{2}$ is in its ground state $M_J = -\frac{1}{2}$ at time t_0, the probability [5–7] for it to occur in upper state $M_J = +\frac{1}{2}$ is

$$P = \left(\frac{\nu_1}{\nu_r} \right)^2 \sin^2 \{ \pi \nu_r(t - t_0) \} \tag{11.1}$$

where the frequency $\nu_r = [\nu_1{}^2 + (\nu - \nu_B)^2]^{1/2}$ is named after the magnetic resonance pioneer, I. I. Rabi. The probability P is maximal at times $t - t_0 = \ell/2\nu_r$, for odd values of integer ℓ. At resonance, $\nu \equiv \nu_B$, the probability for the system to be found in its ground state is unity for ℓ even. At timepoints between these, the probability for each state $M_J = \pm\frac{1}{2}$ to occur is non-zero. The cone axis is to be visualized (Fig. 11.2) as moving in a screw-like motion back and forth between the bottom pole $(-\mathbf{z})$ and the top pole $(+\mathbf{z})$ of a sphere, where $\mathbf{B} \parallel \mathbf{z}$. The rate of spin 'flipping' is slow compared to the rotation frequency ν_B when $B \gg B_1$. This view of magnetic resonance is of course, consistent with the previous descriptions found in earlier chapters, that is, the above-mentioned flips correspond to transfer of photons $h\nu$ between the spin and excitation-field systems. Spin-lattice and spin-spin effects have been ignored here.

Another viewpoint is that the motion of the magnetic dipole ('precession' of the magnetic-moment vector) causes a detectable instantaneous power transfer (plus a frequency shift) in a resonator holding the spin system; for example, the precession induces a voltage in a pick-up coil linked to the energy source or sink (Appendixes E and F).

When $\nu \neq \nu_B$, then the Rabi oscillations (Eq. 11.1) in the probability amplitudes occur at frequencies ν_r higher than ν_1, and the probability for attaining the upper state never reaches unity. The transition probability goes down (quite sharply, dropping roughly as $\nu_1/|\nu - \nu_B|$) as ν departs from resonance, and $\mathbf{B}_1$ rapidly becomes ineffective as far as the spin is concerned. However, as we shall see, the situation $\nu \sim \nu_B$ is an important one. Remember also that the jump probability decreases as $\mathbf{B}_1$ shifts away from being normal to $\mathbf{B}$ (but see Section 1.13).

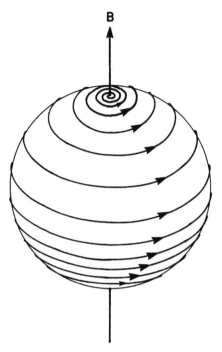

FIGURE 11.2 The locus of the tip of $\langle \mu \rangle$ when the magnetic resonance condition is obeyed. Initially $\langle \mu \rangle$ pointed along $-\mathbf{B}$, that is, the state for which $\alpha g \beta M_J < 0$.

Clearly, if no irreversible energy transfer from or to the total system (spin + radiation) occurs, then there is no net change of any type when averaged over any definite number of oscillation periods. Some effective electromagnetic field other than $\mathbf{B}$ and $\mathbf{B}_1$, that is, spin-lattice relaxation—a coupling to the energy reservoir of the rest of the world—must be present for net EPR power absorption to be observable.

Next, we consider a set of many independent spins in uniform fields $\mathbf{B} \perp \mathbf{B}_1$, which are taken as having magnetic moments identical in magnitude. Such an ensemble can be treated statistically, say, by using the Bloch theory (Chapter 10) or the density-matrix approach.[3] We can benefit by utilizing the net magnetization $\mathbf{M}$ per unit volume, as defined in Eq. 1.8 (see also Note 1.8 and Chapter 10). We take this as being the expectation value describing a spatially averaged macroscopic quantity, a time-dependent vector not subject to quantum-mechanical uncertainties. Initially the phases (positions along the circular 'orbits' about $\mathbf{B}$) of the various individual spin vectors are random, and they remain so after $\mathbf{B}_1$ is switched on. Therefore, in an isotropic material, the time average of any component of $\mathbf{M}$, which is perpendicular to $\mathbf{B}$, is zero in the laboratory frame.

If there were exactly equal numbers of independent spins in states $M_J = -\frac{1}{2}$ and $+\frac{1}{2}$, then the net effect of the upward and downward flips would be to maintain the magnetization as is (i.e., $\mathbf{M}$ would be zero and stay that way). However, for the usual thermal equilibrium situation in a relaxing spin system, there is an excess number ΔN of spins in the lower energy state. These excess spins set the value of $\mathbf{M}$ and hence, as $\mathbf{B}_1$ drives the spins back and forth (in unison) between states $+\frac{1}{2}$, $M_z(t)$ oscillates between extremes for $\mathbf{M}$ parallel and antiparallel to $\mathbf{B}$ [14]. An oscillating magnetic dipole is associated with an oscillating radiation field; that is, an alternating-current (ac) voltage is induced. If $\mathbf{B}_1$ is not turned off during the experiment, then this is a view of cw magnetic resonance spectroscopy (EPR, if $\hat{\mathbf{J}} = \hat{\mathbf{S}}$), as presented in the previous chapters. As discussed, if some of the spin-system energy goes to (say) atomic motions, rather than being returned to $\mathbf{B}_1$, then ΔN is maintained and a net absorption signal is observable. Note that unless the spin-lattice relaxation is adequate, ΔN will go to zero so that eventually no net energy absorption will occur and the spin system will be deemed to have become power-saturated (Section 10.2.3). Note the vanishing of $\mathbf{M}$ in the limit as $\tau_1 \rightarrow \infty$ (Eqs. 10.22).

To understand the physical interactions of the spin moments with the externally applied magnetic fields and with each other, it is important to visualize the time-dependent behavior of the magnetization $\mathbf{M}$. This task can be simplified by choosing the most convenient coordinate system (CS). Very often, use of a 'rotating frame' (Section 10.3.3), which here has its axis $\mathbf{z}_\phi(=\mathbf{z})$ along $\mathbf{B}$ and its axis $\mathbf{x}_\phi$ along the effective rotating (in the lab CS) component of $\mathbf{B}_1$, is called for, since the $\mathbf{B}_1$ rotation and the Larmor precession (when $\nu = \nu_B$) then seemingly are absent [15] (Fig. 10.4). This also greatly simplifies the mathematical detail, as may be seen in Problem 11.1.

One can analyze the situation more completely by means of transient solutions [16] of the Bloch equations (Eqs. 10.17). As seen in the laboratory frame, $\mathbf{M}$ rotates about $\mathbf{B}$ ($\|\mathbf{z}$) at frequency ν, on which motion is superimposed a much slower B_1-dependent nutation (change in angle between $\mathbf{M}$ and $\mathbf{B}$) occurring with the Rabi frequency ν_r. This transient nutation has an initial amplitude dependent on the magnetization present at switch-on of field $\mathbf{B}_1$ ($\perp\mathbf{z}$) and on the proximity of frequency ν to the resonance value. It is damped by the spin-lattice and spin-spin interactions. In the frame rotating about $\mathbf{z}$ at frequency ν, there is slow precession of $\mathbf{M}$ about the direction $\mathbf{B}_1 + (B - h\nu/g\beta_e)\mathbf{z}$ with frequency ν_r. In the spectrometer, a resulting modulation of the absorption (or dispersion) is observable, measured by a detector sensitive to the magnetization in the xy plane [1,16]. This transient on the signal, which is dependent on the presence of a magnetization component $\mathbf{M}_\perp$ in that plane, decays with a nutational relaxation time given (in the absence of field inhomogeneities) by $2(\tau_1^{-1} + \tau_2^{-1})$. The final magnetization is the steady-state one given in the previous chapter (Section 10.3.4).

11.3 THE SINGLE B₁ PULSE

Next, consider the effects of turning off the excitation field $\mathbf{B}_1$, while the transient nutation is still appreciable. Thus a square-wave excitation pulse of temporal length τ is formed. The behavior of $\mathbf{M}$ beginning at pulse-end time is of prime interest in EPR spectroscopy. The observed phenomenon is referred to by the curious name 'free-induction decay' (FID). Here 'free' refers to the absence of $\mathbf{B}_1$ (while the field $\mathbf{B}$ remains on!), in contrast to the situation described above, of the spins being 'driven' by $\mathbf{B}_1$. The FID causes the signal usually dealt with in pulsed EPR.

If $\mathbf{B}_1$ is instantaneously turned off at (any one of the) times $\tau = \ell/(4\nu_1)$ after turn-on, then, at that instant $\mathbf{M}$ is either parallel or antiparallel to $\mathbf{B}$, depending on whether ℓ is an even or odd integer. Here $\nu_1 \equiv g\beta_e B_1/h$. The pulse turning (flip) angle $\Omega(\tau)$ for $\mathbf{M}$ is given (in radians) by the product $2\pi\tau\nu_1$. Thus the π pulse ($\ell = 2$) turns $\mathbf{M}$ by 180° from its initial direction. We note that the π pulse inverts the electron spin population, the higher-energy state becoming more populated than the lower one. Equivalently, this can be said to generate a negative spin temperature to describe the population of the spin system. As time progresses, the spin system recovers and moves toward the normal (positive spin temperature) Boltzmann population (Chapter 10), by radiationless relaxation and the emission of photons $h\nu$. Note that with B_1 turned off, the photon energy density ρ_ν is essentially zero, so that no induced transitions take place. However, the spontaneous photon emission is enhanced when the spin magnetic moments are in phase (superradiant system), that is, while the FID is appreciable. [In cw EPR, the spins are not correlated. Hence most of the energy lost from the spins goes to the 'lattice' of atoms and only some (usually a negligible amount) goes back to $\mathbf{B}_1$ (via incoherent spontaneous emission).] The locus of vector $\mathbf{M}$ is to be visualized as remaining longitudinal (along $\mathbf{B}$) throughout the inversion recovery, that is, shrinking in magnitude along $-\mathbf{z}$, going through zero, and then regaining its initial magnitude along $+\mathbf{z}$. The decay rate is exponential, with a characteristic relaxation time τ_1 (Section 1.5 and Chapter 10). The spin behavior can in many actual cases be described in terms of the Bloch equations for the time dependence of $\mathbf{M}$ (Section 10.3).

We now discuss, in a similar manner, the very important case of the $\pi/2$ pulse: namely, $\tau = 1/(4\nu_1)$. Immediately after cessation of B_1, $\mathbf{M}$ has been rotated, with its magnitude unaltered, to an orientation perpendicular both to $\mathbf{B}$ and (if $\nu = \nu_B$) to the direction 90° from where $\mathbf{B}_1$ was terminated.[4] Note the essential difference between this occurrence of $M_z = 0$, in which the spins are coherently dispersed (see Section 11.8) transverse to the field $\mathbf{B}$ (i.e., in the xy plane), and the random-phase cases when the polarization field $\mathbf{B}$ is first turned on or when there is complete saturation. As with the π pulse, M_z begins to grow back toward its equilibrium value (exponentially, with spin-lattice relaxation time τ_1) while the set of individual spins starts losing phase coherence so that the transverse magnetization $M_\perp \to 0$ (with relaxation time τ_m).[5] This temporal behavior is shown in Fig. 11.3. It is useful to realize that dephasing is of two types: reversible and irreversible (stochastic). The former is given by (real) parameter τ_2^* (includes magnetic-field inhomogeneity) and the latter by τ_m.

The transverse magnetization direction $\mathbf{y}_\phi$, that is, that of $\mathbf{M}_\perp$, continues to precess around $\mathbf{B}$ at frequency $\nu = \nu_B$. With suitable apparatus (i.e., detection in the plane normal to $\mathbf{B}$), the temporal behavior of $\mathbf{M}_y$ can be followed.[6] The phase coherence of the detected signal(s) relative to $\mathbf{B}_1$ can be measured, since the sinusoidal (frequency ν) supply voltages for field $\mathbf{B}_1$ are maintained even when $\mathbf{B}_1$ is not on. The detected signal is an 'interferogram'. That is, the superposition at each instant of

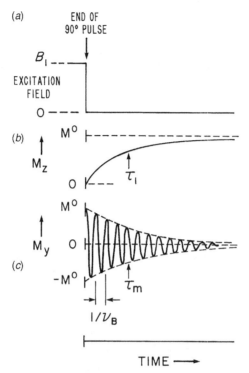

FIGURE 11.3 The magnetization behavior after the end of a $90°$ B_1 pulse shown in (a). Drawings (b) and (c) depict $M_z(t)$ and $M_y(t)$. The situation $\tau_1 = \tau_m$ is depicted.

positive-going and negative-going contributions (induced voltages) from the various spins in the ensemble defines the total signal, which varies rapidly and dramatically with time. In practice, the FID signal must last sufficiently long to be recorded reliably.[7]

Finally, we can now consider more realistic spin systems, where various interactions are at play, so that interpretation of the observed FID signals yields precious chemical and structural information. In other words, all the spin-hamiltonian parameters discussed so far affect the FID and in principle are extractable by measuring it, as are the operative relaxation times. Use of single $\mathbf{B}_1$ pulses and, as we shall see later, of cleverly designed sequences of $\mathbf{B}_1$ pulses, make these goals attainable. These considerations are equally valid for nuclei and electrons; for our purposes, we deal with electron spin, that is, $\hat{\mathbf{J}} = \hat{\mathbf{S}}$.[8]

11.4 FOURIER-TRANSFORM EPR AND FID ANALYSIS

Detection after a pulse gives two separate and complicated functions of time, namely, coherent signals in-phase and out-of-phase with the reference signal at frequency ν ($\mathbf{B}_1$ is now turned off, but the memory of it lingers on in the spin system). These can be repetitively and separately measured, maintaining the above-cited phase coherence, and can be computer-stored.[9] The subsequent analysis of the time evolution of these signals (which begin at the end of the pulse) usually is carried out by the mathematical technique called 'discrete Fourier transformation' [19], a procedure that is immensely assisted by use of modern computer technology, using the fast Fourier-transform algorithm. In essence, the result is conversion of digital time-domain data to digital frequency-domain data containing the line positions (ν_i or B_i) and relative peak intensities identical to those of the cw EPR spectrum (Fig. 11.4). It is possible to obtain analytical expressions for the frequency-domain spectra for suitably simple spin systems, for example, for a disordered system with a single $I = \frac{1}{2}$ nucleus [20]. For best frequency-domain resolution, it is important to follow the time development of the FID as long and completely as possible.

In any actual chemical system with unpaired electrons, numerous electron spin 'precession' frequencies ν_B occur, arising from the many types of local magnetic fields present (e.g., hyperfine effects, electron-electron interactions, inhomogeneities in field $\mathbf{B}$). Therefore, a given frequency ν of excitation field $\mathbf{B}_1$ rarely coincides with an actual ν_B. Thus the off-resonance case referred to earlier is the rule and not the exception. It turns out that to measure maximal portions of the total extent of the EPR spectrum, τ should be as short as practicable. At the same time, the product of $B_1\tau$ is set by the condition that a $\pi/2$ pulse is desired. Thus B_1 must be as large as possible. In practice, a spectral region of ~ 2 mT can be excited by pulse EPR, and τ generally is very short compared to τ_1 and τ_m.

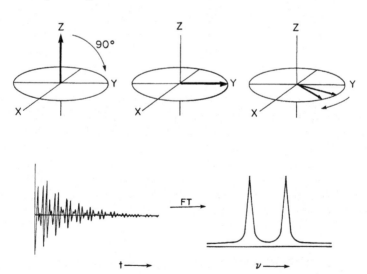

FIGURE 11.4 (*a*) Evolution of the magnetization during a $\pi/2$ pulse EPR experiment on a system with $S = I = \frac{1}{2}$; (*b*) FID spectrum and corresponding FT-EPR spectrum. [After A. Schweiger, *Angew. Chem. Int. Ed. Engl.*, **30**, 265 (1991).]

Let us now list some important advantages of doing pulse EPR rather than cw EPR:

1. *Efficient Data Collection.* This means that all lines are simultaneously detected and the scan time required to do so is independent of the extent covered rather than wasting time, as in cw work, on slowly scanning the various base-line regions between peaks. It also means that one can repetitively apply the pulse and computer-store very many, say, n, such scans. Since the noise is random, the $+$ and $-$ contributions cancel so that the signal-to-noise ratio increases with n (roughly as $n^{1/2}$). Here n easily can be as large as 10^5.

2. *Time Resolution.* A single pulse spectrum can be recorded in ~ 1 ms. A spin system evolving in time can easily be sampled. Thus a given B_1 pulse can be locked to a source of chemical energy, say, a laser pulse, so that reaction products can be sampled, for instance, every microsecond after this primary

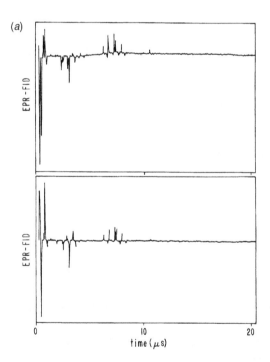

FIGURE 11.5 (*a*) Accumulated FIDs of the fluorenone ketyl radical anion at 220 K. The radical anion was produced by reduction with potassium in tetrahydrofuran. The output of the two phase-orthogonal detection channels is shown. (*b*) Quadrature Fourier spectrum of the data of (*a*). The spectral center at 25 MHz corresponds to the spectrometer frequency 9.271 GHz. (*c*) The stick-spectrum reconstruction based on hyperfine couplings (A/h in MHz) (1) -5.77, (2) $+0.27$, (3) -8.80, (4) $+1.84$. [After O. Dobbert, T. Prisner, K. P. Dinse, *J. Magn. Reson.*, **70**, 173 (1986).]

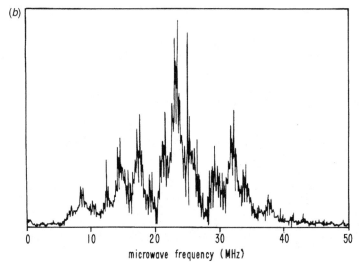

microwave frequency (MHz)

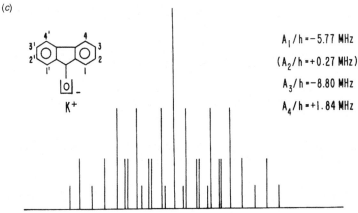

$A_1/h = -5.77$ MHz

$(A_2/h = +0.27$ MHz$)$

$A_3/h = -8.80$ MHz

$A_4/h = +1.84$ MHz

FIGURE 11.5 *Continued.*

event. Many other types of dynamic situations (e.g., chemical and optical creation and decay, diffusion, phase changes, energy transfer between molecules) are also open to study [21, Chapter 1]. Thus time-resolved EPR is seen to be a powerful tool in investigating kinetics.

3. *Efficient Relaxation Measurement.* τ_1 and τ_m can be measured directly from the responses to the pulse(s), rather than extracting them via deconvolution of lineshapes or analysis of cw saturation behavior [22, Chapter 8].

Potential disadvantages include the inability of pulse EPR to scan spectra extending over wide ($\gg 2$ mT) field ranges and the limitations posed by (too rapid) relaxation times.

In Fig. 11.5a, we display the two quadrature-detected (see Appendix E) FID signals arising from the fluorenone ketyl anion radical (**I**) [23].

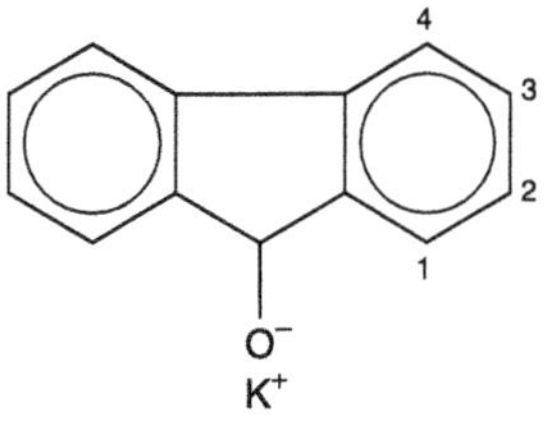

(I) fluorenone ketyl anion radical

Fourier transformation of the spectrum from the time domain to the frequency domain yields the 'ordinary' EPR spectrum consistent with a stick diagram, as shown in Fig. 11.5c.

As a second example of the application of pulse EPR, we consider a short-lived organic free radical created in a reversible photoinduced electron-transfer reaction. Using a pulsed dye laser, electron transfer occurs between excited zinc tetraphenyl-porphyrin (ZnTPP = **II**) and duroquinone (DQ = **III**), both dilute in liquid ethanol, to form the corresponding positive and negative ions [24].

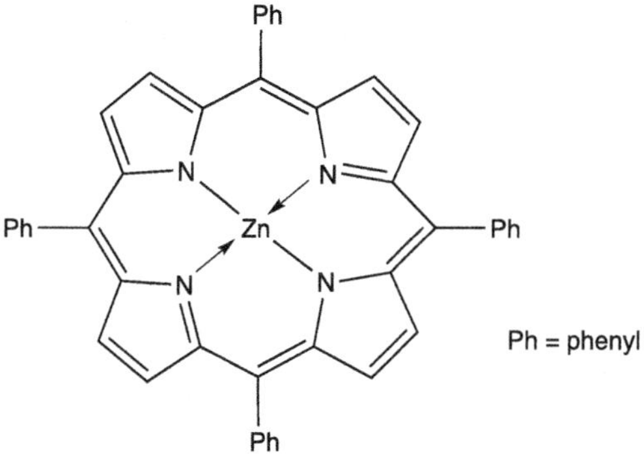

(II) ZnTPP

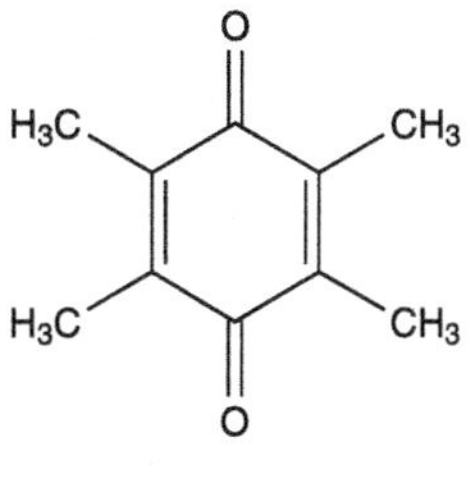

Ph = phenyl

(III) duroquinone

The primary EPR spectrum of DQ$^-$ consists of 13 almost equally spaced narrow lines (0.19 mT apart), arising from the 12 equivalent methyl protons, with relative

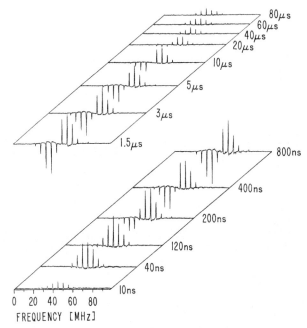

FIGURE 11.6   Quadrature-detected FT-EPR spectra[9] of photogenerated DQ⁻ in liquid ethanol at 245 K, as a function of time delay between the exciting laser pulse (600 nm) and the $\pi/2$ (20 ns) X-band microwave pulse. EPR absorption peaks point up; emission peaks point down. [After T. Prisner, O. Dobbert, K. P. Dinse, H. van Willigen, *J. Am. Chem. Soc.*, **110**, 1622 (1988).]

intensity ratios nominally of $1:12:66:220:495:792:924:$ (etc.). Note that the FID signals of necessity arise from free radicals that were present during the $\pi/2$ pulse. The signal from ZnTPP⁺ is not observed here because its τ_m is too short. Figure 11.6 depicts the FT-EPR spectrum of DQ⁻; not all 13 peaks are visible. The intensities display the kinetics of the electron transfer, as well as the spin-polarization effects (Sections 10.5.3 and 10.7.2) affecting the relative intensities, as discussed in Ref. 24.

It is important to note that straight Fourier-transform analysis of the time-domain experimental spectrum is by no means the only technique for attaining the corresponding frequency-domain spectrum. In NMR spectroscopy, 'linear prediction' and 'maximum-entropy reconstruction' are both used to good effect, and have some specific advantages [25].

11.5   MULTIPLE PULSES

The application of suitable sequences of pulses allows the observer to derive selected spectral parameters cleanly and quickly, removing complexities present in single-shot FIDs and in cw EPR spectra. Thus a multi-line spectrum can often

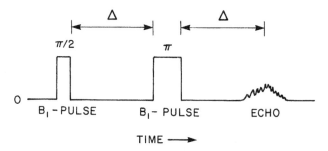

FIGURE 11.7 The Hahn B_1-pulse sequence: $\pi/2 - \Delta - \pi - \Delta -$echo.

be simplified, and both spectral and relaxation effects can be measured in the same experiment.

We now consider trains of n square-wave B_1 pulses (labeled $i, j = 1, 2, \ldots, n$), each of duration τ_i and separated by time intervals Δ_{ij} (where $j > i$). Usually, $n < 4$ in present-day EPR work. These labels are used only when necessary. Field B is kept constant.

Consider the two-pulse sequence $\pi - \Delta - \pi/2$. The first pulse inverts the magnetization **M**. The second pulse places whatever M_z magnetization there is after interval Δ into the xy plane, for FID detection. After a suitable interval, the sequence is repeated, say, with a different Δ. Clearly this choice offers a neat way of measuring τ_1. Other choices can also do so [1, pp. 20ff].

By contrast, consider next the sequence $\pi/2 - \Delta - \pi - \Delta$. By following the progress of **M** in the rotating frame defined earlier, we see that the first pulse rotates **M** from **z** $(=\mathbf{z}_\phi)$ to $\mathbf{y}_\phi$. During interval Δ, the de-phasing process for the individual spin moments progresses. When τ_1 is relatively long, so that $|M_z|$ remains small during this period, the π pulse takes M_y from $\mathbf{y}_\phi$ to $-\mathbf{y}_\phi$. At this point, the coherent de-phasing motions have all been suddenly reversed and, seemingly miraculously, the spins now move toward equal phasing (a maximum in $M_\perp$), which they attain after a (second) period Δ. Near this time, a signal builds up and then afterward decays, so as to yield an 'echo' that can be considered as the juxtaposition of two FIDs back to back [11, pp. 42,43], at least in the absence of envelope modulation. The whole set of events is depicted in Fig. 11.7, and sometimes is called the Hahn sequence, after its discoverer (1950) [26]. The first pulse is usually called the 'excitation pulse' and the second, the 'refocusing pulse'. The electron spin echo often is referred to by the acronym ESE.

11.6 ELECTRON SPIN-ECHO ENVELOPE MODULATION

The spin-echo signal amplitude as a function of the pulse-interval time (Fig. 11.8) can be stored in a computer. The decrease in height of (any point of) the echo signal, measured as a function of increasing Δ, conveniently yields the

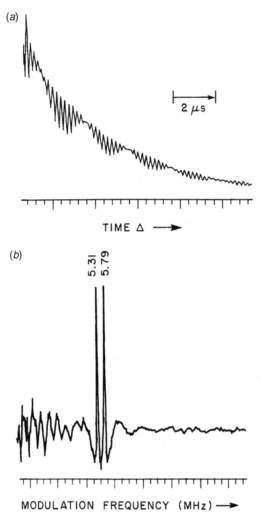

FIGURE 11.8 (*a*) The dependence of the echo amplitude ε as a function of the interval Δ between pulses in a $\pi/2-\Delta-\pi-\Delta-\varepsilon$ experiment. The modulation (ESEEM) visible, after Fourier transformation, is seen to arise from two nuclear-spin (ENDOR) transitions occurring as a narrow-line doublet, depicted in (*b*).

phase-memory relaxation time τ_m. Furthermore, in selected systems, the plot of the intensity of any convenient small region (e.g., the center) of the echo versus Δ reveals a complex superposition, as shown in Fig. 11.8*a*; that is, there is 'modulation' superimposed onto the envelope (decay curve) set by τ_m [27]. As we shall see, this effect arises from hyperfine interaction with (at least) one nuclear-spin moment in solid-state systems. It can divulge accurate and exciting information, especially about neighboring nuclei weakly coupled (due to remote location and/or small g_n) to some center bearing electron-spin density.

Once again, analysis by theory and mathematics is indispensable, and yields very valuable physical information. Using the detailed form of the spin-hamiltonian $\hat{\mathcal{H}}$ and magnetic-moment operators $\hat{\boldsymbol{\mu}}$, and with application of statistical mechanics generally and conveniently carried out via the density-matrix formalism [8–11],[3,10] one can arrive at equations for the magnetization **M** as a function of time (spacings and lengths of pulses, and of measurement periods). With an $S = I = \frac{1}{2}$ system, for example, one finds for a $\Omega_I - \Delta - \Omega_{II} - \Delta - \varepsilon$ scheme that the echo amplitude $\varepsilon(\Delta)$ after the second free-precession interval Δ is given [30] by

$$\varepsilon = (i/2)K[N_{\alpha(e)} - N_{\beta(e)}]\sin\Omega_I \sin^2(\Omega_{II}/2)\times$$

$$\left[\sum_{i,k}|\mu_{ik}|^4 + 2\sum_{i<j,k}\cos(\omega_{ij}\Delta)|\mu_{jk}|^2|\mu_{ik}|^2 + 2\sum_{i,k<n}\cos(\omega_{kn}\Delta)|\mu_{in}|^2|\mu_{ik}|^2 + \right.$$

$$2\sum_{i<j,k<n}re\left\{\cos[(\omega_{ij}+\omega_{kn})\Delta]\,\mu_{jk}^*\,\mu_{jn}\,\mu_{in}^*\,\mu_{ik}+\right.$$

$$\left.\left.\cos[(\omega_{ij}-\omega_{kn})\Delta]\,\mu_{ik}^*\,\mu_{in}\,\mu_{jn}^*\,\mu_{jk}\right\}\right] \tag{11.2}$$

Here the angles Ω are expressed in radians. Superscript * denotes complex conjugate. The second sum is a triple one, with $i < j$, and so on, with the last sum a quadruple one. The subscripts i and j refer to nuclear eigenstates associated with $M_S = +\frac{1}{2}$, while k and n are associated with $-\frac{1}{2}$. Thus the summation indices all run over the (same) range that covers the spin eigenstates: i and j in the $\alpha(e)$ nuclear-spin manifold, k and n in the $\beta(e)$ nuclear-spin manifold.[11] Angular frequencies $\omega_{ij} = (U_i - U_j)/\hbar$ are signed quantities. The proportionality factor K includes the phase factor; detection of the echo signal is done out of phase relative to the pulse sinusoids. As in cw EPR, the Boltzmann population difference ΔN is seen to be crucial.

We note from Eq. 11.2 that the echo is maximal when the turning angles are $\Omega_I = \pi/2$ and $\Omega_{II} = \pi$. Clearly and most important, the echo signal carries information about the nuclear-transition frequencies as well as sums and differences thereof. This means that hyperfine data can be obtained, with the transition frequencies available from the Fourier transform of $\varepsilon(\Delta)$. This is displayed in Fig. 11.8b.

Expressions similar to Eq. 11.2 are available [30,31] for $S = \frac{1}{2}, I = 1$ systems (in which nuclear-quadrupole effects can enter), as well as for three-pulse schemes. For instance, the so-called stimulated echo induced by the pulse sequence $\pi/2 - \Delta_1 - \pi/2 - \Delta_2 - \pi/2 - \Delta_1$ contains modulations at only the nuclear transition frequencies [i.e., in which the sum and difference frequencies (ENDOR frequencies; see Chapter 12) are absent]. Here time interval Δ_1 is held constant and Δ_2 is varied.

If several (n) nuclei contribute to the echo, one can utilize the product rule [32]

$$\varepsilon = \prod_{i=1}^{n}\varepsilon^{(i)} \tag{11.3}$$

valid since the nuclei enter independently to a very good approximation.

It is worth mentioning that one can measure the echo amplitudes at various different fixed fields **B** (or scan B slowly), and assemble from this set what can be called an 'echo-modulated' ESE-detected EPR spectrum [31].

Vector models continue to offer very useful means of visualizing angular momentum and magnetization behavior. We have so far herein used these only sparingly (see Figs. 1.8, 2.2, 11.1, 11.2 and 11.4). In the more recent past, the vector model elucidating ESE and ESEEM signals has been brought forward [33]. Basically, a quantitative description of the effects of nuclei with negligible quadrupole couplings on pulsed EPR signals was developed in terms of simple products of vectors moving in a 3D space. These vectors correspond to nuclear magnetization precessing in the combined effective B fields of the external magnet and the hyperfine interaction.

An example of the Fourier-transformed signal from a $\pi/2 - \Delta - \pi - \Delta - \varepsilon$ experiment [34] is shown in Fig. 11.9, which displays ENDOR-like peaks, including their sums and differences as predicted by Eq. 11.2. These signals arise from interstitial Li^+ ions adjacent to paramagnetic ($S = \frac{1}{2}$) Ti^{3+} ions located at Si^{4+} sites in an α-quartz single crystal. The energy-level labeling for ^7Li as well as the stick

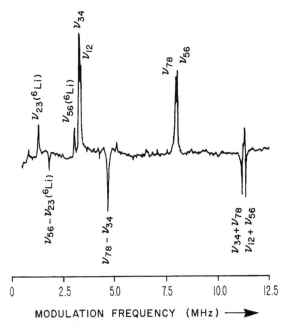

FIGURE 11.9 The Fourier transform of the ESEEM signal (9.25 GHz, at 18 K) from the Ti^{3+} defect center $[TiO_4/Li]^0$ in rose quartz, showing ENDOR-like peaks within the $S = \frac{1}{2}, I = \frac{3}{2}$ manifold for ^7Li, as well as the $S = \frac{1}{2}, I = 1$ manifold for ^6Li. The subindices are energy-level labels (Fig. 11.10). Note the presence of sum and of difference signals, predicted by Eq. 11.2. [After J. Isoya, M. K. Bowman, J. R. Norris, J. A. Weil, *J. Chem. Phys.*, **78**, 1735 (1983).]

representation of the cw hyperfine spectrum are given in Fig. 11.10. By doing a crystal rotation, the ESEEM measurements yielded the parameter matrices $\mathbf{g}$, $\mathbf{A}(^7\text{Li})$ and $\mathbf{P}(^7\text{Li})$, the latter two more accurately than those available from cw EPR. A typical angular dependence of the modulation frequencies is displayed in Fig. 11.11.

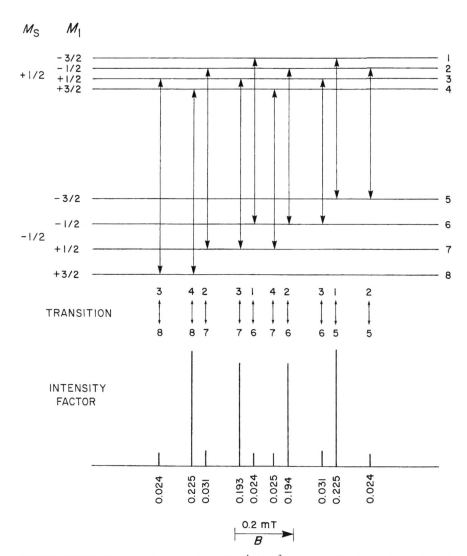

FIGURE 11.10 Energy diagram of an $S = \frac{1}{2}$, $I = \frac{3}{2}$ system with (hyperfine splitting) $A/h = +4.7$ MHz, nuclear Zeeman term $g_n\beta_nB/h = +5.7$ MHz, nuclear-quadrupole matrix $\mathbf{P}$. The levels are labeled in order of decreasing energy. The EPR line positions and intensities for $\Delta M_I = 0$, ± 1 were calculated from the spin hamiltonian. [After J. Isoya, M. K. Bowman, J. R. Norris, J. A. Weil, *J. Chem. Phys.*, **78**, 1735 (1983).]

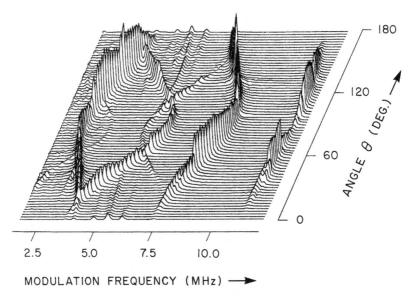

MODULATION FREQUENCY (MHz) ⟶

FIGURE 11.11 Stacked plot of ESEEM spectra (Figs. 11.9 and 11.10) displaying the angular and frequency dependence of the modulation intensity (sites 2,3; **B** ⊥ **a**$_1$; transitions 4 ↔ 8 and 2 ↔ 7). In this crystal rotation, symmetry-related sites 2 and 3 are magnetically equivalent. Near the z axis ($\theta = 0°$), the modulation is weak. The four sets of peaks with major intensities, given in increasing frequency, are the nearly superimposed ν_{23} and ν_{34} ENDOR frequencies, difference frequency $\nu_- = \nu_{78} - \nu_{34}$, ENDOR frequencies ν_{67} and ν_{78}, and sum frequency $\nu_+ = \nu_{23} + \nu_{67}$, with positive hyperfine splitting assumed. Also visible, in the range 3–8 MHz, are weak peaks ascribed to double-quantum transitions. Nuclear (^{7}Li) quadrupole splittings are visible in the θ range 25–45°. [After J. Isoya, M. K. Bowman, J. R. Norris, J. A. Weil, *J. Chem. Phys.*, **78**, 1735 (1983).]

Our second example features ESEEM spectra taken at several frequencies. The stable free radical 2,2′-diphenyl-1-picrylhydrazyl (DPPH, see Section F.2.2) in frozen solution exhibits envelope modulation [35–37], which is attributed to the ^{14}N nuclei ($I = 1$) of the nitro groups. Measurements at several microwave C-band frequencies (4–7 GHz), using loop-gap resonators, were combined with X-band (9–10 GHz) data to extract the isotropic (dominant) part [$A_{iso}/h = -1.12(8)$ MHz] of the hyperfine matrix and accurate nuclear-quadrupole parameters [$P/h = 0.280(2)$ MHz and $\eta = 0.37(3)$, defined in Section 5.6] for the *ortho*-nitro groups. The *para*-nitro group, more remote from the primary spin-density region, gives no observable modulation. The hydrazine nitrogen nuclei give no modulation because they are too strongly coupled; the nuclear Zeeman and quadrupole energies are small compared to the hyperfine term.

As the third example of the ESEEM technique, we mention the single-crystal study of perdeuteropyridine in its lowest triplet state (T_1), using d_6-benzene as the solvent [38]. This molecule, planar in its ground state, is distorted to a boat-like structure in the T_1 state, as deduced from the ^{14}N hyperfine data obtained at 1.2 K by ESE-detected EPR. ESEEM yields the matrices $\mathbf{A}(^2\mathrm{H})$ and $\mathbf{P}(^2\mathrm{H})$ and hence the spin-density distribution.

For ESEEM in systems with $S > \frac{1}{2}$, the valid analysis has become available [39], based on experiments on S-state ions Gd^{3+} and Mn^{2+} in glassy water-ethanol solutions. One notable feature here is the extremely low intensity of the proton sum combination line, now suitably explained.

The relative advantages of measuring ESEEM at S band (1.5–3.9 GHz) as compared to X band have been discussed [40]. The former offers enhancement of the modulation depth arising from weakly coupled nuclei and quadrupolar contributions.

11.7 ADVANCED TECHNIQUES

We cannot expound on the details of the exciting and rapidly developing advanced techniques utilizing various pulsed excitation schemes [34]. Many of these are well known from NMR spectroscopy [15]. Thus two-dimensional (2D) correlation spectroscopy (COSY and HETEROCOSY) and its spin-echo variant (SECSY), as well as exchange spectroscopy (EXSY), are now practiced with electron spins [41–43]. Incorporation of sudden changes in the orientation of field $\mathbf{B}$ is also proving to be very advantageous. For instance, in a uniaxial-g powder spectrum (Fig. 4.7), one can suppress the regions between the principal-axis contributions ($g_\perp$ and $g_\parallel$), allowing resolution of other co-existing spectra and hyperfine splittings [44].

A new class of pulsed EPR experiments, FID-detected hole burning, yields excellent spectral sensitivity and resolution [45]. Here a transient narrow saturation region in an inhomogeneously broadened line is created by a selective microwave pulse, and its behavior (spreading and shifting) is followed via an FID.

The simultaneous application of cw and pulsed microwaves has been found to yield coherent Raman beats (i.e., oscillating absorption and dispersion signals) capable of yielding (for example) detailed hyperfine information [46,47].

Various pulsed ENDOR and ELDOR techniques (Chapter 12) have also become important. Thus the whole realm of pulsed EPR techniques is now becoming an equal partner to cw EPR spectroscopy.

11.8 SPIN COHERENCE AND CORRELATION

The term 'coherence' is widely used in magnetic resonance spectroscopy, unfortunately in several different contexts.[12]

For one, it can mean the phase coherence between the temporal components of a set of sinusoidal excitation fields $\mathbf{B}_1$, say, in a pulse. It can also describe radiation emitted by a spin system.

In the most major context, coherence is used as a concept valuable especially in the pulsed magnetic resonance investigations of spin packets, that is, ensembles of equivalent spins in a field **B** ∥ **z**. More concisely, this term here denotes 'phase coherence' between state functions, spin states in our case [11], Chapters 7 and 9; [18], p. 215. In the present context, coherence is a generalization of polarization ($\langle \bar{J_z} \rangle \neq 0$, where the over-bar refers to the average over the ensemble). It is more appropriate to use the term 'n-quantum coherence' ($n = 0, 1, 2, \ldots$) denoting presence of spin-phase coherence(s) between specific pairs of energy eigenstates, that is, non-zero off-diagonal elements of the density matrix.[3] For example, in a two-spin system, where one spin may be nuclear, coherence between $|+,-\rangle$ and $|-,+\rangle$ (related by a flip-flop transition) is of type $n = 0$; it can be detected by pulse magnetic resonance, and even by a static magnetic susceptibility measurement. Coherence between states $|+,-\rangle$ and $|-,-\rangle$, related by the usual type of EPR transition (spin flip $\Delta M_S = \pm 1$), is of type $n = 1$ and implies the presence of transverse magnetization detectable in the usual way. Coherences of type $n \neq 1$ are not directly detectable by magnetic resonance but become so when such a coherence is transformed into an $n = 1$ type. Coherence transfers between spin states can be accomplished by use of appropriate $\mathbf{B}_1$ pulses or a sufficiently strong cw $\mathbf{B}_1$ field.

We cannot herein do full justice to the usage and importance of coherences, which became crucial first in NMR spectroscopy [48,49], and now have taken firm hold in EPR spectroscopy [22]. It is a language that EPR practitioners will have to learn.

We shall have to content ourselves with merely sketching some aspects of one, now very popular, EPR pulse technique: HYSCORE (hyperfine sublevel correlation spectroscopy). This sequence (Fig. 11.12) was first introduced by Höfer et al. in 1986 [50], and various variations of it have come to light since. This technique

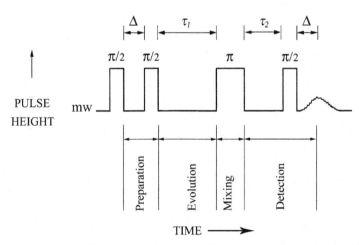

FIGURE 11.12 Idealized pulse scheme for the HYSCORE technique. The $\pi/2$ pulses could last ~ 10 ns, and the π pulse twice as long. Interval Δ could be 30–500 ns. The pulse separation τ_1 and τ_2, being stepped in ~ 20-ns increments, constitute the axes of the 2D experiment. [After J. J. Shane, P. Höfer, E. J. Reijerse, E. de Boer, *J. Magn. Reson.*, **99**, 596 (1992).]

can correlate nuclear frequencies of different M_S manifolds (same S), specifically, two ENDOR frequencies associated with a particular hyperfine interaction. The HYSCORE experiment generally consists of a two-dimensional four-pulse sequence, based on a three-pulse $(\pi/2)$ stimulated echo sequence with a fourth 'mixing' pulse inserted between the second and third ones. For instance, the theory of the nuclear coherence created by the sequence $(\pi/2)-\Delta-(\pi/2)$ has been delineated for a spin system consisting of one electron spin $S = \frac{1}{2}$ and two nuclei with spin $I = \frac{1}{2}$ [51].

To obtain the 2D time-domain modulation signal, the stimulated echo envelope ε is observed as a function of pulse separations τ_1 and τ_2 (Fig. 11.12), with fixed interval Δ. Then $\varepsilon(t)$ is double-Fourier-transformed (with respect to τ_1 and τ_2), and the resulting frequency-domain 2D plot is displayed for analysis. A third dimension in such a plot can represent the correlation peak heights.

HYSCORE can unveil presence of otherwise hard-to-detect spin-bearing nuclei existing near to the primary paramagnetic center (e.g., see Ref. 52). Single-correlation peaks (and, under some conditions, double-quantum correlation peaks from presence simultaneously of two ^{29}Si nuclides in the same magnetic center) can be observed (Fig. 11.13) in the 2D display, for example due to ^{29}Si nuclei

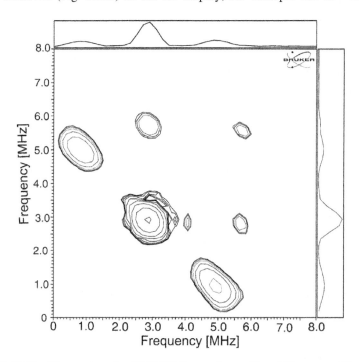

FIGURE 11.13 The results of a HYSCORE experiment ($\Delta = 470$ ns; note Fig. 11.13), showing via a 2D display the presence of two ^{29}Si nuclides (present at natural abundance: 4.7%) in γ-irradiated fused quartz, measured at 9.568 GHz (342.0 mT) and room temperature. Two minor artifacts also are present. (Courtesy of P. Höfer, Bruker Bio-Spin, Rheinstetten, Germany.)

(4.7% natural abundance) near the E_1' center created in γ-irradiated SiO_2 (fused quartz) [53,54]. This capability includes the situations in which two $S > 0$ centers are in close proximity, as for $S = \frac{1}{2}$ radicals [50] in electron-irradiated deuterated crystals of squaric acid, carbon-cyclic $C_4H_2O_4$, which tends to form numerous inter-molecular hydrogen bonds.

REFERENCES

1. T. C. Farrar, E. D. Becker, *Pulse, Fourier Transform NMR*, Academic Press, New York, NY, U.S.A., 1971.

2. M. E. Starzak, *Mathematical Methods in Chemistry and Physics*, Plenum Press, New York, NY, U.S.A., 1989, Chapter 2.

3. L. Allen, J. H. Eberly, *Optical Resonance and Two-Level Atoms*, Dover, New York, NY, U.S.A., 1987.

4. W. H. Louisell, *Quantum Statistical Properties of Radiation*, Wiley, New York, NY, U.S.A., 1973.

5. I. I. Rabi, *Phys. Rev.*, **51**, 652 (1937).

6. J. A. Weil, "On the Intensity of Magnetic Resonance Absorption by Anisotropic Spin Systems", in *Electronic Magnetic Resonance of the Solid State*, J. A. Weil, Ed., Canadian Society for Chemistry, Ottawa, ON, Canada, 1987, Chapter 1.

7. W. J. Archibald, *Am. J. Phys.*, **20**, 368 (1952).

8. U. Fano, *Rev. Mod. Phys.*, **29**, 74 (1957).

9. R. H. Dicke, J. P. Wittke, *Introduction to Quantum Mechanics*, Addison-Wesley, Reading, MA, U.S.A., 1960.

10. K. Blum, *Density Matrix Theory and Applications*, Plenum Press, New York, NY, U.S.A., 1981.

11. C. P. Slichter, *Principles of Magnetic Resonance*, 3rd ed., Springer, Berlin, Germany, 1990.

12. C. S. Johnson Jr., *Adv. Magn. Reson.*, **1**, 33 (1965).

13. R. M. Lynden-Bell, "The Calculation of Lineshapes by Density Matrix Methods", in *Progress in NMR Spectroscopy*, J. W. Emsley, J. Feeney, L. H. Sutcliffe, Eds., Pergamon, London, U.K., 1967, Chapter 2.

14. R. K. Harris, *Nuclear Magnetic Resonance Spectroscopy*, Pitman, London, U.K., 1983.

15. G. E. Pake, T. L. Estle, *The Physical Principles of Electron Paramagnetic Resonance*, 2nd ed., Benjamin, New York, NY, U.S.A., 1973, Chapter 2.

16. H. C. Torrey, *Phys. Rev.*, **76**, 1059 (1949).

17. H. Cho, S. Pfenninger, C. Gemperle, A. Schweiger, R. R. Ernst, *Chem. Phys. Lett.*, **130**, 391 (1989).

18. A. E. Derome, *Modern NMR Techniques for Chemistry Research*, Pergamon, Oxford, U.K., 1987, Chapter 8.

19. D. Shaw, *Fourier Transform NMR Spectroscopy*, 2nd ed., Elsevier, Amsterdam, Netherlands, 1984.

20. A. G. Maryasov, *Appl. Magn. Reson.*, **21**, 79 (2001).

21. M. K. Bowman, "Fourier Transform Electron Spin Resonance", in *Modern Pulsed and Continuous-Wave Electron Spin Resonance*, L. Kevan, M. K. Bowman, Eds., Wiley, New York, NY, U.S.A., 1990, Chapter 1.

22. A. Schweiger, G. Jeschke, *Principles of Pulse Electron Paramagnetic Resonance*, Oxford University Press, Oxford, U.K., 2001.

23. O. Dobbert, T. Prisner, K. P. Dinse, *J. Magn. Reson.*, **70**, 173 (1986).

24. T. Prisner, O. Dobbert, K. P. Dinse, H. van Willigen, *J. Am. Chem. Soc.*, **110**, 1622 (1988); see also A. Angerhofer, M. Toporowicz, M. K. Bowman, J. R. Norris, H. Levanon, *J. Phys. Chem.*, **92**, 7164 (1988); H. van Willigen, M. Vuolle, K. P. Dinse, *J. Phys. Chem.*, **93**, 2441 (1989); M. Pluschau, A. Zahl, K. P. Dinse, H. van Willigen, *J. Chem. Phys.*, **90**, 3153 (1989).

25. J. C. Hoch, A. S. Stern, *NMR Data Processing*, Wiley-Liss, New York, NY, U.S.A., 1996.

26. E. L. Hahn, *Phys. Rev.*, **77**, 297 (1950).

27. W. B. Mims, "Electron Spin Echoes", in *Electron Paramagnetic Resonance*, S. Geschwind, Ed., Plenum Press, New York, NY, U.S.A., 1972, Chapter 4.

28. O. W. Sørensen, G. W. Eich, M. H. Levitt, G. Bodenhauser, R. R. Ernst, *Prog. Nucl. Magn. Reson. Spectrosc.*, **16**, 163 (1983).

29. A. Rahman, *One- and Two-dimensional NMR Spectroscopy*, Elsevier, Amsterdam, Netherlands, 1989, Chapter 14.

30. M. K. Bowman, R. J. Massoth, "Nuclear Spin Eigenvalues, Eigenvectors in Electron Spin Echo Modulation", in *Electronic Magnetic Resonance in the Solid State*, J. A. Weil, Ed., Canadian Society for Chemistry, Ottawa, ON, Canada, 1987, p. 103.

31. L. Kevan, M. K. Bowman, Eds., *Modern Pulsed and Continuous-wave Electron Spin Resonance*, Wiley, New York, NY, U.S.A., 1990.

32. W. B. Mims, *Phys. Rev. B*, **5**(7), 2409 (1972).

33. A. G. Maryasov, M. K. Bowman, Yu. D. Tsvetkov, *Appl. Magn. Reson.*, **23**, 211 (2002).

34. J. Isoya, M. K. Bowman, J. R. Norris, J. A. Weil, *J. Chem. Phys.*, **78**, 1735 (1983).

35. S. A. Dikanov, A. V. Astashkin, Yu. D. Tsvetkov, *J. Struct. Chem.*, **25**, 200 (1984).

36. H. L. Flanagan, D. J. Singel, *Chem. Phys. Lett.*, **137**, 391 (1987).

37. H. L. Flanagan, G. J. Gerfen, D. J. Singel, *J. Chem. Phys.*, **88**, 20 (1988).

38. W. J. Buma, E. J. J. Groenen, J. Schmidt, R. de Beer, *J. Chem. Phys.*, **91**, 6549 (1989).

39. A. V. Astashkin, A. M. Raitsimring, *J. Chem. Phys.*, **117**(13), 6121 (2002).

40. R. B. Clarkson, D. R. Brown, J. B. Cornelius, H. C. Crookham, W.-J. Shi, R. L. Belford, *Pure Appl. Chem.*, **64**, 893 (1992).

41. J. Gorcester, G. L. Millhauser, J. H. Freed, "Two-Dimensional and Fourier-transform EPR", in *Advanced EPR—Applications in Biology and Biochemistry*, A. J. Hoff, Ed., Elsevier, Amsterdam, Netherlands, 1989, Chapter 5.

42. A. Angerhofer, R. J. Massoth, M. K. Bowman, *Isr. J. Chem.*, **28**, 227 (1988).

43. J.-M. Fauth, S. Kababya, D. Goldfarb, *J. Magn. Reson.*, **92**, 203 (1991).

44. S. Pfenninger, A. Schweiger, J. Forrer, R. R. Ernst, *Chem. Phys. Lett.*, **151**, 199 (1988).

45. T. Wacker, G. A. Sierra, A. Schweiger, *Isr. J. Chem*, **32**, 305 (1992).

46. M. K. Bowman, R. J. Massoth, C. S. Yannoni, "Coherent Raman Beats in Electron Paramagnetic Resonance", in *Pulsed Magnetic Resonance: NMR, ESR, Optics*, D. M. S. Bagguley, Ed., Clarendon, Oxford, U.K., 1992, pp. 423–445.

47. M. K. Bowman, *Isr. J. Chem.*, **32**, 339 (1992).

48. A. Bax, *Two-dimensional Nuclear Magnetic Resonance in Liquids*, Reidel, Dordrecht, Netherlands, 1984.

49. R. R. Ernst, G. Bodenhausen, A. Wokaun, *Principles of Nuclear Magnetic Resonance in One and Two Dimensions*, Clarendon Press, Oxford, U.K., 1987.

50. P. Höfer, A. Grupp, H. Nebenführ, M. Mehring, *Chem. Phys. Lett.*, **132**(3), 279 (1986).

51. L. Liesum, A. Schweiger, *J. Chem. Phys.*, **114**(21), 9478 (2001).

52. G. Korda, *J. Non-Cryst. Solids*, **281**, 133 (2001).

53. I. Gromov, B. Glass, J. Keller, J. Shane, J. Forrer, R. Tschaggelar, A. Schweiger, *Concepts Magn. Reson. Part B*, **21B**(1), 1 (2004).

54. A. V. Astashkin, A. Raitsimring, *J. Magn. Reson.*, **148**, 379 (2001).

55. G. Whitfield, A. G. Redfield, *Phys. Rev.*, **106**, 918 (1957).

NOTES

1. Sometimes it is convenient to express the amplitude B_1 in terms of an angular frequency $\omega_1 = 2\pi\nu_1 = |\gamma|B_1 = g\beta B_1/$. Frequency ν_1 should not be confused with the frequency ν of the oscillating field $\mathbf{B}_1$.

2. For $J = \frac{1}{2}$, $(h\nu_1)^2 = 4|<+1/2|\hat{\mathcal{H}}'| - 1/2 >|^2$, where $\hat{\mathcal{H}}'$ is the time-independent amplitude of $\hat{\mathcal{H}}_1 = -\mathbf{B}_1{}^T \cdot \hat{\boldsymbol{\mu}}$ (Eq. 4.21).

3. The density matrix for a state ψ is $n \times n$ hermitian, where n is the total number of energy eigenstates $|k\rangle$. Its ij element is the product $\overline{c_i c_j{}^*}$ of expansion coefficients of the basis states making up the state $\psi = \sum_{k=1}^{n} c_k |k\rangle$ considered. The over-bar (vinculum) denotes averaging over a statistically adequate ensemble. The trace $\sum_k \overline{|c_k|^2}$ is unity. The individual numbers in this sum are the relative populations of the spins in the individual states $|k\rangle$. The matrix is very useful in evaluating average properties (e.g., magnetization components of a spin system) of the ensemble, including their temporal behavior. In magnetic resonance, for instance, dynamic lineshapes and magnetization responses to $\mathbf{B}_1$ pulses are tractable with this mathematical approach (see Refs. 8–10; 11, Chapter 5; 12,13 for details].

4. Fourier analysis reveals that the cut-off step for $\mathbf{B}_1$ creates field components at frequencies spread according to τ^{-1}, centered about the nominal frequency ν [14, p. 74].

5. The phase-memory (spin-coherence) relaxation time τ_m may be defined as the time, as measured from the pulse end, needed for the FID or echo to fall to $1/e$ of its maximum value. This process is not necessarily exponential. Time τ_m need not be equal to the spin-spin relaxation time τ_2 because it may be affected by differences in the local field, including inhomogeneities in field $\mathbf{B}$. Typically for electrons, τ_m is of the order of microseconds. Note that $\tau_m < \tau_1$; that is, the observability of $M_\perp$ is limited by τ_1.

6. There is no necessity for the pulse angle $\Omega(\tau)$ to be exactly $\pi/2$; that is, this angle is optimal for our purposes but the physics described is the same for $0 < \Omega < \pi$.

7. Because of instrumental effects, detection can only begin after a dead-time interval following the end of the pulse. However, special techniques can overcome this problem [17].

8. A special pulse-related jargon finds much use when several magnetic resonance signals are at hand. Thus, a 'hard' pulse is one sufficiently strong to move all the spins away from their previous equilibrium situations, that is, it affects all the spins and their various transitions equally. Such a pulse must have relatively high microwave or radio frequency (rf) power and a relatively short duration. By contrast, a 'soft' pulse (alias 'selective' pulse) is weaker (has lower power and lasts longer) and thus is set to excite only one signal from within the spectrum.

9. The simultaneous (two-mixer) or alternating measurement of both signals is called 'quadrature detection' (Sections E.1.8 and E.2; also Ref. 11, p. 195). Both data sets are needed to obtain complete information, that is, a faithful representation of the absorption and dispersion signals.

10. Another type of analysis, using the product (spin) operator approach, has also been developed [28,29]. It is applicable for systems of not too strongly coupled spins, for pulses equally affecting all transitions ('hard' pulses).

11. The magnetic-moment matrix elements present are those connecting nuclear-spin states for opposite electron-spin configurations.

12. Not least, one must distinguish between spatial and temporal coherence. Spatial coherence described the correlation or predictable relationship between signals at different points in 3-space, whereas temporal coherence does the analogous for signals at different moments in time.

FURTHER READING

1. D. M. S. Bagguley, Ed., *Pulsed Magnetic Resonance: NMR, ESR and Optics*, Oxford University Press, Oxford, U.K., 1992.

2. E. O. Brigham, *The Fast Fourier Transform*, Prentice-Hall, Englewood Cliffs, NJ, U.S.A., 1974.

3. D. C. Champeney, *Fourier Transforms in Physics*, Hilger, Bristol, U.K., 1983.

4. S. A. Dikanov, Yu. D. Tsvetkov, *Electron Spin Echo Envelope Modulation (ESEEM) Spectroscopy*, CRC Press, Boca Raton, FL, U.S.A. (1992).

5. C. P. Keijzers, E. J. Reijerse, J. Schmidt, Eds., *Pulsed EPR: A New Field of Applications*, North Holland, Amsterdam, Netherlands, 1989.

6. L. Kevan, M. K. Bowman, Eds., *Modern Pulsed and Continuous-wave Electron Spin Resonance*, Wiley, New York, NY, U.S.A., 1990.

7. L. Kevan, R. N. Schwartz, Eds., *Time-Domain Electron Spin Resonance*, Wiley, New York, NY, U.S.A., 1979.

8. T.-S. Lin, Electron Spin Echo Spectroscopy of Organic Triplets, *Chem. Rev.*, **84**, 1 (1984).

9. J. D. Macomber, *The Dynamics of Spectroscopic Transitions*, Wiley, New York, NY, 1976.

10. W. B. Mims, "ENDOR Spectroscopy by Fourier Transformation of the Electron Spin Echo Envelope", in *Fourier, Hadamard and Hilbert Transforms in Chemistry*, A. G. Marshall, Ed., Plenum Press, New York, NY, U.S.A., 1982, pp. 307–322.

11. A. Schweiger, "Pulsed Electron Spin Resonance: Basic Principles, Techniques, and Applications", *Angew. Chemie. (Int. Ed. Engl.)* **30**, 265 (1991) (a fine, vintage review).

12. M. Weissbluth, *Photon-Atom Interactions*, Academic Press, Boston, MA, U.S.A., 1989, Chapter 3.

PROBLEMS

11.1   The expectation value $\langle \mathbf{\mu} \rangle = \langle M | \hat{\mathbf{\mu}} | M \rangle$ of the magnetic moment for a spin in a state M obeys the differential equation

$$\frac{d\langle \mathbf{\mu} \rangle}{dt} = \gamma \langle \mathbf{\mu} \rangle \wedge \mathbf{B} \tag{11.4}$$

in the laboratory coordinate system (cf. Eqs. 10.14 and B.75). Consider an applied magnetic field (in terms of cartesian unit vectors $\mathbf{x}$, $\mathbf{y}$, $\mathbf{z}$)

$$\mathbf{B} = B_1 \cos \omega t \, \mathbf{x} + B_1 \sin \omega t \, \mathbf{y} + B_0 \, \mathbf{z} \tag{11.5}$$

composed of a field $\mathbf{B}_1$ rotating about a static field $\mathbf{B}_0$ and normal to it, with $B_1 \ll B_0$. Let us treat the case of resonance, that is, $\omega = \omega_B = |\gamma| B_0$. Transform to the rotating coordinate system

$$\mathbf{x}_\phi = \cos \, \omega t \, \mathbf{x} + \sin \, \omega t \, \mathbf{y} \tag{11.6a}$$

$$\mathbf{y}_\phi = - \sin \, \omega t \, \mathbf{x} + \cos \, \omega t \, \mathbf{y} \tag{11.6b}$$

Show that

(*a*) $\mathbf{B}_1$ is parallel to $\mathbf{x}_\phi$

(*b*)   Equation 11.4 now has the form

$$\frac{d\langle \mathbf{\mu} \rangle}{dt} = \langle \mathbf{\mu} \rangle \wedge (\gamma \mathbf{B} - \omega \mathbf{z}) \tag{11.7}$$

(*c*)   Equation 11.7 has the simple solution

$$\langle \mathbf{\mu} \rangle = \langle \mathbf{\mu} \rangle_0 \left\{ \sin[\omega_1(t - t_0)] \, \mathbf{y} - \cos[\omega_1(t - t_0)] \, \mathbf{z} \right\} \tag{11.8}$$

when $\langle \mathbf{\mu} \rangle$ is antiparallel to the static field at time $t = t_0$. Here $\omega_1 = |\gamma| B_1$. Compare this view of the motion with that for $\langle \mathbf{\mu} \rangle$ in the laboratory frame (Fig. 11.2). If you wish, solve for $\langle \mathbf{\mu} \rangle$ in this coordinate system.

11.2   Calculate the length (in seconds) needed to generate a $90°$ pulse when $B_1 = 1.0$ mT, for (*a*) proton and (*b*) an unpaired free electron.

11.3 Consider the two functions

$$f(x) = \frac{1}{\sqrt{2\pi}} \int_{-\infty}^{\infty} g(k)e^{ikx}\, dk \tag{11.9a}$$

$$g(k) = \frac{1}{\sqrt{2\pi}} \int_{-\infty}^{\infty} f(x)e^{-ikx}\, dx \tag{11.9b}$$

which are assumed, because of this interrelation, to be Fourier transforms of each other. What is the Fourier transform of the following?

(*a*) The gaussian function $f(x) = A \exp(-c^2 x^2)$. Note that

$$\int_{-\infty}^{\infty} \exp(-\lambda x^2 + i\mu x)\, dx = \left(\frac{\pi}{\lambda}\right)^{1/2} \exp(-\mu^2/4\lambda) \tag{11.10}$$

(*b*) A finite cosine pulse

$$f(t) = \begin{cases} f_0 \cos 2\pi vt & \text{for } |t| < t_0, f_0 \neq 0 \qquad (11.11a) \\ 0 & \text{for } |t| > t_0. \qquad\qquad (11.11b) \end{cases}$$

11.4 Translate the expression (e.g., Eq. 10.26*b*)

$$\chi'' = \frac{(\chi'')_0}{1 + [B_0 - B)/\Gamma]^2} \tag{11.12}$$

into angular-frequency units, assuming linearity between B and ω. Then obtain the transverse magnetization via the Fourier transform

$$M_\perp \propto \int_{-\infty}^{\infty} \chi'' \exp[i\omega(t - t_0)]\, d\omega \tag{11.13a}$$

Assume that $M_\perp(t) = M_{0\perp}$ at time t_0. The relevant integral is

$$\int_{-\infty}^{\infty} \frac{\exp(-ipx)}{a^2 + x^2}\, dx = \frac{\pi}{a} \exp(-|ap|) \tag{11.13b}$$

11.5 Derive a value for the longitudinal relaxation time τ_1 (in seconds) from the magnetization (M_z) recovery curve given in Fig. 11.14. Time t_0 marks the end of a 90° pulse.

11.6 The magnetization M_z (of a paramagnetic sample) perpendicular to the plane of a circularly polarized magnetic field $\mathbf{B}_1$ (of angular frequency ω)

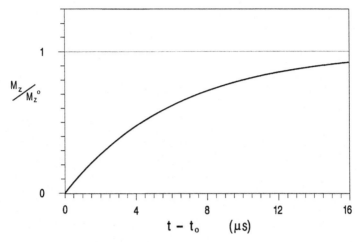

FIGURE 11.14 Recovery of the longitudinal magnetization M_z toward its equilibrium value M_z^0. Here $t = t_0$ indicates the end time of a $\pi/2$ pulse.

is given [55] by

$$M_z = \kappa\mu_0\chi^0 \frac{(\tau_1\Delta\omega)(\gamma B_1\tau_1)B_1 + [1 + (\tau_1\Delta\omega)^2]B_z}{1 + (\gamma B_1\tau_1)^2 + (\tau_1\Delta\omega)^2} \qquad (11.14)$$

where χ^0 is the static magnetic susceptibility (Eqs. 1.16), B_z is an arbitrary static magnetic field, and $\Delta\omega = \gamma B_z - \omega$. We assume that $\tau_1 = \tau_m$ for the sample.

(a) Derive a formula for the difference $\Delta M_z = \pm(\kappa\mu_0\chi^0 B_z - M_z)$ for the magnetizations when $\mathbf{B}_1$ is absent and when it is present (i.e., the amplitude modulation of the field, inducing a voltage in a pick-up coil placed along $\mathbf{z}$).

(b) For the stable free radical 2,2′-diphenyl-1-picrylhydrazyl (DPPH): $g = 2.0035$ and $\tau_1 = \tau_m = 6.2 \times 10^{-8}$ s: Plot $\Delta M_z/\chi^0$ versus B_1 when $B_z = 0$ and also when B_z has a resonant value, say, at $\nu = 20$ MHz and at 10 GHz.

CHAPTER 12

DOUBLE-RESONANCE TECHNIQUES

12.1 INTRODUCTION

In this chapter, we briefly introduce several important techniques that represent elaborations of standard cw and pulse EPR spectroscopy. The experimental aspects of these techniques are described, for instance, by Poole [1].

It was noted in Section 3.3 that EPR may at times be used to characterize the nucleus responsible for hyperfine splitting. However, if in the same system there are two or more nuclei of the same spin, there is ambiguity in the assignment of hyperfine multiplets. Indeed, some hyperfine multiplets in spectra reproduced in this book were originally assigned to the wrong nucleus. Furthermore, if the spacings within a set of hyperfine lines do not exceed their individual widths, one fails to detect this splitting, except perhaps for a broadening. For this reason, splittings arising from the more remote nuclei are rarely observed directly. It would seem that in such EPR spectra one must be resigned to the loss of details of hyperfine interaction. This indeed appeared to be the case until 1956, when Feher [2] proposed and demonstrated the technique of *electron-nuclear double resonance* (ENDOR).[1] His brilliant contribution makes it possible in some cases to obtain otherwise missing details of hyperfine interaction. In many systems, the ENDOR technique completely removes ambiguities. It may provide a wealth of detail about the wavefunction of the unpaired electron. In one favorable case, a distinctive interaction of an unpaired electron with the 23rd nearest-neighbor set of nuclei was established by an ENDOR experiment [4]. The ENDOR technique is applicable whenever any of the spin-bearing nuclides (Table H.4) are present in the paramagnetic sample.

Electron Paramagnetic Resonance, Second Edition, by John A. Weil and James R. Bolton
Copyright © 2007 John Wiley & Sons, Inc.

12.2 A CONTINUOUS-WAVE ENDOR EXPERIMENT

Before undertaking a more detailed description of ENDOR processes, we present a brief phenomenological account of a simple continuous-wave ENDOR experiment on a solid-state system with $S = I = \frac{1}{2}$. Suppose that from the resonant field positions B_k and B_m (Fig. 2.4) of the two hyperfine lines we have established a g factor and have calculated the hyperfine splitting A/h to be, for example, 10 MHz. We now undertake an ENDOR experiment as follows:

1. The sample is placed in a special cavity (one type is shown in Fig. 12.1). At low microwave power, the magnetic field is scanned to find and be set on one transition, say, that at B_k. The spectrometer parameters are now optimized to maximize the EPR signal.

2. The microwave power level (B_{1e}) is then increased to several-fold its level set in part 1, to achieve partial saturation.

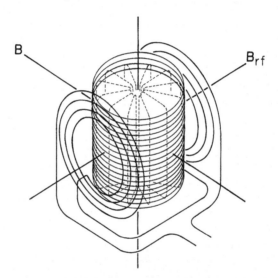

FIGURE 12.1 Schematic reproduction of a TE_{011} cylindrical cavity designed for ENDOR studies. The sample is placed along the axis of the cavity. The side wall is a helix of spaced turns, with interstices filled by a plastic material of low dielectric loss. This design allows for penetration into the cavity of the radiofrequency field and by a modulating magnetic field. The microwave magnetic-field contours are shown by dashed lines. Since it is the component perpendicular to the static field **B** of the microwave field or of the rf field that induces the electron-spin and nuclear-spin transitions, one seeks to keep **B** perpendicular to both. The relative orientation of the microwave and rf fields is in principle arbitrary; that shown here is the most efficient for a set of external coils and involves the least eddy-current loss. In other cavities, the rf field is introduced by a coil (at least partly) inside the cavity. To avoid coupling out microwave energy, the plane of the coil should be parallel to the microwave field. This automatically puts the microwave and the rf magnetic field at right angles to one another. It may be of crucial importance to align the rf field appropriately with respect to a crystal axis (Section 12.4). [After J. S. Hyde, *J. Chem. Phys.*, **43**, 1806 (1965).]

3. A radiofrequency (*rf*) generator of sufficiently wide range and large power output is connected to the side coils[2] of the cavity of Fig. 12.1, so that the sample also experiences an oscillating rf magnetic field $\mathbf{B}_{1n}$. The generator is set to scan the region 2–30 MHz, while the oscilloscope or recorder of the spectrometer is operating. The base line, which, apart from noise, is constant and indicates a constant EPR absorption. Although it may be non-linear, the horizontal axis is a measure of the frequency of the rf generator.

At two frequencies of the rf generator, which we call ν_{n1} and ν_{n2}, the recorder or scope traces out lines such as those shown in Fig. 12.2. This plot of changes in the EPR absorption intensity is called the *ENDOR spectrum*. If the frequencies of these lines are carefully measured as the peak of each line is traversed, it should be noted that the difference $\nu_{n2} - \nu_{n1}$ is numerically just equal to the hyperfine coupling A/h (i.e., 10 MHz), assuming that the higher-order corrections are negligible. This parameter can now be determined with greater accuracy. Furthermore, the mean value of the frequencies ν_{n1} and ν_{n2} is close to $\nu_n = g_n\beta_nB_k/h$, the NMR frequency of the bare nucleus in the magnetic field B_k. If the nuclide responsible for the hyperfine splitting had previously been uncertain, its identity is now available from the value of g_n (Table H.4). If the experiment is repeated, but with the magnetic field set at B_m, the ENDOR spectrum would again consist of two lines, separated in frequency by the hyperfine coupling and symmetrically disposed (in first order) about the NMR frequency of the nucleus in the field B_m. However, the relative intensities of the two lines may not be the same in the two ENDOR spectra.[3]

Since the ENDOR lines typically represent a change in EPR line intensity of $\sim$1% of the EPR line under non-saturated conditions, one requires a spectrometer of high sensitivity. There are also various complexities of the ENDOR spectrometer that we have not enumerated. However, the method is well justified for the following cases:

1. Hyperfine lines are not resolved in the EPR spectrum, or the spectrum is complicated, with many lines.

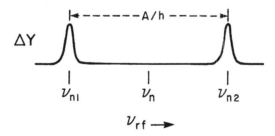

FIGURE 12.2   Change ΔY in the EPR signal amplitude, that is, 'ENDOR lines' for a system with $S = I = \frac{1}{2}$ as the radiofrequency generator is scanned through the region including the frequencies ν_{n1} and ν_{n2}. These are separated by the hyperfine coupling $|A|$ (to first order), and are symmetrically spaced about the nuclear magnetic resonance frequency ν_n of the nucleus for the magnetic field at which the microwave saturation is being carried out.

2. Hyperfine lines are resolved, but more accurate values of the hyperfine couplings are desired.

3. The identity of an interacting nucleus is to be established.

4. Nuclear-quadrupole couplings are to be measured (in a system with $I \geq 1$).

The so-called steady-state ENDOR experiment outlined briefly here is considered in more detail in Section 12.4, after considering the energy levels and possible transitions of this system.

12.3   ENERGY LEVELS AND ENDOR TRANSITIONS

In ENDOR experiments there is no attempt to observe directly the absorption of rf power (ordinary NMR) at these frequencies. Rather, one observes the enhancement of the EPR transition intensity resulting from the redistributions of populations of the various states. A description of the ENDOR lines of Fig. 12.2 (or of more complicated ENDOR spectra) requires

1. Use of the complete spin hamiltonian, including the nuclear Zeeman term (and a quadrupole term if $I > 1$).

2. Consideration of the populations of each state at low microwave power, also under microwave saturation conditions, and during (or immediately after) passage through one of the frequencies ν_{n1} or ν_{n2} at a high rf power level. The relative populations (and thus the ENDOR line behavior) depend on the dominant relaxation mechanisms in the system. The relaxation aspects are considered in Section 12.4.

The spin hamiltonian is (Eqs. 5.7 and 5.50)

$$\hat{\mathcal{H}} = \beta_e \mathbf{B}^T \mathbf{g} \hat{\mathbf{S}} + \sum_i (\hat{\mathbf{S}}^T \mathbf{A}_i \, \hat{\mathbf{I}}_i - g_{xi}\beta_n \mathbf{B}^T \hat{\mathbf{I}}_i + \hat{\mathbf{I}}_i^T \mathbf{P}_i \, \hat{\mathbf{I}}_i) \tag{12.1}$$

where the sum is over all spin-bearing nuclei in the sample. The terms describing interactions with the excitation fields $\mathbf{B}_{1e}$ and $\mathbf{B}_{1n}$ are not needed, for our purposes.

It is convenient to begin with a fixed magnetic field and a fixed orientation of a single-crystal sample for a single nuclide (assuming that the hyperfine coupling is not too extensive, and temporarily neglecting any nuclear-quadrupole term present) such that the effective values g and A may be used in the simplified spin hamiltonian of Eq. C.1:

$$\hat{\mathcal{H}} = g\beta_e \mathbf{B}^T \hat{\mathbf{S}} + A\hat{\mathbf{S}}^T \hat{\mathbf{I}} - g_n \beta_n \mathbf{B}^T \hat{\mathbf{I}} \tag{12.2}$$

The energy levels are (Eqs. C.33)

$$U_{+(1/2),M_I} = +\tfrac{1}{2}g\beta_e B + \left(\tfrac{1}{2}A - g_n\beta_n B\right)M_I \qquad (12.3a)$$

$$U_{-(1/2),M_I} = -\tfrac{1}{2}g\beta_e B - \left(\tfrac{1}{2}A + g_n\beta_n B\right)M_I \qquad (12.3b)$$

These first-order levels for $I = \tfrac{1}{2}$ are shown in Figs. 2.4 and 12.3a, b; in the latter figure, the nuclear transitions at frequencies ν_{n1} and ν_{n2} corresponding to the selection rules $\Delta M_S = 0$, $\Delta M_I = \pm 1$, are also shown. Figure 12.3c shows the levels and the transitions at constant microwave frequency. From Eqs. 12.3, one has the rf frequencies[4]

$$\nu_{n_1} = h^{-1}\left|U_{+(1/2),M_I} - U_{+(1/2),M_I-1}\right| \qquad (12.4a)$$

$$= h^{-1}\left|\tfrac{1}{2}A - g_n\beta_n B\right| \qquad (12.4b)$$

Likewise

$$\nu_{n_2} = h^{-1}\left|U_{-(1/2),\,M_I} - U_{-(1/2),\,M_I-1}\right| \qquad (12.5a)$$

$$= h^{-1}\left|\tfrac{1}{2}A + g_n\beta_n B\right| \qquad (12.5b)$$

Here $|g_n|\beta_n B/h = \nu_n$ is the magnetic resonance frequency of nucleus n in the fixed magnetic field at which the ENDOR spectrum is being observed. The two terms on the right in Eqs. 12.4b and 12.5b represent the two magnetic-field contributions seen by the nucleus: A arises from the unpaired-electron distribution and ν_n arises from the externally applied field. It is important to note that the latter is shiftable, that is, via a change of the EPR frequency (say, from X band to Q band), allowing improvements in resolution. Note that Eqs. 12.4b and 12.5b do not depend on I or M_I.

The two principal results derivable from the magnitudes ν_{n1} and ν_{n2} are as follows:

1. Determination of the hyperfine coupling parameter A. To first order, one obtains

$$\left|\nu_{n_1} \mp \nu_{n_2}\right| = h^{-1}|A| \qquad (12.6)$$

The upper sign applies when $|A| < 2\nu_n$, and the lower sign applies when $|A| > 2\nu_n$. However, unless A is very small, one must use at least a second-order correction. The results cited in Problem 12.2 required fourth-order corrections to match the accuracy of the data. For large hyperfine couplings, one generally resorts to a more general computer solution for the spin-hamiltonian parameters. The greatly increased accuracy of measurement of hyperfine couplings from ENDOR frequencies in

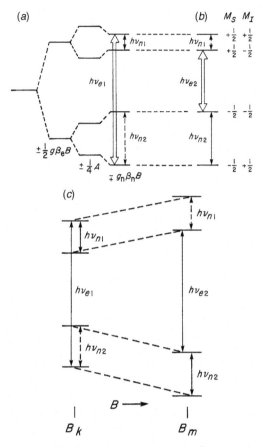

FIGURE 12.3 Energy levels of a system with $S = I = \frac{1}{2}$ in a constant magnetic field. The usual EPR transitions corresponding to the selection rules $\Delta M_S = \pm 1$, $\Delta M_I = 0$ are shown with wide arrows to symbolize the application of higher than usual microwave power. The transitions at the frequencies ν_{n1} and ν_{n2} correspond to the selection rules $\Delta M_S = 0$, $\Delta M_I = \pm 1$. The solid lines represent nuclear transitions that give rise to ENDOR lines if there is only one cross-relaxation process, represented by τ_x (Section 12.4). The dashed transitions also result in ENDOR lines if a second cross-relaxation process is operative. (a) Microwave saturation of the EPR transition $M_I = +\frac{1}{2}$ ($h\nu_{e1}$). (b) Microwave saturation of the EPR transition $M_I = -\frac{1}{2}$ ($h\nu_{e2}$). (c) Energy levels at constant microwave frequency. For the simplest assumptions about relaxation paths in steady-state ENDOR, the partially saturated transition at the field B_k is enhanced by simultaneous irradiation with high rf power at the frequency ν_{n1}. The line at the field B_m is enhanced if the second frequency is ν_{n2}. In some systems, precisely this behavior is observed; however, more typically, enhancement of either line occurs both at ν_{n1} and at ν_{n2}. Since one observes the enhancement as the rf field is scanned, the recorder traces out 'ENDOR lines'.

inhomogeneously broadened spectra is possible by virtue of the (usually) relatively much narrower ENDOR lines. The latter often have widths of $\sim$10 kHz, although they have been observed to range from 3 kHz to $\geq$1 MHz. EPR lines in solids are considered to be narrow if their width ΔB is $\leq$0.1 mT; for $g = 2.00$, this implies a linewidth

$$\Delta \nu = h^{-1} g_e \beta_e \Delta B \cong 2.80 \text{ MHz} \tag{12.7}$$

If an NMR line of a proton ($g_n = 5.5857$) has a width of 0.1 mT, the corresponding frequency width is 4.26 kHz. It is apparent that the smallest observed ENDOR line-widths correspond approximately to typical NMR linewidths. Hence it is not unusual to obtain hyperfine couplings to about 10^{-3} % accuracy, from ENDOR frequencies.

2. Determination of an approximate value of $|g_n|$ from the relation

$$\nu_{n_1} \pm \nu_{n_2} = 2\nu_n = 2 h^{-1} |g_n| \beta_n B \tag{12.8}$$

The upper sign applies when $|A| < 2\nu_n$, and the lower sign applies when $|A| > 2\nu_n$. Even a low-accuracy measurement of the ENDOR frequencies generally permits identification of the nucleus responsible for the hyperfine splitting. In favorable cases, $|g_n|$ may be determined with a precision of 0.1%; however, even with the use of higher-order corrections or computer solution of the spin hamiltonian, one may note a discrepancy between the calculated value of $|g_n|$ and that derived from a table of nuclear moments (Table H.4). The discrepancy may arise from a *pseudo*-nuclear Zeeman interaction; in some cases (e.g., for ions with low-lying excited states) the contribution to g_n from this source is appreciable [3, p. 38].

It may occur to the reader to inquire as to why an elaborate ENDOR experimental system is used to detect transitions between nuclear-spin levels, instead of performing an ordinary NMR experiment. The answer is that the concentration of the nuclei present in most EPR or ENDOR experiments is much too low to permit their NMR detection. The far greater ENDOR sensitivity arises from the following reasons:

1. The energy of the EPR quantum is much greater than that of the NMR quantum. Hence one may have much greater population differences for the more widely spaced levels.
2. The rate of energy absorption is far greater at microwave frequencies. (See Section F.3 for a discussion of sensitivity versus frequency.)
3. The effectiveness of a nucleus in altering the intensity of an EPR line during an ENDOR experiment arises from the fact that it is acting not merely in the applied magnetic field but also in the magnetic field of the electron, which is typically of the order of $10^3 - 10^5$ mT at a nucleus (Problem 12.4). One may thus generate greater population differences than would be possible if the

external magnetic field governed these differences. This phenomenon is referred to as 'enhancement' of the effective magnetic field.

12.4   RELAXATION PROCESSES IN STEADY-STATE ENDOR[5]

The phenomenon here referred to as steady-state ENDOR was termed 'stationary' ENDOR and was first described in detail by Seidel [5]. The usual EPR experiment involves only the one spin-lattice relaxation time τ_{1e} (denoted as τ_1 in Section 10.3.3; if this is very short at 300 or 77 K, one is compelled to make EPR observations at 20 K or even at 4 K). However, even in the simplest four-level system on which ENDOR observations are to be made, there are at least three spin-lattice relaxation times that govern the distribution of population in the several levels. These dictate not only the temperature range in which ENDOR experiments may be performed successfully but possibly also other experimental conditions and hence determine the nature of the observed spectrum. This sensitive dependence on temperature is a disadvantage of the ENDOR technique.

Besides τ_{1e}, one is concerned with relaxation times τ_{1n} and τ_x. Here τ_{1n} is the nuclear spin-lattice relaxation time, associated with the transitions $\Delta M_S = 0$, $\Delta M_I = \pm 1$, and τ_x is a 'cross-relaxation' time associated with mutual 'spin flips', that is, processes for which $\Delta(M_S + M_I) = 0$. Usually, $\tau_{1e} \ll \tau_x \ll \tau_{1n}$. In the absence of microwave or radiofrequency fields, the reciprocals of these times represent the rates of transition between the levels that they connect (Fig. 12.4a).

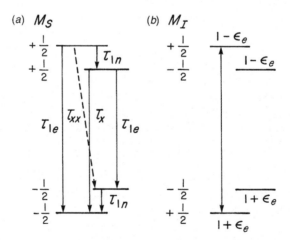

FIGURE 12.4   (a) Relaxation paths for a system with $S = I = \frac{1}{2}$, which are indicated by arrows and labeled by their relaxation times as follows: electron spin-lattice relaxation time τ_{1e}; nuclear spin-lattice relaxation time τ_{1n}; cross-relaxation time τ_x for $\Delta(M_S + M_I) = 0$ processes; cross-relaxation time τ_{xx} for $\Delta(M_S + M_I) = \pm 2$ processes. (b) Relative state populations in the absence of a microwave magnetic field (or in the presence of a very weak microwave field).

For most solid-state systems, one requires temperatures of the order of 4 K to do a successful ENDOR experiment. At these temperatures, one may achieve microwave saturation with modest power because τ_{1e} is relatively long. A lengthened value of τ_{1e} also makes it possible for the nuclear ($\Delta M_I = \pm 1$) transitions to compete with the $\Delta M_S = \pm 1$ transitions. In extreme cases, for example, phosphorus-doped silicon, τ_{1e} is of the order of hours; however, more typically, it is a small fraction of a second. Indeed, if the ENDOR linewidths are of the order of 10 kHz, corresponding to $\tau_2 = 10^{-5}$ s, the value of τ_{1e} can be no shorter if it is not to contribute to broadening from spin-lattice relaxation. With τ_{1e} and τ_2 of this order of magnitude, and if τ_x is not too long, one may hope to do a steady-state ENDOR experiment, that is, to observe ENDOR lines that may be traversed at an arbitrarily slow rate and re-traversed an indefinite number of times. (By contrast, in rapid-passage types of ENDOR experiments, one observes an ENDOR signal only during a rapid traverse, and in the process, there is an equalization of populations so that an immediate re-traverse gives no evidence of an ENDOR line.) This designation 'steady state' may not be fully accurate, if the relaxation time τ_{1n} is very long.

It is now profitable to consider in greater detail the steady-state ENDOR experiment outlined in Section 12.1 for a system with $S = I = \frac{1}{2}$. One commences by optimizing the intensity of an inhomogeneously broadened EPR line, after which one sets the field at the center of the EPR absorption line, for example, the line at B_k in Fig. 12.3c.[5] The microwave power is increased somewhat beyond the value at which the intensity of a homogeneously broadened line would be a maximum (Fig. F.8a). The optimum value of the microwave magnetic field B_{1e} for steady-state ENDOR observations should be such that $\gamma_e^2 B_{1e}^2 \tau_1 \tau_2 \cong 3$ [3, p. 244]. Here g_e is the magnetogyric ratio of the electron, and B_{1e} is the amplitude of the microwave magnetic field. The value of τ_2 to be used here is that appropriate to the single spin packet (Section 10.4.2) being saturated [3, p. 264]. Now the power level of the radiofrequency generator (and thus B_{1n}) is set sufficiently high so that the rate $(dN/dt)\uparrow$ of induced upward transitions at frequency ν_{n1} is large in comparison with τ_x^{-1}, that is, $(dN/dt)\uparrow \tau_x \geq 1$. Stated another way, one requires a large value of B_{1n} because the $\Delta M_S = 0$, $\Delta M_I = \pm 1$ transitions must be able to compete with the $\Delta(M_S + M_I)$ transitions corresponding to the cross-relaxation path measured by τ_x. When the rf frequency passes through the value ν_{n_1}, an ENDOR line is observed. In many four-level systems, one also observes a second ENDOR line, when the frequency ν_{n2} is traversed. If the only effective relaxation paths were those thus far assumed, it would be necessary to saturate the line at the field B_m, after traversing the frequency ν_{n1}, before the ENDOR line at ν_{n2} can be detected.

We now turn to consideration of the population of the levels ($S = I = \frac{1}{2}$) under various conditions. In the absence of a magnetic field, the population of each of the four (almost) degenerate levels would be (almost) $N/4$, where N is the total number of unpaired electrons. In the presence of a magnetic field and ignoring hyperfine effects, the populations of the states are

$$M_S = +\tfrac{1}{2}: \quad N_{+1/2} = \tfrac{1}{4}N \exp[-g_e\beta_e B/(2k_bT)] \cong \tfrac{1}{4}N(1 - \varepsilon_e) \qquad (12.9a)$$

and

$$M_S = -\tfrac{1}{2}: \quad N_{-1/2} = \tfrac{1}{4}N \exp[+g_e\beta_e B/(2k_bT)] \cong \tfrac{1}{4}N(1 + \varepsilon_e) \qquad (12.9b)$$

Here $\varepsilon_e = g_e\beta_e B/2k_bT$.[6] If the $M_I = +\tfrac{1}{2}$ EPR transition (Fig. 12.4) is activated by the microwave field, the only effective relaxation path is that indicated by τ_{1e}; the path via τ_x is ineffective, since the much longer τ_{1n} is in series with it.

At this point, although our steady-state ENDOR experiment involves only *partial* saturation of the electron-spin transition, the discussion is simplified by assuming equalization of the populations of the states $M_I = +\tfrac{1}{2}$, as indicated in Fig. 12.5a. For complete saturation of the transition between the states with $M_I = -\tfrac{1}{2}$, the populations are those given in Fig. 12.5b. It is still true that for $\tau_{1e} \ll \tau_x$, very little cross-relaxation occurs. For saturation of the $M_I = +\tfrac{1}{2}$ transition, it is to be noted that the $|+\tfrac{1}{2},+\tfrac{1}{2}\rangle$ and $|+\tfrac{1}{2},-\tfrac{1}{2}\rangle$ states differ in population by ε_e, whereas in the absence of microwave saturation they would have differed by $\varepsilon_n = g_n\beta_n B/2k_bT$. Thus, if a short-circuiting path is provided between these two states, there can be a significant reduction in the population of the $|+\tfrac{1}{2}, +\tfrac{1}{2}\rangle$ state as compared with that of the $|-\tfrac{1}{2}, +\tfrac{1}{2}\rangle$ state. The intense rf field at the frequency ν_{n1} provides such a path. The rate of transition between the $|+\tfrac{1}{2}, +\tfrac{1}{2}\rangle$ and the $|+\tfrac{1}{2}, -\tfrac{1}{2}\rangle$ states must at least equal τ_x^{-1}. If the $M_I = -\tfrac{1}{2}$ transition is saturated, it is the $|-\tfrac{1}{2}, -\tfrac{1}{2}\rangle$ and the $|-\tfrac{1}{2}, +\tfrac{1}{2}\rangle$ states that have the large population difference ε_e, and hence an intense rf field at frequency ν_{n2} gives rise to an ENDOR line. The equality in population of the other pair of levels, if only these relaxation processes are operative, would not result in an ENDOR line at the frequency ν_{n1}. The same applies to saturation of the $M_I = +\tfrac{1}{2}$ states, for which no ENDOR line at ν_{n2} should be expected. It is apparent that in many systems there must be at least one additional path of relaxation if both ENDOR lines are to be observed on saturating either microwave transition.

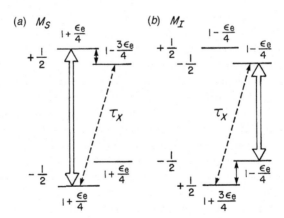

FIGURE 12.5 Relative populations of levels in an $S = I = \tfrac{1}{2}$ system in which the ENDOR behavior is governed by the combined effects of τ_{1e} and τ_x. (a) On saturation of the low-field EPR line, only the lower-frequency ENDOR line is observed. (b) On saturation of the high-field EPR line, only the higher-frequency ENDOR line is observed.

We note that the operators $\hat{S}_+\hat{I}_- + \hat{S}_-\hat{I}_+$ in the hyperfine term (expansion of the second term on the right of Eq. 12.2; see Eq. C.32) admix the states $|M_S - 1, M_I + 1\rangle$ or $|M_S + 1, M_I - 1\rangle$ with the state $| M_S, M_I \rangle$. It is such mixing that makes partially allowed the $\Delta(M_S + M_I) = 0$ transitions associated with the relaxation path τ_x. The alternative path labeled τ_{xx} in Fig. 12.4a could be described as involving $\Delta(M_S + M_I) = \pm 2$ transitions. For the latter transitions, one requires mixing induced by the term $\hat{S}_+\hat{I}_+ + \hat{S}_-\hat{I}_-$ of the states $|M_S + 1, M_I + 1\rangle$ or $|M_S - 1, M_I - 1\rangle$ with the state $|M_S, M_I\rangle$. The mixing coefficient has the form $(A_X - A_Y)/4h\nu_e$, where A_X and A_Y are two of the principal components of the hyperfine matrix [3, p. 247]. It is thus apparent that this mixing is non-zero and τ_{xx} is finite only if the hyperfine interaction is not isotropic.[7] There are now two alternative relaxation pathways (other than τ_{1e}) to reach the lowest-lying state $|-\frac{1}{2}, +\frac{1}{2}\rangle$ from the uppermost state $|+\frac{1}{2}, +\frac{1}{2}\rangle$; one involves $\tau_{xx} + \tau_{1n}$, and the other $\tau_{1n} + \tau_x$ (Fig. 12.5). In either case, the relaxation rate is controlled by τ_{1n} (since τ_{xx} is generally much shorter than τ_{1n}). Application of saturating rf power at either nuclear frequency sufficiently enhances the rate of the transitions $\Delta M_S = 0$, $\Delta M_I = \pm 1$ that the effective value of τ_{1e} is reduced because of the competing relaxation path, independent of which microwave transition is saturated. *This is indeed the essential characteristic of steady-state ENDOR.*

If one or more cross-relaxation times and nuclear spin-lattice relaxation times are of favorable magnitude, the application of saturating rf power reduces the effective value of the electron spin-lattice relaxation time sufficiently so that ENDOR lines may be observed continuously, as long as the exciting frequency is within the ENDOR linewidth.[8]

In many cases the intensities of pairs of ENDOR lines are similar; in others, they may be so unequal that one line is not even detected. This phenomenon is particularly marked when the nuclear Zeeman and the hyperfine interactions are comparable in magnitude [3, p. 221; 6, 7]. For systems with uniaxial symmetry, one can calculate the differences in intensity arising from differences in enhancement (see end of Section 12.2) of the effective rf field by the hyperfine field of the electron at the nucleus, in good agreement with observation. The orientation of the rf field relative to the axis of symmetry can be extremely important in determining the intensity of a particular ENDOR line, even though the rf field is always maintained perpendicular to the static field **B**.

Our discussion has thus far been limited to the response of a four-level system. In the case in which one has a system of general spin S and general nuclear spin I, the maximum possible number of ENDOR lines is $16SI$. This allows for occurrence of all EPR transitions that are forbidden in first order. Thus for $S = I = \frac{1}{2}$, there will be 4 ENDOR lines (see Fig. 5.4).[9]

If any nucleus has spin $I > 1$, it may be essential to consider the term for the nuclear quadrupole interaction (fourth term on the right in Eq. 12.1). Here there are additional relaxation paths, so that it is even more difficult to predict the intensity of the ENDOR lines or to specify in detail the ENDOR (relaxation) mechanism.

We now consider very briefly two other types of ENDOR mechanisms.[10] The first of these, 'packet-shifting' ENDOR, can be discussed by considering the phosphorus-doped silicon system [4,8]. Here the electron wavefunction of the donor extends over a large number of nuclei, some of which are ^{29}Si $(I = \frac{1}{2})$ and, of course, ^{31}P. Since all relaxation times τ_{1e}, τ_{1n} and τ_x are all of the order of hours for this system, one can easily 'burn a hole' in one of the inhomogeneously broadened EPR lines by saturating those spin packets that experience a particular local field. Since the 'hole' recovers with characteristic time τ_{1e}, the redistribution of population of nuclear states associated with this microwave-saturated transition is achieved in leisurely fashion. The most effective means of providing a favorable population redistribution in the nuclear levels is the 'rapid-passage' technique, which allows the populations of a pair of levels to be inverted by the use of an appropriately intense rf pulse.[11] It is expedient to observe the signal in dispersion (rather than absorption) for observation of the transient signals that arise.

A particular spin packet of the inhomogeneously broadened line, which is the envelope of all such packets, represents one particular value of the local field, to which many nuclei contribute. The redistribution of populations by the rapid-passage inversion changes the local field at neighboring nuclei, which in turn may have as neighbors electrons not involved in the microwave saturation. The changed local field means that some spin packets are shifted to other regions of the inhomogeneously broadened line, while other packets now find themselves in just the local field defining the region of the line that was saturated. The net changes of nuclear populations allow transient ENDOR signals to be observed both for ^{29}Si and for ^{31}P nuclear transitions. Thus the term 'packet-shifting' is very appropriate for this type of ENDOR.

Another of the important ENDOR mechanisms is that of 'distant' ENDOR. In the course of investigations on ruby (Cr^{3+} in Al_2O_3), ENDOR lines were observed from ^{27}Al transitions (not surprisingly), but the NMR frequency of the Al nuclei that were involved was not affected by the Cr^{3+} ions [10]. Hence these Al nuclei must have been located so far away from the unpaired electrons that the dipolar interaction was negligible. It can be shown [3, p. 74; 11] that there is marked polarization of nuclei in the vicinity of a paramagnetic ion in a magnetic field, if high microwave power is applied at a frequency on the shoulder of a resonance line. The term 'polarization' implies preferential population of spin levels. Instead of being confined to the vicinity of the paramagnetic ion, these population differences are transmitted throughout the sample by mutual spin flips of the nuclei, at the eventual expense of the energy of the microwave field. This 'spin diffusion' is thus the mode of communication of the paramagnetic ion with distant nuclei. When rf power corresponding to Al nuclear transitions is applied, the change in spin orientation of the distant nuclei is transmitted back to the Cr^{3+} ions; the change in their EPR signal level indicates that energy is absorbed. In consonance with this mechanism, it was noted that when the rf power was removed, the EPR signal recovered with a characteristic time of about 10 s, the ^{27}Al nuclear spin-lattice relaxation time τ_{1n}. For both packet-shifting and for steady-state ENDOR, the recovery rate is of the order of τ_{1e}, which here is about 10^{-1} s.

12.5 CW ENDOR: SINGLE-CRYSTAL EXAMPLES

12.5.1 The *F* Centers in the Alkali Halides

Perhaps the most spectacular successes of the ENDOR method have been in its application to systems that give inhomogeneously broadened lines that are the envelope of large numbers (in some cases, literally hundreds) of overlapping hyperfine components. An example is the electron in an anion vacancy (*F* center) in KBr, for which the width of the (gaussian) EPR line is about 12.5 mT. The six first-shell neighbors in the face-centered structure are either ^{39}K (relative abundance 93.26%) or ^{41}K (relative abundance 6.73%). These nuclides also comprise all other odd-numbered shells. The even-numbered (second, fourth, etc.) shells are composed of ^{79}Br (abundance 50.69%) and ^{81}Br (abundance 49.31%). All four of these nuclides have $I = \frac{3}{2}$. Taking, for a moment, all these nuclides to be equivalent, one may compute (Section 3.5) the maximum number $\Pi_j (2n_j I + 1) = 19 \times 37 = 703$ of hyperfine lines arising just from the six first-shell and the 12 second-shell neighbors. To be sure, many of these lines have low relative intensities. For example, the outermost line of a set of 19 has an intensity only $\frac{1}{580}$th that of the central line (Problem 3.2). Considering the extra lines arising from interaction with additional shells of nuclei, as well as the non-identity of nuclear moments, anisotropic interactions and the effects of the nuclear-quadrupole moments, it is understandable that ordinarily the EPR line of the *F* centers in KBr gives no indication of any structure.

From Eqs. 12.4 and 12.5 one expects that the very wide range of hyperfine interactions of the unpaired electron with nuclei in the various shells ensures that the ENDOR spectrum is spread over a considerable range of frequency.

Both the isotropic and anisotropic parts of the hyperfine interactions fall off with distance from the anion vacancy, in accordance with the ideas discussed in Chapter 5. Thus valuable information about the unpaired-electron distribution of the *F* center becomes available by measuring these effects.

Looking now at the ENDOR spectrum of crystalline KBr in Fig. 12.6, one sees that the frequencies at which the lines are observed vary from roughly 0.5 to 26 MHz. Clearly, there is also a considerable variation in linewidths, the narrowest lines being of the order of 10 kHz. Especially in the 3–4 MHz region, such a small width makes it possible to resolve the pairs expected from Eqs. 12.4 and 12.5, separated by $2\nu_{Br}$, for each of the bromine nuclides. Various line pairs from ^{39}K, ^{41}K, ^{79}Br and ^{81}Br are identified with brackets above the lines (Fig. 12.6).

Identification of the shell numbers (indicated as a subscript to a symbol or bracketed below it) is accomplished by a study of the angular dependence of the lines. In Fig. 12.6, the field **B** is oriented along a ⟨100⟩ axis of KBr. When the hyperfine interaction has uniaxial symmetry (shells I, III and IV), the angular dependence of a line is similar to that shown in Fig. 12.7*a* for the first-shell nuclei of the *F* center in LiF. For nuclei in this and other shells, one can measure and also predict the angular dependence of the dipolar hyperfine interaction. These angular dependences are given in Fig. 12.7 out to the eighth shell [12]. Where the angular dependence of

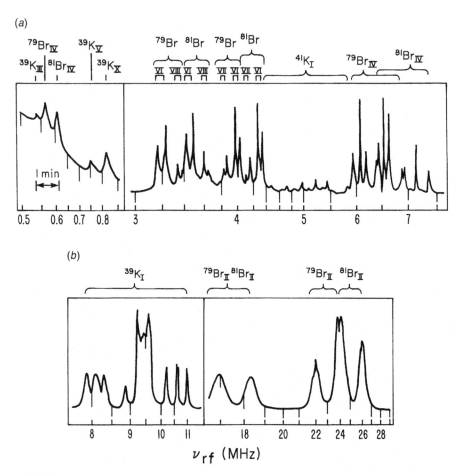

FIGURE 12.6 ENDOR spectrum of the F center in KBr at 90 K for $\mathbf{B} \parallel \langle 100 \rangle$. The line pairs corresponding to ^{39}K, ^{79}Br and ^{81}Br for the various shells are indicated. The triplet lines, illustrative of quadrupole interaction, are most prominent in the 8–11 MHz region. [After H. Seidel, Ref. 5; also see Ref. 4, Section 4.2.3 in FURTHER READING.

two shells is similar, the magnitudes of the hyperfine coupling are usually very different, and the line pairs are usually assignable without ambiguity. If the hyperfine couplings are large, the higher-order terms (Sections 3.6 and 5.3.1) must be taken into account.

In KBr, an additional complication arises from the quadrupolar contributions (Section 5.6) from the nuclides present [$Q = 0.054$ (^{39}K), 0.060 (^{41}K), 0.29 (^{79}Br), 0.27 (^{81}Br), in units of 10^{-28} m^2]. For example, for shell I, there are actually three ^{39}K lines that show the angular dependence corresponding to Fig. 12.7a. These, which also show the effects of second-order hyperfine coupling,

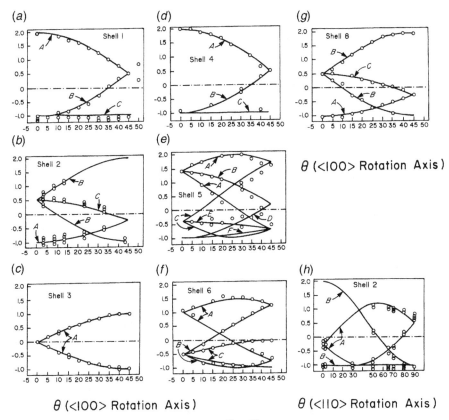

θ (<100> Rotation Axis) θ (<110> Rotation Axis)

FIGURE 12.7 Angular dependence of the (^{7}Li, ^{19}F) ENDOR frequencies for shells 1–8 of nuclei about the trapped electron from the F center in LiF: (a–g) the rotation axis is $\langle 100 \rangle$; (h) for nuclei of shell 2, with $\langle 110 \rangle$ as the rotation axis. The frequency scale has its zero at $a_0/2$ and is divided into units of $b_0/2$, that is, relative to the appropriate hyperfine parameters. Labels $A, B, C, \ldots$ refer to lines from sets of ions at positions equivalent in the absence of the external magnetic field. [After W. C. Holton, H. Blum, *Phys. Rev.*, **125**, 89 (1962).]

are clearly seen in the range 10–12 MHz (Fig. 12.6). These triplets can be understood by consideration of the spin hamiltonian (Eq. 12.1) containing the quadrupole interaction term (Eq. 5.51b). For nuclei in which the quadrupole interaction matrix **P** has uniaxial symmetry ($\eta = 0$), this reduces to

$$\hat{\mathcal{H}}_Q = P[3\hat{I}_z^2 - \hat{I}^2] \tag{12.10}$$

where P is given by Eq. 5.50. At sufficiently high field B and when g and A are close to isotropic, the additive transition-energy term can be obtained by using first-order

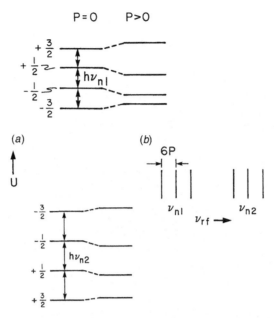

FIGURE 12.8 (a) First-order energy-level splitting arising from nuclei with $I = \frac{1}{2}$, showing the coinciding ENDOR transitions when $A > |g_n|\beta_n B$ and $P = 0$ (upper set for $M_S = +\frac{1}{2}$, lower set for $M_S = -\frac{1}{2}$) as well as the effect of $P > 0$. (b) First-order ENDOR spectrum showing quadrupole splitting, when $P > 0$ and $\theta = 0$.

perturbation theory [3, pp. 225–228; 13] and is given by

$$h\nu_Q = 3|P[3\cos^2\theta - 1](M_I - \tfrac{1}{2})| \tag{12.11}$$

where θ is the angle between **B** and **Z**, and $\Delta M_I = \pm 1$. Here M_I is the greater of the two such quantum numbers. Including the nuclear Zeeman term and the hyperfine

TABLE 12.1 Hyperfine and Quadrupole Couplings (in MHz) in KBr at 90 K, from ENDOR Spectra [a]

Shell	Nuclide	$A_\parallel/h$	$A_\perp/h$	P/h
1	^{39}K	18.33	0.77	0.067
2	^{81}Br	42.85[b]	2.81[b]	0.077
3	^{39}K	0.27	0.022	—
4	^{81}Br	5.70	0.41	0.035
5	^{39}K	0.16[b]	0.021[b]	—
6	^{81}Br	0.84[b]	0.086[b]	—
7	^{81}Br	0.54	0.07	—

[a] H. Seidel, Z. Phys., **165**, 218 (1961).
[b] Detectable departures from uniaxial symmetry.

term, the ENDOR frequencies are given by

$$\nu_{n1} = |+\tfrac{1}{2}[A_\parallel + A_\perp(3\cos^2\theta - 1)] - g_n\beta_n B + 3P(3\cos^2\theta - 1)(M_I - \tfrac{1}{2})|/h \quad (12.12a)$$

$$\nu_{n2} = |-\tfrac{1}{2}[A_\parallel + A_\perp(3\cos^2\theta - 1)] - g_n\beta_n B + 3P(3\cos^2\theta - 1)(M_I - \tfrac{1}{2})|/h \quad (12.12b)$$

Here $\mathbf{A}$ and $\mathbf{P}$ have been taken to be coaxial. The energy levels and the ENDOR spectrum for a nucleus with $I = \tfrac{3}{2}$ are given in Fig. 12.8. The hyperfine and some quadrupole couplings in successive shells for KBr are given in Table 12.1.

12.5.2 Metal-Ion Tetraphenylporphyrins

Detailed studies of ENDOR of $^{63}Cu^{2+}$ and $^{107,109}Ag^{2+}$, of ^{14}N and of ^1H in magnetically dilute single-crystal complexes M(II)TPP have been reported and discussed in detail [14]. The various hyperfine matrices (and nuclear quadrupole matrices for nitrogen) were obtained, and used to obtain a picture of the electronic distribution (d character on M, spin populations on the ligand atoms in specific orbitals, etc.).

12.6 CW ENDOR IN POWDERS AND NON-CRYSTALLINE SOLIDS

For many systems consisting of a paramagnetic guest in a host matrix, it is extremely difficult or impractical to grow a single crystal of a size sufficient for EPR studies. This is especially the case for biologically important systems, for example, metalloenzymes [15–17].

We have noted earlier that for crystalline powders and glassy solid solutions one may see distinct EPR lines, especially from those molecules that have an axis of dominant interaction at right angles to the static magnetic field $\mathbf{B}$. For example, with the V^- center in powdered MgO (Fig. 4.7) where the predominant effect is g anisotropy, the most intense feature corresponds to those defect centers whose tetragonal electric field axis lie perpendicular to $\mathbf{B}$; that is, the position of this line corresponds to $g_\perp$. Figure 5.12 illustrates the case in which there is a marked anisotropy of g and also of a hyperfine coupling A, where the matrices $\mathbf{g}$ and $\mathbf{A}$ have the same principal axes. By contrast, consider systems in which electron spin-spin interaction is dominant (e.g., triplet states), with the g anisotropy small (and $A = 0$), as illustrated in Figs. 6.5 and 6.7. Here the line pairs arise from molecules having magnetic field $\mathbf{B}$ directions parallel to principal axes of the matrix $\mathbf{D}$.

In general, every ENDOR spectrum of a paramagnetic species in a powder or glass arises from molecules occurring in a very limited range of orientations, selected by fixing the EPR absorption at a chosen location within the envelope. Here too, especially informative ENDOR spectra occur when $\mathbf{B}$ is along one of the above-mentioned canonical directions. Provided that the signal-to-noise ratio at the EPR powder spectral setting is adequate, one may obtain ENDOR spectra and gather much information about the paramagnetic species, including geometric detail. Thus relative orientations of the principal axes of matrices $\mathbf{g}$, $\mathbf{D}$ and $\mathbf{A}$ may be obtainable. By suitable analysis, one may be able to extract a number of parameters otherwise obtainable only from single-crystal measurements [18,19].

The hyperfine couplings arising from each type of nuclide (e.g., ^1H) give distinctive contributions that can be related to (in simple cases, centered on) the nuclear resonance frequency ν_n of that nuclide at the magnetic field value B where the ENDOR spectrum is taken. We note (Eqs. 12.4b and 12.5b) that for nuclei with $I = \frac{1}{2}$, each hyperfine coupling, to first order, gives a line pair at radiofrequencies $|\nu_n \pm K/2h|$, centered about ν_n when hyperfine parameter $K(\theta,\phi)$ in Eq. 6.55 is sufficiently small compared to $h\nu_n$. Similarly, for nuclei with $I \geq 1$ (e.g., ^{14}N; see Ref. 19), both lines of such a hyperfine pair may split into $2I$ lines, the set occurring at radiofrequencies $|\nu_n \pm K/2h \pm (3p/2h)(2M_I - 1)|$, where parameter $p \equiv (\mathbf{n}^T\mathbf{g}^T\mathbf{A}^T\mathbf{P}\mathbf{A}\mathbf{g}.\mathbf{n})/g^2K^2$ wherein $\mathbf{P}$ is the nuclear quadrupole coupling matrix (Eq. 5.51a), $\mathbf{n}(\theta,\phi) = \mathbf{B}/B$, and where $-I + 1 \leq M_I \leq I$.

We see then that powder ENDOR spectra can yield information about quadrupole couplings, which is never available from first-order EPR spectra, and that the complications due to the presence of sums, differences and combinations of hyperfine lines in EPR and ESEEM are absent in ENDOR. A brief review of the structural information available from powder ENDOR spectroscopy has been published [20].

As an interesting example of ENDOR in glassy media, we can cite trapped electrons in γ-irradiated aqueous NaOH, methanol and 2-methyltetrahydrofuran at 77 K [21]. Here the EPR line is broadened as a result of dipolar interactions with surrounding matrix protons. Analysis of the ^1H ENDOR lineshape gives information about the extent of delocalization of the unpaired electron.

As a second example, we cite proton ENDOR powder work done on nitrosyl horse-heart myoglobin [22]. Here, detailed analysis using computer simulation, led to identification of the protons in the heme pocket.

12.7 CW ENDOR IN LIQUID SOLUTIONS

The possibility of detecting ENDOR of substances in liquid solution was first demonstrated by Hyde and Maki [23]. Subsequent experimental and theoretical works have shown this to be a very valuable technique [24–27].

The discussion of Section 12.5 has emphasized that for solids one often is able to resolve far more lines in the ENDOR than in the EPR spectrum. For free radicals in liquid solution, this may still be true if inhomogeneous broadening limits the resolution. However, even in these cases, the number of possible lines in the ENDOR spectrum is less than that for the EPR spectrum. Consider a radical with one set of four equivalent protons, which gives the familiar $1:4:6:4:1$ EPR

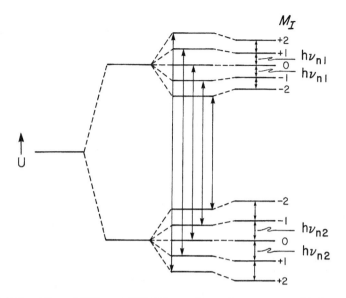

FIGURE 12.9 Allowed EPR and ENDOR transitions for a system with four equivalent protons in a magnetic field. On saturation of any of the five EPR transitions one observes ENDOR transitions only at the frequencies ν_{n1} and ν_{n2} (compare with Fig. 12.3).

spectrum. The full spin hamiltonian is given by Eq. 12.2, where now $\hat{\mathbf{I}} = \Sigma_i \hat{I}_i$, yielding $'I = 2$.[12] That the number of lines in the ENDOR spectrum is less than that for the EPR spectrum may readily be shown by application of Eq. 12.2 to the four-proton system. If the hyperfine coupling constant is small enough so that second-order effects (Section 3.6) are negligible, then the full spin hamiltonian gives a set of five equally spaced levels with $M_S = -\frac{1}{2}$ (Fig. 12.9) and also with $M_S = +\frac{1}{2}$. The five levels corresponding to $M_S = -\frac{1}{2}$ have a uniform spacing greater than that of the $M_S = +\frac{1}{2}$ levels. Hence only *two* ENDOR transitions are observed; these occur at $\nu_{ENDOR} = \nu_p + A/2h$. Here ν_p is the proton NMR frequency at the *constant* magnetic field used for the ENDOR experiment. For radicals in which there are n sets of m non-equivalent protons, there are only $2n$ lines in the ENDOR spectrum, irrespective of the number of protons in any set.[13] In the EPR spectrum there are $(m + 1)^n$ lines.

There are special problems in constructing equipment for studies of ENDOR spectra of liquids. The difficulties in this case arise from the necessity of using an intense rf field, which causes heating if applied continuously. Hence pulse rf ENDOR systems have been developed. The requirement of a large rf field arises because the nuclear transitions must be saturated. Relaxation times τ_{1e} are of the order of $10^{-5}-10^{-6}$ s for free radicals at room temperature, whereas the relaxation times τ_{1n} of protons in these are typically several orders of magnitude longer than this. The application of an intense rf field B_{1n} at the nuclear-resonance frequency

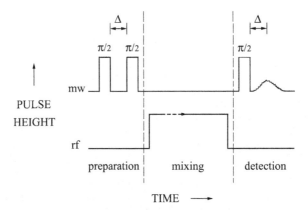

FIGURE 12.10   Mims ENDOR pulse sequence, consisting of a stimulated echo sequence involving three non-selective microwave $\pi/2$ pulses, and a selective rf π pulse of variable frequency applied during the mixing period. Compare with Fig. 11.7. Typically, the three microwave pulses (9.42 GHz, 1 W) last 0.2 μs, the spacing Δ is ~ 10 μs, and the RF pulse (200–1000 kHz) lasts ~ 1 ms. The stimulated echo received of course needs to be Fourier-transformed (FT) to produce the desired spectrum, for instance, to reveal hyperfine splittings.

greatly reduces the lifetime of a nuclear-spin orientation in an excited state because of emission stimulated by this field.

12.8   PULSE DOUBLE-RESONANCE EXPERIMENTS

With the advent of pulse ENDOR spectrometers, all the benefits of time-resolved spectroscopy, as outlined in Chapters 1, 10 and 11, have become available. Various sequences of pulse microwave and RF excitations can be used, with much scope for imaginative research. At present, Mims ENDOR (Fig. 12.10) [29] and Davies ENDOR [30] have a favorite ranking.

The various techniques include ENDOR with circularly polarized $\mathbf{B}_{1n}$, polarization-modulated ENDOR, double and triple ENDOR, stochastic ENDOR, multiple-quantum ENDOR and ENDOR-induced EPR. For example, we can cite a spectrometer operating at 140 GHz, demonstrating orientation-selective Davies ENDOR [31]. These techniques are yielding new levels of understanding of the structural and relaxation properties of unpaired-electron species. Unfortunately, it is not possible to give the details in this book; happily, various excellent descriptions have become available [32–37].

12.9   ELECTRON-ELECTRON DOUBLE RESONANCE (ELDOR)

In an ENDOR experiment one observes a change in intensity of a partially saturated EPR signal when one establishes a connection to an energy level belonging to a *different* hyperfine transition. A very different experiment, termed *electron–electron double resonance* (ELDOR), consists of the observation of a reduction in the intensity of one hyperfine transition when a second hyperfine transition is simultaneously being saturated [38]. Simultaneous electron-spin resonance in one magnetic field for two different transitions requires irradiation simultaneously at two microwave frequencies; that is, one requires a bimodal resonator tunable to two frequencies separated by a multiple of the hyperfine coupling. In principle, the simplest case is that of a single nucleus of spin $\frac{1}{2}$, illustrated in Fig. 12.11. Although the two transitions have no level in common, they may be coupled by two mechanisms:

1. Rapid nuclear relaxation that may be induced by dipolar coupling of electrons and nuclei. The flipping of an electron spin under appropriate conditions can cause a simultaneous flip of a coupled nuclear spin. This mechanism is predominant at low concentrations and at low temperatures.

2. At high concentrations or at sufficiently high temperatures, spin exchange or chemical exchange (Section 10.5.3) tend to equalize the populations of all spin levels.

The ELDOR technique is very sensitive to the various relaxation mechanisms involved. For example, it was used (as an alternative to ENDOR measurements)

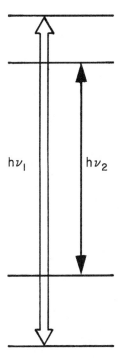

FIGURE 12.11   Electron-electron double-resonance experiment. The intensity of the EPR line observed in a spectrometer operating at a low microwave power level at frequency ν_2 is recorded as a function of the (high) microwave power applied to the same bimodal cavity by a separate source operating at a frequency ν_1.

to distinguish between the two nearly identical large nitrogen hyperfine splittings in DPPH, making use of the differences in the relaxation times of the various nuclei present. This study produced accurate values of the ^{14}N hyperfine coupling constants [39]. A careful ELDOR study of the effect of the Heisenberg spin exchange between pairs of paramagnets (see Section 6.2.1) on their saturation behavior was studied for peroxylamine disulfonate (PADS) radical anions in de-oxygenated water, and agreed with the existing theory [40].

Pulse ELDOR, as applied to a CH fragment ($S = I = \frac{1}{2}$) [i.e., CH(COOH)$_2$ in irradiated malonic acid], yielded both spin-relaxation and cross-relaxation times, the latter between the $M_I = +\frac{1}{2}$ manifolds [41].

Another example is the now-familiar V^- center in MgO (Sections 4.2, 4.8 and 12.6). Here, pulse ELDOR has revealed that the tetragonally distorted hole centers jump between the three possible orientations [42]. One can sit at fixed magnetic field with the pumping microwave excitation (frequency ν_1) on the line from one such site, and the observed signal (frequency ν_2) at another site. Power saturating the first manifests itself on the second, due to the pumping. This saturation transfer probably proceeds, at least at 4.2 K, via tunneling of the hole, taking place without reorientation.

Special 2D ELDOR experiments have given valuable information about magnetization-transfer rates between spin-label nitroxide radicals in disordered solid polymers [43], as well as about local motional effects extending right down to the rigid-limit region in such systems [44]. The pulse sequences are shown explicitly [43]. A review of these techniques is available [45].

Multi-quantum jumps excited in ELDOR spectroscopy, observed initially in pyrrol black powder (oxypyrrol free radical), have been known for some decades now [46].

12.10   OPTICALLY DETECTED MAGNETIC RESONANCE

The intensity of every EPR transition is proportional to the difference ΔN in the populations of the two $|M_S\rangle$ states spanned, that is, to the net spin polarization (Section 4.6). It follows that any other process affecting ΔN influences the EPR intensity, and vice versa, unless the spin-lattice relaxation is too efficient. The above-mentioned concepts are the essence of how ENDOR works. Under the right circumstances, optical transitions induced by polarized light have intensities proportional to ΔN, so that electronic transitions and EPR transitions are linked. Detection of such effects is not easy, generally requiring very low temperatures to maximize ΔN. Such optically detected magnetic resonance (ODMR) experiments make it possible to characterize those portions of broad optical bands that arise from a specific unpaired-electron species. References 47–52 cite applicable reviews of ODMR.

ODMR of the [AlO$_4$]^0 'point' defect (one unpaired electron, primarily on an oxygen anion positioned between the Al and a Si cation) in irradiated crystalline quartz has disclosed which of several optical absorption peaks (1–5 eV) is linked to the paramagnetic center [53].

As second example, we point to crystals of luminescent gold(I) complexes, with halo (triphenyl)phosphine and arsine ligands, which exhibit phosphorescences from phenyl-localized triplet states. These have yielded zero-field ODMR data allowing determination of the electronic quadrupole parameters |D| and |E| (see Section 6.3.1) [54].

An elegant method for measuring relative populating rates of photoexcited triplet-state sublevels during optical pumping, using ODMR in an indole chromophore, has been developed quite recently [55], and points to the potential of ODMR in the future.

12.11 FLUORESCENCE-DETECTED MAGNETIC RESONANCE

A newly developed optical technique enables far better sensitivity and time resolution than does ordinary (cw) EPR, in some circumstances. It uses detection of fluorescence from the recombination product of short-lived free radicals in liquid or solid solution to display the EPR spectrum of one or both of these. For instance, the fluorescence-detected magnetic resonance (FDMR) can be utilized to observe primary radical cations created by ionizing or photoionizing radiation (e.g., pulses

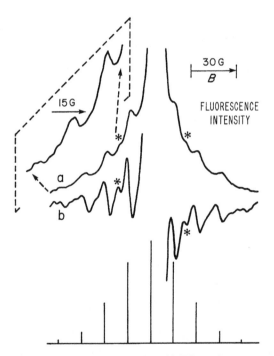

FIGURE 12.12 (*a*) FDMR spectrum observed at 190 K in cyclopentane containing 10^{-3} M cubane and 10^{-4} M perdeuterated anthracene. The asterisks indicate signals from the cyclopentane solvent. The insertion is the enlargement of the left outermost three peaks. (*b*) First-derivative FDMR spectrum of part (*a*). [After X.-Z. Qin, A. D. Trifunac, P. E. Eaton, Y. Xiong, *J. Am. Chem. Soc.*, **113**, 669 (1991).]

of 5–15 ns duration) [56]. A microwave pulse applied immediately thereafter alters the fluorescence intensity, by an amount sensitive to the EPR intensity appropriate to the field **B** present. The 'EPR' spectrum, including hyperfine structure, can be obtained by varying B step-wise.

Thus, for example, the spin-correlated radical pair cubane$^+$

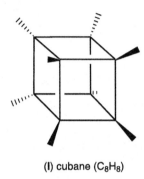

(I) cubane (C_8H_8)

and perdeuteroanthracene$^-$ recombine, within picoseconds after an ionizing 3 MeV electron beam pulse, to yield an excited singlet-state anthracene scintillator molecule [57]. The X-band FDMR spectrum observed at 190 K is shown in Fig. 12.12 and displays the proton hyperfine septet (1:6:15:20:15:6:1 with spacing 1.61 mT) expected for cubane$^+$.

As another example of FDMR, we select the creation of naphthalene$^+$/ naphthalene$^-$ in dilute solution (>0.01 M) at room temperature [58]. The detection of field(B)-dependent fluorescence from such radical pairs, created by electron irradiation, provides a very sensitive technique, using triplet-state EPR at (say) X-band, for observing such short-lived species.

REFERENCES

1. C. P. Poole Jr., *Electron Spin Resonance. A Comprehensive Treatise on Experimental Techniques*, 2nd ed., Wiley-Interscience, New York, NY, U.S.A., 1983.

2. G. Feher, *Phys. Rev.*, **103**, 834 (1956).

3. A. Abragam, B. Bleaney, *Electron Paramagnetic Resonance of Transition Ions*, Oxford University Press, London, U.K., 1970.

4. G. Feher, *Phys. Rev.*, **114**, 1219 (1959).

5. H. Seidel, *Z. Phys.*, **165**, 218, 239 (1961).

6. E. R. Davies, T. Rs. Reddy, *Phys. Lett.*, **31A**, 398 (1970).

7. S. Geschwind, "Special Topics in Hyperfine Structure in EPR", in *Hyperfine Interactions*, A. J. Freeman, R. B. Frankel, Eds., Academic Press, New York, NY, U.S.A., 1967, p. 225.

8. G. Feher, E. A. Gere, *Phys. Rev.*, **114**, 1245 (1959).

9. F. Bloch, *Phys. Rev.*, **70**, 460 (1946).

10. J. Lambe, N. Laurance, E. C. McIrvine, R. W. Terhune, *Phys. Rev.*, **122**, 1161 (1961).

11. B. M. Hoffman, *Proc. Natl. Acad. Sci. U.S.A.*, **100**(7), 3575 (2003).

12. W. C. Holton, H. Blum, *Phys. Rev.*, **125**, 89 (1962).

13. J. A. Weil, *J. Magn. Reson.*, **18**, 113 (1975).

14. T. G. Brown, B. M. Hoffman, *Mol. Phys.*, **39**(5), 1073 (1980).

15. B. M. Hoffman, V. J. DeRose, P. E. Doan, R. J. Gurbiel, A. L. P. Houseman, J. Telser, "Metalloenzyme Active-Site Structure and Function through Multifrequency CW and Pulse ENDOR", in *EMR of Paramagnetic Molecules, Biological Magnetic Resonance*, Vol. 13, L. J. Berliner, J. Reuben, Eds., Plenum Press, New York, NY, U.S.A., 1993.

16. H. Thomann, W. B. Mims, "Pulse Electron-Nuclear Spectroscopy and the Study of Metalloprotein Active Sites", in *Pulsed Magnetic Resonance: NMR, ESR and Optics*, D. M. S. Bagguley, Ed., Oxford University Press, Oxford, U.K., 1992.

17. B. M. Hoffman, *Acc. Chem. Res.*, **24**, 164 (1991).

18. W. T. Doyle, *Phys. Rev.*, **126**, 1421 (1962).

19. G. H. Rist, J. S. Hyde, *J. Chem. Phys.*, **52**, 4633 (1970).

20. D. Attanasio, *J. Chem. Soc. Faraday Trans. I*, **85**, 3927 (1989).

21. J. Helbert, L. Kevan, B. L. Bales, *J. Chem. Phys.*, **57**, 723 (1972).

22. M. Flores, E. Wajnberg, G. Bemski, *J. Biophys.*, **78**(4), 2107 (2000).

23. J. S. Hyde, A. H. Maki, *J. Chem. Phys.*, **40**, 3117 (1964).

24. J. S. Hyde, *J. Chem. Phys.*, **43**, 1806 (1965).

25. J. H. Freed, *J. Chem. Phys.*, **43**, 2312 (1965); **71**, 38 (1967).

26. J. H. Freed, D. S. Leniart, J. S. Hyde, *J. Chem. Phys.*, **47**, 2762 (1967).

27. R. D. Allendoerfer, A. H. Maki, *J. Magn. Reson.*, **3**, 396 (1970).

28. L. C. Kravitz, W. W. Piper, *Phys. Rev.*, **146**, 322 (1966).

29. W. B. Mims, *Proc. Roy. Soc. (London)*, **283**, 452 (1965).

30. E. R. Davies, *Phys. Lett.*, **47A**, 1 (1974).

31. M. Bennati, C. T. Farrar, J. A. Bryant, S. J. Inati, V. Weis, G. J. Gerfen, P. Riggs-Gelasco, J. Stubbe, R. G. Griffin, *J. Magn. Reson.*, **138**, 232 (1999).

32. A. Schweiger, *Struct. Bonding*, **51**, 1 (1982).

33. A. Schweiger, *Angew. Chem., Int. Ed. Engl.*, **30**, 265 (1991).

34. C. Gemperle, A. Schweiger, *Chem. Rev.*, **91**, 1481 (1991).

35. A. Grupp, M. Mehring, "Pulsed ENDOR Spectroscopy in Solids", in *Modern Pulsed and Continuous-wave Electron Spin Resonance*, L. Kevan, M. K. Bowman, Eds., Wiley, New York, NY, U.S.A., 1990, Chapter 4, pp. 195ff.

36. H. Thomann, M. Bernardo, *Spectrosc. Int. J.*, **8**, 119 (1990).

37. K. P. Dinse, "Pulsed ENDOR", in *Advanced EPR—Applications in Biology and Biochemistry*, A. J. Hoff, Ed., Elsevier, Amsterdam, Netherlands, 1989, Chapter 17, pp. 615–631.

38. J. S. Hyde, J. C. W. Chien, J. H. Freed, *J. Chem. Phys.*, **48**, 4211 (1968).

39. J. S. Hyde, R. C. Sneed Jr., G. H. Rist, *J. Chem. Phys.*, **51**, 1404 (1969).

40. M. P. Eastman, G. V. Bruno, J. H. Freed, *J. Chem. Phys.*, **52**(1), 321 (1970).

41. M. Nechtschein, J. S. Hyde, *Phys. Rev. Lett.*, **24**, 672 (1970).

42. G. Rius, A. Hervé, *Solid State Commun.*, **15**, 421 (1974).

43. G. G. Maresch, M. Weber, A. A. Dubinskii, H. W. Spiess, *Chem. Phys. Lett.*, **193**, 134 (1992).

44. B. R. Patyal, R. H. Crepeau, D. Gamliel, J. H. Freed, *Chem. Phys. Lett.*, **175**, 445, 453 (1990).

45. J. Gorcester, G. L. Millhauser, J. H. Freed, "Two-dimensional Electron Spin Resonance", in *Modern Pulsed and Continuous-wave Electron Spin Resonance*, L. Kevan, M. K. Bowman, Eds., Wiley, New York, NY, U.S.A., 1990, Chapter 3.

46. F. Chiarini, M. Martinelli, P. A. Rolla, S. Santucci, *Lett. Nuovo Cimento*, **5**(2), 197 (1972).

47. S. Geschwind, "Optical Techniques in EPR in Solids", in *Electron Paramagnetic Resonance*, S. Geschwind, Ed., Plenum Press, New York, NY, U.S.A., 1972, Chapter 5.

48. B. C. Cavenett, *Adv. Phys.*, **30**, 475 (1981).

49. R. H. Clarke, *Triplet-State ODMR Spectroscopy*, Wiley, New York, NY, U.S.A., 1982.

50. J.-M. Spaeth, "Application of Magnetic Multiple Resonance Techniques to the Study of Point Defects in Solids", in *Electronic Magnetic Resonance of the Solid State*, J. A. Weil, Ed., Canadian Society for Chemistry, Ottawa, ON, Canada, 1987, Chapter 34.

51. J. Hoff, "Optically Detected Magnetic Resonance of Triplet States", in *Advanced EPR: Applications in Biology and Biochemistry*, A. J. Hoff, Ed., Elsevier, Amsterdam, Netherlands, 1989, Chapter 18.

52. J.-M. Spaeth, F. Lohse, *J. Phys. Chem. Solids*, **51**, 861 (1990).

53. B. K. Meyer, F. Lohse, J.-M. Spaeth, J. A. Weil, *J. Phys. C: Solid State Phys.*, **17**, L31 (1984).

54. L. J. Larson, E. M. McCauley, B. Weissbart, D. S. Tinti, *J. Phys. Chem.*, **99**, 7218 (1995).

55. A. Ozarowski, A. H. Maki, *J. Magn. Reson.*, **148**, 419 (2001).

56. D. W. Werst, A. D. Trifunac, *J. Phys. Chem.*, **95**, 3466 (1991).

57. X.-Z. Qin, A. D. Trifunac, P. E. Eaton, Y. Xiong, *J. Am. Chem. Soc.*, **113**, 669 (1991).

58. Yu. N. Molin, O. A. Anisimov, V. M. Grigoryants, V. K. Molchanov, K. M. Salikhov, *J. Phys. Chem.*, **84**, 1853 (1980).

NOTES

1. A detailed account of ENDOR theory, including applications, is given by Abragam and Bleaney [3, Section 1.3 and Chapter 4].

2. Generally, special rf modulation coils or loops are installed, often inside the resonator.

3. In some systems, only the ENDOR line at ν_{n1} would be observed when the magnetic field is set at B_k; furthermore, only the ENDOR line at ν_{n2} would be observed when the magnetic field is set at B_m. We shall consider these cases in Section 12.4.

4. The reader is reminded that the sign of A may be negative. In the normal experiment in which an oscillating radiofrequency field is used, one is unable to establish the order of energy levels; hence it is appropriate to indicate absolute magnitudes ('moduli') where differences are involved. The use of a rotating radiofrequency field allows the order of the energy levels to be determined.

5. In some types of ENDOR experiments, in which one monitors the dispersion (Section F.3.4) instead of the absorption signal, the field is set on one side of the EPR absorption maximum.

6. To avoid repetitive use of the factor $N/4$, we shall divide all population numbers by it; hence the relative populations in Eqs. 12.9 are taken as $1 - \varepsilon_e$ and $1 + \varepsilon_e$, in the absence of a microwave field or in the presence of a very weak one. These populations are shown in Fig. 12.4b.

7. An alternative mechanism that may give rise to the cross-relaxation path τ_{xx} and that may affect the values of the other relaxation times as well, even with isotropic hyperfine interaction, is thermal modulation of this interaction [3, p. 248].

8. This is to be contrasted with either 'packet-shifting' or 'distant' ENDOR, which constitute two other important mechanisms; these are discussed briefly at the end of this section.

9. However, saturation of the forbidden lines of Fig. 5.4 would give rise to ENDOR lines of the same frequency as saturation of the allowed lines. These would be difficult to detect, especially since one would require a very intense rf field to compete with the very fast relaxation rate τ_{1e}.

10. It is not to be inferred that these ENDOR mechanisms are mutually exclusive. The order of magnitude of τ_{1e}, for systems in which packet shifting predominates, is typically much longer than those for steady-state ENDOR; a typical value for the latter case may be taken as $\tau_{1e} < 1$ ms.

11. Rapid passage represents a traverse of the resonant absorption line in a time short compared with τ_1 and τ_2, with a radiofrequency field B_{1n} large enough so that $\tau_2 \gg |g_n B_{1n}|^{-1}$ and $|g_n B_{1n} \tau_1| \gg 1$, where g_n is the nuclear magnetogyric ratio [9].

12. In the discussion of energy levels of free radicals in Chapter 3, the nuclear Zeeman term was omitted since, just as for the hydrogen atom (Appendix C), the *energies* of the EPR transitions are hardly affected by its inclusion.

13. This is true provided that terms of the order of magnitude of $A^2/g_e\beta_e B$ can be neglected [28].

FURTHER READING

1. H. C. Box, *Radiation Effects: ESR and ENDOR Analysis*, Academic Press, New York, NY, U.S.A., 1977.

2. M. M. Dorio, J. H. Freed, Eds., *Multiple Electron Resonance Spectroscopy*, Plenum Press, New York, NY, U.S.A., 1979.

3. C. Gemperle, A. Schweiger, Pulse Electron-Nuclear Double Resonance Technology, *Chem Rev.*, **91**, 1481 (1991).

4. L. Kevan, L. D. Kispert, *Electron Spin Double Resonance Spectroscopy*, Wiley, New York, NY, U.S.A., 1976.

5. A. D. Milov, A. G. Maryasov, A. G. Tvetkov, "Pulsed Electron Double Resonance (PELDOR) and its Applications in Free-Radical Research—A Review", *Appl. Magn. Reson.*, **15**(1), 107 (1998).

6. K. Möbius, M. Plato, W. Lubitz, "Radicals in Solution Studied by ENDOR and Triple Resonance Spectroscopy", *Phys. Rep.*, **87**, 171 (1982).

7. G. E. Pake, T. L. Estle, *The Physical Principles of Electron Paramagnetic Resonance*, Benjamin, Reading, MA, U.S.A., 1973, Chapter 11.

8. J.-M. Spaeth, J. R. Niklas, R. H. Bartram, *Structural Analysis of Point Defects in Solids— an Introduction to Multiple Magnetic Resonance Spectroscopy*, Springer, Berlin, Germany, 1982.

PROBLEMS

12.1 The second-derivative EPR spectrum of Cu-α-picolinate in Zn- α-picolinate powder (Cu^{2+}, 3d^9 in covalent system) is shown in Fig. 12.13a. The

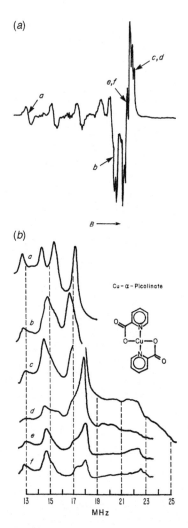

FIGURE 12.13   (*a*) Second-derivative EPR spectrum of Cu(pic)$_2$·4H$_2$O powder. The arrows indicate magnetic-field values at which the ENDOR spectra in (*b*) are taken. (*b*) Powder-type nitrogen ENDOR spectra corresponding to (*a*). The spectrum *a* is a single-crystal-type spectrum, whereas the remaining spectra have contributions from molecules in many orientations. [After G. H. Rist, J. S. Hyde, *J. Chem. Phys.*, **52**, 4633 (1970).]

corresponding ENDOR spectrum is shown in Fig. 12.13*b*. Determine $A_\perp$, ν_n and the quadrupole splitting for ^{14}N.

12.2   The ENDOR spectrum of Co^{2+} (3d^7) in MgO (Fig. 12.14) is observed when the transition $\Delta M_S = \pm 1$, $M_I = +\frac{1}{2}$ is partially saturated at a frequency $\nu = 9.563$ GHz, for $B = 156.1$ mT. From an EPR experiment, one knows that $g = 4.2785$ (Fig. 1.11).

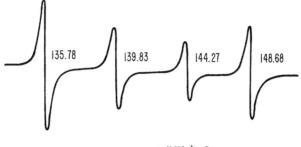

ν (MHz) $\longrightarrow$

FIGURE 12.14 ENDOR spectrum of Co^{2+} in MgO at 4.2 K. The ENDOR frequencies given correspond to the centers of the derivative lines obtained by scanning the radiofrequency while saturating the EPR transition $\Delta M_S = \pm 1$, $M_I = +\frac{1}{2}$ at a field of 1561 G, with $\nu = 9.563$ GHz. Here $g = 4.280$. [After D. J. I. Fry, P. M. Llewellyn, *Proc. Roy. Soc. (London)*, **A266**, 84 (1962).]

(*a*) Why are four lines observed? Assign each transition.

(*b*) Assuming that only second-order hyperfine corrections are necessary, estimate the hyperfine coupling A_o. (Anisotropic effects are here very small.) (The value obtained by using corrections to fourth order is 290.55 MHz.)

(*c*) From the data, estimate the nuclear g factor of ^{59}Co.

12.3 (*a*) Justify the number of separate curves in Figs. 12.7*a*–*g*. (*b*) Explain the angular variation for each curve for shells 1–4.

12.4 At the nucleus, the magnitude B_{hf} of the magnetic field caused by the unpaired electron is (Eq. 5.14*a*) derivable from the relation

$$AM_S = -g_n\beta_n B_{hf} \tag{12.12}$$

(*a*) Calculate B_{hf} at a proton for which the coupling A/h is 142 MHz (one-tenth that of the free hydrogen atom).

(*b*) Calculate B_{hf} at a ^{55}Mn ion ($M_S = +\frac{1}{2}$) for which the coupling $A/hc = -9.10 \times 10^{-3}$ cm^{-1}. The values of g_n can be found in Table H.4.

CHAPTER 13

OTHER TOPICS

13.1 APOLOGIA

The following selections represent sample literature, enabling the interested reader to dig for more. Thus, no claim to completeness is made for the citations listed in the subsequent topics. Obviously also, the number of topics selected is quite limited.

13.2 BIOLOGICAL SYSTEMS

I. Bertini, R. S. Drago, *ESR and NMR of Paramagnetic Species in Biological and Related Systems*, Kluwer, Norwell, MA, U.S.A., 1980.

J. S. Cohen, *Magnetic Resonance in Biology*, Vols. 1 and 2, Wiley-Interscience, New York, NY, U.S.A., 1980, 1982.

L. R. Dalton, Ed., *EPR and Advanced EPR Studies of Biological Systems*, CRC Press, Boca Raton, FL, U.S.A., 1985.

G. Feher, *Electron Paramagnetic Resonance with Applications to Selected Problems in Biology*, Gordon & Breach, New York, NY, U.S.A., 1970.

M. A. Foster, *Magnetic Resonance in Medicine and Biology*, Pergamon Press, Oxford, U.K., 1984.

R. G. Shulman, Ed., *Biological Applications of Magnetic Resonance*, Academic Press, New York, NY, U.S.A., 1979.

13.3 CLUSTERS

B. Mile, J. A. Howard, M. Histed, H. Morris, C. A. Hampson, "Matrix-isolation Studies of the Structures and Reactions of Small Metal Particles", *Faraday Discuss.*, **92**, 129 (1991).

H. Oshio, M. Nakano, "High-spin Molecules with Magnetic Anisotropy towards Single-molecule Magnets", *Chem. Eur. J.*, **11**, 5178 (2005).

13.4 CHARCOAL, COAL, GRAPHITE AND SOOT

R. Botto, Y. Sanada, Eds., *Magnetic Resonance in Carbonaceous Solids*, ACS Advances in Chemistry Series, Vol. 229, American Chemical Society, Washington, DC, U.S.A., 1992.

R. B. Clarkson, P. Ceroke, S.-W. Norby, B. M. Odintsov, "Stable Particulate Paramagnetic Materials as Oxygen Sensors in EPR Oximetry: Coals, Lithium Phthalocyanine and Carbon Chars", *Biol. Magn. Reson.*, **18**, 233 (2003).

N. S. Dalal, A. I. Smirnov, R. L. Belford, "Diesel Particulate Mixtures by W-Band (94 GHz) Electron Spin Resonance", *Appl. Spectrosc. (Commun.)*, **51**, 1429 (1997).

C. C. Jones, A. R. Chughtai, B. Murugaverl, D. M. Smith, "Effects of Air/Fuel Combustion Ratio on the Polycyclic Aromatic Hydrcarbon Content of Carbonaceous Soot from Selected Fuels", *Carbon*, **42**, 2471 (2004).

L. Petrakis, J. P. Fraissard, Eds., *Magnetic Resonance: Introduction, Advanced Topics and Applications to Fossil Energy*, NATO ASI Series C124, Reidel, Dordrecht Netherlands, 1984.

A. M. Ziatdinov, V. V. Kainara, The Nature of Conduction ESR Linewidth Temperature Dependence in Graphite, in *EPR in the 21st Century*, A. Kawamori, J. Yamauchi, H. Ohta, Eds., Elsevier, Amsterdam, Netherlands, 2002, pp. 293–297.

13.5 COLLOIDS

J. P. Fraissard, H. A. Resing, Eds., *Magnetic Resonance in Colloid and Interface Science*, Reidel, Hingham, MA, U.S.A., 1980.

13.6 ELECTROCHEMICAL EPR

I. B. Goldberg, T. M. McKinney, "Principles and Techniques of Electrochemical Electron Spin Resonance Experiments", in *Laboratory Techniques in Electroanalytical Chemistry*, P. T. Kissinger, W. R. Heineman, Eds., Marcel Dekker, New York, NY, U.S.A., 1984, Chapter 24.

A. V. Il'yasov, Yu. M. Kargin, "ESR of Electrochemically Generated Radical Ions and Complexes", *Magn. Reson. Rev.*, **16**, 135 (1993).

J. D. Wadhawan, R. G. Compton, "EPR Spectroscopy in Electrochemistry", in *Encyclopedia of Electrochemistry*, Vol. 2, *Interfacial Kinetics and Mass Transport*, A. J. Bard, M. Stratmann, E. J. Calvo, Eds., Wiley-VCH, Weinheim, Germany 2002, Section 3.2, pp. 170–220.

A. M. Waller, R. G. Compton, "In-situ Electrochemical ESR", in *Comprehensive Chemical Kinetics*, R. G. Compton, A. Hamnet, Eds., Vol. 29, Elsevier, Amsterdam, Netherlands, 1989, Chapter 7.

13.7 EPR IMAGING

A. Blank, C. R. Dunnam, P. B. Borbat, J. H. Freed, "A Three-dimensional Electron Spin Resonance Microscope", *Rev. Sci. Instrum.*, **75**(9), 3050 (2004).

G. R. Eaton, S. S. Eaton, K. Ohno, Eds., *EPR Imaging and In-Vivo EPR*, CRC Press, Boca Raton, FL, U.S.A., 1991.

S. S. Eaton, G. R. Eaton, "EPR Imaging", *Electron Paramagn. Reson.*, **17**, 109 (2000).

C. A. Johnson, D. McGarry, J. A. Cook, N. Devasahayam, J. B. Mitchell, S. Subramanian, M. C. Krishna, "Maximum Entropy Reconstruction in Electron Paramagnetic Resonance Imaging", *Ann. Opt. Res.*, **119**(1), 101 (2003).

K. Ohno, "ESR Imaging", *Magn. Reson. Rev.*, **11**, 275 (1987).

S. Subramanian, K.-I. Matsumoto, J. B. Mitchell, M. C. Krishna, "Radio-frequency Continuous-wave and Time-domain EPR Imaging and Overhauser-enhanced Magnetic Resonance Imaging of Small Animals: Instrumental Developments and Comparison of Relative Merits for Functional Imaging", *NMR Biomed.*, **17**, 263 (2004).

13.8 FERROMAGNETS, ANTIFERROMAGNETS AND SUPERPARAMAGNETS

R. Berger, J.-C. Bissey, J. Kliava, "Lineshapes in Magnetic Resonance Spectra", *J. Phys. Condens. Matter*, **12**, 9347 (2000).

A. S. Borovik-Romanov, N. M. Kreines, V. G. Zhotikov, R. Laiho, T. Levola, "Optical Detection of Ferromagnetic Resonance in K_2CuF_4", *J. Phys. C: Solid State Phys.*, **13**, 879 (1980).

Z. Celinski, K. B. Urquhart, B. Heinrich, "Using Ferromagnetic Resonance to Measure the Magnetic Moments of Ultrathin Films", *J. Magn. Magn. Mater.*, **166**(1), 6 (1997).

E. Chappel, M. D. Núñez-Reguiro, F. Dupont, G. Chouteau, C. Darie, A. Sulpice, "Antiferromagnetic Resonance and High-Magnetic-Field Properties of $NaNiO_2$", *Eur. Phys. J. B*, **17**, 609 (2000).

S. Chikazumi, S. H. Charap, *Physics of Magnetism*, Krieger, Malibar, FL, U.S.A., 1978.

M. Hase, K. Katsumata, "Antiferromagnetic Resonance in the Spin-Peierls Compound $CuGeO_3$ doped with Zn", *RIKEN Rev.* (24), 17 (1999).

A. Layadi, J. O. Artman, "A Ferromagnetic Resonance Investigation of Ferromagnetic Coupling", *J. Phys. D: Appl. Phys.*, **30**, 3312 (1997).

L. Li, Q. Shi, M. Mino, Y. Yamazaki, I. Yamada, "Experimental Observation of the Antiferromagnetic Resonance Linewidth in $KCuF_3$", *J. Phys. Condens. Matter*, **17**, 2749 (2005).

V. K. Sharma, A. Baiker, "Superparamagnetic Effects in the Ferromagnetic Resonance of Silica-supported Nickel Particles", *J. Chem. Phys.*, **75**(12), 5596 (1981).

J. Smit, Ed., *Magnetic Properties of Materials*, McGraw-Hill, New York, NY, U.S.A., 1971.

R. F. Soohoo, *Microwave Magnetics*, Harper & Row, New York, NY, U.S.A., 1985.

13.9 GLASSES

K. H. Fischer, J. A. Hertz, *Spin Glasses*, Cambridge University Press, Cambridge, U.K., 1991.

D. L. Griscom, "Electron Spin Resonance in Glasses", *J. Non-Cryst. Solids*, **40**, 211 (1980); "Electron Spin Resonance", *Glass: Sci. Technol.* **4B**, 151 (1990).

13.10 GEOLOGIC/MINERALOGIC SYSTEMS AND SELECTED GEMS

G. Calas, "Electron Paramagnetic Resonance", *Rev. Mineral.* (*Spectrosc. Methods Mineral. Geol.*), **18**, 513 (1988).

J. M. Cubitt, C. V. Burek, "A Bibliography of Electron Spin Resonance Applications in the Earth Sciences", *Geo Abstracts*, Norwich, U.K., 1980.

A. S. Marfunin, *Advanced Mineralogy*, Springer-Verlag, Berlin, Germany 1994.

A. S. Marfunin, *Spectroscopy, Luminescence and Radiation Centers in Minerals*, Engl. transl. by V. Schiffer, Springer Verlag, Berlin, Germany, 1979.

C. P. Poole Jr., H. A. Farach, T. P. Bishop, "Electron Spin Resonance of Minerals, Parts I and II", *Magn. Reson. Rev.*, **4**, 137 (1977); **4**, 225 (1978).

A. B. Vassilikou-Dova, G. Lehmann, "Investigations of Minerals by Electron Paramagnetic Resonance", *Fortschr. Mineral.*, **65**, 173 (1987).

13.10.1 Amethyst

R. T. Cox, "ESR of an $S = 2$ Centre in Amethyst Quartz and Its Possible Identification as the d^4 Ion Fe^{4+} ", *J. Phys. C: Solid State Phys.*, **9**, 3355 (1976); **10**, 4631 (1977).

13.10.2 Beryl and Chrysoberyl

J. S. Ayala-Arenas, R. L. Andrioli Jr., S. Watanabe, M. Matsuoka, B. C. Bhatt, "Radiation Effect on Thermoluminescence and Electronic Paramagnetic Resonance (EPR) of Pink Beryl", *Rad. Phys. Chem.*, **61**, 417 (2001).

S. Isotani, W. W. Furtado, R. Antonini, A. R. Blak, W. M. Pontuschka, T. Tomé, S. R. Rabbani, "Decay-Kinetics Study of Atomic Hydrogen in *a*-Si:(H, O, N) and Natural Beryl", *Phys. Rev. B*, **42**(10), 5966 (1990).

V. P. Solntsev, E. G. Tsvetkov, A. I. Alimpiev, R. I. Mashkovtsev, "Valent State and Coordination of Cobalt Ions in Beryl and Chrysoberyl Crystals", *Phys. Chem. Minerals*, **31**, 1 (2004).

13.10.3 Diamond

J. Isoya, H. Kanda, Y. Uchida, S. C. Lawson, S. Yamasaki, H. Itoh, Y. Morita, "EPR Identication of the Negatively Charged Vacancy in Diamond", *Phys. Rev. B*, **45**(3), 1436 (1992).

J. Isoya, H. Kanda, J. R. Norris, J. Tang, M. K. Bowman, "Fourier-Transform and Continuous-Wave EPR Studies of Nickel in Synthetic Diamond; Site and Spin Multiplicity", *Phys. Rev. B*, **41**(7), 3905 (1990).

J. H. N. Loubser, J. A. van Wyk, "Electron Spin Resonance in the Study of Diamond", *Rep. Prog. Phys.*, **41**, 1201 (1978).

M. E. Newton, J. M. Baker, "^{14}N ENDOR of the OK1 Centre in Natural Type Ib Diamond", *J. Phys. Condens. Matter*, **1**, 10549 (1989).

13.10.4 Emerald

A. Edgar, D. R. Hutton, "Exchange-Coupled Pairs of Cr^{3+} Ions in Emerald", *J. Phys. C.: Solid State Phys.*, **11**, 5051 (1978).

13.10.5 Opal

D. R. Hutton, G. J. Troup, M. Young, "EPR/ESR Spectra of Natural and Synthetic Opals: A Pilot Study", *Austral. Gemol.*, **19**(9), 365 (1997).

13.10.6 Rock Crystal (α-Quartz)

J. A. Weil, "A Review of Electron Spin Spectroscopy and Its Application to the Study of Paramagnetic Defects in Crystalline Quartz", *Phys. Chem. Minerals*, **10**, 149 (1984).

Also see Section 13.12.1.2 below.

13.10.7 Ruby

G. Kido, N. Miura, "Far-Infrared Electron Spin Resonance of Ruby in Very High Magnetic Fields", *Appl. Phys. Lett.*, **41**(6), 569 (1982).

Z. Min-Guang, X. Ji-An, B. Gui-Ru, X. Huong-Sen, "d-Orbital Theory and High-Pressure Effects upon the EPR Spectrum of Ruby", *Phys. Rev. B*, **27**(3), 1516 (1983); erratum; **36**(13), 7184 (1987).

H. M. Nelson, D. B. Larson, J. H. Gardner, "Very High Pressure Effects upon the EPR Spectrum of Ruby", *J. Chem. Phys.*, **47**(6), 1994 (1967).

J. R. Pilbrow, G. R. Sinclair, D. R. Hutton, G. J. Troup, "Asymmetric Lines in Field-Swept EPR: Cr^{3+} Looping Transitions in Ruby", *J. Magn. Reson.*, **52**, 386 (1983).

E. O. Schulz-DuBois, "Paramagnetic Spectra of Substituted Sapphires—Part I: Ruby", *Bell Syst. Techn. J.*, **38**, 271 (1959).

13.10.8 Sapphire

D. A. Schwartz, E. D. Walter, S. J. McIlwain, V. N. Krymov, D. J. Singel, "High-Frequency (94.9 GHz) EPR Spectroscopy of Paramagnetic Centers in a Neutron-Irradiated Sapphire Single-Crystal Fiber", *Appl. Magn. Reson.*, **16**, 223 (1999).

G. J. Troup, D. R. Hutton, J. R. Pilbrow, "Electron Spin Resonance Spectra of Natural and Synthetic Sapphires at S-Band (2-4 GHz)", *J. Gemol.*, **19**(5), 431 (1985).

13.10.9 Topaz

D. N. da Silva, K. J. Duedes, M. V. B. Pinheiro, S. Schweizer, J.-M. Spaeth, K. Krambrock, "The $O^-(Al_2)$ Centre in Topaz and Its Relation to the Blue Color", *Phys. Stat. Sol. C*, **2**(1), 397 (2005).

A. C. Dickinson, W. J. Moore, "Paramagnetic Resonance of Metal Ions and Defect Centers in Topaz", *J. Phys. Chem.*, **71**, 231 (1967).

V. Priest, D. L. Cowan, D. G. Reichel, F. K. Ross, "A Dangling Silicon-Bond Defect in Topaz", *J. Appl. Phys.*, **68**(6), 3035 (1990).

J. R. Thyer, S. M. Quick, F. Holuj, "ESR Spectrum of Fe^{3+} in Topaz. I. Fine Structure", *Can. J. Phys.*, **45**, 3597 (1967).

13.10.10 Tourmaline

K. Krambrock, M. V. B. Pinheiro, K. J. Guedes, S. M. Medeiros, S. Schweizer, J.-M. Spaeth, "Correlation of Irradiation-Induced Yellow Color with the O^- Hole Center in Tourmaline", *Phys. Chem. Minerals*, **31**, 168 (2004).

13.10.11 Turquoise

J. Diaz, H. A. Farach, C. P. Poole Jr., "An Electron Spin Resonance and Optical Study of Turquoise", *Am. Miner.*, **56**, 773 (1971).

K. B. N. Sharma, L. R. Moorthy, B. J. Reddy, S. Vedanand, "EPR and Electronic Absorption Spectra of Copper-bearing Turquoise Mineral", *Phys. Lett.*, **132**(5), 293 (1988).

13.10.12 Zircon

W. C. Tennant, R. F. C. Claridge, C. J. Walsby, N. S. Lees, "Point Defects in Crystalline Zircon (Zirconium Silicate), $ZrSiO_4$: Electron Paramagnetic Resonance Studies", *Phys. Chem. Minerals*, **31**, 203 (2004).

13.11 LIQUID CRYSTALS

J. Bulthuis, C. W. Hilbers, C. MacLean, *NMR and ESR in Liquid Crystals, Magnetic Resonance*, MTP International Review of Science, Series 1, Vol. 4, C. A. MacDowell, Ed., Butterworths, London, U.K., 1972, p. 201.

13.12 "POINT" DEFECTS

13.12.1 Insulators

13.12.1.1 Alkali Halides J. H. Crawford Jr., L. M. Slifkin, *Point Defects in Solids*, Vol. 1, Plenum Press, New York, NY, U.S.A., 1972.

J. J. Markham, "F-Centers in Alkali Halides", Suppl. 8 of *Solid State Physics*, F. Seitz, D. Turnbull, Eds., Academic, New York, NY, U.S.A., 1966, Chapters 6–8.

H. Seidel, H. C. Wolf, "ESR and ENDOR Spectroscopy of Color Centers in Alkali Halide Crystals", in *Physics of Color Centers*, W. B. Fowler, Ed., Academic Press, New York, NY, U.S.A., 1968, Chapter 8.

A. M. Stoneham, *Theory of Defects in Solids*, Clarendon, Oxford, U.K., 1975.

13.12.1.2 Oxides J. H. Crawford Jr., L. M. Slifkin, *Point Defects in Solids*, Vol. 1, Plenum Press, New York, NY, U.S.A., 1972.

B. Henderson, J. E. Wertz, "Defects in the Alkaline-Earth Oxides", *Adv. Phys.*, **17**, 749 (1968); also Halsted-Wiley, New York, NY, U.S.A., 1977.

J. A. Weil, "A Demi-Century of Magnetic Defects in α-Quartz", in *Defects in SiO$_2$ and Related Dielectrics: Science and Technology*, G. Pacchioni, L. Skuja, D. L. Griscom, Eds., Kluwer Academic, Dordrecht Netherlands, 2000, pp. 197 ff.

13.12.2 Semiconductors

J. Bourgoin, M. Lannoo, *Point Defects in Semiconductors*, Springer, Berlin, Germany, 1983, Chapter 3.

G. Feher, "Review of Electron Spin Resonance Experiments in Semiconductors", in *Paramagnetic Resonance*, Vol. II, W. Low, Ed., Academic Press, New York, NY, U.S.A., 1963, p. 715.

B. Henderson, *Defects in Crystalline Solids*, Arnold, London, U.K., 1972, Chapter 5.

T. A. Kennedy, "EPR in Semiconductors", *Magn. Reson. Rev.*, **7**(1), 41 (1981).

G. Lancaster, *ESR in Semiconductors*, Hilger & Watts, London, U.K., 1972, Chapter 5.

G. W. Ludwig, H. H. Woodbury, "Electron Spin Resonance in Semiconductors", in *Solid State Physics*, Vol. 13, F. Seitz, D. Turnbull, Eds., Academic Press, New York, NY, U.S.A., 1962, pp. 223–304.

13.13 POLYMERS

P.-O. Kinell, B. G. Ränby, V. Runnström-Reio, Eds., *ESR Applications to Polymer Research*, Halsted-Wiley, New York, NY, U.S.A., 1973.

B. G. Ränby, J. F. Rabek, *ESR Spectroscopy in Polymer Research*, Springer, Berlin, Germany, 1977.

S. Schlick, G. Jeschke, "Electron Spin Resonance", in *Encyclopedia of Polymer Science and Engineering*, J. I. Kroschwitz, Ed., Wiley-Interscience, New York, NY, U.S.A., 2004.

13.14 RADIATION DOSAGE AND DATING

K. Beerten, A. Stesmans, "Single-Quartz-Grain ESR Dating of a Contemporary Desert Surface Deposit (Eastern Desert, Egypt)", *Quart. Sci. Rev.*, **24**, 223 (2005).

R. Grün, "Electron Spin Resonance (ESR) Dating", *Quart. Int.*, **1**, 65 (1989).

M. Ikeya, "Use of Electron Spin Resonance Spectrometry in Microscopy, Dating and Dosimetry", *Anal. Sci.*, **5**, 5 (1989).

H. K. Lee, H. P. Schwarcz, "Criteria for Complete Zeroing of ESR Signals during Faulting of the San Gabriel Fault Zone, Southern California", *Tectonophysics*, **235**, 317 (1994).

W. L. McLaughlin, "ESR Dosimetry", *Radiat. Protect. Dosim.*, **47**, 255 (1993).

S. Toyoda, "Formation and Decay of the E_1' Center and its Precursors in Natural Quartz: Basics and Applications", *Appl. Rad. Isot.*, **62**(2), 325 (2005).

A. Wieser, D. F. Regulla, "Ultra-high Level Dosimetry by ESR Spectroscopy of Crystalline Quartz and Fused Silicate", *Rad. Protect. Doc.*, **34**, 291 (1990).

N. D. Yordanov, V. Gancheva, in *EPR of Free Radicals in Solids*, A. Lund, M. Shiotane, Eds., Kluwer, Dordrecht Netherlands, 2003, Chapter 14.

13.15 SPIN LABELS

L. J. Berliner, Ed., *Spin Labeling—Theory and Applications*, Academic Press, New York, NY, U.S.A., 1976.

L. J. Berliner, Ed., *Spin Labeling II—Theory and Applications*, Academic Press, New York, NY, U.S.A., 1979.

J. L. Holtzman, *Spin Labeling in Pharmacology*, Academic Press, New York, NY, U.S.A., 1984.

G. I. Likhtenshtein, *Spin Labeling Methods in Molecular Biology*, Books Demand UMI, Ann Arbor, MI, U.S.A., 1976.

13.16 SPIN TRAPS

G. R. Buettner, "Spin Trapping: ESR Parameters of Spin Adducts", *Free Radical Biol. Med.*, **3**, 259 (1987).

E. G. Janzen, D. L. Haire, "Two Decades of Spin Trapping", in *Advances in Free Radical Chemistry*, Vol. 1, D. D. Tanner, Ed., Greenwich, CT, U.S.A., 1990, pp. 253–295.

D. Rehorek, "Spin Trapping of Inorganic Radicals", *Chem. Soc. Rev.*, **20**, 341 (1991).

13.17 TRAPPED ATOMS AND MOLECULES

K.-P. Dinse, "FT-EPR and Pulsed ENDOR Studies of Encapsulated Atoms and Ions", *Electron Paramagn. Reson.*, **17**, 78 (2000).

W. Weltner Jr., *Magnetic Atoms and Molecules*, Van Nostrand Reinhold, New York, NY, U.S.A., 1983.

APPENDIX A

MATHEMATICAL OPERATIONS

Obviously, there is plentiful use of mathematics in this text, used to describe EPR spectroscopy. This really needs no justification beyond 'It works superbly', and enables rationalization and predictivity. This appendix and the next present a number of mathematical techniques and equations for the convenience of the reader. Although we have attempted to summarize accurately some of the most useful relations, we make no attempt at rigor. A bibliography is included at the end of this appendix for further reading.

A.1 COMPLEX NUMBERS

A complex scalar quantity may be represented as follows

$$u = x + iy = re^{+i\phi} \tag{A.1}$$

where x, y, r and ϕ are real numbers, $e^{i\phi} = \cos \phi + i \sin \phi$ and $i^2 = -1$. One refers to x and y as the real and imaginary components of u [$re(u)$ and $im(u)$], whereas r is the absolute magnitude of u, that is, $r = |u|$. ϕ is called the *phase angle*. The complex conjugate of u, namely, u^*, is obtained by changing the sign of i; that is, $u^* = x - iy$. For functions of complex numbers, the complex-conjugate function is obtained by changing the sign in front of i wherever it appears. The relation between complex numbers and their conjugates is clarified by representing them

Electron Paramagnetic Resonance, Second Edition, by John A. Weil and James R. Bolton
Copyright © 2007 John Wiley & Sons, Inc.

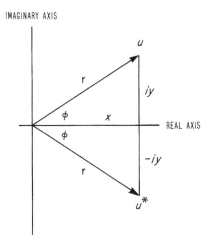

FIGURE A.1   Representation of a point u and of its complex conjugate u^* in complex space (Argand diagram).

as points in the 'complex plane' (Argand diagram; see Fig. A.1). The abscissa is chosen to represent the real axis (x) and the ordinate the imaginary axis (y). Note that re(u) is equal to one-half the sum of u and u^*; the product of u and its complex conjugate is the square of the absolute magnitude, that is,

$$u\,u^* = u^*u = |u|^2 = e^{-i\phi}r\,e^{i\phi} = r^2 \tag{A.2}$$

In this book, in taking square roots, we imply non-negative quantities, for example, $|u| = (u^*u)^{1/2}$. To standardize the format of a complex number of form $1/(x+iy)$, multiply its numerator and denominator by $x - iy$.

A.2   OPERATOR ALGEBRA

A.2.1   Properties of Operators

An operator $\hat{A}$ is a symbolic instruction to carry out a stipulated mathematical operation on some function called the *operand*. Unless its form is explicitly indicated, an operator is designated by a circumflex (^). One of the simplest (and trivial) operators is a constant multiplier; for example, $\hat{k}a = ka$.

An operator $\hat{\Omega}$ is said to be linear if the result of operation on a sum of functions is the same as that obtained by operating on each function separately; that is, if $\hat{\Omega}\alpha = \beta$, then

$$\hat{\Omega}(\alpha_1 + \alpha_2) = \hat{\Omega}\alpha_1 + \hat{\Omega}\alpha_2 = \beta_1 + \beta_2 \tag{A.3}$$

Also, if c is a constant, then

$$\hat{\Omega}(c\alpha) = c\hat{\Omega}\alpha = c\beta \tag{A.4}$$

If $\alpha_1 = f(q_i)$ for some continuous variable q_i, then $\partial/\partial q_i$ is an example of a linear operator. An example of a non-linear operator is $\sqrt{\;}$.

The reader should be familiar with such operators as the summation operator Σ taken to mean

$$\sum_{i=1}^{n} a_i \equiv a_1 + a_2 + a_3 + \cdots + a_n \tag{A.5a}$$

Its use permits a concise representation of a series. Similarly, the product operator Π

$$\prod_{i=1}^{n} a_i = a_1 a_2 a_3 \cdots a_n \tag{A.5b}$$

is at times useful.

Frequently one wishes to summarize a set of equations with constant coefficients such as

$$\psi_1 = c_{11}\phi_1 + c_{12}\phi_2 + c_{13}\phi_3 + \cdots + c_{1n}\phi_n \tag{A.6a}$$
$$\psi_2 = c_{21}\phi_1 + c_{22}\phi_2 + c_{23}\phi_3 + \cdots + c_{2n}\phi_n \tag{A.6b}$$
$$\psi_3 = c_{31}\phi_1 + c_{32}\phi_2 + c_{33}\phi_3 + \cdots + c_{3n}\phi_n \tag{A.6c}$$
$$= \cdots\cdots\cdots\cdots\cdots\cdots\cdots\cdots\cdots$$
$$= \cdots\cdots\cdots\cdots\cdots\cdots\cdots\cdots\cdots$$

The sum of the functions ψ_j can be considered to be consecutive operations on the functions ϕ_k, represented by a double summation

$$\sum_{j} \psi_j = \sum_{j}\sum_{k} c_{jk}\phi_k \tag{A.7}$$

Here one encounters a juxtaposition of two operators, which in the general case are represented as $\hat{A}\hat{B}$. It is understood that $\hat{A}\hat{B}$ implies operation first toward the right with $\hat{B}$ and then with $\hat{A}$. Interchange of the order of the two operators *may* give a different result; for example

$$\hat{x}\frac{\hat{d}}{dx}(x^2) = 2x^2 \quad \text{but} \quad \frac{\hat{d}}{dx}\hat{x}(x^2) = 3x^2$$

where the quantity in parentheses on the left is the operand.[1] If $\hat{A}\hat{B} = \hat{B}\hat{A}$, then $\hat{A}$ and $\hat{B}$ are said to be commuting operators. The difference $\hat{A}\hat{B} - \hat{B}\hat{A}$ is called the *commutator* of $\hat{A}$ and $\hat{B}$ and is represented[2] by the symbol $[\hat{A}, \hat{B}]$. The commutator of two operators is of profound significance in quantum-mechanical systems. The commutators of angular-momentum operators are treated in Appendix B.

A spatial operator $\hat{\Omega}$ is said to be *hermitian* if it obeys the following relation

$$\int_\tau \psi_j^* \hat{\Omega} \psi_k \, d\tau = \int_\tau (\hat{\Omega}^* \psi_j^*) \psi_k \, d\tau \tag{A.8}$$

where ψ_j and ψ_k are continuous and well-behaved functions of τ. Here τ can stand for any positional variable (length or angle), or some multiple such variables (e.g., $d\tau = dx \, dy \, dz$) so that $\int$ represents a multiple integral. A useful aspect of the hermitian property is that (with care!) one may operate 'backward', that is, to the left, when the operator occurs between two operands, as in Eq. A.8. An example of operation to the left is found in Section B.4. Hermitian operators have the important property that if the result of operating on a function is the function itself multiplied by a constant, one is assured that the constant is *real* (Section A.2.2).

Some of the most important operators of quantum mechanics are those associated with observable properties of a physical system, that is, the 'dynamical variables'. A few important linear operators are listed in Table A.1. Some of these operators are identical with the variable itself, whereas others involve derivatives.

TABLE A.1   Classical and Quantum-Mechanical Variables

Variable	Classical Quantity	Quantum-Mechanical Quantity
Mass	m	m
Position	$q \, (=x, y, z)$	q
Time	t	t
Linear momentum	$p_q = m\dfrac{\partial q}{\partial t}$	$\hat{p}_q = -i\hbar\dfrac{\partial}{\partial q}$
Angular momentum [a] about axis z	$\ell_z = xp_y - yp_x$	$\hat{\ell}_z = -i\hbar\left(x\dfrac{\partial}{\partial y} - y\dfrac{\partial}{\partial x}\right)$ $= -i\hbar\dfrac{\partial^b}{\partial\phi}$
Kinetic energy associated with coordinate q	$T = \dfrac{p_q^2}{2m}$	$\hat{\mathcal{H}} = \dfrac{\hat{p}_q^2}{2m} = -\dfrac{\hbar^2}{2m}\dfrac{\hat{\partial}^2}{\partial q^2}\,^c$
Potential energy [d]	$V(\mathbf{r})$	$V(\mathbf{r})$

[a] See Eqs. B.5–B.7.
[b] The angle ϕ measures rotation about the z axis.
[c] This form of the kinetic-energy hamiltonian is valid only for cartesian coordinates.
[d] Here $\mathbf{r} = x\mathbf{i} + y\mathbf{j} + z\mathbf{k}$ $[= r(\sin\theta\cos\phi\mathbf{i} + \sin\theta\sin\phi\mathbf{j} + \cos\theta\mathbf{k})$ in polar coordinates]. See Section A.4.

A.2.2 Eigenvalues and Eigenfunctions

If the result of the application of an operator $\hat{\Lambda}$ to a function ψ_k is

$$\hat{\Lambda}\psi_k = \lambda_k\psi_k \tag{A.9}$$

where λ_k is a constant, the ψ_k is said to be an *eigenfunction* of $\hat{\Lambda}$ with *eigenvalue* λ_k. Note that $c\psi_k$, where c is any non-zero scalar, is also an eigenfunction. When ψ_k is a function of spatial variables, it is also called a 'wavefunction'. When ψ_k deals with spin variables, there is no such functional dependence. The term 'eigenstate' may be used interchangeably with 'eigenfunction'.

The set of all eigenfunctions ψ_k is often called a 'basis set'. The electron-spin functions $\psi_{\alpha(e)}$ and $\psi_{\beta(e)}$ introduced in Section 2.3.1 are examples of eigenfunctions, in this case of the operator $\hat{S}_z$:

$$\hat{S}_z\psi_{\alpha(e)} = +\tfrac{1}{2}\psi_{\alpha(e)} \tag{A.10a}$$

$$\hat{S}_z\psi_{\beta(e)} = -\tfrac{1}{2}\psi_{\beta(e)} \tag{A.10b}$$

(Angular-momentum operator expressions are considered in detail in Sections B.1–B.4.)

A given set of eigenfunctions ψ_k may simultaneously be eigenfunctions of several operators. Operators having the same set of eigenfunctions have the very useful property that *the operators must commute*. In the case of the particle of mass m in a ring of radius r considered in Section 1.7, the wavefunctions ψ_k are eigenfunctions of both the angular-momentum operator $\hat{\ell}_z$ and the hamiltonian operator $\hat{\mathcal{H}}$. The eigenvalue equations are

$$\hat{\ell}_z\psi = P\psi \tag{A.11}$$

$$\hat{\mathcal{H}}\psi = U\psi \tag{A.12}$$

Table A.1 gives $\hat{\ell}_z = -i\hbar\, d/d\phi$, where ϕ measures the angular position of the particle. The kinetic energy of a classical particle having an angular momentum P (along $\mathbf{z}$, which is normal to the ring) and moment of inertia $I_o\ (=mr^2)$ is

$$U = \frac{P^2}{2I_o} \tag{A.13}$$

The hamiltonian operator for a system with potential energy $V = 0$ is

$$\hat{\mathcal{H}} = \frac{\hat{\ell}_z^{\,2}}{2I_o} = \frac{(-i\hbar)^2}{2I_o}\frac{\hat{d}^2}{d\phi^2} = \frac{-\hbar^2}{2I_o}\frac{\hat{d}^2}{d\phi^2} \tag{A.14}$$

Substitution of $\hat{\mathcal{H}}$ from Eq. A.14 into Eq. A.12 gives

$$\frac{-\hbar^2}{2I_o}\frac{\hat{d}^2\psi}{d\phi^2} = U\psi \tag{A.15}$$

Rearranging, one finds

$$\frac{\hat{d}^2\psi}{d\phi^2} = \frac{-2I_oU\psi}{\hbar^2} = -M^2\psi \tag{A.16}$$

Here the constant $2I_oU/\hbar^2$ has been denoted by M^2. Two solutions of Eq. A.16 are

$$\psi_1 = Ae^{+iM\phi} \tag{A.17a}$$

$$\psi_2 = Ae^{-iM\phi} \tag{A.17b}$$

as is evident by substitution. From the requirement that the functions ψ be normalized, that is, that

$$\int_0^{2\pi} \psi^*\psi \, d\phi = 1 \tag{A.18}$$

one finds the amplitude $A = (2\pi)^{-1/2}$. Hence

$$\psi_1 = \frac{1}{\sqrt{2\pi}}e^{+iM\phi} \tag{A.19a}$$

$$\psi_2 = \frac{1}{\sqrt{2\pi}}e^{-iM\phi} \tag{A.19b}$$

Insertion of ψ_1 into Eq. A.12 gives

$$\frac{-\hbar^2}{2I_o}\frac{\hat{d}^2}{d\phi^2}\left(\frac{1}{\sqrt{2\pi}}e^{+iM\phi}\right) = \frac{M^2\hbar^2}{2I_o}\left(\frac{1}{\sqrt{2\pi}}e^{+iM\phi}\right) \tag{A.20}$$

Hence the eigenvalue U of the operator $\hat{\mathcal{H}}$, corresponding to the eigenfunction ψ_1, is $M^2\hbar^2/2I_o$. Use of ψ_2 gives an identical energy value.

Operation by $\hat{\ell}_z$ on ψ_1 and ψ_2 gives the following results:

$$-i\hbar\frac{\hat{d}}{d\phi}\left(\frac{1}{\sqrt{2\pi}}e^{+iM\phi}\right) = +M\hbar\left(\frac{1}{\sqrt{2\pi}}e^{+iM\phi}\right) \tag{A.21a}$$

$$-i\hbar\frac{\hat{d}}{d\phi}\left(\frac{1}{\sqrt{2\pi}}e^{-iM\phi}\right) = -M\hbar\left(\frac{1}{\sqrt{2\pi}}e^{-iM\phi}\right) \tag{A.21b}$$

Hence the eigenvalues P of $\hat{\ell}$, corresponding to the eigenfunctions ψ_1 and ψ_2, are $+M\hbar$ and $-M\hbar$. These correspond to the two directions of rotation around the ring.

The wavefunctions can be eliminated from Eq. A.20 or Eq. A.21 by multiplication on the left by the corresponding complex-conjugate function ψ^*, followed by integration. This yields the expressions for the energy and for the angular momentum of a particle moving in a circle (Eq. 1.2 and Problem A.3).

A.3   DETERMINANTS

A determinant is a scalar quantity that represents a linear combination of products of terms. It may be represented by a square array, for example

$$det[\mathbf{A}^{(2)}] = \begin{vmatrix} A_{11} & A_{12} \\ A_{21} & A_{22} \end{vmatrix} = A_{11}A_{22} - A_{21}A_{12} \tag{A.22a}$$

For evaluation of $det[\mathbf{A}^{(\zeta)}]$, with $\zeta > 2$, see the technique described below. At times, determinants are denoted by a pair of double vertical lines (e.g., Eq. A.28b). Generally a determinant of order k is represented as

$$det[\mathbf{A}^{(k)}] = \begin{vmatrix} A_{11} & A_{12} & \cdots & A_{1k} \\ A_{21} & A_{22} & \cdots & A_{2k} \\ \cdot & & & \\ \cdot & \cdots & \cdots & \cdots \\ \cdot & & & \\ A_{k1} & A_{k2} & \cdots & A_{kk} \end{vmatrix} \tag{A.22b}$$

A determinant may be expanded by the 'method of minors'. The minor of any element A_{ij} is the determinant remaining after the row and column containing the element A_{ij} are removed. The expansion is carried out by multiplying the elements of a specific row or column by their corresponding minors as follows:

$$det[\mathbf{A}^{(k)}] = \sum_{i \text{ or } j} (-1)^{(i+j)} A_{ij} \{det[\mathbf{A}^{(k-1)}]_{ij}\} \tag{A.23}$$

Here $det[\mathbf{A}^{(k-1)}]_{ij}$ is the minor corresponding to the element A_{ij}.

For a determinant of order 3 this expansion may be carried out as follows:

$$\begin{vmatrix} A_{11} & A_{12} & A_{13} \\ A_{21} & A_{22} & A_{23} \\ A_{31} & A_{32} & A_{33} \end{vmatrix} = A_{11} \begin{vmatrix} A_{22} & A_{23} \\ A_{32} & A_{33} \end{vmatrix} - A_{12} \begin{vmatrix} A_{21} & A_{23} \\ A_{31} & A_{33} \end{vmatrix} + A_{13} \begin{vmatrix} A_{21} & A_{22} \\ A_{31} & A_{32} \end{vmatrix}$$

$$= A_{11}A_{22}A_{33} - A_{11}A_{23}A_{32} - A_{12}A_{21}A_{33} +$$
$$A_{12}A_{23}A_{31} + A_{13}A_{21}A_{32} - A_{13}A_{22}A_{31} \tag{A.24}$$

Here the elements of the first row have been used. One could equally well have used the elements of any other row or column. The method of minors is a valuable technique for step-wise reduction of the order of a determinant. For example, a determinant of order 4 can be reduced in one step to a linear combination of four determinants of order 3.

The value of a determinant is not affected by the addition (or subtraction) of the elements of one row to those of another. This is also true for columns. Thus, if any row or column is a multiple of another, the value of the determinant is zero.

Determinants are most frequently used for obtaining solutions to sets of simultaneous equations. Consider the following set of simultaneous equations relating the dependent variables y_1, y_2, y_3 to the independent variables x_1, x_2, x_3:

$$y_1 = c_{11}x_1 + c_{12}x_2 + c_{13}x_3 \tag{A.25a}$$

$$y_2 = c_{21}x_1 + c_{22}x_2 + c_{23}x_3 \tag{A.25b}$$

$$y_3 = c_{31}x_1 + c_{32}x_2 + c_{33}x_3 \tag{A.25c}$$

The solutions may be represented as follows:

$$x_1 = \frac{det(\Delta_1)}{det(\Delta)} \quad x_2 = \frac{det(\Delta_2)}{det(\Delta)} \quad x_3 = \frac{det(\Delta_3)}{det(\Delta)} \tag{A.26}$$

Here

$$det(\Delta) = \begin{vmatrix} c_{11} & c_{12} & c_{13} \\ c_{21} & c_{22} & c_{23} \\ c_{31} & c_{32} & c_{33} \end{vmatrix} \tag{A.27a}$$

and

$$det(\Delta_1) = \begin{vmatrix} y_1 & c_{12} & c_{13} \\ y_2 & c_{22} & c_{23} \\ y_3 & c_{32} & c_{33} \end{vmatrix} \tag{A.27b}$$

$det(\Delta_2)$ or $det(\Delta_3)$ is obtained in an analogous fashion by replacing column 2 or column 3 of $det(\Delta)$ by

$$\begin{matrix} y_1 \\ y_2 \\ y_3 \end{matrix}$$

The signal that the simultaneous equations are not independent is that the value of $det(\Delta)$ is found to be zero. One of the most important of the applications of determinants is the solution of secular equations (Sections 6.3.1, 9.A and C.1.3) for the energies of a quantum-mechanical system.

Determinants are often used to represent antisymmetrized wavefunctions because interchange of two electrons corresponds to interchange of two rows of the determinant. This changes the sign of the wavefunction as required by the Pauli exclusion principle. For example, a two-electron wavefunction is written

$$\Psi = \frac{1}{\sqrt{2!}} \begin{vmatrix} \psi(1)\alpha(1) & \psi(1)\beta(1) \\ \psi(2)\alpha(2) & \psi(2)\beta(2) \end{vmatrix}$$

$$= \psi(1)\psi(2)\frac{1}{\sqrt{2!}}[\alpha(1)\beta(2) - \alpha(2)\beta(1)] \tag{A.28a}$$

Equation A.28a sometimes is written in abbreviated form:

$$\Psi = \frac{1}{\sqrt{2!}} \left\| \psi(1)\alpha(1) \quad \psi(2)\beta(2) \right\| \tag{A.28b}$$

A.4 VECTORS: SCALAR, VECTOR AND OUTER PRODUCTS

Mathematicians define vectors as arrays of scalars, arranged in single rows or columns. The scalars defining the vector, in a given basis, are called its *components*. In ordinary three-dimensional space, for a vector $\mathbf{r}$, there are three components (r_x, r_y and r_z) defined relative to an orthogonal basis set of unit vectors $\mathbf{i}$, $\mathbf{j}$ and $\mathbf{k}$ in the positive x, y and z directions. Such vectors $\mathbf{r}$ are entities with magnitudes and directions. Vectors are indicated in this book by boldface type, for example, $\mathbf{B}$. The *magnitude r* of $\mathbf{r}$ is the non-negative quantity

$$r = [r_x^2 + r_y^2 + r_z^2]^{1/2} \tag{A.29}$$

If the coordinates of a point with respect to fixed axes x, y and z are 7, -3 and 4, the vector $\mathbf{r}$ locating the point is written as

$$\mathbf{r} = 7\mathbf{i} - 3\mathbf{j} + 4\mathbf{k} \tag{A.30}$$

Very often, vectors are denoted as rows or columns setting out their components in brackets or parentheses. Thus

$$\mathbf{r} = \begin{bmatrix} 7 \\ -3 \\ 4 \end{bmatrix} \tag{A.31a}$$

$$\mathbf{r}^T = \begin{bmatrix} 7 & -3 & 4 \end{bmatrix} \tag{A.31b}$$

Both refer to the *same* vector. Here superscript 'T' implies transposition from column format to row format. A definite choice of basis vectors **i**, **j**, **k** is implicit in this notation.

Addition of vectors such as **r** implies summation of their x, y and z components. Suppose that a second vector is given by

$$\mathbf{s} = 3\mathbf{i} + 4\mathbf{j} - 6\mathbf{k} \tag{A.32}$$

then the sum and difference are

$$\mathbf{r} + \mathbf{s} = (7+3)\mathbf{i} + (-3+4)\mathbf{j} + (4-6)\mathbf{k}$$
$$= 10\mathbf{i} + \mathbf{j} - 2\mathbf{k} \tag{A.33a}$$
$$\mathbf{r} - \mathbf{s} = 4\mathbf{i} - 7\mathbf{j} + 10\mathbf{k} \tag{A.33b}$$

Three types of multiplication of two vector quantities are defined:

1. The *scalar* product (also called an *inner* product) $\mathbf{a}^T \cdot \mathbf{b}$ of vectors **a** and **b** is defined to be $ab \cos \theta_{a,b}$, where $\theta_{a,b}$ is the angle between **a** and **b**. If one writes $\mathbf{a} = a_x\mathbf{i} + a_y\mathbf{j} + a_z\mathbf{k}$ and $\mathbf{b} = b_x\mathbf{i} + b_y\mathbf{j} + b_z\mathbf{k}$, then

$$\mathbf{a}^T \cdot \mathbf{b} = \begin{bmatrix} a_x & a_y & a_z \end{bmatrix} \cdot \begin{bmatrix} b_x \\ b_y \\ b_z \end{bmatrix}$$
$$= a_x b_x + a_y b_y + a_z b_z \tag{A.34}$$

since $\mathbf{i}^T \cdot \mathbf{i} = \mathbf{j}^T \cdot \mathbf{j} = \mathbf{k}^T \cdot \mathbf{k} = 1$ and $\mathbf{i}^T \cdot \mathbf{j} = \mathbf{i}^T \cdot \mathbf{k} = \mathbf{j}^T \cdot \mathbf{k} = 0$. The vectors **i**, **j** and **k** are said to be *orthonormal*. Note that $\mathbf{a}^T \cdot \mathbf{b} = \mathbf{b}^T \cdot \mathbf{a}$. If a and b are complex quantities, the scalar product is taken as $(\mathbf{a}^*)^T \cdot \mathbf{b}$.

2. The *vector* product $\mathbf{c} = \mathbf{a} \wedge \mathbf{b}$ of vectors **a** and **b** is a vector **c** perpendicular to the plane containing **a** and **b**; it is drawn from the origin of **a** and **b** and is of length $ab \sin \theta_{a,b}$. The sense of the vector is obtained from the right-hand rule; if the right forefinger is parallel to **a** and the middle finger parallel to **b**, then the thumb indicates the direction of the vector product **c**. Considering the unit vectors **i**, **j** and **k**, one notes that $\mathbf{i} \wedge \mathbf{i} = \mathbf{j} \wedge \mathbf{j} = \mathbf{k} \wedge \mathbf{k} = 0$; $\mathbf{i} \wedge \mathbf{j} = \mathbf{k}$, $\mathbf{j} \wedge \mathbf{k} = \mathbf{i}$, $\mathbf{i} \wedge \mathbf{k} = -\mathbf{j}$ and so on. Expansion of $\mathbf{a} \wedge \mathbf{b}$ in terms of its components yields

$$\mathbf{c} = \mathbf{a} \wedge \mathbf{b} = (a_x\mathbf{i} + a_y\mathbf{j} + a_z\mathbf{k}) \wedge (b_x\mathbf{i} + b_y\mathbf{j} + b_z\mathbf{k})$$
$$= a_x b_y \mathbf{k} - a_x b_z \mathbf{j} - a_y b_x \mathbf{k} + a_y b_z \mathbf{i} + a_z b_x \mathbf{j} - a_z b_y \mathbf{i}$$
$$= (a_y b_z - a_z b_y)\mathbf{i} + (a_z b_x - a_x b_z)\mathbf{j} + (a_x b_y - a_y b_x)\mathbf{k} \tag{A.35a}$$

Note that the result in Eq. A.35a could have been obtained directly by writing a mnemonic 'determinant'

$$\mathbf{a} \wedge \mathbf{b} = \begin{vmatrix} \mathbf{i} & \mathbf{j} & \mathbf{k} \\ a_x & a_y & a_z \\ b_x & b_y & b_z \end{vmatrix} \tag{A.35b}$$

with the unit coordinate vectors in the first row and the components of $\mathbf{a}$ and $\mathbf{b}$ in the second and third rows. One important use of vector products is in the description of the components of angular momentum (see Table A.1 and Appendix B).

3. A third type of product of two three-dimensional vectors is called the *outer product* (Note Section 4.8). The result is a second-rank tensor. The results of such a multiplication are as follows

$$\mathbf{a}\mathbf{b} = \mathbf{C} \tag{A.36a}$$

to be interpreted as

$$\begin{pmatrix} a_1 & a_2 & a_3 \end{pmatrix} \begin{pmatrix} b_1 & b_2 & b_3 \end{pmatrix} = \begin{bmatrix} C_{11} & C_{12} & C_{13} \\ C_{21} & C_{22} & C_{23} \\ C_{31} & C_{32} & C_{33} \end{bmatrix} \tag{A.36b}$$

where $C_{ij} \equiv a_i b_j$.

A.5 MATRICES

A *matrix* is defined as any rectangular array of $n \times m$ numbers or symbols ('matrix elements'), where n is the number of rows and m the number of columns. Note that no rule for doing mathematical operations with the matrix elements is included; that is, a matrix is just an ordered set of numbers. At times, instructions for doing things with the elements may be given, in addition to the matrix or matrices (e.g., Eqs. A.22) to produce the single number called the determinant $det(\mathbf{A})$ of the matrix $\mathbf{A}$. Similarly, $tr(\mathbf{A})$ means sum the diagonal elements. Every matrix is symbolized using a boldface capital letter, for example, $\mathbf{C}$. If $n = m = 1$, then the matrix is a representation of a scalar quantity. If $n = 1$ and $m > 1$, then the resulting row matrix $\mathbf{R}$ may be regarded as one representation of the row vector $\mathbf{r}^T$. If $n > 1$ and $m = 1$, column matrix $\mathbf{C}$ may similarly be regarded as a representation of a column vector $\mathbf{c}$. Thus

$$\mathbf{C} = \mathbf{c} = \begin{bmatrix} c_1 \\ c_2 \\ \cdot \\ \cdot \\ \cdot \\ c_n \end{bmatrix} \qquad \mathbf{R} = \mathbf{r}^T = \begin{bmatrix} r_1 & r_2 & \cdots & r_n \end{bmatrix} \tag{A.37}$$

Generally we use the superscript 'T' to indicate a row vector, that is, the transpose of the corresponding column vector (see Table A.2 for a definition of the transpose operation). Such representations of vectors are a common practice. The notation used for transposition of any general matrix is the same as that for a vector.

A square matrix is one in which $n = m$. This special type of matrix is said to be an nth-order matrix. The square matrix $\mathbf{B}$ may be written

$$\mathbf{B} = \begin{bmatrix} b_{11} & b_{12} & \cdots & b_{1n} \\ b_{21} & b_{22} & \cdots & b_{2n} \\ \cdots & \cdots & \cdots & \cdots \\ b_{n1} & b_{n2} & \cdots & b_{nn} \end{bmatrix} \tag{A.38}$$

If $b_{ij} = b_{ji}$, the matrix is said to be symmetric. The transpose $\mathbf{B}^T$ of matrix $\mathbf{B}$ is obtained from $\mathbf{B}$ by interchanging elements b_{ij} and b_{ji} at all off-diagonal positions.

TABLE A.2   Matrix Transformations

Matrix Symbol	Components	Example
$\mathbf{A}$	A_{ij}	$\begin{bmatrix} 2 & 3+i \\ 4i & 5 \end{bmatrix}$
Transpose($\mathbf{A}$) = $\mathbf{A}^T$	$(\mathbf{A}^T)_{ij} = A_{ji}$	$\begin{bmatrix} 2 & 4i \\ 3+i & 5 \end{bmatrix}$
Complex conjugate($\mathbf{A}$) = $\mathbf{A}^*$	$(\mathbf{A}^*)_{ij} = A_{ij}{}^*$	$\begin{bmatrix} 2 & 3-i \\ -4i & 5 \end{bmatrix}$
Adjoint($\mathbf{A}$) = $\mathbf{A}^\dagger$	$(\mathbf{A}^\dagger)_{ij} = A_{ij}{}^*$	$\begin{bmatrix} 2 & -4i \\ 3-i & 5 \end{bmatrix}$
Inverse($\mathbf{A}$) = $\mathbf{A}^{-1}$	$(\mathbf{A}^{-1})_{ij} = \dfrac{1}{det(\mathbf{A})} \dfrac{\partial\, det(\mathbf{A})}{\partial A_{ij}}$	$\dfrac{14+12i}{340}\begin{bmatrix} 5 & -3-i \\ -4i & 2 \end{bmatrix}$

[a]In this table, $det(\mathbf{A})$ refers to the determinant with elements identical with those of the matrix $\mathbf{A}$. The inverse exists only for a square matrix, the determinant of which is non-zero. The symbolism '$det(\mathbf{A})/\partial A_{ji}$' implies that in taking the derivative with respect to A_{ji}, all other elements are kept fixed. For a 2×2 determinant

$$det(\mathbf{A}) = |\mathbf{A}| = \begin{vmatrix} A_{11} & A_{12} \\ A_{21} & A_{22} \end{vmatrix} = A_{11}A_{22} - A_{21}A_{12}$$

Hence

$$\frac{\partial |A|}{\partial A_{21}} = -A_{12}$$

At the right, for inverse ($\mathbf{A}$), the numerator and denominator have been multiplied by $14 + 12i$ to rationalize the demoninator.

A.5.1   Addition and Subtraction of Matrices

The operation

$$\mathbf{D} = \mathbf{A} + \mathbf{B} \tag{A.39a}$$

or

$$\mathbf{E} = \mathbf{A} - \mathbf{B} \tag{A.39b}$$

is accomplished by adding or subtracting corresponding matrix elements of $\mathbf{A}$ and $\mathbf{B}$; for example, the element d_{ij} is equal to $a_{ij} + b_{ij}$, and the element e_{ij} is equal to $a_{ij} - b_{ij}$. The following numerical examples illustrate the procedure:

$$\begin{bmatrix} 3 & -2 & 7 \\ -2 & 5 & -4 \\ 7 & -4 & 8 \end{bmatrix} + \begin{bmatrix} 6 & 4 & -2 \\ 4 & 2 & 3 \\ -2 & 3 & -5 \end{bmatrix} = \begin{bmatrix} 9 & 2 & 5 \\ 2 & 7 & -1 \\ 5 & -1 & 3 \end{bmatrix} \tag{A.40a}$$

$$\begin{bmatrix} 3 & -2 & 7 \\ -2 & 5 & -4 \\ 7 & -4 & 8 \end{bmatrix} - \begin{bmatrix} 6 & 4 & -2 \\ 4 & 2 & 3 \\ -2 & 3 & -5 \end{bmatrix} = \begin{bmatrix} -3 & -6 & 9 \\ -6 & 3 & -7 \\ 9 & -7 & 13 \end{bmatrix} \tag{A.40b}$$

Note that only matrices of the same dimensions can be added or subtracted.

A.5.2   Multiplication of Matrices

The multiplication of a matrix by a scalar is accomplished by multiplying each element by the scalar, for example

$$6 \begin{bmatrix} 3 & -2 & 7 \\ -2 & 5 & -4 \\ 7 & -4 & 8 \end{bmatrix} = \begin{bmatrix} 18 & -12 & 42 \\ -12 & 30 & -24 \\ 42 & -24 & 48 \end{bmatrix} \tag{A.41}$$

The rules for multiplication of two matrices can be summarized as follows:

1. Two matrices can be multiplied only if the first is a $z \times n$ matrix and the second an $n \times y$ matrix; that is, the number of columns in the first matrix must equal the number of rows in the second matrix. The resulting product matrix has dimensions $z \times y$.

2. Each element of the product matrix is obtained as follows

$$(ab)_{jk} = \sum_{q=1}^{n} a_{jq} b_{qk} \tag{A.42}$$

As a first example, consider the multiplication of a column matrix by a row matrix

$$\begin{bmatrix} 3 & 5 & -4 \end{bmatrix} \cdot \begin{bmatrix} 2 \\ -1 \\ 1 \end{bmatrix} = [3 \times 2] + [5 \times (-1)] + [-4 \times 1] = -3 \qquad (A.43)$$

Note that in this case the answer is a scalar. The result of this type of multiplication is called a 'scalar product', since it corresponds to the scalar product of vectors (Section A.4). For example, the scalar product $\mathbf{B}^{\mathrm{T}} \cdot \hat{\mathbf{S}}$ of the two vectors with the components B_x, B_y, B_z and $\hat{S}_x$, $\hat{S}_y$, $\hat{S}_z$ is

$$\begin{bmatrix} B_x & B_y & B_z \end{bmatrix} \cdot \begin{bmatrix} \hat{S}_x \\ \hat{S}_y \\ \hat{S}_z \end{bmatrix} = B_x\hat{S}_x + B_y\hat{S}_y + B_z\hat{S}_z \qquad (A.44)$$

Next consider the product of a 1×3 matrix and a 3×3 matrix:

$$\begin{bmatrix} 3 & 5 & -4 \end{bmatrix} \cdot \begin{bmatrix} 3 & -2 & 7 \\ -2 & 5 & -4 \\ 5 & -3 & 8 \end{bmatrix} = \begin{bmatrix} -21 & 31 & -31 \end{bmatrix} \qquad (A.45)$$

The product of a 3×3 matrix and a 3×3 matrix results in a 3×3 product matrix; for example, with two symmetrical matrices, one obtains

$$\begin{bmatrix} 3 & -2 & 7 \\ (-2) & \cdots & (5) & \cdots & (-4) \\ 7 & -4 & 8 \end{bmatrix} \cdot \begin{bmatrix} 6 & 4 & (-2) \\ (4) & \cdots & (2) & \cdots \rightarrow & (3) \\ -2 & 3 & (-5) \end{bmatrix} = \begin{bmatrix} -4 & 29 & -47 \\ 16 & -10 & 39 \\ 10 & 44 & -66 \end{bmatrix}$$

$$(A.46)$$

It perhaps is clearer if one calculates a few elements of the product presented above. For example, the element a_{23} in the product matrix is $(-2)(-2) + (5)(3) + (-4)(-5) = 39$. The location of an element resulting from multiplication of a particular row and column corresponds to that obtained by mentally (or actually) drawing lines through the row and column being multiplied; the intersection of the lines locates the product element. This is shown in Eq. A.46 for the element a_{23}.

In general, $\mathbf{A} \cdot \mathbf{B} \neq \mathbf{B} \cdot \mathbf{A}$. If $\mathbf{A} \cdot \mathbf{B} = \mathbf{B} \cdot \mathbf{A}$, matrices $\mathbf{A}$ and $\mathbf{B}$ are said to commute. For example, if the order of multiplication of the matrices in Eq. A.46 is reversed, the result is quite different:

$$\begin{bmatrix} 6 & 4 & -2 \\ 4 & 2 & 3 \\ -2 & 3 & -5 \end{bmatrix} \cdot \begin{bmatrix} 3 & -2 & 7 \\ -2 & 5 & -4 \\ 7 & -4 & 8 \end{bmatrix} = \begin{bmatrix} -4 & 16 & 10 \\ 29 & -10 & 44 \\ -47 & 39 & -66 \end{bmatrix} \tag{A.47}$$

Each *row* of the product matrix in Eq. A.46 is the same as a *column* of the product matrix in Eq. A.47. This is a result of the fact that each original matrix is symmetric. Had the original matrices not been symmetric, the product matrices would, in general, have been dissimilar.

The multiplication of matrices is associative:

$$\mathbf{A} \cdot \mathbf{B} \cdot \mathbf{C} = (\mathbf{A} \cdot \mathbf{B}) \cdot \mathbf{C} = \mathbf{A} \cdot (\mathbf{B} \cdot \mathbf{C}) \tag{A.48}$$

The student may wish to check the validity of this statement.

In this book there will be occasion to perform the following type of multiplication:

$$\begin{bmatrix} a_1 & a_2 & a_3 \end{bmatrix} \cdot \begin{bmatrix} g_{11} & g_{12} & g_{13} \\ g_{21} & g_{22} & g_{23} \\ g_{31} & g_{32} & g_{33} \end{bmatrix} \cdot \begin{bmatrix} b_1 \\ b_2 \\ b_3 \end{bmatrix} = ? \tag{A.49}$$

The result is a scalar. Multiplication of the first two matrices yields a row matrix, and a row matrix times a column matrix is a scalar (Eq. A.43). It is left to the reader to work out the algebraic result.

A further example of matrix multiplication is the operation of rotation about $\mathbf{z}$ of coordinate axes in the xy plane through an arbitrary angle ϕ. After counterclockwise rotation of the x and y axes through the angle ϕ, the coordinates of a point in a rigid body change from (x_1, y_1) to (x_2, y_2). Reference to Fig. A.2 yields the relations between the new and old coordinates

$$x_2 = +x_1 \cos \phi + y_1 \sin \phi \tag{A.50a}$$
$$y_2 = -x_1 \sin \phi + y_1 \cos \phi \tag{A.50b}$$

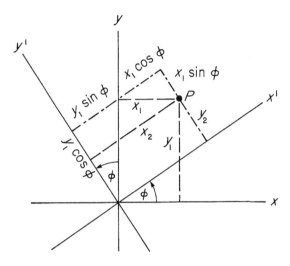

FIGURE A.2 Coordinates of a point P in a plane, before (x_1, y_1) and after (x_2, y_2) counterclockwise rotation of the coordinate axes through an angle ϕ.

Alternatively, both the initial and final coordinates may be expressed as column vectors:

$$\begin{bmatrix} x_2 \\ y_2 \end{bmatrix} = \begin{bmatrix} \cos\phi & \sin\phi \\ -\sin\phi & \cos\phi \end{bmatrix} \cdot \begin{bmatrix} x_1 \\ y_1 \end{bmatrix} = \begin{bmatrix} x_1\cos\phi + y_1\sin\phi \\ -x_1\sin\phi + y_1\cos\phi \end{bmatrix} \tag{A.51}$$

It is clear that Eq. A.51 is equivalent to Eqs. A.50. The square matrix in Eq. A.51 is called a *coordinate-rotation matrix*.

In various instances in this book, we need to deal with matrix projections of the type $\mathbf{n}^T \cdot \mathbf{Y} \cdot \mathbf{n}$, where $\mathbf{n}^T = \begin{bmatrix} \sin\theta\cos\phi & \sin\theta\sin\phi & \cos\theta \end{bmatrix}$ is the unit magnitude vector in three dimensions (expressed in spherical polar coordinates) and where 3×3 matrix $\mathbf{Y}$ is not necessarily symmetric across its main diagonal. We then have

$$Y = \mathbf{n}^T \cdot \mathbf{Y} \cdot \mathbf{n} \tag{A.52a}$$

$$Y = Y_{11}\sin^2\theta\cos^2\phi + (Y_{12} + Y_{21})\sin^2\theta\cos\phi\sin\phi +$$

$$Y_{22}\sin^2\theta\sin^2\phi + (Y_{13} + Y_{31})\cos\theta\sin\theta\cos\phi +$$

$$(Y_{23} + Y_{32})\cos\theta\sin\theta\sin\phi + Y_{33}\cos^2\theta \tag{A.52b}$$

which is not distinguishable from $\mathbf{n}^T \cdot \mathbf{Y}^T \cdot \mathbf{n}$ or from $\mathbf{n}^T \cdot [(\mathbf{Y}^T + \mathbf{Y})/2] \cdot \mathbf{n}$. Generally, scalar $Y(\theta, \phi)$ is the quantity measurable. Thus, for example, setting $\theta = 90°$ and $\phi = 0°$ yields Y_{11}. It should now be clear that knowledge of $Y(\theta, \phi)$ at six suitable points suffices to determine the six parameters in Eq. A.52b.

TABLE A.3 Properties of Special Matrices

Matrix	Alternative Definitions	
Unit[a]	$\mathbf{1}_n$	$A_{ii} = 1, A_{ij} = 0$ if $i \neq j$. Here n is the dimension of the matrix
Diagonal	$^d\mathbf{A}$	$A_{ij} = 0$ if $i \neq j$
Symmetric	$\mathbf{A}^T = \mathbf{A}$	$A_{ij} = A_{ji}$
Antisymmetric	$\mathbf{A}^T = -\mathbf{A}$	$A_{ij} = -A_{ji}$
Real	$\mathbf{A}^* = \mathbf{A}$	$A_{ij}^* = A_{ij}$
Hermitian	$\mathbf{A}^T = \mathbf{A}$	$A_{ij}^* = A_{ji}$
Unitary	$\mathbf{A}^{-1} = \mathbf{A}^\dagger$	Generally, not all elements A_{ij} real
Real orthogonal (real unitary)	$\mathbf{A}^{-1} = \mathbf{A}^T$	All elements A_{ij} real

^{a} of course, $\mathbf{1}_n = {}^d\mathbf{1}_n$.

A.5.3 Special Matrices and Matrix Properties

A given matrix may be transformed into various related matrices, some of which have especially useful properties. Table A.2 defines several matrices derived from a matrix $\mathbf{A}$; various examples illustrating the relations among matrix components are given in the third column.

The properties of some matrices of special importance are defined in Table A.3.

A.5.4 Dirac Notation for Eigenfunctions and Matrix Elements

A shorthand notation, introduced by Dirac [1] for eigenfunctions, is at times employed in this book; for example, the wavefunction ψ_k may be represented by $|k\rangle$, where k is an identifying label, usually a set of one or more quantum numbers. The function $|k\rangle$ is called a 'ket'; for example, Eq. A.9 can be written

$$\hat{\Lambda}|k\rangle = \lambda_k|k\rangle \tag{A.53}$$

The ket $|k\rangle$ is an eigenfunction of $\hat{\Lambda}$ and is labeled with the quantum number k, since the eigenvalue λ_k is labeled by k. Spin functions corresponding to $M_S = +\frac{1}{2}$ and $M_S = -\frac{1}{2}$ are conventionally represented by $|\alpha\rangle$ and $|\beta\rangle$. Corresponding to each ket, there is a function called a 'bra', written as $\langle k|$. A bra can be combined with a ket yielding a scalar. For the bra $\langle j|$ and the ket $|k\rangle$, the notation $\langle j|k\rangle$ implies integration over the full range of all spatial variables, that is

$$\langle j|k\rangle = \int_\tau \psi_j^* \psi_k \, d\tau \tag{A.54}$$

where ψ_j and ψ_k are spatial functions rather than spin functions. The 'bracket' nota-tion remains useful even when integration is not defined, that is, with spin functions. In Eq. A.54, if $j \neq k$ and $\langle j|k \rangle = 0$, the wavefunctions are said to be *orthogonal*. If $j = k$ and $\langle j|j \rangle = \langle k|k \rangle = 1$, the wavefunctions are said to be *normalized*. It is always possible and usually convenient to choose angular-momentum wavefunctions to be orthogonal and normalized, that is, *orthonormal*. Frequently, integrals of the form $\int \psi_j^* \hat{B} \psi_k d\tau$ are encountered as elements of a matrix (Eq. A.62). In the Dirac notation such integrals are represented by

$$\int_\tau \psi_j^* \hat{B} \psi_k \, d\tau = \langle j|\hat{B}|k \rangle \qquad (A.55a)$$

$$= B_{jk} \qquad (A.55b)$$

The expression $\langle j|B^\wedge|k \rangle$ is called a 'matrix element'. If $j = k$, the function is called a 'diagonal matrix element', and if $j \neq k$, then it is called an 'off-diagonal matrix element'. Note that j labels the row, and k labels the column.

The hermitian condition introduced in Eq. A.8 is transformed, in our bracket notation, to

$$\langle j|\hat{B}|k \rangle = \langle k|\hat{B}|j \rangle^* \qquad (A.56)$$

and thus is now defined also for non-spatial operators, such as spin angular momenta, for which spatial integrations are not meaningful.

The average value (or expectation value) $\langle B \rangle$ of any observable B for a state described by the orthonormal wavefunction ψ_k is obtained using the operation

$$\langle B \rangle = \int_\tau \psi_k^* \hat{B} \psi_k \, d\tau = \langle k|\hat{B}|k \rangle \qquad (A.57)$$

where $\hat{B}$ is the operator corresponding to the observable B.

An important property of the bra and ket functions is given by the relation

$$\langle j|k \rangle = \langle k|j \rangle^* \qquad (A.58)$$

This relation is evident from Eq. A.56 by taking $\hat{B}$ to be the unit operator.

In the matrix element $\langle j|\hat{B}|k \rangle$ it is assumed that $\hat{B}$ is to operate in the forward direction $(\rightarrow)$ on the ket $|k \rangle$. To operate backward $(\leftarrow)$ on the bra $\langle j|$, one must take the adjoint of the matrix element (Table A.2), using Eq. A.58, to obtain

$$\underset{\rightarrow}{\langle j|\hat{B}|k \rangle} = \underset{\leftarrow}{\langle j|\hat{B}|k \rangle^\dagger} = \underset{\rightarrow}{\langle k|\hat{B}|j \rangle^*} = x_j^* \langle k\,|j' \rangle^* = x_j^* \langle j'|k \rangle \qquad (A.59)$$

Here x_j can be a complex number.

When operator $\hat{B}$ is the hamiltonian operator $\hat{\mathcal{H}}$, which is hermitian, then the corresponding matrix $\mathcal{H}$ is called the 'hamiltonian matrix'. The spin hamiltonian has primary importance throughout this book, since its eigenvalues are the energies of the spin states.

If $\hat{B}$ is a non-hermitian operator (e.g., the ladder operators in Appendix B), the effect of $\hat{B}$ is to alter the wavefunction ($|j\rangle \rightarrow |j'\rangle$), as well as to multiply it by a constant x_j. If the operator $\hat{B}$ corresponds to an observable quantity, and is therefore hermitian, then x_j must be real and $j = j'$. Hence in this case it does not matter whether $\hat{B}$ operates forward or backward in the matrix element of Eq. A.59.

At times, when there are several independent particles in the system being considered, it is necessary to deal with a composite ket (or bra) made up of individual-particle kets; for example, see Eqs. 2.42.

A.5.5   Diagonalization of Matrices

Every normal matrix of interest herein can be transformed into diagonal form. The resulting diagonal elements are called the *eigenvalues* (or alternatively, the 'principal values') of the matrix. The associated eigenvectors (principal axes or directions) need be specified only to within an arbitrary multiplicative constant, which is usually chosen to yield unit 'lengths' (this is called 'normalization'); thus each vector orientation, but not its direction along this line, is meaningful. The principal axes (this set is called the 'principal-axis system') usually are orthogonal to each other.

As an example, consider the two-dimensional case offered by the spin operators $\hat{S}_x$, $\hat{S}_y$ and $\hat{S}_z$ for $S = \frac{1}{2}$. (We could equally well consider a nuclear spin $I = \frac{1}{2}$.) The eigenfunctions of $\hat{S}_z$ are usually taken as a basis set for the spin functions (Section 2.3.3):

$$\hat{S}_z|\alpha\rangle = +\tfrac{1}{2}\,|\alpha\rangle \tag{A.60a}$$

$$\hat{S}_z|\beta\rangle = -\tfrac{1}{2}\,|\beta\rangle \tag{A.60b}$$

Here we omit the electron label (i.e., in $|\alpha(e)\rangle$ and $|\beta(e)\rangle$). Multiplication of Eq. A.60a from the left by $\langle\alpha|$ gives

$$\langle\alpha|\hat{S}_z|\alpha\rangle = +\tfrac{1}{2}\langle\alpha|\alpha\rangle = +\tfrac{1}{2} \tag{A.61a}$$

Similarly

$$\langle\beta|\hat{S}_z|\alpha\rangle = +\tfrac{1}{2}\langle\beta|\alpha\rangle = 0 \tag{A.61b}$$

$$\langle\alpha|\hat{S}_z|\beta\rangle = -\tfrac{1}{2}\langle\alpha|\beta\rangle = 0 \tag{A.61c}$$

$$\langle\beta|\hat{S}_z|\beta\rangle = -\tfrac{1}{2}\langle\beta|\beta\rangle = -\tfrac{1}{2} \tag{A.61d}$$

It proves to be very convenient to deal with a single mathematical entity, matrix $\bar{\mathbf{S}}_z$, instead of numerous operator equations such as Eqs. A.61. These equations may be combined into the following matrix form:

$$\bar{S}_z = \begin{bmatrix} \langle\alpha|\hat{S}_z|\alpha\rangle & \langle\alpha|\hat{S}_z|\beta\rangle \\ \langle\beta|\hat{S}_z|\alpha\rangle & \langle\beta|\hat{S}_z|\beta\rangle \end{bmatrix} = \begin{bmatrix} +\frac{1}{2} & 0 \\ 0 & -\frac{1}{2} \end{bmatrix} \quad (A.62)$$

Here $\bar{S}_z$ is a matrix that includes all possible matrix elements of the operator $\hat{S}_z$ between the states α and β (for rotation, see Table I.1). We see then how a quantum-mechanical operator can be represented by a square matrix, whose dimension k is just the number of basis states. If this matrix is diagonal, then the basis wavefunctions must be eigenfunctions of the operator. Thus $|\alpha\rangle$ and $|\beta\rangle$ are eigenfunctions of $\hat{S}_z$. Examples of spin operators and spin matrices are given in Sections B.5 and B.10.

Operation by $\hat{S}_x$ on $|\alpha\rangle$ and $|\beta\rangle$ gives the following non-zero results:

$$\hat{S}_x|\alpha\rangle = \tfrac{1}{2}|\beta\rangle \quad (A.63a)$$

$$\hat{S}_x|\beta\rangle = \tfrac{1}{2}|\alpha\rangle \quad (A.63b)$$

The corresponding matrix then is

$$\bar{S}_x = \begin{bmatrix} \langle\alpha|\hat{S}_x|\alpha\rangle & \langle\alpha|\hat{S}_x|\beta\rangle \\ \langle\beta|\hat{S}_x|\alpha\rangle & \langle\beta|\hat{S}_x|\beta\rangle \end{bmatrix} = \begin{bmatrix} 0 & +\frac{1}{2} \\ +\frac{1}{2} & 0 \end{bmatrix} \quad (A.64)$$

$\bar{S}_x$ is not a diagonal matrix since the basis functions are *not* eigenfunctions of $\hat{S}_x$.

Assume that the unknown eigenfunctions of operator $\hat{S}_x$ can be represented as a linear combination of $|\alpha\rangle$ and $|\beta\rangle$, that is

$$|\phi_1\rangle = c_{11}|\alpha\rangle + c_{12}|\beta\rangle \quad (A.65a)$$

$$|\phi_2\rangle = c_{21}|\alpha\rangle + c_{22}|\beta\rangle \quad (A.65b)$$

or in matrix form

$$\begin{bmatrix} |\phi_1\rangle \\ |\phi_2\rangle \end{bmatrix} = \underbrace{\begin{bmatrix} c_{11} & c_{12} \\ c_{21} & c_{22} \end{bmatrix}}_{\bar{C}} \cdot \begin{bmatrix} |\alpha\rangle \\ |\beta\rangle \end{bmatrix} \quad (A.66a)$$

Alternatively (see Table A.3)

$$[\langle\phi_1|\langle\phi_2|] = [\langle\alpha|\langle\beta|]\, C^\dagger \quad (A.66b)$$

Each row of the square matrix $\overline{C}$ in Eq. A.66a (For notation, see Table I.1) may be considered as a row vector $\mathbf{c}^T$ in spin space. These two vectors have the property that

$$\overline{\mathbf{S}}_x \cdot \mathbf{c}_i = \lambda \mathbf{c}_i \quad i = 1, 2 \tag{A.67}$$

Hence the vectors $\mathbf{c}_i$ (and $\mathbf{c}_i{}^T$) are called *eigenvectors* of the matrix $\overline{\mathbf{S}}_x$ with eigenvalues λ.

Insertion of the unit matrix $\mathbf{1}_2$ into Eq. A.67 and rearrangement gives

$$(\overline{\mathbf{S}}_x - \lambda \mathbf{1}_2) \cdot \mathbf{c}_i = \begin{pmatrix} 0 \\ 0 \end{pmatrix} \tag{A.68}$$

Equation A.68 represents two simultaneous equations (called the 'secular equation'; see Note C.2), which are not independent. If $\mathbf{c} \neq 0$, then these equations may be solved [2, Chapter 4]. Expansion of the secular determinant $|\overline{\mathbf{S}}_x - \lambda \mathbf{1}_2|$ yields

$$\begin{vmatrix} 0 - \lambda & +\frac{1}{2} \\ +\frac{1}{2} & 0 - \lambda \end{vmatrix} = \lambda^2 - \frac{1}{4} = 0 \tag{A.69a}$$

or

$$\lambda = \pm \frac{1}{2} \tag{A.69b}$$

For $\lambda = \frac{1}{2}$, the two simultaneous equations corresponding to Eq. A.68 are

$$\left(0 - \tfrac{1}{2}\right)c_{11} + \tfrac{1}{2}c_{12} = 0 \tag{A.70a}$$

$$\tfrac{1}{2}c_{11} + \left(0 - \tfrac{1}{2}\right)c_{12} = 0 \tag{A.70b}$$

yielding

$$c_{11} = c_{12}$$

With $\lambda = -\frac{1}{2}$, one obtains $c_{21} = -c_{22}$. If $|\phi_1\rangle$ and $|\phi_2\rangle$ are to be normalized, $|c_{11}|^2 + |c_{12}|^2 = 1$ and $|c_{21}|^2 + |c_{22}|^2 = 1$. Hence the final eigenfunctions of matrix $\overline{S}_x$ can be represented by

$$\begin{bmatrix} |\phi_1\rangle \\ |\phi_2\rangle \end{bmatrix} = \begin{bmatrix} \dfrac{1}{\sqrt{2}} & \dfrac{1}{\sqrt{2}} \\ \dfrac{1}{\sqrt{2}} & -\dfrac{1}{\sqrt{2}} \end{bmatrix} \cdot \begin{bmatrix} |\alpha\rangle \\ |\beta\rangle \end{bmatrix} \tag{A.71}$$

C

where square matrix $\mathbf{C}$ contains the coefficients c_{ij} and is a unitary matrix (in fact, real orthogonal; see Table A.3). We note that the transformation

$$\mathbf{C} \cdot \bar{\mathbf{S}}_x \cdot \mathbf{C}^\dagger = \begin{bmatrix} \dfrac{1}{\sqrt{2}} & \dfrac{1}{\sqrt{2}} \\ \dfrac{1}{\sqrt{2}} & -\dfrac{1}{\sqrt{2}} \end{bmatrix} \cdot \begin{bmatrix} 0 & +\dfrac{1}{2} \\ +\dfrac{1}{2} & 0 \end{bmatrix} \begin{bmatrix} \dfrac{1}{\sqrt{2}} & \dfrac{1}{\sqrt{2}} \\ \dfrac{1}{\sqrt{2}} & -\dfrac{1}{\sqrt{2}} \end{bmatrix} \tag{A.72a}$$

$$= \begin{bmatrix} +\dfrac{1}{2} & 0 \\ 0 & -\dfrac{1}{2} \end{bmatrix} \tag{A.72b}$$

yields $^d \bar{\mathbf{S}}_x$, that is, the diagonal form of $\bar{\mathbf{S}}_x$. Matrix $\mathbf{C}$ is not unique: there are three others that accomplish this (Problem A.10).

When we consider operator $\hat{\mathbf{S}}_y$ (i.e., matrix $\bar{\mathbf{S}}_y$), a new feature develops, in that the operations analogous to those in Eq. A.63 yield pure imaginary numbers. Here the corresponding matrices $\mathbf{C}$ are again unitary, but not real orthogonal. However, the eigenvalues of $\bar{\mathbf{S}}_y$ are once again $\pm\frac{1}{2}$.

A general method for the diagonalization of any 2×2 hermitian matrix (Table A.3) may be developed by taking $\mathbf{C}$ to be the two-dimensional coordinate rotation matrix (Eq. A.51), which is unitary (Table A.3). The appropriate diagonalization procedure is

$$\overset{\mathbf{C}}{\begin{bmatrix} \cos\omega & \sin\omega \\ -\sin\omega & \cos\omega \end{bmatrix}} \cdot \begin{bmatrix} a & c \\ c^* & b \end{bmatrix} \cdot \overset{\mathbf{C}^{-1}}{\begin{bmatrix} \cos\omega & -\sin\omega \\ \sin\omega & \cos\omega \end{bmatrix}} = \begin{bmatrix} u & 0 \\ 0 & v \end{bmatrix} \tag{A.73}$$

The auxiliary angle ω is defined by

$$\tan 2\omega = -\frac{c + c^*}{a - b} \tag{A.74a}$$

$$\cos^2 \omega = \frac{1}{2}\left[1 + \left(1 + \frac{(c + c^*)^2}{(a - b)^2}\right)^{-1/2}\right] \tag{A.74b}$$

$$\sin^2 \omega = \frac{1}{2}\left[1 - \left(1 + \frac{(c + c^*)^2}{(a - b)^2}\right)^{-1/2}\right] \tag{A.74c}$$

$$\sin\omega\cos\omega = \frac{(c + c^*)}{2[(a - b)^2 + (c + c^*)^2]^{1/2}} \tag{A.74d}$$

The choice of $+$ or $-$ sign for the square-root quantities in Eqs. A.74 must be made with cognizance of the sign of $b - a$, consistent with Eq. A.74a.

The general solutions for u and v in Eq. A.73 are

$$u = a\cos^2\omega + b\sin^2\omega + (c + c^*)\sin\omega\cos\omega \qquad (A.75a)$$

$$v = a\sin^2\omega + b\cos^2\omega - (c + c^*)\sin\omega\cos\omega \qquad (A.75b)$$

These can also be written as eigenvalue equations

$$\begin{pmatrix} a & c \\ c^* & b \end{pmatrix} \cdot \begin{pmatrix} \cos\omega \\ \sin\omega \end{pmatrix} = u \begin{pmatrix} \cos\omega \\ \sin\omega \end{pmatrix} \qquad (A.76a)$$

$$\begin{pmatrix} a & c \\ c^* & b \end{pmatrix} \cdot \begin{pmatrix} -\sin\omega \\ \cos\omega \end{pmatrix} = v \begin{pmatrix} -\sin\omega \\ \cos\omega \end{pmatrix} \qquad (A.76b)$$

The advantage of this method is that, with ω given by Eqs. A.74, the rotation matrix on the left of Eq. A.73 becomes the eigenvector matrix; that is, the two elements in each row are the coefficients of one of the two eigenvectors.

If the 2×2 hermitian matrix to be diagonalized is not real (i.e., off-diagnol element $c \equiv p + iq$ with p, q real has an imaginary part ($q \neq 0$), then one must first, before employing the above procedure, use the similarity transformation

$$\begin{pmatrix} \xi^* & 0 \\ 0 & \xi \end{pmatrix} \cdot \begin{pmatrix} a & p+iq \\ p-iq & b \end{pmatrix} \cdot \begin{pmatrix} \xi & 0 \\ 0 & \xi^* \end{pmatrix} = \begin{pmatrix} a & \gamma \\ \gamma & b \end{pmatrix} \qquad (A.77)$$

performed with the unitary matrix depicted. Here $\xi \equiv e^{i\alpha/2}$, $\tan\alpha \equiv q/p$, and $\gamma \equiv |(p^2 + q^2)^{1/2}|$ real. Thus the basis function ket $|\phi_1, \phi_2\rangle$ is replaced by ket $|\phi_1 e^{i\alpha/2}, \phi_2 e^{-i\alpha/2}\rangle$, and Eqs. A74–A76 apply again, but with c replaced by γ.[3]

Exactly the same ideas prevail when dealing with matrices of arbitrary dimensions. Their eigenvalues can be obtained by solving the corresponding secular determinants. In general, any change of basis (coordinate system) causes the components of an operator matrix (or of a parameter matrix) to change in accordance with a similarity transformation. Thus, for example, transformations

$$\mathbf{U} \cdot \mathbf{P} \cdot \mathbf{U}^{-1} = \mathbf{U} \cdot \mathbf{P} \cdot \mathbf{U}^\dagger = {}^D\mathbf{P} \qquad (A.78)$$

diagonalize $n \times n$ hermitian matrices $\mathbf{P}$. Here $n \times n$ unitary matrix $\mathbf{U} = \mathbf{C}^*$ is derived from the n eigenvectors of $\mathbf{P}$, as shown above for $n = 2$ and $\mathbf{S}_x$. The most frequent examples encountered in this book are the diagonalization of spin-hamiltonian matrices (e.g., Appendix C) and the diagonalization of parameter (g-factor, hyperfine and electronic quadrupole) matrices (Chapters 4–6). Discussion of procedures for diagonalization of such a matrix can be found elsewhere [3–8]. Large matrices can be diagonalized numerically by computer.

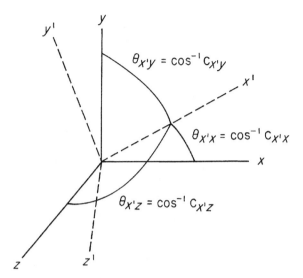

FIGURE A.3 A set of new Cartesian axes x', y', z' derived from the old Cartesian axes x, y, z by a rotation about an arbitrary direction through the origin.

We are now equipped to deal with any n-dimensional matrix, whether this is a quantum-mechanical operator for spin greater than $\frac{1}{2}$ or a parameter matrix referring to the usual three-dimensional space. We consider the latter case as our final example.

For a 3×3 parameter matrix $\mathbf{Y}$ (e.g., $\mathbf{g}$ or $\mathbf{gg}$ in Chapter 4, $\mathbf{A}$ or $\mathbf{AA}$ in Chapter 5, $\mathbf{D}$ in Chapter 6), we wish to find its principal parameters. In general, $\mathbf{Y}$ is expressed in the crystal coordinate system and is not diagonal. Its principal values can be obtained from its secular determinant. Alternatively, $\mathbf{Y}$ can be diagonalized using the appropriate similarity transformation (Eq. A.78) that is, rotating the coordinate system (Fig. A.3) to the principal-axis system. The rotation matrix $\mathbf{C}$ is real orthogonal, containing sines and cosines of several rotation angles. Its form is usually not simple [3–8].

Physical properties (e.g., energies and hence spin hamiltonians) cannot depend on the choice of coordinate system. For instance, consider the measurable property $Y = \mathbf{n}^T \cdot \mathbf{Y} \cdot \mathbf{n}$ (e.g., Eq. A.52a), where $\mathbf{Y}$ is a 3×3 parameter matrix and $\mathbf{n}$ is a unit vector (e.g., the direction of magnetic field $\mathbf{B}$, as in Eqs. 4.11 or Eq. 5.15b). In changing coordinates, $\mathbf{Y} = \mathbf{C} \cdot \mathbf{Y} \cdot \mathbf{C}^T$, $\mathbf{n}_{new}^T = \mathbf{n}^T \cdot \mathbf{C}^T$ and $\mathbf{n}_{new} = \mathbf{C} \cdot \mathbf{n}$, and hence $\mathbf{Y}$ is left unaltered.

The effect of rotating the coordinate system on a vector in 3-space is shown in Fig. A.4. We see that the vector is unaltered in direction and length, but that its components change.

Finally, we note that matrices have components bearing two indices (row and column) that are of two types. In one, these indices refer to axes of two different coordinate systems (e.g., rotation matrices where the indices refer to new and old axis sets). In the other, the indices refer to axes of the same coordinate systems. Operator matrices as well as parameter matrices are of the latter type.[4]

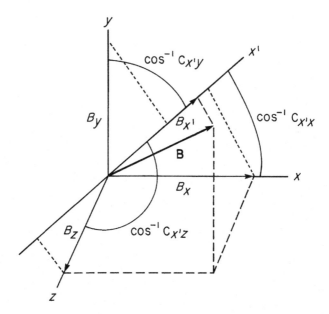

FIGURE A.4 Representation for a vector **B** of its component $B_{x'}$ and the contributions to $B_{x'}$ from B_x, B_y and B_z. Consult Fig. A.3.

A.5.6 Matrix Invariants

It can be shown that every $d \times d$ matrix **Y** (where d is any positive integer) has d invariants s_j ($j = 1, 2, \ldots, d$) [9, Section 1.20]. The latter are functions $f_j(\mathbf{Y})$ not affected when **Y** is changed to another equivalent $d \times d$ matrix **Y**′ via a similarity transformation (i.e., a change of basis, change of coordinate system): **Y** and **Y**′ have the same invariants. The set of invariants contain the d numbers $tr(\mathbf{Y}^j)$; the two most important are $s_1 = -tr(\mathbf{Y})$ and

$$s_d = det(\mathbf{Y}) = (-1)^{d-1}[s_{d-1}tr(\mathbf{Y}) + s_{d-2}tr(\mathbf{Y}^2) + \cdots + tr(\mathbf{Y}^d)]$$

We may note that the functions $\mathbf{n}^T \cdot \mathbf{Y} \cdot \mathbf{n}$ (here **n** is a $1 \times d$ unit vector) also are invariant to such transformations, as explained in Section A.5.5. The importance of this will be evident from Eqs. 6.55.

A.6 PERTURBATION THEORY

For sufficiently simple quantum-mechanical problems, exact solutions in terms of eigenvalues and eigenfunctions are available. In most real applications, the relevant hamiltonian is too complicated to allow solution of the problem. Often, however, some part of the hamiltonian is simple in the sense implied above, and the remaining part can be treated approximately. As an example of a well-known situation of this kind, we know that the rotational motion of a diatomic molecule is well approximated by the exact solutions to the quantum-mechanical rigid-rotor problem, while centrifugal distortion must be added as a perturbation and handled inexactly. Similar situations occur with spin hamiltonians.

The hamiltonian of a perturbed system, as described above, can usually be written as

$$\hat{\mathcal{H}} = \hat{\mathcal{H}}_0 + \lambda \hat{\mathcal{H}}' \tag{A.79}$$

where $\hat{\mathcal{H}}_0$ is the hamiltonian for which solutions are known and $\hat{\mathcal{H}}'$ is the perturbation operator. λ is a convenient (but not physical) perturbation magnitude parameter, the value of which can be set to unity at the end of the calculation.

Suppose that the eigenfunctions and eigenvalues of $\hat{\mathcal{H}}_0$ are given by

$$\hat{\mathcal{H}}_0 \psi_i^{(0)} = U_i^{(0)} \psi_i^{(0)} \tag{A.80a}$$

or equivalently in Dirac notation by

$$\hat{\mathcal{H}}_0 |i\rangle^{(0)} = U_i^{(0)} |i\rangle^{(0)} \tag{A.80b}$$

The index $i \, (=1, 2, 3, \ldots, n)$ ranges over the full set of 'zero-order' eigenfunctions $|1\rangle^{(0)}, |2\rangle^{(0)}, \ldots, |n\rangle^{(0)}$. The $U_i^{(0)}$ values are called 'zero-order' energies.

The unknown eigenfunctions and eigenvalues of $\hat{\mathcal{H}}$ are given by

$$\hat{\mathcal{H}} |i\rangle = U_i |i\rangle \tag{A.81}$$

Since the solutions of Eq. A.81 must go continuously into those of Eq. A.80b as $\lambda \to 0$, it is assumed that $|i\rangle$ and U_i can be expanded as a power series in λ; that is

$$|i\rangle = |i\rangle^{(0)} + \lambda |i\rangle^{(1)} + \lambda^2 |i\rangle^{(2)} + \cdots \tag{A.82a}$$

$$U_i = U_i^{(0)} + \lambda U_i^{(1)} + \lambda^2 U_i^{(2)} + \cdots \tag{A.82b}$$

where the numbers in parentheses give the order of the perturbation terms. Substitution of Eqs. A.79 and A.80 into Eq. A.81 yields

$$\begin{aligned}
[\hat{\mathcal{H}}_0 + \lambda \hat{\mathcal{H}}'] \, & [|i\rangle^{(0)} + \lambda |i\rangle^{(1)} + \lambda^2 |i\rangle^{(2)} + \cdots] \\
&= [U_i^{(0)} + \lambda U_i^{(1)} + \lambda^2 U_i^{(2)} + \cdots] \times \\
&\quad [|i\rangle^{(0)} + \lambda |i\rangle^{(1)} + \lambda^2 |i\rangle^{(2)} + \cdots]
\end{aligned} \tag{A.83}$$

or

$$\hat{\mathcal{H}}_0|i\rangle^{(0)} + \lambda[\hat{\mathcal{H}}'|i\rangle^{(0)} + \hat{\mathcal{H}}_0|i\rangle^{(1)}] + \lambda^2[\hat{\mathcal{H}}'|i\rangle^{(1)} + \hat{\mathcal{H}}_0|i\rangle^{(2)}] + \cdots$$
$$= U_i^{(0)}|i\rangle^{(0)} + \lambda[U_i^{(1)}|i\rangle^{(0)} + U_i^{(0)}|i\rangle^{(1)}]$$
$$+ \lambda^2[U_i^{(2)}|i\rangle^{(0)} + U_i^{(1)}|i\rangle^{(1)} + U_i^{(0)}|i\rangle^{(2)}] + \cdots \tag{A.84}$$

Equation A.84 must be valid for all possible values of λ. This is possible only if the coefficients of a given power of λ are equal on both sides of the equation. Thus one may write

$$\hat{\mathcal{H}}_0|i\rangle^{(0)} = U_i^{(0)}|i\rangle^{(0)} \tag{A.85a}$$

$$\hat{\mathcal{H}}'|i\rangle^{(0)} + \hat{\mathcal{H}}_0|i\rangle^{(1)} = U_i^{(1)}|i\rangle^{(0)} + U_i^{(1)}|i\rangle^{(1)} \tag{A.85b}$$

$$\hat{\mathcal{H}}'|i\rangle^{(1)} + \hat{\mathcal{H}}_0|i\rangle^{(2)} = U_i^{(2)}|i\rangle^{(0)} + U_i^{(1)}|i\rangle^{(1)} + U_i^{(0)}|i\rangle^{(2)} \tag{A.85c}$$

It was assumed that the solutions to Eq. A.85a are known.

In Eq. A.85b the functions $|i\rangle^{(1)}$ are unknown; similarly the effect of $\hat{\mathcal{H}}'$ on $|i\rangle^{(0)}$ is unknown. Assume that these functions of the zero-order basis set may be expanded as follows:

$$|i\rangle^{(1)} = A_1|1\rangle^{(0)} + A_2|2\rangle^{(0)} + \cdots + A_j|j\rangle^{(0)} + \cdots \tag{A.86}$$

$$\hat{\mathcal{H}}'|i\rangle^{(0)} = H_{1i}|1\rangle^{(0)} + H_{2i}|2\rangle^{(0)} + \cdots + A_{ji}|j\rangle^{(0)} + \cdots \tag{A.87}$$

Multiplication of Eq. A.87 from the left by $^{(0)}\langle j|$ gives

$$H_{ji} = {}^{(0)}\langle j|\hat{\mathcal{H}}'|i\rangle^{(0)} \tag{A.88}$$

since the kets are assumed to be orthonormal. Rearrangement of Eq. A.85b gives

$$(\hat{\mathcal{H}}_0 - U_i^{(0)})|i\rangle^{(1)} = (U_i^{(1)} - \hat{\mathcal{H}}')|i\rangle^{(0)} \tag{A.89}$$

Substitution of Eqs. A.86 and A.87 into Eq. A.89 gives

$$\sum_j (U_j^{(0)} - U_i^{(0)})A_j|j\rangle^{(0)}$$
$$= U_i^{(1)}|i\rangle^{(0)} - H_{1i}|1\rangle^{(0)} - H_{2i}|2\rangle^{(0)} - \cdots - H_{ji}|j\rangle^{(0)} - \cdots \tag{A.90}$$

Multiplication from the left by $^{(0)}\langle i|$ gives

$$U_i^{(1)} = H_{ii} = {}^{(0)}\langle i|\hat{\mathcal{H}}'|i\rangle^{(0)} \tag{A.91}$$

The quantity $U_i^{(1)}$ is the first-order correction to the energy. Multiplication of Eq. A.90 from the left by $^{(0)}\langle j|$ gives

$$A_j = \frac{-H_{ji}}{U_j^{(0)} - U_i^{(0)}} \quad i \neq j \tag{A.92}$$

The coefficient A_i remains to be determined (since for $i = j$, Eq. A.92 is not valid). By requiring $|i\rangle$ to be normalized, it is readily shown that $A_i = 0$. Hence the first-order correction to the wavefunction is

$$|i\rangle^{(1)} = -\sum_{j \neq i} \frac{H_{ji}}{U_j^{(0)} - U_i^{(0)}} |j\rangle^{(0)} \tag{A.93}$$

where $H_{ij} = H_{ji}^*$.

An approach similar to the solution of Eq. A.85c yields the second-order corrections to the energies and wavefunctions

$$U_i^{(2)} = -\sum_{j \neq i} \frac{H_{ij}H_{ji}}{U_j^{(0)} - U_i^{(0)}} \tag{A.94}$$

$$|i\rangle^{(2)} = \sum_{k \neq i} \left[\sum_{j \neq i} \frac{H_{kj}H_{ji}}{(U_i^{(0)} - U_k^{(0)})(U_i^{(0)} - U_j^{(0)})} - \frac{H_{ii}H_{ki}}{(U_i^{(0)} - U_k^{(0)})^2} \right] |k\rangle^{(0)} -$$

$$\frac{1}{2} \left[\sum_{k \neq i} \frac{H_{ki}H_{ik}}{(U_i^{(0)} - U_k^{(0)})^2} \right] |i\rangle^{(0)} \tag{A.95}$$

Here i represents an arbitrary particular state. Examples of first- and second-order corrections to energies and wavefunctions are given in Section C.1.7.

Owing to the occurrence of energy differences in the denominator of Eqs. A.92–A.95, it is necessary to modify the procedure above presented when applying perturbation methods to systems with degenerate energy levels [10].

A.7 DIRAC DELTA FUNCTION

This mathematical 'function' furnishes a convenient notation for evaluation of functions at specific points (e.g., Sections 4.6 and 9.2.4). It often also serves to represent the conceptual lineshape function in the limit when the width is shrunk to zero. The mathematical properties of δ can be found in many sources

[9, pp. 55–59, 156; 11–13]. Its primary one is

$$\int_{-\infty}^{\infty} f(x)\delta(x - x_{\text{o}})\, dx = f|_{x=x_0} \qquad (A.96)$$

A.8   GROUP THEORY

It seems likely that there really is no such thing in nature as symmetry (i.e., only the identity operation in group theory exists in practice), due to the perpetual motions and internal structure of all 'particles', to the wave aspects impinging on each other, and indeed the very laws of nature.

On the other hand, it is certainly helpful in EPR spectroscopy (see Chapter 4) to utilize approximate symmetry, thus to classify physical situations, and to simplify dealing with them. Group theory, the mathematical framework developed to quantify symmetry aspects, is highly developed, and is applied equally to spatial symmetry and to quantum-mechanical concepts (permutation groups). We cannot herein cover details, but can point to some excellent introductory texts on the subject (early gems as well as more recent ones) [2, Chapter 7; 14–17].

REFERENCES

1. P. A. M. Dirac, *The Principles of Quantum Mechanics*. 3rd ed., Oxford University Press, London, U.K., 1947, p. 18.

2. See, for example, G. G. Hall, *Matrices and Tensors*, Pergamon Press, Oxford, U.K., 1963.

3. H. E. Zimmerman, *Quantum Mechanics for Organic Chemists*, Academic Press, New York, NY, U.S.A., 1975.

4. G. B. Arfken, H. J. Weber, *Mathematical Methods for Physicists*, 6th ed., Academic Press, Orlando, FL, U.S.A., 2005.

5. H. Margenau, G. M. Murphy, *The Mathematics of Physics and Chemistry*, 2nd ed., Van Nostrand, Princeton, NJ, U.S.A., 1956.

6. H. Goldstein, C. P. Poole Jr., J. L. Safko, *Classical Mechanics*, 3rd ed., Addison-Wesley, Reading MA, U.S.A., 2002, Chapter 4 and Appendix A.

7. M. E. Starzak, *Mathematical Methods in Chemistry and Physics*, Plenum Press, New York, NY, U.S.A., 1989.

8. B. Kolman, *Introductory Linear Algebra with Applications*, 3rd ed., Macmillan, New York, NY, U.S.A., 1984.

9. F. B. Hildebrand, *Methods of Applied Mathematics*, 2nd ed., Prentice-Hall, Englewood Cliffs, NJ, U.S.A. 1965.

10. L. I. Schiff, *Quantum Mechanics*, 3rd ed., McGraw-Hill, New York, NY, U.S.A., 1968, pp. 248–251.

11. A. S. Davydov, *Quantum Mechanics*, 2nd ed., Pergamon Press, New York, NY, U.S.A., 1976, p. 14 and Appendix A.

12. E. Merzbacher, *Quantum Mechanics*, 2nd ed., Wiley, New York, NY, U.S.A., 1970, pp. 82–85, 140.

13. R. Skinner, J. A. Weil, *Am. J. Phys.*, **57**, 777 (1989).

14. J. F. Nye, *Physical Properties of Crystals*, 2nd ed., Oxford University Press, Oxford, U.K., 1985, Chapter 15.

15. G. E. Pake, T. L. Estle, "Ligand or Crystal Fields", in *The Physical Principles of Electron Paramagnetic Resonance*, 2nd ed., Benjamin, Reading, MA, U.S.A., 1973, Chapter 3.

16. S. A. Al'tshuler, B. M. Kozyrev, *Electron Paramagnetic Resonance in Compounds of Transition Elements*, 2nd ed. (revised) (Engl. transl.), Wiley, New York, NY, U.S.A., 1974.

17. H. Eyring, J. Walter, G. E. Kimball, *Quantum Chemistry*, Wiley, New York, NY, U.S.A., 1944, Chapter 10.

NOTES

1. Strictly speaking, multiplication by x always implies application of an operator $\hat{x}$, but we do not adhere to labeling such operators with a superscript $\wedge$; the same is true for other elementary operations such as addition or subtraction.

2. There also exists an *anticommutator*: $[\hat{A}, \hat{B}]_{+} \equiv \hat{A}\hat{B} + \hat{B}\hat{A}$, useful for various purposes.

3. The authors thank Prof. B. M. Hoffman for working out the corrected and elaborated version of the diagonalization procedure appearing in the present edition.

4. Such matrices are representative of second-rank tensors. Vectors are tensors of first rank, whereas scalars are tensors of rank zero. Higher-rank tensors can be of importance in EPR when spin hamiltonians for spins $S > \frac{3}{2}$ are dealt with.

FURTHER READING

1. J. M. Anderson, *Mathematics for Quantum Chemistry*, Benjamin, New York, NY, U.S.A., 1966 (good for operator algebra and perturbation theory).

2. P. W. Atkins, R. S. Friedman, *Molecular Quantum Mechanics*, 3rd ed., Oxford University Press, Oxford, U.K., 1997.

3. T. Bak, J. Lichtenberg, *Mathematics for Scientists*, 3 vols., Benjamin, New York, NY, U.S.A., 1966. (Volume 1, *Vectors, Tensors and Groups*, is recommended for present purposes.)

4. A. G. Hamilton, *Linear Algebra*, Cambridge University Press, Cambridge, U.K., 1989.

5. J. F. Nye, *Physical Properties of Crystals*, 2nd ed., Oxford University Press, Oxford, U.K., 1985, Chapters 1 and 2 (excellent discussion of second-rank tensors).

6. W. A. Wooster, *Tensors and Group Theory for the Physical Properties of Crystals*, Clarendon, Oxford, U.K., 1972.

7. A. Yariv, *An Introduction to Theory and Application of Quantum Mechanics*, Wiley, New York, NY, U.S.A., 1982.

PROBLEMS

A.1 Write the sum $\mathbf{S} = \mathbf{A} + \mathbf{B}$, difference $\mathbf{D} = \mathbf{A} - \mathbf{B}$, and scalar products $\mathbf{P} = \mathbf{A} \cdot \mathbf{B}$, $\mathbf{Q} = \mathbf{B} \cdot \mathbf{A}$ of the matrices

$$\mathbf{A} = \begin{bmatrix} 3 & 1 & -2 \\ 4 & -2 & 3 \\ -2 & 1 & -1 \end{bmatrix} \quad \text{and} \quad \mathbf{B} = \begin{bmatrix} 2 & 0 & -1 \\ -4 & 1 & 2 \\ 1 & -1 & 0 \end{bmatrix}$$

A.2   Given

$$\mathbf{A} = \begin{bmatrix} 2 & 1 \\ -1 & 3 \end{bmatrix} \quad \text{and} \quad \mathbf{B} = \begin{bmatrix} -1 & 2 \\ 3 & -2 \end{bmatrix}$$

show that $(\mathbf{A} + \mathbf{B}) \cdot (\mathbf{A} - \mathbf{B}) \neq \mathbf{A} \cdot \mathbf{A} - \mathbf{B} \cdot \mathbf{B}$. Refer to Fig. 12.12.

A.3   Multiply both sides in Eqs. A.20 and A.21 from the left by the complex-conjugate wavefunctions. Then integrate the resulting expressions over the appropriate range of ϕ to obtain expressions for the energy and for the angular momentum for a particle moving in a circle of fixed radius.

A.4   Find x, y and z such that

$$\begin{bmatrix} 2 & -1 & 3 \\ 1 & 2 & -4 \\ -1 & 3 & -2 \end{bmatrix} \begin{bmatrix} x \\ y \\ z \end{bmatrix} = \begin{bmatrix} 1 \\ -3 \\ 6 \end{bmatrix}$$

For instance, calculate and apply the inverse of the preceding 3×3 matrix to do so.

A.5   Obtain the eigenvalues (principal values) of the matrix

$$\begin{bmatrix} 5 & 0 & -2 \\ 0 & -3 & 0 \\ -2 & 0 & 2 \end{bmatrix}$$

by the secular determinant method.

A.6   Diagonalize the matrix

$$\mathbf{A} = \begin{bmatrix} 5 & 1 & -1 \\ 1 & 3 & -1 \\ -1 & -1 & 3 \end{bmatrix}$$

using the corresponding eigenvector matrix

$$\mathbf{C} = \begin{bmatrix} 0 & \dfrac{1}{\sqrt{2}} & \dfrac{1}{\sqrt{2}} \\ \dfrac{1}{\sqrt{3}} & -\dfrac{1}{\sqrt{3}} & \dfrac{1}{\sqrt{3}} \\ -\dfrac{2}{\sqrt{6}} & \dfrac{1}{\sqrt{6}} & -\dfrac{1}{\sqrt{6}} \end{bmatrix}$$

A.7 Perform the following matrix multiplication

$$\begin{bmatrix} B & 0 & 0 \end{bmatrix} \cdot \begin{bmatrix} g_{xx} & g_{xy} & g_{xz} \\ g_{yx} & g_{yy} & g_{yz} \\ g_{zx} & g_{zy} & g_{zz} \end{bmatrix} \cdot \begin{bmatrix} \hat{S}_x \\ \hat{S}_y \\ \hat{S}_z \end{bmatrix}$$

to obtain another expression for the Zeeman spin hamiltonian.

A.8 The coordinates of the corners of the rectangular base of a parallelepiped are (2.191 0.448 0); (−2.191 −0.448 0); (1.673 −1.484 0); (−1.673 1.484 0).

(a) Determine the angle between the x axis and the long side of the rectangle.

(b) Using the appropriate rotation matrix (see Eq. A.51), transform the coordinates to an axis system in which the long side of the rectangle is parallel to the x axis.

A.9 A radical R is produced in a single crystal of RH by irradiation. Cartesian axes **x**, **y** and **z** are chosen for the crystal (say, utilizing faces of a single crystal). The elements of the (symmetric) g matrix determined by EPR are found to be $g_{xx} = 2.0039$, $g_{xy} = -0.0007$, $g_{xz} = -0.0001$, $g_{yy} = 2.0030$, $g_{yz} = 0$, $g_{zz} = 2.0035$.

(a) Find the principal elements g_X, g_Y and g_Z of this matrix;

(b) Confirm that the direction cosines of the principal axes relative to the crystal axes are (0.87386, −0.47349, 0.11035), (−0.07082, 0.10058, 0.99241) and (0.48099, 0.87504, −0.05436).

A.10 Derive the other three possible matrices **C** satisfying the conditions imposed in deriving the one shown in Eq. A.71.

A.11 Verify by matrix multiplication of Eq. A.73 that Eqs. A.74 and A.75 represent the requirement for diagonalization of a symmetric 2 × 2 matrix, and prove that parameter c in Eq. A.73 must be real for this equation to hold. What does this imply for matrix **P** in Eq. A.78?

A.12 Prove that the coordinate-rotation matrix

$$\begin{bmatrix} \cos\phi & \sin\phi & 0 \\ -\sin\phi & \cos\phi & 0 \\ 0 & 0 & 1 \end{bmatrix}$$

is real orthogonal.

A.13 Prove for an asymmetric parameter matrix, say

$$Y = \begin{bmatrix} Y_{11} & 0 & 0 \\ 0 & Y_{22} & Y_{23} \\ 0 & Y_{32} & Y_{33} \end{bmatrix}$$

that the principal axes do not form an orthogonal (cartesian) set.

A.14 Note that quantum-mechanical matrices can be spatial vectors. Thus, for example, take the three-component matrices $\bar{J}_x, \bar{J}_y, \bar{J}_z$ (in Section B.10.3) for angular momentum $j = \frac{1}{2}$ and combine them into a single 2×2 vector matrix $\bar{J}$, making use of the spatial unit vectors $\mathbf{i}, \mathbf{j}, \mathbf{k}$.

A.15 Convince yourself that every matrix Y obeys its own characteristic (secular) equation: $Y^d + s_1 Y^{d-1} + \cdots + s_{d-1} Y + s_d 1_d = 0 1_d$, where d is a positive integer (see Section A.5.6). This is the famous Cayley-Hamilton theorem. Perhaps, try it out on some 3×3 matrix of your own manufacture.

APPENDIX B

QUANTUM MECHANICS OF ANGULAR MOMENTUM

B.1 INTRODUCTION

Those properties (such as the total energy or angular momentum) that are conserved in an isolated system are called 'constants of motion'. These constants and their corresponding operators play an important role in quantum mechanics. Consider the operator $\hat{\mathcal{H}}$ for the total energy, that is, the hamiltonian operator. When $\hat{\mathcal{H}}$ operates on an energy eigenfunction $\psi_n(x, y, z)$ of a particular system (Eq. A.12), the result (Schrödinger's equation) is

$$\hat{\mathcal{H}}\psi_n = U_n\psi_n \tag{B.1}$$

where n denotes the state and U_n is the *exact* total energy of the nth state of the system. Equation B.1 is the time-independent Schrödinger equation. Such exact values U_n are called 'eigenvalues'. Operators on functions that do not correspond to constants of motion do not yield eigenvalues. Suppose now that λ_n is another constant of motion, arising from some other operator $\hat{\Lambda}$. It is an important result of quantum mechanics that $\psi_n(x, y, z)$, the eigenfunctions of $\hat{\mathcal{H}}$, can always be chosen in such a way that they are also eigenfunctions of $\hat{\Lambda}$.

Angular momentum offers some constants of motion for an isolated system. It is not the vector operator $\hat{\mathbf{J}}$ for the total angular momentum, but rather the square of this total angular momentum, which can produce an exact result when applied to

---

Electron Paramagnetic Resonance, Second Edition, by John A. Weil and James R. Bolton
Copyright © 2007 John Wiley & Sons, Inc.

an appropriately chosen $\psi_n(x, y, z)$. If the latter operator is designated by

$$\hat{J}^2 = \hat{\mathbf{J}}^{\mathrm{T}}\hat{\mathbf{J}} \tag{B.2}$$

then

$$\hat{J}^2\psi_n = \lambda_j\psi_n \tag{B.3}$$

Here λ_j is an eigenvalue of $\hat{J}^2$ and is expressed in units of $[h/2\pi]^2 = \hbar^2$. By an appropriate choice of an axis system, ψ_n can be made to be an eigenfunction of any *one* of the components of $\hat{\mathbf{J}}$, that is, $\hat{J}_x$, $\hat{J}_y$ or $\hat{J}_z$, in addition to being an eigenfunction of $\hat{J}^2$ and of $\hat{\mathcal{H}}$. However, ψ_n can never be a simultaneous eigenfunction of two or more of the components of $\hat{J}^2$. If ψ_n is to be a simultaneous eigenfunction of any group of operators, each of which corresponds to a constant of motion, all the operators must commute among themselves. For instance, if ψ_n is to be simultaneously an eigenfunction of $\hat{\mathcal{H}}$, $\hat{J}^2$ and $\hat{J}_z$, then the following relations must hold

$$(\hat{\mathcal{H}}\hat{J}^2 - \hat{J}^2\hat{\mathcal{H}})\psi_n = 0 \tag{B.4a}$$

$$(\hat{\mathcal{H}}\hat{J}_z - \hat{J}_z\hat{\mathcal{H}})\psi_n = 0 \tag{B.4b}$$

and

$$(\hat{J}^2\hat{J}_z - \hat{J}_z\hat{J}^2)\psi_n = 0 \tag{B.4c}$$

The eigenvalue λ_j of $\hat{J}^2$ is linked closely to the *symmetry* properties of $\psi_n(x, y, z)$. Thus, although the values of U_n vary according to the wavefunction used, the eigenvalue λ_j is the same for all wavefunctions that have the same symmetry characteristics. For example, all spherically symmetric wavefunctions have $\lambda_j = 0$. Hence, if one can obtain a set of solutions to Eq. B.3, these eigenvalues are generally associated with all wavefunctions of the same symmetry.

So far $\psi_n(x, y, z)$ has been presented as a three-dimensional spatial wavefunction. When $\hat{\mathbf{J}}$ represents a spatial angular momentum (e.g., orbital motion), λ_j is found to be a function of the quantum number j, which is an integer. For instance, j could be the electron orbital angular-momentum quantum number L or the molecular rotation quantum number N. However, electrons and many nuclei possess an intrinsic angular momentum (called 'spin angular momentum') that cannot be described in terms of a spatial wavefunction $\psi_n(x, y, z)$. In order to deal with this spin angular momentum, a 'spin coordinate' must be included in the wavefunction.[1] It is found that in these cases the quantum number j can then take on half-integer values as well (i.e., $j = 0, \frac{1}{2}, 1, \frac{3}{2}, \ldots$). This spin angular momentum has no classical counterpart; however, it can be accommodated in the quantum mechanics if a generalized angular momentum is defined (Section B.3). Henceforth the symbol $\hat{\mathbf{J}}$ stands for the operator of such a generalized angular momentum (which could be $\hat{\mathbf{L}}$, $\hat{\mathbf{S}}$, $\hat{\mathbf{I}}$, ... or a vector sum of any of them). It often is very useful to visualize the time development of a spin set using 'classical' pictures of the appropriate vectors in 3D space (e.g., see Chapter 11).

B.2 ANGULAR-MOMENTUM OPERATORS

In order to obtain the quantum-mechanical operators for angular momentum, one must first consider the classical expression

$$\ell = \mathbf{r} \wedge \mathbf{p} \tag{B.5}$$

for the orbital angular momentum ℓ of a particle orbiting about an origin $\mathbf{O}$. Here $\mathbf{r}$ represents the position vector of the particle, and $\mathbf{p}$ is its linear-momentum vector. The vector product indicated in Eq. B.5 can be computed as in Eq. A.35b. The components of ℓ are

$$\ell_x = yp_z - zp_y \tag{B.6a}$$
$$\ell_y = zp_x - xp_z \tag{B.6b}$$
$$\ell_z = xp_y - yp_x \tag{B.6c}$$

The operator $\hat{\mathbf{r}}$ is the same (just multiplication) in quantum mechanics as in classical mechanics, so that the symbol $^\wedge$ need not be attached. However, $\mathbf{p}$ must be replaced by $-i\hbar\,\hat{\Delta}$, where $\hat{\Delta} = \mathbf{i}\partial/\partial x + \mathbf{j}\partial/\partial y + \mathbf{k}\partial/\partial z$. Thus in quantum mechanics the components of $\hat{\ell}$ become

$$\hat{\ell}_x = -i\hbar\left(y\frac{\hat{\partial}}{\partial z} - z\frac{\hat{\partial}}{\partial y}\right) \tag{B.7a}$$

$$\hat{\ell}_y = -i\hbar\left(z\frac{\hat{\partial}}{\partial x} - x\frac{\hat{\partial}}{\partial z}\right) \tag{B.7b}$$

$$\hat{\ell}_z = -i\hbar\left(x\frac{\hat{\partial}}{\partial y} - y\frac{\hat{\partial}}{\partial x}\right) \tag{B.7c}$$

The spatial angular-momentum operator is more conveniently represented in terms of the operator $\hat{\mathbf{L}}$, that is, $\hat{\ell}$ is measured in units of $\hbar$, and hence

$$\hat{L}_x = -i\left(y\frac{\hat{\partial}}{\partial z} - z\frac{\hat{\partial}}{\partial y}\right) \tag{B.8a}$$

$$\hat{L}_y = -i\left(z\frac{\hat{\partial}}{\partial x} - x\frac{\hat{\partial}}{\partial z}\right) \tag{B.8b}$$

$$\hat{L}_z = -i\left(x\frac{\hat{\partial}}{\partial y} - y\frac{\hat{\partial}}{\partial x}\right) \tag{B.8c}$$

As in classical mechanics of angular momentum, the square of a vector operator is equivalent to the sum of the squares of the three component operators:

$$\hat{L}^2 = \hat{\mathbf{L}}^{\mathrm{T}}\hat{\mathbf{L}} = \hat{L}_x{}^2 + \hat{L}_y{}^2 + \hat{L}_z{}^2 \tag{B.9}$$

In classical mechanics the magnitude and direction of an angular momentum vector are well defined. In quantum mechanics only the magnitude of the *total* angular-momentum vector and any *one* of its components are exactly and simultaneously measurable. It is possible to determine two observables exactly and simultaneously only if the operators corresponding to them commute (Section B.1).

B.3 COMMUTATION RELATIONS FOR GENERAL ANGULAR-MOMENTUM OPERATORS

We find it convenient to deal here with the commutator $\hat{A}\hat{B} - \hat{B}\hat{A}$ denoted by $[\hat{A},\hat{B}]_-$ for any two operators. By expanding $\hat{\mathbf{J}}^{\mathrm{T}}\hat{\mathbf{J}} = \hat{J}^2$, it can be shown that $\hat{J}^2$ commutes with all of the components $\hat{J}_x$, $\hat{J}_y$, and $\hat{J}_z$:

$$[\hat{J}^2,\hat{J}_x]_- = [\hat{J}^2,\hat{J}_y]_- = [\hat{J}^2,\hat{J}_z]_- = 0 \tag{B.10}$$

However, no two components commute among themselves. For example

$$\hat{J}_x\hat{J}_y = -\left(y\frac{\hat{\partial}}{\partial z} - z\frac{\hat{\partial}}{\partial y}\right)\left(z\frac{\hat{\partial}}{\partial x} - x\frac{\hat{\partial}}{\partial z}\right)$$

$$= -\left[y\frac{\hat{\partial}}{\partial x} + yz\frac{\hat{\partial}^2}{\partial z\partial x} - xy\frac{\hat{\partial}^2}{\partial z^2} - z^2\frac{\hat{\partial}^2}{\partial x\partial y} + xz\frac{\hat{\partial}^2}{\partial y\partial z}\right] \tag{B.11a}$$

$$\hat{J}_y\hat{J}_x = -\left(z\frac{\hat{\partial}}{\partial x} - x\frac{\hat{\partial}}{\partial z}\right)\left(y\frac{\hat{\partial}}{\partial z} - z\frac{\hat{\partial}}{\partial y}\right)$$

$$= -\left[yz\frac{\hat{\partial}^2}{\partial x\partial z} - z^2\frac{\hat{\partial}^2}{\partial x\partial y} - xy\frac{\hat{\partial}^2}{\partial z^2} + yz\frac{\hat{\partial}^2}{\partial z\partial x} + x\frac{\hat{\partial}}{\partial y}\right] \tag{B.11b}$$

$$[\hat{J}_x,\hat{J}_y]_- = \hat{J}_x\hat{J}_y - \hat{J}_y\hat{J}_x = x\frac{\hat{\partial}}{\partial y} - y\frac{\hat{\partial}}{\partial x} = i\hat{J}_z \tag{B.11c}$$

Similar expressions hold for the commutators of the other components. Consequently, it is not possible simultaneously to determine three or even two components of the angular-momentum operator. However, it is possible to determine the square of the magnitude of $\hat{\mathbf{J}}$ and *one* of the components of $\hat{\mathbf{J}}$, which is usually taken as $\hat{J}_z$.

These important commutation relations are summarized as follows:

$$[\hat{J}_x, \hat{J}_y]_- = i\hat{J}_z \qquad (B.12a)$$

$$[\hat{J}_y, \hat{J}_z]_- = i\hat{J}_x \qquad (B.12b)$$

$$[\hat{J}_z, \hat{J}_x]_- = i\hat{J}_y \qquad (B.12c)$$

$$[\hat{J}^2, \hat{J}_x]_- = [\hat{J}^2, \hat{J}_y]_- = [\hat{J}^2, \hat{J}_z]_- = 0 \qquad (B.10)$$

A generalized angular momentum (i.e., one that may include spin) *is defined as any vector operator whose components obey the commutation relations of Eqs. B.10 and B.12.*

At this point it is convenient to introduce the so-called *ladder* operators, which are linear combinations of $\hat{J}_x$ and $\hat{J}_y$, defined by

$$\hat{J}_+ \equiv \hat{J}_x + i\hat{J}_y \qquad (B.13a)$$

$$\hat{J}_- \equiv \hat{J}_x - i\hat{J}_y \qquad (B.13b)$$

where $\hat{J}_+$ is the *raising* operator and $\hat{J}_-$, the *lowering* operator. The significance of these operators becomes apparent later. As can readily be verified by substitution of their definitions, they obey the commutation relations

$$[\hat{J}^2, \hat{J}_+]_- = [\hat{J}^2, \hat{J}_-]_- = 0 \qquad (B.14a)$$

$$[\hat{J}_z, \hat{J}_+]_- = \hat{J}_+ \qquad (B.14b)$$

$$[\hat{J}_z, \hat{J}_-]_- = \hat{J}_- \qquad (B.14c)$$

$$[\hat{J}_+, \hat{J}_-]_- = 2\hat{J}_z \qquad (B.14d)$$

B.4 EIGENVALUES OF $\hat{J}^2$ AND $\hat{J}_Z$

Let the eigenvalues of $\hat{J}^2$ and $\hat{J}_z$ be λ_j and λ_m, respectively. The angular-momentum eigenvalues depend only on the primary and secondary quantum numbers j and m.[2] As we shall see, quantum number j is characteristic of the total angular momentum magnitude, and m is characteristic of the z component of the angular momentum. The angular-momentum eigenfunctions are completely specified by j and m. Hence the functions can be represented by kets written as $|j, m\rangle$ (Section A.5.4).[3] The eigenvalue equations for $\hat{J}^2$ and $\hat{J}_z$ can then be written

$$\hat{J}^2|j, m\rangle = \lambda_j|j, m\rangle \qquad (B.15a)$$

$$\hat{J}_Z|j, m\rangle = \lambda_m|j, m\rangle \qquad (B.15b)$$

where λ_j and λ_m are the eigenvalues of $\hat{J}^2$ and $\hat{J}_z$. Note that the kets $|j, m\rangle$ are orthonormal. This means that

$$\langle j', m' \mid j, m \rangle = 1 \quad \text{for} \quad j' = j \text{ and } m' = m \qquad (B.16a)$$

$$\langle j', m' \mid j, m \rangle = 0 \quad \text{for} \quad j' \neq j \text{ and/or } m' \neq m \qquad (B.16b)$$

The operator $\hat{J}^2$ may be expanded as

$$\hat{J}^2 = \hat{J}_x^2 + \hat{J}_y^2 + \hat{J}_z^2 \qquad (B.17a)$$

$$\hat{J}_x^2 + \hat{J}_y^2 = \hat{J}^2 - \hat{J}_z^2 \qquad (B.17b)$$

Thus the operator $\hat{J}_x^2 + \hat{J}_y^2$ also has the discrete eigenvalues

$$(\hat{J}_x^2 + \hat{J}_y^2) \mid j, m\rangle = (\hat{J}^2 - \hat{J}_z^2) \mid j, m\rangle = (\lambda_j - \lambda_m^2) \mid j, m\rangle \qquad (B.18)$$

The operators $\hat{J}_x$ and $\hat{J}_y$ correspond individually to experimental observables (and hence are hermitian; see Section A.2). When they are applied to $|j, m\rangle$, they must give real numbers. Hence the eigenvalues of $\hat{J}_x^2 + \hat{J}_y^2$ must be *real and non-negative*, which implies (see Eqs. B.15) that

$$\lambda_j - \lambda_m^2 \geq 0 \qquad (B.19)$$

To establish the exact form of the eigenvalues λ_m, it is convenient to examine the matrix elements of the commutator $[\hat{J}_z, \hat{J}_+]_- = \hat{J}_+$ (Eq. B.14b), that is

$$\langle j, m' \mid \hat{J}_z \hat{J}_+ - \hat{J}_+ \hat{J}_z \mid j, m \rangle = \langle j, m' \mid \hat{J}_+ \mid j, m \rangle \qquad (B.20)$$

where the bra and the ket have the same value of j but where m and m' may differ. Evaluation of the left-hand side of Eq. B.20 demonstrates the effect of $\hat{J}_+$ on the wavefunctions $|j, m\rangle$. The left-hand matrix element can be expanded into two matrix elements

$$\langle j, m' \mid \hat{J}_z \hat{J}_+ \mid j, m \rangle - \langle j, m' \mid \hat{J}_+ \hat{J}_z \mid j, m \rangle$$

Use of Eq. B.15b allows the second matrix element to be reduced to

$$\langle j, m' \mid \hat{J}_+ \hat{J}_z \mid j, m \rangle = \lambda_m \langle j, m' \mid \hat{J}_+ \mid j, m \rangle \qquad (B.21a)$$

The first matrix element can be reduced with the help of Eq. A.54

$$\langle j, m' \mid \hat{J}_z \hat{J}_+ \mid j, m \rangle, = \lambda_{m'}^* \langle j, m' \mid \hat{J}_+ \mid j, m \rangle \qquad (B.21b)$$

where the arrow implies operation toward the left. Since $\hat{J}_z$ is a hermitian operator, $\lambda_{m'}$ must be real, so that $\lambda_{m'}^* = \lambda_{m'}$. Thus Eq. B.20 reduces to

$$(\lambda_{m'} - \lambda_m)\langle j, m' | \hat{J}_+ | j, m\rangle = \langle j, m' | \hat{J}_+ | j, m\rangle \qquad (B.22)$$

This means that the only non-zero matrix elements of $\hat{J}_+$ are those for which $\lambda_{m'} - \lambda_m = +1$. Consequently, the only nonzero result of the operator $\hat{J}_+$ acting on $| j, m\rangle$ is

$$\hat{J}_+ | j, m\rangle = \xi_m | j, m + 1\rangle \qquad (B.23)$$

that is, $m' = m + 1$. Similarly, an examination of the matrix element of $\hat{J}_-$ shows that the only non-zero matrix elements of $\hat{J}_-$ are those for which $\lambda_{m'} - \lambda_m = -1$:

$$\hat{J}_- | j, m\rangle = \zeta_m | j, m - 1\rangle \qquad (B.24)$$

Here scalars ξ_m and ζ_m of Eqs. B.23 and B.24 are not eigenvalues and may be complex numbers, so that a factor $e^{i\phi}$, where ϕ is a real phase angle, may appear in ξ_m and ζ_m (Section A.1). It is now apparent, from Eqs. B.23 and B.24, why $\hat{J}_+$ and $\hat{J}_-$ are called *raising* and *lowering* operators.

This analysis shows that for a given value of λ_j, one may obtain a whole set of states $| j, m\rangle$ having the eigenvalues

$$\ldots \lambda_{m-2}, \lambda_{m-1}, \lambda_m, \lambda_{m+1}, \lambda_{m+2} \ldots$$

This series must terminate at both ends, since from Eq. B.19, $\lambda_m^2 \leq \lambda_j$. The λ_m values differ by integers, and the m quantum number is assumed to increase in integral steps for a given value of j; hence one may equate λ_m and m.

Within the series given above, the highest eigenvalue of $\hat{J}_z$ is designated by $\overline{m}$ and the lowest eigenvalue of $\hat{J}_z$ by $\underline{m}$. Therefore

$$\hat{J}_+ | j, \overline{m}\rangle = 0 \qquad (B.25a)$$

$$\hat{J}_- | j, \underline{m}\rangle = 0 \qquad (B.25b)$$

Otherwise, there would be a value of $\overline{m}$ higher than $|\lambda_j^{1/2}|$ and a value of $\underline{m}$ less than $-|\lambda_j^{1/2}|$; this is contrary to the limitation imposed by Eq. B.19.

Next, apply $\hat{J}_-$ to Eq. B.25a. When $\hat{J}_- \hat{J}_+$ is expanded, one obtains

$$\begin{aligned}
\hat{J}_- \hat{J}_+ &= (\hat{J}_x - i\hat{J}_y)(\hat{J}_x + i\hat{J}_y) \\
&= \hat{J}_x^2 + \hat{J}_y^2 + i[\hat{J}_x, \hat{J}_y]_- \\
&= \hat{J}^2 - \hat{J}_z^2 - \hat{J}_z \qquad (B.26)
\end{aligned}$$

Therefore

$$\hat{J}_-\hat{J}_+|j, \overline{m}\rangle = (\lambda_j - \overline{m}^2 - \overline{m})|j, \overline{m}\rangle = 0 \qquad (\text{B.27})$$

and it follows that

$$\lambda_j = \overline{m}(\overline{m} + 1) \qquad (\text{B.28})$$

Similarly, by applying $\hat{J}_+$ to Eq. B.25b, one obtains

$$\lambda_j = \underline{m}(\underline{m} - 1) \qquad (\text{B.29})$$

Equations B.28 and B.29 are compatible only if $\overline{m} = -\underline{m}$.

Since successive values of m differ by unity, $\overline{m} - \underline{m}$ is a non-negative integer. It thus can denoted by $2j$.[4] Hence j can have the values

$$j = 0, \tfrac{1}{2}, 1, \tfrac{3}{2}, \ldots$$

Then from $\overline{m} - \underline{m} = 2j$ and $\overline{m} = -\underline{m}$

$$\overline{m} = +j \quad \text{and} \quad \underline{m} = -j \qquad (\text{B.30})$$

Hence $m = -j, -j + 1, \ldots, +j - 1, +j$. There are thus $2j + 1$ permissible values of m for each value of j. From Eq. B.28 (or Eq. B.29)

$$\lambda_j = \overline{m}(\overline{m} + 1) = j(j + 1) \qquad (\text{B.31})$$

In summary, then, the eigenvalues of $\hat{J}^2$ and $\hat{J}_z$ are

$$\hat{J}^2|j, m\rangle = j(j + 1)|j, m\rangle \qquad (\text{B.32}a)$$

and

$$\hat{J}_z|j, m\rangle = m|j, m\rangle \qquad (\text{B.32}b)$$

Note that by convention $\hat{\mathbf{J}}$ is unitless so that the units ($\hbar^2$ in Eq. B.32a and $\hbar$ in Eq. B.32b) are assumed to have been divided out.

It is important to note that eigenvalue m of the projection $\hat{J}_z$ changes when one changes from one quantization axis $\mathbf{z}$ to some other $\mathbf{z}'$. Thus when a value of m is given, one should always keep in mind what spatial direction is associated with it. When changing axes, each state $|j, m'\rangle$ in the new system is a linear combination of the old states $|j, m\rangle$.

The quantities ξ_m and ζ_m of Eqs. B.23 and B.24 remain to be evaluated. Consider the diagonal matrix element

$$\langle j, m \mid \hat{J}_- \hat{J}_+ \mid j, m \rangle = \xi_m \langle j, m \mid \hat{J}_- \mid j, m + 1 \rangle$$
$$= \xi_m \zeta_{m+1} \langle j, m \mid j, m \rangle = \xi_m \zeta_{m+1} \tag{B.33}$$

Using Eq. B.26, one has

$$\langle j, m \mid \hat{J}_- \hat{J}_+ \mid j, m \rangle = j(j+1) - m^2 - m \tag{B.34}$$

Thus

$$\xi_m \zeta_{m+1} = j(j+1) - m(m+1) \tag{B.35}$$

The matrix element of Eq. B.33 may be evaluated in a third way

$$\langle j, m \mid \hat{J}_- \hat{J}_+ \mid j, m \rangle = \xi_m [\langle j, m \mid \hat{J}_- \mid j, m + 1 \rangle]^\dagger \\ \qquad\qquad\qquad \overleftarrow{}$$
$$= \xi_m [\langle j, m + 1 \mid \hat{J}_+ \mid j, m \rangle]^* \\ \qquad\qquad\qquad \overrightarrow{}$$
$$= \xi_m \xi_m^* [\langle j, m + 1 \mid j, m + 1 \rangle]$$
$$= \xi_m \xi_m^* \tag{B.36}$$

where the symbol † means adjoint. Equation A.54 has again been used, together with the fact that $\hat{J}_-^* = \hat{J}_+$. Thus from Eqs. B.33 and B.34, we obtain

$$\xi_m^* = \zeta_{m+1} \tag{B.37}$$

and

$$\xi_m \xi_m^* = |\xi_m|^2 = j(j+1) - m(m+1) \tag{B.38}$$

or

$$\xi_m = [j(j+1) - m(m+1)]^{1/2} \tag{B.39}$$

The value of ξ_m in Eq. B.39 should have been multiplied by $e^{i\phi}$, where ϕ is a phase angle. We note that $(e^{i\phi})(e^{i\phi})^* = 1$. By convention, ϕ is chosen to be zero, so that ξ_m is real and non-negative. If this convention is applied consistently, then this choice has no effect on the final results, since all experimental observables correspond to real numbers.

Similarly, from the matrix element of $\hat{J}_+ \hat{J}_-$, one obtains

$$\zeta_m = [j(j+1) - m(m-1)]^{1/2} \tag{B.40}$$

Hence the operation of $\hat{J}_+$ and $\hat{J}_-$ on $|j, m\rangle$ gives the results

$$\hat{J}_+|j, m\rangle = [j(j + 1) - m(m + 1)]^{1/2} \, | \, j, m + 1\rangle \qquad (B.41a)$$

$$\hat{J}_-|j, m\rangle = [j(j + 1) - m(m - 1)]^{1/2} \, |j, m - 1\rangle \qquad (B.41b)$$

B.5 SUPERPOSITION OF STATES

In quantum mechanics [1, 2], every particle has eigenstates of various observables, and in general its state is a linear sum of the eigenstates, say, the kets for the electron (or proton) spin

$$|\rangle = c_+ \, | \, \tfrac{1}{2}, +\tfrac{1}{2}\rangle + c_- \, | \, \tfrac{1}{2}, -\tfrac{1}{2}\rangle \qquad (B.42)$$

where the coefficients are complex numbers obeying the normalization condition $|c_+|^2 + |c_-|^2 = 1$. Measurement theory states that when the spin component of that electron is measured sufficiently many times, for the electron always in the state cited above, that the fractional result will be $|c_+|$ occurrences of $+\tfrac{1}{2}$ and $|c_-|$ occurrences of $-\tfrac{1}{2}$. Analogously, for the photon, the kets will be $|1, \pm 1\rangle$.

B.6 ANGULAR-MOMENTUM MATRICES

At this point one may summarize the non-zero matrix elements of $\hat{J}_x, \hat{J}_y, \hat{J}_z, \hat{J}_+, \hat{J}_-$ and $\hat{J}^2 = \hat{\mathbf{J}}^T \hat{\mathbf{J}}$, as follows:

$$\langle j, m + 1|\hat{J}_x|j, m\rangle = (1/2)[j(j + 1) - m(m + 1)]^{1/2} \qquad (B.43a)$$

$$\langle j, m - 1|\hat{J}_x|j, m\rangle = (1/2)[j(j + 1) - m(m - 1)]^{1/2} \qquad (B.43b)$$

$$\langle j, m + 1|\hat{J}_y|j, m\rangle = (-i/2)[j(j + 1) - m(m + 1)]^{1/2} \qquad (B.43c)$$

$$\langle j, m - 1|\hat{J}_y|j, m\rangle = (+i/2)[j(j + 1) - m(m - 1)]^{1/2} \qquad (B.43d)$$

$$\langle j, m|\hat{J}_z|j, m\rangle = m \qquad (B.43e)$$

$$\langle j, m + 1|\hat{J}_+|j, m\rangle = [j(j + 1) - m(m + 1)]^{1/2} \qquad (B.43f)$$

$$\langle j, m - 1|\hat{J}_-|j, m\rangle = [j(j + 1) - m(m - 1)]^{1/2} \qquad (B.43g)$$

$$\langle j, m|\hat{J}^2|j, m\rangle = j(j + 1) \qquad (B.43h)$$

Equations B.43a to B.43d follow from Eqs. B.13; that is, $\hat{J}_x = (\hat{J}_+ + \hat{J}_-)/2$ and $\hat{J}_y = (\hat{J}_+ - \hat{J}_-)/2i$. Equation B.43e follows from Eq. B.32b.

For any given value of j, the matrix elements such as those in Eqs. B.43 form a square matrix. The matrix is said to represent the operator in question. It is depicted by the same boldfaced symbol, but without the ^. The order of the matrix is $2j + 1$, corresponding to the possible values of m. Consider the spin matrices for $j = \frac{1}{2}$. These are directly applicable to the electron-spin case, where $S = \frac{1}{2}$, and to the nuclear-spin cases with $I = \frac{1}{2}$. The matrix representation of the operator $\hat{J}_x$ is

$$
\begin{array}{c}
\left|\frac{1}{2}, +\frac{1}{2}\right\rangle \left|\frac{1}{2}, -\frac{1}{2}\right\rangle \\
\mathbf{J}_x = \begin{array}{c} \langle\frac{1}{2}, +\frac{1}{2}| \\ \langle\frac{1}{2}, -\frac{1}{2}| \end{array} \begin{bmatrix} 0 & +\frac{1}{2} \\ +\frac{1}{2} & 0 \end{bmatrix} = +\frac{1}{2}\begin{bmatrix} 0 & +1 \\ +1 & 0 \end{bmatrix}
\end{array}
\tag{B.44a}
$$

where the rows and columns are labeled by j, m_j and the bras and kets contain the sets j, m. The elements appearing in the $\mathbf{J}_x$ matrix B.44a were obtained by inserting the operator $\hat{J}_x$ between the corresponding bra (indicated to the left of a given matrix element) and ket (above that element). For example, the 1,2 element $\langle\frac{1}{2}, +\frac{1}{2}|\hat{J}_x|\frac{1}{2}, -\frac{1}{2}\rangle$ of Eq. B.44a is computed (via use of Eq. B.43a) to be $[(\frac{1}{2})(\frac{3}{2}) - (-\frac{1}{2})(\frac{1}{2})]^{1/2}/2 = \frac{1}{2}$. In a similar fashion, $\mathbf{J}_y$ and $\mathbf{J}_z$ are written as

$$
\mathbf{J}_y = \begin{bmatrix} 0 & -\frac{1}{2}i \\ +\frac{1}{2}i & 0 \end{bmatrix} = +\frac{1}{2}\begin{bmatrix} 0 & -i \\ +i & 0 \end{bmatrix}
\tag{B.44b}
$$

$$
\mathbf{J}_z = \begin{bmatrix} +\frac{1}{2} & 0 \\ 0 & -\frac{1}{2} \end{bmatrix} = +\frac{1}{2}\begin{bmatrix} +1 & 0 \\ 0 & -1 \end{bmatrix}
\tag{B.44c}
$$

The matrices on the right of Eqs. B.44 are often called the *Pauli spin matrices*, symbolized by $\boldsymbol{\sigma}_x$, $\boldsymbol{\sigma}_y$ and $\boldsymbol{\sigma}_z$. Hence $\mathbf{J}_k = \boldsymbol{\sigma}_k/2$, where $k = x$, y, or z. One can obtain matrices $\mathbf{J}_+$ and $\mathbf{J}_-$ either from Eqs. B.43e and B.43f or from matrix addition:

$$
\mathbf{J}_+ = \mathbf{J}_x + i\mathbf{J}_y = \begin{bmatrix} 0 & +1 \\ 0 & 0 \end{bmatrix}
\tag{B.44d}
$$

$$
\mathbf{J}_- = \mathbf{J}_x - i\mathbf{J}_y = \begin{bmatrix} 0 & 0 \\ +1 & 0 \end{bmatrix}
\tag{B.44e}
$$

We note that the matrices representing the various quantum-mechanical operators can in a sense be themselves considered to be operators, operating on a column vector (Section A.5.5).

Since the eigenvalue of $\hat{J}^2$ for each spin function in the case of $j = \frac{1}{2}$ must be $\frac{1}{2}(\frac{1}{2} + 1) = \frac{3}{4}$, the matrix $\mathbf{J}^2$ is

$$
\mathbf{J}^2 = \begin{bmatrix} +\frac{3}{4} & 0 \\ 0 & +\frac{3}{4} \end{bmatrix} = \frac{3}{4}\mathbf{1}_2
\tag{B.45}
$$

This can be verified by computing

$$
\mathbf{J}^2 = \mathbf{J}_x^2 + \mathbf{J}_y^2 + \mathbf{J}_z^2
\tag{B.46}
$$

For instance, matrix multiplication of $\mathbf{J}_x$ by itself gives

$$
\begin{bmatrix} 0 & +\frac{1}{2} \\ +\frac{1}{2} & 0 \end{bmatrix} \cdot \begin{bmatrix} 0 & +\frac{1}{2} \\ +\frac{1}{2} & 0 \end{bmatrix} = \begin{bmatrix} +\frac{1}{4} & 0 \\ 0 & +\frac{1}{4} \end{bmatrix}
$$

with identical results for $\mathbf{J}_y^2$ and $\mathbf{J}_z^2$. Addition of the matrices in Eq. B.46 yields the result in Eq. B.45. Similar to the Pauli matrices, correspondingly larger $[(2j+1) \times (2j+1)]$ matrices can easily be written for $j > \frac{1}{2}$ (e.g., Section B.10).

B.7 ADDITION OF ANGULAR MOMENTA

One often encounters problems in which there are two angular momenta that may or may not be 'coupled by an interaction'. More accurately, interaction terms in the hamiltonian often can be visualized by use of vector models, in which various angular-momentum vectors are added. This situation arises in the following cases:

1. Coupling of spin and orbital angular momenta of some particle (e.g., an electron).
2. Coupling of the angular momenta of two different particles (e.g., the spins of an electron and a proton).

We begin by considering two angular-momentum operators $\hat{\mathbf{J}}_1$ and $\hat{\mathbf{J}}_2$. We may construct two equivalent representations of these two, an *uncoupled* representation and a *coupled* representation. We find situations in which one or the other of these representations is the more convenient.

We start with the uncoupled representation, where $\hat{\mathbf{J}}_1$ and $\hat{\mathbf{J}}_2$ are independent, that is, assumed not to be coupled initially. The eigenfunctions associated with operators $\hat{\mathbf{J}}_1$ and $\hat{\mathbf{J}}_2$ are taken as $|j_1, m_1\rangle$ and $|j_2, m_2\rangle$. Thus

$$\hat{J}_1^{\,2}|j_1, m_1\rangle = j_1(j_1 + 1)|j_1, m_1\rangle \tag{B.47a}$$

$$\hat{J}_2^{\,2}|j_2, m_2\rangle = j_2(j_2 + 1)|j_2, m_2\rangle \tag{B.47b}$$

$$\hat{J}_{1z}|j_1, m_1\rangle = m_1|j_1, m_1\rangle \tag{B.47c}$$

$$\hat{J}_{2z}|j_2, m_2\rangle = m_2|j_2, m_2\rangle \tag{B.47d}$$

The direct-product kets $|j_1, m_1\rangle|j_2, m_2\rangle = |j_1, j_2, m_1, m_2\rangle$ form a set that define the uncoupled representation.

We now turn to the coupled representation. The total angular momentum $\hat{\mathbf{J}}$ is defined by

$$\hat{\mathbf{J}} = \hat{\mathbf{J}}_1 + \hat{\mathbf{J}}_2 \tag{B.48}$$

Note that from Section A.4 and the law of cosines, one obtains

$$\hat{\mathbf{J}}_1{}^T\hat{\mathbf{J}}_2 = \tfrac{1}{2}(\hat{J}^2 - \hat{J}_1{}^2 - \hat{J}_2{}^2) \tag{B.49}$$

Since $\hat{\mathbf{J}}$ is an angular momentum, its components must satisfy the commutation relations Eqs. B.10 and B.12. For example

$$[\hat{J}_x, \hat{J}_y]_- = [\hat{J}_{1x} + \hat{J}_{2x}, \hat{J}_{1y} + \hat{J}_{2y}]_-$$
$$= [\hat{J}_{1x}, \hat{J}_{1y}]_- + [\hat{J}_{1x}, \hat{J}_{2y}]_- + [\hat{J}_{2x}, \hat{J}_{1y}]_- + [\hat{J}_{2x}, \hat{J}_{2y}]_- \tag{B.50}$$

Each of the middle two commutators is zero, since independent angular momenta in different sets (e.g., $\hat{\mathbf{S}}$ and $\hat{\mathbf{I}}$) commute. Thus

$$[\hat{J}_x, \hat{J}_y]_- = i\hat{J}_{1z} + i\hat{J}_{2z} = i\hat{J}_z \tag{B.51}$$

The eigenfunctions $|j_1, j_2, j, m\rangle$ of the operators $\hat{J}_1{}^2, \hat{J}_2{}^2, \hat{J}^2$ and $\hat{J}_z$ and $\hat{J}_z$ form the set defining the coupled representation. Thus

$$\hat{J}_1{}^2 |j_1, j_2, j, m\rangle = j_1(j_1 + 1)|j_1, j_2, j, m\rangle \tag{B.52a}$$

$$\hat{J}_2{}^2 |j_1, j_2, j, m\rangle = j_2(j_2 + 1)|j_1, j_2, j, m\rangle \tag{B.52b}$$

$$\hat{J}^2 |j_1, j_2, j, m\rangle = j(j + 1)|j_1, j_2, j, m\rangle \tag{B.52c}$$

$$\hat{J}_z |j_1, j_2, j, m\rangle = m|j_1, j_2, j, m\rangle \tag{B.52d}$$

The coupled and uncoupled representations are connected by the transformation[5]

$$|j_1, j_2, j, m\rangle = \sum_{m_1, m_2} C(j_1, j_2, j; m_1 m_2 m) |j_1, j_2, m_1, m_2\rangle \tag{B.53}$$

The numerical factors $C(j_1 j_2 j; m_1 m_2 m)$ are variously called *vector-coupling*, *Clebsch-Gordan* or *Wigner coefficients* and are to be found in various tabulations [3–5]. They are described further (Eqs. B.68–B.71).

If the operator $\hat{J}_z = \hat{J}_{1z} + \hat{J}_{2z}$ is applied to Eq. B.53, one obtains

$$m|j_1, j_2, j, m\rangle = \sum_{m_1, m_2} (m_1 + m_2) C(j_1 j_2 j; m_1 m_2 m)|j_1, j_2, m_1, m_2\rangle$$

and hence (using Eq. B.53)

$$\sum_{m_1, m_2} (m - m_1 - m_2) C(j_1 j_2 j; m_1 m_2 m)|j_1, j_2, m_1, m_2\rangle = 0 \tag{B.54}$$

Since the kets $|j_1, j_2, m_1, m_2\rangle$ are linearly independent, this sum can vanish only if the coefficient of each term is identically zero. Hence

$$(m - m_1 - m_2) \, C(j_1 j_2 j; m_1 m_2 m) = 0 \tag{B.55}$$

Thus $C(j_1 j_2 j; m_1 m_2 m) = 0$ unless $m = m_1 + m_2$. Consequently, m, m_1 and m_2 are not independent, and the double sum in Eq. B.53 can be replaced by a single sum over m_1, since $m_2 = m - m_1$. Hence, with simplified notation, we obtain

$$|j_1, j_2, j, m\rangle = \sum_{m_1} C(j_1 j_2 j; m_1 m - m_1) |j_1, j_2, m_1, m - m_1\rangle \tag{B.56}$$

Further restrictions on the vector-coupling (VC) coefficients can be derived from the orthonormal properties of the $|j_1, j_2, j, m\rangle$ eigenfunctions:

$$\langle j_1, j_2, j', m' | j_1, j_2, j, m\rangle = \delta_{jj'} \, \delta_{mm'}$$

$$= \sum_{m_1'} \sum_{m_1} \Big[C(j_1 j_2 j; m_1 m - m_1) \, C(j_1 j_2 j'; m_1' m' - m_1') \times \tag{B.57}$$

$$\langle j_1, j_2, m_1, m - m_1 | j_1, j_2, m_1', m' - m_1'\rangle \Big]$$

The VC coefficients have been assumed to be real in Eq. B.57. Note that j and j' are obtained from the same values of j_1 and j_2.

Equation B.57 thus restricts the sum in Eq. B.56 to functions that have the same values of j and m. Since the values of j_1 and j_2 must also be the same on both sides, the notation may be further simplified to

$$|j, m\rangle = \sum_{m_1} C(j_1 j_2 j; m_1 m - m_1) |m_1, m_2\rangle \tag{B.58}$$

where $m = m_1 + m_2$.

Nothing has yet been said about the ranges of j and m. Since $\hat{\mathbf{J}}$ is a generalized angular momentum, one invokes the restrictions $\overline{m} = j$ and $\underline{m} = -j$ found in Eq. B.30. Here $\overline{m}$ and $\underline{m}$ are the maximum and minimum values of m. Since $m = m_1 + m_2$, the maximum value of m for all values of j is $j_1 + j_2$. This must also be the maximum value of j; otherwise there would exist a larger value of m. Thus

$$\overline{j} = j_1 + j_2 \tag{B.59a}$$

where $\overline{j}$ is the maximum value of j. Similarly

$$\underline{j} = |j_1 - j_2| \tag{B.59b}$$

where $\underline{j}$ is the minimum value of j. A detailed proof of these relations is given below (Eqs. B.64–B.66). When j and m have their maximum values, the relation between the coupled and uncoupled representations is especially simple, since there is only one permissible value of m_1 (and hence of m_2). Thus from Eq. B.58

$$\begin{aligned}
|\overline{j}, \overline{m}\rangle &= C(j_1 j_2 j_1 + j_2; j_1 j_2)|\overline{m}_1, \overline{m}_2\rangle \\
&= |\overline{m}_1, \overline{m}_2\rangle
\end{aligned} \tag{B.60a}$$

since the standard convention takes $C(j_1 j_2 j_1 + j_2; j_1 j_2) = 1$. Similarly

$$|\overline{j}, \underline{m}\rangle = |\underline{m}_1, \underline{m}_2\rangle \tag{B.60b}$$

Equations B.60 generate two of the $2j+1$ kets in the set with $j = \overline{j}$. The other members of this set can be obtained by applying the lowering operator $\hat{J}_- = \hat{J}_{1-} + \hat{J}_{2-}$ to Eq. B.60a or by either applying the raising operator $\hat{J}_+ = \hat{J}_{1+} + \hat{J}_{2+}$ to Eq. B.60b; for example

$$\hat{J}_-|\overline{j}, \overline{m}\rangle = (\hat{J}_{1-} + \hat{J}_{2-})|\overline{m}_1, \overline{m}_2\rangle \tag{B.61a}$$

and

$$\zeta_m|\overline{j}, \overline{m} - 1\rangle = \zeta_{m_1}|\overline{m} - 1_1, \overline{m}_2\rangle + \zeta_{m_2}|\overline{m}_1, \overline{m}_2 - 1\rangle \tag{B.61b}$$

The coefficients ζ are obtained from Eq. B.40. Hence

$$C(j_1 j_2 \ j_1 + j_2; \overline{m}_1 - 1 \ m_2) = \zeta_{m_1}/\zeta_m \tag{B.62a}$$

and

$$C(j_1 j_2 \ j_1 + j_2; \overline{m}_1 \ m_2 - 1) = \zeta_{m_2}/\zeta_m \tag{B.62b}$$

Sequential addition of $\hat{J}_-$ to Eq. B.61 generates all the $2j+1$ kets in the set with $j = \overline{j}$.

The set of kets corresponding to $j = \overline{j} - 1$ are fewer in number by two than the set with $j = \overline{j}$ and are bounded by the kets $|\overline{j} - 1, \overline{m} - 1\rangle$ and $|\overline{j} - 1, \underline{m} + 1\rangle$. The use of Eq. B.58 demonstrates that $|\overline{j} - 1, \overline{m} - 1\rangle$ must be related to the same uncoupled kets as $|\overline{j}, \overline{m} - 1\rangle$. Furthermore, there can be no other functions with $m = \overline{m} - 1$. If one writes

$$|\overline{j}, \overline{m} - 1\rangle = c_1|\overline{m}_1 - 1, \overline{m}_2\rangle + c_2|\overline{m}_1, \overline{m}_2 - 1\rangle \tag{B.63a}$$

$$|\overline{j} - 1, \overline{m} - 1\rangle = c_1'|\overline{m}_1 - 1, \overline{m}_2\rangle + c_2'|\overline{m}_1, \overline{m}_2 - 1\rangle \tag{B.63b}$$

then the orthonormal properties of the kets require that $c_1' = c_2$ and $c_2' = -c_1$.

Now that the ket $|\overline{j} - 1, \overline{m} - 1\rangle$ has been defined, all the other members of that set can be obtained by the application of $\hat{J}_-$. The ket $|\overline{j} - 2, \overline{m} - 2\rangle$ is obtained from the condition that it must be orthonormal to $|\overline{j} - 1, \overline{m} - 2\rangle$ and to

$|\overline{j}, \overline{m} - 2\rangle$. The preceding sequence of processes is continued until all the kets have been generated.

The number of uncoupled states must be the same as the number of coupled states:

$$\sum_{\underline{j}}^{\overline{j}} (2j + 1) = (2j_1 + 1)(2J_2 + 1) \tag{B.64}$$

This counting procedure determines $\underline{j}$, since $\overline{j} = j_1 + j_2$. The left-hand side of Eq. B.64 can be evaluated using

$$\sum_{\alpha}^{\beta} j = \tfrac{1}{2}[\beta(\beta + 1) - \alpha(\alpha - 1)] \tag{B.65a}$$

and

$$\sum_{\alpha}^{\beta} 1 = \beta - \alpha + 1 \tag{B.65b}$$

where j, α and β must be integers or (for j) half-integers. Thus Eq. B.64 becomes

$$[(j_1 + j_2)(j_1 + j_2 + 1) - \underline{j}\,(\underline{j} - 1)] + [j_1 + j_2 - \underline{j} + 1]$$
$$= (2j_1 + 1)(2j_2 + 1) \tag{B.66a}$$

or

$$\underline{j}^{\,2} = (j_1 - j_2)^2 \tag{B.66b}$$

Since $j \geq 0$ and real, one has

$$\underline{j} = |j_1 - j_2| \tag{B.59b}$$

This j is restricted to the values

$$j = |j_1 - j_2|, |j_1 - j_2| + 1, \ldots, j_1 + j_2 - 1, j_1 + j_2 \tag{B.67}$$

A number of symmetry relations exist among the VC coefficients; also a general relation may be derived for these coefficients.[6] However, it is often easier to evaluate the VC coefficients from relations such as Eqs. B.62. The following example illustrates the method.

Consider two angular momenta such that $j_1 = j_2 = 1$. From Eq. B.60a, we obtain

$$|2, +2\rangle_c = |+1, +1\rangle_u \tag{B.68a}$$

where subscripts c on $|j, m\rangle$ and u on $|m_1, m_2\rangle$ indicate the coupled and uncoupled representations. Application of $\hat{J}_-$ gives

$$|2, +1\rangle_c = \frac{1}{\sqrt{2}}[|+1, 0\rangle_u + |0, +1\rangle_u] \tag{B.68b}$$

Equations B.62 have been used to obtain the VC coefficients. A second application of $\hat{J}_-$ gives

$$|2,0\rangle_c = \frac{1}{\sqrt{6}}[2|0,0\rangle_u + |+1,-1\rangle_u + |-1,+1\rangle_u] \qquad (B.68c)$$

Further application of $\hat{J}_-$ (or the use of $\hat{J}_+$ on $|2,-2\rangle_c$) gives

$$|2,-1\rangle_c = \frac{1}{\sqrt{2}}[|-1,0\rangle_u + |0,-1\rangle_u] \qquad (B.68d)$$

and

$$|2, -2\rangle_c = |-1,-1\rangle_u \qquad (B.68e)$$

In general

$$|j,\pm m\rangle = \sum_{m_1} C(j_1 j_2 j; m_1 m_2) | \pm m_1, \pm m_2\rangle \qquad (B.69)$$

Thus only the first $j+1$ members of a set for given j need be evaluated.

The members of the set with $j = 1$ are evaluated by using the condition that $|2, +1\rangle_c$ and $|1, +1\rangle_c$ must be orthogonal. This requires that

$$|1, +1\rangle_c = \frac{1}{\sqrt{2}}[|+1,0\rangle_u - |0, +1\rangle_u] \qquad (B.70a)$$

Application of $\hat{J}_-$ to Eq. B.70a gives

$$|1, 0\rangle_c = \frac{1}{\sqrt{2}}[|+1, -1\rangle_u - |-1, +1\rangle_u] \qquad (B.70b)$$

The third member of the set is

$$|1, -1\rangle_c = \frac{1}{\sqrt{2}}[|-1, 0\rangle_u - |0, -1\rangle_u] \qquad (B.70c)$$

The single function $|0, 0\rangle_c$ of the set with $j = 0$ can be obtained from its orthogonality with $|2, 0\rangle_c$ and $|1, 0\rangle_c$. The result is

$$|0, 0\rangle_c = \frac{1}{\sqrt{3}}[|0, 0\rangle_u - |+1, -1\rangle_u - |-1, +1\rangle_u] \qquad (B.71)$$

The coupled-representation kets $|j, m\rangle_c$ are still eigenfunctions of $\hat{J}^2$, $\hat{J}_z$, $\hat{J}_1^2$ and $\hat{J}_2^2$, but generally are no longer eigenfunctions of either $\hat{J}_{1z}$ or $\hat{J}_{2z}$. The use of the coupled representation becomes especially convenient when the angular momenta $\hat{\mathbf{J}}_1$ and $\hat{\mathbf{J}}_2$

are coupled by an interaction term in the hamiltonian. For example, $\hat{\mathbf{L}}$ and $\hat{\mathbf{S}}$ are coupled through the hamiltonian term $\lambda \hat{\mathbf{L}}^T \hat{\mathbf{S}}$. In Section 6.2 we deal with two electron spins. An example of the application of the methods outlined in this section is found, for $\hat{\mathbf{J}}_1 = \hat{\mathbf{S}}$ and $\hat{\mathbf{J}}_2 = \hat{\mathbf{I}}$, in Appendix C.

We now turn to consideration of important other representations of the spin matrices. Thus, when one deals with a set of spins (e.g., one unpaired electron and one proton), then the dimension n of the spin hamiltonian is $n = (2j_1 + 1)(2j_2 + 1)(j_3 + 1)\ldots$. To build the spin-hamiltonian matrix and also to enable computation of the EPR line intensities, one needs to construct spin angular momentum operators designed for the n-dimensional space. Thus, for the simple case $S = I = \frac{1}{2}$, one needs the six 4×4 matrices $\mathbf{S}_x$, $\mathbf{S}_y$, $\mathbf{S}_z$, $\mathbf{I}_x$, $\mathbf{I}_y$ and $\mathbf{I}_z$. For the electron in $^1H^0$, the Pauli matrices (Sections B5 and B10) must be extended. Thus, for example

$$
\mathbf{S_x} = \begin{array}{c} \\ \langle \alpha_e \alpha_n | \\ \langle \beta_e \alpha_n | \\ \langle \alpha_e \beta_n | \\ \langle \beta_e \beta_n | \end{array}
\begin{array}{cccc} |\alpha_e \alpha_n\rangle & |\beta_e \alpha_n\rangle & |\alpha_e \beta_n\rangle & |\beta_e \beta_n\rangle \\ \left[\begin{array}{cccc} 0 & +1 & 0 & 0 \\ +1 & 0 & 0 & 0 \\ 0 & 0 & 0 & +1 \\ 0 & 0 & +1 & 0 \end{array}\right] \end{array}
\tag{B.72}
$$

All matrix elements connecting rows and columns differing in M_I are seen to vanish. When $\mathbf{B}_1 \| \mathbf{x}$, this matrix is the key to obtaining (see Eqs. C.22) the relative intensities of the four possible EPR transitions. Similarly, for the four NMR-type transitions of $^1H^0$, extended matrix $\mathbf{I}_x$ is required to compute intensities.

B.8 NOTATION FOR ATOMIC AND MOLECULAR STATES

It is very helpful to use the various angular-momentum quantum numbers associated with a given energy state of an atomic system to label that state [6]. Thus for atoms or atomic ions, it is customary to specify the net electron spin S, the net electronic orbital angular momentum L, and the total electronic angular momentum J, arising from $\hat{\mathbf{J}} = \hat{\mathbf{L}} + \hat{\mathbf{S}}$. Often when the nuclear spin is non-zero, it is necessary to append the nuclear-spin quantum number I. The total angular-momentum operator may be denoted by $\hat{\mathbf{F}} = \hat{\mathbf{J}} + \hat{\mathbf{I}}$. The three electronic quantum numbers cited above are exactly valid only when relativistic effects are absent (the so-called Russell-Saunders case [7, Sections 11.1–11.2]), but the latter are generally small for the usual systems studied and the three labels can still be useful even when this is not the case.

The notation for a given electronic state is given by

$$
^{2S+1}\underset{\sim}{L}_J
$$

where the pre-superscript $2S + 1$ is the electron-spin multiplicity; L gives the orbital state according to the usual spectroscopic notation

L	0	1	2	3	4	$\cdots$
$\underset{\sim}{L}$	S	P	D	F	G	$\cdots$

and the subscript J is any integer in the range $|L - S|, |L - S| + 1, \ldots, L + S - 1,$ $L + S$. The component quantum numbers M_J arising from $\hat{J}_z$ are needed only when an external field splits the $2J + 1$ states associated with each J manifold. In many cases, label J is omitted (e.g., ^2S, ^3P), implying additional degeneracy.

Usually, for single-electron states, lower-case letters are used instead of capitals, and the number of electrons is appended as a post-superscript (e.g., $2s$, $2p^2$, $3d^5$).

Analogously, for the electronic states of diatomic (and all other linear) molecules, one uses [7, p. 320] the notation

$$^{2s+1}\underset{\sim}{\Lambda}_{|\Omega|}$$

with $\Lambda = M_L$ and $\Sigma = M_S$, both measured along the internuclear direction Z. The molecular analog for post-subscript J in atoms is post-subscript $|\Omega|$. Here $\Omega = \Lambda + \Sigma$ describes the total electronic angular-momentum projection along Z. The Greek-letter spectroscopic notation used for the orbital part is

$|\Lambda|$	0	1	2	3	4	$\cdots$
$\underset{\sim}{\Lambda}$	Σ	Π	Δ	Φ	Λ	$\cdots$

For given $|\Lambda|$, there is degeneracy of $2S + 1$ if $\Lambda = 0$, and double that degeneracy if $|\Lambda| > 0$ (resulting from $\pm\Lambda$). Note that the state label Σ here should not be confused with the quantum number Σ used above. The angular-momentum components normal to **Z** affect the energy only in higher-order approximation. Only the component of angular momentum along the internuclear axis remains meaningful for each of the three electronic angular momenta, since the loss of spherical symmetry implies 'quenching' of the other two components [7, p. 379]. For Σ states, a post-superscript + or − is added to describe the behavior of the given wavefunction under reflection in any plane containing the axis **Z**. When there is a center of inversion (e.g., homonuclear diatomic molecules), the postsubscript g or u is added to describe the effects of inversion on the wavefunction; here g stands for gerade and u for ungerade (respectively, *even* and *odd* in German).

B.9   ANGULAR MOMENTUM AND DEGENERACY OF STATES

The orbital angular-momentum wavefunctions of an isolated atom (neutral or ionized) are usually taken to be eigenfunctions of $\hat{L}^2$ and of $\hat{L}_z$ (i.e., the complex functions called *spherical harmonics* Y_{L,M_L} [7, p. 66]). However, in an external electric field, some or all of the orbital angular momentum may be 'quenched'; that is, the diagonal matrix element (mean value) of $\hat{L}^2$ and of $\hat{L}_z$ may be zero. Concomitantly, the orbital degeneracy in each energy level is removed. If some of the orbital degeneracy remains, the wavefunctions of the degenerate states may be chosen to be complex, that is, of the form $\psi = Ae^{-iM_L\phi}$. However, if all orbital degeneracy is removed, one can use the real form of the orbital angular-momentum wavefunctions given in the third column of Table B.1 to describe the perturbed state.

That an orbitally non-degenerate level should have zero associated angular momentum (and hence zero orbital magnetic moment) is shown as follows. Suppose that an energy eigenfunction (corresponding to some eigenstate of the system) is complex. In this case there always exists at least one other independent eigenfunction having the same energy, namely, the complex conjugate. Consequently any state that can be represented by a complex eigenfunction must be at least *doubly* degenerate. Conversely, if the state is non-degenerate, the eigenfunction must be real (at least in the case where the potential energy is purely electrostatic).

The operator $\hat{L}_z$ for the z component of the orbital angular momentum is the pure imaginary operator (Eq. B.8c)

$$\hat{L}_z = -i\left(x\frac{\partial}{\partial y} - y\frac{\partial}{\partial x}\right) \tag{B.73}$$

$\hat{L}_z$ is a hermitian operator; this means that its eigenvalues must be real. If the wavefunction of an orbitally nondegenerate state of a system is designated by $|n\rangle$,

TABLE B.1   The Angular-Momentum Eigenfunctions Y_{L,M_L} (for $L = 0, 1$), and their Real Linear Combinations [a]

L	Spherical Harmonics Y_{L,M_L}	Orbitals (Real)
0	$Y_{0,0} = \dfrac{1}{\sqrt{4\pi}}$	$s = \dfrac{1}{\sqrt{4\pi}} = Y_{0,0}$
1	$Y_{1,-1} = \left(\dfrac{3}{8\pi}\right)^{1/2} \sin\theta\, e^{-i\phi}$	$p_x = \left(\dfrac{3}{4\pi}\right)^{1/2} \sin\theta\cos\phi = \dfrac{Y_{1,-1} - Y_{1,+1}}{\sqrt{2}}$
	$Y_{1,0} = \left(\dfrac{3}{4\pi}\right)^{1/2} \cos\theta$	$p_y = \left(\dfrac{3}{4\pi}\right)^{1/2} \sin\theta\sin\phi = \dfrac{-Y_{1,-1} - Y_{1,+1}}{\sqrt{2}}$
	$Y_{1,+1} = -\left(\dfrac{3}{8\pi}\right)^{1/2} \sin\theta\, e^{i\phi}$	$p_z = \left(\dfrac{3}{4\pi}\right)^{1/2} \cos\theta = Y_{1,0}$

[a] We may denote the functions $Y_{L,M_L}(\theta, \phi)$ by $|L, M_L\rangle$.

operation by $\hat{L}_z$ gives

$$\hat{L}_z|n\rangle = M_L|n\rangle \tag{B.74}$$

where M_L must be real. However, if $|n\rangle$ corresponds to an orbitally non-degenerate state, then $|n\rangle$ must also be real. Operator $\hat{L}_z$ is pure imaginary, and hence the result of operating with an imaginary operator on a real eigenfunction must be imaginary or zero. Since M_L must be real, the only possibility here is to have $M_L = 0$. In a similar fashion, it can be shown that $\hat{L}_x|n\rangle = \hat{L}_y|n\rangle = 0$. Thus the expectation values of all the components of $\hat{L}$ are zero for this state.

B.10   TIME DEPENDENCE

In quantum mechanics, time t is to be considered as a parameter, rather than as an operator [8]. Thus, in this aspect, it differs from the spatial coordinates.

The lifetime of any quantum-mechanical state is linked with the range of energies it can possess. A true eigenstate of some hamiltonian has the oscillatory form $\psi_0 \exp(-iUt/\hbar)$, with an exact energy eigenvalue U. When the state evolves with time in any other manner (a transition may occur), then there is a spread dU in its energy, and the relation [7, p. 202; 8,9]

$$\tau\,\delta U \approx h/4\pi \tag{B.75}$$

holds, where τ is the probable lifetime of the state. This spread in energy shows up as broadening of the spectral line associated with the transition(s) at hand.

The time dependence of the angular momentum $\hat{\mathbf{J}}$ and hence (Eq. 1.9) of the magnetic momentum operator $\hat{\boldsymbol{\mu}} = \alpha g\beta\hat{\mathbf{J}}$ is of crucial importance for magnetic resonance. From the fundamentals of quantum mechanics it follows [10,11] that

$$\frac{d\hat{\boldsymbol{\mu}}}{dt} = \gamma\,\hat{\boldsymbol{\mu}} \wedge \mathbf{B} \tag{B.76}$$

Analysis of the consequences of Eq. B.76 yields the ideas concerning the dynamics, including transition moments, of spin systems. For instance, ensemble averaging leads to the fundamental Bloch equations (Chapter 10). In this book, however, we cannot present a detailed description of this broad topic.

B.11   PRECESSION

The angular momentum $\mathbf{J}$ of a particle can be said to 'precess' about any magnetic field $\mathbf{B}$ (also $\mathbf{H}$, if the two are co-linear) present at the particle's location, if there is a magnetic moment 'attached' to it, associated with $\mathbf{J}$. This is so for every electron, as well as each nucleus that has a spin, since each has a magnetic dipole moment $\boldsymbol{\mu}$. Usually, the latter can be deemed to be proportional to its $\mathbf{J}$ vector (see Eq. 1.9):[8]

$$\boldsymbol{\mu} = \alpha g \beta \mathbf{J} \tag{B.77}$$

These definitions hold for particles at rest (or nearly so) in the laboratory frame.

With photons, too, one can visualize precession (see Appendix D). For the massive particles cited above, one can consider the origin to be at the particle center, and in vacuum the axis of the precession is field $\mathbf{B}$. For the photon, the location is largely indeterminate [12] and its translational speed is always, of course, the speed of light.

The 'gyroscopic' concept of 'precession' is handy because of the easy visualization of the physical effect (see Fig. B.1), but is not trivial [13]. Thus most of the key symbols above are quantum-mechanical operators, so that it is actually their expectation values that must be dealt with. Because the Heisenberg uncertainty principle forbids simultaneous exact knowledge of the three spatial component eigenvalues of operator $\mathbf{J}$; strictly speaking, one cannot consider any individual particle as precessing. However, for their expectation values [14], one can prove that

$$\langle J_x \rangle = (\sin\theta\cos\phi)/2 \tag{B.78a}$$
$$\langle J_y \rangle = (\sin\theta\sin\phi)/2 \tag{B.78b}$$
$$\langle J_z \rangle = (\cos\theta)/2 \tag{B.78c}$$

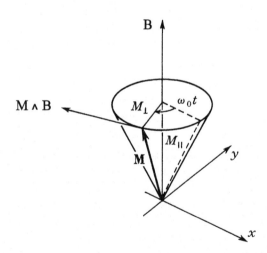

FIGURE B.1 Precession of the macroscopic magnetization $\mathbf{M}$ about a uniform steady magnetic induction $\mathbf{B}$ externally applied, in an isotropic medium. Note that the torque (rotary force) vector is normal to both $\mathbf{B}$ and $\mathbf{M}$ [15].[7] The clockwise sense of rotation (looking 'down', along $-\mathbf{B}$) is consistent with precession of proton spins and not electron spins.

where $\mathbf{z} \parallel \mathbf{B}$. Angle θ is the polar angle between the chosen quantization axis for $\mathbf{J}$ and the field $\mathbf{B}$ (and $\mathbf{H}$ as well, in an isotropic medium), and ϕ is the corresponding azimuthal angle. Here the two conditions $\theta = \arccos(2a^2 - 1)$ and $\phi = \beta - \alpha - \omega t$ must hold, wherein the complex coefficients for the spin ket are $c_{+1/2} = a \exp(i\alpha)$ and $c_{-1/2} = b \exp(i\beta)$ with normalization of the real coefficients: $a^2 + b^2 = 1$. We see that $\langle J_z \rangle$ is independent of time (t). Happily, the same relations hold for the components of the magnetization $\mathbf{M}$ of a statistically sufficiently large set of such particles in a uniform magnetic field ($\mathbf{B}$, as well as $\mathbf{H}$). Vector $\mathbf{M}$ can well and truly be said to precess about $\mathbf{H}$, as depicted in Fig. B.1.

The angular precession frequency ω, called the *Larmor frequency* (see Section 10.3 and also Problem 1.5), is given by the relation written in vector form as

$$\boldsymbol{\omega} = -\gamma \mu_m \mathbf{H} \tag{B.79}$$

valid in isotropic media. Equivalently and more simply (see Eq. 1.4)

$$\boldsymbol{\omega} = -\gamma \mathbf{B} \tag{B.80}$$

The sign of $-\gamma$ gives the sense of rotation.

Thus, for each and every electron, the spin magnetic moment vector $\langle \boldsymbol{\mu} \rangle$ precesses counterclockwise, when looking from the electron along vector $\mathbf{B}$. The precession of the spin $\langle \mathbf{S} \rangle$ is opposite in direction. For each and every proton, the $\langle \boldsymbol{\mu} \rangle$ rotation is clockwise.

With photons, the situation differs but presents some analogies to that shown above (see Appendix D). Here the spin angular momentum is measured along the direction ($\equiv \mathbf{z}$) of the photon's travel (at the speed of light c). Photons in 'pure' states can be classified as being of two types, σ_+ and σ_-, that is, with $\langle J_z \rangle = +1$ and -1, respectively. The precession is measured about this linear momentum axis, and not about any external field. The speed of precession is the frequency associated with the photon $\omega = U/(h/2\pi)$, where U is the photon energy.

B.12 MAGNETIC FLUX QUANTIZATION

It is, of course, common knowledge that electromagnetic fields (light) is quantized in photons. Less well known [16,17] is that changes in magnetic induction B also occur in quantized steps, which can be discussed in terms of the basic magnetic flux unit

$$\Phi_0 \equiv h/2|e| \tag{B.81a}$$

$$= 2.0678336 \times 10^{-15} \, \text{T m}^2 \, [= \text{weber (Wb)}] \tag{B.81b}$$

The flux $\Phi = \int \mathbf{B} \, d\mathcal{A}$ is caught by a superconducting loop. The area ($\mathcal{A}$) in question here is that of the (mostly) superconducting loop. Here $\Phi = n\Phi_0$, where n is a positive integer giving the number of flux-unit quanta present. The 'superconducting quantum interference device' (SQUID) is capable of measuring such a quantum (10^{-14} T threshold). Note that one period of sinusoidal voltage variation produced by (say) a SQUID can be represented by

$$\Delta\Phi = -\Phi_0 \tag{B.82}$$

which corresponds to an observed increase of one flux quantum. In terms of frequency, one can write

$$\nu = \Delta V/\Phi_0 \tag{B.83}$$

where ΔV is the voltage induced by the flux change. In fact, 'volt' as an electrical unit (JC^{-1}) is defined thereby.

B.13 SUMMARY

For convenience in reference, the essential results of this appendix are summarized below. For orbital, electronic-spin, or nuclear-spin angular momenta, the appropriate expressions are obtained from those below by substituting L, S or I, respectively, for j; the quantum number expressed as m is then M_L, M_S or M_I, respectively:

1. Operations giving generally non-zero values

$$\hat{J}^2 \, |j, m\rangle = j(j+1) \, |j, m\rangle \tag{B.32a}$$

$$\hat{J}_z \, |j, m\rangle = m \, |j, m\rangle \tag{B.32b}$$

$$\hat{J}_+ \, |j, m\rangle = [j(j+1) - m(m+1)]^{1/2} \, |j, m+1\rangle \tag{B.41a}$$

$$\hat{J}_- \, |j, m\rangle = [j(j+1) - m(m-1)]^{1/2} \, |j, m-1\rangle \tag{B.41b}$$

(see also Problem B.4).
2. Matrix elements (see Eqs. B.16 and B.43).
3. Angular momentum matrices in the $|j, m\rangle = |J, M_J\rangle$ basis.
$j = \tfrac{1}{2}$:

$$\mathbf{J}_x = \frac{1}{2}\begin{bmatrix} 0 & +1 \\ +1 & 0 \end{bmatrix} \quad \mathbf{J}_y = \frac{1}{2}\begin{bmatrix} 0 & -i \\ +i & 0 \end{bmatrix} \quad \mathbf{J}_z = \frac{1}{2}\begin{bmatrix} +1 & 0 \\ 0 & -1 \end{bmatrix}$$

$$\mathbf{J}_+ = \begin{bmatrix} 0 & +1 \\ 0 & 0 \end{bmatrix} \quad \mathbf{J}_- = \begin{bmatrix} 0 & 0 \\ +1 & 0 \end{bmatrix} \tag{B.84}$$

$j = 1$:

$$\mathbf{J}_x = \frac{1}{\sqrt{2}} \begin{bmatrix} 0 & +1 & 0 \\ +1 & 0 & +1 \\ 0 & +1 & 0 \end{bmatrix} \quad \mathbf{J}_y = \frac{1}{\sqrt{2}} \begin{bmatrix} 0 & -i & 0 \\ +i & 0 & -i \\ 0 & +i & 0 \end{bmatrix}$$

$$\mathbf{J}_z = \begin{bmatrix} +1 & 0 & 0 \\ 0 & 0 & 0 \\ 0 & 0 & -1 \end{bmatrix}$$

$$\mathbf{J}_+ = \sqrt{2} \begin{bmatrix} 0 & +1 & 0 \\ 0 & 0 & +1 \\ 0 & 0 & 0 \end{bmatrix} \quad \mathbf{J}_- = \sqrt{2} \begin{bmatrix} 0 & 0 & 0 \\ +1 & 0 & 0 \\ 0 & +1 & 0 \end{bmatrix} \quad \text{(B.85)}$$

For the matrices up to $j = \frac{7}{2}$, see Orton [18].

REFERENCES

1. P. A. M. Dirac, *The Principles of Quantum Mechanics*, 3rd ed., Clarendon Press, Oxford, U.K. 1947, pp. 18, 36, 46.

2. G. Baym, *Lectures on Quantum Mechanics*, Benjamin, New York, NY, U.S.A., 1969. Chapter 1.

3. A. R. Edmonds, *Angular Momentum in Quantum Mechanics*, 2nd ed., Princeton University Press, Princeton, NJ, U.S.A., 1974.

4. E. U. Condon, G. H. Shortley, *The Theory of Atomic Spectra*, Cambridge University Press, Cambridge, U.K., 1970.

5. M. Rotenberg, R. Bivins, N. Metropolis, J. K. Wooten Jr., *The 3−j and 6−j Symbols*, Technology Press (MIT), Cambridge, MA, U.S.A., 1959; also T. Inoue, *Table of the Clebsch-Gordan Coefficients*, Tokyo Tosho, Tokyo, Japan, 1966.

6. M. Gerloch, *Orbitals, Terms and States*, Wiley, Chichester, U.K., 1986.

7. P. W. Atkins, R. S. Friedman, *Molecular Quantum Mechanics*, 4th ed., Oxford University Press, Oxford, U.K., 2005.

8. M. Bunge, Can. *J. Phys.*, **48**, 1410 (1970).

9. L. D. Landau, E. M. Lifshitz, *Quantum Mechanics*, Addison-Wesley, Reading, MA, U.S.A., 1958, Section 44.

10. A. Abragam, B. Bleaney, *Electron Paramagnetic Resonance of Transition Ions*, Oxford University Press, Oxford, U.K., 1970, Chapter 1.

11. G. E. Pake, T. L. Estle, *The Physical Principles of Electron Paramagnetic Resonance*, 2nd ed., Benjamin, Reading, MA, U.S.A., 1973, Chapter 2.

12. L. Mandel, E. Wolf, *Optical Coherence and Quantum Optics*, Cambridge University Press, Cambridge, U.K. 1995, Section 12.11.5.

13. H. Goldstein, C. P. Poole Jr., J. L. Safko, *Classical Mechanics*, 3rd ed., Addison-Wesley, Cambridge, MA, U.S.A., 1980, Chapter 5.

14. C. P. Slichter, *Principles of Magnetic Resonance*, 3rd ed., Springer-Verlag, Berlin, Germany. 1990, pp. 15–16.

15. J. A. Stratton, *Electromagnetic Theory*, McGraw-Hill, New York, NY, U.S.A., 1941, Sections 4.8–4.9.

16. J. J. Sakurai, with S.-F. Tuan, *Modern Quantum Mechanics*, revised ed., Addison-Wesley, Reading, MA, U.S.A., 1994.

17. M.-A. Park, Y.-J. Kim, *Supercond. Sci. Technol.*, **17**, L10 (2004).

18. J. W. Orton, *Electron Paramagnetic Resonance*, Gordon & Breach, New York, NY, U.S.A., 1969, Appendix 1.

19. A. Shadowitz, *The Electromagnetic Field*, 2nd ed., McGraw-Hill, New York, NY, U.S.A., 1988.

NOTES

1. Spin eigenfunctions are not really functions in the true mathematical sense (and do not depend on space or time). They are better called 'eigenvectors in an abstract spin space'. Nevertheless, we continue herein to use the term 'eigenfunctions' in connection with relations such as Eqs. B.3 and B.4, even when spin operators are used.

2. For convenience, in this appendix we use j and m, rather than J and M_J.

3. For spatial angular momenta, they are the well-known spherical harmonic angular functions.

4. Note that if the integer $\overline{m} - \underline{m}$ had been set equal to j, then the eigenvalue of $\mathbf{J}^{\mathrm{T}} \cdot \mathbf{J}$ would have been $(j/2)[(j/2) + 1]$. Hence half-integral quantum numbers appear quite naturally in this treatment. Half-integral values of j occur for electronic-spin and nuclear-spin angular momenta.

5. The function on the left-hand side of this equation is in the coupled representation, whereas the functions on the right-hand side are in the uncoupled representation. The 'equals' sign here should be taken to mean equivalence.

6. See FURTHER READING section in this appendix.

7. A torque $\mathbf{N}$ when acting on an individual particle is $\mathbf{r} \wedge \mathbf{F}$ in classical mechanics (with $\mathbf{F}$ an applied mechanical force), and is $\boldsymbol{\mu} \wedge \mathbf{B}$ in our magnetic dipole situation. It has SI units of joules and is, of course, an axial vector.

8. More generally (see Eq. 4.8), in some systems, the relationship is tensorial, and not simply proportional.

FURTHER READING

1. A. R. Edmonds, *Angular Momentum in Quantum Mechanics*, 2nd ed., Princeton University Press, Princeton, NJ, U.S.A., 1960.

2. E. U. Condon, H. Odabasi, *Atomic Structure*, Cambridge University Press, Cambridge, U.K., 1980.

3. M. E. Rose, *Elementary Theory of Angular Momentum*, Wiley, New York, NY, U.S.A., 1974.

4. A. Yariv, *An Introduction to Theory and Application of Quantum Mechanics*, Wiley, New York, NY, U.S.A., 1982.

5. R. N. Zare, *Angular Momentum*, Wiley, New York, NY, U.S.A., 1988.

PROBLEMS

B.1   Show that $[\hat{J}_y, \hat{J}_z]_- = i\hat{J}_x$.

B.2   Derive the commutation relations in Eqs. B.14.

B.3   Show that the matrix element $\langle j, m'|\hat{J}_-|j, m\rangle$ is non-zero only if $m' = m - 1$.

B.4   Derive the relations for $\hat{J}_x|j, m\rangle$ and $\hat{J}_y|j, m\rangle$ from those for $\hat{J}_+|j, m\rangle$ (Eqs. B.41), as well as expressions for their matrix elements.

B.5   Establish the angular-momentum matrices $\mathbf{J}_x$, $\mathbf{J}_y$, $\mathbf{J}_z$, $\mathbf{J}_+$, $\mathbf{J}_-$ and $\mathbf{J}^2$ for $j = \frac{3}{2}$.

B.6   For $j = \frac{3}{2}$, show that the matrix relation $\mathbf{J}_x^2 + \mathbf{J}_y^2 + \mathbf{J}_z^2 = \mathbf{J}^2$ holds.

B.7   By matrix addition and multiplication, find the commutators $[\mathbf{J}_+, \mathbf{J}_-]_-$, $[\mathbf{J}_+, \mathbf{J}^2]_-$ and $[\mathbf{J}_-, \mathbf{J}^2]_-$ for $j = 1$.

B.8   For $j = 1/2$, the matrix equations $\mathbf{J}_x^2 = \mathbf{J}_y^2 = \mathbf{J}_z^2$ hold. By calculation of the appropriate matrices for $j = 1$ and $j = \frac{3}{2}$, show that these relations are not satisfied for $j > 1$. On the other hand, show that the trace (sum of the diagonal elements) is the same for these three matrices for a given value of j.

B.9   Verify Eq. 6.12 by matrix multiplication. Similarly, derive expressions for $\hat{S}_{1y}\hat{S}_{2y}$, $\hat{S}_{1z}\hat{S}_{2z}$, $\hat{S}_{1x}\hat{S}_{2y}$, as well as $\hat{S}_x\hat{S}_y$ and $\hat{S}_y\hat{S}_x$.

B.10   For $j_1 = 2, j_2 = 1$, show that the kets for the coupled representation, in terms of the uncoupled kets, are given by

$$|3, \pm 3\rangle_c = |\pm 2, \pm 1\rangle_u$$

$$|3, \pm 2\rangle_c = \frac{1}{\sqrt{6}}\left[2|\pm 1, \pm 1\rangle_u + \sqrt{2}|\pm 2, 0\rangle_u\right]$$

$$|3, \pm 1\rangle_c = \frac{1}{\sqrt{30}}\left[2\sqrt{3}|0, \pm 1\rangle_u + 4|\pm 1, 0\rangle_u + \sqrt{2}|\pm 2, \mp 1\rangle_u\right]$$

$$|3, 0\rangle_c = \frac{1}{\sqrt{15}}\left[\sqrt{3}|+1, -1\rangle_u + 3|0, 0\rangle_u + \sqrt{3}|-1, +1\rangle_u\right]$$

$$|2, \pm 2\rangle_c = \frac{1}{\sqrt{6}}\left[\sqrt{2}|\pm 1, \pm 1\rangle_u - 2|\pm 2, 0\rangle_u\right]$$

$$|2, \pm 1\rangle_c = \frac{1}{\sqrt{6}}\left[\sqrt{3}|0, \pm 1\rangle_u - |\pm 1, 0\rangle_u - \sqrt{2}|\pm 2, \mp 1\rangle_u\right]$$

$$|2, 0\rangle_c = \frac{1}{\sqrt{2}}[|-1, +1\rangle_u - |+1, -1\rangle_u]$$

$$|1, \pm 1\rangle_c = \frac{1}{\sqrt{10}}\left[|0, \pm 1\rangle_u - \sqrt{3}|\pm 1, 0\rangle_u + \sqrt{6}|\pm 2, \mp 1\rangle_u\right]$$

$$|1, 0\rangle_c = \frac{1}{\sqrt{10}}\left[\sqrt{3}|+1, -1\rangle_u - 2|0,0\rangle_u + \sqrt{3}|-1, +1\rangle_u\right]$$

Note that the absolute sign of any ket (or bra) occurring alone has no physical significance.

B.11 Obtain all six extended Pauli matrices for the hydrogen-atom problem.

B.12 Various vectors can occur in the spin-hamiltonian operator $\hat{\mathcal{H}}_s$. These include:

1. Externally applied magnetic (Zeeman) field **B**.
2. Externally applied magnetic (excitation) field **B**$_1$.
3. Externally applied electric (Stark) field **E**.
4. Externally applied electric (excitation) field **E**$_1$.
5. Spin angular-momentum operators **J** (i.e., **S**, **I**, . . .).

What are their spatial properties of importance in magnetic resonance spectroscopy?

Consider:

a. Magnetic-Field inversion

   (1) What happens to $\hat{\mathcal{H}}_s$ when **B** → −**B** ?

   (2) What happens to the energy eigenvalue and the M_J label(s) of any given spin-state when **B** → −**B** ?

   (3) What happens to the complete spin-energy manifold when **B** → −**B** ?

   (4) What happens to the magnetic resonance spectrum (line positions and intensities) when **B** → −**B** ?

   (5–8) Repeat the questions above four considering **B**$_1$ instead of **B**. One can also repeat the questions above with **E**$_1$, **E**, and **J**.

b. Crystal-position inversion

   (9 *a,b*) What happens to the magnetic resonance spectrum when a single-crystal sample is turned upside-down, taking **B** to be

      (i) Horizontal?

      (ii) Vertical (up–down)?

   (10 *a–c*) Compare

      (i) The set of magnetic resonance spectra obtained when an anisotropic crystal sample has been rotated in steps (say, clockwise looking into the magnet) about some axis **n,** with

      (ii) The set obtained when the crystal is turned upside-down and rotated the same way again. Take **B** to be ⊥**n** and to be horizontal.

         (a) Is there conceivably any difference between the two sets?

         (b) If the sets of spectra can be different, will this be so for all possible crystal symmetries? If crystal symmetry does matter, how so?

         (c) How is the determination of parameter matrices, say, **g** and **D**, in EPR work affected by the above maneuvers?

c. Coordinate inversion

   (11) What is the importance (if any) of the spatial inversion operation **I** in magnetic resonance theory?

Notes for Problem B.12

1. Herein, we consider Euclidean 3-space exclusively. Time will not be included.

2. Obviously it is important always to state clearly what coordinate-axis system is being considered: one fixed in the lab, fixed relative to **B**, fixed in the crystal, or ??

3. The inversion operator **I** acting on a proper vector takes the three vector components into their negatives. Vectors can be considered to be objects that are independent of the coordinate system used, whereas their components are not. Completely generally, the inversion can be regarded as the product of two commuting operations: reflection in any plane containing the origin and rotation by π about the axis **p** passing through the origin and normal to the plane. Is the choice of direction **p** here arbitrary?

4. Proper vectors **V** originating from the origin go into their negatives under operation of the inversion operator **I**, that is, $\mathbf{IV} = -\mathbf{V}$. By contrast, **I** operating on a pseudo-vector **V** stemming from the origin automatically yields that vector back again, that is, $\mathbf{IV} = \mathbf{V}$. For more details, see Section 14–2 of Ref. 19. The vectors **V** and $-\mathbf{V}$ both do conceptually exist and are distinct.

5. As an example of a proper vector, take any radius vector **r** (one starting at the coordinate system's origin). The inversion operator takes **r** into its negative: $\mathbf{Ir} = -\mathbf{r}$, that is, its coordinate set $\{\mathbf{x}, \mathbf{y}, \mathbf{z}\}$ under inversion goes into $\rightarrow \{-\mathbf{x}, -\mathbf{y}, -\mathbf{z}\}$, but its magnitude $|\mathbf{r}| = r$ remains positive and unchanged. Are the unit vectors $(\mathbf{x}, \mathbf{y}, \mathbf{z})$ of the coordinate axes used affected by action of I?

6. For any proper vector **V**, inversion $\mathbf{IV} = \mathbf{V}'$ implies that $\mathbf{V} \parallel \mathbf{V}'$. If **V** is located away from the origin, then so is **V**'. For any pseudo-vector **V**, relation $\mathbf{IV} = \mathbf{V}'$ implies $\mathbf{V} \neq \mathbf{V}'$ unless **V** stems from the origin, in which case $\mathbf{V} = \mathbf{V}'$.

7. Vectors **B**, $\mathbf{B}_1$ and **J** are pseudo-vectors (alias 'axial' vectors). This implies that each can be written as a vector product ($\wedge$) of two proper vectors (alias 'polar' vectors), or of two pseudo-vectors. We have $\mathbf{B} = \mathbf{DEL} \wedge \mathbf{A}$ (= curl **A**) where **A** is the electromagnetic vector potential, and $\mathbf{J} = \mathbf{J} \wedge \mathbf{J}$. Operator **DEL** is a proper vector, whereas **A** is a pseudo-vector. Also, note that magnitudes $|\mathbf{B}| = B$ as well as $|\mathbf{E}| = E$ are non-negative by definition. Is **E** a proper vector? Explain and discuss.

8. Explain why physicists deem the concept of the inversion operation to be so informative and useful? Consider the action of **I** on spherical harmonics (e.g., hydrogenic orbitals), and the parity classification into g and u states, as well as the resulting ease in arriving at a basic spectroscopic selection rule.

APPENDIX C

THE HYDROGEN ATOM AND SELECTED RADICALS RH$_n$

The hyperfine interaction in the hydrogen atom was treated in an approximate manner in Chapter 2. An exact calculation, as well as approximate methods, are presented in this appendix [1]. An extension to a system containing one unpaired electron and one effective or two equivalent protons (the RH$_2$ free radical) is briefly discussed.

C.1 HYDROGEN ATOM

We consider the hydrogen atom in its electronic ground state, located in an externally applied magnetic field **B** not sufficiently intense to perturb appreciably the electronic orbital wavefunctions. The relatively minor effects [2] on the ground-state orbital of the hyperfine interaction can be ignored herein. Thus only the spin systems in the atom are of interest here.

C.1.1 Spin Hamiltonian

A more exact spin hamiltonian than that of Eq. 2.40 for an isotropic system of one electron ($S = \frac{1}{2}$) and one proton ($I = \frac{1}{2}$) in a magnetic field **B** is[1]

$$\hat{\mathcal{H}} = g\beta_e \mathbf{B}^\mathrm{T} \cdot \hat{\mathbf{S}} + A_0 \hat{\mathbf{S}}^\mathrm{T} \cdot \hat{\mathbf{I}} - g_n \beta_n \mathbf{B}^\mathrm{T} \cdot \hat{\mathbf{I}} \tag{C.1}$$

Electron Paramagnetic Resonance, Second Edition, by John A. Weil and James R. Bolton
Copyright © 2007 John Wiley & Sons, Inc.

For the z axis chosen to be along **B**, Eq. C.1 becomes

$$\hat{\mathcal{H}} = g\beta_e B \hat{S}_z + A_0(\hat{S}_z\hat{I}_z + \hat{S}_x\hat{I}_x + \hat{S}_y\hat{I}_y) - g_n\beta_n B\hat{I}_z \tag{C.2a}$$

Using the operators $\hat{S}_+, \hat{S}_-, \hat{I}_+$ and $\hat{I}_-$, as defined by Eqs. B.13, one finds that $\hat{S}_x\hat{I}_x + \hat{S}_y\hat{I}_y$ equals $\frac{1}{2}(\hat{S}_+\hat{I}_- + \hat{S}_-\hat{I}_+)$. Hence Eq. C.2a can be rearranged to

$$\hat{\mathcal{H}} = g\beta_e B \hat{S}_z + A_0[\hat{S}_z\hat{I}_z + \tfrac{1}{2}(\hat{S}_+\hat{I}_- + \hat{S}_-\hat{I}_+) - g_n\beta_n B\hat{I}_z \tag{C.2b}$$

This form is of use in Section C.1.2.

C.1.2   The Spin Eigenkets and Energy Matrix

The bra and ket notation (Section A.5.4) will be used for the spin 'eigenfunctions', for example, $|M_S, M_I\rangle$. There are four independent spin eigenkets, as given in Section 2.4.

The energy matrix consists of the matrix elements between all the spin eigenkets (i.e., $\langle M_S, M_I | \hat{\mathcal{H}} | M_S', M_I'\rangle$). It is thus a 4×4 matrix. Use of the angular-momentum matrices for $S = \frac{1}{2}$ (introduced in Section B.5) permits computation of these matrix elements. The matrices $\bar{\mathbf{I}}_z$, $\bar{\mathbf{I}}_+$ and $\bar{\mathbf{I}}_-$ have the analogous form (since $I = \frac{1}{2}$) as the corresponding electron spin matrices, since such matrices apply to any system with *total angular-momentum magnitude* $\langle J\rangle = [J(J+1)]^{1/2}$ with $J = \frac{1}{2}$.

The matrix elements are divided into two classes:

1. *Diagonal Matrix Elements.* A diagonal matrix element is one in which the bra and ket have the same labels. Inspection of the spin matrices (Eqs. B.76) shows that only $\bar{\mathbf{S}}_z$ and $\bar{\mathbf{I}}_z$ have non-zero diagonal elements; hence the only non-zero diagonal matrix elements of $\hat{\mathcal{H}}$ (Eq. C.2b) arise from

$$\langle M_S, M_I | \hat{S}_z | M_S, M_I\rangle, \langle M_S, M_I | \hat{I}_z | M_S, M_I\rangle$$

and

$$\langle M_S, M_I | \hat{S}_z\hat{I}_z | M_S, M_I\rangle$$

A typical diagonal element of the matrix associated with $\hat{\mathcal{H}}$ is

$$\langle \alpha(e), \alpha(n) | g\beta_e B \hat{S}_z + A_0\hat{S}_z\hat{I}_z - g_n\beta_n B\hat{I}_z | \alpha(e), \alpha(n)\rangle$$
$$= \tfrac{1}{2}g\beta_e B + \tfrac{1}{4}A_0 - \tfrac{1}{2}g_n\beta_n B \tag{C.3}$$

As detailed in Chapter 2 (see also Sections A.5.4 and B.5), spin functions $\alpha(e)$ and $\alpha(n)$ correspond to $M_S = +\frac{1}{2}$ and $M_I = +\frac{1}{2}$; $\beta(e)$ and $\beta(n)$ correspond to $M_S = -\frac{1}{2}$ and $M_I = -\frac{1}{2}$.

2. *Off-Diagonal Matrix Elements.* Inspection of the spin matrices shows that $\bar{S}_+$, $\bar{S}_-$, $\bar{I}_+$ and $\bar{I}_-$ have only off-diagonal non-zero elements. Hence for the operators $\hat{S}_+\hat{I}_-$ and $\hat{S}_-\hat{I}_+$, the non-zero off-diagonal elements of the spin hamiltonian in Eq. C.2*b* are of the type

$$\langle M_S + 1, M_I - 1|\hat{S}_+\hat{I}_-| M_S, M_I\rangle$$

and

$$\langle M_S - 1, M_I + 1|\hat{S}_-\hat{I}_+| M_S, M_I\rangle$$

For example

$$\langle \alpha(e), \beta(n)|\hat{S}_+\hat{I}_-|\beta(e), \alpha(n)\rangle = 1 \tag{C.4}$$

The spin-hamiltonian matrix $\tilde{\mathcal{H}}$ in the uncoupled representation (see Section B.6) then is

$$\mathcal{H} = \begin{array}{c} \\ \langle\alpha_e\alpha_n| \\ \\ \langle\beta_e\alpha_n| \\ \\ \langle\alpha_e\beta_n| \\ \\ \langle\beta_e\beta_n| \\ \end{array} \begin{array}{cccc} |\alpha_e\alpha_n\rangle & |\alpha_e\beta_n\rangle & |\beta_e\alpha_n\rangle & |\beta_e\beta_n\rangle \\ \begin{bmatrix} \frac{1}{2}g\beta_e B + \frac{1}{4}A_0 + \\ -\frac{1}{2}g_n\beta_n B \end{bmatrix} & 0 & 0 & 0 \\ 0 & \begin{array}{c}\frac{1}{2}g\beta_e B - \frac{1}{4}A_0 + \\ +\frac{1}{2}g_n\beta_n B\end{array} & \frac{1}{2}A_0 & 0 \\ 0 & \frac{1}{2}A_0 & \begin{array}{c}-\frac{1}{2}g\beta_e B - \frac{1}{4}A_0 + \\ -\frac{1}{2}g_n\beta_n B\end{array} & 0 \\ 0 & 0 & 0 & \begin{array}{c}-\frac{1}{2}g\beta_e B + \frac{1}{4}A_0 + \\ +\frac{1}{2}g_n\beta_n B\end{array} \end{array} \tag{C.5}$$

Note that this energy matrix occurs already factored into blocks along the principal diagonal, with all other elements zero. When such factorization is present, one may deal in succession with each of the central submatrices in turn. This procedure results in a considerable simplification of the calculations. Similar considerations apply to the corresponding determinants.

C.1.3   Exact Solution for the Energy Eigenvalues

To obtain the energies of the four states, one can proceed by subtracting a variable (say, U) from each diagonal element and setting the resulting determinant (called the 'secular determinant')^2 equal to zero. The four roots of the quartic equation in U are the energies of the allowed states [3,4]. By inspection, one can see that two of the state energies are

$$U_{\alpha(e)\alpha(n)} = +\tfrac{1}{2}g\beta_e B + \tfrac{1}{4}A_0 - \tfrac{1}{2}g_n\beta_n B \qquad (C.6a)$$

and

$$U_{\beta(e)\beta(n)} = -\tfrac{1}{2}g\beta_e B + \tfrac{1}{4}A_0 + \tfrac{1}{2}g_n\beta_n B \qquad (C.6b)$$

Clearly, these two energies vary linearly with field B.

The other two state energies can be obtained by expanding the remaining (2×2) determinant

$$\begin{vmatrix} \tfrac{1}{2}g\beta_e B - \tfrac{1}{4}A_0 + \tfrac{1}{2}g_n\beta_n B - U & \tfrac{1}{2}A_0 \\ \tfrac{1}{2}A_0 & -\tfrac{1}{2}g\beta_e B - \tfrac{1}{4}A_0 - \tfrac{1}{2}g_n\beta_n B - U \end{vmatrix} = 0 \qquad (C.7)$$

Solving for U, one obtains the energies

$$U_{\{\alpha(e)\beta(n)\}} = -\tfrac{1}{4}A_0 + \tfrac{1}{2}\left[(g\beta_e + g_n\beta_n)^2 B^2 + A_0{}^2\right]^{1/2} \qquad (C.8a)$$

$$U_{\{\beta(e)\alpha(n)\}} = -\tfrac{1}{4}A_0 - \tfrac{1}{2}\left[(g\beta_e + g_n\beta_n)^2 B^2 + A_0{}^2\right]^{1/2} \qquad (C.8b)$$

The braces around the $\alpha(e)\beta(n)$ and $\beta(e)\alpha(n)$ subscripts are meant to show that the corresponding two states are *mixtures* of $\alpha(e)\beta(n)$ and $\beta(e)\alpha(n)$ [5]. The two coefficients in each such state depend on B, one tending to zero as B increases. The eigenvalues $U_{\{\alpha(e)\beta(n)\}}$ and $U_{\{\beta(e)\alpha(n)\}}$ are subscripted as shown, since $\alpha(e)\beta(n)$ and $\beta(e)\alpha(n)$ become the correct energy eigenstates at a sufficiently high magnetic field. Note the choice of notations for the kets, for example, $|1,0\rangle = \{\alpha, \beta\} = \{\alpha(e), \beta(n)\}$.

One can best label the states (and energies) with the total spin angular momentum (Sections B.4 and B.7). Thus, taking $\hat{\mathbf{F}} = \hat{\mathbf{S}} + \hat{\mathbf{I}}$ as the appropriate vector operator, one may denote the spin states by kets $|F, M_F\rangle$, where the total angular-momentum quantum numbers are

$$F = |S - I|, |S - I| + 1, \ldots, S + 1 \qquad (C.9)$$

and where $M_F = M_S + M_I$. For the hydrogen atom, $F = 0$ or 1. If $F = 0$, then $M_F = 0$; if $F = 1$, then $M_F = 0, \pm 1$. The four sets F, M_F furnish unique labels for the four spin states.

Equations C.6 and C.8 are called the 'Breit-Rabi formulas' (and can, in fact, be incorporated into a single equation) [3–5]. The Breit-Rabi energies are plotted as a function of magnetic field in Fig. C.1. The original application was to the character-ization of hydrogen (^{1}H) neutral atoms exhibiting the four allowed low-field transitions occurring in a beam of these atoms [6].

We note (Fig. C.1a) the triple degeneracy of the $F = 1$ state at zero magnetic field, with the 'spin-paired' $F = 0$ ground state lying $A_0/h = 1.42040575 \times 10^9$ Hz below

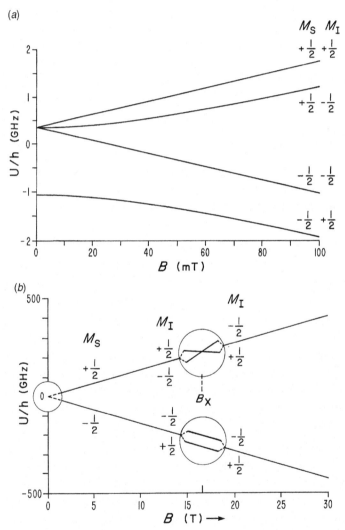

FIGURE C.1 Spin energy levels of the ground-state hydrogen atom (^{1}H) at (a) at low and moderate magnetic fields (Breit-Rabi diagram) and (b) high magnetic fields. The doublet-level structure is unresolved on this scale. Here the left-most circle contains part (a) of the figure.

this level set [7,8]. In the intermediate-field range (0.2–1.5 T), which has routinely been used for EPR, the lower hyperfine levels, both labeled with $M_S = -\frac{1}{2}$, run approximately parallel with increasing B, as do the upper such doublet levels labeled with $M_S = +\frac{1}{2}$. At $B = (g\beta_e - g_n\beta_n) A_o/(2g\beta_e g_n\beta_n) = 16.655$ T, the upper levels cross, with vanishing of the intensity of the (NMR) transition between them, as is evident from Eq. C.17d. At fields above the crossing value, the two level pairs continue to diverge as B increases.

It has been customary, but not really correct, to label the four states with the pair M_S, M_I (rather than F, M_F). We note also that there is possible uncertainty in the literature in using the label M_I for the members of the upper doublet, as a consequence of the existence of the cross-over. In what follows, we shall at times adhere to the M_S, M_I practice for discussion of the intermediate-field region.

We emphasize that, throughout the entire field range, F and M_F remain good quantum numbers, with $M_F = M_S + M_I$; M_S remains an unambiguous label for both the lower and the upper pairs of states. We note, however, that F can change in some of the allowed transitions.

C.1.4 Energy Eigenstates and Allowed Transitions

The fact that non-zero off-diagonal elements appear in the energy matrix C.5 means that the four basis spin 'functions' are not all eigenkets of the spin hamiltonian given by Eqs. C.2. It is desirable to find a set of spin functions that truly are eigenfunctions of $\hat{\mathcal{H}}$.

The desired four eigenkets of $\hat{\mathcal{H}}$ should be expressed in the coupled representation (Section B.6). One notes that $|\alpha(e), \alpha(n)\rangle$ and $|\beta(e), \beta(n)\rangle$ already are eigenkets of $\hat{\mathcal{H}}$ and, in the new basis, these are denoted by $|F, M_F\rangle = |1, +1\rangle$ and $|1, -1\rangle$. As already seen (Fig. C.1), their energies vary linearly with B.

The remaining two eigenkets $|F, M_F\rangle = |1, 0\rangle$ and $|0, 0\rangle$ are obtained by diagonalization of the 2×2 matrix given previously in Eqs. C.5 and C.7. This is best accomplished using the coordinate-rotation matrix method outlined in Section A.5.2. The two $M_F = 0$ eigenkets are expressed as normalized linear combinations (Eqs. A.65):

$$|1, 0\rangle = \cos \omega \, |\alpha(e), \beta(n)\rangle + \sin \omega \, |\beta(e), \alpha(n)\rangle \tag{C.10a}$$

$$|0, 0\rangle = -\sin \omega \, |\alpha(e), \beta(n)\rangle + \cos \omega \, |\beta(e), \alpha(n)\rangle \tag{C.10b}$$

The 'mixing' angle ω, which measures the relative importance of the Zeeman and hyperfine energies, is obtainable from

$$\sin 2\omega = \left|(1 + \xi^2)^{-1/2}\right| \tag{C.11a}$$

$$\cos 2\omega = \xi \left|(1 + \xi^2)^{-1/2}\right| \tag{C.11b}$$

where $\xi \equiv (g\beta_e + g_n\beta_n)B/A_0$; ω is restricted to the range $0 - \pi$ radians. Thus all four eigenstates (kets; bras as well, of course) in the coupled representation (see Section B.6) are readily available.

At $B = 0$, one has $\omega = \pi/4$ and the two spin states occur equally admixed to yield

$$|1, 0\rangle = \frac{1}{\sqrt{2}}(|\alpha(e), \beta(n)\rangle + |\beta(e), \alpha(n)\rangle) \tag{C.12a}$$

$$|0, 0\rangle = \frac{1}{\sqrt{2}}(|\alpha(e), \beta(n)\rangle - |\beta(e), \alpha(n)\rangle) \tag{C.12b}$$

They are separated in energy by A_0.

It is important to realize that, as $B \to \infty$, angle $\omega \to 0$ and hence $|1, 0\rangle \to |\alpha(e), \beta(n)\rangle$, while $|0, 0\rangle \to |\beta(e), \alpha(n)\rangle$. At the crossing of the upper two levels of ^{1}H, referred to earlier (Fig. C.1b), $\tan \omega = g_n\beta_n/g_e\beta_e$, so that $\omega = 0.087°$.

The interaction of electromagnetic radiation with the hydrogen-atom system can lead to transitions between certain of the energy levels. We consider only the electronic ground state, in a static magnetic field **B**.

The transition probability between spin states |initial⟩ and |final⟩ is proportional (Section 4.6) to $|\langle\text{initial}| \hat{\mathcal{H}}_1 |\text{final}\rangle|^2$. This applies under the resonance condition $h\nu = U_{\text{final}} - U_{\text{initial}}$ for absorption of a photon, where ν is the frequency of the electromagnetic radiation. Here $\hat{\mathcal{H}}_1$ is the amplitude (time-independent part) of the spin-hamiltonian operator, representing the effect of the applied exciting electromagnetic field $\mathbf{B}_1$, which for magnetic-dipole transitions is given (cf. Eq. 1.14a) by

$$\hat{\mathcal{H}}_1 = -\hat{\boldsymbol{\mu}}^{\mathsf{T}} \cdot \mathbf{B}_1 \tag{C.13}$$

Column vector $\hat{\boldsymbol{\mu}}$ is the total magnetic-dipole operator (Eq. 1.9 extended) for the given system, and $\mathbf{B}_1$ is the linearly polarized amplitude column vector of the oscillating magnetic field.[3] Thus Eq. C.13 becomes

$$\hat{\mathcal{H}}_1 = (g\beta_e\hat{\mathbf{S}} - g_n\beta_n\hat{\mathbf{I}})^{\mathsf{T}} \cdot \mathbf{B}_1 \tag{C.14}$$

We note that it is the matrix element between |initial⟩ and |final⟩ of the transition-moment matrix of operator $\hat{\boldsymbol{\mu}}^{\mathsf{T}} \cdot \mathbf{B}_1/B_1$, a property of the magnetic species, which is important in determining the magnetic resonance intensities.

Consider the important case $\mathbf{B}_1 \perp \mathbf{B}$, choosing $\mathbf{B}_1 \| \mathbf{x}$ and $\mathbf{B} \| \mathbf{z}$. Then one has

$$\hat{\mathcal{H}}_1 = g\beta_e B_1\hat{S}_x - g_n\beta_n B_1\hat{I}_x \tag{C.15}$$

and hence we shall require the off-diagonal elements of the matrices for $\hat{S}_x$ and $\hat{I}_x$.

Because of the mixing of states (Eqs. C.10), four transitions of type $\Delta M_F = \pm 1$ are in principle detectable. The relative intensities can be computed by evaluating the appropriate matrix elements of the matrix representing $\hat{\mathcal{H}}_1$.

The four allowed transitions (see Fig. C.1, and also Fig. C.2) can be written in terms of the kets $|F, M_F\rangle$. Their relative intensities are found to be

$$|0,0\rangle \rightarrow |1,-1\rangle \text{ intensity} \propto (g\beta_e \sin\omega + g_n\beta_n \cos\omega)^2 \qquad (C.16a)$$

$$|0,0\rangle \rightarrow |1,+1\rangle \text{ intensity} \propto (g\beta_e \cos\omega + g_n\beta_n \sin\omega)^2 \qquad (C.16b)$$

$$|1,-1\rangle \rightarrow |1,0\rangle \text{ intensity} \propto (g\beta_e \cos\omega - g_n\beta_n \sin\omega)^2 \qquad (C.16c)$$

$$|1,0\rangle \rightarrow |1,+1\rangle \text{ intensity} \propto (g\beta_e \sin\omega - g_n\beta_n \cos\omega)^2 \qquad (C.16d)$$

Note from Eqs. C.16 that resonance at zero magnetic field (i.e., Eqs. C.11 of Ref. 10: $\omega = \pi/4$) is observable, and well known. Remember also that as $B \rightarrow \infty$, angle $\omega \rightarrow 0$. Thus Eqs. C.16a,d represent NMR transitions, while C.16b, c represent EPR. The sum of all the right-hand terms in (Eqs. C.16) simply is $g^2\beta_e^2 + g_n^2\beta_n^2$, for frequency-swept (with one fixed field B) conditions. The transitions with $\Delta M_F = 0$ and ± 2 are forbidden when $\mathbf{B}_1 \propto \mathbf{B}$.

We now turn to further examination of the usual EPR field range, above the initial region of curvature, where the levels split into M_S doublets, each pair running 'parallel' with changing B, but far below the cross-over field (Figs. C.1 and C.2).

With the usually good approximation of neglecting the nuclear Zeeman term, Eq. C.14 is given by

$$\hat{\mathcal{H}}_1 = g\beta_e B_1 \hat{S}_z \qquad (C.17)$$

A general matrix element in the uncoupled representation of this $\hat{\mathcal{H}}_1$ is

$$\langle M_S, M_I | \hat{\mathcal{H}}_1 | M_S', M_I' \rangle = g\beta_e B_1 \langle M_S, M_I | \hat{S}_z | M_S', M_I' \rangle \qquad (C.18a)$$

$$= g\beta_e B_1 \langle M_S | \hat{S}_z | M_S' \rangle \langle M_I | M_I' \rangle \qquad (C.18b)$$

$$= g\beta_e B_1 M_S \langle M_S | M_S' \rangle \langle M_I | M_I' \rangle \qquad (C.18c)$$

From the orthogonality of the state functions this matrix element is non-zero only if $M_S' = M_S$ and $M_I' = M_I$.

With $\mathbf{B}_1 \parallel \mathbf{x}$ (i.e., $\mathbf{B}_1 \perp \mathbf{B}$), one has the excitation spin hamiltonian C.17, for which the general element of its matrix form in the uncoupled representation is

$$\langle M_S, M_I | \hat{\mathcal{H}}_1 | M_S', M_I' \rangle = g\beta_e B_1 \langle M_S, M_I | \hat{S}_x | M_S', M_I' \rangle \qquad (C.19a)$$

$$= g\beta_e B_1 \langle M_S | \hat{S}_x | M_S' \rangle \langle M_I | M_I' \rangle \qquad (C.19b)$$

From Eqs. B.42a,b one notes that the matrix element of $\bar{S}_x$ is zero unless $M_S' = M_S \pm 1$. Hence the selection rules when $\mathbf{B}_1 \parallel \mathbf{x}$ are

$$\Delta M_S = \pm 1 \qquad \text{and} \qquad \Delta M_I = 0 \qquad\qquad \text{(C.20)}$$

This clearly represents EPR transitions. Identical results are obtained with $\mathbf{B}_1$ anywhere in the plane perpendicular to $\mathbf{B}$.

As we now see, for $\mathbf{B}_1 \parallel \mathbf{x}$, the relative intensities of possible EPR transitions are approximately given by

$$|\langle M_S, M_I | \hat{S}_x | M_S', M_I' \rangle|^2$$

Alternatively, since $\hat{S}_x = \frac{1}{2}(\hat{S}_+ + \hat{S}_-)$, the matrix elements of $\hat{S}_+$ and $\hat{S}_-$ (Eqs. B.42f,g) can also be used. In a specific instance, only one of these is effective in causing a transition; if $\Delta M_S = +1$, then

$$|\langle M_S + 1, M_I | \hat{S}_+ | M_S', M_I' \rangle|^2$$

governs the intensity; if $\Delta M_S = -1$, then

$$|\langle M_S + 1, M_I | \hat{S}_- | M_S', M_I' \rangle|^2$$

determines the intensity.

These selection rules are exact only when the kets $|M_S, M_I\rangle$ are eigenkets of the spin hamiltonian (Eq. C.2), that is, when the term

$$\tfrac{1}{2} A_0 (\hat{S}_+ \hat{I}_- + \hat{S}_- \hat{I}_+)$$

is neglected. This approximation is valid when $g\beta_e B \gg |A_0|$.

Consider next the situation with $\mathbf{B}_1$ along $\mathbf{z}$, the direction of the static magnetic field $\mathbf{B}$. Here, since $\cos \omega \approx 1$, the allowed transitions are those given in Eqs. C.16b, c (Fig. 1.4). With $\mathbf{B}_1 \parallel \mathbf{B}$, the selection rule set is

$$\Delta M_S = 0 \qquad \text{and} \qquad \Delta M_I = 0 \qquad\qquad \text{(C.21)}$$

and thus, under these conditions, one cannot expect to see absorption, since no single spin flips are allowed. However, simultaneous electronic and nuclear spin flips (e.g., $|-\frac{1}{2}, +\frac{1}{2}\rangle \leftrightarrow |+\frac{1}{2}, -\frac{1}{2}\rangle$) do occur when $\mathbf{B}_1 \parallel \mathbf{B}$ and give rise to a spectral line of relatively low intensity.[4]. The transition can be written in terms of the kets

$|F, M_F\rangle$, and the relative intensity as

$$|1, 0\rangle \rightarrow |0, 0\rangle \text{ intensity} \propto (g\beta_e + g_n\beta_n)^2 \sin^2 2\omega \tag{C.22}$$

The subject of parallel-field EPR spectroscopy is summarized in Section 1.13. Further details for atomic hydrogen can be found in a more recent publication [9].

C.1.5 Resonant Frequencies in Constant Magnetic Field

When B is constant, the separation between the energy levels is fixed. On scanning the microwave frequency, one observes resonance whenever $\Delta U \approx h\nu$ and the selection rules are obeyed. Considering $\Delta M_S = \pm 1$ with $\Delta M_I = 0$, resonance is observed at the following two frequencies, derived using Eqs. C.6 and C.8:

$$\nu_k = \frac{U_{\{\alpha(e),\alpha(n)\}} - U_{\{\beta(e),\alpha(n)\}}}{h}$$

$$= \frac{1}{2h} \left\{ (g\beta_e - g_n\beta_n)B + A_0 + [(g\beta_e + g_n\beta_n)^2 B^2 + A_0^2]^{1/2} \right\} \tag{C.23a}$$

$$\nu_m = \frac{U_{\{\alpha(e),\beta(n)\}} - U_{\{\beta(e),\beta(n)\}}}{h}$$

$$= \frac{1}{2h} \left\{ (g\beta_e - g_n\beta_n)B - A_0 + [(g\beta_e + g_n\beta_n)^2 B^2 + A_0^2]^{1/2} \right\} \tag{C.23b}$$

The difference $h(\nu_k - \nu_m)$ is exactly the hyperfine *coupling parameter* A_0.

C.1.6 Resonant Magnetic Fields at Constant Excitation Frequency

Use of a constant excitation frequency ν and a varying magnetic field is the typical experimental arrangement. This situation is more complicated, since the magnetic field is *not* the same for the two transitions.

The solution is given to excellent approximation [5] by

$$B = \frac{A_0}{g\beta_e} \frac{1}{1 - \left(\frac{A_0}{2h\nu}\right)^2} \left[-M_I \pm \left\{ M_I^2 + \left[1 - \left(\frac{A_0}{2h\nu}\right)^2 \right] \left[\left(\frac{h\nu}{A_0}\right)^2 - \left(I + \frac{1}{2}\right)^2 \right] \right\}^{1/2} \right] \tag{C.24}$$

Here $g_n\beta_n$ has been taken as negligible compared to $g\beta_e$, since $g_n\beta_n \approx 10^{-3} g\beta_e$. Equation C.24 is in fact valid for any nuclear spin I. It follows that

$$\frac{B_{-|M_I|} - B_{|M_I|}}{2|M_I|} = \frac{\frac{A_0}{g\beta_e}}{1 - \left(\frac{A_0}{2h\nu}\right)^2} \tag{C.25}$$

which affords a simple means of obtaining A_0, with all higher-order approximations included, when g and ν are known. Note that for $A_0/g > 0$, the EPR line labeled with fixed $M_I = -\frac{1}{2}$ occurs at fields above that with $M_I = +\frac{1}{2}$ (Fig. 1.4); the opposite is true for $A_0/g < 0$.

An alternative treatment, valid when $|(g\beta_e + g_n\beta_n)B| \gg |A_0|$, is now examined in some detail for ^{1}H. Here one can carry out a binomial expansion of the square-root term in energy Eqs. C.8:[5]

$$[(g\beta_e + g_n\beta_n)^2 B^2 + A_0^2]^{1/2} = (g\beta_e + g_n\beta_n)B + \frac{1}{2}\frac{A_0^2}{(g\beta_e + g_n\beta_n)B}$$
$$- \frac{1}{8}\frac{A_0^4}{(g\beta_e + g_n\beta_n)^3 B^3} + \cdots \qquad (C.26)$$

Only the first two right-hand terms are retained. The resonant field B_k corresponds to the transition $|\beta(e),\ \alpha(n)\rangle \rightarrow |\alpha(e),\ \alpha(n)\rangle$ and B_m, to the transition $|\beta(e),\ \beta(n)\rangle \rightarrow |\alpha(e),\ \beta(n)\rangle$. For $\nu_k = \nu_m = \nu$ (the fixed microwave frequency), Eqs. C.23 can be transformed by substitution of Eq. C.26 and multiplication by B_k or B_m to gives

$$g\beta_e B_k^2 - (h\nu - \tfrac{1}{2}A_0)B_k + \frac{A_0^2}{4g\beta_e} = 0 \qquad (C.27a)$$

and

$$g\beta_e B_m^2 - (h\nu + \tfrac{1}{2}A_0)B_m + \frac{A_0^2}{4g\beta_e} = 0 \qquad (C.27b)$$

Here the approximation $g\beta_e \gg |g_n\beta_n|$ has again been utilized. Solution of Eqs. C.27a,b gives

$$B_k = \frac{1}{4\,g\beta_e}\left[2\,h\nu - A_0 + (4\,h^2\nu^2 - 4\,h\nu A_0 - 3A_0^2)^{1/2}\right] = 0 \qquad (C.28a)$$

and

$$B_m = \frac{1}{4\,g\beta_e}\left[2\,h\nu + A_0 + (4\,h^2\nu^2 + 4\,h\,\nu A_0 - 3A_0^2)^{1/2}\right] = 0 \qquad (C.28b)$$

It is clear that $B_m - B_k \neq A_0/g\beta_e$, consistent with Eq. C.25.

For the H atom, $A_0/h = 1420.40575$ MHz and $g = 2.0022838$ [7,8]. The latter is very close to the free-electron g factor (Section 1.8). With $\nu = 9.500$ GHz (X-band microwaves), substitution in Eqs. C.28 gives

$$B_k = 311.586 \text{ mT} \qquad \text{and} \qquad B_m = 362.561 \text{ mT}$$

Hence $B_m - B_k = 50.975$ mT. Since $A_0/g_e\beta_e = 50.685$ mT, one sees that the actual splitting a_0 differs from $A_0/g_e\beta_e$ by 0.290 mT. The difference is not large but is significant. It is of interest to note the average value $(B_m + B_k)/2 = 337.074$ mT. This is 1.916 mT lower than the field $h\nu/g\beta_e$ (B_ℓ in Fig. 2.4b). There would thus be a considerable error in determining the g factor from the mean position. For most organic radicals the difference between proton hyperfine components is much smaller.

C.1.7   Calculation of Spin Energy Levels by Perturbation Theory

Here we wish to discuss the most utilized approximate mathematical technique (Section A.6) for estimating energies, which is needed since for most spin systems one cannot obtain exact solutions. We continue to consider the one-electron atom. The spin hamiltonian (Eq. C.2) may be separated into two parts

$$\hat{\mathcal{H}} = \hat{\mathcal{H}}_0 + \hat{\mathcal{H}}' \tag{C.29}$$

where

$$\hat{\mathcal{H}}_0 = g\beta_e B\hat{S}_z + A_0\hat{S}_z\hat{I}_z - g_n\beta_n B\hat{I}_z \tag{C.30}$$

and the 'perturbation' is

$$\hat{\mathcal{H}}' = \tfrac{1}{2}A_0(\hat{S}_+\hat{I}_- + \hat{S}_-\hat{I}_+) \tag{C.31}$$

If terms arising from $\hat{\mathcal{H}}'$ are much smaller than those arising from $\hat{\mathcal{H}}_0$ (this implies sufficiently high magnetic fields), then one may use the eigenkets of $\hat{\mathcal{H}}_0$ as a basis set for determining the energy corrections arising from $\hat{\mathcal{H}}'$.

The zero-order energies $U^{(0)}$ are just the diagonal matrix elements of $\hat{\mathcal{H}}_0$ (Eq. C.5) (Section A.6), namely

$$U_{\alpha(e),\alpha(n)}^{(0)} = +\tfrac{1}{2}g\beta_e B + \tfrac{1}{4}A_0 - \tfrac{1}{2}g_n\beta_n B \tag{C.32a}$$

$$U_{\alpha(e),\beta(n)}^{(0)} = +\tfrac{1}{2}g\beta_e B - \tfrac{1}{4}A_0 + \tfrac{1}{2}g_n\beta_n B \tag{C.32b}$$

$$U_{\beta(e),\alpha(n)}^{(0)} = -\tfrac{1}{2}g\beta_e B - \tfrac{1}{4}A_0 - \tfrac{1}{2}g_n\beta_n B \tag{C.32c}$$

$$U_{\beta(e),\beta(n)}^{(0)} = -\tfrac{1}{2}g\beta_e B + \tfrac{1}{4}A_0 + \tfrac{1}{2}g_n\beta_n B \tag{C.32d}$$

as defined the energy matrix C.5. The effects of each of the terms in Eqs. C.32 are indicated sequentially in Fig. C.2. Note that the nuclear Zeeman interaction in the hydrogen atom does *not* affect the first-order EPR transition energies.

Regardless of the form of $\hat{\mathcal{H}}'$ in Eq. C.29, one may use the general expression Eq. A.81b with Eqs. A.90 and A.93 for the energy arising from the perturbation as

$$U_i = U_i^{(0)} + \langle i|\hat{\mathcal{H}}'|i\rangle + \sum_{j \neq i} \frac{\langle i|\hat{\mathcal{H}}'|j\rangle\langle j|\hat{\mathcal{H}}'|i\rangle}{U_j^{(0)} - U_i^{(0)}} + \cdots \tag{C.33}$$

where the sum goes over all spin states. Here $U_i^{(0)}$ and $U_j^{(0)}$ are the zero-order energies for the spin states $|i\rangle$ and $|j\rangle$. The second term on the right-hand side

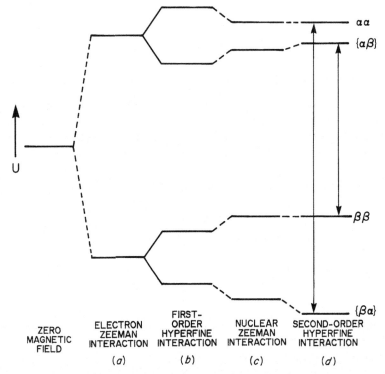

FIGURE C.2 Spin energy levels and allowed EPR transitions for the hydrogen atom (at moderate constant magnetic field) showing the effects of successive terms in the spin hamiltonian (Eq. C.2): (*a*) electron Zeeman interaction $g\beta_e B\hat{S}_z$. (*b*) addition of the first-order proton hyperfine interaction $A_0\hat{S}_z\hat{I}_z$; (*c*) addition of the nuclear Zeeman interaction $-g_n\beta_n B\hat{I}_z$; (*d*) addition of the second-order proton hyperfine interaction $\frac{1}{2}A_0(\hat{S}_+\hat{I}_- + \hat{S}_-\hat{I}_+)$.

of Eq. C.33 is the first-order correction $U^{(1)}$ to the energy of the *i*th state; it is given by the *i*th diagonal matrix element of $\hat{\mathcal{H}}'$, evaluated for the *i*th zero-order 'wavefunction'. However, since $\hat{\mathcal{H}}'$ involves only the raising and lowering operators, all diagonal matrix elements of $\hat{\mathcal{H}}'$ are zero. Thus the state energies previously calculated with the use of $\hat{\mathcal{H}}_0$ are correct to first order. Hence it is necessary to utilize the third right-hand term of Eq. C.33 to obtain additional (second-order) corrections $U^{(2)}$.

Writing the states in Eq. C.33 in terms of the M_S and M_I values, one finds the second-order correction to be given by

$$U_{M_S M_I}^{(2)} = \sum_{M_S' \neq M_S,\, M_I' \neq M_I} \frac{\langle M_S, M_I | \hat{\mathcal{H}}' | M_S', M_I' \rangle \langle M_S', M_I' | \hat{\mathcal{H}}' | M_S, M_I \rangle}{U_{M_S, M_I}^{(0)} - U_{M_S', M_I'}^{(0)}} \qquad (C.34)$$

As in the exact treatment (Section C.3), the only non-zero off-diagonal matrix elements of $\hat{\mathcal{H}}'$ are

$$\langle \beta(e), \alpha(n) | \hat{S}_-\hat{I}_+ | \alpha(e), \beta(n) \rangle = 1 \tag{C.35a}$$

and

$$\langle \alpha(e), \beta(n) | \hat{S}_+\hat{I}_- | \beta(e), \alpha(n) \rangle = 1 \tag{C.35b}$$

Therefore only the energies of the states $|\alpha(e), \beta(n)\rangle$ and $|\beta(e), \alpha(n)\rangle$ are affected; the second-order energy corrections to these states are

$$U_{\{\alpha(e),\beta(n)\}}^{(2)} = -U_{\{\beta(e),\alpha(n)\}}^{(2)} = \frac{A_0{}^2}{4\, g\beta_e B} \tag{C.36}$$

The energies to second order are included in Fig. C.2. The transition frequencies (Eq. C.23) at *constant field* are

$$v_k = \frac{1}{h}\left(g\beta_e B + \tfrac{1}{2}A_0 + \frac{A_0{}^2}{4\, g\beta_e B} \right) \tag{C.37a}$$

and

$$v_m = \frac{1}{h}\left(g\beta_e B - \tfrac{1}{2}A_0 + \frac{A_0{}^2}{4\, g\beta_e B} \right) \tag{C.37b}$$

The term $g_n\beta_n B$ has been neglected in comparison with $g\beta_e B$. As before, $v_k - v_m = A_0/h$ when B is held constant and v is scanned, but not when the field is swept. Comparison with the exact solutions (Eqs. C.24) discloses (after a series expansion of these) that these become identical with Eqs. C.37 as $B \to \infty$. Note that we can use Eqs. C.37 to calculate the fields B_k and B_m at which resonance occurs in a fixed-frequency experiment $(v = v_k = v_m)$ only to second-order approximation.

Discussion as to how to obtain perturbation-theoretic corrections to the kets can be found in Section A.6.

C.2 RH RADICALS

The literature includes descriptions of EPR studies of various free radicals that can be termed 'RH species', in that these $S = \frac{1}{2}$ molecules show hyperfine structure primarily with one proton. We shall neglect here consideration of any hyperfine structure arising from other nuclides in these molecules, which are usually organic species. A prime example is RH = malonic acid π radical HOOC—CH—COOH

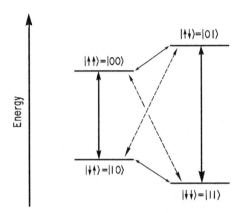

FIGURE C.3   The energy-level scheme for a two-spin system, say, $S = I = \frac{1}{2}$. The solid arrows denote allowed transitions, whereas the dashed arrows indicate 'forbidden' transitions, taking place between 'entangled' states. The phase dependence, under rotation about quantization axis **z**, of the four quantum states also is indicated.

(see Sections 9.2 and 12.9). Clearly all the theory in Section C.1 is applicable here; only the quantities of the parameters differ.

The latter free radical is of special interest since it has been used for combined pulsed EPR and NMR spectroscopy to establish existence of 'entangled' spin states [11], creating normally highly forbidden transitions to mix states, as depicted in Fig. C.3. As is evident, the 2-spin description can be recast to a qubit description (with $\uparrow = 0$, $\downarrow = 1$). These ideas may well assist in developing quantum computing.

C.3   RH$_2$ RADICALS

When more than one magnetic nucleus is interacting with a single unpaired electron, the calculation of the state energies is best preceded by a careful inspection of the nuclear-spin wavefunctions. If the nuclei are equivalent, it is usually convenient to use a 'coupled' representation for the nuclear-spin states (Section B.6). The nuclear spins of the two protons of RH$_2$ (R is any suitable molecular group) are added vectorially to obtain one set of wavefunctions with total nuclear spin $^tI = 1$ and another set with $^tI = 0$. The new nuclear-spin functions are represented by $|^tI, ^tM\rangle$ and are related to the kets $|M_{I_1}, M_{I_2}\rangle$ by

$$|1, +1\rangle \equiv |+\tfrac{1}{2}, +\tfrac{1}{2}\rangle \qquad\qquad\qquad\qquad\quad (C.38a)$$

$$\left.\begin{array}{l} |1, 0\rangle \ \ \equiv \tfrac{1}{\sqrt{2}}\left(|+\tfrac{1}{2}, -\tfrac{1}{2}\rangle + |-\tfrac{1}{2}, +\tfrac{1}{2}\rangle\right) \\[2mm] \end{array}\right\} \ \ ^tI = 1 \qquad (C.38b)$$

$$|1, -1\rangle \equiv |-\tfrac{1}{2}, -\tfrac{1}{2}\rangle \qquad\qquad\qquad\qquad\ \ (C.38c)$$

$$|0, 0\rangle \ \ \equiv \tfrac{1}{\sqrt{2}}\left(|+\tfrac{1}{2}, -\tfrac{1}{2}\rangle - |-\tfrac{1}{2}, +\tfrac{1}{2}\rangle\right) \qquad ^tI = 0 \qquad (C.38d)$$

Because the *square* of the total nuclear angular momentum [characterized by $'I('I+1)$] is a constant of the motion, states with different values of $'I$ do not mix; that is, there are no non-zero matrix elements between $'I = 0$ and $'I = 1$ states. Including the electron spin, the total spin wavefunctions are designated by $|M_S, 'I, 'M_I\rangle$.

C.3.1   Spin Hamiltonian and Energy Levels

The spin hamiltonian for the RH_2 fragment, which has $S = \frac{1}{2}$, is again separated into two parts so that perturbation theory may be used:

$$\hat{\mathcal{H}}_0 = g\beta_e B\hat{S}_z + A_0\hat{S}_z\,{}'\hat{I}_z \tag{C.39a}$$

$$\hat{\mathcal{H}}' = \tfrac{1}{2}A_0(\hat{S}_+\,{}'\hat{I}_- + \hat{S}_-\,{}'\hat{I}_+) \tag{C.39b}$$

The nuclear Zeeman terms have been omitted since they do not appreciably affect the transition energies in our high-field approximation. Equations C.39 are analogous to Eqs. C.30 and C.31, with the total nuclear-spin operator $'\hat{I}$ replacing the individual nuclear-spin operators $\hat{I}$.

The zero-order energies $U^{(0)}_{M_S, 'I, 'M_I}$ for the eight spin states are then the diagonal matrix elements of $\hat{\mathcal{H}}_0$, yielding

$$
\begin{array}{ll}
U_{1/2,1,+1}{}^{(0)} = \tfrac{1}{2}g\beta_e B + \tfrac{1}{2}A_0 & U_{-1/2,1,-1}{}^{(0)} = -\tfrac{1}{2}g\beta_e B + \tfrac{1}{2}A_0 \\[4pt]
U_{1/2,0,0}{}^{(0)} = \tfrac{1}{2}g\beta_e B & U_{-1/2,1,0}{}^{(0)} = -\tfrac{1}{2}g\beta_e B \\[4pt]
U_{1/2,1,0}{}^{(0)} = \tfrac{1}{2}g\beta_e B & U_{1/2,1,-1}{}^{(0)} = \tfrac{1}{2}g\beta_e B - \tfrac{1}{2}A_0 \\[4pt]
U_{-1/2,0,0}{}^{(0)} = -\tfrac{1}{2}g\beta_e B & U_{-1/2,1,+1}{}^{(0)} = -\tfrac{1}{2}g\beta_e B - \tfrac{1}{2}A_0
\end{array}
\tag{C.40a–h}
$$

C.3.2   EPR Transitions

Since the selection rules are $\Delta M_S = \pm 1$, $\Delta'M_I = 0$, the spectrum at constant microwave frequency consists to first approximation of three lines occurring at the resonant fields

$$B_k = \frac{h\nu}{g\beta_e} - \frac{A_0}{g\beta_e} \tag{C.41a}$$

$$B_\ell = \frac{h\nu}{g\beta_e} \tag{C.41b}$$

$$B_m = \frac{h\nu}{g\beta_e} + \frac{A_0}{g\beta_e} \tag{C.41c}$$

The line at B_ℓ is twice as intense as the lines at B_k and B_m because the states contributing to the line at B_ℓ are doubly degenerate (Fig. 3.1a).

As before (Eq. C.31), the second-order energy corrections involve only off-diagonal matrix elements; from Eqs. B.42f,g, only the following four are non-zero:

$$\langle +\tfrac{1}{2}, 1, -1|\hat{S}_+{}^t\hat{I}_-| -\tfrac{1}{2}, 1, 0\rangle = \sqrt{2} \qquad (C.42a)$$

$$\langle +\tfrac{1}{2}, 1, 0|\hat{S}_+{}^t\hat{I}_-| -\tfrac{1}{2}, 1, +1\rangle = \sqrt{2} \qquad (C.42b)$$

$$\langle -\tfrac{1}{2}, 1, 0|\hat{S}_-{}^t\hat{I}_+| +\tfrac{1}{2}, 1, -1\rangle = \sqrt{2} \qquad (C.42c)$$

$$\langle -\tfrac{1}{2}, 1, +1|\hat{S}_-{}^t\hat{I}_+| +\tfrac{1}{2}, 1, 0\rangle = \sqrt{2} \qquad (C.42d)$$

The energies to second order are given in Fig. C.4. The transition frequencies at *constant field* are

$$\nu_k = \frac{1}{h}\left(g\beta_e B + A_0 + \frac{1}{2}\frac{A_0{}^2}{g\beta_e B} \right) \qquad (C.43a)$$

$$\nu_{\ell'} = \frac{1}{h}\left(g\beta_e B + \frac{A_0{}^2}{g\beta_e B} \right) \qquad (C.43b)$$

$$\nu_{\ell''} = \frac{1}{h}\left(g\beta_e B \right) \qquad (C.43c)$$

$$\nu_m = \frac{1}{h}\left(g\beta_e B - A_0 + \frac{1}{2}\frac{A_0{}^2}{g\beta_e B} \right) \qquad (C.43d)$$

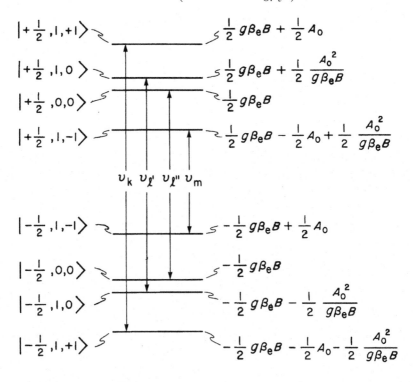

FIGURE C.4   Spin energy levels of the RH$_2$ fragment, to second order. The kets $|M_S, {}^t I, {}^t M_I\rangle$ for the eight states are indicated. Here A_0 has been neglected in comparison with $g\beta_e B$, in the second-order terms.

If A_0 is small compared to $g\beta_e B$, the latter can be set equal to $h\nu$ in the correction terms, and hence the resonant fields at constant ν are approximately

$$B_k = \frac{h\nu}{g\beta_e} - \frac{A_0}{g\beta_e} - \frac{1}{2}\frac{A_0^2}{g\beta_e\, h\nu} \tag{C.44a}$$

$$B_{\ell'} = \frac{h\nu}{g\beta_e} - \frac{A_0^2}{g\beta_e\, h\nu} \tag{C.44b}$$

$$B_{\ell''} = \frac{h\nu}{g\beta_e} \tag{C.44c}$$

$$B_m = \frac{h\nu}{g\beta_e} + \frac{A_0}{g\beta_e} - \frac{1}{2}\frac{A_0^2}{g\beta_e\, h\nu} \tag{C.44d}$$

Thus the spectrum now consists of four lines, with the
lines, except that at $B_{\ell''}$, shifted downfield from the zero-order positions. These are illustrated in Fig. 3.15a.

REFERENCES

1. R. S. Dickson, J. A. Weil, *Am. J. Phys.*, **59**, 125 (1991).

2. J. A. Weil, *Can. J. Phys.*, **59**, 841 (1981).

3. G. Breit, I. I. Rabi, *Phys. Rev.*, **38**, 2082L (1931).

4. J. E. Nafe, E. B. Nelson, *Phys. Rev.*, **73**, 718 (1948).

5. J. A. Weil, *J. Magn. Reson.*, **4**, 394 (1971).

6. N. F. Ramsey, *Molecular Beams*, Oxford University Press, London, U.K., 1956, Chapters 3 and 9.

7. L. Essen, R. W. Donaldson, E. G. Hope, M. J. Bangham, *Metrologia*, **9**, 128 (1973).

8. J. S. Tiedeman, H. G. Robinson, *Phys. Rev. Lett.*, **39**, 602 (1977).

9. J. A. Weil, *Concepts Magn. Reson., Part A*, **28A**(5), 331 (2006).

10. J. A. Weil, J. R. Bolton, J. E. Wertz, *Electron Paramagnetic Resonance, Elementary Theory and Practical Applications*, Wiley, New York, NY, U.S.A., 1992.

11. M. Mehring, J. Mende, W. Scherer, *Phys. Rev. Lett.*, **90**(15), 153001 (2003).

NOTES

1. It is worthwhile to note that when we replace $\hat{\mathbf{S}}$ with $\hat{\mathbf{I}}_1$ and $\hat{\mathbf{I}}$ with $\hat{\mathbf{I}}_2$ (and $g\beta_e$ by $-g_n\beta_n$), we attain the well-known AB case encountered in liquid-phase NMR.

2. The terms *secular* ('continuing through long ages') and *non-secular* derive from celestial mechanics, from the perturbation techniques (Section A.6) treating small disturbances to the trajectories of planets and satellites. In EPR, as we shall see, spin-hamiltonian elements (e.g., $\hat{S}_z$ and $\hat{S}_z\,\hat{I}_z$) not causing transitions are secular, while those (e.g., $\hat{S}_x$ and

$\hat{S}_+\hat{I}_-$) associated with transitions are non-secular. This terminology is also applied to relaxation (τ_2 and τ_1; see Chapter 10) and associated line-broadening effects.

3. Note that only half of B_1 is effective (Fig. 10.5).

4. There are appropriate off-diagonal elements, dependent on A_0^{-1}, in the matrix $\mathbf{S}_z$ (expressed in the basis in which the spin-hamiltonian matrix is diagonal) to yield non-zero transition probabilities.

5. $(x^2 + y^2)^{1/2} = x + \frac{1}{2}x^{-1}y^2 - \frac{1}{8}x^{-3}y^4 + \cdots +$, where $x = (g\beta_e + g_n\beta_n)B$ and $y = A_0$, when $|x| > |y|$.

FURTHER READING

1. H. A. Bethe, E. E. Salpeter, *Quantum Mechanics of One- and Two-Electron Atoms*, Springer, Berlin, Germany, 1957.

2. A. Carrington, A. D. McLachlan, *Introduction to Magnetic Resonance*, Harper & Row, New York, NY, U.S.A., 1967.

3. R. Erickson, A. Lund, "Analytical Expressions of Magnetic Energies and Wave Functions of Paramagnetic Systems with $S = \frac{1}{2}$ and $I = 1$ or $\frac{3}{2}$", *J. Magn. Reson.*, **92**, 146 (1991).

4. A. P. French, E. F. Taylor, *An Introduction to Quantum Physics*, Norton, New York, NY, U.S.A., 1978.

5. L. B. Knight Jr., W. E. Rice, L. Moore, E. R. Davidson, R. S. Dailey, "Theoretical and Electron Spin Resonance Studies of the H...H, H...D and D...D Spin-Pair Radicals in Rare-Gas Matrices: A Case of Extreme Singlet-Triplet Mixing", *J. Chem. Phys.*, **109**(4), 1409 (1998).

PROBLEMS

C.1 Consider the spectrum of the hydrogen atom shown in Fig. 2.6. Use the expressions developed in this appendix to compute an accurate value of the hyperfine coupling constant A_0/h (MHz) and the g factor. Comment on the differences from the corresponding values for the free gaseous hydrogen atom.

C.2 Compare the numerical results of the frequencies ν_k and ν_m at fixed fields B, for the free hydrogen atom ^{1}H, as given by the exact formulas C.23 and the second-order approximate formulas C.37, at fields $B = 0.1$, 1, and 10 T.

C.3 Derive expressions C.16 for the relative intensities of the hydrogenic EPR lines, as well as the quasi-forbidden double-spin-flip line $|0,0\rangle \rightarrow |1,0\rangle$.

C.4 Consider the hydrogen atom trapped in an anisotropic medium, so that its hyperfine energy must be treated in terms of a rhombic hyperfine matrix $\mathbf{A}$.

    (*a*) For the zero-field case ($B = 0$), set up the appropriate spin hamiltonian and find the energy eigenvalues and spin eigenstates. Take g and g_n to be isotropic and utilize the principal-axis system of $\mathbf{A}$.

(**b**) Set up the matrices $\mathbf{S}_i$ ($i = X, Y, Z$) in the energy-diagonal basis and predict the zero-field EPR spectra when $\mathbf{B}_1$ is along each principal axis.

C.5 The effects of the nuclear quadrupole moment on the energy levels of a nucleus with $I = 1$ in an inhomogeneous electric field may be seen by examining the shifts of levels of the deuterium atom. The six unperturbed levels are given by

$$U_{M_S M_I} = g\beta_e B M_S + A_0 M_S M_I \tag{C.45}$$

The nuclear quadrupole spin hamiltonian (Eqs. 5.50 and 5.51) for a uniaxial electric field along $\mathbf{Z}$ is

$$\hat{\mathcal{H}}_Q = Q'\left[\hat{I}_z^2 - \frac{I(I+1)}{3}\right] \tag{C.46}$$

where

$$Q' = \frac{3|e|Q}{4I(2I-1)} \left.\frac{\partial^2 \phi}{\partial Z^2}\right|_n \tag{C.47}$$

Here e is the electronic charge, Q is the nuclear quadrupole moment (in m^2), $-\partial^2 \phi/\partial Z^2$ is the gradient of the electric field $-\partial \phi/\partial Z$ in which the nuclear quadrupole is located, and ϕ is the potential describing the charge distribution around the nucleus n.

(**a**) Plot the energy of the six levels versus magnetic field.

(**b**) Show the shift of each of these levels caused by $\hat{\mathcal{H}}_Q$, expressing the shift in terms of Q' ($=3P$; see Eq. 5.51b).

(**c**) Show the allowed transitions (still $\Delta M_S = \pm 1$, $\Delta M_I = 0$).

(**d**) Can one detect a nuclear quadrupole interaction from the EPR spectrum? [These shifts must be taken into account when interpreting electron-nuclear double resonance (ENDOR) spectra (Chapter 12).]

C.6 The matrix operator representing the x component of the electron-spin angular momentum is given by

$$\bar{S}_x = \frac{1}{2}\begin{pmatrix} 0 & S & C & 0 \\ S & 0 & 0 & C \\ C & 0 & 0 & -S \\ 0 & C & -S & 0 \end{pmatrix} \begin{matrix} \alpha,\alpha \\ \{\alpha,\beta\}_{F=1} \\ \{\alpha,\beta\}_{F=0} \\ \beta,\beta \end{matrix} \tag{C.48}$$

In the coupled (F, M_F) representation. Here $C = \cos \omega$ and $S = \sin \omega$ (See Eqs. C.8 and C.10, and Fig. C.2). What predictions of observable spectroscopic phenomena can you make using this matrix? Provide detail.

C.7   Derive the value (Section C.1.3) for the field B at which the upper two spin levels of the ground-state hydrogen atom cross. What will this field be for the tritium atom?

C.8   Explore whether the spacings between all four spin levels of the ground-state hydrogen atom diverge without limit as $B \to \infty$. In nature, what phenomena other than the spin Zeeman effects enter as the external magnetic field becomes really large?

C.9   Derive the relative intensities for all possible magnetic resonance transitions for the ground-state hydrogen atom when (*a*) $\mathbf{B}_1 \perp \mathbf{B}$ (i.e., derive Eqs. C.16) and (*b*) when $\mathbf{B}_1 \parallel \mathbf{B}$. Which are NMR transitions?

C.10   Consider Eq. C.25:

   (*a*)   Calculate $B_m - B_k$ at 9.500 GHz for the H atom and compare with the result obtained by use of Eqs. C.28 as discussed in Section C.1.6.

   (*b*)   Expand the denominator term in the equation, arriving at an expression for $B_m - B_k$ as a power series a_o^{n+1}/B^n. Rationalize the absence of an $n = 1$ term, such as occurs in Eq. 3.2.

C.11   Use the methods developed for RH_2 to calculate to second order the energies of the states for a free radical containing three equivalent protons. For $A_0/h = 100$ MHz, $g = 2.00232$, and microwave frequency 9.500 GHz, calculate the field positions and relative intensities of all allowed transitions.

APPENDIX D

PHOTONS

D.1 INTRODUCTION

A fair view of EPR spectroscopy admits that while unpaired electrons constitute a most essential ingredient, there is another species that constitutes an equally important component, namely, *photons*. Both types of particles demand that, in some circumstances, they must be considered as 'wave packets', but there are major differences between them. This appendix focuses on photons, since the rest of the book deals primarily with electrons. The understanding of photons continues to develop rapidly (e.g., see Ref. 1).

D.2 THE PHYSICAL ASPECTS OF PHOTONS

Electromagnetic radiation is absorbed and emitted by atomic systems in separate portions (quanta of energy) called *photons*. This phenomenon occurs in all types of spectroscopy, and thus certainly so in magnetic resonance. Electromagnetic radiation, which generally is considered to be a bulk property, arises from and disappears into sets of massive electric charges and magnetic poles, located on massive particles.[1] Both the latter can be classified mathematically in terms of multipoles [2], and, via parentage, so can the radiation and its photons. Thus electric, as well as magnetic dipolar and quadrupolar, radiation is frequently encountered. In magnetic resonance, it is magnetic dipolar radiation that is of central importance.

Electron Paramagnetic Resonance, Second Edition, by John A. Weil and James R. Bolton
Copyright © 2007 John Wiley & Sons, Inc.

The photon is one type of *vector boson* and, of course, is the carrier of every electromagnetic interaction between massive particles; that is, it transmits the Coulomb force between charged species (as well as magnetic dipole-dipole contacts, etc.). It couples to the electric charge q of each such particle in the system, with a strength proportional to this charge.

The cosmic blackbody radiation can be considered to be due to free gaseous photons in our Universe, with an energy (frequency) distribution of $h\nu$ values corresponding to a temperature of 2.7 K. When impinging on material surfaces (sets of atoms), photons can be assumed to exert pressure, and they can bounce off (reflection: re-emission?). Photons can transfer energy, linear momentum as well as angular momentum when encountering massive particles.

The photon is something of a physics monstrosity, but this 'particle' concept, when properly used, works wonderfully well. However, in some circumstances, the concept fails to describe observed facts, and the wave picture must be invoked and works (the same is true of the electron [3]). The photon in all probability has zero mass [4], unlike the electron, and one cannot assign an extent to it, so that one cannot ever give a position for it. A photon does not have a measurable spatial wavefunction, and considering the result of confining it in some small volume reveals that other photons (i.e., radiation) are then created elsewhere [5]. Nevertheless, spatial parity (eigenstates of the 3D inversion operator) can be assigned to it [6, pp. 578 ff; 7, pp. 29 ff], basically because the linear momentum changes sign under transformation with the parity operator.

The photon of course always travels at the speed c of light, the numerical value of which, however, depends on the medium traversed.[2] Photons having exactly the same properties are indistinguishable.[3]

The photon has two properties that are continuous variables and intimately related to each other. One is its radiational energy $h\nu$, and the other is its linear momentum $h\nu/c$. It also has angular momentum: its *spin*, with discrete (quantized) values; the photon spin angular momentum is always present. For some discussion of photonic *orbital angular momentum*, a questionable concept since there is no center of mass [1,9].

It is evident that the frequency ν of the photon can approach zero. Thus, in the limiting sense, the energy and momentum can display zero values while, nevertheless, such a photon travels at the speed of light. There seems, at this time, to be no high-energy upper limit in theory to the frequency ν (values of ν up to $\sim 10^{35}$ Hz are known [10]).

No photon can ever be said to have the exact energy $h\nu$: that is, in reality (despite common usage) there always is some uncertainty in the photon oscillation frequency ν. This has to do with uncertainty in the source emission of photons. Thus there never is exactly monochromatic light, but we usually ignore this fact.

Photons have no electric or magnetic dipoles, or higher electromagnetic moments. However, one can talk about the electric field **E** and magnetic field **H** associated with each photon. These are related via $\mathbf{H} = (\mu_0 c)^{-1}\mathbf{z} \wedge \mathbf{E}$ [i.e., $\mathbf{H} = (c)^{-1} \mathbf{z} \wedge \mathbf{E}$, in vacuum].[2] Thus, knowing one field vector, one knows the other. These are both perpendicular to the direction **z** of photon propagation, and to each other. If one wishes, one can define $\mathbf{x} \equiv \mathbf{E}/|\mathbf{E}|$ and $\mathbf{y} \equiv \mathbf{H}/|\mathbf{H}|$.

It appears, however, that these fields are observable only posthumously when the photon meets a massive particle. For instance, a photon beam impinging onto an appropriate solid surface perpendicular to its path **z** induces a photocurrent in the xy plane, whose direction depends on the predominant photon spin direction (helicity) [8, pp. 101–102]. The type of photon encountered is determined by its source and possibly also by devices that it passes through. The circumstances of its creation determine its properties. While 'in flight', the photon cannot be typed. However, when it meets its eventual annihilation, such a determination is feasible. We have noted that photons can transfer both energy and angular momentum to material particles encountered. Note, too, that the latter survive detection, while photons do not. Thus, the properties of any given photon can be established (inferred) only by measuring the properties of its source (post-natal) and/or its sink (post-mortem). It is clear then that the birth and death of a photon are linked, in the sense of yielding information about its properties. The actual properties of any given photon usually are determined after its death, by examining its entrails, that is, the electromagnetic effects of the atom into which it has disappeared. In other words, when a photon is absorbed, it reports its properties, which actually are properties of its source. It is in that sense only that we can speak (below) of σ_+ and σ_- photons.

In the latter case, a photon traveling in some direction (usually defined to be **z**; thus its travel runs from $-$ to $+$ along **z**) can occur in either of two helicity states: ± 1. Here $+1$ means that its spin angular-momentum z component (in units of $\hbar = h/2\pi$) is along the direction of its travel (i.e., forward), while -1 means that its angular momentum z component is along the negative of its travel direction (i.e., backward). Thus a photon can occur in either of these two states. There is no state of helicity 0 (this follows from the fact that the photon always travels at the speed of light: relativity and zero mass).

One can visualize the photon's spin angular momentum, which is always present, via a precession vector circulating in a cone about the z axis, as seen by an observer looking from a point on the photon trajectory in the direction of its forward travel. Then helicity $+1$ means clockwise rotation (with z projection $+1$); this implies counterclockwise rotation looking from the opposite cone expanding backward from the photon. Obviously, -1 means counterclockwise rotation looking forward. Thus its helicity defines a *unique* sense of rotation (angular momentum) for each photon. Experimentally, it is known how helicity can be switched from one state to the other (photon spin flipping: presumably there is a switch of photon identity during this process), for instance [10], by using a quartz-crystal $\lambda/4$ polarizer plate. We note that if the z angular-momentum component is well defined, then the z and y components are necessarily uncertain.

We can denote the states of photons as σ_+ and σ_- [11]. The two symbols can be thought of as describing the individual photon's possible bras and kets (state functions). The terms 'right-hand' and 'left-hand' circularly polarized are to be associated with σ_+ and σ_- photons, respectively.

The terms 'linearly polarized' and 'circularly polarized' (and, more generally, 'elliptically polarized') really apply only to statistically sufficiently large *sets* of photons, not to individual ones, and imply presence of an angular-momentum phase correlation between all pairs of photons in the set. We can have both $+1$ and -1 types of photons in the set, that is, representing elliptical polarization of the light. Admixture of equal amounts of left and right circularly 'light' of the same frequency and traveling within the same ray **c** is equivalent to presence of linearly polarized light polarized normal to **c**, and vice versa (see Fig. 10.5).[4]

It is known [12] that there is a relationship, $\delta n \, \delta \phi \sim 1$, between the uncertainties in the total number n of monochromatic photons present and the phases ϕ of the (well-defined sinusoidal components of the) corresponding electromagnetic fields. In typical magnetic resonance work, n can be considered to be very large.

Efforts to visualize the wave-like and corpuscular nature of light beams continue. Thus, in one proposed highly conjectural model [13], an array of equally spaced electric dipoles (say, sets of charges $\pm e$ separated along direction **x** while moving steadily along direction **z**, with the dipoles oscillating sinusoidally along **x**), generate oscillating fields $\mathbf{E} \parallel \mathbf{x}$ and $\mathbf{B} \parallel \mathbf{y}$. The fields **E** and **B** are interrelated by Maxwell's equations. A closer approach to modeling actual light is attained if the pairs of charges are anti-particles of each other,[3] and are fermions, since in that case the dipoles can have spin of 1 and magnetic moments of 0, as is the case for photons. Clearly, such a model demands coherence in location and vibration.

Nothing seems to be known, even qualitatively, about the length of time required for a photon to be absorbed or emitted by a massive system.

D.3 MAGNETIC-RESONANCE ASPECTS

When a single photon interacts with an electron, it can transfer its spin angular momentum to the electron spin, under the right conditions (e.g., to change M_S from $-\frac{1}{2}$ to $+\frac{1}{2}$). This is a magnetic-dipole transition, and there is no change in parity. Since the electron spin has a unique sense of precession, it is only σ_+ that can accomplish this resonance action: EPR. After the event, the photon is gone (absorbed). The opposite change in M_S occurs when a photon is created, deriving its spin angular momentum from the electron spin.

Similarly, on meeting a proton (^{1}H), a photon can transfer its spin angular momentum to the nuclear spin (e.g., to change M_I from $+\frac{1}{2}$ to $-\frac{1}{2}$) only if it is of type σ_-: NMR, since the proton magnetic moment (and hence precession sense) is of opposite sign to that of the electron-spin moment.

We can benefit by next considering the impact of a beam of many photons of equal frequency traveling parallel to each other, say, along a direction defined to be **z**. Let the beam impinge on an atom (say, ^{1}H^0). Consider the photons to be linearly polarized, so that there equal numbers of σ_+ and σ_- types. First, we consider the two possible situations:

1. *Zeeman Field* **B** *Perpendicular to* **z,** *say, along* **x.** Then excitation magnetic field $\mathbf{B}_1$ is in the xy plane, and the direction of this polarization vector does not affect the physical situation. As we stated, one can resolve $\mathbf{B}_1$ into two equal-amplitude counter-rotating circularly polarized fields, that is, both σ_+ and σ_- photons are available. One or the other type of the photons can flip either spin in the atom, that is, can cause a $|\Delta M| = 1$ alias σ-type transition. Its angular momentum is transferred to the spin system.

2. **B** *Parallel to* **z.** Here there are two subcases to be considered. In both of these, the photon angular-momentum component is well defined along **z.** Then

    a. $\mathbf{B}_1 \parallel \mathbf{x}$, *and also* $\mathbf{B}_1 \parallel \mathbf{B}$. No photon angular momentum can be transferred to the spin system, *but* energy can be transferred by flipping both spins simultaneously. This is an $\Delta M = 0$ alias π-type transition.

    b. $\mathbf{B}_1 \parallel \mathbf{y}$, *and also* $\mathbf{B}_1 \perp \mathbf{B}$. No spin flips can be accomplished by the photons.[5] Nothing happens! We can consider emission of photons in the same manner, and we can consider the situation when $\mathbf{B}_1$ is produced in an AC current-carrying coil, or in a resonant metal cavity. The ideas described above still apply.

Besides these single-photon effects, there can be multiple-photon action, in which combinations of several σ_+ and σ_- photons enter or leave a material system (see Fig. D.1). Thus, for example, $2\sigma_+ + 1\sigma_-$ photons of equal energy may enter an electron-spin system, each with one-third of the spin Zeeman splitting energy, $h\nu/3$. In all such multi-photon absorption events, there must always be one 'extra' σ_+ photon, in the case of electron spin systems. With protons, it is σ_- which plays the special role.

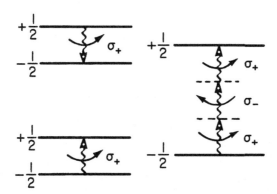

FIGURE D.1   Single-photon $S = \frac{1}{2}$ EPR transitions: absorption (lower left) and emission (upper left), and an example of a multi-photon absorption (at right). Here, linearly polarized excitation magnetic field $\mathbf{B}_1$ is perpendicular to the static Zeeman field **B**.

The π-type of transitions referred to above have been shown [14] to occur when modulation of the magnetic field **B** is applied, as often is the case in EPR cw spectroscopy.

REFERENCES

1. J. Leach, M. J. Padgett, S. M. Barnett, S. Franke-Arnold, J. Courtial, *Phys. Rev. Lett.*, **88**(25), 257901(1–4) (2002).

2. J. D. Jackson, *Classical Electrodynamics*, 3rd ed., Wiley, New York, NY, U.S.A., 1999, Chapter 9.

3. A. Hobson, *Am. J. Phys.*, **73**(7), 630–634 (2005).

4. J. D. Jackson, *Classical Electrodynamics*, 3rd ed., Wiley, New York, NY, U.S.A., 1999, Section I.2.

5. L. Mandel, E. Wolf, *Optical Coherence and Quantum Optics*, Cambridge University Press, Cambridge, U.K., 1995, Section 12.11.5.

6. A. S. Davydov, *Quantum Mechanics* (transl. D. ter Haar), Pergamon Press, Oxford, U.K., 1965.

7. A. I. Akhiezer, V. B. Berestetskii, *Quantum Electrodynamics*, Interscience (Wiley), New York, NY, U.S.A., 1965, pp. 11ff.

8. C. Cohen-Tannoudji, J. Dupont-Roc, G. Grynberg, *Atom-Photon Interactions—Basic Processes and Applications*, Wiley, New York, NY, U.S.A., 1992.

9. A. T. O'Neill, I. MacVicar, L. Allen, M. J. Padgett, *Phys. Rev. Lett.*, **88**(5), 053601(1–4) (2002).

10. J. W. Elbers, P. Sommers, *Astrophys. J.*, **441**, 151–161 (1995).

11. S. D. Ganichev, E. L. Ivchenko, W. Prettl, *Physica E*, **14**, 166–171 (2002).

12. A. Abragam, *The Principles of Nuclear Magnetism*, Oxford University Press, Oxford, U.K., 1961, p. 4.

13. J. P. Wesley, *Phys. Essays*, **16**(4), 499–502 (2003)

14. M. Kälin, I. Gromov, A. Schweiger, *J. Magn. Reson.*, **160**(2), 166–182 (2003).

NOTES

1. The term 'massive' here means that the particle mass is non-zero.

2. Technically speaking, the word 'vacuum' implies absence of all massive particles *and* absence of electromagnetic radiation (non-virtual photons) [8]. Thus the phrase 'speed of light in the vacuum' presents a conundrum. For our purposes, we'll adjust the meaning to allow one and only one photon into the 'vacuum'.

3. The set of photons, unlike electrons, obeys Bose-Einstein statistics. Photons, as stated above, are one of the set of *vector bosons*, and are deemed to be their own antiparticles, and hence are called *truly neutral*; they are assigned negative "charge parity" [6, p. 200], as determined 'experimentally' (the potentials of the electromagnetic fields change sign under charge conjugation, and so must all electrical charges within the system). The latter property corresponds to an eigenvalue of the charge-conjugation

operator $\hat{\Lambda}$ (which commutes with the hamiltonian, linear momentum and angular momentum operators) [8, p. 283].

4. The oscillating linearly polarized magnetic excitation field $\mathbf{B}_1$ commonly used in magnetic resonance work is made up of large sets of photons of these two types.

5. One may ask what happens to the photon angular momentum (helicity) in this case. Here we note that the photon angular component along $\mathbf{z}$ is effectively zero, with certainty, and hence we can be sure that the other two components cannot be well defined.

APPENDIX E

INSTRUMENTATION AND
TECHNICAL PERFORMANCE

This appendix deals with the technical aspects of EPR spectroscopy. We outline the important parts of modern spectrometers, relegating discussion of factors affecting use of these machines to Appendix F. There is a great deal of detail that cannot be covered here, but happily such material is readily available, and can be found (as usual in this text) via the FURTHER READING section at the end of this appendix.

E.1 INSTRUMENTAL: BACKGROUND

EPR absorption has been detected from zero magnetic field[1] to fields as large as 100 T ($\nu = 3.0 \times 10^{12}$ Hz for g or z). At one extreme, a sensitive magnetometer used by geologists utilizes EPR absorption to measure the Earth's magnetic field, which is only ~ 0.05 mT. At the other end, pulsed electromagnets, designed to avoid destructive power dissipation, have been used at ultra-high fields; laser sources in the far-infrared are used to provide the required excitation fields $\mathbf{B}_1$.

There are various considerations that limit the choice of the radiation frequency ν. One primary concern is that of sensitivity. This requirement, at first sight, appears theoretically to dictate that the excitation frequency ν be as high as possible, since the sensitivity of an EPR spectrometer has routinely been thought to increase approximately as ν^2.

Happily, the Eatons research group has pursued this complicated question during the last decade (see Ref. 3 and papers cited therein). As it turns out, if one assumes

Electron Paramagnetic Resonance, Second Edition, by John A. Weil and James R. Bolton
Copyright © 2007 John Wiley & Sons, Inc.

that resonator resistance, not sample loss, dominates the signal-to-noise ratio (S/N) and that all amplifiers in the signal detection system have the same noise figures, and if the sample size and the resonator size are constant, then the EPR S/N increases as $v^{7/4}$. On the other hand, if the sample and resonator both scale inversely with frequency, then the S/N scales $v^{\frac{1}{4}}$; that is, the S/N is better at lower frequency. If the sample size is constant and the resonator size scales inversely with frequency, then the S/N increases as $v^{11/4}$.

However, several factors place a limit on the frequency employed, so that frequencies in the microwave region remain favored:

1. One limitation is the size of the sample; at frequencies of $\sim$30–40 GHz, typical microwave resonant-cavity dimensions are of the order of a few milli-meters.[2] Thus, although the sensitivity *per unit volume* is high, the sample volume is limited to about 0.02 mL.

2. High frequencies require high magnetic fields, homogeneous over the sample volume. With electromagnets, sufficiently homogeneous magnetic fields in excess of 2.5 T are difficult to produce. Superconducting magnets may produce fields of the order of 10 T, but they are expensive.

3. The small size of microwave components for high frequencies makes their fabrication a costly process; small imperfections can give rise to relatively high dielectric losses. Dielectric losses are seldom the problem with the micro-wave components. There are reflection problems and conductive losses as the frequency increases. For example, often oversized or corrugated waveguide (e.g., see Ref. 5 for such technical details, regarding a far-IR EPR spec-trometer) is used in what is sometimes called 'quasioptical' designs. Also, one of the key points is that noise figures of amplifiers generally is poorer at higher frequencies. In addition, a non-trivial matter is that tube wall thick-ness makes the filling factor decrease even if the dimensions nominally scale with inverse frequency. Sample size, resonator size and waveguide size happen to be very convenient at X band. These and other factors have resulted in a choice of $\sim$10 GHz as the working frequency of most commercial spectrometers.

Several quantum-theoretic factors also affect the choice of v [6]. First and fore-most, the various terms in the spin hamiltonian (Sections 6.6 and 6.7) are either field-dependent or not, that is, their relative importance in contributing to the EPR spectrum varies with B and hence with v. Thus it is advantageous to measure some parameters at relatively low fields and others at high fields. Resolution between peaks arising from different chemical species increases as $B(v)$ increases.

Some materials, such as aqueous systems, including many of biological import-ance, absorb microwaves via dielectric (rather than magnetic) processes. Such dielectric losses are usually highly frequency dependent and may lower the resona-tor quality factor Q and thus decrease the capability of detecting the very small losses of energy caused by EPR absorption. Figure E.1 illustrates the frequency

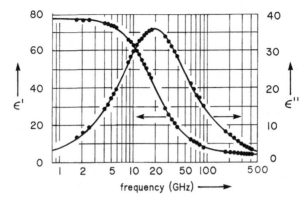

FIGURE E.1 Frequency dependence of the relative permittivity (dielectric constant) of liquid water at 25°C. Here ε' (v) is the dielectric dispersion and ε'' (v) is the loss. [After J. Barthel, K. Bachhuber, R. Buchner, H. Hetzenauer, *Chem. Phys. Lett.*, **165**, 369 (1990).]

dependence of the dielectric properties of water. Here ε' is the real part (dispersion) and ε'' is the imaginary part (absorption) of the complex relative permitivity $\varepsilon/\varepsilon_0$, that is, of the dielectric constant (Section F.3.4).

In early cw NMR spectroscopy it was a matter of preference as to whether one fixed the field and varied the frequency, or vice versa. This was a consequence of the ease of continuously and precisely varying a low-noise radiofrequency source over the approximately 20 ppm frequency range required for most ^{1}H NMR spectroscopy. Modern NMR spectrometers almost exclusively use pulsed Fourier-transform methods, with fixed field and frequency. For most cw EPR, the field/frequency spectral width is much too large relative to the bandwidth of the EPR resonator to permit frequency-swept spectra. Consequently, the microwave frequency usually is kept constant, at the resonant frequency of a high-Q resonator, and the magnetic field is swept. Older EPR spectrometers, especially those operating in the X band and Q band, used klystrons as the stable frequency source. Many of these spectrometers are still in use, and the terminology developed for klystron sources (such as the mode pattern) still dominates the description of spectrometer operation. [See Ref. 7, Section 3C and Ref. 8 for some discussion of klystrons and some other aspects of this generation of spectrometers (also, see Section E.2.2).] The most commonly used microwave sources in modern commercial spectrometers are stabilized Gunn oscillators, operating at fixed frequencies. Owing to the increased use of resonators with high Q values, a microwave source with lower phase noise than klystrons is required in order to fully benefit from the increased sensitivity offered by such resonators. Carefully designed Gunn oscillators typically deliver about two orders of magnitude improvement in phase noise compared to klystrons.

Modern spectrometers achieve a sensitivity approaching the theoretical limit imposed by thermal noise and the noise figures of amplifiers and detectors (Section F.3.1). There exist both continuous-wave (cw) and pulsed EPR spectrometers. We first examine a traditional cw EPR instrument.

E.2 CW EPR SPECTROMETERS

A typical cw spectrometer is shown blocked out in Fig. E.2 according to the function of groups of components. First, the magnet system provides a stable, homogeneous and magnetic field **B** linearly variable within the desired range. The region labeled 'source' contains those components that produce the excitation electromagnetic waves (i.e., **B**$_1$). The cavity system shown holds the paramagnetic sample, and directs and controls the microwave beam to and from the sample. The modulation (taken as functioning at 100 kHz) and detection systems monitor, amplify and record the signal. A central concept is that the magnetic-field modulation encodes spectral information on the microwave signal. The phase-sensitive detection system extracts this encoded information Also shown in this figure is a useful feature, the reference arm (sometimes called 'bypass arm'), present in many spectrometers. This takes microwave power from the waveguide ahead of the circulator and restores it, with adjusted power level and phase, after the circulator (and

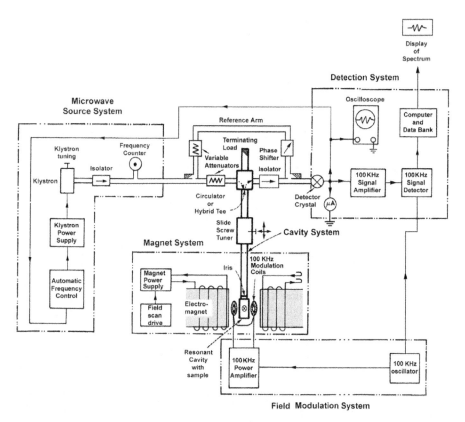

FIGURE E.2   Block diagram of a typical X-band EPR spectrometer employing 100 kHz phase-sensitive detection. In a modern spectrometer the computer is used not only for data acquisition and analysis but also to control essential parameters of the spectrometer.

resonator). With suitable settings, this reference arm allows not only for appropriate biasing of the power level at the detector (even when the power reflected from the resonator is very small) but also the choice of whether the absorption or dispersion (Sections E.2.5 and F.3.5) signal from the spin system is detected. The functions of the individual components within each of the blocks are considered first.

E.2.1 Magnet System

The usual source of the static magnetic field **B** causing the desired splittings of the spin energy levels is an electromagnet (or, for low fields, an air-core magnet [9], which can be a solenoid). This field should be stable and uniform over the sample volume; field variations should be kept within ± 1 μT to resolve very narrow EPR lines and to obtain the correct lineshape [10]. The north-south polarity (direction) of **B** between the pole caps is generally of no importance. Usually, one desires to have **B** at right angles to the excitation field $\mathbf{B}_1$ (Sections 4.6 and C.1.4; see also Ref. 11) at the sample, hence the resonator must be oriented in the magnet so as to achieve this configuration. However, note that parallel-field ($\mathbf{B} \parallel \mathbf{B}_1$) spectroscopy is becoming more practiced (see Section 1.13).

Stability is achieved by energizing the magnet with a highly stable regulated power supply. Since the dependence of magnetic field on the magnet current of an iron-core magnet is inherently non-linear and subject to hysteresis, any serious EPR studies *must* rely on measurement and control of the field to a high degree. The exceptions to this general statement are the air-core magnets used for low-frequency EPR. The simplest procedure is to place a Hall-effect probe, which gives an output voltage proportional to field B, at a fixed location in the gap. It should be thermostatted or temperature-compensated to avoid drift with changing probe temperature. A feedback system adjusts the magnet current to maintain a fixed Hall-probe voltage. Typically, provision is made in the Hall-probe control system for linear variation (scanning) of the field over the region of EPR absorption. The scan should be very reproducible and linear to allow accurate measurement of spectral parameters. Modern EPR spectrometers use digital-to-analog converters with as many as 24 bits to control the magnet current.

Field homogeneity is, of course, essential in that linewidths of some samples are less than 0.01 mT. Manufacturer's specification for magnets with pole diameters of 30 cm, designed for EPR usage, cite a homogeneity within ~ 1 μT within a central volume of ~ 1 mL. The magnetic field at the sample may be measured by means of a proton NMR probe placed beside (or even inside) the microwave cavity.[3] Detection of the NMR signal and measurement of the corresponding resonance frequency with a counter permit a measurement of the magnetic field to about one part in 10^8. Measurement of B at two or three field points calibrates the entire spectrum if the field scan is truly linear. If only field differences between lines are required, one may use a dual cavity: a single cavity resonating in an appropriate mode to enable two sample ports [7, p. 188]. One of the samples can then be a standard that gives a known multiple-line spectrum (Section F.4.3).

Typical EPR absorption lines are shown in Figs. F.1a and F.2a; the corresponding first-derivative presentations are given in Figs. F.1b and F.2b. The derivative signal displayed may be positive or negative on the left-hand side of the scan, depending on the modulation phase adjustment of the phase-sensitive detector or the polarity of the recorder connection. *The sense of the recorded signal is irrelevant to its interpretation.* (One is the mirror image of the other.) However, it is present practice to present EPR spectra such that the lowest-field line goes in the positive direction at its low-field side. Recorded spectra of either phase are found in the literature. Further details may be found in Appendix F.

Development of very high-field magnets is continuing. Direct-current (dc) fields to ~30 T have been produced, and pulsed fields to ~100 T are a reality [12,13]. Very high magnetic fields and the associated EPR spectrometer components are especially expensive specialty items and hence are often housed in national laboratories, such as those in Grenoble (France), Tallahassee (Florida, U.S.A.) and Sendai (Japan). EPR studies at such fields and concomitant high-excitation frequencies have been widely reviewed [6,12,14–21].

With the advent of such high B fields, special effects, such as field-induced phase transitions in crystal structure, are now known. Clearly, investigations of systems with transition frequencies greater than 500 GHz are becoming feasible.

For some purposes, installation of quantitative and adjustable B-field gradients is required: that is, for static magnetic susceptibility measurements by the Faraday method and for EPR imaging work (Section 13.12).

E.2.2   Radiation Source

At low frequencies ($\nu < 1$ GHz), standard rf oscillators can provide the excitation fields $\mathbf{B}_1$. In the microwave region ($1 \leq \nu \leq 100$ GHz), various special tubes are used. These include backward-wave oscillators, special diodes and triodes (some solid-state devices, such as Gunn diodes), klystrons, magnetrons and traveling-wave tubes. Of these, klystrons were the most commonly used sources for many years, but they have now been largely replaced by stabilized Gunn diode sources in commercial spectrometers. In the far-infrared region ($100 \leq \nu \leq 1000$ GHz), harmonics generated from klystrons used in conjunction with backward-wave oscillators have been used, as well as certain gas lasers [6,12,14,15]. In this book, we discuss in some detail only the klystron.

We note that these sources are all coherent ones; that is, their sinusoidal output is in principle uninterrupted by random changes. This property is especially important in both cw and pulsed experiments (Chapter 11).

Turning now to more detail, a *klystron* is a vacuum tube that can produce microwave oscillations centered on a small range of frequency; the output as a function of frequency is referred to as the *klystron mode*. It is frequency-stable at the resonant frequency of a built-in evacuated metal cavity (which, however, is temperature-sensitive). A klystron may operate in any of several modes; the mode corresponding to the highest-power output is usually the one utilized. The mode (power output vs. frequency) may be displayed on an oscilloscope if one ramps the klystron output frequency over the range of the mode.

The frequency of the nearly monochromatic output radiation is determined by mechanical tuning of the klystron cavity and by the voltages applied to the klystron. These adjustments allow one to vary the center frequency of a given mode over a limited range. It is desirable that the klystron frequency be very stable; hence fluctuations of the klystron temperature or of applied voltages must be minimized and mechanical vibration suppressed. For accurate EPR work, the output frequency must be measured to at least six decimal places. Fortunately, electronic counters are now readily available for frequencies in excess of 100 GHz. For some purposes, the phase of the outgoing sinusoidal radiation should also be constant.

As stated, modern EPR spectrometers have Gunn diode microwave sources whose power output is flatter with frequency than is the output of a klystron, so that the operator no longer observes a 'mode pattern'. There is some variation in the flatness of the power display with frequency in the tune mode because of reflections from small impedance mismatches at connectors, and so on. The absorption of power by the resonator is still a 'dip' in the display.

Modern pulsed EPR (discussed in Section F.2) requires phase-coherent (i.e., constant frequency) microwave pulses, very much shorter than those used in NMR. Consequently, fast switches are needed to shape the $\sim 10-100$ ns pulses. To excite the maximum extent of the EPR spectrum, it is common to use 1 kW traveling-wave-tube (TWT) amplifiers. For some applications, even higher-power amplifiers are needed. The detection system has to be protected from the high-power pulses, using either a limiter or a switch. Fast phase shifters are needed to cancel unwanted signals. The bandwidth of the detection system has to be large enough to encompass all frequency components of the signal. Current commercial pulsed EPR spectrometers achieve 4 ns resolution and use 250 MHz analog to digital converters. The large bandwidth in the detection system results in more noise than in continuous-wave (cw) spectrometers.

One other source of microwaves, the gyrotron, is a cyclotron-resonance maser that has been operated at 140 GHz for EPR spectroscopy [22] at $B = 5$ T. It can work at high power (90 dBm, i.e., 1 MW) and short pulse output (microseconds).

Despite much research on solid-state microwave generators, their frequency stability and inherent noise make them inferior to the best present-day klystrons. However, for pulsed EPR spectroscopy, their characteristics are often adequate over the required short periods, making them preferred sources. Over such brief pulses, one is not too concerned with random frequency variations of the source. When required, very short (ns) pulses can be amplified by a traveling-wave tube, the resulting power level is sufficient to excite a substantial part of the entire EPR spectrum. Pulsed EPR is discussed in Section E.3.

E.2.3 Microwave Transmission

Microwaves can be transmitted effectively by use of conduits, which include coaxial cables, striplines or metal waveguides; modern microwave systems employ all three

TABLE E.1   Traditional Microwave Frequency Bands and Magnetic Fields for $g = 2$ EPR

Band Designation	Band Range	Typical EPR Frequency ν (GHz)	Typical EPR Field B (mT)
L	0.390–1.550	1.5	54
S	1.550–3.900	3.0	110
C	3.900–6.200	6.0	220
X	6.200–10.900	9.5	340
K	10.900–36.000	23	820
Q	36.000–46.000	36	1300
V	46.000–56.000	50	1800
W	56.000–100.000	95	3400

Sources: Adapted from S. Y. Liao, *Microwave Devices and Circuits*, 3rd ed., Prentice-Hall, Englewood Cliffs, NJ, U.S.A., 1990; elaborated in S. S. Eaton, G. R. Eaton, *Magn. Reson. Rev.*, **16**, 157 (1993), Table 1.

types. However, waveguides are usually used to transmit microwaves to and from the cavity resonators [7, Chapter 2], and coaxial cable is used for most loop-gap resonators and other lumped-circuit resonators, including the Bruker FlexLine split-ring resonators. Waveguides are manufactured in distinct sizes. The popular 'X band' uses 12.7×25.4 mm-OD rectangular brass pipe (these are, in fact, defined in inches). This allows transmission with small losses for frequencies of 8–12 GHz. Two other waveguide sizes are used less frequently: K band for the range 10.9–36.0 GHz and Q band for the range 36.0–46.0 GHz. Table E.1 lists the traditional names for the microwave frequency bands along with the magnetic-field values appropriate for $g = 2$ EPR detection. The choice of $\sim$10 GHz operating frequency represents a generally good compromise between sensitivity, convenience and cost of microwave components.

E.2.4   Coupling of the Source to the Resonator

Three important components are involved in transmission of the microwave source power output to the resonator holding the sample. These are the isolator, the attenuator and the circulator. We first consider the case of but a single source (however, see Section E.2.8).

Serious perturbations of the microwave source frequency may occur if there are significant backward reflections of microwave energy from the rest of the system. These can be minimized by using an *isolator*, which is a non-reciprocal device; that is, its properties are not identical for waves traveling in the forward and in the reverse directions. It readily passes microwave energy in the forward direction, while it strongly attenuates any reflections back to the microwave source. Isolators commonly used in modern spectrometers are three-port circulators with one port terminated with a matched load. The circulator uses a ferrite rod situated in a magnetic field produced by magnets in the device to steer microwaves in the direction 1 to 2 to

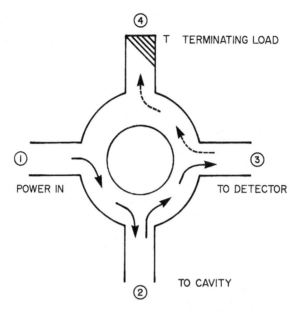

FIGURE E.3   A four-port microwave circulator showing the directions of microwave transmission among the several arms.

3 to 4 to 1 [7, Section 4C]. Thus, it is a non-reciprocal device in that energy incident on port 2 does not go to port 1, but rather goes to port 3. If port 3 is efficiently terminated, then reflections at port 2 are isolated from port 1.

The *attenuator* adjusts the level of the microwave power incident on the sample. It contains an absorptive element, and corresponds to a neutral-density filter in light-absorption measurements.

A *circulator*, like the isolator, is a non-reciprocal device; it serves to direct the microwave power to the cavity and simultaneously to direct the power reflected from the cavity to the detector. The operation of a four-port circulator is indicated in Fig. E.3. A four-port circulator is composed of two 3-port circulators combined in one unit to achieve greater directional behavior. A terminating load on the fourth arm serves to absorb any power that might be reflected back from the detector arm. Because a magnetic field is inherent in the operation of a circulator, it is affected by fringe fields, so that the circulator and isolator have to be kept away from the main magnet.

E.2.5   Resonator System

The heart of a typical EPR spectrometer is a device called a 'resonator', which contains the sample. Achievement of resonance in an EPR experiment does not *require* the use of a resonant cavity. For sufficiently strong EPR samples, a transmission system can be used. However, in practice, only a few very high frequency EPR spectrometers use transmission systems, and they usually do so because of temporary constraints of the magnet geometry during prototype construction phases. Although it has been conventional in EPR to refer to the resonator as the cavity, many non-cavity resonators are becoming common, especially for frequencies other than X band (and for pulsed EPR, even at X band). Numerous other resonant devices also permit EPR detection. A few of these (helical resonator, loop-gap resonator, etc.) are mentioned at the end of this section. For very high frequencies, Fabry-Perot resonators [23,24] have been utilized.

The reader may well be familiar with the phenomenon of acoustic resonance in organ pipes or in nearly enclosed vessels. The latter too are referred to as 'cavity resonators'; their properties were described in detail more than a century ago by Helmholtz [25]. Reflection of sound waves from the walls leads to destructive interference at wavelengths that are not submultiples of one dimension of such resonant cavities. The frequency at which one-half of a wavelength corresponds to a cavity dimension is called the *fundamental resonant frequency*; this frequency increases with decreasing dimensions of the cavity. A cavity *may* be excited to produce more than one type of standing-wave pattern or resonant 'mode'. The energy density associated with a traveling wave (say, in a waveguide) is usually small; by comparison, considerable acoustical energy may be stored in the *standing* waves of a resonant cavity. One can readily observe acoustical resonance because the wavelengths of sound are in the range of a few centimeters to a few meters.

The same phenomenon also occurs with electromagnetic waves. For microwaves, the wavelength is also typically of the order of centimeters (3 cm at X band). Hence the dimensions of a resonant cavity are convenient. The cavity walls are of highly conducting metal, and the effective electric current (associated with the electromagnetic waves in the cavity) flows within only a thin portion (skin effect) on the inner surface adjacent to this space [26,27]. The useful cavity modes, which are selected at the design stage, usually are those in which the microwave field B_1 is perpendicular to the external magnetic field B. Field B_1 is perpendicular to the current flow in the walls of the resonator. As stated, mathematical expressions and visual images of the simultaneous electric (E_1) and magnetic (B_1) standing-wave patterns for a given resonator can be derived from Maxwell's equations with suitable boundary conditions (Figs. E.4 and E.5) [7, Chapter 5; 28,29]. The locations of the maxima of the E_1 and B_1 fields in the cavity are different; their relative location depends on the mode at hand. To be useful for EPR, a cavity mode should (1) permit a sufficiently high energy density, (2) allow placement of the sample at a maximum of B_1 (a minimum of E_1) and (3) usually have B_1 perpendicular to the static magnetic field B.

The nomenclature for the cavity resonant modes (see below) is easily accessible, for example [30,31], and is intimately associated with the solutions for Maxwell's equations giving the dynamic electric and magnetic fields within volumes contained via various possible boundary conditions (metal walls).

In paramagnetic resonance, it is, of course, the field $B_1(t)$, oscillating at frequency ν, that causes the spin transitions. Generally, the direction of B_1 is along one direction (linearly polarized), perpendicular to the magnetic field B. The actual value of

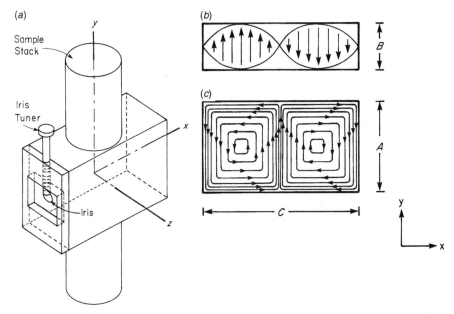

FIGURE E.4 A rectangular parallelepiped TE_{102} microwave cavity. (*a*) Cylindrical extensions above and below the cavity prevent excessive leakage of microwave radiation out of the cavity and act as positioning guides for the sample. The microwave energy is coupled into the cavity through the iris hole at the left. This coupling may be varied by means of the iris screw. (*b*) The electric-field ($\mathbf{E}_1$) contours lie in the *xy* planes. One half-wavelength in the **x** direction corresponds to the shortest distance between points of equal field intensity but of opposite phase. (*c*) Magnetic-field ($\mathbf{B}_1$) contours in the *xz* plane. Length *a* is approximately one half-wavelength and *c* is exactly two half-wavelengths. The length *b* is not critical but should be less than one half-wavelength.

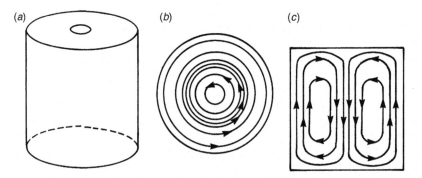

FIGURE E.5 Cylindrical cavity operating in the TE_{011} mode. (*a*) Cavity; the height and diameter of the cylinder govern the resonant frequency. Cylindrical extensions above the cavity prevent excessive leakage of microwave radiation out of the cavity and act as positioning guides for sample insertion. (*b*) Electric-field ($\mathbf{E}_1$) contours. (*c*) Magnetic-field ($\mathbf{B}_1$) contours.

B_1 at the sample can be determined by various methods, as discussed in Section F.3.5.

Cavities commonly employed for EPR spectrometers are shown in Figs. E.4a and E.5a. The modes are referred to as *transverse electric* (TE) and *transverse magnetic* (TM); the subscripts designate the number of half-wavelengths along the several dimensions (see Problem E.3).

For the rectangular-parallelepiped cavity of Fig. E.4a, the mode is TE_{102}, since there are one and two half-waves along **a** and **c**; the fields do not vary along **b**. The spatial distribution of electric and of magnetic fields for the TE_{102} cavity may be seen from Figs. E.4b and E.4c. Such a cavity permits the insertion of large samples (as long as these have a low dielectric constant) without drastic reduction in the energy density. It is especially useful for liquid samples, which may extend through the entire height of the cavity.

A most common and convenient general-purpose EPR resonator in fact is the TE_{102} resonant cavity, because of its high efficiency (Q), adaptability and ease of sample access. Such a cavity usually has smaller background signal than the other resonators mentioned here, because the B_1 is lower at the walls where many background signals originate.

The cylindrical cavity of Fig. E.5a can have a relatively high energy density when operated in the TE_{011} mode shown in Figs. E.5b and E.5c. This energy density is typically higher by a factor of at least 3 compared to that in a TE_{102} cavity under comparable conditions. This cavity is especially useful for observing transitions in gaseous systems since very large (25 mm at X band) diameter sample tubes may be used.

The frequency characteristics of the resonator can be displayed by temporarily sweeping across the resonator bandwidth with, say, a sawtooth voltage, and detecting the microwave power reflected from the resonator. If the narrow resonant frequency range of the latter is within the frequency range swept, then the resonator absorption shows up on an oscilloscope (Fig. E.2) as a sharp dip representing the resonator's power absorption. When the microwave source is a klystron, the resonator dip will be within a non-linear display of power versus frequency that is called a 'klystron mode pattern'. The output of a Gunn diode is more nearly flat with frequency.

The sharpness of response of any resonant system is commonly described by a figure of merit, a quality factor, universally represented by the symbol Q, defined in Appendix F (Eqs. F.7). That definition implies that for a fixed frequency, efficient energy storage (and hence Q) increases with cavity volume (i.e., going to higher modes, rarely used). With the bandwidth definition $Q \equiv \nu/\Delta\nu$ in mind, one can say that the narrower the bandwidth (the higher the Q), the sharper the dip in the cavity display. This definition also implies that Q may be increased by decreasing energy losses from currents flowing in the cavity walls or sample. Heavy silver plating is effective in maximizing Q; a final thin gold plate prevents deterioration of the silver. The Q is lowered if a sample having a high dielectric constant extends into regions of appreciable microwave electric field. The various factors (including the EPR energy absorption itself) that affect Q are dealt with in Section F.2.6.

In any transmission line, the reflection is a minimum if the line is terminated with a device having the characteristic impedance of the line. When this requirement is met, the energy transmission is a maximum. The coupling of microwaves into (and out of) a reflection cavity is achieved by means of a small hole, called the *iris*, in the cavity wall. Transmission cavities (Fig. 1.3a) have two irises, one at each end. An adjustable screw with a conductive tip adjacent to the iris permits optimal impedance matching (Fig. E.4a). The setting of the iris screw is determined empirically and depends on the size and nature of the sample in the cavity. The iris serves a function analogous to that of an impedance-matching transformer in a typical electrical circuit. Most irises are used in conjunction with an adjacent tuning device, say, a dielectric screw. Its adjustment, to provide optimal coupling and highest EPR sensitivity, has been amply discussed [7, pp. 143–151; 32].

One can think in terms of either the power going to the cavity and the power reflected back from it, or of the incident and reflected microwave magnetic fields B_1 (or electric field E_1). The reflected B_1 carries with it the information about the sample's EPR behavior (dispersion χ' and absorption χ''; see Fig. 10.6) [33].

The frequency output of the microwave source needs to match the narrow-band resonator frequency to attain optimal energy storage in the latter, and thus achieve adequate power to excite the EPR transitions of the sample. To avoid drifting of the frequency, it is useful to make use of an automatic frequency-control (AFC) system utilizing an electronic feedback loop to keep the source frequency locked to the center frequency of the resonator. Furthermore, the power loss to the sample arising from the EPR energy absorption is accompanied by a cavity frequency shift (dispersion; see Section F.3.5), which is removed by the AFC.

A useful elaboration of the usual resonator is the bimodal cavity, which can be used, for example, to measure the EPR dispersion [34]. The design and use of a $TM_{110} \times TM_{110}$, featuring orthogonal modes, is well described, and yields a better signal-to-noise ratio (S/N) than do analogous single-mode cavities.

The 'dielectric cavity' consists of a diamagnetic material, such as sapphire, to raise the filling factor by concentrating the excitation field B_1 over the volume of the sample. Just as electrical currents in metal surfaces of a cavity or loop-gap resonator result in the B_1 field at the sample, the dielectric properties of sapphire or other material with high dielectric constant result in a B_1 field directed along the cylindrical axis of the dielectric [35].

Development of a simple ferroelectric ($KTaO_3$ single crystal) cavity insert, placed around the sample to be studied, has been reported [36]. This enhances the EPR sensitivity by an order of magnitude, for various types of resonators.

An alternative to the usual EPR cavity is the *loop-gap* resonator, alias split-ring resonator [37–39]. This device basically consists of a series of gaps in the inner of two concentric metal cylinders the outer cylinder being the shield. Careful comparison between the performance characteristics of such a resonator with those of a TE_{102} cavity disclose that the three-loop two-gap resonator is a feasible alternative at X band [4]. The loop-gap devices have excellent filling factors, valuable for non-saturable samples and in pulse work. Their low Q is an advantage for dispersion studies. Such split-ring resonators can also be used to detect spins at surfaces of a sample, for instance, to perform in vivo EPR at 1 GHz [40]. This use is analogous to the surface coils commonly used in magnetic resonance imaging. The EPR signal is detected in the microwave field that is 'outside' the resonator.

The microwave helical resonator (slow-wave structure [41]), although lacking somewhat in sensitivity (low Q), is useful when a sharp frequency response is to be avoided [7, Chapter 14].

For EPR imaging, there is the developing need to encompass especially large 'samples', some of them living. For example, parallel-coil resonators (11 single loops) for time-domain imaging of biological objects are now a reality [42].

E.2.6   Field-Modulation System

A serious disadvantage of the spectrometer outlined in Figs. 1.3a and E.2 is the contribution of the noise components (principally at low frequencies) to the output signal. The phase-sensitive amplitude detection technique employed in the spectrometer of Fig. E.2 utilizes small-amplitude sinusoidal magnetic-field modulation (see Note F.2) superimposed on the dc field **B**. The effect of the modulation is depicted in Fig. E.6. This technique allows (1) the amplification of the EPR signal using ac techniques, (2) elimination of most of the noise-contributing components and (3) enhanced spectral resolution. Modulation of the magnetic field **B** at the commonly used frequency of 100 kHz is achieved by placing small Helmholtz coils on each side of the cavity along the axis of the static field. The cavity walls in the region of the modulation coils must consist of very thin conductive layers the cavity walls to allow penetration of the 100 kHz magnetic field, yet these layers must be thick (several skin depths) enough to retain the microwaves, and must not vibrate. Usually only the component at modulation frequency ν_m (e.g., 100 kHz) of the absorption signal is detected, even though higher harmonics are present and can give improved spectral resolution of multi-line spectra, and have other uses in the study of spin dynamics. See the discussion in Section F.2.1 regarding Ref. 14 cited therein. Note also the occurence of sidebands, as treated in Section F.3.7.

The magnitude B_m of the modulation magnetic field at the sample may be measured by inserting a suitably calibrated pick-up coil into the appropriate region in the resonator.

The relative advantages and disadvantages of using field modulation for detection of EPR have been discussed [43,44]. The magnetic field modulation and phase-sensitive detection at the modulation frequency improves the S/N and stabilizes the baseline. However, other problems can be created by field modulation. For instance, this technique can cause lineshape distortion, as discussed in Appendix F. One approach to an alternative detection scheme is the newly developed electron paramagnetic rotary resonance scheme, using two incident microwave sources whose frequency difference matches the precession frequency of the magnetization in the rotating frame (Chapters 10 and 11).

E.2.7   Coupling of the Resonator to the Detector

As shown in Fig. E.2, the signal from the resonator generally uses the same waveguide or coaxial cable as is used to feed the resonator. The exit wave is separated

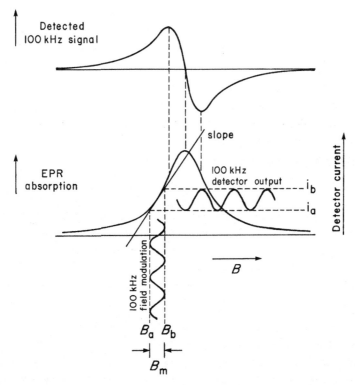

FIGURE E.6 Effect of small-amplitude 100 kHz field modulation on the detector output current. The static magnetic field B is modulated between the limits B_a and B_b. The corresponding detector current varies between the limits i_a and i_b. The upper diagram shows the recorded rectified 100 kHz signal as a function of B.

from the incident wave by using a circulator as shown in Fig. E.3 (or an equivalent device) and is sent to the detector.

Discontinuities in the waveguide and/or imperfect matching of microwave elements results in reflection of some fraction of the incident energy; hence standing waves arise. The power appearing at the detector should arise solely from reflections *originating at the microwave cavity*, since reflections from other sources tend to decrease the sensitivity of the instrument. The *slide-screw tuner* is a device used in early EPR spectrometers that allowed one to set up standing waves of amplitude and phase such as to cancel standing waves that may exist in the waveguide. It consists of a small metallic finger with adjustable depth and variable position along the waveguide, which sets the magnitude and phase of the reflected wave. Strictly speaking, the reference arm in modern spectrometers renders the slide-screw tuner unnecessary.

E.2.8 Detection System

The most commonly used detector for cw EPR is a Schottky diode crystal, which acts as a rectifier of the microwave radiation, maintaining ac signals at lower frequencies that may be modulating the microwave carrier wave. Generally such rectifier devices are composed of silicon in contact with a tungsten 'cat whisker'. They tend to be delicate, subject to damage caused by static electricity (sparks).

An array of crystals called a 'double-balanced mixer' (DBM) is also commonly used, especially in pulsed EPR, where use is made of the fact that the DBM can be constructed to give outputs [in phase, and $90°$ out of phase with $B_1(t)$]. For optimal detection, the crystal is biased via the reference arm with approximately 1 mW of power to achieve a 'detector current' of approximately 200 μA. Under these conditions, the detector current varies as the square root of the power, that is, linearly with B_1. If the sample is undergoing saturation (explained in Chapter 10), one must use an attenuator to reduce the power incident at the resonator. This could result in a detector current too small for optimum sensitivity if it were biased only with power reflected from the resonator. In the simplest possible spectrometer, this presents a problem. However, this deficiency in the simple spectrometer of can be eliminated by providing the reference arm (see Fig. E.2) referred to previously (Section E.1), which allows constant bias on the detector crystal.

A crystal detector produces an inherent noise that is inversely proportional to the frequency f of the detected signal ('$1/f$ noise', where f is the relevant frequency). The widespread use of 100 kHz as the field-modulation frequency is based on the fact that at this frequency the $1/f$ detector noise is less than that from other causes [7, Section 7C]. Cooling the detector can decrease the noise level.

The introduction of 100 kHz modulation was a major advance in early EPR spectrometers. A possible disadvantage of 100 kHz as a modulation frequency is the presence of modulation sidebands at ± 100 kHz (± 3.6 μT) from the center of an absorption line (Section F.2.3). Whereas normally these are not resolved, they do prevent resolution of very narrow EPR lines (~ 5.0 μT). In modern spectrometers, the modulation frequency can be reduced with little increase in the detector noise, so that improved resolution should be possible without loss of sensitivity. Use of modulation frequencies lower than 100 kHz is becoming increasingly common for this reason.

After rectification, the 100-kHz component undergoes amplification. At this point one can achieve still further reduction of noise by use of a mixer [45], for example, a phase-sensitive detector. This noise reduction is achieved by rejection of all frequency components except those in a very narrow band ($\sim \pm 1$ Hz) about 100 kHz. The amplified EPR signal is mixed with the output of the field-modulating 100 kHz oscillator. If the two signals are opposite in phase, the output from the system is a minimum; if they are exactly in phase, the output voltage has a maximum amplitude. Thus a phase shifter is installed to allow optimization of the output. The operation of the phase-sensitive detector is readily understood with the aid of Fig. E.7, which shows a schematic diagram for a phase-sensitive detector. An alternative to the rectifier shown is a four-diode bridge.

If the amplitude B_m of the 100-kHz field modulation is kept small ($<25\%$; 10% is good in practice if the lineshape is important) compared to the linewidth, the amplitude of the detected 100 kHz signal as the field B is scanned is approximately proportional to the slope of the absorption curve evaluated at B (i.e., at the null of the sinusoidal modulating field B_m). This can be seen from Fig. E.6. The tangent to the

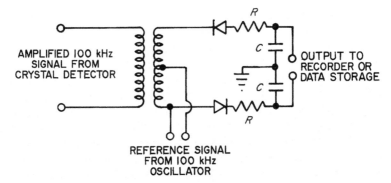

AMPLIFIED 100 kHz
SIGNAL FROM
CRYSTAL DETECTOR

OUTPUT TO
RECORDER OR
DATA STORAGE

REFERENCE SIGNAL
FROM 100 kHz
OSCILLATOR

FIGURE E.7 Schematic diagram of a phase-sensitive detector. A transformer with a tapped secondary coil can be used to combine the amplified output of the crystal detector with a fraction of the output of the oscillator driving the modulation coils. The combined signal is rectified, filtered and recorded. The output signal depends markedly on the amplitudes and relative phase of the signal and reference voltages. The time constant of the device is set by varying the value of RC, that is, the product of the resistance and capacitance.

absorption curve at a chosen point is taken to approximate the small portion of the curve traversed by the modulating magnetic field. As the total field varies between the limits B_a and B_b, the detector current varies sinusoidally between the limits i_a and i_b. When the slope of the absorption curve is zero (i.e., *at* the resonant field value and at values far from resonance), the 100-kHz component at the detector is zero. At the inflection points, where the slope has a maximum magnitude, the output signal is also a maximum. The output polarity of the phase-sensitive detector is governed by the sign of the slope; hence for small modulation amplitudes, the output signal appears approximately as the first derivative of the absorption signal. Use of modulation amplitudes approaching the linewidth leads to distortion in the observed lineshapes (Section F.2.2). Usually the output signal is filtered by a selected resistor-capacitor combination. The filter time constant (in seconds) is given by the product RC, where R is the resistance in MΩ (megohms) and C the capacitance in μF (microfarads). Modern spectrometers use multiple-stage filters so as to more sharply suppress high-frequency noise, and they increasingly use digital filters.

In some cases, modulation with more than one signal (i.e., with several frequencies) is advantageous [7, Section 6K]. Alternatively, detection at a harmonic of a single modulation frequency can be carried out. With some EPR spectra, enhanced resolution can thus be attained, say, by using third-derivative presentation. Other methods of monitoring spins use a variety of modulation methods to encode spectral information.

Another type of detection system makes use of the superheterodyne principle, that is, the mixing of a signal with the output of a local oscillator so as to produce an intermediate frequency, which is then amplified and detected. This was used for a while to improve S/N by side-stepping the $1/f$ problem. More recently, such mixing techniques have been used again for the purpose of lowering frequencies to the range where they can be directly digitized.

Hyde and his colleagues introduced impetus into the field of working with multiple (n) excitation microwave frequencies, in a series of beautiful experiments, cogently described (see Ref. 46 and preceding papers cited there as well as Ref. 47). For the most important case, $i = 2$, the validity of the 'Anderson' equation

$$(\omega_2 - \omega_1)^2 = (\gamma B - \omega_1)^2 + \gamma^2 B_1{}^2 \tag{E.1}$$

was explored, wherein the stronger cw transverse field is labeled 1. Technical details for such multi-quantum EPR spectroscopy were cogently presented and the multiple implications of such techniques explained.

In early EPR measurements, the microwave signal reflected from the sample cavity was mixed with the output of a 'local-oscillator' klystron; the latter generally operating at a frequency 60 MHz above or below that of the signal klystron. The resulting 60-MHz difference frequency produced at the output of the mixing device contains all the information of interest. At such high frequencies the $1/f$ detector noise is negligible. A low-frequency field modulation can be used without loss of sensitivity, since the noise added at the detection frequency of 60 MHz is negligible. Some samples require the use of very low levels of microwave power.

The advantage(s) of using frequency modulation, rather than field modulation, has also discussed in detail elsewhere [48], and tests using DPPH powder have been presented.

The superheterodyne system has unparalleled sensitivity for these cases. However, the complexity and cost of such a system make its use less popular for normal EPR, although, as stated above, there is renewed interest in related mixing techniques to avoid frequency limitation on other spectrometer features.

E.3 PULSED EPR SPECTROMETERS

The key aspect of doing pulsed EPR is the reliable generation of sets of microwave pulses of suitable shape. The important parameters are the power of the pulse, the accuracy of the time that each pulse is on, and the intervals between pulses. Pulse shape per se is less important. Typically, each pulse is on for tens to hundreds of nanoseconds, that is, much shorter periods than with NMR. The difficulties in achieving short pulses explain why pulsed EPR techniques have taken longer to develop. The intervals between pulses range from nanoseconds to several microseconds.

The microwave cw source, say, at ~ 10 GHz, can be a klystron or a solid-state (Gunn) diode. Its output in part goes to an on/off switch, computer-controlled by a pulse programmer. The microwave pulse transmitted through the switch is then amplified by a pulsed traveling-wave-tube amplifier capable of giving the requisite high power (1 kW is typical).

The amplified pulses are fed to a resonant cavity (located in the magnet) and excite the EPR transitions of the spin system. No field modulation is needed for detection, but the same magnet system as for cw work can be used. Microwave signals, produced by the samples as a result of stimulation by the microwave pulses, emanate from the cavity and are led to a suitable low-noise amplifier and detector diode, or to a DBM. As in cw spectrometers, various attenuators, isolators and circulators are installed. Descriptions of pulsed EPR spectrometers can be found in the literature [49–53].

Two major types of pulsed EPR experiments are distinguished by whether microwaves are incident on the sample during the signal detection period. In saturation recovery measurements, a high-power pulse saturates the spin system, ideally for a time long relative to any relaxation times, and then the signal is detected with microwave power low enough to avoid saturation, during recovery of the spin system to thermal equilibrium. In a spin echo or FID experiment, in their many variations, short high-power pulses are used to excite the spins, and then the response of the spins (echo or FID signal) is monitored—with no microwaves incident on the sample. The time-dependence of the echo or FID response is digitized, sometimes after using a boxcar integrator[4] (as a function of time points within the interval) to improve the S/N.

Special spectrometers have been designed and used for saturation-recovery work, and have yielded relaxation-time data, and insight into spin diffusion and spin-spin distances in various materials [54–56].

The whole pulse sequence can be repeated numerous times, at intervals dictated by how fast the spin system can recover to its original pre-pulse state. Thus the response signal can be superimposed onto the algebraic sum of all previous signals stored in a computer, canceling random ($\pm$) voltages due to noise and enhancing the signal-to-noise ratio of the EPR signals. In most experiments, it is also important to sum the response to pulses of different phases.

With the use of two mixers, often combined into one unit called a 'quadrature mixer', it is possible to measure two distinct time-dependent signals, that is, one out-of-phase by 90° and the other in phase with the pulse (i.e., quadrature detection). These represent dispersion and absorption, that is, the real and imaginary parts of the EPR signal function (see Chapter 10).

E.4 COMPUTER INTERFACING WITH EPR SPECTROMETERS

A modern spectrometer makes effective use of computer technology. At first computers were employed only to record and analyze the EPR signals coming from the detector of the spectrometer; however, today the computer is a central component of the spectrometer, setting and controlling many of the spectrometer components, such as the magnetic field, the modulation amplitude and the microwave power and optimizing the spectrometer performance through control of amplitude and phase of key signals as they pass through the electronic circuits. A review of developments in the 1970s is available [57], as well as a more recent one [58]. However, this aspect of EPR spectrometry continues to undergo explosive development. Two references for programs to operate pulse EPR spectrometers [59,60] will suffice.

Following detection and amplification, the EPR signal is usually digitized in an analog-to-digital converter and stored temporarily in a digital oscilloscope (also known as a 'transient recorder'), before being transferred to the computer's memory. Observation of the EPR trace (as a function of field or time) on the digital oscilloscope allows a visual control of the spectrometer performance.

Often a number of scans of the EPR trace (vs. field or time) are added together in the computer. This process serves to increase the signal-to-noise ratio because the true signal always adds linearly, whereas noise components increase only as the square root of the number n of scans. Hence the signal-to-noise ratio increases as $\sqrt{n}$. This technique is particularly valuable in flash-photolysis studies where the spectrometer must be operated in a broad-band mode in order to achieve a fast time resolution; this mode results in a much higher noise level, which can be reduced only by using the time-averaging process from repetitive scans. Computer software for acquisition and analysis of EPR spectra has become elegant and is a significant part of a commercial spectrometer (and its cost), but also becomes somewhat limiting of local modifications. Users are dependent on the instrument manufacturers for software of sufficient flexibility to permit as-yet-unthought-of experiments.

E.5 TECHNIQUES FOR TEMPERATURE VARIATION AND CONTROL

To all of the above technical requirements, we must add the demands for operation over extremes of temperature, carefully regulated and accurately measured. Temperature-control systems can be divided into those in which the entire cavity is heated or cooled and those in which the sample is heated or cooled in a jacket inserted into the cavity. The latter system is by far the most widely used and is described in some detail here.

In the temperature range (90–400 K), a system employing heated or cooled nitrogen flowing over the sample in a quartz dewar may be used. For temperatures below room temperature, cold nitrogen gas (boiled off from liquid nitrogen or cooled in a metal coil immersed in liquid nitrogen) flows to the cavity dewar through a dewar transfer line. A heater and temperature sensor (often a platinum-wire thermometer) are placed near the bottom of the cavity dewar to heat the nitrogen gas to the desired temperature. The temperature sensor is connected to a closed feedback loop to control the current in the heater so as to maintain the desired temperature. For temperatures above room temperature, the dry nitrogen gas (without cooling) is heated by the heater in the bottom of the cavity dewar. For temperatures in the range 6–100 K, a similar system is employed, except using cold helium gas. A transfer line delivers liquid helium (4.2 K) to the bottom of the cavity dewar, where this refrigerant evaporates and, if necessary, is heated to the desired temperature.

The development of small and efficient closed-cycle helium-gas refrigerators operating in the range 12 K to room temperature provides an alternative method for achieving controlled low temperature; this system is more complex, and inherently transmits vibrations to the resonator (which can effectively be removed by installation, say, of notch filters); however, such a device does have the attraction that it eliminates the need for dewars and for liquid refrigerants [61].

One option that has been used is to cool the complete cavity, including the sample. The cylindrical TE_{011} cavity (cooled by liquid nitrogen or liquid helium, if needed) is an excellent choice for versatility, allowing the insertion of various modules (e.g., for irradiation). The chilled cavity has a higher Q, for maximum sensitivity. A heater is wound around the cavity for temperature control, using direct current.

For operation above room temperature, one may use a cool TE_{011} cavity with internal heating by parallel platinum strips in an insulated sample jacket inserted along the axis of the cavity. Gas-flow systems for this purpose are also feasible. Alternatively, with considerable care to minimize oxidation, one may heat the entire cavity.

Examples of EPR work done at very high temperatures (e.g., ≤ 1000 K) exist [62–64], but are not plentiful. Sample stability and line broadening mitigate against such work in many instances.

Numerous thermometric devices for low or high temperatures have been reported. Some of these, such as germanium resistors or silicon diodes, unfortunately tend to be affected by magnetic fields. Others, such as carbon resistors, have an excessive resistance at low temperatures. Thermistors are useful over only a small range. A carbon/glass resistance thermometer allows rapid cycling of temperature without damage, but its sensitivity near room temperature is low. It becomes the preferred thermometer below 77 K. Above that temperature a platinum resistance thermometer is best chosen. A 'preferred' sensor at low temperatures is now the Cernox sensor. It is insensitive to magnetic-field changes and can be acquired with an accurate calibration.

Above room temperature, one is impelled to use a thermocouple (where one can use wire diameters as small as 0.03 mm); the heat capacity of the heated section is too low for alternatives. A preferred couple is Pt : Pt(10%Rh) with the reference junction in an ice-water bath.

Finally, one needs a highly stable temperature measurement and control system, but which is also capable of rapid variations. The output of either of the thermometric sensors (carbon resistor or platinum thermometer) is compared with a value preset on an accurate measuring bridge. The difference is amplified and fed back to the heater being controlled. Alternatively, one may program a desired temperature sequence.

E.6 TECHNIQUES FOR PRESSURE VARIATION

The pressure dependence of the EPR parameters has had only a modest amount of attention. The primary ways of applying pressure to a sample are via hydrostatic or elevated atmospheric means, and via application of uniaxial mechanical stresses (see Fig. 9.11). As another example of the latter, the anisotropic effects on the D and E parameters of Cr^{3+} in ruby have been studied (see Section 13.9.5). Various publications decribing appropriate resonators for pressure variation have been described. From among these, we cite Refs. 65–67; see also Ref. 7, Section 9D.

REFERENCES

1. R. Bramley, S. J. Strach, *Chem. Rev.*, **83**, 49 (1983).
2. R. Bramley, *Int. Rev. Phys. Chem.*, **5**, 211 (1986).
3. G. A. Rinard, R. W. Quine, S. S. Eaton, G. R. Eaton, *J. Magn. Reson.*, **154**, 80 (2002); 156, 113 (2002).
4. J. S. Hyde, W. Froncisz, T. Oles, *J. Magn. Reson.*, **82**, 223 (1989).
5. K. A. Earle, D. S. Tipikin, J. H. Freed, *Rev. Sci. Instrum.*, **67**(7), 2502 (1996).
6. R. L. Belford, R. B. Clarkson, J. B. Cornelius, K. S. Rothenberger, M. J. Nilges, M. D. Timken, "EPR over Three Decades of Frequency; Radiofrequency to Infrared", in *Electronic Magnetic Resonance of the Solid State*, J. A. Weil, Ed., Canadian Society for Chemistry, Ottawa, ON, Canada, 1987, p. 21.
7. C. P. Poole Jr., *Electron Spin Resonance—a Comprehensive Treatise on Experimental Techniques*, 2nd ed., Interscience, New York, NY, U.S.A., 1983.
8. C. J. Bender, "EPR: Instrumental Methods", in *Biological Magnetic Resonance*, Vol. 21, Kluwer Academic/Plenum Press, New York, NY, U.S.A., 2004, Chapter 1.
9. G. A. Rinard, R. W. Quine, S. S. Eaton, G. R. Eaton, E. D. Barth, C. A. Pelizzari, H. J. Halpern, *Magn. Reson. Eng.*, **15**, 51 (2002).
10. J. A. Burt, *J. Magn. Reson.*, **37**, 129 (1980).
11. G. R. Sinclair, J. R. Pilbrow, D. R. Hutton, G. J. Troup, *J. Magn. Reson.*, **57**, 228 (1984).
12. J. Witters, F. Herlach, *Bull. Magn. Reson.*, **9**, 132 (1987).
13. O. Y. Grinberg, L. J. Berliner, Eds., "Very High-Frequency (VHF) ESR/EPR", *Biological Magnetic Resonance*, Vol. 22, Kluwer Academic/Plenum Press, New York, NY, U.S.A., 2004.
14. M. Date, *J. Phys. Soc. Jpn.*, **39**, 892 (1975).
15. M. Date, M. Motokawa, A. Seki, H. Mollymoto, *J. Phys. Soc. Jpn.*, **39**, 898 (1975).
16. Ya. S. Lebedev, "Very-high-field EPR and its Applications", *Appl. Magn. Reson.*, **7**, 339–362 (1994).
17. K. A. Earle, D. E. Budil, J. H. Freed, *Adv. Magn. Reson. Opt. Reson.*, **19**, 253–323 (1996).
18. T. Prisner, *Adv. Magn. Opt. Reson.*, **20**, 245–299 (1997).
19. G. R. Eaton, S. S. Eaton, *Appl. Magn. Reson.*, **16**(2), 161–166 (1999).
20. S. S. Eaton, G. R. Eaton, "High Magnetic Fields and High Frequencies", in *Handbook of Electron Spin Resonance*, C. P. Poole Jr., H. A. Farach, Eds., Vol. 2, Springer-Verlag, AIP Press, New York, NY, U.S.A., 1999, pp. 345–370.
21. A. K. Hassan, L. A. Pardi, J. Krzystek, A. Sienkiewicz, P. Goy, M. Rohrer, L.-C. Brunel, *J. Magn. Reson.*, **142**, 300–312 (2000).
22. L. R. Becerra, G. J. Gerfen, B. F. Bellew, J. A. Bryant, D. A. Hall, S. J. Inati, R. T. Weber, S. Un, T. F. Prisner, A. E. McDermott, K. W. Fishbein, K. E. Kreischer, R. J. Temkin, D. J. Singel, R. G. Griffin, *J. Magn. Reson., A*, **117**, 28 (1995).

23. I. Amity, *Rev. Sci. Instrum.*, **41**(10), 1492 (1970).

24. J. P. Barnes, J. H. Freed, *Rev. Sci. Instrum.*, **69**(8), 3022 (1998).

25. H. L. F. Helmholtz, *On the Sensations of Tone*, transl. by A. J. Ellis, Longmans, Green & Co., London, U.K., 1875, p. 68, 579.

26. G. S. Smith, *Am. J. Phys.*, **58**, 996 (1990).

27. M. A. Plonus, *Applied Electromagnetics*, McGraw-Hill, New York, NY, U.S.A., 1978, pp. 488–490.

28. S. Y. Liao, *Microwave Devices and Circuits*, 3rd ed., Prentice-Hall, Englewood Cliffs, NJ, U.S.A., 1990, Chapter 4.

29. R. R. Mett, W. Froncisz, J. S. Hyde, "Axially Uniform Resonant Cavity Modes for Potential Use in Electron Paramagnetic Resonance Spectroscopy", *Rev. Sci. Instrum.*, **72**(11), 4188–4200 (2001).

30. C. G. Montgomery, R. H. Dicke, E. M. Purcell, Eds., *Principles of Microwave Circuits*, Radiation Lab Series, Vol. 8, McGraw-Hill, New York, NY, U.S.A., 1948.

31. E. Argence, T. Kahan, *Theory of Waveguides and Cavity Resonators*, Engl. transl., Hart, New York, NY, U.S.A., 1967.

32. G. Feher, *Bell Syst. Tech. J.*, **36**, 449 (1957).

33. A. Yariv, *Quantum Electronics*, Wiley, New York, NY, U.S.A., 1967, p. 135.

34. C. Mailer, H. Thomann, B. H. Robinson, L. R. Dalton, *Rev. Sci. Instrum.*, 51(12), 1714 (1980).

35. R. D. Richtmyer, *J. Appl. Phys.*, **10**, 391 (1939).

36. Y. E. Nesmelov, J. T. Surek, D. D. Thomas, *J. Magn. Reson.*, **153**, 7 (2001).

37. J. S. Hyde, W. Froncisz, *J. Magn. Reson.*, **47**, 515 (1982); also J. S. Hyde, W. Froncisz, "Loop-Gap Resonators", in *Advanced EPR—Applications in Biology and Biochemistry*, A. J. Hoff, Ed., Elsevier, Amsterdam, Netherlands, 1989, Chapter 7.

38. S. Pfenninger, J. Forrer, A. Schweiger, Th. Weiland, *Rev. Sci. Instrum.*, **59**, 752 (1988).

39. W. N. Hardy, L. A. Whitehead, *Rev. Sci. Instrum.*, **52**, 213 (1981).

40. M. J. Nilges, T. Walczak, H. M. Swartz, *Phys. Med.*, **2–4**, 195 (1989).

41. R. M. Bevensee, *Electromagnetic Slow-Wave Systems*, Wiley, New York, NY, U.S.A., 1964.

42. N. Devasahayam, S. Subramanian, R. Murugesam, J. A. Cook, M. Afeworki, R. G. Tschudin, J. B. Mitchell, M. C. Krishna, *J. Magn. Reson.*, **142**, 168 (2000).

43. J. S. Hyde, P. W. Sczaniecki, W. Froncisz, *J. Chem. Soc. Faraday Trans. I*, **85**, 3901 (1989).

44. H. Hirata, M. Ueda, M. Ono, Y. Shimoyama, *J. Magn. Reson.*, **155**, 140 (2002).

45. C. P. Slichter, *Principles of Magnetic Resonance*, 3rd ed., Springer, Berlin, Germany, 1990, p. 184.

46. H. S. Mchaourab, J. S. Hyde, *J. Chem. Phys.*, **98**(3), 1786 (1993).

47. C. S. Klug, T. G. Camenisch, W. L. Hubbell, J. S. Hyde, *Biophys. J.*, **88**, 3641 (2005).

48. H. Hirata, T. Kuyama, M. Ono, Y. Shimoyama, *J. Magn. Reson.*, **164**, 233 (2003).

49. M. K. Bowman, "Fourier Transform Electron Spin Resonance", in *Modern Pulsed and Continuous-wave Electron Spin Resonance*, L. Kevan, M. K. Bowman, Eds., Wiley, New York, NY, U.S.A., 1990, Chapter 1.

50. J.-M. Fauth, A. Schweiger, L. Braunschweiler, J. Forrer, R. R. Ernst, *J. Magn. Reson.*, **66**, 74 (1986).

51. J. Gorcester, J. H. Freed, *J. Chem. Phys.*, **88**, 4678 (1988).

52. T. F. Prisner, S. Un, R. G. Griffin, *Isr. J. Chem.*, **32**, 357 (1992).

53. R. W. Quine, G. A. Rinard, S. S. Eaton, G. R. Eaton, "A Pulsed and Continuous Wave 250 MHz Electron Paramagnetic Resonance Spectrometer", *Magn. Reson. Eng.*, **15**, 59 (2002), and various papers by the same group cited therein.

54. R. W. Quine, S. S. Eaton, G. R. Eaton, *Rev. Sci. Instrum.* **63**, 4251–4262 (1992).

55. W. Froncisz, T. G. Camenisch, J. J. Ratke, J. S. Hyde, *Rev. Sci. Instrum.*, **72**(3), 1837 (2001).

56. J. S. Hyde, J.-J. Yin, W. K. Subczynski, T. C. Camenisch, J. R. Ratke, W. Froncisz, *J. Phys. Chem. B*, **108**, 9524 (2004).

57. H. L. Vancamp, A. H. Heiss, *Magn. Reson. Rev.*, **7**, 1 (1981).

58. B. Kirste, in *Handbook of Electron Spin Resonance: Data Sources,* Vol. 1, *Computer Technology, Relaxation and ENDOR*, C. P. Poole Jr., H. A. Farach, Eds., AIP, New York, NY, U.S.A., 1994, pp. 27–50.

59. I. Gromov, B. Glass, J. Keller, J. Shane, J. Forrer, B. Tschaggelar, A. Schweiger, *Concepts Magn. Reson., Part B*, **21B**(1), 1 (2004).

60. B. Epel, I. Gromov, S. Stoll, A. Schweiger, D. Goldfarb, *Concepts Magn. Reson., Part B*, **26B**(1), 36 (2005).

61. B. D. Perlson, J. A. Weil, *Rev. Sci. Instrum.*, **46**, 874 (1975).

62. J. S. van Wieringen, J. G. Rensen, *Proc. XIIth Colloque Ampère*, North-Holland Publ., Amsterdam, Netherlands, 1964, pp. 229–231.

63. J. S. van Wieringen, J. G. Rensen, *Philips Res. Rep.*, **20**, 659 (1965).

64. T. A. Yager, W. D. Kingery, *Rev. Sci. Instrum.*, **51**(4), 464 (1980).

65. S. E. Bromberg, I. Y. Chen., *Rev. Sci. Instrum.*, **63**(7), 3670 (1992).

66. M. Krupski, *Rev. Sci. Instrum.*, **67**(8), 2894 (1996).

67. A. Sienkiewicz, B. Vileno, S. Garaj, M. Jaworski, L. Forró, *J. Magn. Reson.*, **177**, 278 (2005).

NOTES

1. If there is an internal magnetic field, as from nuclei, one may detect resonance at zero applied field [1,2].

2. Special designs, such as the loop-gap resonator, permit the use of much larger samples [4].

3. It is usually necessary to make a correction for a difference between the magnetic field at a probe outside the cavity and at the sample inside the cavity.

4. A boxcar integrator is an analog device that enables the storage of repeated transient signals. It consists of a series of capacitors that store charge, each in turn triggered on and off as dictated by some external program (e.g., following a laser flash or a microwave pulse). The final integrated signals are read out as the voltages on the capacitors and represent a time-based input signal.

FURTHER READING

1. R. S. Alger, *Electron Paramagnetic Resonance: Techniques and Applications*, Wiley, New York, NY, U.S.A., 1967. (This book contains numerous drawings of experimental equipment and details of construction.)

2. L. J. Berliner, C. J. Bender, Eds., "EPR: Instrumental Methods", *Biol. Magn. Reson.*, 21 (2004).

3. A. Colligiani, P. Guillan, I. Longo, M. Martinelli, L. Pardi, "An X-band ESR Apparatus Using a Whispering-Gallery Mode Dielectric Resonator", *Appl. Magn. Reson.*, **3**, 827 (1992).

4. D. Duret, M. Beranger, M. Moussavi, P. Turek, J. J. Andre, "A New Ultra Low-Field ESR Spectrometer", *Rev. Sci. Instrum.*, 62, 685 (1991); see also *IEEE Trans. Magn.*, **27**, 5405 (1991).

5. S. S. Eaton, G. R. Eaton, "Electron Paramagnetic Resonance", in *Analytical Instrumentation Handbook*, 3rd ed., J. Cazes, Ed., Marcel Dekker, NY, U.S.A., Chapter 12, 2005, pp. 349–398.

6. J. H. Freed, "1-mm Wave ESR Spectrometer", *Rev. Sci. Instrum.*, **59**, 1345 (1988).

7. A. F. Gullá, D. E. Budil, "Engineering and Design Concepts for Quasi-optical High-Field Electron Paramagnetic Resonance", *Concepts Magn. Reson. B*, **22B**(1), 15 (2005).

8. S. Y. Liao, *Microwave Devices and Circuits*, 3rd ed., Prentice-Hall, Englewood Cliffs, NJ, U.S.A., 1990.

9. C. P. Poole Jr., *Electron Spin Resonance. A Comprehensive Treatise on Experimental Techniques*, 2nd ed., Wiley-Interscience, New York, NY, U.S.A., 1983.

10. C. P. Poole Jr., H. A. Farach, *Handbook of Electron Spin Resonance*, 2 Vols., American Institute of Physics, New York, NY, U.S.A., 1994, 1999.

11. T. L. Squires, *An Introduction to Electron Spin Resonance*, Academic Press, New York, NY, U.S.A., 1963.

PROBLEMS

E.1 Why is it that when a sample tube filled with water is placed inside a microwave cavity, the Q factor is greatly decreased, whereas when the same tube contains benzene it has little effect on Q?

E.2 What variables in the EPR spectrometer would you expect to affect the signal-to-noise ratio?

E.3 The resonance frequency of a metallic air-filled rectangular cavity is given approximately by

$$\nu = \frac{c}{2}\left[\left(\frac{n_x}{C}\right)^2 + \left(\frac{n_y}{A}\right)^2 + \left(\frac{n_z}{B}\right)^2\right]^{1/2}$$

where C, A and B are the x, y and z inner dimensions and c is the vacuum speed of electromagnetic radiation. Design an X-band cavity to resonate in the $TE_{n_x n_y n_z}$ ($n_x = 1$, $n_y = 0$, $n_z = 2$) mode (Fig. E.4), using standard waveguide (1.016 × 2.286 cm ID).

APPENDIX F

EXPERIMENTAL CONSIDERATIONS

Now that we have examined the details of EPR instrumentation in Appendix E, we turn to various analysis methods and techniques that allow the experimenter to assess and optimize the spectra produced from the EPR spectrometer.

F.1 TECHNIQUES FOR GENERATION OF PARAMAGNETIC SPECIES

Only a very small fraction of substances are found naturally in a paramagnetic state (e.g., O_2, NO, nitroxide radicals, coal, soot, charred organic materials, numerous transition ions and certain point defects such as those caused by irradiation of teeth, bones, quartz, foodstuffs). Fortunately it is possible to excite literally every substance into a paramagnetic state by the use of various procedures that include the following:

1. *One-Electron Chemical Reduction* For example, reaction of certain poly-acene hydrocarbons (e.g., anthracene) with alkali metals in ether solvents in vacuo produces stable solutions of the corresponding radical anions.

2. *One-Electron Chemical Oxidation* Reaction of these same polyacenes with a strong oxidant, such as concentrated sulfuric acid or $AlCl_3$, results in stable solutions of the corresponding cation radicals.

Electron Paramagnetic Resonance, Second Edition, by John A. Weil and James R. Bolton
Copyright © 2007 John Wiley & Sons, Inc.

3. *Hydrogen-Atom Abstraction and Addition* Hydroxyl radicals (e.g., formed by photolysis of hydrogen peroxide) readily abstract hydrogen atoms from many organic molecules, leaving behind a free radical. The OH radicals, generated by chemical means or by photolysis, are usually reacted with the substrate molecules (e.g., Fig. 1.5) in a flow system. Photolysis of organic peroxides yields alkoxy radicals (e.g., *tert*-butyloxy), which act as hydrogen-atom abstractors [1]. Bombardment of organic single crystals with hydrogen atoms produces some interesting free radicals, especially where the H atoms can add to double bonds [2]. In fact, it can be said that almost any kind of high-energy irradiation of single crystals produces free radicals.

4. *Photolysis* On irradiation (usually in the UV region) of a suitable sample in the cavity, photochemical processes often produce radical intermediates. If these have moderate lifetimes, they may be observed by steady-state photolysis; however, for short-lived intermediates and where kinetic information is desired, flash-photolysis techniques must be employed [3]. Clearly, the laser has become an important tool in this area. Photolysis has also been used very successfully to excite molecules to metastable triplet states ($S = 1$) as discussed in Chapter 6.

5. *Radiolysis* High-energy radiation (x rays, γ rays and high-energy electrons from an accelerator) almost always produces radical intermediates and products. Products that may be stable (e.g., at low temperature) can be observed after radiolysis; however, a wealth of information has been obtained on short-lived radicals by means of either steady-state or pulsed radiolysis using high-energy electron beams (e.g., from a Van der Graaff accelerator). Some paramagnetic products are stable for long times at room temperature. Among these are irradiated crystalline amino acids such as alanine, which is used as a dosimeter, teeth, bones, quartz and almost any material in which the defect is prevented from further reaction.

6. *Heat* Thermal decomposition of species (e.g., $[\eta\text{-}C_3H_5)Fe(CO)_3]_2$ in solution [4], and Cl_2 (by a hot platinum wire in the cavity [5]) lead to free radicals. Almost every charred organic material contains free radicals.

7. *Spin-Trapping* Certain substances (e.g., nitrones) can react with very short-lived radicals (e.g., OH) and produce relatively stable radical products, which can be detected and analyzed by EPR using spin traps (see Section 13.14). The characteristics of the product radical often give information as to the identity and reactivity of the original short-lived radical.

8. *Electrochemical Methods* By constructing a small electrochemical cell capable of insertion into the cavity, it is possible to carry out in situ one-electron (or three-electron) oxidations and reductions of a wide variety of molecules [6].[1]

9. *Discharges* For gas-phase systems, installation of a suitable microwave or dc electric-discharge chamber adjacent to the EPR resonator allows the production of gaseous free radicals that flow into the resonator (Chapter 7).

10. *Matrix Isolation* Many highly reactive radicals have been trapped at low temperatures by the matrix-isolation technique [8]. The host is usually an inert substance such as nitrogen, argon or methane. Even though the radicals are randomly oriented, much information may be obtained.

This list is by no means complete; however, it serves to give the reader a flavor of the rich harvest of results that may be obtained from almost any substance by applying the appropriate technique for generation of paramagnetic species. Each spin species has its own EPR characteristics, such as line positions, lineshapes, relaxation pattern, field and frequency behavior and temperature dependence. We now turn to some of these factors.

F.2   LINESHAPES AND INTENSITIES

F.2.1   Lineshapes

Individual EPR absorption lines tend to be symmetric about their central position. A single such line can be characterized by its shape, $Y(x - x_r)$, where $x = B$ for fixed-frequency spectroscopy, and $x = \nu$ for work at constant magnetic field. The subscript r denotes the central resonance position. Function Y can be tabulated digitally, or approximated using analytical lineshape formulas (gaussian, lorentzian or other functions as discussed below). Each must be multiplied by the factor determining the actual intensity, that is, by a dimensionless factor $\mathcal{I}$, which is virtually independent of x (Section 4.6) but is dependent on B_1.

The observed *amplitude* of the line, defined as the maximum height $\mathcal{I}Y(0)$, depends on experimental factors such as power level, detector sensitivity and amplifier settings, as well as on the sample composition and temperature. Relative amplitudes, when several lines occur in a given spectrum, are, of course, of greater fundamental interest since the factors mentioned above then tend to cancel.

It is the integral of $\mathcal{I}Y$ over the whole range of x (i.e., the area $\mathcal{A}$ under the curve of $\mathcal{I}Y$ vs. x) that is of particular interest, since it is proportional to the number of spins in the EPR sample. By *intensity* of a line we mean the area $\mathcal{A}$.

When two or more lines occur, we call the distance δx between the centers of the first and last lines the *extent* of the spectrum. The *width* of each line can be parameterized in several related ways. One type of width is the full-width $\Delta x_{1/2}$ at half-height $[\mathcal{I}Y(0)/2]$, that is, the spread between absorption points half-way down from the line maximum. The area $\mathcal{A}$ depends linearly on $Y(0)$; for a series of lines with a given $\mathcal{A}$, $Y(0)$ and $\Delta x_{1/2}$ are inversely proportional. Thus, for a given spin concentration, broad lines imply small amplitudes and hence lower detectability compared to narrow lines.

The shapes of EPR lines are usually described by either lorentzian or gaussian lineshapes (Figs F.1a and F.2a). Analytical expressions for these are

$$\text{Lorentzian:} \qquad Y_\ell(x - x_r) = Y_\ell(0) \, [1 + c^2(x - x_r)^2]^{-1} \qquad \text{(F.1}a\text{)}$$

$$\text{Gaussian:} \qquad Y_g(x - x_r) = Y_g(0) \, \exp[-c^2(x - x_r)^2] \qquad \text{(F.1}b\text{)}$$

Detailed expressions in terms of measurable experimental parameters are given in Tables F.1. The parameters describing these lines are the maximum amplitude $Y(0)$ and the inverse width factor $c = 1/\Gamma$ (see Tables F.1). For convenience these parameters often are chosen such that the integrated area $\mathcal{A}$ under each curve is unity; that is, the expressions Y are normalized. Thus

$$\mathcal{A} = \int_{-\infty}^{\infty} Y \, d(x - x_r) = 1 \tag{F.2}$$

Tables F.1 include expressions for the first and second derivatives for normalized lorentzian and gaussian lines. Expressions for the usual linewidth parameters, $\Delta x_{1/2}$ and Δx_{pp}, are given in Tables F.1, and they are illustrated in Figs. F.1 and F.2.

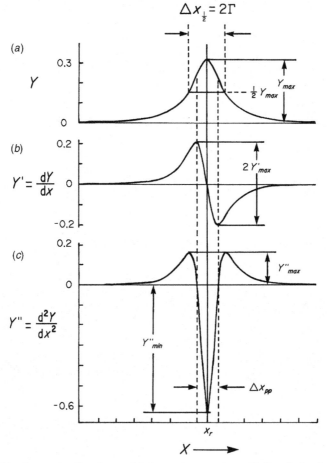

FIGURE F.1 Lorentzian lineshapes: (a) absorption spectrum; (b) first-derivative spectrum; (c) second-derivative spectrum. For an explanation of the symbols, see Table F.1a.

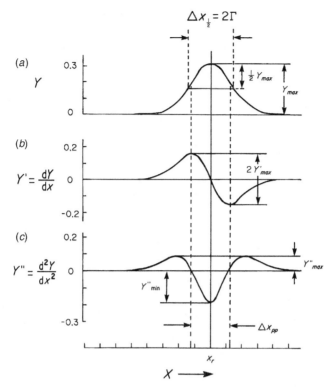

FIGURE F.2 Gaussian lineshapes: (*a*) absorption spectrum; (*b*) first-derivative spectrum; (*c*) second-derivative spectrum. For an explanation of the symbols, see Table F.1*b*.

Note that it follows from Eq. F.2 that the intensity factor $\mathscr{I}$ equals the area below the fitted curve portraying the actual experimental absorption line.

Lorentzian lineshapes are usually observed for EPR lines of systems if the lines are relaxation-determined. This situation occurs if there is no hyperfine broadening, if the concentration of paramagnetic centers is low and if there is dynamic averaging (Chapter 10), say, by spin exchange in liquid solution. These are called 'homogeneously broadened lines'. Lines often approach the gaussian shape if the line is a superposition of many components. One usually refers to such composites as 'inhomogeneously broadened' (Section 10.4.2). When both types of contributions are appreciable, more complex functions are useful. One such function, developed for magnetic-resonance application [9], has been called the 'Tsallian function', after its originator (C. Tsallis). It features one extra adjustable parameter, and is given in detail herein in Table F.1*c*. Another, called the 'Voigt profile', also is deemed useful [10, p. 490; 11, pp. 188–189]. A rational-function approximation for a generalized lineshape, especially useful for uniaxial powder patterns, also has been developed [12].

All absorption lineshape functions, including the ones described above, can be related to stick-spectral functions, that is, to Dirac delta (δ) functions, by considering

TABLE F1a   Properties of Normalized Lorentzian Lines [a,b]

Property	Equation
Absorption $Y(x)$	$Y = Y_{max} \dfrac{\Gamma^2}{\Gamma^2 + (x - x_r)^2}$
Full-width at half-height of Y	$\Delta x_{1/2} = 2\Gamma$, where Γ is the half-width at half-height.
Peak amplitude of Y	$Y_{max} = Y\big|_{x=x_r} = \dfrac{1}{\pi\Gamma}$
First-derivative $\dfrac{dY}{dx} \equiv Y'(x)$	$Y' = -Y_{max}\dfrac{2\Gamma^2(x - x_r)}{\left[\Gamma^2 + (x - x_r)^2\right]^2}$
Peak-to-peak amplitude of Y'	$2Y'_{max} = \dfrac{3\sqrt{3}}{4\pi}\dfrac{1}{\Gamma^2} \equiv A_{pp}$
Peak-to-peak width of Y'	$\Delta x_{pp} = \dfrac{2}{\sqrt{3}}\Gamma$
Second-derivative $\dfrac{d^2Y}{dx^2} \equiv Y''(x)$	$Y'' = -Y_{max}2\Gamma^2\dfrac{\Gamma^2 - 3(x - x_r)^2}{\left[\Gamma^2 + (x - x_r)^2\right]^3}$
Max-to-max width of Y''	$\Delta x_{++} = 2\Gamma$
Peak amplitude of the positive lobe(s) of Y''	$Y''_{min} = -Y_{max}\dfrac{1}{\Gamma^2}$
Peak amplitude of the negative lobe of Y''	$Y''_{min} = -Y_{max}\dfrac{2}{\Gamma^2}$, which occurs at $x = x_r$.

[a] Normalized as per Eq. F.2.
[b] Case $q = 2$ in Table F.1c.

their limits as their linewidth goes to zero [13]. As an example, for a lorentzian, one can write

$$\lim_{\Gamma \to 0} Y_\Gamma(x) = \lim_{\Gamma \to 0} \frac{1}{\pi} \frac{\Gamma}{\Gamma^2 + (x - x_r)^2} \tag{F.3a}$$

$$= \delta(x - x_r) \tag{F.3b}$$

where $\Gamma = \frac{1}{2}\Delta x_{1/2}$ (see Table F.1c).

When field modulation together with phase-sensitive detection is used, as explained previously (Sections E.1.6 and E.1.8), it is the first derivative $\mathscr{I} dY/dB$ for each line that is recorded as a function of B (Figs. E.6, F.1b and F.2b). Here the line center is denoted by B_r, and, as stated, $\mathscr{I}$ is the field-independent multiplier of interest. It is convenient to use the distance ΔB_{pp} between the maximum peak and the minimum peak of a first derivative as a measure of linewidth; this is closely related to $\Delta B_{1/2}$ (Tables F.1). Clearly, two integrations [a 'running' one to obtain the absorption function $\mathscr{I} Y(B - B_r)$ and a 'numerical' one between set limits to yield simply a number] are needed to obtain the desired area from the experimentally obtained first-derivative profile. Note (Tables F.1) that it is the quantity $\mathscr{I} Y'_{max} \Gamma^2$ (for fixed experimental conditions) that is proportional to the area.

TABLE F.1b   Properties of Normalized Gaussian Lines [a,b]

Property	Equation
Absorption $Y(x)$	$Y = Y_{\max} \exp\left[\dfrac{-(\ln 2)(x - x_r)^2}{\Gamma^2}\right]$
Full-width at half-height of Y	$\Delta x_{1/2} = 2\Gamma$, where Γ is the half-width at half-height.
Peak amplitude of Y	$Y_{\max} = Y\big|_{x=x_r} = \left(\dfrac{\ln 2}{\pi}\right)^{1/2}\dfrac{1}{\Gamma}$
First-derivative $\dfrac{dY}{dx} \equiv Y'(x)$	$Y' = -Y_{\max}\dfrac{2(\ln 2)(x - x_r)}{\Gamma^2}\exp\left[\dfrac{-(\ln 2)(x - x_r)^2}{\Gamma^2}\right]$
Peak-to-peak width of Y'	$\Delta x_{pp} = \left(\dfrac{2}{\ln 2}\right)^{1/2}\Gamma$
Peak-to-peak amplitude of Y'	$2Y'_{\max} = 2\left(\dfrac{2}{\pi e}\right)^{1/2}\dfrac{\ln 2}{\Gamma^2} \equiv A_{pp}$
Second-derivative $\dfrac{d^2 Y}{dx^2} \equiv Y''(x)$	$Y'' = -Y_{\max}\dfrac{2(\ln 2)}{\Gamma^4}\left\{\Gamma^2 - 2(\ln 2)(x - x_r)^2\right\} \times$ $\exp\left[\dfrac{-(\ln 2)(x - x_r)^2}{\Gamma^2}\right]$
Max-to-max width of Y''	$\Delta x_{++} = 2\sqrt{6}\,\Gamma$
Peak amplitude of the positive lobe(s) of Y''	$Y''_{\max} = Y_{\max}\dfrac{4e^{-3/2}\ln 2}{\Gamma^2}$
Peak amplitude of the negative lobe of Y''	$Y''_{\min} = -Y_{\max}\dfrac{2\ln 2}{\Gamma^2}$, which occurs at $x = x_r$.

[a] Normalized as per Eq. F.2.
[b] Case $q = 1$ in Table F.1c.

One obtains a trace that is approximately the second derivative of the absorption line by tuning the lock-in amplifier to twice the modulation frequency (e.g., 2×100 kHz) or by using two modulation frequencies and detection systems. It can be seen from Figs F.1c and F.2c that the shape of the second-derivative presentation is very sensitive to the nature of the absorption line.

Hyde and coworkers [14] have demonstrated the power of a mathematical technique that they call 'pseudo-modulation' as a signal-detection method to obtain more information (higher derivatives) without a large increase in noise. The technique involves simultaneous acquisition of multiple in- and out-of-phase harmonics of the EPR response when field modulation is applied. This capability is now one of the routine data-analysis functions in commercial software.

It is worth re-emphasizing here (see Section 4.6) that there are subtle differences between B-field scans at constant frequency ν and ν-scans at constant B. Even in the simplest case $(S = \frac{1}{2})$ when $h\nu = g\beta_e B$, one has $\Delta B_{1/2} = \Delta\nu_{1/2}h/g\beta_e$ and $\mathcal{A}_B = \mathcal{A}_\nu g_e/g$, and the function Y in the field domain and the corresponding one in the frequency domain are not simultaneously normalized [15,16].

TABLE F.1c Properties of Normalized Tsallian Lines, [a] for $q > 1$. Parameter $q = 1$ Produces the Gaussian Shape, While $q = 2$ yields the Lorentzian[b]

Property	Equation
Absorption $Y(x)$	$Y = Y_{max}\left[1 + (2^{q-1} - 1)\left(\dfrac{x - x_r}{\Gamma}\right)^2\right]^{-1/(q-1)}$ where $q \in (1, \infty)$.
Full-width at half-height of Y	$\Delta x_{1/2} = 2\Gamma$, where Γ is the half-width at half-height.
Peak amplitude of Y	$Y_{max} = Y\big|_{x=x_r} = \dfrac{(2^{q-1} - 1)^{1/2}}{\Gamma\beta(\frac{1}{2}, \frac{1}{q-1} - \frac{1}{2})}$ where β is the Euler's integral of the first kind (β function).
First-derivative $\dfrac{dY}{dx} \equiv Y'(x)$	$Y' = -Y_{max}\dfrac{2^{q-1} - 1}{q - 1}\dfrac{2}{\Gamma^2}(x - x_r) \times$ $\left[1 + (2^{q-1} - 1)\left(\dfrac{x - x_r}{2\Gamma}\right)^2\right]^{-q/(q-1)}$
Peak-to-peak width of Y'	$\Delta x_{pp} = 2\Gamma\left(\dfrac{q - 1}{q + 1}\dfrac{1}{2^{q-1} - 1}\right)^{1/2}$
Peak-to-peak amplitude of Y'	$2|Y'_{max}| = \dfrac{2}{\Gamma}Y_{max}\dfrac{2^{q-1} - 1}{q - 1}\Delta x_{pp} \times$ $\left[1 + (2^{q-1} - 1)\left(\dfrac{\Delta x_{pp}}{2\Gamma}\right)^2\right]^{-q/(q-1)} \equiv A_{pp}$
Second-derivative $\dfrac{d^2Y}{dx^2} \equiv Y''(x)$	$Y'' = -Y_{max}\dfrac{2^{q-1} - 1}{q - 1}\dfrac{2}{\Gamma^2}\left[1 - \dfrac{q + 1}{q - 1}(2^{q-1} - 1)\left(\dfrac{x - x_r}{\Gamma}\right)^2\right] \times$ $\left[1 + (2^{q-1} - 1)\left(\dfrac{x - x_r}{\Gamma}\right)^2\right]^{-(2q-1)/(q-1)}$
Max-to-max width of Y''	$\Delta x_{++} = 2\left(\dfrac{3(q - 1)}{(2^{q-1} - 1)(q + 1)}\right)^{1/2}\Gamma$
Peak amplitude of the positive lobe(s) of Y''	$Y''_{max} = Y_{max}\dfrac{2^{q-1} - 1}{q - 1}\dfrac{4}{\Gamma^2}\left(1 + 3\dfrac{q - 1}{q + 1}\right)^{-(2q-1)/(q-1)}$
Peak amplitude of the negative lobe of $Y''(x)$	$Y''_{min} = -Y_{max}\dfrac{2^{q-1} - 1}{q - 1}\dfrac{2}{\Gamma^2}$, which occurs at $x = x_r$.

[a] Normalized as per Eq. F. 2.
[b] When $q = 1$, see Table F.1b; $q = 2$ yields Table F.1a.

It should be emphasized that in modern EPR, both cw and pulsed, excellent simulations of spectra are feasible (Section 1.14, and e.g., see Figs. 1.4, 3.3, 5.2, 6.5b and 6.7b). Knowledge of the spin hamiltonian suffices to yield exact line positions and relative intensities, say, by exact diagonalization of $\mathcal{H}$ using computer techniques. Attachment to each line of a suitable shape function, of appropriate width, is less quantitative, but feasible and much practiced (e.g., see Ref. 17). Several computer program packages are obtainable (e.g., see Ref. 18), one of which is commercially available [19].

F.2.2   Signal Intensities and Spin-Concentration Standards

The following factors determine the intensity of the EPR signal:

1. The number of relevant paramagnetic centers in the sample.
2. The spin S and g factor of each such species.
3. The transition probability per second per spin (Section 4.6).
4. The number of lines (e.g., fine-structure and hyperfine-structure components) in the spectrum.
5. The sample temperature.
6. The microwave frequency ν and/or the applied magnetic field magnitude B.
7. The relative orientations of **B** and $\mathbf{B}_1$.
8. The amplitude B_1 of the microwave magnetic field effective at the sample.[2]
9. The field modulation amplitude B_m (and phase) at the sample. As mentioned in Appendix E, an alternative technique employs modulation of the excitation frequency.
10. The overall spectrometer gain, including multi-scanning spectral superposition capability.

In any chemico-analytical applications, one desires knowledge of the number of paramagnetic centers giving rise to the observed signal. In practice, determinations of absolute concentration (number of spins of a given type per unit volume of sample) determinations are rarely carried out since so many errors can enter. Instead, relative intensities of pairs of lines are preferable since many of the above-listed variables cancel. Comparison standards are employed to minimize some of the errors and are of two types: (1) concentration standards and (2) absolute spin-number standards.

If one requires only the concentration of paramagnetic species in a solid, liquid solution or gas, then a concentration standard can be employed. The following conditions should apply to both standard and unknown:

1. The same solvent or host and the same sample geometry should be employed to ensure that B_1 is the same for the unknown and the standard samples.
2. The amplitude of the EPR signals should be proportional to $P_0^{1/2}$ ($\propto B_1$); that is, there should be no power saturation for either sample or standard. Ideally, P_0 should be the same for sample and standard.
3. The field modulation amplitude B_m can be large [i.e., $\approx (2-4)\Delta B_{pp}$] provided that it is the *area* under the absorption curve that is to be determined. This is especially important for weak signals where overmodulation must be employed to achieve adequate sensitivity.
4. The unknown and standard samples should be at the same temperature.
5. Identical holders for both samples must be provided.

Under these conditions the concentration of the unknown paramagnetic species X is given by

$$[X] = \frac{\mathcal{A}_X \, \mathcal{R}_X (\text{Scan}_X)^2 G_{\text{Std}} (B_m)_{\text{Std}} (g_{\text{Std}})^2 [S(S+1)]_{\text{Std}}}{\mathcal{A}_{\text{Std}} \, \mathcal{R}_{\text{Std}} (\text{Scan}_{\text{Std}})^2 G_X (B_m)_X (g_X)^2 [S(S+1)]_X} [\text{Std}] \qquad (\text{F.4})$$

Here 'Scan' is the horizontal scale in mT per unit length on the scope or chart paper, G is the relative gain of the signal amplifier, B_m is the modulation amplitude in mT, and $\mathcal{R}$ is defined by Eq. F.6. Parameter $\mathcal{A}$ is the measured area under the absorption curve. This may be in arbitrary units as long as they are the same for unknown and standard.

Area $\mathcal{A}$ can be obtained by computing the first moment [10, p. 461],[4] by analog integration [20] or by digital double integration, or even (in the 'good old days') by weighing a cut-out absorption curve displayed on paper. It is now so routine to integrate spectra acquired, stored in a computer, that the first and second integrals of an experimental derivative spectrum are commonly used to judge the quality (phasing, etc.) of the experimental data. Meaningful integration requires a constant base line or a base-line correction after each integration, so there has to be good base line on the low-field and high-field sides of the spectrum. One must use extreme care in evaluating the area under an absorption curve. Serious errors may result from failure to extend measurements sufficiently far from the center of the line [21]. The percentage error resulting from finite truncation of the first-derivative curve is shown in Fig. F.3, which may be used to apply corrections. The errors are especially large for lorentzian lines because of their extensive wings. Similar considerations hold for dispersion signals and their integration.

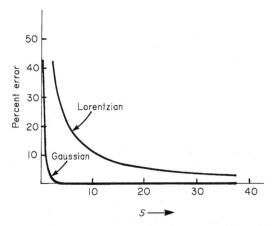

FIGURE F.3 Percent error in determining the area under an absorption line when the first-derivative lorentzian or gaussian curve is truncated at the finite limits $\pm S$, where S is measured in units of ΔB_{pp}. (Data taken from M. L. Randolph, "Quantitative Considerations in the ESR Studies of Biological Materials", in *Biological Applications of Electron Spin Resonance*, H. M. Swartz, J. R. Bolton, D. C. Borg, Eds., Wiley, New York, NY, U.S.A., 1972, Chapter 3, Tables 3-2 and 3-3.)

Clearly, sensitivity for anisotropic signals decreases when going from liquid solutions and single-crystal systems to glasses and powders. In the latter, the extent of the spectrum may be much enlarged, and it may be difficult to tell how far out to measure to attain accurate absorption areas.

The following concentration standards have proved useful:

1. *2,2′-Diphenyl-1-picrylhydrazyl (DPPH)* (**I**). This free radical can be analytically determined [22] and dissolves readily in (say) benzene; however, its solutions are not stable over long periods of time, especially in bright light. In powder form, DPPH is seldom pure; for example, it may contain unoxidized 2,2′-diphenyl-1-picrylhydrazine and entrapped solvent molecules. The best lineshapes and linewidths for DPPH are obtained when the sample is solvent-free, recrystallized from CS_2 [23].

(I) DPPH

2. *Potassium peroxylamine disulfonate* $[K_2(SO_3)_2NO]$ (= Fremy's salt). This is a good standard (also for g and magnetic field) for aqueous solutions, since concentrations can be determined optically [24,25]. Aqueous solutions should be prepared in 10% Na_2CO_3; even these are stable for only about one day, and the stability is sensitive to the preparation method.

3. $MnSO_4 \cdot H_2O$ and $CuSO_4 \cdot 5H_2O$. These are good intensity standards since they are readily available in pure form; however, their lines are rather broad and asymmetric (note that for $^{55}Mn^{2+}$, $S = I = \frac{5}{2}$). The $CuSO_45H_2O$ must be kept carefully hydrated to serve as a quantitative standard.

4. The nitroxide free-radical compounds 2,2,6,6-tetramethyl-1-piperidinyl-1-yl oxyl (=TEMPO) (**II**) and 4-maleimido-TEMPO (**III**) have proved to be very versatile standards.

(II) 2,2,6,6-tetramethyl-1-piperidinyl-1-yl oxyl   (III) 4-maleimido TEMPO

They can be obtained in very pure form, and they dissolve readily in a variety of solvents, including water. In addition, such solutions (when stored in a refrigerator) are stable over a period of months. There exist many variants of these: X-TEMPO, where X can be OH (TEMPOL, alias TANOL), NH_2 and so on, many of which are used as spin labels (e.g., see Refs. 26–29), increasingly in biomedical circumstances, to measure concentrations of specific active sites.

5. The EPR lines of Cr^{3+} in Al_2O_3 (synthetic ruby) provide a good secondary standard [30]. The crystal is chemically stable and exhibits no lines in the $g = 2$ region when suitably oriented.

6. The EPR line of S_3^- in ultramarine pigment (powder) also has been found to be useful as a secondary standard [31]. This signal has the advantage in some situations of having a strong signal at a g value of 2.030, significantly higher than g_e [32].

7. The set of lines (any one sufficiently strong and conveniently located) from ^{55}Mn present in plasticine putty [33] can be very useful.

The determination of the absolute number of spins is difficult, and its discussion is postponed to Section F.4.3.

F.3   SENSITIVITY AND RESOLUTION

The optimization of the signal-to-noise ratio of a spectrometer is frequently a prerequisite to the successful performance of an EPR experiment. Such optimization requires familiarity with the various factors that affect the *signal* level and the *noise* level.

F.3.1   Optimum Sensitivity

The minimum detectable number N_{min} of paramagnetic centers in an EPR cavity, assuming a signal-to-noise ratio of unity, is given to a good approximation by

$$N_{min} = \frac{12\pi V_c k_b T_s \Gamma}{\mu_0 g^2 \beta_e^2 S(S + 1)B_r Q_U} \left(\frac{F k_b T_d b}{P_0}\right)^{1/2} \tag{F.5}$$

Here

V_c = the volume (m^3) of the cavity (assumed to be operated in the TE_{102} mode)
k_b = Boltzmann constant ($J\ K^{-1}$)
T_s = temperature of the sample (K)
Γ = half-width (in mT) at half-height of the single absorption line (Tables F.1)
B_r = magnetic field (mT) at the center of the absorption line
Q_U = the effective unloaded Q factor of the cavity (see Eq. F.11)
T_d = microwave-detector temperature (K)
b = bandwidth in s^{-1} of the entire detecting and amplifying system
P_0 = microwave power ($J\ s^{-1}$) incident on the cavity

F = a noise factor (>1) attributable to sources other than thermal detector
noise (an ideal spectrometer would have $F = 1$)

μ_0 = the magnetic constant ($J^{-1} m^3 T^2$)

All quantities here are in SI: mksC units. The derivation of Eq. F.5 [10, Chapter 11; 34] assumes that the absorption shape is lorentzian (Table F.1a), that the Curie law (Eqs. 1.17c,d) applies, and that microwave saturation does not occur. The meaning of 'saturation' in this context is that the power is in the region in which the EPR signal no longer increases linearly with increase in B_1.

An estimate of N_{min} is obtained by inserting the following typical values:

$V_c = 1.1 \times 10^{-5} m^3$ (for a TE_{102} cavity at X band)
$T_s = T_d = 300$ k, $g = 2.00$
$\Gamma = 0.1$ mT, $S = \frac{1}{2}$
$B_r = 340.0$ mT
$Q_U = 5000$
$b = 1$ s^{-1}
$P_0 = 10^{-1}$ J s$^{-1} = 100$ mW
$F = 100$

These factors give $N_{min} \approx 10^{11}$ spins. For a typical sample, the minimum detectable concentration of paramagnetic centers is $\sim 10^{-9}$ M. Figures such as these are typically quoted by manufacturers of EPR spectrometers. However, unless the conditions of measurement are given, a quoted N_{min} value may be misleading. In particular, many samples are so readily saturated that power levels in excess of 1 mW are out of the question. Hence N_{min} would be effectively larger by a factor of 10 relative to the calculation above, which assumed 100 mW. In this calculation, several potentially relevant effects of the field modulation have been ignored.

When an EPR spectrum contains hyperfine lines, the intensity of a given line is only a fraction of the total intensity. Hyperfine splitting increases N_{min} by the factor

$$\mathcal{R} = \frac{\sum_j d_j}{d_k} \tag{F.6}$$

where d_k is the degeneracy of the most intense line and $\sum_j d_j$ is the sum of the degeneracies of all N lines in the spectrum ($j = 1, 2, \ldots, N$). For the 2,5-dioxy-p-benzosemiquinone trianion (Fig. 3.20), $\mathcal{R} = 2.00$, and for the naphthalene anion radical (Fig. 3.8), $\mathcal{R} = 7.11$. Similar considerations hold when fine-structure (matrix **D** non-zero) or other splittings are present.

The statistical methods for analyzing spectra are continuing to be developed. For instance, the principle of maximum entropy states that given certain testable information about a probability distribution that is not in itself sufficient to determine that distribution uniquely, one should prefer to choose as best that distribution which maximizes the Shannon information entropy [35]. This principle is much applied in analysis of both the time and frequency domains of magnetic-resonance spectra [36], mostly in NMR so far, but also increasingly in EPR spectroscopy and imaging. Herein, we content ourselves with mention of one especially informative paper [37] that analyzes a set of one-shot field-swept cw EPR spectra [X band, at room temperature (RT)] of the TEMPO free radical at various concentrations in methanol. The techniques used, which are thoroughly explained, lead to significant improvement in sensitivity, compared to conventional spectral enhancement techniques, and yield more reliable hyperfine splittings and relative peak intensities.

The present section deals primarily with sensitivity. However, many spectra may be so rich in hyperfine components that resolution becomes an additional factor to be optimized. Increased resolution often results in a decreased sensitivity. It may then be necessary to sacrifice some sensitivity in order to gain the requisite resolution. Various factors affecting sensitivity and resolution will be treated in subsequent sections.

F.3.2 Sample Temperature

Even when the temperature of the sample being studied does not markedly affect the EPR linewidth, for maximum sensitivity one should work at as low a temperature as feasible, since the signal amplitude and area are generally inversely proportional to the absolute temperature (Curie's law; see Eqs. 1.17c,d as well as Eqs. 10.27–10.29). However, in many cases, the temperature also has an effect on the width.

If the linewidth is determined by a short relaxation time τ_1, as in transition ions, a significant decrease in sample temperature can have a dramatic effect on the EPR spectrum. The reason is that τ_1 generally is a strong function of the sample temperature (e.g., in some cases τ_1 is proportional to the inverse seventh power of the sample temperature as the temperature is lowered). For some samples, liquid-helium temperatures (4 K or lower) are necessary to obtain sufficiently narrow lines. This is especially true of many of the rare-earth and actinide ions. (For general coverage of this topic, see Ref. 10, Chapter 13, and Refs. 11, 38 and 39.)

F.3.3 Microwave Frequency

The microwave frequency is a parameter that is varied only slightly in most EPR work. The principal reason is that most spectrometers permit frequency variations of no more than $\pm 10\%$ of the center frequency of the source. However, there are numerous situations for which a major change (usually an increase) in the microwave frequency can result in a very significant improvement in sensitivity, despite the inconvenience (and expense!) involved. Several aspects are considered:

1. *Filling Factor, Microwave Power and Dielectric Loss.* As the microwave frequency increases, the size of the resonator (for the same mode) must decrease. If the sample size is scaled in the same proportion as the resonator dimension, then the filling factor η remains constant. In this case of scaling both the sample and the resonator with the inverse of frequency, the result is $\nu^{-1/4}$. If the dielectric loss is significant, then this factor is $\nu^{1/2}\rho^{1/2}$ where ρ is the sample resistivity. Since the manipulation of a sample in a small resonator is difficult, there is little advantage of an increase in frequency *in this case*. Indeed, if the sample is readily saturated, then the improvement is even more minimal; for constant B_1 at the sample, the sensitivity increases only as $\nu^{3/4}$. For the case of constant incident power for a

non-saturable sample, the factors are (a) $\nu^{1/2}$ for the case in which resonator resistance dominates, and (b) $\nu^2 \rho$ for the case in which sample loss dominates.

2. *Sample Size.* If the sample size is fixed, as is often true for single crystals, then an increase in the microwave frequency may result in a dramatic improvement in sensitivity. In this case, the sensitivity increases as $\nu^{9/2}$. The important factor here is that for constant sample volume, the filling factor increases as ν^3. All other factors being equal, a change from X band to Q band would result in a 500-fold increase in sensitivity!

3. *Aqueous Samples.* Unfortunately, the dielectric loss of water increases with frequency from X band to Q band (Fig. E.1). The resultant reduction in the Q factor largely cancels any gain in sensitivity that would otherwise be achieved. It is for this reason that almost all aqueous solution studies are carried out at X band or lower frequencies.

F.3.4 *Q* Factor of the Resonator

For our use, the quality factor (also known as the 'figure of merit') for the resonant system may be defined as

$$Q = \frac{2\pi \text{ (maximum microwave energy stored in the resonator)}}{\text{energy dissipated by the resonator per microwave cycle}} \quad \text{(F.7a)}$$

$$Q = \frac{\text{resonant frequency } \nu_0}{\text{bandwidth } \Delta\nu} \quad \text{(F.7b)}$$

and is seen to be a dimensionless parameter. Thus, for a resonator (say, a metal cavity), the value of Q under the condition that only losses within the empty resonator are considered (i.e., resistive losses in the walls) is called the *unloaded Q factor*: Q_u. However, when the resonator is a cavity, an iris is usually used to couple to the waveguide system. Since this entails additional losses, there is a further lowering of Q. This coupling loss is measured by

$$Q_r = \frac{2\pi \text{ (maximum microwave energy stored in the resonator)}}{\text{energy lost through openings, per microwave cycle}} \quad \text{(F.8)}$$

The ratio $\beta_Q = Q_u/Q_r$ is called the *coupling parameter*. For optimum coupling (i.e., maximum transmission of power into the resonator), $\beta_Q = 1$. The overall or loaded quality factor Q_L is defined by

$$\frac{1}{Q_L} = \frac{1}{Q_u} + \frac{1}{Q_r} \quad \text{(F.9)}$$

Hence, when $\beta_Q = 1$, $Q_L = Q_u/2$.

If within the resonator there are materials that have a non-vanishing imaginary part ε'' of the complex dielectric constant $\varepsilon' - i\varepsilon''$ [10, p. 61], additional losses can occur. To account for these losses, the dielectric Q factor is defined as

$$Q_\varepsilon = \frac{2\pi \text{ (maximum microwave energy stored in the resonator)}}{\text{energy lost in the dielectric, per microwave cycle}} \qquad \text{(F.10)}$$

An effective unloaded factor Q_U is defined such that

$$\frac{1}{Q_U} = \frac{1}{Q_u} + \frac{1}{Q_\varepsilon} \qquad \text{(F.11)}$$

and is the factor that should be used in sensitivity calculations (e.g., Eq. F.5). The loaded Q changes accordingly, replacing Q_u by Q_U in Eq. F.9.

Most of the dielectric losses usually occur within the sample or the sample tube. For instance, the effect of lossy solvents on quantitative EPR studies have been examined in some detail [40]. The glass or quartz sample tube has a smaller effect. However, it is important to position the tube in a region of the resonator for which the microwave electric field is a minimum. Positioning is extremely critical for samples (such as aqueous solutions) that exhibit a high degree of dielectric loss. For these, the quite satisfactory container is a flat high-purity silica cell that can be accurately oriented along the nodal plane of the field $\mathbf{E}_1$. It is found that optimum sensitivity occurs when $Q_u = Q_\varepsilon$, that is, for a reduction in Q_U to one-half of its value in the absence of dielectric loss ($Q_L = \infty$) [10, p. 439]. For aqueous solutions this requires that the silica cell walls be separated by $\sim$0.3 mm for X-band frequencies. At higher frequencies, the dielectric loss for water is even more serious (Fig. E.1). For organic solvents with $\varepsilon'' < 10$, cylindrical sample tubes with an internal diameter of $\sim$3 mm are permissible.

The TM_{102} cavity is very useful for aqueous solutions because it enables use of a larger sample, in a special silica flat cell, and hence attains achievement of a higher sensitivity. Much effort has been expanded to optimize the design of resonators for the very important use of aqueous EPR samples [41–43], with striking results. Furthermore, a special multi-tube aqueous sample container (AquaX cell) is now commercially available.

Since most glass and ordinary silica materials give EPR signals, many are not suitable as sample-tube materials. Use of high-purity fused silica alleviates these difficulties, if thermally annealed (and not exposed to high-energy radiation; see Sections 11.8 and 13.13). This material is also advantageous since it has a very low dielectric loss and hence is a desirable material for the fabrication of dewar inserts.

Cylindrical cavities operated in the TE_{011} mode (Fig. E.5) generally have a significantly larger Q factor than does a rectangular cavity in the TE_{102} mode. Thus if the sample causes only a low dielectric loss, use of such a cylindrical cavity may be advantageous for cw EPR.

Loop-gap resonators (see Section E.1.5) usually have a lower Q value, but much higher filling factor, relative to cavity resonators. They are usually the resonator of choice for spin-echo and FID measurements, and for saturation recovery if the relaxation time is too short relative to the Q value of a cavity resonator.

F.3.5   Microwave Power Level and Measurements of B_1

To measure the mean magnitude B_1 (averaged over the sample volume) is not a trivial assignment. This requires a knowledge of P_0, Q_U (Eq. F.11), and the distribution (filling factor η) of $\mathbf{B}_1$ within the resonator. For a small sample in a TE$_{102}$ cavity, $\eta \sim 2V_s/V_c$,[5] where V_s is the sample volume and V_c is the cavity volume.

Among other techniques, B_1 at the sample can be determined by inserting a small metallic sphere into the appropriate region and measuring the resulting shift in resonator frequency [44]. In pulsed EPR spectrometers, B_1 can be measured by nulling FID and echo signals and by nutation experiments [45–48]. Happily, discussion of a series of techniques to obtain B_1 has been furnished by Poole [10, Section 5I], and we can refer the reader to this.

At power levels in excess of 10^{-4} W, the signal output voltage from the detector crystal of a reflection-cavity EPR spectrometer is proportional to $P_0^{1/2}$. In a modern cw spectrometer, the user realistically assumes that the signal response is linear with square root of power. The user does not know what the power incident on the crystal is. Often, this can be easily obtained from microwave power meters.

The microwave power P_0 at the sample should be low enough that saturation effects are negligible. Note use of the bypass arm discussed in Section E.1, which allows the user to vary the power without worrying about the detector-crystal bias.

When power saturation occurs, the area under the absorption curve is no longer a valid measure of spin concentration in the sample, and distortion of the lineshape may occur. This depends on the spin relaxation times, on spectral diffusion rates, on the field (B) sweep rate and on the modulation field amplitude (B_m) and modulation frequency (ν_m). The various *passage* cases have been classified [49; 50, Section 2.4; 51], which is useful for the analysis of such effects.

It is instructive to note that when the first-derivative amplitude is at its maximum with respect to P_0 (Fig. F.8a), the linewidth has risen to only ~1.2 times the width in the absence of saturation (Fig. F.8b). Hence, at this maximum, microwave saturation does not increase width as much as does excessive modulation amplitude. It is good practice to obtain first a maximum derivative amplitude. If resolution is important, then P_0 should be reduced to about 75% of its value at the maximum amplitude.

For inhomogeneously broadened single lines (which tend to be gaussian in shape), the derivative amplitude theoretically increases monotonically to a limiting value with increasing power (P_0). This behavior is indicated by the dashed curve in Fig. F.8a. In practice, it is observed that even for lines that one classifies as inhomogeneously broadened, the amplitude goes through a maximum with increasing power. This implies some measure of homogeneous broadening, arising perhaps from mutual spin flips with the surroundings of the spin.

Some spectrometers are capable of detecting the dispersion, that is, the real part χ' of the dynamic magnetic susceptibility (Eq. 10.27a), which accompanies absorption. The dispersion signal does not saturate as readily as the absorption signal. Hence it is desirable to detect dispersion when dealing with signals that saturate readily. This is especially important when working at liquid helium temperatures, where τ_1 may be very long. The usefulness of dispersion measurements in EPR has been discussed [52]; for instance, the static magnetic susceptibility $\chi°$ for DPPH has been obtained from such data. We can note the work of Hoffman and coworkers [53], who routinely use dispersion spectra at low temperatures to avoid relaxation-time problems. Dispersion-versus-absorption (DISPA) diagrams are useful in EPR for analysis of modulation broadening and instrumental distortion [54,55]. Dispersion spectra are noisy in a high-Q resonator. Hyde and co-workers [56] showed that the decreased demodulation of phase noise in a low-Q loop-gap resonator made it superior to a high-Q cavity for dispersion spectra. An even greater improvement for dispersion was found in a crossed-loop resonator [57].

F.3.6   Modulation Amplitude

It was noted previously that field-modulation techniques may be employed to improve the sensitivity of a spectrometer. However, an excessive modulation amplitude or an excessively high modulation frequency can lead to line distortion.

The observed peak-to-peak first-derivative linewidth is represented by $\Delta B_{pp}(B_m)$. The modulation amplitude B_m should be a small fraction of $\Delta B_{pp}(0)$, since the portion of an absorption line scanned during any half-cycle of field modulation must be nearly linear in field B to obtain an output that is close to being the first derivative of the absorption line (Fig. E.6). As B_m approaches and exceeds $\Delta B_{pp}(0)$, the derivative line amplitude first increases linearly with B_m, then reaches a maximum, and finally decreases slowly (Fig. F.4). However, long before the line amplitude reaches a maximum, the observed linewidth $\Delta B_{pp}(B_m)$ increases significantly (Fig. F.5). This phenomenon has been analyzed for lorentzian [58] and for gaussian [59] lines (Section F.2). The results were given[3] in Table E.2 of the earlier edition of the present book [61], and are displayed in Figs. F.4 and F.5 herein. It follows that a maximum in the first-derivative amplitude $A_{pp}(B_m) = 2Y_{max}'$ is obtained when $B_m \approx 3.5\,\Delta B_{pp}(0)$ for lorentzian lines and $\sim 1.8\,\Delta B_{pp}(0)$ for gaussian lines. At these settings the lines are considerably broadened (by a factor of 3 for lorentzian lines and of 1.6 for gaussian lines).

The optimum setting of B_m depends on how much sensitivity can be sacrificed for faithful reproduction of the lineshape or vice versa. If resolution and true lineshape are important, then the modulation amplitude should satisfy the condition $B_m \leq 0.2\,\Delta B_{pp}(0)$. However, if sensitivity is the prime concern and some line distortion can be tolerated, then B_m should be increased until a maximum derivative amplitude is obtained. A reasonable compromise between sensitivity and resolution is then achieved by reducing B_m by a factor of 4–5 from the value that makes A_{pp} a maximum. In practice, this might lead to excessively high modulation amplitudes for broad lines, and the practical limit may be heating of the resonator or

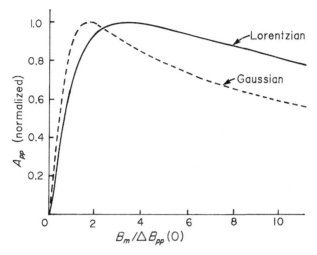

FIGURE F.4   Normalized peak-to-peak amplitudes (A_{pp}) for first derivatives of lorentzian and gaussian lines as a function of modulation amplitude (B_m). $\Delta B_{pp}(0)$ is the limiting peak-to-peak first-derivative linewidth as B_m goes to zero.

microphonics caused by the large eddy currents. The effect of increasing B_m for a very narrow line can be seen in Fig. F.6 and, if $B_m \geq \Delta B_{pp}(0)$, then B_m can be determined directly from the peak separation, say, using that figure.

Computational approaches have been developed to correct for the lineshape distortion due to overmodulation [62–64].

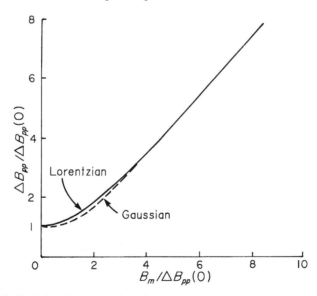

FIGURE F.5   Relative first-derivative linewidths at increasing values of the relative modulation amplitude $B_m/\Delta B_{pp}(0)$.

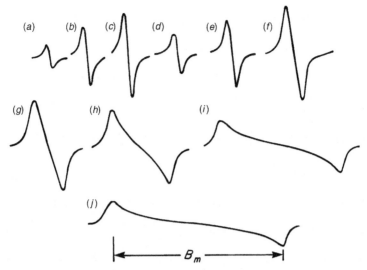

FIGURE F.6 NMR signals of protons in an aqueous solution of $Cr(NO_3)_3$ ($\Delta B_{pp}(0) = 19$ mT; $\nu_m = 40$ Hz) as a function of modulation amplitude B_m. The field scan is the same for each trace. Values of $B_m/\Delta B_{pp}(0)$ are as follows: (a) 0.150; (b) 0.398; (c) 0.552; (d) 0.552; (e) 1.052; (f) 2.48; (g) 4.94; (h) 10.14; (i) 20.6; (j) 28.8. The gain for traces (a)–(c) is twice that for traces (d)–(j). [From G. W. Smith, *J. Appl. Phys.*, **35**, 1217 (1964).]

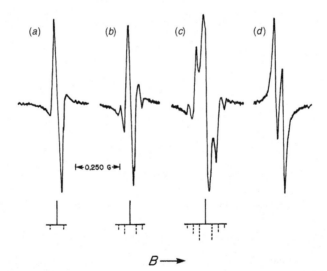

FIGURE F.7 Modulation sidebands on the EPR spectrum of the *F* center in CaO, for which the linewidth is less than 2.0 μT. Modulation amplitude: (a) 0.4 μT; (b) 2.0 μT; (c) 5.0 μT. Spectrum (d): Phase adjusted so that the central line is not seen. The two lines are the first modulation sidebands. The modulation sidebands are opposite in phase to the central line; their positions are indicated by dashed lines in (a), (b) and (c). The separation of the sidebands corresponds to $h\nu_m/g_e\beta_e$, which, for $\nu_m = 100$ kHz, is 3.6 μT. (Unpublished spectra supplied by J. E. Wertz.)

F.3.7 Modulation Frequency

The observed line also is distorted if the magnitude of the modulation frequency approaches that of the linewidth (say, in Hz), that is, if $v_m \sim (g_e\beta_e/h)\Delta B_{pp}$. Since the crystal detector is a non-linear device, its output contains the sum and the difference of the microwave and the modulation frequencies. This results in the production of sideband resonance lines spaced $h v_m/g_e\beta_e$ apart and extending over a range B_m. For a modulation frequency of 100 kHz, $h v_m/g_e\beta_e = 3.6\,\text{mT}$. The development of these sidebands as the amplitude of the modulation is increased is shown in Fig. F.7 for a line having a width less than 2.0 µT. The sidebands have a phase difference of 180° with respect to the central line; the latter can be made to vanish (as in Fig. F.7d) by appropriate adjustment of the phase-sensitive detector. As expected, the sidebands are separated by about 3.6 µT. For a description of the relevant theory, see the article by Kälin et al. [65]. The phase of the sidebands relative to the centerband depend on depth of modulation and the phases of the RF and audiofrequencies [66].

F.4 MEASUREMENTS

F.4.1 *g* Factors and Hyperfine Splittings

The absolute determination of a *g* factor implies the simultaneous measurement of a frequency v (of $\mathbf{B}_1$) and a magnetic-field magnitude B. There are various problems in practice [25,67], including the following:

1. *Decision as to Where on the Lineshape to Measure the Magnetic Field* The EPR line must be symmetric (no admixture of absorption and dispersion) and imposed on a flat horizontal base line. The narrower the line, the easier it is to ascertain its center (zero of the first derivative). The choice of line presentation (e.g., absorption curve or its first derivative) does affect the accuracy attained for *g*. Obviously, also the signal-to-noise ratio is an additional relevant factor.

2. *Sample Considerations* The EPR line position is affected by the sample's temperature, spin concentration, bulk diamagnetic contribution (the demagnetizing field depends on the sample shape and can even be anisotropic) and its container.

3. *Signal Distortions* These usually arise from field modulation, power saturation, scan rate and field inhomogeneity.

4. *Magnetic-Field Measurement* If an NMR gaussmeter is used, the probe should be very close to the sample to minimize positional field differences. An appropriate correction for this factor is to apply the field difference method. The NMR *g* factor must be corrected for medium effects, including those caused by paramagnetic relaxation additives. Since this measurement of *B* amounts to measuring an NMR (e.g., proton) frequency, it may be possible to utilize a single counter, that is, reduce attaining *g* basically

to a frequency-ratio measurement. Modern calibrated Hall probes in commercial spectrometers measure the magnetic field about as accurately as an NMR gaussmeter. A species with known g factor has to be used to calibrate the offset between the field in the resonator and the field at the Hall probe.

5. *Frequency Counter Accuracy*
6. *Stability of v and B, and of Other Instrumental Factors*
7. *Limits on the Accuracy of the Natural Constants (h, β_e, β_n) Used*
8. *Special Considerations Regarding Determination of the Sign of g.* See text below

Most of these factors are also relevant to measurement of hyperfine and other zero-field splittings that involve differences of line positions. We note that the g factor may well depend on sample temperature and composition.

As in the case of spin-concentration measurements, it is useful to use a relative measurement of g factors against well-established standards. Table F.2 lists some

TABLE F.2   Some Radicals for Which the g Factors Are Accurately Known

Radical	Solvent[a]	g Factor[b]	Reference
Naphthalene⁻	DME/Na −58°C	2.002743 ± 0.000006	d
Perylene⁻ ⎱(IV)	DME/Na	2.002657 ± 0.000003	d
Perylene⁺ ⎰	Concentrated H_2SO_4	2.002569 ± 0.000006	d
Tetracene⁺	Concentrated H_2SO_4	2.002590 ± 0.000007	d
p-Benzosemiquinone⁻	Butanol/KOH at 23°C	2.004665 ± 0.000006[e]	72
Wurster's blue cation	Absolute ethanol	2.003051 ± 0.000012	76
DPPH ⎱(I)	Benzene	2.00354 ± 0.00003	74
DPPH ⎰	None (powder)	2.0037 ± 0.0002	75
Pitch (C)	In KCl powder	2.0028	73

[a]Deoxygenated.
[b]Not corrected for any second-order hyperfine shifts present.
[c]DME is dimethoxyethane.
[d]Temperature-dependent.

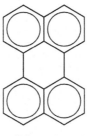

(IV) perylene

of the commonly used materials. We consider a constant-frequency experiment. The unknown factor g_X can then be obtained from

$$g_X - g_{Std} = -\frac{B_X - B_{Std}}{B_{Std}} g_{Std} \qquad (F.12)$$

where g_{Std} is the g factor of the standard, B_{Std} is the center-field position of the EPR spectrum of the standard and B_X is the center field of the unknown EPR spectrum. In using Eq. F.12 it is important to use the same medium and sample configuration to avoid spurious field shifts.

The magnetic-field sweep may also be calibrated conveniently by the use of a dual-sample cavity (see Section F.4.3). Here it is convenient to use as a standard a substance giving a many-line spectrum for which an accurate determination of the hyperfine splittings has been made.

One commonly used spin concentration and sensitivity standard is pitch (containing carbonaceous radicals) mixed into powdered KCl. Typically, for a weak-pitch sample (0.00033%; 10^{13} spins per centimeter of length), $g = 2.0028$, $\Delta B_{pp} = 0.17$ mT and $\mathcal{A}/[Y'_{max} (\Delta B_{pp})^2] = 5.46$ [10, p. 445]. The single line is not exactly lorentzian.

It should be noted that, while DPPH is a good standard for many purposes [68], there are problems in making it really quantitative. As mentioned, its purity varies quite widely, with solvent molecules and the parent hydrazine (2,2′-diphenyl-1-picrylhydrazine) as typical inclusions. Furthermore, single crystals show EPR anisotropy. One way of coping with this problem is to use appreciable amounts of the powder, but moving it nearly (basically) 'out' of the resonant cavity to adjust the signal intensity to be not excessive [69].

One useful standard is Wurster's blue perchlorate (see Section 9.2.8 and Ref. 70). Table E.4 in the previous (1994) edition of this book lists, for ethanol solutions, the line positions and relative intensities of some of the strong lines in this spectrum (some of the weak outermost lines are also included).

A review of the calibration methods and the reference materials that have been used or recommended for EPR spectroscopy is available [71].

The sign of the various g values is a vexatious topic, and determination of the sign is rarely considered (however, see Sections 4.4 and 8.7; also Ref. 10, Section 5O).

F.4.2 Relaxation Time

The transverse relaxation time τ_2 is obtainable in appropriate cases from measurement of the linewidth (Section 10.4). To obtain the longitudinal relaxation time τ_1, the spin system must be perturbed from thermal equilibrium, so as to change the relative populations of its energy levels. Suitable measurements as B_1 is increased to achieve saturation can yield an estimate for τ_1. For instance, from

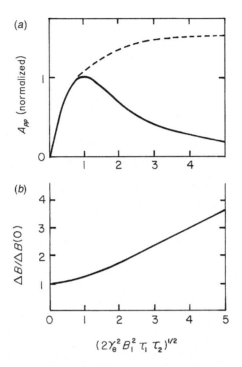

FIGURE F.8 (*a*) The normalized first-derivative amplitude as a function of B_1 (proportional to $P_0^{1/2}$) for a homogeneously broadened EPR line. The dashed curve denotes an inhomogeneously broadened line. (*b*) The normalized peak-to-peak linewidth as a function of B_1 for a homogeneously broadened line.

Eq. 10.27*b*, one can derive the expression

$$A_{pp} = 2Y_{max}' = \frac{9}{4\sqrt{3}\pi} \frac{|\gamma_e| B_1 \tau_2^2}{(1 + \gamma_e^2 B_1^2 \tau_1 \tau_2)^{3/2}} \tag{F.13}$$

for the peak-to-peak first-derivative amplitude (Fig. F.8*a*). By differentiating this expression with respect to B_1 to obtain the value of B_1^2 (i.e., P_0) that gives a maximum in this amplitude, one finds the spin-lattice relaxation time to be given by

$$\tau_1 = \frac{1}{2\tau_2 \gamma_e^2 B_1^2|_{max}} \tag{F.14}$$

where B_1 is the excitation field amplitude at the sample. This relation offers one technique for evaluating τ_1 when τ_2 and the maximal B_1 is known. This method is useful if all hyperfine lines are well resolved, and other conditions are chosen such that the linewidth is a measure of τ_2. Other methods include scanning repeatedly through resonance in time short relative to the relaxation time [72]. A technique utilizing double transverse excitation and longitudinal detection also allows measurement of τ_1 without the need to measure B_1, provided that τ_2 is available [73].

General discussions about techniques to obtain relaxation can be found in abundance. Thus, many references to various techniques using cw spectrometers exist [10, Chapter 13; 50, Section 5.4; 74–76]. Pulse EPR techniques for measurement of τ_1 and τ_2 also abound. Some aspects of these are evident from the discussion in Chapter 11. As an example, we cite the free-precession technique. Here a $180°$ pulse inverts the direction of the magnetization component $\mathbf{M}^T \cdot \hat{\mathbf{B}}$, which returns to its equilibrium value with the time constant τ_1 (Section 10.3). Monitoring the recovery at suitable intervals by either spin-echo or FID measurements yields τ_1, unless there is spectral diffusion. A $90°–\tau–180°$-τ echo sequence, incrementing τ, will yield τ_2 under the condition that there is no instantaneous diffusion. For details of measurement of relaxation times, consult [10, Chapter 13; 77–79].

F.4.3 Spin-Number Determinations

This type of measurement may well be among the most difficult in the realm of EPR spectroscopy, as is the determination of the absolute total intensity (transition probability) for a known sample. A thorough discussion of this topic can be found in Alger's book [50, Section 5.3].

If a sample containing a known number of spins is to be used as a comparison standard, then filling-factor aspects are all-important. Accurate knowledge of the spatial variations of both the microwave magnetic field and the modulation field is required, and ignoring these distributions can cause serious errors. The sample size, and its effect on the microwave field distribution, should be minimal. Mitigating against this is the need for good signal-to-noise ratio. Saturation effects, too, are deleterious.

If a standard is used, it and the unknown must be placed in equivalent positions and must have the same sample volume. Weighed samples of pure DPPH, $CuSO_4 \cdot 5H_2O$, or the nitroxide compounds mentioned above can serve as satisfactory standards. The resonator quality factor Q (see below) should be measured for accurate work, unless the samples and standards are known to have identical effects on Q.

The use of a dual-sample cavity (TE_{104}) is advantageous when making absolute-intensity measurements since both standard and unknown can be run simultaneously [10, p. 188]. The standard and unknown samples should be interchanged to ensure that differences between the two sample positions are accounted for. One can also record the sum and difference spectra of two samples by using this device [80].

Clearly, minimizing the linewidth as much as possible is beneficial. However, broadening is not important to the problem of determining spin concentration, unless the lines are so broad that they are difficult to integrate.

REFERENCES

1. D. Griller, *Magn. Reson. Rev.*, **5**, 1 (1979).
2. H. C. Heller, S. Schlick, T. Cole, *J. Phys. Chem.*, **71**, 97 (1967).
3. K. A. McLauchlan, D. G. Stevens, *Acc. Chem. Res.*, **21**, 54 (1988).
4. M. C. Baird, *Chem. Rev.*, **88**, 1217 (1988) and references cited therein.
5. M. S. de Groot, C. A. de Lange, A. A. Monster, *J. Magn. Reson.*, **10**, 51 (1973).
6. A. M. Waller, R. G. Compton, "In-situ Electrochemical ESR", in *Comprehensive Chemical Kinetics*, R. G. Compton, A. Hamnett, Eds., Vol. 29, Elsevier, Amsterdam, Netherlands, 1989, Chapter 7.
7. E. Solon, A. J. Bard, *J. Am. Chem. Soc.*, **86**, 1926 (1964).
8. A. M. Bass, H. P. Broida, *Formation and Trapping of Free Radicals*, Academic Press, New York, NY, U.S.A., 1960.
9. D. F. Howarth, J. A. Weil, Z. Zimpel, *J. Magn. Reson.*, **161**, 215 (2003).
10. C. P. Poole Jr., *Electron Spin Resonance—a Comprehensive Treatise on Experimental Techniques*, 2nd ed., Wiley Interscience, New York, NY, U.S.A., 1983.
11. C. P. Poole, Jr., H. A. Farach, *Theory of Magnetic Resonance*, 2nd ed., Wiley, New York, NY, U.S.A., 1987.
12. M. M. Maltempo, *J. Magn. Reson.*, **68**, 102 (1986).
13. G. B. Arfken, H. J. Weber, *Mathematical Methods for Physicists*, 6th ed., Academic Press, New York, U.S.A., 2005, Section 1.15.
14. J. S. Hyde, H. S. Mchaourab, T. G. Camenisch, J. J. Ratke, R. W. Cox, W. Froncisz, *Rev. Sci. Instrum.*, **69**(7), 2622 (1998).
15. R. Aasa, T. Vänngård, *J. Magn. Reson.*, **19**, 308 (1975).
16. J. R. Pilbrow, *J. Magn. Reson.*, **58**, 186 (1984).
17. S. Brumby, *Magn. Reson. Rev.*, **8**, 1 (1983).
18. M. J. Mombourquette, J. A. Weil, FORTRAN Program *EPR-NMR*, Department of Chemistry, University of Saskatchewan, Saskatoon, SK, S7N 5C9, Canada (1996, and still evolving)].
19. G. R. Hanson, K. E. Gates, C. J. Noble, A. Mitchell, S. Benson, M. Griffin, K. Burrage, in *EPR of Free Radicals in Solids*, A. Lund, M. Shiotane, Eds., Kluwer, Dordrecht, Netherlands, 2003, Chapter 5.
20. M. L. Randolph, *Rev. Sci. Instrum.*, **31**, 949 (1960).
21. M. L. Randolph, in *Biological Applications of Electron Spin Resonance*, H. M. Swartz, J. R. Bolton, D. C. Borg, Eds., Wiley, New York, NY, U.S.A., 1972, pp. 119 ff.
22. J. A. Weil, J. K. Anderson, *J. Chem. Soc.*, 5567 (1965).
23. S. V. Kolaczkowski, J. T. Cardin, D. E. Budil, *Appl. Magn. Reson.*, **16**, 293 (1999).
24. M. T. Jones, *J. Chem. Phys.*, **38**, 2592 (1963).
25. M. T. Jones, R. Ahmed, R. Kastrup, V. Rapini, *J. Phys. Chem.*, **83**, 1327 (1979).
26. K. Saito, K. Takeshita, J.-I. Ueda, T. Ozawa, *J. Pharm. Sci.*, **92**(2), 275 (2003).
27. T.-H. Kwon, K. Malloy, D. Sun, B. Alessandri, M. R. Bullock, *J. Neurotraum.*, **20**(4), 337 (2003).
28. K. Itoh, T. Kimizuka, K. Matsui, *Anal. Sci.*, **17** Suppl., i1157 (2001).

29. C. Kroll, H. H. Borchert, *Eur. J. Pharm. Sci.*, **8**(1), 5 (1999).

30. L. S. Singer, *J. Appl. Phys.*, **30**, 1463 (1959).

31. N. Gobeltz-Hautecoeur, A. Demortier, B. Lede, J. P. Lelieur, C. Duhayon, *Inorg. Chem.*, **41**(11), 2848 (2002).

32. G. R. Eaton, S. S. Eaton, J. W. Stoner, R. W. Quine, G. A. Rinard, A. I. Smirnov, R. T. Weber, J. Krzystek, A. K. Hassan, L.-C. Brunel, A. Demortier, *Appl. Magn. Reson.*, **21**, 563 (2001).

33. P. Rahimi-Moghaddam, Y. Upadrashta, M. J. Nilges, J. A. Weil, *Appl. Magn. Reson.*, **24**, 113 (2003).

34. G. Feher, *Bell. Syst. Tech. J.*, **36**, 449 (1957).

35. http://en.wikipedia.org/wiki/MaxEnt_thermodynamics.

36. L. Vahamme, T. Sundin, P. Van Hecke, S. Van Huffel, *NMR Biomed.*, **14**(4), 233 (2001).

37. B. A. Goodman, S. M. Glidewell, J. Skilling, *Free Rad. Res.*, **22**(4), 337 (1995).

38. K. J. Standley, R. A. Vaughan, *Electron Spin Relaxation Phenomena in Solids*, Adam Hilger, London, U.K., 1969, Chapters 5–8.

39. S. S. Eaton, G. R. Eaton, in *Distance Measurements in Biological Systems by EPR*, G. R. Eaton, S. S. Eaton, L. J. Berliner, Eds.; reprinted in *Biol. Magn. Reson.*, **19**, 29 (2000).

40. D. P. Dalal, S. S. Eaton, G. R. Eaton, *J. Magn. Reson.*, **44**, 415 (1981).

41. J. S. Hyde, R. R. Mett, *Curr. Top. Biophys.*, **26**(1), 7 (2002).

42. R. R. Mett, J. S. Hyde, *J. Magn. Reson.*, **165**, 137 (2003).

43. J. W. Sidabras, R. R. Mett, J. S. Hyde, *J. Magn. Reson.*, **172**, 333 (2005).

44. E. L. Ginzton, *Microwave Measurements*, McGraw-Hill, New York, NY, U.S.A., 1957, p. 445.

45. W. H. Perman, M. A. Bernstein, J. C. Sandstrom, *Magn. Reson. Med.*, **9**, 16 (1989).

46. M. Huisjen, J. S. Hyde, *J. Chem. Phys.*, **60**, 1682 (1974).

47. P. W. Percival, J. S. Hyde, *Rev. Sci. Instrum.*, **46**, 1522 (1975).

48. H. C. Torrey, *Phys. Rev.*, **76**, 1059 (1949).

49. M. Weger, *Bell Syst. Tech. J.*, **39**, 1013 (1960).

50. R. S. Alger, *Electron Paramagnetic Resonance: Techniques and Applications*, Wiley-Interscience, New York, NY, U.S.A., 1968.

51. G. E. Pake, T. L. Estle, *The Physical Principles of Electron Paramagnetic Resonance*, 2nd ed., Benjamin, New York, NY, U.S.A., 1973, Section 2.2.

52. J. Talpe, L. van Gerven, *Phys. Rev.*, **145**, 718 (1966).

53. P. D. Christie, H.-I. Lee, L. M. Cameron, B. J. Hales, W. H. Orme-Johnson, B. M. Hoffman, *J. Am. Chem. Soc.*, **118**, 8707 (1996).

54. F. G. Herring, A. G. Marshall, P. S. Phillips, D. C. Roe, *J. Magn. Reson.*, **37**, 293 (1980).

55. H. A. Buckmaster, T. Duczmal, in *Electronic Magnetic Resonance of the Solid State*, J. A. Weil, Ed., Canadian Society for Chemistry, Ottawa, ON, Canada, 1987, pp. 57 ff.

56. J. S. Hyde, W. Froncisz, A. Kusumi, *Rev. Sci. Instrum.*, **53**(12), 1934 (1982).

57. G. A. Rinard, R. W. Quine, B. T. Ghim, S. S. Eaton, G. R. Eaton, *J. Magn. Reson.*, **A122**, 58 (1996).

58. H. Wahlquist, *J. Chem. Phys.*, **35**, 1708 (1961).

59. G. W. Smith, *J. Appl. Phys.*, **35**, 1217 (1964).

60. J. E. Wertz, J. R. Bolton, *Electron Spin Resonance: Elementary Theory and Practical Applications*, McGraw-Hill, New York, NY, U.S.A., 1972.

61. J. A. Weil, J. R. Bolton, J. E. Wertz, *Electron Paramagnetic Resonance: Elementary Theory and Practical Applications*, Wiley, New York, NY, U.S.A., 1994.

62. V. A. Bikineev, E. A. Zavatski, V. V. Isaev-Ivanov, V. V. Lavrov, A. V. Lomakin, V. N. Fomichev, K. A. Shabalin, *Tech. Phys.*, **40**, 619 (1995).

63. B. H. Robinson, C. Mailer, A. W. Reese, *J. Magn. Reson.*, **138**, 199, 210 (1999).

64. M. Fedin, I. Gromov, A. Schweiger, *J. Magn. Reson.*, **171**, 80 (2004).

65. M. K. Kälin, I. Gromov, A. Schweiger, *J. Magn. Reson.*, **160**(2), 166 (2003).

66. O. Haworth, R. E. Richards, in *Progress in Nuclear Magnetic Resonance Spectroscopy*, J. W. Emsley, J. Feeney, L. H. Sutcliffe, Eds., Vol. 1, 1966, Chapter 1.

67. B. G. Segal, M. Kaplan, G. K. Fraenkel, *J. Chem. Phys.*, **43**, 4191 (1965), as corrected by R. D. Allendoerfer, *J. Chem. Phys.*, **55**, 3615 (1971).

68. M. S. Blois Jr., H. W. Brown, J. E. Maling, *Arch. Sci. (Genève)*, **13**, 243 (1960).

69. N. Chen, Y. Pan, J. A. Weil, *Newsl. Int. EPR Soc.*, **10**(3), 15 (1999).

70. W. R. Knolle, Ph.D. thesis, University of Minnesota, Minneapolis, MN, U.S.A., 1970.

71. T. Chang, *Magn. Reson. Rev.*, **9**, 65 (1984).

72. D. R. Locker, D. C. Look, *J. Appl. Phys.*, **39**, 6119 (1968).

73. M. Giordano, M. Martinelli, L. Pardi, S. Santucci, *Mol. Phys.*, **42**, 523 (1981).

74. C. P. Poole Jr., H. A. Farach, *Relaxation in Magnetic Resonance*, Academic Press, New York, NY, U.S.A., 1971.

75. J. Pescia, *J. Phys.*, **27**, 782 (1966) (in French).

76. K. J. Standley, R. A. Vaughan, *Electron Spin Relaxation Phenomena in Solids*, Adam Hilger, London, U.K., 1969, Chapters 5–8.

77. M. Sahlin, A. Gräslund, A. Ehrenberg, *J. Magn. Reson.*, **67**, 135 (1986).

78. S. S. Eaton, G. R. Eaton, in *Distance Measurements in Biological Systems by EPR*, G. R. Eaton, S. S. Eaton, L. J. Berliner, Eds.; reprinted in *Biol. Magn. Reson.* **19**, 29 (2000).

79. A. Schweiger, G. Jeschke, *Principles of Pulse Electron Paramagnetic Resonance*, Oxford University Press, Oxford, U.K., 2001, Chapter 8.

80. R. C. Smith, T. H. Wilmshurst, *J. Sci. Instrum.*, **40**, 371 (1963).

NOTES

1. For example, the 'stable' free radical 2,2'-diphenyl-1-picrylhydrazyl (DPPH) in solutions can be transformed into its hydrazine (2,2'-diphenyl-1-picrylhydrazine) electrolytically, reversibly, with appearance and disappearance of its purple color and EPR signal [7].

2. Note that field magnitude B_1 is defined by the expression $B_1(\nu) = 2B_1 \cos 2\pi\nu t$, where ν is the microwave frequency. Field $\mathbf{B}_1$ is assumed to be linearly polarized (Eqs. 10.23).

3. The modulation amplitude B_m at the modulation frequency ν_m was defined in [60, p. 453] in a non-standard way via the relation $B = B_0 + \frac{1}{2}B_m \sin(2\pi\nu_m t + \phi_m)$. In Refs. 58 and 59, the factor $\frac{1}{2}$ is absent, as is usual. The phase angle ϕ_m is important in the signal detection scheme (Sections E.1.6 and E.1.8).

4. The first moment of linewidth computed by evaluating the integral

$$\int_{-\infty}^{\infty} (B - B_r)^n Y' \, dB$$

for $n = 1$. Here $Y'(B - B_r)$ is the first-derivative lineshape function, B is the magnetic-field variable, B_r is the field at the center of the derivative curve, and $n = 0, 1, 2, \ldots$.

5. Actually, the filling factor η equals $V_s\langle B_1^2\rangle_s / V_c\langle B_1^2\rangle_c$, which leads to the approximation given, assuming a sufficiently small sample in a uniform B_1 field that is not changed by the sample [10, pp. 157, 439].

FURTHER READING

1. R. S. Alger, *Electron Paramagnetic Resonance: Techniques and Applications*, Wiley, New York, NY, U.S.A., 1967. (This book contains numerous drawings of experimental equipment and details of construction.)

2. R. G. Kooser, E. Kirschmann, T. Matkov, "Measurements of Spin Concentrations in Electron Paramagnetic Resonance Spectroscopy", *Concepts Magn. Reson.*, **4**, 145 (1992).

3. T. M. McKinney, I. B. Goldberg, "Electron Spin Resonance", in *Physical Methods of Chemistry*, Vol. 3B, *Determination of Chemical Compounds and Molecular Structure*, B. W. Rossiter, J. F. Hamilton, Eds., Wiley, New York, NY, U.S.A., 1989, pp. 383–584.

4. C. P. Poole Jr., *Electron Spin Resonance—A Comprehensive Treatise on Experimental Techniques*, 2nd ed., Wiley-Interscience, New York, NY, U.S.A., 1983.

5. A. A. Westenberg, "Use of ESR for the Quantitative Determination of Gas-Phase Atom and Radical Concentrations", *Prog. React. Kinet.*, **7**, 23 (1973).

PROBLEMS

F.1 (*a*) Plot on the same piece of graph paper a lorentzian line and a gaussian line, each with the same value of $Y_{\max}$ and Γ.

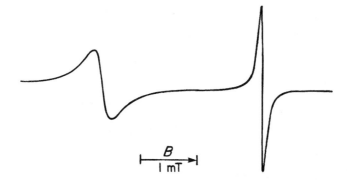

FIGURE F.9 An EPR spectrum showing two lorentzian first-derivative lines of different amplitude and width.

(*b*) On a second piece of graph paper plot the first-derivative curves for these two lines.

(*c*) On the basis of (*a*) and (*b*), what criteria would you use to determine the lineshape of an experimental line?

F.2   What methods could be employed to determine the intensity of an EPR line?

F.3   Calculate the relative intensities of the two lines shown in Fig. F.9.

F.4   Analyze the EPR spectrum of the cobalt dioxygen carrier presented in Fig. 3.14, drawing its eight first-derivative components and their superposition. Estimate the linewidth ΔB_{pp} (see Ref. 19 in Chapter 4), $A_0(^{59}Co)$ and g. Are the second-order hyperfine corrections appreciable? Also, draw the integrated absorption envelope. What effect would increase of ΔB_{pp} have on the inner part of the spectrum?

F.5   Suppose that kinetic studies are to be conducted on a radical for which the half-life is $\sim 10^{-3}$ s at $0°C$. With an active sample volume of ~ 0.5 cm^3, Q_u of the cavity is ~ 3000. The radical gives rise to a single lorentzian line with a peak-to-peak linewidth of 0.5 mT. The g factor is 2.01. The first-derivative signal amplitude reaches a maximum when the microwave power is ~ 10 mW. The measurements are made at X band, and a TE$_{102}$ cavity of volume 11 cm^3 is used. Assume $F = 200$, $T_d = 300$ K and $b = 10^{-4}$ s. With the conditions as given above, determine the appropriate minimum radical concentration detectable in this system for a signal-to-noise ratio of 10.

F.6   Determine the value of R in Eq. F.6 for the following radicals (see Chapter 3 and Table 9.3 for structures):
(*a*) Anthracene anion.
(*b*) Pyrazine anion.
(*c*) $^{13}CD_2H$.

F.7   For a certain radical, $\tau_1 = 10^{-5}$ s. Estimate the value of B_1 that would give a maximum first-derivative amplitude. The peak-to-peak linewidth is 0.1 mT.

F.8   For a field-modulation frequency of 15 kHz, compute the positions (in mT) of the modulation sidebands relative to the central line.

F.9   Derive Eq. F.14 from Eq. F.13, and show that the extremum given is for a *maximum* value of B_1. How would you measure this value, and also τ_2, experimentally?

APPENDIX G

EPR-RELATED BOOKS AND SELECTED CHAPTERS

The following is an alphabetical list (by first author) of significant books and monographs that feature electron paramagnetic resonance spectroscopy and its applications.

A. Abragam, B. Bleaney, *Electron Paramagnetic Resonance of Transition Ions*, Clarendon, Oxford, U.K., 1970.

A. Abragam, *The Principles of Nuclear Magnetism*, Clarendon, Oxford, U.K., 1961.

AIP, *NMR and EPR; Selected Reprints*, American Institute of Physics, New York, NY, U.S.A., 1965.

I. G. Aleksandrov, *The Theory of Nuclear Magnetic Resonance*, Academic Press, New York, NY, U.S.A., 1966.

B. F. Alekseev, Yu. V. Bogachev, V. Z. Drapkin, A. S. Serdjuk, N. B. Strakhov, S. G. Fedin, *Radiospectroscopy of Natural Substances (by EPR and NMR)*, Norell, Mays Landing, NJ, U.S.A., 1991.

R. S. Alger, *Electron Paramagnetic Resonance: Techniques and Applications*, Wiley-Interscience, New York, NY, U.S.A., 1968.

P. S. Allen, E. R. Andrew, C. A. Bates, Eds., *Magnetic Resonance and Related Phenomena, Proc. 18th Congress Ampère*, 2 volumes, Nottingham, U.K., 1974, North-Holland and American Elsevier, New York, NY, U.S.A., 1975.

S. A. Al'tshuler, B. M. Kozyrev, *Electron Paramagnetic Resonance*, C. P. Poole Jr., Ed., Academic Press, New York, NY, U.S.A., 1964.

S. A. Al'tshuler, B. M. Kozyrev, *Electron Paramagnetic Resonance in Compounds of Transition Elements*, 2nd ed., Halsted-Wiley, New York, NY, U.S.A., 1974.

Electron Paramagnetic Resonance, Second Edition, by John A. Weil and James R. Bolton
Copyright © 2007 John Wiley & Sons, Inc.

E. R. Andrew, *Nuclear Magnetic Resonance*, Cambridge University Press, Cambridge, U.K., 1954.

Annual Review of Physical Chemistry, **1** (1950) through **53** (2002) (Miscellaneous articles on EPR).

H. M. Assenheim, *Introduction to Electron Spin Resonance*, Hilger & Watts, London, U.K., 1966.

N. M. Atherton, *Electron Spin Resonance*, Halsted-Wiley, London, U.K., 1973.

N. M. Atherton, *Principles of Electron Spin Resonance*, Prentice-Hall, New York, NY, U.S.A., 1993.

P. W. Atkins, M. C. R. Symons, *The Structure of Inorganic Radicals*, Elsevier, Amsterdam, Netherlands, 1967.

P. Averbuch, Ed., *Magnetic Resonance and Radio-frequency Spectroscopy, Proc. 15th Colloque Ampère*, Grenoble, France, North-Holland, Amsterdam, Netherlands, 1969.

P. B. Ayscough, *Electron Spin Resonance in Chemistry*, Methuen, London, U.K., 1967.

P. B. Ayscough, Ed., *Electron Spin Resonance*, Vols. 1–5, American Chemical Society, Washington, DC, U.S.A., 1973–1979.

D. M. S. Bagguley, Ed., *Pulsed Magnetic Resonance: NMR, ESR and Optics*, Oxford University Press, Oxford, U.K., 1992.

A. M. Bass, H. P. Broida, Eds., *Formation and Trapping of Free Radicals*, Academic Press, New York, NY, U.S.A., 1960.

A. Bencini, D. Gatteschi, *Electron Paramagnetic Resonance of Exchange-Coupled Systems*, Springer, Berlin, Germany, 1990.

G. B. Benedek, *Magnetic Resonance at High Pressure*, Interscience, New York, NY, U.S.A., 1963.

L. J. Berliner, Ed., *Spin Labeling—Theory and Applications*, Academic Press, New York, NY, U.S.A., 1976.

L. J. Berliner, Ed., *Spin Labeling II—Theory and Applications*, Academic Press, New York, NY, U.S.A., 1979.

L. J. Berliner, S. S. Eaton, G. R. Eaton, et al., Eds., *Biological Magnetic Resonance*, Vols. 1–27 etc., Plenum Press, New York, NY, U.S.A., 1976–.

R. Bernheim, *An Introduction to Optical Pumping*, Benjamin, New York, NY, U.S.A., 1965.

M. Bersohn, J. C. Baird, *An Introduction to Electron Paramagnetic Resonance*, Benjamin, New York, NY, U.S.A., 1966.

I. Bertini, R. S. Drago, *ESR & NMR of Paramagnetic Species in Biological and Related Systems*, NATO Advanced Studies Institute, Kluwer, Boston, MA, U.S.A., 1980.

B. H. J. Bielski, J. M. Gebicki, *Atlas of Electron Spin Resonance Spectra*, Academic Press, New York, NY, U.S.A., 1967.

R. Blinc, *Magnetic Resonance and Relaxation*, North-Holland, Amsterdam, Netherlands, 1967.

F. Bloch, Ed., *Spectroscopic and Group-theoretical Methods in Physics*, North-Holland, Amsterdam, Netherlands, 1968.

M. S. Blois, Jr., H. W. Brown, R. M. Lemmon, R. O. Lindblom, M. Weissbluth, Eds., *Free Radicals in Biological Systems*, Symposium Proceedings, Academic Press, New York, NY, U.S.A., 1961.

N. Bloembergen, *Nuclear Magnetic Relaxation*, Drukkery fa. Schotanus and Jens, Utrecht, Netherlands, 1948.

L. A. Blumenfel'd, V. V. Voevodskii, A. G. Semenov, *Applications of ESR in Chemistry*, Akademii Nauk SSSR, Sibirsk, Otd., U.S.S.R., 1962.

L. A. Blumenfel'd, W. W. Wojewolski, A. G. Semenov, *Die Anwendung der Paramagnetischen Elektronen Resonanz in der Chemie*, Akademische Verlagsgesellschaft, Leipzig, Germany, 1966.

J. Bourgoin, M. Lannoo, *Point Defects in Semiconductors*, Springer, Berlin, Germany, 1983, Chapter 3.

H. C. Box, *Radiation Effects: ESR and ENDOR Analysis*, Academic Press, New York, NY, U.S.A., 1977.

R. F. Boyer, S. E. Keinath, Eds., *Molecular Motion in Polymers by ESR*, Harwood-Academic Press, New York, NY, U.S.A., 1980.

A. L. Buchachenko, *Stable Radicals*, Consultants Bureau, New York, NY, U.S.A., 1965.

A. Carrington, A. D. McLachlan, *Introduction to Magnetic Resonance*, Harper & Row, New York, NY, U.S.A., 1967.

W. J. Caspers, *Theory of Spin Relaxation*, Interscience, New York, NY, U.S.A., 1964.

Chemical Society (London), *Electron Spin Resonance*, Specialist Periodical Reports, Vol. 1–, London, U.K., 1973–.

B. Catoire, Ed., *Electron Spin Resonance (ESR) Applications in Organic and Bioorganic Materials*, 1990 Conference Proceedings, Springer Verlag, Berlin, Germany, 1992.

R. H. Clarke, Ed., *Triplet-state ODMR Spectroscopy*, Wiley, New York, NY, U.S.A., 1982.

J. S. Cohen, *Magnetic Resonance in Biology*, Vols. 1 and 2, Wiley-Interscience, New York, NY, U.S.A., 1980, 1982.

G. Cohen, B. Giovannini, Eds., *EPR of Magnetic Ions in Metals*, Conference Proceedings, Haute-Nendez, Switzerland, 1973, Université de Geneve, Geneva, Switzerland, 1974.

Colloques/Congresses Ampére: European Meetings covering Magnetic Resonance and Related Phenomena. Proceedings yearly or so, from 1952–present, with various editors.

Colloque du C.N.R.S., *La Structure Hyperfine Magnétique des Atomes et des Molécules*, C.N.R.S., Paris, France, 1966.

C. K. Coogan, Ed., *International Symposium on Electron and Nuclear Magnetic Resonance*, Melbourne, Australia, 1963.

C. K. Coogan, N. S. Ham, S. N. Stuart, J. R. Pilbrow, G. V. H. Wilson, Eds., *International Symposium on Electron and Nuclear Magnetic Resonance*, Melbourne, Australia, 1969, Plenum Press, New York, NY, U.S.A., 1970.

R. C. Cross, Ed., *Molecular Relaxation Processes*, Academic Press, New York, NY, U.S.A., 1966.

R. Czoch, A. Francik, *Instrumental Effects in Homodyne EPR Spectrometers*, Horwood, Chichester, U.K., 1989.

L. R. Dalton, Ed., *EPR and Advanced EPR Studies of Biological Systems*, CRC Press, Boca Raton, FL, U.S.A., 1985.

D. W. Davies, *The Theory of the Electric and Magnetic Properties of Molecules*, Wiley, New York, NY, U.S.A., 1967.

C. DeWitt, B. Dreyfus, P. G. de Gennes, Eds., *Low-temperature Physics—Les Houches Lectures*, Grenoble, 1961, Gordon & Breach, New York, NY, U.S.A., 1962.

S. A. Dikanov, Y. D. Tzvetkov, *Electron Spin Echo Envelope Modulation (ESEEM) Spectroscopy*, CRC, Boca Raton, FL, U.S.A., 1992.

W. T. Dixon, *Theory and Interpretation of Magnetic Resonance Spectra*, Plenum Press, New York, NY, U.S.A., 1972.

M. M. Dorio, J. H. Freed, Eds., *Multiple Electron Resonance Spectroscopy*, Plenum Press, New York, NY, U.S.A., 1979.

R. S. Drago, *Physical Methods for Chemists*, Saunders College Publishing, Philadelphia, PA, U.S.A., 1992. Chapter 9.

G. R. Eaton, S. S. Eaton, K. Ohno, Eds., *EPR Imaging and In-vitro EPR*, CRC Press, Boca Raton, FL, U.S.A., 1991.

G. R. Eaton, S. S. Eaton, K. M. Salikhov, *Foundations of Modern EPR*, World Scientific, Singapore, 1997.

S. S. Eaton, G. R. Eaton, L. J. Berliner, Eds., *Biomedical EPR*, Vols. 23–24 in series Biological Magnetic Resonance, Kluwer, New York, NY, U.S.A., 2004.

S. S. Eaton, G. R. Eaton, "Electron Paramagnetic Resonance", in *Ewing's Analytical Instrumentation Handbook*, 3rd ed., J. Cazes, Ed., Marcel Dekker, New York, NY, U.S.A., 2005, Chapter 12.

A. Ehrenberg, B. G. Malmström, T. Vänngärd, Eds., *Proc. Int. Conf. Magnetic Resonance in Biological Systems*, Stockholm, Sweden, 1966, Pergamon Press, London, U.K., 1967.

A. Erbeia, Ed., *Resonances Magnetique*, Centre d'Actualisation Scientifique et Technique Monographies, No. 4, Recueil de Travaux des Sessions de Perfectionement, Institut National des Sciences Appliquées, Lyon, France, 1967, Masson, Paris, France, 1969.

Faraday Society, *Microwave and Radiofrequency Spectroscopy*, General Discussion of the Faraday Society, Aberdeen University Press, Aberdeen, U.K., 1955.

G. Feher, *Electron Paramagnetic Resonance with Applications to Selected Problems in Biology*. Les Houches Lectures, 1969, Gordon & Breach, New York, NY, U.S.A., 1970.

A. R. Forrester, J. M. Hay, R. H. Thomson, *Organic Chemistry of Free Radicals*, Academic Press, New York, NY, U.S.A., 1968.

M. A. Foster, *Magnetic Resonance in Medicine and Biology*, Pergamon Press, Oxford, U.K., 1984.

G. K. Fraenkel, "Paramagnetic Resonance Absorption", in *Techniques of Organic Chemistry*, Vol. 1, A. Weissberger, Ed., Interscience, New York, NY, U.S.A., 1960.

J. P. Fraissard, H. A. Resing, Eds., *Magnetic Resonance in Colloid and Interface Science*, Reidel, Hingham, MA, U.S.A., 1980.

C. Franconi, *Magnetic Resonance of Biological Systems*, Gordon & Breach, New York, NY, U.S.A., 1971.

A. J. Freeman, R. B. Frankel, *Hyperfine Interactions*, Academic Press, New York, NY, U.S.A., 1967.

S. Fujiwara, Ed., *Recent Developments of Magnetic Resonance in Biological Systems*, Hirokawa, Tokyo, Japan, 1968.

F. Gerson, *High Resolution E.S.R. Spectroscopy*, Wiley, London, U.K., 1970.

F. Gerson, W. Huber, *Electron Spin Resonance Spectroscopy for Organic Chemists*, VCH, Weinheim, Germany, 2003.

S. Geschwind, Ed., *Electron Paramagnetic Resonance*, Plenum Press, New York, NY, U.S.A., 1972.

I. B. Goldberg, A. J. Bard, "Electron Spin Resonance Spectroscopy", in *Treatise on Analytical Chemistry*, Vol. 10, 2nd ed., P. J. Elving, Ed., Wiley, New York, NY, U.S.A., 1983, pp. 225ff.

R. M. Golding, *Applied Wave Mechanics*, Van Nostrand, London, U.K., 1969, Chapter 7.

W. Gordy, W. V. Smith, R. F. Trambarulo, *Microwave Spectroscopy*, Wiley, New York, NY, U.S.A., 1953.

W. Gordy, "Theory and Applications of Electron Spin Resonance", in *Techniques of Chemistry*, Vol. 15, W. End, Ed., Wiley, New York, NY, U.S.A., 1980.

C. J. Gorter, *Paramagnetic Relaxation*, Elsevier, New York, NY, U.S.A., Amsterdam, Netherlands, 1947.

C. J. Gorter, Ed., *Progress in Low-temperature Physics*, Annual Series, Interscience, New York, NY, U.S.A., 1957–.

O. Y. Grinberg, L. J. Berliner, *Very High Frequency (VHF) ESR/EPR*, Kluwer Academic/Plenum, New York, NY, U.S.A., Vol. 22 in series Biological Magnetic Resonance, 2004.

P. Gütlich, H. A. Goodwin, Eds., *Spin Crossover in Transition Metal Compounds III*, Springer-Verlag, Berlin, Germany, 2004.

D. ter Haar, Ed., *Fluctuation, Relaxation and Resonance in Magnetic Systems*, Oliver and Boyd, London, U.K., 1961.

J. E. Harriman, *Theoretical Foundations of Electron Spin Resonance*, Academic Press, New York, NY, U.S.A., 1978.

H. G. Hecht, *Magnetic Resonance Spectroscopy*, Wiley, New York, NY, U.S.A., 1967.

K. H. Hellwege, A. M. Hellwege, Eds., *Magnetic Properties of Free Radicals*, Springer, Berlin, Germany, 1967.

J. N. Herak, K. J. Adamic, Eds., *Magnetic Resonance in Chemistry and Biology* (lectures at the Ampére International Summer School, Basko Polje, Yugoslavia, June 1971), Marcel Dekker, New York, NY, U.S.A., 1975.

H. M. Hershenson, *Nuclear Magnetic Resonance and Electron Spin Resonance Spectra Index, 1958–63*, Academic Press, New York, NY, U.S.A., 1965.

H. A. O. Hill, P. Day, Eds., *Physical Methods in Advanced Inorganic Chemistry (ESR, NMR, Mössbauer)*, Wiley, New York, NY, U.S.A., 1968.

A. J. Hoff, Ed., *Advanced EPR: Applications in Biology and Biochemistry*, Elsevier, Amsterdam, Netherlands, 1989.

J. L. Holtzman, *Spin Labeling in Pharmacology*, Academic Press, New York, NY, U.S.A., 1984.

V. Hovi, Ed., *Magnetic Resonance and Related Phenomena, Proc. 17th Collogue Ampère*, Turku, Finland, 1972; North-Holland, Amsterdam, The Netherlands, 1973.

R. P. Hudson, *Principles and Applications of Magnetic Cooling*, North-Holland, Amsterdam, Netherlands and American Elsevier, New York, NY, U.S.A., 1972.

C. A. Hutchison Jr., "Magnetic Susceptibilities", in *Determination of Organic Structures by Physical Methods*, E. A. Braude, F. C. Nachod, Eds., Academic Press, New York, NY, U.S.A., 1955, Chapter 7.

M. Ikeya, *New Applications of Electron Spin Resonance: Dating, Dosimetry and Microscopy*, World Scientific, Singapore, 1993.

D. J. E. Ingram, *Free Radicals as Studied by Electron Spin Resonance*, Butterworths, London, U.K., 1958.

D. J. E. Ingram, *Biological and Biochemical Application of Electron Spin Resonance*, Adam Hilger, London, U.K., 1969.

D. J. E. Ingram, *Spectroscopy at Radio and Microwave Frequencies*, Butterworths, London, U.K., 1967.

D. J. E. Ingram, *Radio and Microwave Spectroscopy*, Butterworths, London, U.K., 1975.

C. D. Jeffries, *Dynamic Nuclear Orientation*, Wiley-Interscience, New York, NY, U.S.A., 1963.

R. A. Y. Jones et al., *Techniques of NMR and ESR*, United Travel Press, London, U.K., 1965.

E. T. Kaiser, L. Kevan, Eds., *Radical Ions*, Wiley-Interscience, New York, NY, U.S.A., 1968.

M. Kaupp, M. Bühl, V. G. Malkin, Eds., *Calculation of NMR and EPR Parameters*, Wiley VCH, Weinheim, Germany, 2004.

A. Kawamori, J. Yamauchi, H. Ohta, Eds., *EPR in the 21st Century: Basics and Applications to Material, Life and Earth Sciences*, Elsevier, Amsterdam, Netherlands, 2002.

C. P. Keijzers, E. J. Reijerse, J. Schmidt, Eds., *Pulsed EPR—A New Field of Applications*, North-Holland, Amsterdam, Netherlands, 1989.

L. Kevan, M. K. Bowman, Eds., *Modern Pulsed and Continuous-wave Electron Spin Resonance*, Wiley, New York, NY, U.S.A., 1990.

L. Kevan, L. D. Kispert, *Electron Spin Double Resonance Spectroscopy*, Wiley, New York, NY, U.S.A., 1976.

L. Kevan, R. N. Schwartz, Eds., *Time-Domain Electron Spin Resonance*, Wiley, New York, NY, U.S.A., 1979.

R. Kirmse, J. Stach, *ESR Spektroskopie*, Akademie, Berlin, Germany, 1985.

Ia. G. Kliava, *EPR Spectroscopy of Disordered Solids*, Zinatne, Riga, Latvia, 1988 (in Russian).

P. F. Knowles, D. Marsh, H. W. E. Rattle, *Magnetic Resonance of Biomolecules*, Wiley-Interscience, New York, NY, U.S.A., 1976.

B. I. Kochelaev, Y. Y. Yablokov, *The Beginning of Paramagnetic Resonance*, World Scientific, Singapore, 1995.

E. Kundla, E. Lippmaa, T. Saluvere, Eds., *Magnetic Resonance and Related Phenomena*, Springer, Berlin, Germany, 1979.

H. Kurreck, B. Kirste, W. Lubitz, *Electron Nuclear Double Resonance Spectroscopy of Radicals in Solution*, VCH, Weinheim, Germany, 1988.

G. Lancaster, *Electron Spin Resonance in Semiconductors*, Hilger and Watts, London, U.K., 1966.

Ya. S. Lebedev, *Atlas of Electron Spin Resonance Spectra*, Consultants Bureau, New York, NY, U.S.A., 1963–1964.

J. E. Leffler, *An Introduction to Free Radicals*, Wiley, New York, NY, U.S.A., 1993.

W. Low, *Paramagnetic Resonance in Solids*, Solid State Physics, Vol. 2 Supplement, Academic Press, New York, NY, U.S.A., 1960.

W. Low, Ed., *Paramagnetic Resonance*, Proc. 1st Int. Conf., Jerusalem, Israel, 1962, 2 volumes, Academic Press, New York, NY, U.S.A., 1963.

A. Lund, M. Shiotani, Eds., *EPR of Free Radicals in Solids*, Luwer, Dordrecht, Netherlands, 2003.

F. E. Mabbs, D. Collison, "Electron Paramagnetic Resonance of *d*-transition Metal Compounds", in *Studies in Inorganic Chemistry Series*, Vol. 16, Elsevier Science, Amsterdam, Netherlands, 1992.

C. A. McDowell, Ed., *Magnetic Resonance*, MIP International Review of Science, Vol. 4, Butterworths, London, U.K., University Park Press, Baltimore, MD, U.S.A., 1972.

S. P. McGlynn, T. Azumi, M. Kinoshita, *Molecular Spectroscopy of the Triplet State*, Prentice-Hall, New York, NY, U.S.A., 1969.

A. D. McLachlan, *Electron Spin Resonance*, Harper & Row, New York, NY, U.S.A., 1969.

K. A. McLauchlan, *Magnetic Resonance*, Oxford University Press, Don Mills, ON, Canada, 1972.

J. McMillan, *Electron Paramagnetism*, Reinhold, New York, NY, U.S.A., 1968.

R. McWeeny, *Spins in Chemistry*, Academic Press, New York, NY, U.S.A., 1970.

A. A. Manenkov, R. Orbach, Eds., *Spin Lattice Relaxation in Ionic Solids*, Harper & Row, New York, NY, U.S.A., 1966.

J. D. Memory, *Quantum Theory of Magnetic Resonance Parameters*, McGraw-Hill, New York, NY, U.S.A., 1968.

W. B. Mims, *The Linear Electric Field Effect in Paramagnetic Resonance*, Oxford University Press, Oxford, U.K., 1976.

G. J. Minkoff, *Frozen Free Radicals*, Interscience, New York, NY, U.S.A., 1960.

Yu. N. Molin, K. M. Salikhov, K. I. Zamaraev, *Spin Exchange—Principles and Applications in Chemistry and Biology*, Springer, Berlin, Germany, 1980.

L. T. Muus, P. W. Atkins, *Electron Spin Relaxation in Liquids*, Plenum Press, New York, NY, U.S.A., 1972.

R. J. Myers, *Molecular Magnetism and Magnetic Resonance Spectroscopy*, Prentice-Hall, Englewood Cliffs, NJ, U.S.A., 1973.

J. W. Orton, *Electron Paramagnetic Resonance. An Introduction to Transition-Group Ions in Crystals*, Iliffe, London, U.K., 1968.

D. E. O'Reilly, J. H. Anderson, "Magnetic Properties", in *Physics and Chemistry of the Organic Solid State*, Vol. 2, D. Fox, M. Labes, A. Weissberger, Eds., Interscience, New York, NY, U.S.A., 1963.

F. J. Owens, C. P. Poole Jr., H. A. Farach, Eds., *Magnetic Resonance of Phase Transitions*, Academic Press, New York, NY, U.S.A., 1979.

G. E. Pake, *Paramagnetic Resonance*, Benjamin, New York, NY, U.S.A., 1962.

G. E. Pake, T. L. Estle, *The Physical Principles of Electron Paramagnetic Resonance*, 2nd ed., Benjamin, New York, NY, U.S.A., 1973.

R. V. Parish, *NMR, NQR, EPR and Mössbauer Spectroscopy in Inorganic Chemistry*, Ellis Horwood, New York, NY, U.S.A., 1990.

J. Peisach, W. E. Blumberg, *Electron Spin Resonance of Metal Complexes*, Plenum Press, New York, NY, U.S.A., 1969.

L. Petrakis, J. P. Fraissard, Eds., *Magnetic Resonance: Introduction, Advanced Topics and Applications to Fossil Energy*, NATO ASI Series C124, Reidel, Dordrecht, Netherlands, 1984.

J. R. Pilbrow, *Transition Ion Electron Paramagnetic Resonance*, Clarendon Press, Oxford, U.K., 1990.

C. P. Poole Jr., *Electron Spin Resonance. A Comprehensive Treatise on Experimental Techniques*, Interscience, New York, NY, 1967; 2nd ed., Wiley-Interscience, New York, NY, U.S.A., 1983; reprinted by Dover, New York, NY, U.S.A. 1996, with minor corrections but no repagination.

C. P. Poole Jr., H. A. Farach, *Handbook of Electron Spin Resonance: Data Sources, Computer Technology, Relaxation, and ENDOR*, Vol. 1, American Institute of Physics, Mineola, NY, U.S.A., 1994.

C. P. Poole Jr., H. A. Farach, *Handbook of Electron Spin Resonance*, Vol. 2, Springer-Verlag, AIP Press, New York, NY, U.S.A., 1999.

C. P. Poole Jr., H. A. Farach, *Relaxation in Magnetic Resonance, Dielectric and Mössbauer Applications*, Academic Press, New York, NY, U.S.A., 1971.

C. P. Poole Jr., Ed., *Magnetic Resonance Reviews*, Vols. 1–, Gordon & Breach, New York, NY, U.S.A., 1971–.

C. P. Poole Jr., H. A. Farach, *The Theory of Magnetic Resonance*, Wiley-Interscience, New York, NY, U.S.A., 1972; 2nd ed., 1987.

G. T. Rado, H. Suhl, Eds., *Magnetism*, Vol. II, Part A, Academic Press, New York, NY, U.S.A., 1965.

N. F. Ramsey, *Nuclear Moments*, Wiley, New York, NY, U.S.A., 1953.

B. G. Ränby, J. F. Rabek, *ESR Spectroscopy in Polymer Research*, Springer, New York, 1977.

G. M. Rosen, B. E. Britigan, H. J. Halpern, S. Pou, Eds., *Free Radicals: Biology and Detection by Spin Trapping*, Oxford Press, New York, NY, U.S.A., 1999.

K. M. Salikhov, Yu. N. Molin, R. Z. Sagdeev, A. L. Buchachenko, *Spin Polarization and Magnetic Effects in Radical Reactions*, Elsevier, Amsterdam, Netherlands, 1984.

K. M. Salikhov, A. G. Semenov, Yu. D. Tsvetkov, *Electron Spin Echoes and Their Applications*, Nauka, Novosibirsk, U.S.S.R., 1976 (in Russian).

G. Schoffa, *Elektronenspinresonanz in der Biologie*, G. Braun, Karlsruhe, Germany, 1964.

R. T. Schumacher, *Introduction to Magnetic Resonance*, Benjamin, New York, NY, U.S.A., 1970.

A. Schweiger, G. Jeschke, *Principles of Pulse Electron Paramagnetic Resonance Spectroscopy*, Oxford University Press, Oxford, U.K., 2001.

F. Seitz, D. Turnbull, Eds., *Solid State Physics*, Vol. 5, Academic Press, New York, NY, U.S.A., 1957.

R. Servant, A. Charru, Eds., *Electronic Magnetic Resonance and Solid Dielectrics, Proc. 12th Colloque Ampère*, North-Holland, Amsterdam, Netherlands, 1964.

H. Sixl, *Festkörperspektroskopie II—Resonanzspektroskopie*, Hochschulverlag, Stuttgart, Germany, 1979.

D. V. Skobel'tsyn, Ed., *Quantum Electronics and Paramagnetic Resonance*; Plenum Press, New York, NY, U.S.A., 1971.

C. P. Slichter, *Principles of Magnetic Resonance*, Harper & Row, New York, NY, U.S.A., 1963; 2nd ed., Springer, Berlin, Germany, 1978; 3rd ed., 1989.

J. Smidt, Ed., *Magnetic and Electric Resonance and Relaxation, Proc. 11th Colloque Ampère*, Eindhoven, Netherlands, North-Holland, Amsterdam, Netherlands, 1963.

W. Snipes, Ed., *Conference on Electron Spin Resonance and the Effects of Radiation on Biological Systems*, Gatlinburg, TN, U.S.A., 1965, National Academy of Science and National Research Council, Washington, DC, U.S.A., 1966.

L. A. Sorin, M. V. Vlasova, *Electron Spin Resonance of Paramagnetic Crystals*, Plenum Press, New York, NY, U.S.A., 1973.

J.-M. Spaeth, J. R. Niklas, R. H. Bartram, *Structural Analysis of Point Defects in Solids—an Introduction to Multiple Magnetic Resonance Spectroscopy*, Springer, Berlin, Germany, 1982.

T. L. Squires, *An Introduction to Electron Spin Resonance*, Academic Press, New York, NY, U.S.A., 1964.

K. J. Standley, R. A. Vaughan, *Electron Spin Relaxation Phenomena in Solids*, Adam Hilger, London, U.K., 1969.

L. D. Stepin, *Quantum Radio-frequency Physics*, H. H. Stroke, Ed., MIT Press, Cambridge, MA, U.S.A., 1965.

A. M. Stoneham, *Theory of Defects in Solids*, Clarendon, Oxford, U.K., 1975, Chapter 13.

M. W. P. Strandberg, *Microwave Spectroscopy*, Wiley, New York, NY, U.S.A., 1954.

S. Sugano, Y. Tanabe, H. Kamimura, *Multiplets of Transition-Metal Ions in Crystals*, Academic Press, New York, NY, U.S.A., 1970.

H. M. Swartz, J. R. Bolton, D. C. Borg, Eds., *Biological Applications of Electron Spin Resonance*, Wiley-Interscience, New York, NY, U.S.A., 1972.

M. C. R. Symons, *Chemical and Biochemical Aspects of Electron Spin Resonance Spectroscopy*, Wiley, Chichester, U.K., 1978.

J. Talpe, *Theory of Experiments in Paramagnetic Resonance*, Pergamon Press, Oxford, U.K., 1971.

C. H. Townes, A. L. Schawlow, *Microwave Spectroscopy*, McGraw-Hill, New York, NY, U.S.A., 1955.

I. Ursu, *Resonanta Electronic de Spin*, Editura Academiee, Bucharest, Romania, 1965.

I. Ursu, Ed., *Magnetic Resonance and Related Phenomena, Proc. 16th Colloque Ampère*, Bucharest, Romania, Academy of Socialist Republic of Romania, Bucharest, 1971.

I. Ursu, *La Resonance Paramagnetique Electronique*, Dunod, Paris, France, 1968.

L. Van Gerven, Ed., *Koninlijke Vlaamse Academie voor Wetenschappen, Proc. 13th Colloque Ampère*, Leuven, Belgium, North-Holland, Amsterdam, Netherlands, 1965.

L. L. Van Reijen, *Electron Spin Resonance Studies of Pentavalent and Trivalent Chromium*, Multitype, Amsterdam, Netherlands, 1964.

Varian Associates, *Workshop on Nuclear Magnetic Resonance and Electron Paramagnetic Resonance*, Pergamon Press, New York, NY, U.S.A., 1959. [See Ch. 1, Further Reading 7.]

S. V. Vonsovskii, Ed., *Ferromagnetic Resonance*, U.S. Department of Commerce, Washington, DC, U.S.A., 1964.

J. S. Waugh, Ed., *Advances in Magnetic Resonance*, Vol. 1–, Academic Press, New York, NY, U.S.A., 1965–.

J. A. Weil, Ed., *Electronic Magnetic Resonance of the Solid State*, Chemical Institute of Canada, Ottawa, ON, Canada, 1987.

J. A. Weil, J. R. Bolton, J. E. Wertz, *Electron Spin Resonance; Elementary Theory and Practical Applications*, Wiley, New York, NY, U.S.A., 1994 (First edition of the present book).

M. Weissbluth, *The Triplet State in Molecular Biophysics*, B. Pullman, M. Weissbluth, Eds., Academic Press, New York, NY, U.S.A.,1965.

M. Weissbluth, *Photon-Atom Interactions*, Academic Press, New York, NY, U.S.A., 1989, Chapter 3.

W. Weltner Jr., *Magnetic Atoms and Molecules*, Scientific and Academic Editions, Van Nostrand Reinhold, New York, NY, U.S.A., 1983.

G. K. Wertheim, A. Hausmann, W. Sander, *The Electronic Structure of Point Defects as Determined by Mössbauer Spectroscopy and by Spin Resonance*, American Elsevier, New York, NY, U.S.A., 1972.

J. E. Wertz, J. R. Bolton, *Electron Spin Resonance; Elementary Theory and Practical Applications*, McGraw-Hill, New York, NY, U.S.A., 1972; reprinted by Chapman & Hall, New York, NY, 1987 (earlier version of the present book).

T. H. Wilmshurst, *Electron Spin Resonance Spectrometers*, Hilger, London, U.K., 1967.

J. Winter, *Magnetic Resonance in Metals*, Oxford University Press, Oxford, U.K., 1971.

S. J. Wyard, *Solid State Biophysics*, McGraw-Hill, New York, NY, U.S.A., 1969.

A. Yariv, *Quantum Electronics*, Wiley, New York, NY, U.S.A., 1967, Chapter 8.

T. F. Yen, Ed., *Electron Spin Resonance of Metal Complexes*, Symposium on Electron Spin Resonance of Metal Chelates, Cleveland, OH, 1968; Plenum Press, New York, NY, U.S.A., 1969.

N. D. Yordanov, Ed., *Electron Magnetic Resonance of Disordered Systems*, World Scientific, Singapore, 1989.

A. B. Zahlan, Ed., *Excitons, Magnons and Phonons in Molecular Crystals*, Beirut Symposium, Beirut, Lebanon, 1968.

A. B. Zahlan et al., Eds., *The Triplet State*, Proceedings of International Symposium, Beirut, Lebanon, 1967; Cambridge University Press, Cambridge, U.K., 1967.

APPENDIX H

FUNDAMENTAL CONSTANTS, CONVERSION FACTORS, AND KEY DATA

In this appendix (Tables H.1–H.4) we have assembled for the convenience of the reader a selection of the most recent compilation (2003) of fundamental constants relevant to the EPR field, as well as conversion equations of general interest and tables of properties of selected nuclides. As stated, we use the SI (rationalized mksC) system.

TABLE H.1 Fundamental Constants[a,b]

1. Speed of light in the electromagnetic vacuum
$$c = 2.99792458 \times 10^8 \text{ m s}^{-1} \text{ (defined)}$$

2. Magnetic constant (permeability of the vacuum)
$$\mu_0 = 4\pi \times 10^{-7} = 12.566370614 \times 10^{-7} \text{ J C}^{-2} \text{ s}^2 \text{ m}^{-1} \text{ (defined)}$$
(also $= \text{T}^2 \text{ J}^{-1} \text{m}^3$, where 1 joule (J) $= 1 \text{ kg m}^2 \text{s}^{-2} \text{C}^0$ and 1 tesla (T) $= 1 \text{ kg m}^0 \text{ s}^{-1} \text{C}^{-1}$)

3. Electric constant (permittivity of the vacuum)
$$\varepsilon_0 = \mu_0^{-1} c^{-2} = 8.854187817 \times 10^{-12} \text{ J}^{-1} \text{ C}^2 \text{ m}^{-1} \text{ (defined)}$$

4. Planck constant
$$h = 6.6260693(11) \times 10^{-34} \text{ J s}$$
$$\equiv h/2\pi = 1.05457168(18) \times 10^{-34} \text{ J s}$$

5. Elementary charge magnitude
$$|e| = 1.60217653(14) \times 10^{-19} \text{ C}$$

(continued)

Electron Paramagnetic Resonance, Second Edition, by John A. Weil and James R. Bolton
Copyright © 2007 John Wiley & Sons, Inc.

TABLE H.1 *Continued*

6. Electron mass
$$m_e = 9.1093826(16) \times 10^{-31} \, \text{kg}$$

7. Proton mass
$$m_p = 1.67262171(29) \times 10^{-27} \, \text{kg}$$

8. Bohr magneton
$$\beta_e = \frac{|e|\hbar}{2\,m_e} = 9.27400949(80) \times 10^{-24} \, \text{J T}^{-1}$$

9. Free-electron g factor
$$g_e = 2.0023193043718(75)$$

Note that herein, g_e is taken to be positive, whereas the CODATA tables issued by NIST take this symbol to denote the negative: $g_e = \alpha_e|g_e|$ (see Section 1.8).

10. Electron magnetic moment
$$\mu_e = -g_e\beta_e S = -9.28476412(80) \times 10^{-24} \, \text{J T}^{-1} \qquad (S = \tfrac{1}{2})$$

11. Free-electron magnetogyric ratio
$$\gamma_e = \mu_e/S = -1.76085974(15) \times 10^{11} \, \text{s}^{-1} \, \text{T}^{-1}$$

12. Nuclear magneton
$$\beta_n = \frac{|e|\hbar}{2m_p} = 5.05078343(43) \times 10^{-27} \, \text{J T}^{-1}$$

13. Proton g factor
$$g_p = 5.585694701(56)$$

14. Proton g factor (corrected for diamagnetism, in a spherical water sample at 298 K)
$$g_p{}' = 5.585551211(56)$$

15. Proton magnetic moment
$$\mu_p = g_p\beta_n I = 1.41060671(12) \times 10^{-26} \, \text{J T}^{-1} \qquad (I = \tfrac{1}{2})$$

16. Proton magnetogyric ratio
$$\gamma_p = \mu_p/I\hbar = 2.67522205(23) \times 10^8 \, \text{s}^{-1} \, T^{-1}$$

17. Proton magnetogyric ratio (corrected for diamagnetism, in a spherical water sample at 298 K)
$$\gamma_p{}' = 2.67515333(23) \times 10^8 \, \text{s}^{-1} \, \text{T}^{-1}$$

18. Bohr radius
$$r_b = \frac{4\pi\varepsilon_o\hbar^2}{m_e e^2} = 5.291772108(18) \times 10^{-11} \, \text{m}$$

19. Boltzmann constant
$$k_b = 1.3806505(24) \times 10^{-23} \, \text{J K}^{-1}$$

20. Avogadro constant
$$N_A = 6.0221415(10) \times 10^{23} \, \text{mol}^{-1}$$

[a]Taken from "CODATA Internationally Recommended Values of the Fundamental Physical Constants", The NIST Reference on Constants, Units and Uncertainty, http://physics.nist.gov/cuu/Constants/.
[b]The figures given in parentheses are the 2003 uncertainties in the last two decimal places.

TABLE H.2   Useful Conversion Factors

1. Magnetic field B (mT) to electron resonant frequency ν_e (MHz) and $\bar{\nu}_e$ (cm^{-1})

$$\nu_e \text{ (MHz)} = \frac{g_e \beta_e B}{h} \frac{g}{g_e} = 28.02495 \frac{g}{g_e} B \text{ (mT)}$$

$$B \text{ (mT)} = 0.03568249 \frac{g}{g_e} \nu_e \text{ (MHz)}$$

$$\nu_e \text{ (MHz)} = c \, [\text{m s}^{-1}] \times 10^{-4} \, \bar{\nu}_e \text{ (cm}^{-1}) = 2.99792458 \times 10^4 \, \bar{\nu}_e \text{ (cm}^{-1})$$

$$\bar{\nu}_e \text{ (cm}^{-1}) = 0.33356410 \times 10^{-4} \, \nu_e \text{ (MHz)}$$

$$= 9.3481139 \times 10^{-4} \frac{g}{g_e} B \text{ (mT)}$$

2. Magnetic field B (mT) to proton resonant frequency ν_p(MHz)

$$\nu_p \text{ (MHz)} = 0.04257748 \, B \text{ (mT); } 0.04257639 \, B \text{ (mT) (in pure water)}$$
$$B \text{ (mT)} = 23.48659 \, \nu_p \text{ (MHz); } 23.48719 \, \nu_p \text{ (MHz) (in pure water)}$$

3. Ratio of proton to electron resonant frequency

$$\frac{\nu_p}{\nu_e} = 1.519270 \times 10^{-3} \frac{g_e}{g}; \frac{\nu_p}{\nu_e} = 1.519231 \times 10^{-3} \frac{g_e}{g} \text{ (proton in pure water)}$$

4. Calculation of g factors

$$g = \frac{h\nu_e}{\beta_e B} = 0.07144773 \frac{\nu_e \text{ (MHz)}}{B \text{ (mT)}}$$

$$= \frac{g_p \beta_n}{\beta_e} \frac{\nu_e}{\nu_p} = 3.042064 \times 10^{-3} \frac{\nu_e}{\nu_p}; 3.041987 \times 10^{-3} \frac{\nu_e}{\nu_p} \text{ (proton in pure water)}$$

5. Hyperfine coupling and hyperfine splitting parameters

$$\frac{A}{h} \text{ (MHz)} = 28.02495 \, a \text{ (mT)}$$

$$a \text{ (mT)} = 0.03568249 \frac{A}{h} \text{ (MHz)}$$

$$\frac{A}{hc} \text{ (cm}^{-1}) = 0.33356410 \times 10^{-4} \frac{A}{h} \text{ (MHz)}$$

$$= 9.3481182 \times 10^{-4} \frac{A}{g_e \beta_e} \text{ (mT)}$$

TABLE H.3 Selected Free-Atom/Ion One-Electron Spin-Orbit Coupling Parameters $\zeta_{n\ell}$ (in cm^{-1})[a]

Element	Charge Number						
	0	+1	+2	+3	+4	+5	+6
For $n = 2$, $\ell = 1$[b]							
B	15						
C	30						
N	50						
O	70						
F	140						
For $n = 3$, $\ell = 2$[c]							
Ti	70	90	123	155			
V	95	135	170	210	250		
Cr	135	185	230	275	355	380	
Mn	190	255	300	355	415	475	540
Fe	275	335	400	460	520	590	665
Co	390	455	515	580	650	715	790
Ni		565	630	705	790	865	950
Cu			830	890	960	1030	1130
For $n = 4$, $\ell = 2$[c]							
Zr		(300)	(400)	(500)			
Nb		(420)	(610)	(800)			
Mo			(670)	800	(850)	(900)	
Tc		(950)	(1200)	(1300)	(1500)		(1700)
Ru			(1250)	(1400)	(1500)		(1700)
Rh					(1700)	(1850)	(2100)
Pd		(1300)	(1600)				
Ag			(1800)				
For $n = 5$, $\ell = 2$[c]							
Hf							
Ta				(1400)			
W			(1500)	(1800)	(2300)	(2700)	
Re			(2100)	(2500)	(3300)	(3700)	(4200)
Os				(3000)	(4000)	(4500)	(5000)
Ir					(5000)	(5500)	(6000)
Pt		(3400)					
Au			(5000)				

For $n = 4$, $\ell = 3$, 3+ ion[b]

Ion:	Ce	Pr	Nd	Pm	Sm	Eu	Gd
ζ:	640	750	900	—	1180	1360	—
Ion:	Tb	Dy	Ho	Er	Tm	Yb	
ζ:	1620	1820	2080	2470	2750	2950	

[a] To obtain the Russell-Saunders parameter λ, insert the total electron spin quantum number S into the relation $\lambda = \pm\zeta/2S$, applying the + sign for shells less than half–full and the – sign for shells more than half–full. For half–full shells (i.e., $L = 0$), the spin-orbit coupling energy is essentially zero. Values in parentheses are estimates.

[b] M. Gerloch, *Orbitals, Terms and States*, Wiley, Chichester, U.K., 1986, p. 73, Table 4.1.

[c] B. N. Figgis, *Introduction to Ligand Fields*, Wiley-Interscience, New York, NY, U.S.A. 1966, p. 60, Table 3.4.

TABLE H.4 Properties of Selected Nuclides and Atoms

Z	Nuclide[a]	Natural Abundance (mol%)	Spin I	g_n	$\dfrac{g_n \beta_n}{g_e \beta_e} \times 10^5$ [b]	Nuclear Electric Quadrupole Moment Q (10^{-24} cm^2)[c]	Isotropic Hyperfine Parameter a_0 (mT)[d]	Uniaxial Hyperfine Parameter b_0 (mT)[e]
0	^{1}n		$\tfrac{1}{2}$	−3.82608544	−104.06688			
1	^{1}H	99.985	$\tfrac{1}{2}$	5.5856948	151.927038		50.6850[e] (50.68377 calc.)	
	^{2}H	0.015	1	0.8574388228	23.32174	0.00286	7.78027	
2	^{3}He	0.000137	$\tfrac{1}{2}$	−4.25499544	−115.7329		226.83	
3	^{6}Li	7.5	1	0.8220473	22.35912	−0.00083	5.43[f]	0.760
	^{7}Li	92.5	$\tfrac{3}{2}$	2.1709750	59.04902	−0.0406	14.34[f]	2.271
4	^{9}Be	100	$\tfrac{3}{2}$	−0.784993	−21.3513	0.053	−16.11	
5	^{10}B	19.9	3	0.600214927	16.32543	0.0847	30.43	
	^{11}B	80.1	$\tfrac{3}{2}$	1.7924326	48.75293	0.0407	90.88	
6	^{13}C	1.07	$\tfrac{1}{2}$	1.404824	38.21024		134.77	3.832
7	^{14}N	99.632	1	0.40376100	10.98202	0.02001	64.62	1.981
	^{15}N	0.368	$\tfrac{1}{2}$	−0.56637768	−15.40508		−90.65	−2.779
8	^{17}O	0.038	$\tfrac{5}{2}$	−0.757516	−20.60391	−0.02578	−187.80	−6.009
9	^{19}F	100	$\tfrac{1}{2}$	5.257736	143.0077		1886.53	62.80
10	^{21}Ne	0.27	$\tfrac{3}{2}$	−0.441198	−12.00034	0.103	−221.02	−7.536
11	^{23}Na	100	$\tfrac{3}{2}$	1.4784371	40.21151	0.104	31.61[f] (33.08 calc.)	
12	^{25}Mg	10.00	$\tfrac{5}{2}$	−0.34218	−9.30713	0.199	17.338	
13	^{27}Al	100	$\tfrac{5}{2}$	1.4566028	39.61883	0.1402	139.55	2.965
14	^{29}Si	4.69	$\tfrac{1}{2}$	−1.1106	−30.20778		−163.93	−4.0750

(continued)

TABLE H.4 Continued

Z	Nuclide[a]	Natural Abundance (mol%)	Spin I	g_n	$\dfrac{g_n\beta_n}{g_e\beta_e} \times 10^{5b}$	Nuclear Electric Quadrupole Moment Q $(10^{-24}\ \mathrm{cm}^2)$[c]	Isotropic Hyperfine Parameter a_0 (mT)[d]	Uniaxial Hyperfine Parameter b_0 (mT)[e]
15	^{31}P	100	$\frac{1}{2}$	2.26320	61.55793		474.79	13.088
16	^{33}S	0.76	$\frac{3}{2}$	0.4292141	11.67158	$-0.084//-0.678$	123.57	3.587
17	^{35}Cl	75.78	$\frac{3}{2}$	0.5479162	14.90304	-0.08249	204.21	6.266
17	^{37}Cl	24.22	$\frac{3}{2}$	0.4560824	12.40521	-0.06493	169.98	5.216
19	^{39}K	93.258	$\frac{3}{2}$	0.26100487	7.09882	0.049	8.238[f] (8.152 calc.)	
19	^{41}K	6.730	$\frac{3}{2}$	0.14326183	3.89644	0.060	4.525	
20	^{43}Ca	0.135	$\frac{7}{2}$	-0.3764694	-10.23828	-0.049	-22.862	
21	^{45}Sc	100	$\frac{7}{2}$	1.358996	36.9807	-0.216	100.73	3.430
22	^{47}Ti	7.44	$\frac{5}{2}$	-0.31539	-8.5784	0.29	-27.904	-1.051
22	^{49}Ti	5.41	$\frac{7}{2}$	-0.315477	-8.58051	0.24	-27.910	-1.051
23	^{50}V	0.250	6	0.55761482	15.13906	0.21	56.335	2.368
23	^{51}V	99.750	$\frac{7}{2}$	1.4710587	39.9387	-0.043	148.62	6.246
24	^{53}Cr	9.50	$\frac{3}{2}$	-0.31636	-8.560	$-0.15, +0.04, -0.028$	-26.698	-1.470
25	^{55}Mn	100	$\frac{5}{2}$	1.38748716	37.587	0.33	179.70	-8.879
26	^{57}Fe	2.119	$\frac{1}{2}$	0.1812460	4.912		26.662	1.395
27	^{59}Co	100	$\frac{7}{2}$	1.322	35.849	0.41	212.20	12.065

28	^{61}Ni	1.140	$\frac{3}{2}$	−0.50001	−13.6000	0.162	−89.171	−5.360
29	^{63}Cu	69.17	$\frac{3}{2}$	1.4848971	40.36	−0.211 // 0.220	213.92	17.085
30	^{65}Cu	30.83	$\frac{3}{2}$	1.5877	43.19	−0.195	228.92	18.283
	^{67}Zn	4.10	$\frac{5}{2}$	0.35008196	9.52831	0.150	74.470	5.021
31	^{69}Ga	60.11	$\frac{3}{2}$	1.34439	36.5668	0.168	435.68	7.274
	^{71}Ga	38.89	$\frac{3}{2}$	1.70818	46.4616	0.106	553.58	9.242
32	^{73}Ge	7.73	$\frac{9}{2}$	−0.1954373	−5.315654	−0.17	−84.32	−1.716
33	^{75}As	100	$\frac{3}{2}$	0.959653	26.10193	0.314	523.11	11.905
34	^{77}Se	7.63	$\frac{1}{2}$	1.070147	29.084		717.93	17.542
35	^{79}Br	50.69	$\frac{3}{2}$	1.404267	38.19535	0.305	1144.34	29.174
	^{81}Br	49.31	$\frac{3}{2}$	1.513708	41.17206	0.254	1233.52	31.448
36	^{83}Kr	11.58	$\frac{9}{2}$	−0.215704	−5.86704	0.253	−211.85	−5.515
37	^{85}Rb	72.17	$\frac{5}{2}$	0.54134060	14.72182	0.273, 0.274	36.11 ^{f} (37.00 calc.)	
	^{87}Rb	27.83	$\frac{3}{2}$	1.834545	49.8913	0.132, 0.127	122.38	
38	^{87}Sr	7.00	$\frac{9}{2}$	−0.24302289	−6.6070	0.34	−30.46	
39	^{89}Y	100	$\frac{1}{2}$	−0.2748308	−7.475408		−44.60	−0.888
40	^{91}Zr	11.22	$\frac{5}{2}$	−0.521448	−14.18313	−0.206	−98.23	−2.221
41	^{93}Nb	100	$\frac{9}{2}$	1.3712	37.296	−0.32 // −0.37	235.15	6.527
42	^{95}Mo	15.92	$\frac{5}{2}$	−0.3657	−9.944	−0.022	−70.79	−2.151
	^{97}Mo	9.55	$\frac{5}{2}$	−0.3734	−10.156	0.255	−72.30	−2.197
44	^{99}Ru	12.74	$\frac{5}{2}$	−0.256	−6.77	0.079	−62.94	−2.279

(continued)

TABLE H.4 *Continued*

Z	Nuclide[a]	Natural Abundance (mol%)	Spin I	g_n	$\frac{g_n\beta_n}{g_e\beta_e} \times 10^5$ [b]	Nuclear Electric Quadrupole Moment Q $(10^{-24}\ \text{cm}^2)$ [c]	Isotropic Hyperfine Parameter a_0 (mT) [d]	Uniaxial Hyperfine Parameter b_0 (mT) [e]
	^{101}Ru	17.05	$\frac{5}{2}$	$-0.288,$ -0.286	-7.59	0.46	-70.52	-2.554
45	^{103}Rh	100	$\frac{1}{2}$	-1.768	-4.809		-43.85	-1.728
46	^{105}Pd	22.33	$\frac{5}{2}$	-0.257	-6.96	0.660	-2.683	
47	^{107}Ag	51.839	$\frac{1}{2}$	-0.22735930	-6.1806		-65.33	-2.924
	^{109}Ag	48.161	$\frac{1}{2}$	-0.26173812	-7.11928		-75.25	-3.368
48	^{111}Cd	12.80	$\frac{1}{2}$	-1.189772	-32.3791		-487.07	-18.41
	^{113}Cd	12.22	$\frac{1}{2}$	-1.244602	-33.8523			
49	^{113}In	4.29	$\frac{9}{2}$	1.2286	33.43	0.80		
	^{115}In	95.71	$\frac{9}{2}$	1.2313	33.4905	0.81	720.07	10.147
50	^{113}Sn	0.34	$\frac{1}{2}$	-1.8377	-49.99			
	^{117}Sn	7.68	$\frac{1}{2}$	-2.00208	-54.555		-1497.98	-24.98
	^{119}Sn	8.59	$\frac{1}{2}$	-2.09456	-56.971		-1567.18	-26.13
51	^{121}Sb	57.21	$\frac{5}{2}$	1.3454	36.597	-0.45	1252.4	22.44
	^{123}Sb	42.79	$\frac{7}{2}$	0.72851	19.8220	-0.49	678.38	12.15
52	^{123}Te		$\frac{1}{2}$	-1.473896				
	^{125}Te	7.07	$\frac{1}{2}$	-1.777010	-48.322		-1983.6	-37.42
53	^{127}I	100	$\frac{5}{2}$	1.12531	30.6076	0.689 // -0.789	1484.40	28.989
54	^{129}Xe	26.44	$\frac{1}{2}$	-1.55595	-42.3211	-0.116	-2418.92	-47.815
	^{131}Xe	21.18	$\frac{3}{2}$	0.461241	12.54550		717.06	14.143
55	^{133}Cs	100	$\frac{7}{2}$	0.73797509	20.06910	-0.00371	82.00[f] (88.03 calc.)	

TABLE H.4 *Continued*

Z	Isotope							
56	^{135}Ba	6.592	3/2	0.559085	15.2001	0.160,...	126.67	
	^{137}Ba	11.232	3/2	0.62491	17.0036	0.246	141.70	
				0.62489				
57	^{138}La		5	0.7427292		0.45 // 0.43		3.384
	^{139}La	99.910	7/2	0.7951586	21.6292	0.20	214.35	12.62
59	^{141}Pr	100	5/2	1.7102	43.5	−0.059	445.7	
60	^{143}Nd	12.18	7/2	−0.3043	−8.367	−0.48 // −0.61	−84.82	−2.268
	^{145}Nd	8.30	7/2	−0.187	−5.17	−0.253,...	−52.39	−1.401
62	^{147}Sm	15.0	7/2	−0.2328	−6.316	−0.261	−71.86	−2.389
	^{149}Sm	13.8	7/2	0.1919	5.209	0.075 // −0.078	59.26	1.970
63	^{151}Eu	47.8	5/2	1.3887	37.78	0.903,...	462.33	15.606
	^{153}Eu	57.2	5/2	0.61296	16.684	2.41,...	204.17	6.892
64	^{155}Gd	14.80	3/2	−0.1715	−4.686	1.30 // −0.44	−69.48	−0.940
	^{157}Gd	15.65	3/2	−0.2247	−6.128	1.36 // −0.46	−90.85	−1.229
65	^{159}Tb	100	3/2	1.343	97.37	1.432	486.35	17.86
66	^{161}Dy	18.9	5/2	−0.192	−5.14	2.51	−75.12	−2.775
	^{163}Dy	24.9	5/2	0.269	7.24	2.65	105.73	3.905
67	^{165}Ho	100	7/2	1.19	32.42	3.58,...	483.86	18.340
68	^{167}Er	22.95	7/2	−0.16110	−4.401	2.827 // 3.57	−69.01	−2.705
69	^{169}Tm	100	1/2	−0.4620	−12.67		−208.21	−8.355
70	^{171}Yb	14.3	1/2	0.98734	26.887		476.02	19.065
	^{173}Yb	16.12	5/2	−0.27196	−7.3968	2.80	−130.96	−5.245
				0.27201				

(continued)

TABLE H.4 *Continued*

Z	Nuclide[a]	Natural Abundance (mol%)	Spin I	g_n	$\dfrac{g_n\beta_n}{g_e\beta_e}\times 10^5$ [b]	Nuclear Electric Quadrupole Moment Q $(10^{-24}\,\text{cm}^2)$ [c]	Isotropic Hyperfine Parameter a_0 (mT) [d]	Uniaxial Hyperfine Parameter b_0 (mT) [e]
71	^{175}Lu	97.41	$\frac{7}{2}$	0.639426	17.3921	3.49	379.31	3.985
	^{176}Lu		7	0.4527		4.92 // 4.97		
72	^{177}Hf	18.606	$\frac{7}{2}$	0.2267	6.166	3.37 // 3.36	157.36	1.771
	^{179}Hf	13.629	$\frac{9}{2}$	−0.1424	−3.873	3.79, …	−98.84	−1.112
73	^{180}Ta		9	0.5361		4.95		
	^{181}Ta	99.988	$\frac{7}{2}$	0.67729	18.4221	3.35 // 3.17	535.95	6.357
74	^{183}W	14.314	$\frac{1}{2}$	0.23556952	6.407372		206.14	2.605
75	^{185}Re	37.40	$\frac{5}{2}$	1.2748	34.674	2.18 // 2.19	1253.60	16.347
	^{187}Re	62.60	$\frac{5}{2}$	1.2879	35.027	2.07	1266.38	16.514
76	^{187}Os	1.6	$\frac{1}{2}$	0.1293038	3.51693			
	^{189}Os	16.1	$\frac{3}{2}$	0.439955	13.27	0.86	471.0	6.437
77	^{191}Ir	37.3	$\frac{3}{2}$	0.09740 / 0.1005	2.64	0.816	112.96	0.907
	^{193}Ir	62.7	$\frac{3}{2}$	0.1061 / 0.1091	2.91	0.751	124.60	1.754
78	^{195}Pt	33.8	$\frac{1}{2}$	1.2190	33.156		1227.84	21.038
79	^{197}Au	100	$\frac{3}{2}$	0.0987720	2.66469	0.594	102.62	1.884
80	^{199}Hg	16.87	$\frac{1}{2}$	1.011771	27.51966		1494.4	22.99
	^{201}Hg	13.18	$\frac{3}{2}$	−0.3734838	10.15856	0.38	−551.6	−8.49
81	^{203}Tl	29.524	$\frac{1}{2}$	3.24451574	88.24919		6496.7	44.54
	^{205}Tl	70.476	$\frac{1}{2}$	3.27642922	89.089		6558.5	44.96

TABLE H.4 *Continued*

| 82 | ^{207}Pb | 22.1 | $\frac{1}{2}$ | 1.18517 | 31.954 | | 2908.49 | 23.208 |
| 83 | ^{209}Bi | 100 | $\frac{9}{2}$ | 0.91347 | 25.51 | $-0.77 \; // -0.55$ | 2766.5 | 23.68 |

[a] Data taken from N. J. Stone, *Table of Nuclear Magnetic Dipole and Electric Quadrupole Moments*, Oxford Physics, Clarendon Laboratory, Parks Road, Oxford, OX1 3PU, U.K., 2001; see: also *Atom Data Nucl. DataTable*, 90, 75 (2005) and *Handbook of Chemistry and Physics*, D. R. Lide, Editor-in-Chief, 86th ed., CRC Press, Boca Raton, FL, U.S.A., 2005–2006, pp. 9.92–9.94, 11.50–11.184. The nuclear magnetic moments appearing in some tables are defined to be $g_n\beta_n I$. The relevant columns in the N.J. Stone table are labeled I, μ, and Q. The values used herein were only from those nuclides which are 'stable' or have half-lives of $>10^{10}$ years. If the table stated more than one value for μ or Q, then the value containing the most significant figures was used. If the same number of significant digits were present within those multiple values, then the value with the least error was used. Symbol I was defined as the nuclear spin. The nuclear magnetic dipole moment μ was given in units of nuclear magnetons whereas Q was defined as the nuclear electric quadrupole 'moment', in units of barns (b).

[b] Corrected for the new (2003) value of the Bohr magneton and nuclear magneton (see Table H.1).

[c] Q is given in 10^{-24} cm^2, units called 'barns', and, of course, equal to 10^{-28} m$^2 = (10^1$ fm$)^2$, in SI. The nuclear quadrupole moment is $|e|Q$.

[d] The isotropic hyperfine couplings are tabulated as

$$a_0 = \frac{2\mu_o}{3} g_n \beta_n |\psi(0)|^2$$

which is Eq. 2.38 with $g = g_e$, in magnetic-field units. Except for the H atom and the alkali-metal atoms, where experimental values are readily available, a_0 has been obtained from calculated values of $|\psi(0)|^2$ using the tables published by J. R. Morton, K. F. Preston, *J. Magn. Reson.*, 30, 577 (1978). It should be noted that these calculated values are only approximate and will probably be increasingly erroneous at higher atomic numbers.

[e] For example, the anisotropic hyperfine couplings are for one unpaired electron in a p orbital centered on the atom. In this case, the expression for $b_0 = \delta A/g_e\beta_e$ (Eq. 5.26) is

$$b_0 = \frac{2}{5}\frac{\mu_o}{4\pi} g_n \beta_n \langle r^{-3}\rangle_p$$

where the angular brackets imply integration over the p orbital. The b_0 values can also be used for d and f orbitals with adjustment factors given in J. R. Morton, K. F. Preston, *J. Magn. Reson.*, **30**, 577 (1978).

[f] Experimental atomic hyperfine couplings as measured using the atomic beam technique [J. R. Morton, K. F. Preston in *Landolt-Börnstein Numerical Data and Functional Relationships in Science and Technology*, New Series, H. Fischer, K.-H. Hellwege, Eds., Group II, Vol. 9a, Springer, Berlin, Germany, 1977, pp. 5ff; J. R. Morton, K. F. Preston, in *Landolt-Börnstein Numerical Data and Functional Relationships in Science and Technology*, New Series, H. Fischer, Ed., Group II, Vol. 17a, Springer, Berlin, Germany 1987, pp. 5ff].

APPENDIX I

MISCELLANEOUS GUIDELINES

I.1 NOTATION FOR SYMBOLS

Basically, there are two types of ingredients leading to the notation needed herein:

1. The aspects involving the 3-space (x,y,z) we live in, leading to scalars, spatial vectors, and two types of 3×3 matrices,
2. The aspects involving the quantum-mechanical spin spaces, once again with scalars (classifiable into two types), vectors (bras and kets) and matrices (spin angular momenta, and spin hamiltonians). These can all occur as quantum-mechanical operators.

The types have been classified in separate sections of Table I.1. Inspection of these and logic has led to the notation allowing us to clearly and quickly distinguish between them, to some extent utilizing context. This led to the right-hand part of the table, utilizing symbols bold, italic, over-bar and over-'hat'. Examples of the notation are included. Note that many situations that at first sight might occur do not arise in practice, and are indicated by crossed-out lines.

Electron Paramagnetic Resonance, Second Edition, by John A. Weil and James R. Bolton
Copyright © 2007 John Wiley & Sons, Inc.

TABLE I.1   Types of Symbols (Dashes in Table imply absence of such usage in this book.)

Type Number	Spatial				QM Entity						Example(s)	Notation				
	Scalar	Vector	Parameter Matrix	Rotation Matrix	Plain Scalar	Eigenvalue Scalar	Non-spatial Vector	Non-matrix Operator	Matrix Operator	'Rotation' Matrix		Non: Ital., Hat, Bold, Bar	Italic*	Over-Hat (^)	Bold	Over-Bar**
1	✓				✓						g, β_e	✓				
2	✓					✓					S, M_I		✓			
3	✓						✓				$\mathbf{c}_i$				✓	
4	✓								✓		$\hat{\ell}_X, \hat{s}_X$			✓		
5	✓									✓	$\bar{H}, \bar{I}_6, \bar{S}_x$					✓
6		✓			✓						$\mathbf{R}, \mathbf{B}$				✓	
—	—	✓	—	—	—	✓	—	—	—	—	—	—	—	—	✓	—
—	—	✓	—	—	—	—	✓	—	—	—	—	—	—	—	✓	—
7		✓						✓			$\hat{\mathbf{S}}$			✓	✓	
8		✓								✓	$\bar{\mathbf{S}}, \bar{\mathbf{I}}$				✓	✓
9			✓		✓						$\mathbf{g}, \mathbf{A}$				✓	
—	—	—	✓	—	—	✓	—	—	—	—	—	—	—	—	✓	—
—	—	—	✓	—	—	—	✓	—	—	—	—	—	—	—	✓	—
—	—	—	✓	—	—	—	—	✓	—	—	—	—	—	—	✓	—
—	—	—	✓	—	—	—	—	—	✓	—	—	—	—	—	✓	—
10			✓		✓						$\mathbf{1}_3, \mathbf{R}$				✓	
—	—	—	—	✓	—	✓	—	—	—	—	—	—	—	—	✓	—
—	—	—	—	✓	—	—	✓	—	—	—	—	—	—	—	✓	—
—	—	—	—	✓	—	—	—	✓	—	—	—	—	—	—	✓	—
11			✓						✓		$\hat{\mathbf{1}}_3$			✓	✓	
12										✓	$\bar{C}$					✓

* Italics also appear in the text in some other usages (e.g., B, U).

** Over-bars also denote certain averaged quantities (e.g., $\bar{\mathbf{B}}$) or quantities represented in cm^{-1} units (e.g. $\bar{D}$).

I.2 GLOSSARY OF SYMBOLS

The list in Table I.2 contains most of the (non-chemical) symbols and abbreviations used in this book. Some special symbols, which are used in only a limited vein, are defined in the relevant chapter. A careful attempt has been made to standardize usage throughout this book; however, it must be realized that usage in the original literature is not standardized. Trivial use of subscripts and superscripts is not always included in the compilation below. In general, vectors and matrices are set in boldface type, and quantum-mechanical operators are indicated by circumflexes (see Table I.1).

TABLE I.2 Glossary of Symbols

Symbol	Unit [a]	Section or Chapter [b]	Description
a	T (mT, G)	2,10.5.1	Hyperfine splitting 'constant' ($\bar{a}$ = value averaged over species)
a	—	9.2.4	Designation of an orbitally non-degenerate level
a, b, c, a', c'	—	5.3.2	Crystallographic axes; also unit-cell dimensions
a_0	T (mT, G)	2.3.3, 2.4, 5.2	Isotropic hyperfine splitting 'constant'
A	—	8.3	Designation of an orbitally non-degenerate state
A	J	5.1	Hyperfine coupling 'constant'
A	—	9.2.1	Antisymmetric benzene π orbital
$\mathcal{A}$	m^2	4.7, F.2	Area
A_0	J	2.3	Isotropic hyperfine coupling 'constant'
A_{pp}	—	F.4	Amplitude of first-derivative spectral line
$A_{u\ell}$	s^{-1}	4.5	Einstein spontaneous emission coefficient
A	J	5	Hyperfine parameter matrix
AA	J^2	5.3.2	$\mathbf{A}^T \cdot \mathbf{A}$
δA	J	5.1,5.3.2	Purely anisotropic part of the hyperfine coupling
b	s^{-1} (Hz, kHz)	F.3.1	Bandwidth
b_0	T (mT, G)	5.2	Uniaxial hyperfine splitting parameter
B	—	9.2.4	Designation of one type of orbitally non-degenerate state
B	T (mT, G)	1.8	Magnetic-field (induction) vector
B	T (mT, G)	1.1, 10.5.1	Magnetic-field (induction) magnitude ($\bar{B}$ = value averaged over species)

(*continued*)

TABLE I.2   *Continued*

Symbol	Unit [a]	Section or Chapter [b]	Description
B_{hf}	T (mT, G)	5.3.2	Magnetic-field amplitude at nucleus, arising from hyperfine interaction
B_{local}	T (mT, G)	1.12, 5.3.2	Local magnetic-field amplitude
B_m	T (mT, G)	F.3.6	Magnetic modulation-field amplitude
$\overline{B}$	T (mT, G)	6.3.3	Average resonant magnetic-field amplitude
B_r, B', B_0	T (mT, G)	2.1, 6.3.3, F.1.1	Resonant magnetic-field amplitude
B_{rot}	J	7.3	Rotational energy coefficient
$\mathbf{B}_1$	T (mT, G)	1.1	Excitation magnetic-field vector at frequency ν_1
$B_{u\ell}$	m^3 J^{-1} s^{-1}	4.6	Einstein induced radiation coefficient
$\Delta B, \Delta B_0$	T (mT, G)	4.7, 6.3, 9.2.5, 10.5	Range, separation or shift of resonant magnetic-field amplitudes
ΔB_{pp}	T (mT, G)	F.2.1, F3.6	Separation in field between the extrema (peak-peak) for a first-derivative EPR line
c	m s^{-1}	H.1	Speed of light
c	—	9.2	Coefficient in linear expansion of wavefunctions
c_0	T (mT, G)	5.2	Rhombic hyperfine splitting parameter
c_p	—	5.3.2.1	p-Orbital coefficient of spatial wavefunction
c_s	—	5.3.2.1	s-Orbital coefficient of spatial wavefunction
C	K	1.10	Curie constant
C	J^{-1} C^2	E.2.8	Capacitance
$\mathbf{C}$	—	4.4	Direction cosine matrix
d_j	—	F.3.1	Degeneracy of jth line
D	—	8.2, B.7, Table 8.1	Designation of a state with total orbital angular momentum $L = 2$
D	J	6.3.1, 9.7.1	Uniaxial electronic quadrupole parameter
D_0	J	9.7.1	Electronic quadrupole parameter in Newman model
D'	T (mT, G)	6.3.1	Electronic quadrupole parameter in magnetic-field units; $D' = D/g_e\beta_e$

(continued)

TABLE I.2 *Continued*

Symbol	Unit ^{a}	Section or Chapter ^{b}	Description
$\overline{D}$	m^{-1} (cm^{-1})	6.3.1	Rhombic electronic quadrupole parameter in wavenumber units; $\overline{D} = D/hc$
D	J	4.8	Electronic quadrupole parameter matrix
det	—	A.5	Determinant
e	C	A.5	Electronic charge magnitude
e	J	6.6	Energy coefficient
E	—	8.2	Designation of a state with two-fold orbital degeneracy
E	J	6.3.1	Rhombic electronic quadrupole parameter
E'	T (mT, G)	6.3.1	Rhombic electronic quadrupole parameter (in magneticfield units; $E' = E/g_e\beta_e$)
$\overline{E}$	m^{-1} (cm^{-1})	6.3.1	Rhombic electronic quadrupole parameter (in wavenumber units; $\overline{E} = E/hc$)
$\mathbf{E}_1$	V m^{-1}	1.1	Alternating electric-field vector
f	—	10.5.1	Fractional population; mole fraction
F	—	Tables 8.1, 8.7	Designation of a state with total orbital angular momentum $L = 3$
F	—	7.2	Quantum number for the combined electron and nuclear angular momenta
F	—	F.3.1	Noise factor
$\hat{\mathbf{F}}$	—	7.2, 8.7	$\hat{\mathbf{J}} + \hat{\mathbf{I}}$
g	—	1 and 4, 4.2	g factor ($g_\parallel, g_\perp$–values, in directions parallel and perpendicular to a unique symmetry axis)
g	—	7.3, 8.2	Gerade (even)
g_e	—	1.7	g factor of the free electron
g_n	—	1.11	Nuclear g factor
g	—	1.6, 4	g parameter matrix
gg	—	4.4	$\mathbf{g}^T \cdot \mathbf{g}$
G	J T^{-1} m^{-3}	10.5.1	Magnetization component $M_{+\phi}$
G	—	4.8, 8.3	Ground state
$\langle G\vert, \vert G\rangle$	—		Bra and ket Dirac notation for ground-state functions
h	J s	H.1	Planck's constant
$\hbar$	J s rad^{-1}	H.1	$h/2\pi$

(*continued*)

TABLE I.2 *Continued*

Symbol	Unit ^{a}	Section or Chapter ^{b}	Description
H	$J\ T^{-1}\ m^{-3}$	1.8	Magnetic field $\mathbf{B}/\mu_m$
H_{ij}	J	Appendix 9A	i, j Element of matrix representation of $\hat{\mathcal{H}}$ $i = j$ (Coulomb integral) $i \neq j$, i adjacent to j (resonance integral)
$\Delta H^{\ddagger}$	J	10.5.6	Enthalpy change of activation
$\hat{\mathcal{H}}$	J	2.3.1	Hamiltonian operator, or spin hamiltonian operator (sub- or superscripts indicate the nature of the operator)
$\bar{\mathcal{H}}$	J	4.4	Spin-hamiltonian matrix
i	$C\ s^{-1}\ (=A)$	1.7, E.1.8	Electric current magnitude
i	—	A.1	$\sqrt{-1}$
im	—	A.1	Imaginary part
$\hat{\mathbf{I}}$	—	1.8	Nuclear-spin vector operator (components $\hat{I}_x, \hat{I}_y, \hat{I}_z$)
$\hat{I}_+, \hat{I}_-$	—	C.1	'Raising' and 'lowering' nuclear-spin operators
I	—	1	Primary quantum number for nuclear spin
^tI	—	3.1	Total nuclear-spin quantum number
I_0	$kg\ m^2$	A.2.2	Moment of inertia
$\mathscr{I}$	—	5.3.1, F.2.1	Intensity of EPR absorption
J, j	—	1.6, B.7	Primary quantum number for general angular momentum
$\hat{\mathbf{J}}$	—	1.6, B.7	General or total electronic angular-momentum vector operator
J	—	6.2.1	General angular-momentum matrix
J_0	J	6.2.1	Electron-exchange isotropic interaction constant
k_b	$J\ K^{-1}$	H.1	Boltzmann constant
$\langle k|, |k\rangle$	—	2.3.2, A.5.4	Bra and ket representation of the state k, for example, of state function ψ_k
$k_{(j)}$	$M^{1-j}\ s^{-1}$	10	Rate constant of order j
K	J	6.7	Hyperfine parameter
K	—	10.5	Equilibrium constant
ℓ	$\mathbf{J}\ s$	B.2	Orbital angular-momentum vector (components $\ell_x, \ell_y, \ell_z) = \mathbf{L}\hbar$

(*continued*)

TABLE I.2 *Continued*

Symbol	Unit ^{a}	Section or Chapter ^{b}	Description
L	—	1.6, 4	Primary quantum number for total orbital angular momentum
$\hat{\mathbf{L}}$	—	4.8	Orbital angular-momentum vector operator
m	kg	1.7	Mass
m_e	kg	1.8	Electron mass
m_p	kg	1.8	Proton mass
M	mol L^{-1}		Molarity
$\mathbf{M}$	J T^{-1} m^{-3}	1.8, 10.3.1	Magnetization vector
^{t}M	—	3.6	Composite secondary quantum number for a set of nuclei
M_I	—	1.4	Secondary quantum number for (the z component of) the nuclear-spin angular momentum
M_J, m	—	B.7	Secondary quantum number for (the z component of) the total (e.g., spin plus orbital) angular momentum
M_L	—	B.7	Secondary quantum number for (the z component of) the orbital angular momentum
M_S	—	1.4, B.7	Secondary quantum number for (the z component of) the electron-spin angular momentum
M^0	J T^{-1} m^{-3}	1.8	Equilibrium macroscopic magnetization of a spin system in the presence of field $\mathbf{B}$
$M_{\pm \phi}$	J T^{-1} m^{-3}	10.3.3	Component of complex transverse magnetization in the rotating frame
$\mathbf{n}$	—	4.4	Unit vector along field $\mathbf{B}$
N	—	4.6, 10.2.2	Population of a state or energy level
N	—	7.3	Quantum number associated with the rotational angular momentum of a linear molecule in the gas phase
$N_{\min}$	—	F3.1	Smallest detectable number of electron spins
N_V	m^{-3}	1.9	Number of magnetic dipoles per unit volume
ΔN	—	10.2.2	Population difference between two states or levels
$\hat{\mathbf{N}}$	—	7.3	Rotational angular-momentum vector operator; also $\hat{\mathbf{\Lambda}} + \hat{\mathbf{R}}$

(continued)

TABLE I.2   *Continued*

Symbol	Unit ^{a}	Section or Chapter ^{b}	Description
p	kg m s^{-1}	B.2	Linear momentum vector (components p_x, p_y, p_z)
P	—	B.7	Designation of a state with total orbital angular momentum $L = 1$
P	— or $\cdots$	4.7, 10.2	Probability or probability function
P_0	J s^{-1} (=W)	F.3	Incident microwave power
P_a	J s^{-1} (=W)	10.3.4	Absorbed microwave power
P	J	5.6	Nuclear-quadrupole factor
P_d	J	8.3	d-Orbital hyperfine parameter
P	J	5.6	Nuclear-quadrupole parameter matrix
q	C	1.7	Electric charge
q_{efg}	J C^{-2} m^{-2}	5.6	Electric-field gradient parameter
Q	T (mT, G)	9.2	Proportionality constant connecting hyperfine parameter a and unpaired-electron population ρ
Q	—	F.3.4	Quality factor for a resonator Note: Q_L, Q_r, Q_u, Q_v, Q_e
Q	m^{-2}	5.6	Nuclear-quadrupole moment
Q'	J	Problem C.5	Nuclear-quadrupole energy factor ($=3P$)
Q band	—	Table E.1	Designation of a microwave frequency band
r	m	1.7, 5.2	Radius, inter-particle distance
r_b	m	2.2, Tables 5.5, H.1	Bohr radius
r	m	5.2	Inter-particle vector
re	—	A.1	Real part
R	V A^{-1} (=J C^{-2} s) = Ω	E.7	Resistance [in ohms (Ω)]
R	m	8.2	Distance
[R]	mol L^{-1} (=M)	10.5, 3.1	Concentration of a chemical species R
R	—	4.5	Rotation matrix
$\hat{\mathbf{R}}$	—	7.3	Molecular rotation operator
R_D	m	9.7.1	Distance reference parameter in Newman model
$\mathcal{R}$	—	F.3.1	Degeneracy factor
S	—	6.2, B.7	Designation of a state with $L = 0$ (orbital singlet state)
S	—	1, B.7	Primary quantum number for electron spin

(*continued*)

TABLE I.2 *Continued*

Symbol	Unit [a]	Section or Chapter [b]	Description
S	—	9.2.1	Symmetric benzene π orbital
$\hat{\mathbf{S}}$	—	1	Electron-spin angular-momentum vector operator (components $\hat{S}_x$, $\hat{S}_y$, $\hat{S}_z$)
$\hat{S}_+$, $\hat{S}_-$	—	C.1	Raising and lowering electron-spin operators
S_{ij}	—	Appendix 9A	Overlap integral
$\Delta S^{\ddagger}$	$J\,mol^{-1}\,K^{-1}$	10.5.6	Entropy change of activation
t	s	10.2	Time
t_d	—	9.7.1	Empirical parameter in Newman model
T	K	—	Absolute temperature
T	J	Table A.1	Kinetic energy
T	—	8.3	Designation of a state with total electron spin $S = 1$ (triplet state)
T_d	K	F.3.1	Microwave-detector temperature
T_s	K	10.2.1, F.3.1	Spin temperature, or sample temperature
$\mathbf{T}$	J	5.2	Purely anisotropic hyperfine interaction matrix
TE_{ijk}	—	E.1.5	Designation of a transverse-electric cavity mode
TM_{ijk}	—	E.1.5	Designation of a transverse-magnetic cavity mode
tr	—	A.5	Trace of a matrix
u	—	7.6	Ungerade (odd)
U, U	J	1	Energy
U_0	J	8.2, Fig. 8.11	Reference energy in Tanabe-Sugano model
$\Delta U^{\ddagger}$	J	5.3.2.2	Energy change of activation
v	—	7.3	Vibrational quantum number
v	$m\,s^{-1}$	1.7	Speed
$\mathbf{v}$	$m\,s^{-1}$	1.7	Velocity vector
V	m^3	1.8	Volume
V	J	Table A.1	Potential energy
v.e.	—	Chapter 8	Valence electron
$W_\uparrow$, $W_\downarrow$	s^{-1}	10.2.3	Transition probabilities for connection to the lattice
$\mathbf{x, y, z}$	m	—	Laboratory-fixed axes
X band	—	Table E.1	Designation of a microwave frequency range

(continued)

TABLE I.2 *Continued*

Symbol	Unit ^{a}	Section or Chapter ^{b}	Description
X, Y, Z	m	—	Molecule-fixed axes, principal axes
Y	s or T^{-1}	F.2.1	Shape curve of an absorption line
Y_{L,M_L}	—	8.8	Spherical harmonic functions (θ, ϕ)
Z	—	5.3.2.3, H.4	Electric-charge number
$Z_\uparrow, Z_\downarrow, Z$	s^{-1}	10.2.2	Transition probabilities
$\mathbf{1}_n$	—	Table A.3	$n \times n$ identity matrix
α	J	Appendix 9A	Coulomb integral
α	—	1.8	$+1/-1$ (nucleus/electron)
$\alpha(e), \alpha(n)$	—	2.3.1	Electron-, nuclear-spin state function with $M_S = +\frac{1}{2}, M_I = +\frac{1}{2}$
α atom	—	5.3.2.2	Atom (e.g., H, F) attached to the carbon atom on which the unpaired electron is primarily localized in an alkyl radical
β	$J\,T^{-1}$	1.8	Magneton
β	J	Appendix 9A	Resonance integral
$\beta(e), \beta(n)$		2.3.1	Electron-, nuclear-spin state function with $M_S = -\frac{1}{2}, M_I = -\frac{1}{2}$
β_e	$J\,T^{-1}$	1.7, H.4	Bohr magneton $= |e|\hbar/2m_e$
β_n	$J\,T^{-1}$	1.8, H.4	Nuclear magneton $= |e|\hbar/2m_p$
β_Q	—	F.3.4	Quality-factor ratio $=$ coupling parameter
β atom	—	5.3.2.2	Atom (e.g., H, F) attached to a carbon atom adjacent to the carbon atom on which the unpaired electron is localized primarily in an alkyl radical
γ	rad $s^{-1}\,T^{-1}$	1.7	Magnetogyric ratio ($\bar{\gamma} =$ value averaged over species)
γ_e, γ_n	rad $s^{-1}\,T^{-1}$	—	Electron, nuclear magnetogyric ratio
δ	J	4.8, 8.2	Separation of orbital energy levels due to the tetragonal component of the crystal field
Γ	T (mT, G)	F.2	Half the linewidth at half-height, in the absence of microwave saturation ($\bar{\Gamma} =$ value averaged over species)
Δ	—	—	Modifier to indicate difference between two quantities (e.g., ΔB, ΔN, ΔM_S)
Δ	s	10.5	Pulse interval

(continued)

TABLE I.2 *Continued*

Symbol	Unit [a]	Section or Chapter [b]	Description
Δ	J	8.2	Separation of orbital energy levels in an octahedral or tetrahedral crystal field
Δ	—	B.7	State of a linear diatomic molecule in which $\Lambda = \pm 2$
$\Delta x_{1/2}$	s^{-1} or T	Tables F.1	Full width at half-height of curve $Y(x)$
Δx_{++}	s^{-1} or T	Tables F.1	Max-to-max width of $Y''(x)$
Δx_{pp}	s^{-1} or T	Tables F.1	Peak-to-peak width of $Y'(x)$
ε	m (Å)	Fig. 8.1	Distance increment
ε	—	11.6	Echo amplitude function
ε', ε''	—	F.3.4	In-phase and out-of-phase relative permittivities (dielectric constants)
ε_0	$C^2\,J^{-1}\,m^{-1}$	H.1	Electric 'constant' (permittivity of the vacuum)
ε_e	—	12.4	$g_e\beta_e B/2k_b T$
ε_n	—	12.4	$g_n\beta_n B/2k_b T$
η	$kg\,m^{-1}\,s^{-1}$	10.5.5	Coefficient of viscosity (1 poise = 1 g cm^{-1} s^{-1})
η	—	5.6	Nuclear-quadrupole asymmetry factor
η	—	F.1.2	Filling factor of a cavity
θ	rad or deg	2.1, 5.7	Angle, for example, that between magnetic-dipole vectors or $\mathbf{B}$; also polar angle
κ	—	10.4.1	Lineshape factor
κ_m, κ	—	1.8	Relative permeability μ_m/μ_0
λ	J	4.8	Spin-orbit coupling parameter
λ	—	9.2.3	g-factor coefficient of resonance integral
λ	—	A.7	Perturbation expansion parameter
Λ	J^{-1}	4.8	Spin-orbit effect perturbation matrix
$\hat{\Lambda}$	—	2.3.1	General quantum-mechanical operator
Λ	—	7.3	Quantum number corresponding to $\hat{\Lambda}$, that is, the projection of $\hat{\mathbf{L}}$ onto the internuclear axis in a diatomic molecule
$\langle\mu\rangle$	$J\,T^{-1}$	11	Expectation value of the magnetic moment
$\boldsymbol{\mu}$	$J\,T^{-1}$	1.8	Magnetic moment vector (with component μ_z, $\mathbf{z}$ along $\mathbf{B}$)
μ_m	$T^2\,J^{-1}\,m^3$	1.8	Magnetic permeability

(continued)

TABLE I.2   *Continued*

Symbol	Unit ^{a}	Section or Chapter ^{b}	Description
μ_0	$T^2\,J^{-1}\,m^3$	Table H.1	Magnetic 'constant' (permeability of the vacuum)
μ_{rot}	J	7.3	Rotational-spin magnetic coupling parameter
ν	$s^{-1}\,(=Hz)$	1.1	Frequency
ν_B	$s^{-1}\,(=Hz)$	11.1	Larmor frequency $(=\omega_B/2\pi)$
π	—	1.11	Photon type
π orbital	—	Appendix 9A, Note 7.5	Molecular orbital composed of a combination of atomic $2p_z$ orbitals
π radical	—	9.1, 9.2	Species with an unpaired electron in an orbital, perpendicular to a planar set of single bonds, of the double-bond type
Π	—	B.7	Designation of a state of a linear molecule in which $\Lambda = \pm 1$
ρ	—	9.2	Unpaired-electron population
ρ_v	$J\,m^{-3}\,s$	4.6, 10.2.3	Radiation density
ρ_s	m^{-3}	2.6, Note 9.1	Spin density at nucleus n
$\hat{\rho}_s$	m^{-3}	2.6	Spin-density operator
σ	—	1.11	Photon type
σ	—	1.12	Chemical-shift parameter
σ orbital	—	Note 7.5	Orbital that is cylindrically symmetric about an internuclear direction
σ radical	—	9.1, 9.3	Species with an unpaired electron in an orbital of the single-bond type
Σ	—	B.7	State of a diatomic molecule in which $\Lambda = 0$
Σ	—	7.3	Quantum number corresponding to the projection of $\mathbf{S}$ onto the internuclear axis of a diatomic molecule
τ	s	10.5.1	Mean lifetime of a state
τ	s	10.3	Time interval between pulses
τ_c	s	10.5.5	Correlation time for molecular tumbling
τ_m	s	10.3	Phase-memory time
τ_x	s	12.4	Cross-relaxation time when $\Delta(M_S + M_I) = 0$
τ_{xx}	s	12.4	Cross-relaxation time when $\Delta(M_S + M_I) \neq 0$
τ_1	s	10	Spin-lattice relaxation time

(*continued*)

TABLE I.2 *Continued*

Symbol	Unit ^{a}	Section or Chapter ^{b}	Description
τ_2	s	10	Spin-spin relaxation time
ϕ	deg or rad	—	Azimuthal angle
ϕ	J C^{-1}	Problem C.5	Electric potential
ϕ	m$^{-p/2}$ rad$^{-q/2}$	2.3.1, 2.3.2	Alternative symbol for wavefunction (in p distance variables and q angular variables) or spin state
Φ_o	T m^2 (=Weber)	B.12	Magnetic flux unit or spin state
χ_m, χ	—	1.9	Volume magnetic susceptibility (χ', χ'' = in-phase, out-of-phase components of the dynamic magnetic susceptibility)
χ	—	1 (Note 14)	Volume magnetic susceptibility matrix
ψ	m$^{-p/2}$ rad$^{-q/2}$	2.3.2	Wavefunction (in p distance variables and q angular variables)
ω	rad s^{-1}	10.2.3, 10.3	Angular frequency = $2\pi\nu$
ω_B	rad s^{-1}	10.3.1	Angular frequency corresponding to field **B**
Ω	sterad	4.7	Solid angle
Ω	—	7.3, B.7	Quantum number corresponding to $\Lambda + \Sigma$
Ω	rad or deg	10.6	Spin-magnetization turning angle in pulsed EPR

^{a} The units in parentheses are commonly used variants.
^{b} Or other element (table, problem, etc.).

I.3 ABBREVIATIONS

ac	Alternating current
ACS	Anti-crossing spectroscopy
AFC	Automatic frequency control
CIDEP	Chemically induced dynamic electron polarization
COSY	Correlation spectroscopy
CS	Coordinate system
cw	Continuous wave
DBM	Double-balanced mixer (not to be confused with dBm, which is a measure of power: 1 dBm = $10^{0.10}$ mW)
dc	Direct current
DFT	Density-functional theoretical
DNP	Dynamic nuclear polarization

efg	Electric-field gradient
EDMR	Electrically detected magnetic resonance
ELDOR	Electron–electron double resonance
EMR	Electron magnetic resonance
ENDOR	Electron-nuclear double resonance
EPR	Electron paramagnetic resonance
ESE	Electron-spin echo
ESEEM	Electron spin echo envelope modulation
ESR	Electron-spin resonance
EXSY	Exchange spectroscopy
FDMR	Fluorescence-detected magnetic resonance
FID	Free-induction decay
FT	Fourier-transform
FTEPR	Fourier-transform EPR
HMO	Hückel molecular orbital
HYSCORE	Hyperfine sublevel correlation spectroscopy
LCAO	Linear combination of atomic orbitals
LCS	Level-crossing spectroscopy
LEPR	Laser electron paramagnetic resonance
LODEPR	Longitudinally detected EPR
MASER	Microwave amplification of stimulated emission of radiation
MBMR	Molecular-beam magnetic resonance
MO	Molecular orbital
MOSFET	Metal-oxide semiconductor field emission transistor
MOMRIE	Microwave optical resonance induced by electrons
MRIE	Magnetic resonance induced by electrons
NMR	Nuclear magnetic resonance
NQR	Nuclear quadrupole resonance
oct	Octahedral
ODMR	Optically detected magnetic resonance
rf	Radiofrequency
SECSY	Spin-echo correlation spectroscopy
SMM	Single-molecule magnet
SOMO	Semi-occupied molecular orbital
SQUID	Superconducting quantum interference device
trg	Trigonal
tth	Tetrahedral
ttg	Tetragonal
TWT	Traveling-wave tube
VC	Vector coupling

I.4   EXPONENT NOMENCLATURE

Refer to Table I.3:

I.5   JOURNAL REFERENCE STYLE

TABLE I.3

Factor	Prefix	Symbol	Factor	Prefix	Symbol
10^{-24}	yocto	y	10^{+24}	yotta	Y
10^{-21}	zepto	z	10^{+21}	zeta	Z
10^{-18}	atto	a	10^{+18}	exa	E
10^{-15}	femto	f	10^{+15}	peta	P
10^{-12}	pico	p	10^{+12}	tera	T
10^{-9}	nano	n	10^{+9}	giga	G
10^{-6}	micro	μ	10^{+6}	mega	M
10^{-3}	milli	m	10^{+3}	kilo	k (or K)
10^{-2}	centi	c	10^{+2}	hecto	h (or H)
10^{-1}	deci	d	10^{+1}	deka	Da (or D)
			10^{0} uni		

Journal names cited in this book have been abbreviated as advised by the Chemical Abstracts Service Source Index (CASSI), American Chemical Society, Washington, DC, U.S.A., 2005.

AUTHOR INDEX

Page numbers in *italics* indicate the page on which an author is listed in a Reference List. Otherwise the page number refers to the page number on which the citation occurs. The superscript numbers indicate the number of times that author occurs on that page.

Electron Paramagnetic Resonance, Second Edition, by John A. Weil and James R. Bolton
Copyright © 2007 John Wiley & Sons, Inc.

SUBJECT INDEX

Page numbers in *italics* indicate main coverage; numbers in **boldface** indicate figures.

Electron Paramagnetic Resonance, Second Edition, by John A. Weil and James R. Bolton
Copyright © 2007 John Wiley & Sons, Inc.

CPSIA information can be obtained
at www.ICGtesting.com
Printed in the USA
BVOW06*0318270617

487888BV00007B/39/P